The Absolute, Ultimate Guide to
Lehninger Principles of Biochemistry
Fifth Edition

The Absolute, Ultimate Guide to
Lehninger Principles of Biochemistry
Fifth Edition

Study Guide and Solutions Manual

Marcy Osgood
University of New Mexico School of Medicine

Karen Ocorr
University of California, San Diego

Solutions Manual based on
a previous edition by

Frederick Wedler
Robert Bernlohr
Ross Hardison
Teh-Hui Kao
Ming Tien

Pennsylvania State University

W. H. FREEMAN AND COMPANY
NEW YORK

ISBN-13: 978-1-4292-1241-0
ISBN-10: 1-4292-1241-1

Printed in the United States of America

Third printing

W. H. Freeman and Company
41 Madison Avenue
New York, NY 10010
Houndmills, Basingstoke RG21 6XS, England

www.whfreeman.com

Study Guide

Contents

Solutions Manual

Contents

Preface

Learning a complex subject, such as biochemistry, is very much like learning a foreign language.

In the study of a foreign language there are several distinct components that must be mastered: the vocabulary, the grammatical rules, and the integration of these words and rules so they can be used to communicate ideas. Similarly, in the study of biochemistry there is a very large (some would say vast) number of new terms and concepts, as well as a complex set of "rules" governing biochemical reactions that must all be memorized. All of this information must be integrated into an interrelated whole that describes biological systems.

Memorizing vocabulary and grammatical rules will not make you fluent in a foreign language; neither will such memorization make you fluent in biochemistry.

Similarly, listening to someone speak a foreign language will not, by itself, make you more capable of producing those same sounds, words, and sentences. The key to mastery of any new subject area, whether it is a foreign language or biochemistry, is the interaction of memorization, practice, and application until the information fits together into a coherent whole.

In this workbook we attempt to guide you through the material presented in Lehninger Principles of Biochemistry, Fifth Edition *by Nelson and Cox. The* **Step-by-Step Guide** *to each chapter includes three parts:*

- A one- to two-page summary of the **Major Concepts** helps you to see the "big picture" for each chapter.
- **What to Review** helps you make direct connections between the current material and related information presented elsewhere in the text.
- **Topics for Discussion** for each section and subsection in the chapter focus your attention on the main points being presented and help you internalize the information by *using* it.
- **Discussion Questions for Study Groups** are questions that are especially suited to Study Groups, either because they pull together several points made in the chapter or because they are more involved questions that would benefit from collaborative insight.

Each chapter also includes a **Self-Test** *for you to assess your progress in mastering biochemical terminology and facts, and learning to integrate and apply that information.*

- **Do You Know the Terms?** asks you to complete a crossword puzzle using the new vocabulary introduced in the chapter.
- **Do You Know the Facts?** tests how well you have learned the "rules" of biochemistry.
- **Applying What You Know** tests how well you "speak the language" of biochemistry, often in an experimental or metabolically relevant context.

Two especially popular features of the Absolute, Ultimate Guide *are:*

- The **Biochemistry on the Internet** problems will expose you to just a few of the analytical resources that are available to scientists on the Internet. The molecular models are fun to play with, and many of the questions provide you with an opportunity to analyze "data" as you might in an actual laboratory setting.

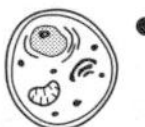

- The **Cell Map** is based on many semesters of use by our students. It is designed to help you place the biochemical pathways that you are learning about into their proper cellular perspective. The Cell Map questions tell you what to include in your Map, but you can be creative!

Students have told us it is a great study aid, that really helped them to make the connections between the various pathways.

In our experience, this material can best be assimilated when it is discussed in a study group.

A study group is nothing more than two to four people who get together on a regular basis (weekly) to "speak biochemistry." This type of interaction is critical to fluency in a foreign language and is no less critical in the successful assimilation of biochemistry. We designed our Step-by-Step Guide to each chapter with study groups in mind. The questions posed for each section can be used as springboards for study group discussions. We purposefully have *not* supplied answers to this section to force you to wrestle with the concepts. It is the struggle that will make you learn the material. In addition, because most of the answers can be readily worked out by a careful reading of the section of the text, the questions will focus your attention on the more important aspects of the material.

Detailed **Solutions** *to all the end-of-chapter textbook problems are included as a separate section in the* Absolute, Ultimate Guide.

We have taken great care to ensure that the solutions are correct, complete, and informative. The final answer to numerical problems has been rounded off to reflect the number of significant figures in the data.

We thank each and every one of our students for their invaluable feedback and input, which have helped to make this study guide its absolute and ultimate best.

About the Authors

Karen Ocorr received her Ph.D. from Wesleyan University, where she studied the physiology and neurochemistry of the lobster cardiac ganglion. She was an NIH postdoctoral research fellow at the University of Texas, examining the roles of enzymes and second messengers in neuronal plasticity in *Aplysia californica.* She continued these investigations at the California Institute of Technology and Stanford University, examining the role of intracellular signaling underlying long-term potentiation in the vertebrate hippocampus. She taught Introductory Biochemistry for 10 years at the University of Michigan, Ann Arbor, where she also taught Animal Physiology, Cell Biology, and Introductory Biology for Non-Majors. She is currently researching the roles of K^+ channels in heart function at the Burnham Institute for Medical Research in La Jolla, California, and is a visiting Lecture Professor at Hunan Normal University in Changsha, China, where she teaches biochemistry.

Marcy Osgood received her Ph.D. from Rensselaer Polytechnic Institute. After two postdoctoral positions, one that involved the investigation of mammalian mitochondrial Complex III, and another that involved an attempt to construct a mercury-detecting microbial biosensor, she became an Assistant Professor in the Department of Physical Sciences at Albany College of Pharmacy, Union University. She was then a Lecturer in the Biology Department at the University of Michigan, Ann Arbor for nine years, where her research interests began to focus on educational issues. Her current position is as Assistant Professor of Biochemistry Education in the Department of Biochemistry and Molecular Biology at the University of New Mexico School of Medicine. She has won numerous teaching awards and has organized and participated in teaching workshops at national meetings. Her current research efforts deal with assessment of peer-assisted learning strategies.

Osgood and Ocorr have collaborated for 10 years to develop effective techniques for teaching biochemistry. This study guide is the result of these efforts and embodies much of what they have found to be effective over years of instruction at many levels and in many areas of the biological sciences. It includes thousands of discussion, quiz, and exam questions from their biochemistry courses. The results of some of their teaching efforts have been published in the journal *Biochemistry and Molecular Biology Education.* Osgood and Ocorr each have won numerous teaching awards, and between them, they have been responsible for teaching biochemistry to more than 10,000 undergraduate students.

Study Guide

chapter

The Foundations of Biochemistry

1

STEP-BY-STEP GUIDE

Major Concepts

All cells have common structural elements.

All cells are defined by a **plasma membrane,** which separates the contents of a cell from its surroundings and is a barrier to diffusion. All cells are divided into two major internal regions, the cytoplasm and a nuclear region. The **cytoplasm** contains soluble enzymes, metabolites, and cellular organelles. It is a very active and organized place, with constantly changing interactions occurring between its components. The contacts between biomolecules are primarily weak, noncovalent interactions, which collectively produce complex structure and functions. The **nuclear region** contains primarily DNA and associated proteins.

Cells can be classified according to the complement of cellular membranes and the complexity of the nuclear region.

There are two **domains** of single-celled microorganisms, **Bacteria** and **Archaea,** which differ in specific biochemical characteristics. **Eukaryotes** generally are larger than prokaryotes and include all protists, fungi, plants, and animals. The nuclear region of the single-celled organisms, the **nucleoid,** has no membrane to separate it from the rest of the cytoplasm. In addition, there are no other internal membranes and no internal **organelles.** Eukaryotic cells have plasma membranes as well as many membrane-bounded intracellular organelles and a membrane-bounded nucleus. Cells can be alternatively classified based upon their sources of energy and carbon.

Living matter is composed of low atomic weight elements.

Hydrogen, oxygen, nitrogen, and carbon are the most abundant elements in biomolecules. The bonding versatility of carbon makes it the most important and defining element in biochemical compounds. Most biomolecules are derivatives of hydrocarbons with a variety of attached functional groups. These functional groups determine the chemical behavior of biomolecules.

Biochemistry is three-dimensional.

In addition to the functional groups, the overall shape of a biomolecule greatly affects the types of interactions in which it can participate. Most biomolecules are asymmetric. Typically, only one of the possible (**chiral**) forms is found in living organisms. Interactions between many biomolecules (e.g., enzymes and their substrates) depend upon their ability to differentiate between **stereoisomers.**

Living organisms are interdependent; they exchange energy and matter with each other and the environment.

The sun is the ultimate source of (almost) all energy used by organisms. The **anabolic** reactions that use energy to form biological macromolecules and the **catabolic** reactions that liberate energy constitute the **metabolic pathways.** Most of these reactions require **enzymes** in order to proceed at useful rates; these biological catalysts lower the **activation energy.** Metabolic reactions can be linked through **ATP,** which can capture or release stored energy as needed by cells.

A thermodynamic system can be a single, simple chemical reaction or an entire organism.

The transfer of energy in a single reaction or in an organism can be described by three thermodynamic quantities: (Gibbs) **free energy,** G**; enthalpy,** H**; and entropy,** S**.** These quantities are related by the equation

$$\Delta G = \Delta H - (T\,\Delta S)$$

where T is the absolute temperature (in degrees Kelvin, or K). ΔG describes the change in free energy that occurs in a chemical reaction.

Reactions proceed spontaneously only if ΔG is negative, meaning energy is released by the reaction.
If ΔG is positive, the reaction requires an input of energy. Reactions are at equilibrium if $\Delta G = 0$.

The standard free-energy change, $\Delta G'^\circ$ is a physical constant and can be calculated from the equilibrium constant, K'_{eq}: $\Delta G'^\circ = -RT \ln K'_{eq}$.
For a system at equilibrium the rates of the forward and reverse reactions are equal; no further net change occurs in the system.

Information needed for the formation and function of all living organisms is contained in their genetic material, which is usually DNA.
The linear sequence of nucleotides in a strand of DNA provides information specifying the linear sequence of amino acids in all the proteins made by an organism. In turn, the complement of proteins determines the repertoire of metabolic reactions and functions that an organism can perform. (In some viruses, the genetic material is RNA.)

Biological systems have a hierarchy of structure.
A relatively small number of **monomeric** subunits are the building blocks used to construct macromolecules. Macromolecules can then be arranged into supramolecular complexes. Different classes of macromolecules have different roles in cells. For example, membranes (supramolecular complexes of lipids, proteins, and carbohydrates) define and compartmentalize cells, whereas chromosomes (supramolecular complexes of DNA and proteins) encode and store genetic information.

Biomolecules first arose by chemical evolution.
The conditions under which life arose on Earth cannot be absolutely determined, but chemical evolution can be simulated in the laboratory. Considering the simplicity of the chemical reactions that probably led to the formation of small biomolecules, and eventually to macromolecules, the evolution of life seems to be possible on other planets.

What to Review

Part of this chapter is a review of organic and inorganic chemical principles. You may find it helpful to consult a good inorganic or organic chemistry text for additional background information. Using this and other texts as sources, make sure that you thoroughly understand the following definitions and concepts.

- Covalent bonds and their relationship to the number of unpaired electrons.
- The significance of the atomic numbers.
- Stereoisomerism; chirality.
- Electronegativity.
- Nucleophilic and electrophilic centers in reactions.

Topics for Discussion

Answering each of the following questions, especially in the context of study group discussions, should help you understand the important points of the chapter.

1.1 Cellular Foundations

1. What distinguishes living organisms from inanimate objects?

Cells Are the Structural and Functional Units of All Living Organisms

2. Cellular dimensions are difficult to imagine on a human scale. To put cells into perspective it is helpful to magnify them. The following table shows the relative sizes of cells and their components magnified 10,000-fold (from micrometer to millimeter scale). A typical eukaryotic cell, when magnified 10,000-fold, is about the size of a large platter. Complete the table below with familiar objects that are approximately the sizes indicated and assemble them on a large serving platter.

Cell or Cellular Component	Size at 10,000 x (mm)	Equivalent Size Object
Ribosome	0.3	
Lysosome	10	
Flagellum, diameter	0.15	
Nucleoid	10	
Mitochondrion	5 x 30	
Prokaryotic cell	10–100	
Chloroplast	20 x 80	
Nucleus	100	
Eukaryotic cell	500 (50 cm)	Large serving platter

Cellular Dimensions Are Limited by Diffusion

3. What sets the lower limit of cell size? The upper limit?

There Are Three Distinct Domains of Life

4. Why are eukaryotes considered to be more closely related to the archaea than to bacteria?

5. Why are cyanobacteria considered among the most self-sufficient groups of organisms?

Escherichia coli *Is the Most-Studied Bacterium*

6. What are the structural features of prokaryotes that distinguish them from eukaryotes, which are discussed in the next section of the text?

Eukaryotic Cells Have a Variety of Membranous Organelles, Which Can Be Isolated for Study

7. What might be an advantage of compartmentalization of cells?

8. What are the differing characteristics of organelles that allow researchers to separate and isolate them from one another?

The Cytoplasm Is Organized by Cytoskeleton and Is Highly Dynamic

9. What are the three classes of cytoskeletal proteins? What are the functions of the cytoskeleton?

10. How do the organelles and cytoskeleton interact?

Cells Build Supramolecular Structures

11. What are the bonds and/or interactions that are important at each of the three levels of cell structure?

In Vitro Studies May Overlook Important Interactions among Molecules

12. What are the advantages, and the disadvantages, of the in vitro approach to studying biomolecules?

1.2 Chemical Foundations

13. What are the four most common elements in living organisms?

14. What is the importance of trace elements to animal life?

Biomolecules Are Compounds of Carbon with a Variety of Functional Groups

15. What is meant by the "bonding versatility" of carbon?

16. What factors influence the strength, length, and rotation of carbon bonds?

17. Review the common families of functional groups or organic compounds that are encountered in biomolecules.

Cells Contain a Universal Set of Small Molecules

18. What is the definition of *metabolome*?

Macromolecules Are the Major Constituents of Cells

19. What are macromolecules and how are they categorized?

20. Proteins and nucleic acids are always considered **informational macromolecules,** while polysaccharides are not always considered so; why is this the case?

Three-Dimensional Structure Is Described by Configuration and Conformation

21. What are the three types of models that are used to represent the three-dimensional structure of molecules? What different kinds of information are provided by each type of model?

22. What is the difference between a diastereomer and an enantiomer?

Louis Pasteur and Optical Activity: In Vino, Veritas

23. Use the structural formulas of Pasteur's two forms of tartaric acid to follow the description of RS nomenclature in the text.

24. How does the high potential energy of the eclipsed conformation affect the dynamics of ethane?

25. What are the advantages and disadvantages of the techniques of x-ray crystallography and NMR spectroscopy?

Interactions between Biomolecules Are Stereospecific

26. Why do cells produce only one form of a chiral compound rather than a racemic mixture?

1.3 Physical Foundations

Living Organisms Exist in a Dynamic Steady State, Never at Equilibrium with Their Surroundings

27. What is the difference between **dynamic steady state** and **equilibrium?** Make sure you understand these terms; the concepts will recur constantly throughout the text.

Organisms Transform Energy and Matter from Their Surroundings

28. Is it possible to design a closed system that incorporates living organisms? Why or why not?

The Flow of Electrons Provides Energy for Organisms

29. What happens (in terms of electron flow) to the reactant that is oxidized?

Creating and Maintaining Order Requires Work and Energy

30. What are the qualities that determine the free energy (G) in a system?

31. How are cells able to synthesize polymers if such reactions are thermodynamically unfavorable?

Energy Coupling Links Reactions in Biology

32. In terms of potential energy, energy transductions, and entropy, explain the following normal human daily activities: eating, moving, excreting. Where do the sun and ATP fit into this scheme?

Entropy: The Advantages of Being Disorganized

33. Does the oxidation of glucose represent an increase or decrease in entropy? Does it have a positive or negative ΔG?

K_{eq} and ΔG° Quantify a Reaction's Tendency to Proceed Spontaneously

34. What are the two ways of expressing a driving force on a reaction?

Enzymes Promote Sequences of Chemical Reactions

35. Why are enzymes essential in biochemical reactions?

36. How do enzymes overcome activation barriers?

37. Why is the option of increasing temperature to overcome these barriers *not* possible in living cells?

Metabolism Is Regulated to Achieve Balance and Economy

38. What would be the negative consequences to the cell of making too much of one metabolite?

1.4 Genetic Foundations

Genetic Continuity Is Vested in Single DNA Molecules

39. Why can we not describe the "average" behavior of a DNA molecule?

The Structure of DNA Allows for Its Replication and Repair with Near-Perfect Fidelity

40. How do complementary strands assure continuity of information?

The Linear Sequence in DNA Encodes Proteins with Three-Dimensional Structures

41. How are the instructions contained in the linear sequence of DNA translated into a three-dimensional enzyme or structural protein?

42. What forces contribute to the three-dimensional structure?

1.5 Evolutionary Foundations

Changes in the Hereditary Instructions Allow Evolution

43. How can a change in the DNA instructions affect the final form of a protein?

Biomolecules First Arose by Chemical Evolution

44. Under what environmental conditions are the first biomolecules thought to have been formed?

45. What energy sources may have been available to drive prebiotic evolution?

RNA or Related Precursors May Have Been the First Genes and Catalysts

46. What are the lines of evidence suggesting that RNA or a similar molecule was both the first gene and the first catalyst?

Biological Evolution Began More Than Three and a Half Billion Years Ago

47. How might it be possible to determine if a fossil carbon compound were of biological origin?

The First Cell Probably Used Inorganic Fuels

48. Why is it likely that autotrophs evolved only after the appearance of heterotrophs?

Eukaryotic Cells Evolved from Simple Precursors in Several Stages

49. How are nuclei thought to have evolved?

Molecular Anatomy Reveals Evolutionary Relationships

50. Compare the definitions of *homologs*, *paralogs*, and *orthologs*

Functional Genomics Shows the Allocations of Genes to Specific Cellular Processes

51. What is the difference between *E. coli* and *H. sapiens* in terms of percentage of genes devoted to membrane transporters?

Genomic Comparisons Have Increasing Importance in Human Biology and Medicine

52. What would you predict would be the difference between a chimpanzee and *H. sapiens* in terms of percentage of genes devoted to membrane transporters?

Discussion Questions for Study Groups

- Why are the six complex characteristics detailed in the first part of this chapter necessary to define "life" completely?
- The fossil record indicates that organisms evolved mechanisms to adapt to dramatic changes in the composition of Earth's atmosphere. How can we use this information in evaluating the consequences of modern global warming?
- Why would the development of impermeable layers of lipid-like compounds have favored continued chemical evolution?

SELF-TEST

Do You Know the Terms?

ACROSS

1. *G*; free ______.
4. Amino acids are the ______ subunits of proteins.
5. The randomness of the components of a chemical system; *S*.
6. Amino acid is to ______ as monomer is to polymer.
8. The complete set of genetic material needed for the growth and development of an organism.
11. Glycine is the only amino acid lacking an ______ asymmetric or carbon.
12. Reactions requiring an input of energy from the surroundings ______ are thermic reactions.
14. In ______ - ______ reactions, electrons are transferred from a more reduced to a more oxidized molecule.
17. A type of weak interaction that stabilizes the native conformation of a biomolecule or supramolecular complex.
19. Describes a reaction for which the free-energy change (ΔG) is negative.
21. The internal components of cells and the aqueous solution in which they are suspended.
22. Molecules having the same composition and order of atomic connections, but different spatial arrangements among the atoms.

DOWN

1. Mitochondria are thought to have evolved from bacteria that formed ______ associations with the ancestors of modern eukaryotes.
2. Proteins encoded by two genes that share similar nucleotide sequences.
3. Enzymes enhance the rate of chemical reactions by lowering the ______ energy that constitutes an energy barrier between reactants and products.
7. An equimolar mixture of the D and L isomers of an optically active compound is a ______ mixture.
9. Describes a reaction for which the free-energy change (ΔG) is positive.
10. Structural components of membranes; energy storage molecules.
13. The energy or heat content of a system; *H*.
15. Organisms that can synthesize most of the molecules necessary for their growth from simple compounds, such as CO_2 and NH_3.
16. A system that exchanges energy and material with its surroundings is said to be ______.
18. Membrane-bounded compartment, present only in eukaryotes, that contains chromosomes.
20. ______ -synthetic organisms convert solar energy into ATP.

ANSWERS

Do You Know the Terms?

ACROSS

1. energy
4. monomeric
5. entropy
6. protein
8. genome
11. chiral
12. endo
14. oxidation-reduction
17. noncovalent
19. exergonic
21. cytoplasm
22. stereoisomers

DOWN

1. endosymbiotic
2. homologs
3. activation
7. racemic
9. endergonic
10. lipids
13. enthalpy
15. autotrophs
16. open
18. nucleus
20. photo

chapter

2

Water

STEP-BY-STEP GUIDE

Major Concepts

The special properties of water, all of which are derived from its polarity and hydrogen-bonding capability, are central to the behavior of biomolecules dissolved in it.

The shape and polarity of the water molecule are responsible for its hydrogen-bonding capability.

This capability is what gives water its internal cohesion and its versatility as a solvent. An increase in entropy (randomness) of the ions dissolving from a crystal lattice drives the dissolution of crystalline substances, and an increase in entropy of surrounding water molecules drives the behavior of hydrophobic and amphipathic molecules in water. Individual hydrogen bonds, ionic interactions, hydrophobic interactions, and van der Waals interactions are weak relative to covalent bonds. However, cumulatively, these small binding forces result in strong associations that are critical to the function of a biomolecule.

The ionization of water is an experimentally measurable quantity, which is expressed as K_w, the ion product of water.

This K_w is the basis for the **pH scale,** which designates the concentration of H^+ in an aqueous solution. The pH of the environment and the pK_a (the negative log of the equilibrium constant) of each of the participating biomolecules are extremely influential in the reactions of biological systems.

Buffers are mixtures of a weak acid (proton donor) and its conjugate base (proton acceptor).

Buffers will resist changes in pH most effectively when the concentrations of the proton donor and proton acceptor are equal. Under these circumstances, the pH of the solution is equal to the pK_a of the weak acid, as illustrated by the **Henderson-Hasselbalch equation:**

$$\mathrm{pH} = \mathrm{p}K_a + \log \frac{[\text{proton acceptor}]}{[\text{proton donor}]}$$

The phosphate and bicarbonate systems are critical buffer systems in cytoplasm and blood.

Water acts as a reactant as well as a solvent in many biochemical reactions.

A molecule of water is eliminated in condensation reactions, added to bonds in hydrolytic reactions, and split in the process of photosynthesis.

What to Review

The use of significant figures in science is a topic covered in most introductory high school chemistry courses. The rules for determining significant figures are simple, but sometimes forgotten. Rule 1: Nonzero digits are always significant. A number such as 34.78 has four significant figures; 2.85 has three. Rule 2: Any and all zeros between any two other significant digits are significant. A number such as 301 has three significant figures; but 24,360 has only four. Rule 3: When experimental or measured quantities are used in addition or subtraction, the answer should contain the same number of digits after the decimal point as does the least precise quantity. If you add 14.08 m, 6.186 m, and 15 m, the actual sum is 35.266, but should be reported as 35 meters. Rule 4: When multiplying and dividing, the number of significant figures in the result is the same as the smallest number of significant figures in the original set of numbers. Rule 5: Conversion factors, integers, and pure numbers are exact and therefore can be considered to have an infinite number of significant figures. They do not affect the number of significant figures in the result.

Topics for Discussion

Answering each of the following questions, especially in the context of a study group discussion, should help you understand the important points of this chapter.

2.1 Weak Interactions in Aqueous Systems

Hydrogen Bonding Gives Water Its Unusual Properties

1. What is the shape of the water molecule? How is its unequal charge distributed over that shape?

2. How do hydrogen bonds contribute to the high melting and boiling points of water?

3. Why is ice less dense than liquid water?

Water Forms Hydrogen Bonds with Polar Solutes

4. What other functional groups of biomolecules can form hydrogen bonds? Why *don't* —CH groups participate in H bonds?

Water Interacts Electrostatically with Charged Solutes

5. What are some of the functional groups of biomolecules that will interact with water electrostatically?

6. What is the force or strength of the ionic attraction between a Na^+ and Cl^- (10 nm apart) in water? In benzene?

Entropy Increases as Crystalline Substances Dissolve

7. Describe what occurs when a crystalline salt dissolves in water in terms of the enthalpy (H) and the entropy (S) of the system.

Nonpolar Gases Are Poorly Soluble in Water

8. What physical property of oxygen could contribute to its concentration being a limiting factor for aquatic animals in deep water?

Nonpolar Compounds Force Energetically Unfavorable Changes in the Structure of Water

9. What is the driving force behind the formation of micelles?

10. Amphipathicity is important to the structure (and therefore the function) of which biomolecules?

Van der Waals Interactions Are Weak Interatomic Attractions

11. Van der Waals interactions are a weak, transient subcategory of which type of noncovalent interaction?

Weak Interactions Are Crucial to Macromolecular Structure and Function

12. How do the four types of weak interactions among biomolecules compare in strength to each other and to covalent bonds?

13. Why would a zipper or a Velcro® strip be an appropriate analogy to weak interactions in biochemical reactions?

14. How are tightly bound water molecules in DNA and proteins different from "free" water molecules?

Solutes Affect the Colligative Properties of Aqueous Solutions

15. Which of the following would have the greatest effect on the freezing point of a liter of water: the addition of 2 mol of NaCl or 2 mol of glucose?

2.2 Ionization of Water, Weak Acids, and Weak Bases

Pure Water Is Slightly Ionized

16. What specific information does the equilibrium constant of a reaction provide?

The Ionization of Water Is Expressed by an Equilibrium Constant

17. Why is the ion product of water (K_w) at 25 °C always equal to 1×10^{-14}?

The pH Scale Designates the H^+ and OH^- Concentrations

18. How does K_w relate to the pH scale?

Weak Acids and Bases Have Characteristic Dissociation Constants

19. Does a strong acid have a greater or lesser tendency to lose its proton than does a weak acid? Does the strong acid have a higher or lower K_a? A higher or lower pK_a?

Titration Curves Reveal the pK_a of Weak Acids

20. At a pH equal to the pK_a of a weak acid, what can be said about the concentrations of the acid and its conjugate base? What point on a titration curve indicates the pK_a of that weak acid?

2.3 Buffering against pH Changes in Biological Systems

Buffers Are Mixtures of Weak Acids and Their Conjugate Bases

21. What are the two equilibrium reactions that are simultaneously adjusting during an experimental titration of a weak acid?

22. What does the relatively flat zone of a titration curve tell you about the pH changes within that zone?

The Henderson-Hasselbalch Equation Relates pH, pK_a, and Buffer Concentration

23. How does the Henderson-Hasselbalch equation prove that the pK_a of a weak acid is equal to the pH of the solution at the midpoint of its titration?

Weak Acids or Bases Buffer Cells and Tissues against pH Changes

24. What is the importance of the functional groups of proteins that act as weak acids or bases in biological systems?

25. What is the extracellular buffering system that is used by animals with lungs? What is the primary intracellular buffering system?

26. Where and how does the phosphate buffer system function?

27. What are the three reversible equilibria involved in the bicarbonate buffer system in animals with lungs?

Untreated Diabetes Produces Life-Threatening Acidosis

28. What is the most sensitive aspect of cell function (mentioned many times in this chapter) in relationship to changes in pH?

2.4 Water as a Reactant

29. What general types of cellular reactions form water, and which consume water? Which of these reactions are endergonic and which exergonic?

2.5 The Fitness of the Aqueous Environment for Living Organisms

30. What are some of the ways that organisms on earth exploit the special properties of water in terms of temperature regulation?

Discussion Questions for Study Groups

- Consider two sets of solutions: one set is composed of a solution of pH 1 and a solution of pH 2; the second set is composed of a solution of pH 11 and a solution of pH 12. Both sets thus comprise solutions differing from one another in pH by one pH unit. Is the difference in H^+ concentration between the solutions in each set equal? Why or why not?

SELF-TEST

Do You Know the Terms?

ACROSS

3. Describes a solution with a $[H^+]$ of 1×10^{-8}.

4. Hydro _______ molecules can form energetically favorable interactions with water molecules.

6. Water is often referred to as the "universal _______" because of its ability to hydrate molecules and screen charges.

7. Denotes the concentration of H^+ (and therefore of OH^-) in an aqueous solution.

8. The ion product of water; it is 1×10^{-14} M in aqueous solutions at 25 °C.

9. The _______-_______ equation; describes the relationship between pH and the pK_a of a buffer.

10. The equilibrium constant for the reaction $HA \rightleftharpoons H^+ + A^-$ is also called the _______ constant, K_a.

14. Hydro _______ molecules decrease the entropy of an aqueous system by causing water molecules to become more ordered.

15. The numbers 1, 10, 100, and 1000 are placed at equal intervals on a _______ scale.

16. Reaction in which two reactants combine to form a single product with the elimination water.

18. A plot of pH vs. OH^- equivalents added is a _______ curve.

19. Weak interactions that are crucial to the structure and function of macromolecules.

21. Describes a solution in which $[OH^-]$ is greater than $[H^+]$.

22. A mixture of a weak acid and its conjugate base.

24. Enzymes show maximum activity at a characteristic pH _______.

25. HA is a proton _______.

26. The point in a reversible chemical reaction at which the rate of product formation equals the rate of product breakdown to the starting reactants.

27. A^- is a proton _______.

DOWN

1. The _______ radius is approximately twice the distance of a covalent radius for a single bond. (3 words)

2. Stable structures formed by lipids in water, which are held together by hydrophobic interactions.

4. $H_2PO_4^- \rightleftharpoons H^+ + PO_4^{2-}$ describes a _______ buffer system.

5. Compound containing both polar and nonpolar regions.

9. The electrostatic interactions between the hydrogen and oxygen atoms on adjacent H_2O molecules constitute a _______. (2 words)

11. Dissolved molecules.

12. Covalent bond breakage by the addition of water.

13. Water molecules readily dissolve compounds such as NaCl because they screen _______ interactions between Na^+ and Cl^-.

14. pH at which $[HAc] = [Ac^-]$.

17. $H_2CO_3 \rightleftharpoons H^+ + HCO_3^-$ describes a _______ buffer system.

20. Describes a solution in which $[H^+]$ is greater than $[OH^-]$.

23. Noncovalent bonds have weaker bond _______ than covalent bonds.

Do You Know the Facts?

In questions 1–4, decide whether the statement is true or false, and explain your answer.

1. The oxygen atom in water has a partial positive charge.

2. Each hydrogen atom of water bears a partial positive charge.

3. H bonds can form only between water molecules.

4. H bonds are relatively weak compared to covalent bonds.

In questions 5–9, choose the one best response.

5. You are running your first marathon on a very warm day. You start to sweat heavily and realize that you may be in danger of dehydration. Why is severe dehydration potentially life-threatening?
   A. Water is a solvent for many biomolecules.
   B. Water is a chemical participant in many biological reactions.
   C. Water is necessary for buffering action in the body.
   D. Water's attraction to itself drives hydrophobic interactions.
   E. All of the above are true.

6. Which of the following is true of hydrogen bonds?
   A. The attraction between the oxygen atom of a water molecule and the hydrogen atom of another molecule constitutes a hydrogen bond.
   B. Hydrogen bonds form as covalent bonds between positively and negatively charged ions.
   C. Hydrogen bonds form between nonpolar portions of biomolecules.
   D. A and B are true.
   E. A, B, and C are true.

7. Which of the following is true of pH?
   A. pH is the negative logarithm of $[OH^-]$.
   B. Lemon juice, which has a pH of 2.0, is 60 times more acidic than ammonia, which has a pH of 12.0.
   C. Varying the pH of a solution will alter the pK_a of an ionizable group in that solution.
   D. Varying the pH of a solution will alter the degree of ionization of an ionizable group in that solution.
   E. All of the above are true.

8. Consider a weak acid in a solution with a pH of 5.0. Which of the following statements is true?
   A. The weak acid is a proton acceptor.
   B. The weak acid has a lower affinity for its proton than does a strong acid.
   C. At its pK_a, the weak acid will be totally dissociated.
   D. The $[H^+]$ is 10^{-5} M.
   E. All of the above are true.

9. Water derives all its special properties from its:
   A. cohesiveness and adhesiveness.
   B. high boiling point and melting point.
   C. small degree of ionization.
   D. polarity and hydrogen-bonding capacity.
   E. high dielectric constant.

10. The pK_a values for the three ionizable groups on tyrosine are pK_a (—COOH) = 2.2, pK_a ($—NH_3^+$) = 9.11, and pK_a (−R) = 10.07. In which pH ranges will this amino acid have the greatest buffering capacity?
A. At all pH's between 2.2 and 10.07
B. At pH's near 7.1
C. At pH's between 9 and 10
D. At pH's near 5.7
E. Amino acids cannot act as buffers.

11. In a typical eukaryotic cell the pH is usually around 7.4. What is the $[H^+]$ in a typical eukaryotic cell?
A. 0.00000074 M
B. 6.6 μM
C. 4×10^{-8}
D. 2.3 nM
E. 7.4×10^{-5} M

12. Carbonic acid has a K_a of 1.70×10^{-4} and acetic acid has a K_a of 1.74×10^{-5}. Which of the following is true?
A. Carbonic acid has the higher K_a of the two and would therefore be the best buffer at pH 6.
B. The acid with the larger K_a is a better proton acceptor.
C. Carbonic acid is the stronger acid and has a lesser tendency to lose its proton compared to acetic acid.
D. Neither carbonic acid nor acetic acid can be effective buffers at any pH.
E. Acetic acid is a weaker acid and has a lesser tendency to lose its proton compared to carbonic acid.

13. As climbers approach the summit of a mountain they usually increase their rate of breathing to compensate for the "thinner air" due to the lower oxygen pressures at higher elevations. This increased ventilation rate results in a reduction in the levels of CO_2 dissolved in the blood. Which of the following accurately describes the effect of lowering the $[CO_2]_{dissolved}$ on blood pH?
A. It will result in an increase in the dissociation of $H_2CO_3 \rightarrow H^+ + HCO_3^-$ and a drop in pH.
B. It will result in an increase in the dissociation of $H_2CO_3 \rightarrow H_2O + CO_2$ and a drop in pH.
C. It will result in an increase in the association of $H^+ + HCO_3^- \rightarrow H_2CO_3$ and an increase in pH.
D. It will result in a decrease in the association of $H^+ + HCO_3^- \rightarrow H_2CO_3$ and a decrease in pH.
E. Lowering $[CO_2]_{dissolved}$ has no effect on blood pH.

14. You mix 100 ml of solution of pH 1 with 100 ml of a solution of pH 3. The pH of the new 200 ml solution will be:
A. 1.0.
B. 2.0.
C. 3.0.
D. between pH 1.0 and pH 2.0.
E. between pH 2.0 and pH 3.0.

15. The pH of a sample of blood is 7.4; the pH of a sample of gastric juice is 1.4. The blood sample has an $[H^+]$:
A. 5.29 times lower than that of the gastric juice.
B. a million times higher than that of the gastric juice.
C. 6,000 times lower than that of the gastric juice.
D. a million times lower than that of the gastric juice.
E. 0.189 times that of the gastric juice.

Applying What You Know

1. Certain insects can "skate" along the top surface of water in ponds and streams. What property of water allows this feat, and what bonds or interactions are involved?

2. When you are very warm, because of high environmental temperature or physical exertion, you perspire. What property(ies) of water is your body exploiting when it sweats?

3. What is the absolute difference in $[H^+]$ between two aqueous solutions, one of pH 2.0 and one of pH 3.0? What is the $[OH^-]$ of the solution of pH 2.0?

4. Formic acid has a pK_a of 3.75; acetic acid has a pK_a of 4.76. Which is the stronger acid? Does the stronger acid have a greater or lesser tendency to lose its proton than the weaker acid?

5. In a solution of pH 4.76, containing both acetic acid and acetate, what can you say about the concentrations of acetic acid (CH_3COOH) and acetate (CH_3COO^-) present?

6. The Henderson-Hasselbalch equation is $\text{pH} = pK_a + \log \frac{[\text{proton acceptor}]}{[\text{proton donor}]}$.
Show how it proves that the pK_a of a weak acid is equal to the pH of the solution at the midpoint of its titration.

7. Define pH and pK_a and explain how they are different.

ANSWERS

Do You Know the Terms?

ACROSS

3. alkaline
4. philic
6. solvent
7. pH
8. K_w
9. Henderson-Hasselbalch
10. dissociation
14. phobic
15. log
16. condensation
18. titration
19. noncovalent
21. basic
22. buffer
24. optimum
25. donor
26. equilibrium
27. acceptor

DOWN

1. van der Waals
2. micelles
4. phosphate
5. amphipathic
9. hydrogen bond
11. solutes
12. hydrolysis
13. ionic
14. pK_a
17. carbonate
20. acidic
23. energy

Do You Know the Facts?

1. False; the oxygen atom has a partial negative charge.

2. True; because the oxygen atom is more electronegative than the two hydrogen atoms, the electrons are more often in the vicinity of the oxygen, giving each of the hydrogen atoms a partial positive charge.

3. False; they can form between any electronegative atom (usually oxygen or nitrogen) and a hydrogen atom covalently bonded to another electronegative atom in the same or another molecule.

4. True; hydrogen bonds have a bond dissociation energy of 23 kJ/mol, whereas covalent single bonds have a stabilization energy of approximately 200–460 kJ/mol.

5. E
6. A
7. D
8. D
9. D
10. C
11. C
12. E
13. C
14. D
15. D

Applying What You Know

1. Extensive hydrogen binding among water molecules accounts for the surface tension that allows some insects to walk on water.

2. The high heat of vaporization of water, which is a measure of the energy required to overcome attractive forces between molecules, allows your body to dissipate excess heat through the evaporation of the water that is perspired.

3. The solution with a pH of 2.0 has a $[H^+]$ of 10^{-2} M, and an $[OH^-]$ of 10^{-12} M; the solution with a pH of 3.0 has a $[H^+]$ of 10^{-3} M. The difference in $[H^+]$ is 10^{-2} M $-$ 10^{-3} M $=$ 0.009 M.

4. Formic acid is the stronger acid. It has a greater tendency to lose its proton than does acetic acid.

5. At pH 4.76, the concentrations of acetic acid and acetate in the solution will be equal.

6. $$\text{pH} = \text{p}K_a + \log \frac{[\text{proton acceptor}]}{[\text{proton donor}]}$$

At the midpoint of the titration, [proton acceptor] = [proton donor]. The log of 1 = 0, so pH = pK_a + 0; pH = pK_a.

7. pH is the negative logarithm of $[H^+]$ in an aqueous solution. It provides a standard way to measure the H^+ concentration in an aqueous solution. pK_a is the negative logarithm of an equilibrium constant. It is equal to the pH at which a weak acid is one-half dissociated; i.e., the pH at which there are equal concentrations of a weak acid and its conjugate base. The pK_a occurs at the midpoint of the titration curve of a weak acid, the center of the range providing the maximum buffering capacity of the conjugate acid-base pair. The pK_a is an integral property of an ionizable group. It is the extent of ionization of an ionizable group that varies with the pH of the solution.

chapter

3

Amino Acids, Peptides, and Proteins

STEP-BY-STEP GUIDE

Major Concepts

Amino acids share a common structure.

There are 20 standard amino acids found in proteins and they are usually referred to by either a three-letter or a one-letter code. Each amino acid contains a **carboxyl group,** an **amino group,** and a hydrogen atom bonded to a central carbon atom, the α **carbon.** The α carbon is **chiral** when the R group is anything other than a hydrogen. Therefore, with the exception of glycine, there are two **enantiomers** of each amino acid, a D-form and an L-form. Proteins contain only the L isomers of amino acids.

The different chemical properties of amino acids are the result of the different properties of their R groups, which are the basis for categorizing amino acids as nonpolar, aromatic, polar, positively charged, or negatively charged.

Amino acids ionize in aqueous solution.

Both the carboxyl and the amino groups can ionize, and amino acids are capable of acting as both weak acids (proton donors) and weak bases (proton acceptors). The R groups of some amino acids can also ionize, affecting overall acid-base behavior. All of these ionizable groups affect the net charge on each amino acid. Consequently, amino acids are often characterized by the pH at which they have no net charge; this characteristic is their **isoelectric point,** or **pI.**

Polymers of amino acids are polypeptides or proteins.

Amino acids in these polymers are covalently linked through a peptide bond formed by a condensation reaction between the carboxyl group of one amino acid and the amino group of a second amino acid. In general, proteins are very large molecules, generally containing more than 100 amino acid residues. Peptides generally contain fewer than 100 amino acid residues. Small peptides perform a number of biological functions such as intercellular signaling. Some proteins contain non-amino acid prosthetic groups, which usually play an important role in the protein's biological function.

Proteins can be studied using a variety of techniques.

Proteins can be purified on the basis of **solubility, size** (size-exclusion chromatography, gel electrophoresis), **shape** or **binding characteristics** (affinity chromatography), or **charge** (anion- or cation-exchange chromatography, isoelectric focusing). Often, purification requires multiple steps, each step taking advantage of a different characteristic of the protein being purified.

A protein's function depends on its primary structure, which can be determined experimentally.

Protein structure is described in terms of four levels: primary (1°), secondary (2°), tertiary (3°), and quaternary (4°). Primary structure is the specific number and sequence of amino acids in a specific protein. The location of disulfide bonds is also a part of the primary structure of a protein. The primary structure determines how a protein will fold into its three-dimensional, functional form. Altering the primary structure in critical areas can therefore affect the protein's function. Primary structure is specified by DNA coding regions; therefore, DNA sequence information provides a fast and accurate method for determining the amino acid sequence of a protein. Alternatively, the amino acid sequence can be determined directly by first breaking disulfide bonds, cleaving the protein into small fragments, sequencing the fragments by the Edman degradation procedure, and then reconstructing the amino acid sequence of the intact protein.

Proteins that have similar functions in different species have similar amino acid sequences.

Typically, stretches of amino acid sequences are invariant in the same proteins from different species. These invariant residues are usually found in functionally critical regions.

What to Review

Answering the following questions and reviewing the relevant concepts, which you have already studied, should make this chapter more understandable.

- Review the chemical properties of the biologically important functional groups (Fig. 1–15), especially $—NH_3^+$, and $—COO^-$. How do these functional groups behave in aqueous solution (pp. 47–50)?
- Be sure you understand the difference between a strong acid or base and a weak acid or base (pp. 56–59). Into which category do $—COO^-$ and $—NH_3^+$ groups fit?
- What are buffers? How do they work? How can a titration curve tell you at what pH a molecule will act as an effective buffer (Fig. 2–17)? Use this information to interpret the titration curve of glycine in this chapter (Fig. 3–10).
- Using models if necessary, satisfy yourself that enantiomers are *not* superimposable and cannot function as identical molecules (Fig. 3–3).
- Review condensation reactions. The formation of a peptide bond is just one of many examples of monomers joining to form polymers through condensation reactions.
- Review what a positive free-energy change means in terms of the spontaneity of a reaction (pp. 22–24). Does this mean that an endergonic reaction cannot occur? If not, what circumstances must prevail in order for such a reaction to occur?

Topics for Discussion

Answering each of the following questions, especially in the context of study group discussions, should help you understand the important points of the chapter.

3.1 Amino Acids

Amino Acids Share Common Structural Features

1. Why is glycine *not* optically active?

Not chiral

2. Both the three-letter and one-letter abbreviations for the 20 standard amino acids are very commonly used in the biochemical literature and should be committed to memory. Use the Worksheet in the Self-Test to help you memorize the various amino acid structures, names, and abbreviations.

Use "amino acid quiz" as a search term to find interactive quizzes on the Internet.

The Amino Acid Residues in Proteins Are L-Stereoisomers

3. Could a life form that evolved on a planet where all proteins were made up of D-amino acids survive on Earth? Why or why not? (More on this topic in Chapter 4.)

Amino Acids Can Be Classified by R Group

4. The structures and R-group classifications of the 20 standard amino acids are essential parts of a working vocabulary for biochemistry, and should be committed to memory. Look at different amino acid structures and be sure you can identify the α carbon, the carboxyl group, the amino group, and the R group in each.

5. Which amino acid R groups are capable of forming hydrogen bonds? Which will promote hydrophobic interactions? Which amino acid fits into both of these categories?

6. Which amino acid allows the most structural flexibility when found in a protein? Which allows the least? Which is the only amino acid that forms disulfide bridges?

7. Does the ability of some amino acids to absorb ultraviolet light translate into a useful technique for the detection of *all* proteins? Why or why not?

8. What are the pK_a values of R groups on the charged amino acids (see Table 3–1)? Which amino acid has an R group with a pK_a value near the pH of most living systems?

Box 3–1 Absorption of Light by Molecules: The Lambert-Beer Law

9. The measurement of protein concentration is a common laboratory procedure. Measurements of light absorption by unknown samples are compared to a standard curve that can be generated by measuring the absorption of solutions with known amounts of protein. Why must these standard solutions be made in the same solvent and at the same pH as that of the unknown protein solutions?

The absorbance, A, is directly proportional to the concentration of the absorbing solute

Uncommon Acids Also Have Important Functions

10. Proteins can contain nonstandard amino acids that are modified versions of the 20 amino acids discussed in this chapter. Do these modifications occur before or after the amino acids are incorporated into the protein molecule? Are amino acids always found incorporated into polypeptide chains in cells?

Amino Acids Can Act as Acids and Bases

11. How does the zwitterionic nature of individual amino acids in solution affect their solubility in water? Why *must* individual amino acids be water soluble? (Note that although R groups vary in water solubility, *individually all* amino acids having a single amino group and a single carboxyl group are zwitterionic in neutral aqueous solution.)

12. Are amino acids weak or strong acids and bases?

Amino Acids Have Characteristic Titration Curves

13. Over what pH range(s) is glycine an effective buffer? Over what pH range(s) is glycine not at all an effective buffer?

Titration Curves Predict the Electric Charge of Amino Acids

14. At approximately what pH would you expect glycine to have a net charge of -1?

Amino Acids Differ in Their Acid-Base Properties

15. Note that the pI values for glutamate and histidine are *not* the average of the pK_a values for each of their three ionizable groups. Can these pI values be calculated as the average of some set of pK_a values? Does this calculation hold for pI values of the other charged amino acids?

3.2 Peptides and Proteins

Peptides Are Chains of Amino Acids

16. What is lost from amino acids in the formation of a peptide bond?

Go to www.rschb.org and type "1 × y2" in the search box. Click on **Display Molecule** and use Rasmol or Swiss-PDB viewer to view the peptide oxytocin. Click on the MDL logo to display a drop-down menu. Try **Color** or **Display** to verify that this is an octapeptide.

Peptides Can Be Distinguished by Their Ionization Behavior

17. Is the tetrapeptide Ala–Glu–Gly–Lys (see Fig. 3–15) a good buffer at pH 7.0? Will it move in an electric field at pH 7.0?

Biologically Active Peptides and Polypeptides Occur in a Vast Range of Sizes and Compositions

18. List some functions of small bioactive peptides.

19. Provide an example of a very large protein and a very small protein.

Investigating Proteins with Mass Spectrometry

20. Why is the partial sequencing of protein fragments by mass spectrometry such a useful tool for modern proteome research?

21. How might the number, sequence, and properties of amino acids in a protein affect its structure and function?

Some Proteins Contain Chemical Groups Other Than Amino Acids

22. Name five or six different types of prosthetic groups that can be attached to proteins.

23. If you divide the molecular weights of the proteins listed in Table 3–2 by the average MW of an amino acid residue, you will find that the number of residues thus calculated does not always agree with that reported. What are two possible explanations for this discrepancy?

3.3 Working with Proteins

Proteins Can Be Separated and Purified

As you read the rest of this chapter you will be able to answer the following questions about protein purification.

24. Which protein purification and separation techniques separate proteins on the basis of size?

25. What technique might you use to separate two proteins that comigrate on an SDS polyacrylamide gel?

26. Which techniques separate proteins on the basis of charge?

27. Which techniques separate proteins on the basis of solubility?

28. Which techniques separate proteins on the basis of binding specificity?

Proteins Can Be Separated and Characterized by Electrophoresis

29. What does a single band on a protein gel represent?

30. What is SDS, and why is it used in electrophoretic procedures?

31. Explain why two-dimensional gel electrophoresis is a more sensitive analytical method than one-dimensional gel electrophoresis.

Unseparated Proteins Can Be Quantified

32. When you finish this chapter, make a list of different methods for quantifying the amount of a protein in a complex solution, such as a blood sample or tissue homogenate.

33. When scientists study the activities of proteins such as the enzyme chymotrypsin it is important to be able to measure the amount of a protein. Why?

3.4 The Structure of Proteins: Primary Structure

The Function of a Protein Depends on Its Amino Acid Sequence

34. How does information about primary structure contribute to an understanding of a protein's function?

The Amino Acid Sequences of Millions of Proteins Have Been Determined

35. Use the amino acid sequence of insulin (Fig. 3–24) to quiz yourself: What are the one-letter abbreviations and R-group classifications for each amino acid listed?

Short Polypeptides Are Sequenced with Automated Procedures

36. Outline the individual steps required to sequence a protein.

Large Proteins Must Be Sequenced in Smaller Segments

37. What general steps must be taken before larger proteins can be sequenced using automated procedures?

38. Placing peptide fragments in their proper order is like piecing together a puzzle. Be sure to understand the example in Figure 3–37; there is a similar puzzle in the Self-Test.

Amino Acid Sequences Can Also Be Deduced by Other Methods

39. What information concerning the primary structure of a protein would *not* be available from the DNA sequence?

Small Peptides and Proteins Can Be Chemically Synthesized

40. Explain how the use of a solid support for chemical synthesis of polypeptides eliminates a step of traditional organic synthesis.

Amino Acid Sequences Provide Important Biochemical Information

41. Return to an earlier question (#34) and now expand on your answer: In what ways does information about primary structure contribute to an understanding of a protein's function?

Protein Sequences Can Elucidate the History of Life on Earth

42. When using the primary sequences of proteins to study evolution, one can often obtain misleading information; why would comparisons of some proteins not provide useful information?

43. What constitutes a conservative substitution for an amino acid residue in a protein?

Box 3–3 Consensus Sequences and Sequence Logos

44. In Figure 1b, what does the presence of the large, red D at position 1 suggest about the characteristics of this end of the protein sequence? What does {ILVFYW} indicate?

Discussion Questions for Study Groups

- Refer to Table 3–1. Note the pK_a values for the —COOH on the R group of aspartate and glutamate. Why are these pK_a values higher than those of —COOH groups bonded to the α carbon of amino acids?
- Why is the ability to synthesize large proteins a useful tool for understanding the structure and function of proteins such as enzymes? How might you use this capability to study the function of an enzyme?
- What are the reasons that proteins with very similar functions might be coded for by very different DNA sequences?
- The consensus sequence C-X(2-4)-C-X(3)-F-X(5)-L-X-(2)-H-X(3)-H represents a portion of a DNA binding protein that forms a finger-like projection. This structure is stabilized by the binding of a zinc ion to four of the amino acids in this sequence. Which amino acids are likely to be crucial to the stability of this loop and why? Is this loop a secondary or tertiary structure? What key term from this chapter describes the zinc ion? Hint: go to www.rcsb.org and look and up "4znf" to see what a "zinc finger" structure looks like.
- Which of the following consensus sequences might you expect to find in a paralog of the zinc finger protein just discussed?
  C-X(4)-C-X(12)-C-X3-C
  or
  C-X4-C-X(2)-H-X3-C

SELF-TEST

Worksheet

The following worksheet helps you memorize the structures, one- and three-letter abbreviations, and R-group category of each amino acid. The first example is filled in as a guide. (Note: many students find it useful to make "flash cards" of the amino acid structures in order to facilitate memorization.)

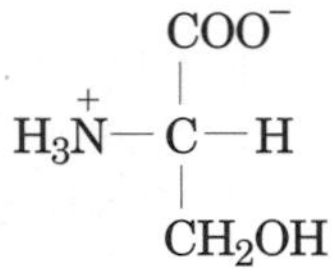

COO^-
$H_3\overset{+}{N}—C—H$
CH_2OH

Name	Serine
1 letter	S
3 letter	Ser
Group	Polar, uncharged

COO^-
$H_3\overset{+}{N}—C—H$
H

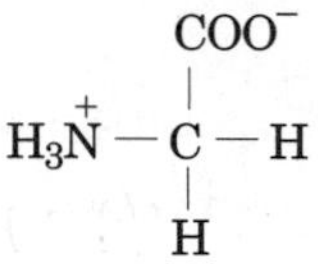

Name	Glycine
1 letter	G
3 letter	Gly
Group	Nonpolar, aliphatic

COO^-
$H_3\overset{+}{N}—C—H$
$H—C—OH$
CH_3

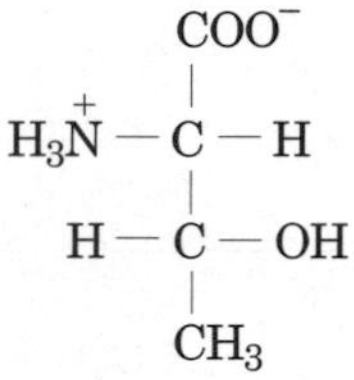

Name	Threonine
1 letter	T
3 letter	thr
Group	Polar, uncharged

COO^-
$H_3\overset{+}{N}—C—H$
CH_2
COO^-

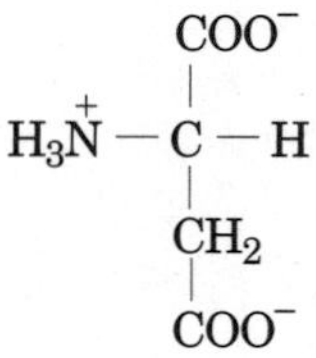

Name	Aspartate
1 letter	D
3 letter	Asp
Group	Negatively charged

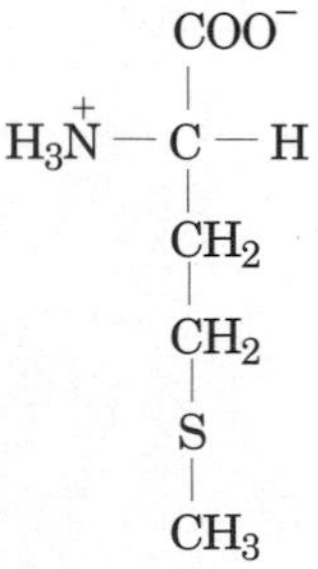

COO^-
$H_3\overset{+}{N}—C—H$
CH_2
CH_2
S
CH_3

Name	Methionine
1 letter	M
3 letter	Met
Group	Nonpolar, aliphatic

COO^-
$H_3\overset{+}{N}—C—H$
CH_2
CH_2
CH_2
CH_2
$^+NH_3$

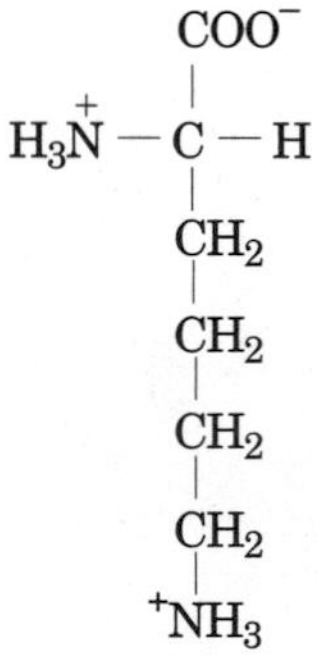

Name	Lysine
1 letter	K
3 letter	Lys
Group	Positively charged

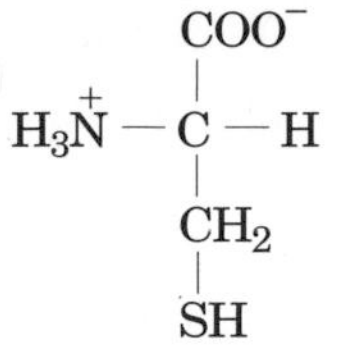

COO^-
$H_3\overset{+}{N}-C-H$
CH_2
SH

Name	Cysteine
1 letter	C
3 letter	Cys
Group	Polar, uncharged

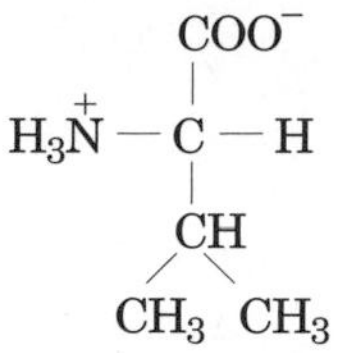

COO^-
$H_3\overset{+}{N}-C-H$
CH
$CH_3\ \ CH_3$

Name	Valine
1 letter	V
3 letter	Val
Group	Nonpolar, aliphatic

COO^-
$H_3\overset{+}{N}-C-H$
CH
$CH_3\ \ CH_3$

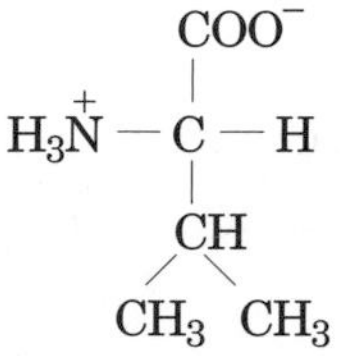

Name	
1 letter	
3 letter	
Group	

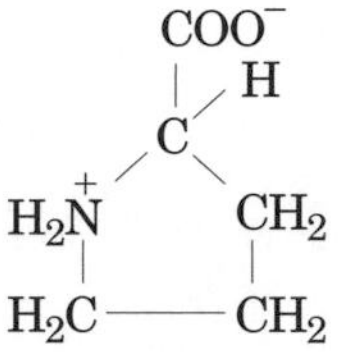

COO^-, H
C
$H_2\overset{+}{N}$ CH_2
H_2C — CH_2

Name	Proline
1 letter	P
3 letter	Pro
Group	Nonpolar, aliphatic

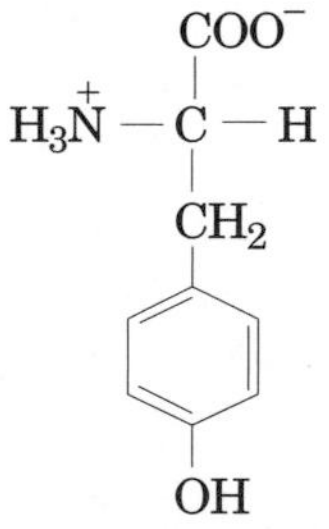

COO^-
$H_3\overset{+}{N}-C-H$
CH_2
OH

Name	Tyrosine
1 letter	Y
3 letter	Tyr
Group	Aromatic

COO^-
$H_3\overset{+}{N}-C-H$
CH_2
$C-NH$
CH
$C-N$
H

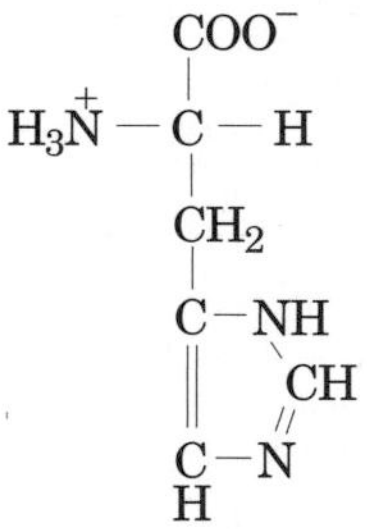

Name	Histidine
1 letter	H
3 letter	His
Group	Positively charged

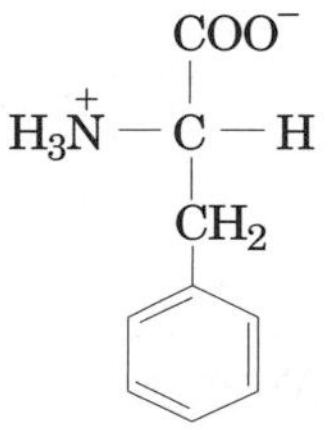

COO^-
$H_3\overset{+}{N}-C-H$
CH_2

Name	Phenylalanine
1 letter	F
3 letter	Phe
Group	Aromatic

COO^-
$H_3\overset{+}{N}-C-H$
CH_2
C
H_2N O

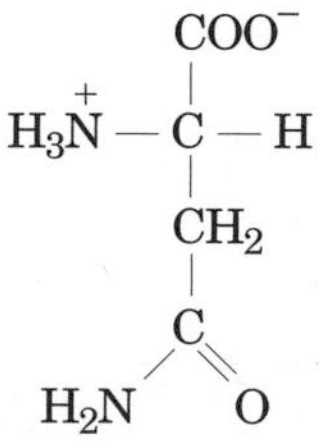

Name	Asparagine
1 letter	N
3 letter	Asn
Group	Polar, uncharged.

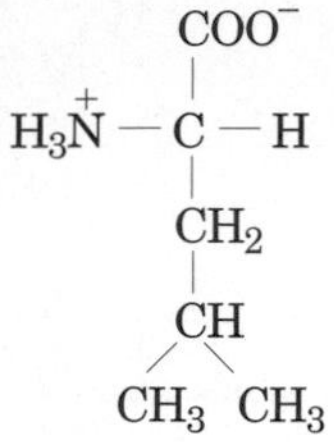

COO^-
$H_3\overset{+}{N}-C-H$
CH_2
CH
CH_3 CH_3

Name	Leucine
1 letter	L
3 letter	Leu
Group	Nonpolar, aliphatic

COO^-
$H_3\overset{+}{N}-C-H$
CH_2
$C=CH$
NH

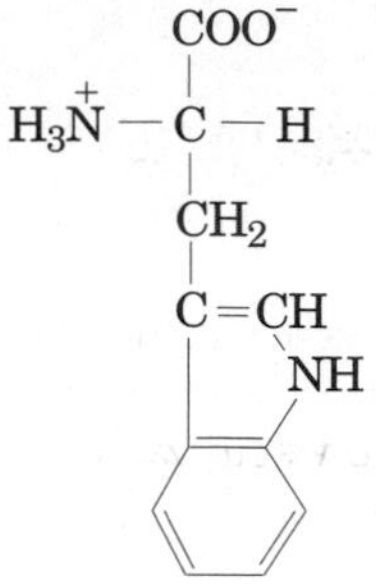

Name	Tryptophan
1 letter	W
3 letter	Trp
Group	Aromatic

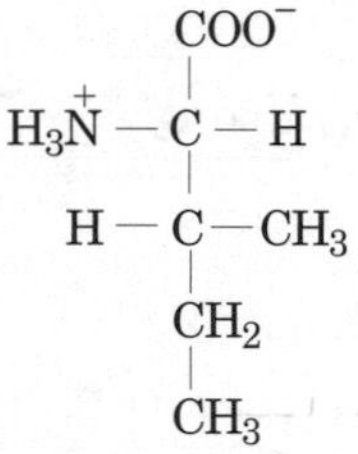

COO^-
$H_3\overset{+}{N}-C-H$
$H-C-CH_3$
CH_2
CH_3

Name	Isoleucine
1 letter	I
3 letter	Ile
Group	Nonpolar, Aliphatic

COO^-
$H_3\overset{+}{N}-C-H$
CH_2
CH_2
COO^-

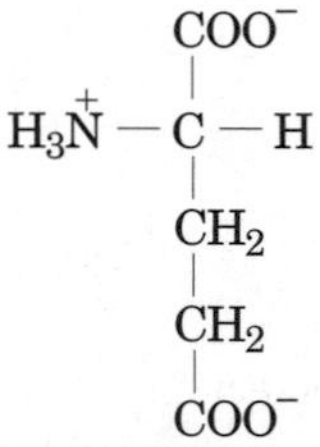

Name	Glutamate
1 letter	E
3 letter	Glu
Group	Negatively charged

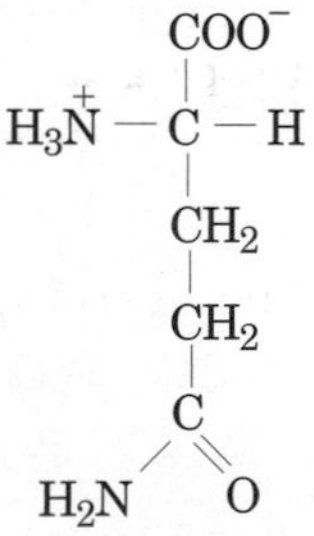

COO^-
$H_3\overset{+}{N}-C-H$
CH_2
CH_2
C
H_2N O

Name	Glutamine
1 letter	Q
3 letter	Gln
Group	Polar, uncharged

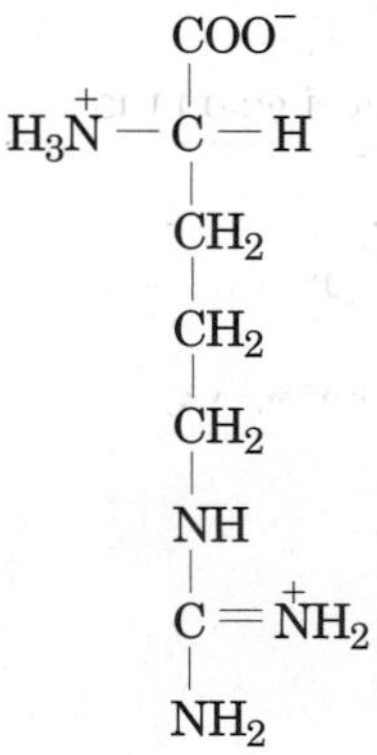

COO^-
$H_3\overset{+}{N}-C-H$
CH_2
CH_2
CH_2
NH
$C=\overset{+}{N}H_2$
NH_2

Name	Arginine
1 letter	R
3 letter	Arg
Group	Positively charged

SELF-TEST

Do You Know the Terms?

ACROSS

1. The lipid portion of a lipoprotein is known as a _______ group.
4. A covalent bond between two nonadjacent cysteines in a polypeptide chain is a _______ bond.
6. An example of a(n) _______ amino acid is histidine, which can either accept protons or donate them at a pH that is close to physiological pH values.
9. All stereoisomers must have at least one _______ center.
11. _______ out is a technique that selectively precipitates some proteins, while others remain in solution. Ammonium sulfate ($(NH_4)_2SO_4$) is often used for this purpose.
12. The bond type that forms the primary structure.
13. The pH at which the numbers of positive and negative charges on an amino acid are equal is referred to as the _______ point or pI.
15. A single unit within a polymer: for example, lysine in a protein molecule.
16. At pH 7, any amino acid with an uncharged R group is a _______.
17. Proteins from different species that have similar amino acid sequences and functions are referred to as _______.

DOWN

2. A reagent used in electrophoresis to separate polypeptides on the basis of mass. (abbr.)
3. Many types of separation can be done using this chromatographic technique; its advantage lies in the reduction of transit time on the column, limiting diffusional spreading of protein bands and improving resolution. (abbr.)
4. After "salting out" proteins, removal of excess ammonium sulfate can be accomplished by _______ of the protein-salt solution overnight against large volumes of buffer.
5. The whole assortment of proteins in an organism; analogous to the genome.
7. A linear chain of amino acid residues that usually has a molecular weight less than 10,000 daltons.
8. The _______ degradation procedure provides information about a protein's primary structure.
10. Insulin obtained from sheep can be used to treat human diabetics because sheep and human insulin are _______ proteins.
14. Hemoglobin, which contains two sets of identical subunits, is often referred to as a(n) _______.

Do You Know the Facts?

1. Draw the chemical structures of the following amino acids. Give their one- and three-letter abbreviations, and R-group classification.

(a) Alanine

(b) Proline

(c) Cysteine

(d) Phenylalanine

(e) Lysine

(f) Aspartate

2. Name the following structures. Give their one- and three-letter abbreviations, and R-group classifications.

(a)

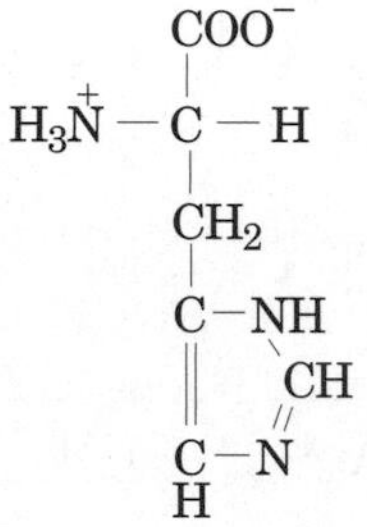

COO^-
$H_3\overset{+}{N}-C-H$
CH_2
$C-NH$
$\|\quad CH$
$C-N$
H

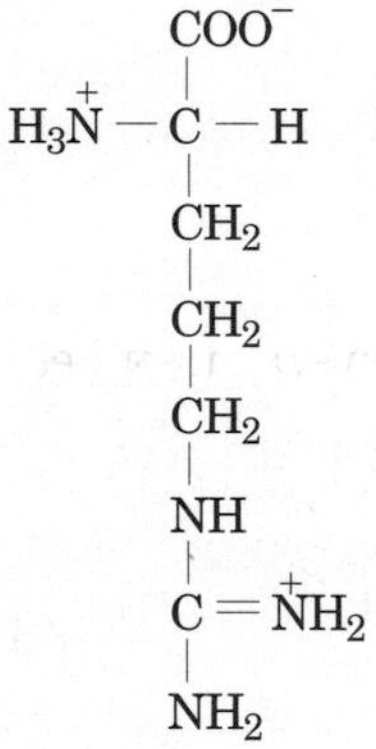

(b)

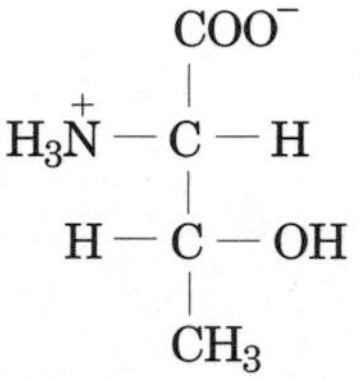

COO^-
$H_3\overset{+}{N}-C-H$
$H-C-OH$
CH_3

(c)

COO^-
$H_3\overset{+}{N}-C-H$
CH_2
CH_2
CH_2
NH
$C=\overset{+}{N}H_2$
NH_2

(d)

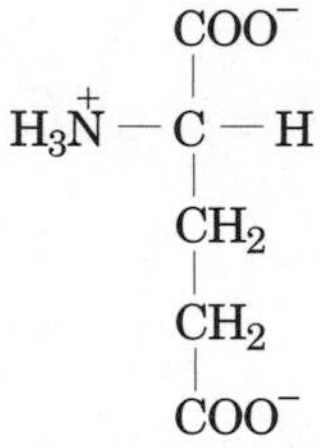

COO^-
$H_3\overset{+}{N}-C-H$
CH_2
CH_2
COO^-

3. Name the amino acid(s) described.

(a) Provides the least amount of steric hindrance in proteins.

(b) Positively charged at physiological pH.

(c) Aromatic R group; hydrophobic and neutral at any pH.

(d) Saturated hydrocarbon R group; important in hydrophobic interactions.

(e) The only amino acid having an R group with a pK_a near 7; important in the active site of some enzymes.

Histidine

(f) The only amino acid with a substituted α-amino group; influences protein folding by forcing a bend in the chain. Proline

4. What is the amino acid sequence of the peptide abbreviated KQNY?

Lysine - Glutamine - Asparagine - Tyrosine

5. What is the amino acid sequence of the peptide abbreviated RLWEQ?

Arginine - Leucine - Tryptophan - Glutamate - Glutamine

6. What is the one-letter abbreviation for the amino acid that has an aromatic R group capable of forming hydrogen bonds?

tryptophan

7. What is the one-letter abbreviation for the amino acid that has a sulfur-containing R group and is able to form a disulfide bridge?

C

8. Which of the following statements describe(s) the peptide bond? (More than one answer may be correct.)

A. It is the only covalent bond between amino acids in polypeptides.
(B) It is a substituted amide linkage.
(C) It is formed in a condensation reaction.
D. It is formed in an exergonic reaction.
E. It is unstable under physiological conditions.

9. Which of the following explains why all individual amino acids are soluble in water but not all peptides are soluble?

(A) Individual amino acids are zwitterions at physiological pHs.
B. All peptides are insoluble in water.
C. The R groups of the amino acid residues in the peptide are charged at physiological pHs.
D. All the amino acid residues in the peptide are zwitterions at physiological pHs.
E. The R groups on all amino acids can interact noncovalently with water at pH 7.4.

10. Name the techniques described for separating cellular proteins.

(a) Taking advantage of unique structural or functional properties of a protein, this technique specifically removes the protein of interest from a solution.

(b) Proteins leave the mobile phase, associating with a negatively charged immobile substrate, such as a bead or resin.

(c) Proteins are separated on the basis of their ability to migrate in an electric field, an indicator of relative size.

(d) Proteins are chromatographically separated solely on the basis of size.

Refer to the table below for questions 11 and 12.

	Molecular Weight	Number of Residues	Number of Polypeptide Chains
Cytochrome *c* (human)	13,000	104	1
Myoglobin (equine heart)	16,890	153	1
Serum albumin (human)	68,500	~550	1
Apolipoprotein B (human)	513,000	4,536	1

11. Which protein is likely to be retained the longest on (i.e., will elute last from) a size-exclusion chromatographic column? Why?

12. Which protein would show up as the band at the top of an SDS polyacrylamide gel after electrophoresis? Why?

Applying What You Know

1. What do the amino acids threonine and tyrosine have in common?

hydroxy group

2. What makes the amino acid cysteine important? Can methionine perform the same function?

disulfide bonds, and no

3. What structural property of amino acids permits the measurement of protein concentration by UV light absorption? Which amino acids have this property?

Aromatics

F, Y, W

280nm

4. Indicate whether the following statements concerning histidine are true or false. Refer to the following titration curve.

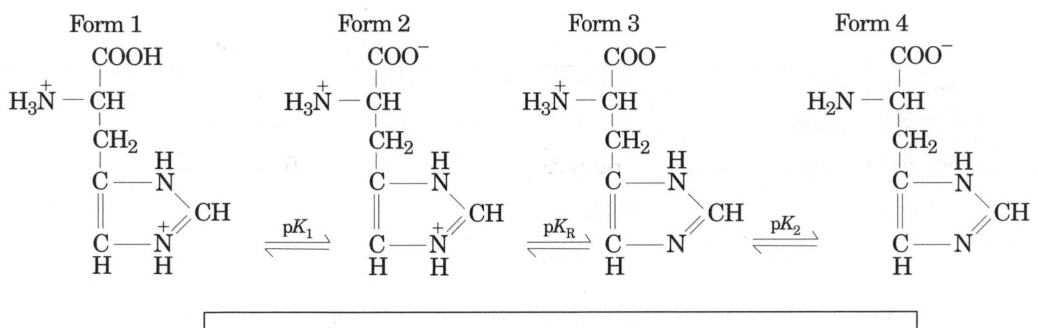

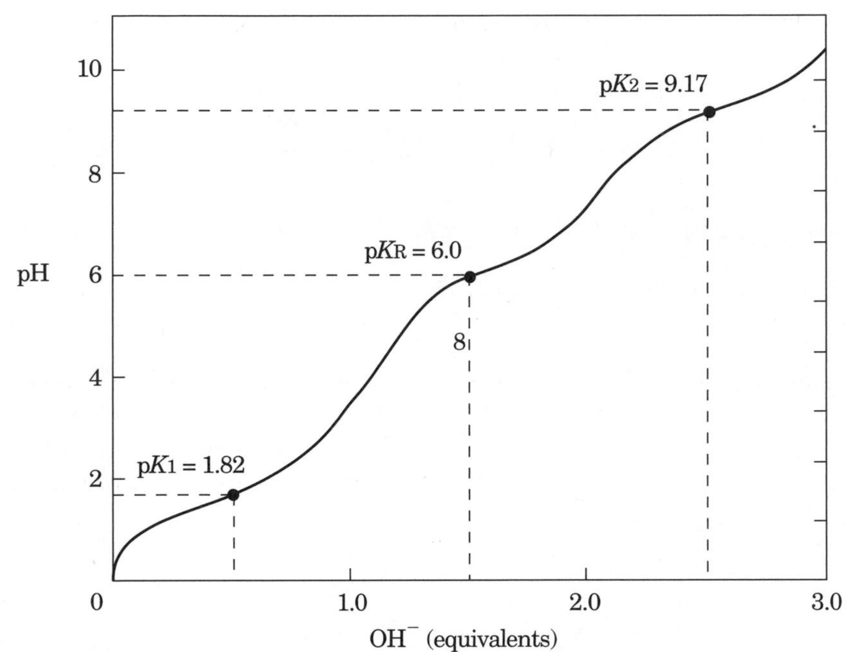

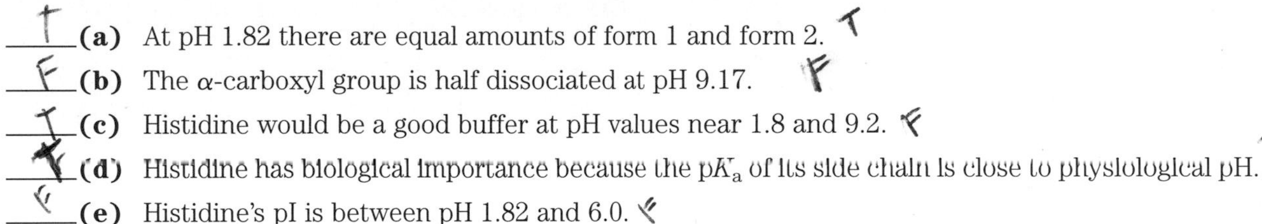

______ **(a)** At pH 1.82 there are equal amounts of form 1 and form 2.

______ **(b)** The α-carboxyl group is half dissociated at pH 9.17.

______ **(c)** Histidine would be a good buffer at pH values near 1.8 and 9.2.

______ **(d)** Histidine has biological importance because the pK_a of its side chain is close to physiological pH.

______ **(e)** Histidine's pI is between pH 1.82 and 6.0.

______ **(f)** Histidine's pI is between pH 6.0 and 9.17.

______ **(g)** Histidine is an aromatic amino acid.

5. Indicate whether the following statements concerning glutamate are true or false. Refer to the following titration curve.

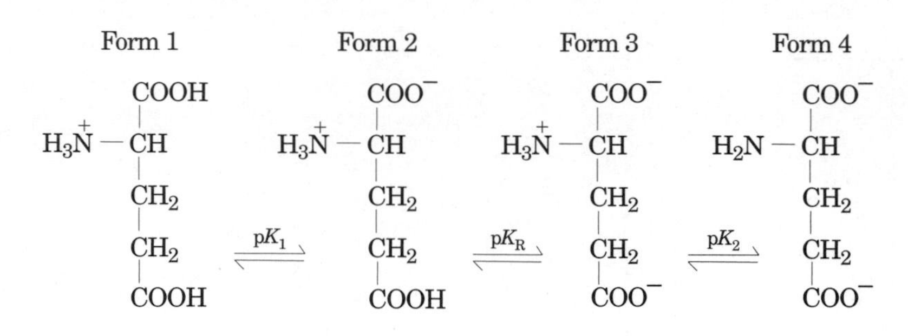

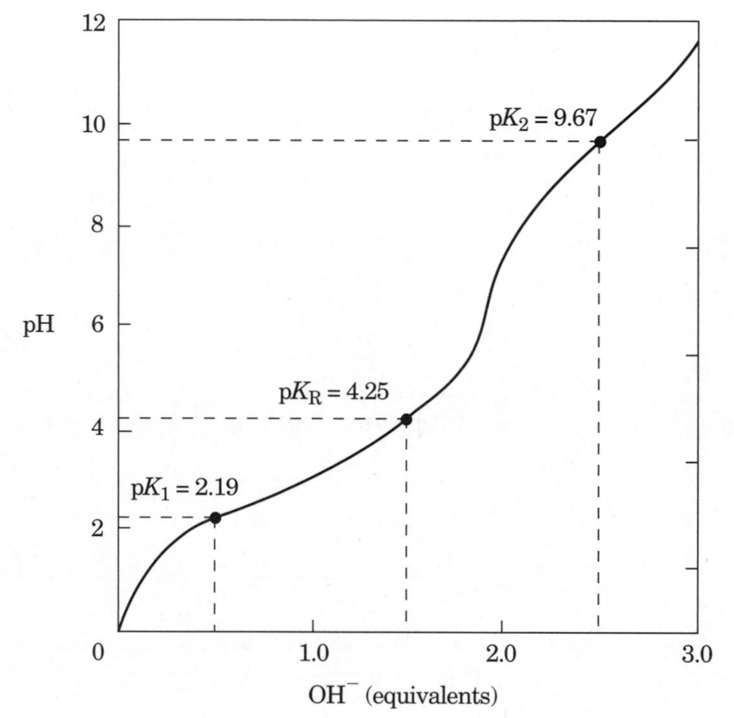

______**(a)** At pH 2.19 there are equal amounts of form 1 and form 2.

______**(b)** The α-carboxyl group is half dissociated at pH 9.67.

______**(c)** Glutamate would be a good buffer at pH values near 2.2 and 9.7.

______**(d)** Glutamate has biological importance because the pK_a of its side chain is close to physiological pH.

______**(e)** Glutamate's pI is between pH 2.19 and 4.25.

______**(f)** Glutamate's pI is between pH 4.25 and 9.67.

______**(g)** Glutamate is an aromatic amino acid.

6. Refer to the following table and titration curve to answer the questions below.

Compound	pK_1	pK_2	pK_R
Lactate	3.86		
Acetate	4.76		
Glycine	2.34	9.60	
Glutamate	2.19	9.67	4.25
Arginine	2.17	9.04	12.48

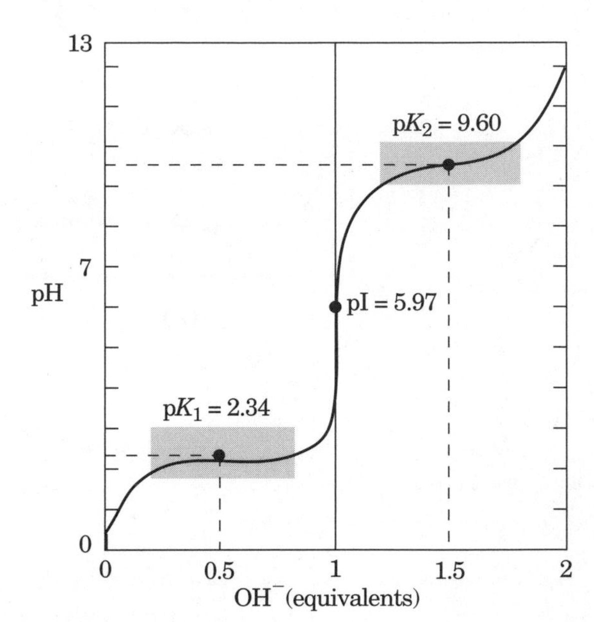

(a) Which compound from the table is illustrated by the titration curve?

(b) Which compound from the table has a net charge of zero nearest to physiological pH?

7. (a) Name the peptide shown below.

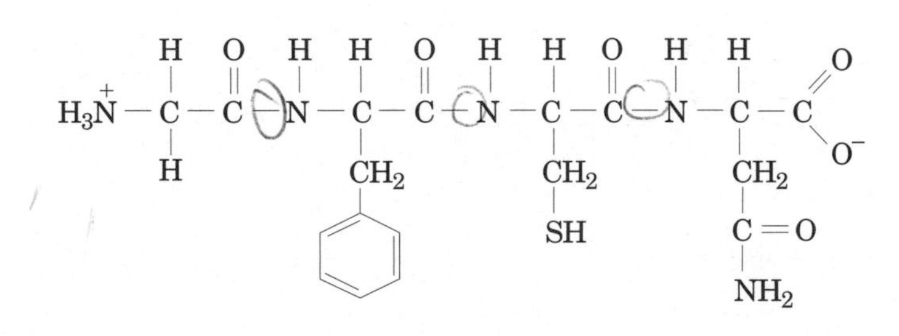

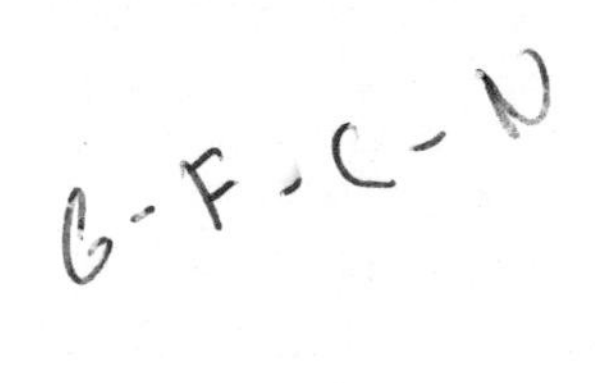

(b) How many peptide bonds are present in this peptide?

(c) How many disulfide bonds are present in this peptide?

8. Calculate the net charge on the following tripeptides at pH 7.0:

(a) Asp-Glu-Ser **(b)** Ser-Gly-Thr **(c)** Gly-Lys-Arg

(d) Which tripeptide will be retained the longest on a cation-exchange chromatographic column in a pH 7.0 buffer?

9. Although gel electrophoresis allows researchers to identify specific protein bands on a gel and provides approximate molecular weights for proteins, it does not provide any information about amino acid sequence. Sequence information is extremely valuable in that it can be used to predict protein structure and function. Assume you have isolated a relatively abundant protein, and you want to obtain the amino acid sequence. You perform the following experiments:

(a) Addition of dithiothreitol to the protein sample followed by gel electrophoresis results in the protein gel shown on the next page. What can you conclude?

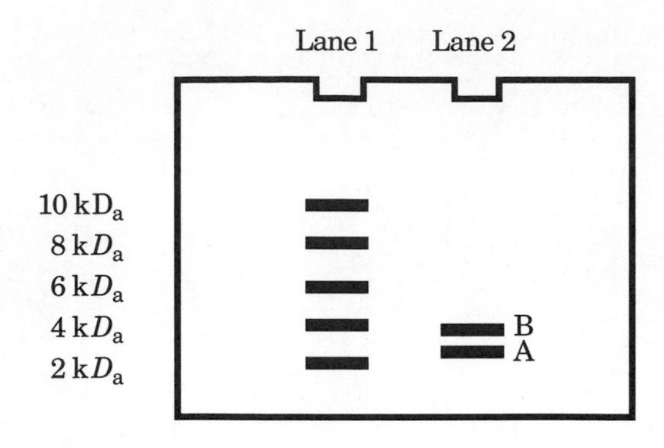

Lane 1 contains molecular weight markers as indicated to the left of the figure.
Lane 2 contains the DTT-treated, purified protein.

(b) Cleavage with chymotrypsin produces the following fragments:

Band A CN
NLQNY
GIVEQCCHKRCSEY

Band B F
Y
DPTKM
IACGVRGF
RTTGHLCGKDLVNALY

Cleavage with *Staphylococcus aureus* V8 protease produces the following fragments:

Band A GIVE
YNLQNYCN
QCCHKRCSE

Band B PTKM
RTTGHLCGKD
LVNALYIACGVRGFFYD

What is the amino acid sequence of your isolated protein?

ANSWERS

Worksheet

1. Refer to Figure 3–5 for the identification and classification of each amino acid structure. Refer to Table 3–1 for the one- and three-letter codes.

Do You Know the Terms?

ACROSS

1. prosthetic
4. disulfide
6. amphoteric
9. chiral
11. salting
12. peptide
13. isoelectric
15. residue
16. zwitterion
17. orthologs

DOWN

2. SDS
3. HPLC
4. dialysis
5. proteome
7. polypeptide
8. Edman
10. homologous
14. oligomer

Do You Know the Facts?

1. **(a)**

$$\begin{array}{c} COO^- \\ | \\ H_3\overset{+}{N}-C-H \\ | \\ CH_3 \end{array}$$

Alanine, Ala, A, nonpolar

(b)

$$\begin{array}{c} COO^- \\ |\ /H \\ C \\ /\quad\backslash \\ H_2\overset{+}{N} \qquad CH_2 \\ | \qquad\qquad | \\ H_2C - CH_2 \end{array}$$

Proline, Pro, P, nonpolar

(c)

$$\begin{array}{c} COO^- \\ | \\ H_3\overset{+}{N}-C-H \\ | \\ CH_2 \\ | \\ SH \end{array}$$

Cysteine, Cys, C, polar, uncharged

(d)

$$\begin{array}{c} COO^- \\ | \\ H_3\overset{+}{N}-C-H \\ | \\ CH_2 \\ | \\ C_6H_5 \end{array}$$

Phenylalanine, Phe, F, aromatic

(e)

$$\begin{array}{c} COO^- \\ | \\ H_3\overset{+}{N}-C-H \\ | \\ CH_2 \\ | \\ CH_2 \\ | \\ CH_2 \\ | \\ CH_2 \\ | \\ ^+NH_3 \end{array}$$

Lysine, Lys, K, positively charged

(f)

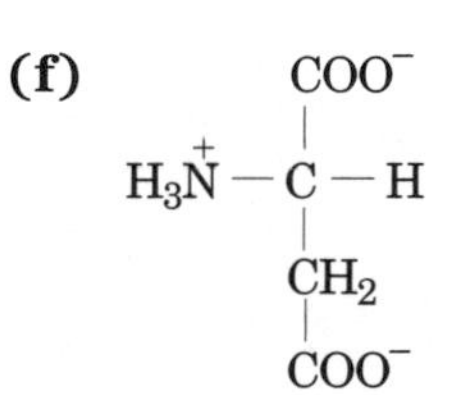

$$\begin{array}{c} COO^- \\ | \\ H_3\overset{+}{N}-C-H \\ | \\ CH_2 \\ | \\ COO^- \end{array}$$

Aspartate, Asp, D, negatively charged

2. **(a)** Histidine, His, H, positively charged
   (b) threonine, Thr, T, polar, uncharged
   (c) arginine, Arg, R, positively charged
   (d) glutamate, Glu, E, negatively charged

3. **(a)** Glycine; **(b)** lysine, arginine, histidine (possible); **(c)** phenylalanine, tryptophan; **(d)** alanine, valine, leucine, isoleucine; **(e)** histidine; **(f)** proline

4. Lysine, glutamine, asparagine, tyrosine

5. Arginine, leucine, tryptophan, glutamate, glutamine

6. Y

7. C

8. B, C

9. A

10. **(a)** affinity chromatography; **(b)** cation exchange chromatography; **(c)** electrophoresis; **(d)** gel filtration.

11. Cytochrome *c*. Size-exclusion chromatography separates molecules on the basis of size. Larger molecules are unable to fit into the tiny pores in the chromatographic beads, and so are excluded based on their size. These larger molecules take a direct route through the column, eluting before smaller molecules, which are detoured into the bead pores. The smallest protein, cytochrome *c*, would be retained the longest and elute last. (This result assumes that all of proteins in the mixture are globular.)

12. Apolipoprotein B. SDS polyacrylamide gel electrophoresis is used to separate proteins on the basis of molecular weight. SDS molecules, which are negatively charged, bind to proteins uniformly along their length, rendering the intrinsic charge of the protein insignificant. Proteins in SDS thus assume a similar shape and have a net negative charge roughly proportional to their mass. Electrophoresis in the presence of SDS separates proteins almost exclusively on the basis of mass, with smaller polypeptides migrating more rapidly to the bottom of the gel when an electric field is applied, and larger proteins moving through the gel much more slowly.

Applying What You Know

1. Both have an —OH group, which contributes polarity and hydrogen-bonding capability.

2. Cysteine's sulfhydryl group readily oxidizes to form a covalently linked disulfide bridge, a dimeric amino acid called cysteine. Cysteine residues stabilize the structures of proteins in which they occur. Methionine cannot form disulfide bridges.

3. The aromatic rings of the side chains of Trp and Tyr, and to a much lesser extent Phe, absorb light strongly around 280 nm.

4. **(a)** T; **(b)** F; **(c)** T; **(d)** T; **(e)** F; **(f)** T; **(g)** F

5. **(a)** T; **(b)** F; **(c)** T; **(d)** F; **(e)** T; **(f)** F; **(g)** F

6. **(a)** Glycine; **(b)** glycine

7. **(a)** Glycylphenylalanylcysteinylasparagine; **(b)** 3; **(c)** 0

8. **(a)** −2; **(b)** 0; **(c)** +2; **(d)** Gly–Lys–Arg

9. **(a)** From the gel results, one can conclude that the protein is composed of two subunits corresponding to the two bands in lane 2. The protein in band A has a molecular mass of approximately 3 kD, and the protein in band B has a molecular mass of approximately 4 kD.

(b) Protease treatment provides the information that the protein in band A has the sequence
GIVEQCC**HKR**CS**E**Y**N**L**Q**NYCN
The protein in band B has the sequence
RTTGHLCG**KD**LV**N**ALY**IA**CGVRGFFY**D**P**TKM**

chapter

4

The Three-Dimensional Structure of Proteins

STEP-BY-STEP GUIDE

Major Concepts

A protein's conformation, or three-dimensional structure, is described by its secondary, tertiary, and quaternary structure.

All of these are dependent on the primary structure. Weak, noncovalent interactions stabilize these levels of architecture, which determine a protein's overall shape.

Peptide bonds connect amino acid residues in proteins.

These C—N bonds have some double-bond character, which restricts rotation, limiting the possible conformations of a polypeptide chain. The N—Cα and Cα—C bonds can rotate, but their possible conformations are limited by steric hindrance. A **Ramachandran plot** depicts the conformations theoretically permitted for peptides.

Secondary structure refers to the arrangement of adjacent amino acids in regular, recurring patterns.

Some common secondary structures are the α **helix,** β **conformation,** β **turn,** and **collagen helix.** These structures are formed, in part, in response to noncovalent interactions between neighboring amino acids. They are stable because hydrogen-bond formation is maximized and steric repulsion is minimized in these conformations. Specific amino acids are more likely to be found in some secondary structures than in others; consequently, information about the amino acid sequence of a protein can be used to predict where these structures are likely to occur.

Tertiary structure refers to the actual three-dimensional arrangement of a single chain of amino acids.

The linear amino acid sequence is a critical determinant in the final shape of a protein. Secondary structure, bend-producing amino acids, and hydrophobic interactions also contribute to the folding of a protein molecule and therefore to its overall shape. Protein folding is a complex process that is still not well understood. In some cases, protein folding occurs spontaneously, determined solely by the primary sequence. Weak, noncovalent interactions between individual amino acid residues play a role in stabilizing a protein's tertiary structure. Most soluble proteins are compact, globular structures that are somewhat flexible. The ability to change conformation is critical to protein function; enzymes, for example, often change shape when they bind substrate. **X-ray diffraction** techniques have provided direct information on the tertiary structure of many proteins.

Quaternary structure is a level of protein architecture found only in proteins having more than one polypeptide chain (subunit).

The three-dimensional arrangement of the separate subunits in the protein is referred to as its quaternary structure. Interactions between the individual subunits are stabilized by the same weak, noncovalent interactions that help stabilize the secondary and tertiary structure. One well-characterized protein with quaternary structure is hemoglobin, which consists of four distinct polypeptide chains. The association of these subunits into an oligomer contributes to the ability of hemoglobin to both bind and release oxygen under the appropriate physiological conditions (see Chapter 5 for more information on hemoglobin function). The quaternary structure of fibrous proteins such as collagen contributes to the tensile strength of these proteins. Very large assemblies of polypeptides have also been described; for example ribosomes are supramolecular assemblies that contain more than 80 proteins.

What to Review

Answering the following questions and reviewing the relevant concepts, which you have already studied, should make this chapter more understandable.

- Protein chemistry takes place in aqueous solution. Renew your appreciation of the role water plays in all biochemistry discussions (pp. 43–49).
- Weak, noncovalent interactions are extremely important in protein architecture. What are the various types of weak interactions common in biomolecules, and what are their relative strengths (pp. 50–51)?
- How are peptide bonds formed? Does a polypeptide have direction (p. 82)?
- Recall that proteins are made using only L-amino acids. Draw a representation of an L- and a D-amino acid (p. 72).
- The amino acid sequence, or primary structure, of a protein influences all other levels of protein architecture. What bonds and interactions are most important in this level of protein structure? How can the sequence of amino acids in a polypeptide be determined (pp. 82, 94–100)?

Topics for Discussion

Answering each of the following questions, especially in the context of a study group discussion, should help you understand the important points of this chapter.

4.1 Overview of Protein Structure

A Protein's Conformation Is Stabilized Largely by Weak Interactions

1. Would the "simple rules" stated at the end of this section (p. 115) be the same if protein chemistry occurred in a nonpolar solvent?

The Peptide Bond Is Rigid and Planar

2. What makes peptide bonds planar?

3. Why can't psi (ψ) and phi (ϕ) both be zero?

4.2 Protein Secondary Structure

The α Helix Is a Common Protein Secondary Structure

4. Find the ψ and ϕ angles for an α helix on the Ramachandran plot shown in Figure 4–3. Are these angles theoretically allowed?

5. Are the side chains of amino acids in an α helix on the outside or inside of the helix?

Box 4–1 Knowing the Right Hand from the Left

6. Attempt to draw for yourself an α helix containing a *mixture* of L- and D-amino acids. What does this attempt tell you about α helices?

Amino Acid Sequence Affects Stability of the α Helix

7. The properties of the various amino acid side chains dictate their interactions in secondary structure elements. In α helices, the properties of the side chains place constraints on the stability of the helix.

Explore an α helix. Go to www.rcsb.org and type "1 AL1" into the search box. Click on **Display Molecule** and select either Rasmol Viewer or Swiss-PDB Viewer to display the structure. Click on the MDL logo (or right click on a PC) to view a menu of display options. Choose **Display → Ball & Stick;** drag molecule to move it around in the window.

- Rotate the helix so you are looking down its central axis. Where are the R groups found?
- What is the amino acid sequence of this α helix? (Hint: **Color → Amino Acid,** then **Options → Labels**)
- Hydrogen bonds are formed at regular intervals between backbone amines and carbonyls; in terms of linear sequence, how many amino acid residues lie between two residues involved in a single hydrogen bond? (Hint: **Options → Display → Hydrogen Bonds**)
- What other forces stabilize the α helix structure?

The β Conformation Organizes Polypeptide Chains into Sheets

8. Where are the R groups located in a β sheet?

9. What noncovalent forces stabilize β-sheet structures?

10. In silk fibroin, which amino acid residues have functional groups participating in interchain H bonds? Which have groups extending above and below the plane of the sheet?

11. What are the limitations on the kinds of amino acids found in β structures? Compare these to the constraints discussed for the α helix.

β Turns Are Common in Proteins

12. What is unusual about peptide bonds involving the imino nitrogen of proline in β turns?

13. How many amino acid residues are typically found in a β turn and how is this structure stabilized?

Common Secondary Structures Have Characteristic Dihedral Angles

14. Note where each of the secondary structures falls on the Ramachandran plot in Figure 4–3.

Common Secondary Structures Can Be Assessed by Circular Dichroism

15. How might a manager of a biological supply company use circular dichroism to ensure the quality of protein-based products?

4.3 Protein Tertiary and Quaternary Structures

16. What distinguishes tertiary from quaternary structure in proteins? Do all proteins have quaternary structure?

17. How could you determine if a protein has quaternary structure?

Fibrous Proteins Are Adapted for a Structural Function

18. What contributes to the insolubility of fibrous proteins such as α-keratin and how does this insolubility contribute to the function of these proteins?

19. What covalent bonds contribute strength in each of these fibrous proteins: α-keratin, collagen, elastin?

20. Wool and silk are both composed of fibrous proteins; wool can stretch and shrink but silk cannot. What is the molecular basis for the different characteristics of these two fibers?

Box 4–2 Permanent Waving Is Biochemical Engineering

21. Why can a badly done perm be so damaging to hair?

Box 4–3 Why Sailors, Explorers, and College Students Should Eat Their Fresh Fruits and Vegetables

22. What is the relationship between vitamin C and scurvy?

23. How does vitamin C participate in collagen formation?

Box 4–4 The Protein Data Bank

24. Go to www.rcsb.org. and look up 1v4f or 1cdg to see a collagen triple helix. Click on **Display Molecule** and use Rasmol or Swiss-PDB viewers to display the structure. Click on the MDL logo and try out different display and coloring options.

Structural Diversity Reflects Functional Diversity in Globular Proteins

25. Try to envision how protein structure contributes to the functions of the various globular proteins mentioned in this section.

Myoglobin Provided Early Clues about the Complexity of Globular Protein Structure

26. Why does it make functional sense that a globular protein, such as myoglobin, is soluble in water?

27. What weak interactions contribute to the close packing and stability of the structure of myoglobin?

Globular Proteins Have a Variety of Tertiary Structures

28. Take a look at Figure 4–21 to see a sampling of the 3-D structures that can be formed from the three basic secondary structures.

29. What is the difference between a domain and a motif?

30. Provide some examples of common protein motifs.

Box 4–5 Methods for Determining the Three-Dimensional Structure of a Protein

31. Why can't you see proteins under a high-power light microscope?

32. Why must x rays be used to generate "images" of proteins?

33. Why is it important to know about protein structure in aqueous solutions?

Protein Motifs Are the Basis for Protein Structural Classification

34. Why are the proteins belonging to a structural "family" likely to be related with respect to their evolution?

Protein Quaternary Structures Range from Simple Dimers to Large Complexes

35. After reading this section, list some supramolecular complexes consisting only of proteins and some consisting of proteins associated with other types of biomolecules.

36. What kinds of bonds or interactions would be important in the assembly of large protein complexes?

4.4 Protein Denaturation and Folding

Loss of Protein Structure Results in Loss of Function

37. Under what conditions are proteins denatured?

38. What bonds or interactions are disrupted during denaturation?

Amino Acid Sequence Determines Tertiary Structure

39. When denatured proteins refold in the presence of low levels of detergent or denaturant, they form many incorrect disulfide bonds. What does this tell you about the process of protein folding?

Polypeptides Fold Rapidly by a Stepwise Process

40. What are the two models of protein folding presented in this section?

41. What are the noncovalent interactions that drive folding in each of these models?

Some Proteins Undergo Assisted Folding

42. Yeast mutants that are deficient in heat shock proteins (some of which are of molecular chaperones) have proteins that do not fold correctly. One role of molecular chaperones is to bind to nascent protein chains (proteins that are being synthesized) and *prevent* associations between newly synthesized regions. Why would this be necessary?

Defects in Protein Folding May Be the Molecular Basis for a Wide Range of Human Genetic Disorders

43. Which type of noncovalent interaction is likely to be responsible for stabilizing amyloid fibrils?

44. Describe two or three cellular consequences of protein misfolding.

Box 4–6 Death by Misfolding: The Prion Diseases

45. The conversion of the normal PrP protein to the PrP^{Sc} form is the result of an alteration that changes the structure of the prion protein. Based on information from this chapter, what other changes in a protein's amino acid composition would be expected to change a protein's structure and therefore its function?

Discussion Questions For Study Groups

- Why is the α helix often referred to as a condensed secondary structure, whereas the β conformation is often called extended? How might this relate to a protein's function?
- Which amino acids will usually be found in the interior of a globular protein and which will be found on the exterior? Why? Under what circumstances might you find exceptions to this "rule"?

- Explore some of the protein structures in this chapter (see list below). Go to www.rcsb.org and type in the protein name or PDB identification number in the search window. Click on **Display Molecule** and use the Rasmol or Swiss-PDB viewer to see the structure. Drag the molecule to rotate it; click on the MDL logo to access additional display commands.

Protein	*PDB ID*
Alpha-hemolysin	7 AHL
Deoxy hemoglobin	2 HHB
Collagen	1 CGD
Globin	1 VRF
GroEL/GroES complex	1 AON
Myoglobin	1 MBO / 2 MBW
Poliovirus	2 PLV
Pyruvate Kinase	1 PKN
Tobacco Mosaic Virus	1 VTM
Troponin	4 TNC

SELF-TEST

Do You Know the Terms?

ACROSS

1. Cellular agents that assist in protein folding at elevated temperatures.
6. Covalently linked amino acids with a single amino terminus and a single carboxyl terminus is called a(n) _______.
8. Bonds that occur between cysteine residues in proteins.
9. Also called a "motif."
11. Hemoglobin is a(n) _______ protein because it has two or more polypeptide chains.
14. They protrude in opposite directions from the zigzag structure of the β conformation. (2 words)
15. GCKKGGLVCAH (with an S—S bond linking C and C) for example; _______ structure.
17. Muscle fibers are an example of a(n) _______ complex.
18. Protein secondary structure that extends 0.35 nm per amino acid residue.
20. Though unrelated based on their amino acid sequences, proteins that belong to a(n) _______ have related structural features.
23. The noncovalent interactions that are thought to be the driving force behind the formation of a "molten globule."
26. An example is the re-formation of disulfide bonds during permanent waving.

DOWN

1. A native protein is in its functional _______.
2. An example of protein misfolding that has lethal consequences.
3. A stable arrangement of a few secondary structures.
4. α helices are stabilized by _______ bonds between the carbonyl oxygen and the amino hydrogen.
5. A β turn is an example of _______ structure.
7. Disrupting the hydrophobic interactions of a single-subunit protein would have the greatest effect on the _______ structure of that protein.
9. An example of a supramolecular assembly is the collagen _______.
10. α-keratin is referred to as a supramolecular complex of protein subunits; hemoglobin with only four subunits is referred to as a(n) _______.
12. The saddle conformation is a(n) _______ structure.
13. Myoglobin is to tertiary as hemoglobin is to _______.
16. Roasting a chicken results in the permanent _______ of myosin and actin proteins in the muscle cells.
19. Individual amino acids when polymerized in a protein.
21. The $\alpha\beta$ subunits in hemoglobin compose a single _______; the intact hemoglobin tetramer contains two of these.
22. Protein secondary structure that extends 0.15 nm per amino acid residue.
24. This class of proteins binds to and shields hydrophobic portions of unfolded polypeptides in cells. These proteins also are denatured by elevated temperatures.
25. Refers to the portion of a protein that is often composed of noncontiguous amino acid sequences and is usually defined on the basis of its contribution to protein function.

Do You Know the Facts?

1. Which of the following statements is/are true concerning peptide bonds?

A. They are the only covalent bond formed between amino acids in polypeptide structures.
B. The angles between the participating C and N atoms are described by the values psi (ψ) and phi (ϕ).
C. They have partial double-bond character.
D. A and C.
E. All of the above are true.

2. In an α helix, the R groups on the amino acid residues:

A. are found on the outside of the helix spiral.
B. participate in the hydrogen bonds that stabilize the helix.
C. allow only right-handed helices to form.
D. A and B are true.
E. A, B, and C are true.

3. Which of the following bonds or interactions is/are possible contributors to the stability of the tertiary structure of a globular protein? (Hint: remember the amino acid categories.)

A. peptide bonds between a metal ion cofactor and a histidine residue
B. hydrophobic interactions between histidine and tryptophan R groups
C. covalent disulfide cross-links between two methionine residues
D. hydrogen bonds between serine residues and the aqueous surroundings
E. All of the above contribute.

Refer to the following numbered statements to answer questions 4–7.

(1) Found in the same percentage in all proteins.
(2) Stabilized by H bonds between —NH and —CO groups.
(3) Found in globular proteins.
(4) Affected by amino acid sequence.
(5) An extended conformation of the polypeptide chain.
(6) Includes all 20 standard amino acids in equal frequencies.
(7) Hydrophobic interactions are responsible for the primary structure.

4. Which statements are true of α helices?

5. Which statements are true of β sheets?

6. Which statements are true of both α helices and β sheets?

7. Which statements are true of neither α helices nor β sheets?

8. Describe and compare the positions of the R groups of amino acids that are in an α helix and in a β conformation.

9. Fibrous proteins, such as α-keratin, collagen, and elastin, have evolved to be strong and/or elastic. They are also insoluble in water. What would you guess this insolubility means about the spatial positioning of the various groups of amino acids that constitute these proteins?

10. What does denaturation mean? Under what conditions are proteins denatured?

11. From the types of bonds and interactions below, identify which is *most* responsible for the structures described in (a)–(j).

ionic interactions *covalent bonds*
hydrophobic interactions *hydrogen bonds*
van der Waals interactions

(a) the association of hemoglobin subunits in its quaternary structure
(b) primary structure of proteins
(c) secondary structure of proteins
(d) the interaction between hemoglobin subunits upon binding oxygen
(e) the binding of iron within the heme group
(f) an α helix
(g) a β sheet
(h) association of the heme group within the myoglobin polypeptide
(i) the hardness of rhinoceros horn
(j) the compactness of the interior of myoglobin

Refer to Table 3–2 in the textbook for questions 12 and 13.

12. How many proteins in the table exhibit quaternary structure? How can you tell?

A. All
B. None
C. Half
D. Five
E. One cannot tell from the data provided.

13. How many proteins in the table exhibit α helix secondary structure? How can you tell?

A. All
B. None
C. Half
D. Two
E. One cannot tell from the data provided.

Applying What You Know

1. Answer the following questions concerning the polypeptides whose amino acid sequences are shown below. It may help to refer to Table 3–1 for the three-letter abbreviations and to Figure 3–5 for amino acid structures.

(1) Gly–Ile–Trp–Leu–Ile–Ile–Phe–Gly–Val–Val–Ala–Gly–Val–Ile–Gly–Trp–Ile–Leu–Leu–Ile

(2) Gly–Pro–Hyp–Gly–Pro–Met–Gly–Pro–Ser–Gly–Pro–Arg–Gly–Pro–Hyp–Gly–Pro–Hyp–Gly

(3) Gly–Met–Trp–Pro–Glu–Met–Cys–Gly–Glu–Pro–Ala–His–Val–Arg–Asp–Tyr–Pro–Leu–Leu

(4) Gl –Met–Trp–Pro–Glu–Met–Cys–Gly–Glu–Pro–Ala–His–Val–Arg–Asp–Tyr–Cys–Leu–Leu

(a) Which would be *most likely* to form an α-helical structure?

(b) Which could have at least one disulfide bond?

(c) Which would be *most likely* to be part of a fibrous protein found in cartilage?

(d) Suppose that the first amino acid residue in peptide (1) were altered from Gly to Pro. Would you expect this change to have an effect on the secondary structure of this peptide? Why or why not?

2. Myoglobin and the individual subunits of hemoglobin are similar, though certainly not identical, in size, overall shape, and function. Would you expect a molecule of myoglobin or a subunit of hemoglobin to have a greater ratio of nonpolar to polar amino acids? Why?

3. How would the following agents and/or procedures interfere with or disrupt the different levels of protein architecture? Why?

(a) Addition of SDS (see p. 92).

(b) Oxidation of cysteine with performic acid to cleave disulfide bonds (Fig. 3–26).

(c) Addition of proteases such as trypsin (p. 99).

Biochemistry on the Internet

Chapters 3 and 4 have introduced you to the structural hierarchy of cellular proteins. The extent to which the linear sequence of amino acids, the primary structure, determines the final three-dimensional shape, or tertiary structure, of proteins is quite remarkable.

As our understanding of the mechanisms underlying protein folding has expanded, it has become possible to identify unknown proteins and predict their shape, and therefore their function, based only on amino acid sequence information. This type of protein analysis relies on information that is available from sequence databases around the world. The key to unlocking this information is some knowledge of the DNA or amino acid sequence of the "mystery" protein.

The problem outlined below provides a key (an amino acid sequence) and directs you to some of the doors (database access sites) to the ever-expanding amount of scientific information available on the Internet. The list of databases presented here is by no means exhaustive; this information is intended only to provide a limited and somewhat structured exposure to protein databases and molecular modeling.

As you proceed through this exercise you will be able to obtain additional information by following any of the links that appear during your search. It will be up to you to sift through all this information to find what is relevant to your research, so don't stray too far in your initial forays and be sure to find your way back.

The Problem

In the course of evaluating a 6-year-old patient with bronchitis, a systematic hematological study of the child's blood was performed. An abnormal protein was detected and additional tests on the patient's blood sample were performed. These tests, which included isoelectric focusing, electrophoresis, and cation-exchange high performance liquid chromatography, confirmed the existence of a novel blood protein. The protein was also observed in blood samples from the child's father and other blood relatives, indicating a genetic basis for the abnormality. The abnormal protein was isolated in a pure form and the following amino acid sequence was determined:

```
  1 VLSPADKTNV KAAWGKVGAH AGEYGAEALE RMFLSFPETT KTYFPHFDLS
 51 HGSAQVKGHG KKVADALTNA VAHVDDMPNA LSALSDLHAH KLRVDPVNFK
101 LLSHCLLVTL AAHLPAEFTP AVHASLDKFL ASVSTVLTSK YR
```

(a) Your job is to try to identify this novel blood protein. There are a number of excellent sites for analyzing amino acid and protein sequences on the Internet. Many of these sites use BLAST (Basic Local Alignment Search Tool), which is a set of search engines designed to query available databases. Choose one of the following sites for your analysis.

- NCBI Blast search at www.ncbi.nlm.nih.gov/blast
- PIR—International Protein Sequence Database at http://pir.georgetown.edu

For either of these sites, follow the internal links to locate a sequence comparison engine. For example, at the NCBI site use the "Protein BLAST" links. Type the entire sequence above into the appropriate search field and submit your sequence for analysis. Be sure to limit your search to *Homo sapiens* when possible.

What does this analysis tell you about the identity of your protein?

(b) The molecular mass of the abnormal blood protein is approximately 100 daltons more than the mass of the "normal" protein. What does this suggest about the abnormal protein, and what might you do to verify your hypothesis?

(c) What information can you find on the secondary, tertiary, and quaternary structure of the "normal" version of this protein? You can start your structural analysis at a variety of Web sites that provide information on the three-dimensional structure of proteins. For this initial analysis try using the RCSB Protein Data Bank at www.rcsb.org/pdb. Perform a keyword search using the protein class name that you identified in part (a).

ANSWERS

Do You Know the Terms?

ACROSS

1. chaperones
6. polypeptide
8. disulfide
9. fold
11. multisubunit
14. R groups
15. primary
17. supramolecular
18. β conformation
20. superfamily
23. hydrophobic
26. renaturation

DOWN

1. conformation
2. prion
3. motif
4. hydrogen
5. secondary
7. tertiary
9. fibril
10. oligomer
12. supersecondary
13. quaternary
16. denaturation
19. residues
21. protomer
22. α helix
24. hsp
25. domain

Do You Know the Facts?

1. C. Disulfide bonds between cysteine residues are also covalent bonds in polypeptides. Psi (ψ) and phi (ϕ) describe the Cα—C and the N—Cα bonds, not those involved in the peptide bond.

2. A. H bonds form between —NH and —CO groups of the polypeptide backbone, not between R groups. Both left- and right-handed helices can occur (though extended left-handed helices have not been observed in proteins).

3. D. The hydroxyl group on the serine side chain can form H bonds with surrounding water molecules. Peptide bonds occur only between amino acids, not between an amino acid and a cofactor. Histidine has no hydrophobic R group. Only cysteines can form disulfide linkages.

4. 2, 3, 4

5. 2, 3, 4, 5

6. 2, 3, 4

7. 1, 6, 7

8. In an α helix, the polypeptide backbone is tightly wound along the long axis of the molecule; the R groups extend outward from the rod, like spokes, in a helical array. In a β conformation, the backbone of the polypeptide chain is extended into a zigzag rather than a helical rod; the R groups of adjacent amino acids protrude in opposite directions from the zigzag structure, creating an alternating pattern when viewed from the side.

9. These fibrous proteins are an exception to the rule that hydrophobic R groups of amino acid residues must be buried in the interior of a protein; they have a high concentration of hydrophobic amino acids in the interior and on the surface of the protein.

10. Denaturation is the total loss or randomization of three-dimensional structure. It can be reversible or irreversible. Extremes of heat and pH, exposure to some organic solvents such as alcohol or acetone, or some other substances such as urea or some detergents, will denature proteins. The covalent peptide bonds of proteins are not broken, but denaturation disrupts the many weak, noncovalent interactions that are most important in maintaining the native three-dimensional conformation.

11. **(a)** Hydrophobic interactions; **(b)** covalent bonds; **(c)** hydrogen bonds; **(d)** ionic interactions; **(e)** covalent bonds; **(f)** hydrogen bonds; **(g)** hydrogen bonds; **(h)** covalent bonds; **(i)** covalent bonds; **(j)** hydrophobic interactions.

12. D. A protein has quaternary structure only if it has more than one polypeptide chain, or subunit. Five of the proteins listed have at least two subunits, and so have quaternary structure.

13. E. None of the data in the table provides any information regarding secondary structure.

Applying What You Know

1. **(a)** 1; **(b) 4;** (c) 2; **(d)** No; peptide (1) could adopt an α structure before the change of the amino-terminal residue to proline. Though proline residues are rarely found in α helices, the amino terminus is one location where the proline R group would have relatively little effect. The amino (imino) group would not interfere with the helical structure.

2. The hemoglobin subunit would have a higher ratio of nonpolar to polar amino acids. In hemoglobin's native conformation, each subunit is in contact with the other subunits; hydrophobic interactions between the subunits are important in the stability of the tetramer, and nonpolar amino acids can be on the surface of the subunit. The external surface of the myoglobin molecule would have more polar amino acid residues, to increase hydrogen bonding with the aqueous surroundings.

3. (a) SDS, a detergent, would interfere with hydrophobic and ionic interactions, and would therefore interfere in secondary, tertiary, and quaternary structure; primary structure would not be affected. (b) For proteins with disulfide bridges, oxidation of cysteine with performic acid to cleave disulfide bonds would alter all levels of protein architecture. Disulfide bonds are part of primary structure, and a loss of these stabilizing covalent cross-links would very likely disrupt a protein's native conformation. Proteins without disulfide bridges would be unaffected. (c) Proteases cleave peptide bonds, destroying primary structure. All other levels depend on the amino acid sequence, and so would be affected as well.

Biochemistry on the Internet

(a) When the unknown sequence is compared to known sequences in protein data bases, search engines detect domains that are conserved between the proteins. Limiting the search to *Homo sapiens* shows a high degree of similarity between the unknown protein and the alpha chain of human hemoglobin.

(b) The higher molecular mass coupled with the sequence similarity to the α chain of hemoglobin suggests that the protein is a mutant form of hemoglobin that has more amino acids than the normal form. Because the average weight of a single amino acid in a polypeptide chain is around 110, it is possible that an extra amino acid residue has been inserted into this abnormal hemoglobin subunit.

This can be verified by directly comparing the amino acid sequence of the abnormal protein and the human hemoglobin α-chain sequence. The abnormal Hb identified in the patient contains 142 residues with an additional glutamate (E) inserted after the proline (P) at position 37.

In fact, the amino acid sequence of the abnormal protein provided in the original question is the complete sequence for an α-chain variant called Hb Cantonsville, named after the town where it was first identified. For the complete reference, see Moo-Penn, W.F., Swan, D.C., Hine, T.K., Baine, R.M., Jue, D.L., Benson, J.M., Johnson, M.H., Virshup, D.M., and Zinkham, W.H. Hb Cantonsville (Glutamic acid inserted between Pro-37 (C2) α and Thr-38 (C3) α) *J. Biol. Chem.*, 264:21454, 1989.

(c) There are a number of good sites for the Internet-based analysis of protein structure. One excellent site is the Protein Data Bank. Using "globin" as a keyword produces more than 400 related structures while using "compound: hemoglobin and source: human" produces only about 100 related structures. Several molecule display options allow you to see the protein in 3-D, but many require the use of plugins that can be downloaded from the Web (e.g., Java). Help with downloads is available on this same page. Still images of the molecule can also be viewed from this page.

In most of the 3-D viewers, you can see the protein's secondary structure by setting the **Display** option to **Ribbons.** This will display the protein's beta conformation and α-helical structures. These models are interactive allowing you to manipulate the movement of the proteins by clicking on the structure and then dragging it with your mouse.

In the hemoglobin model you should see four subunits composed entirely of α helices connected by β turns. Note the heme molecules bound to each of the four subunits; this is where hemoglobin binds oxygen.

chapter

5 Protein Function

STEP-BY-STEP GUIDE

Major Concepts

Several key principles of protein function can be illustrated by the behavior of oxygen-binding proteins, immune system proteins, and muscle proteins: Proteins are not rigid but, in fact, must be flexible. Many proteins bind smaller molecules called ligands (e.g., oxygen, calcium, hormones). Ligand binding can affect protein conformation, which often alters a protein's ability to bind to other ligands, as we will see in the case of oxygen binding by hemoglobin.

Myoglobin is a protein consisting of a single polypeptide chain with a heme prosthetic group that binds oxygen reversibly.

The binding properties of the heme group *within* the globin protein are different from those of the free heme group. The binding of oxygen (the ligand, L) to myoglobin can be described quantitatively; for a monomeric protein, the fraction of binding sites occupied by the ligand is a hyperbolic function of [L].

Hemoglobin is a multisubunit, heme-containing, oxygen-binding protein.

This protein is a tetramer with two α chains and two β chains. Each subunit contains a heme prosthetic group; consequently, each is capable of binding one molecule of oxygen. Hemoglobin displays **positive cooperativity** with respect to its oxygen binding; the binding of oxygen to hemoglobin follows a sigmoid curve. This cooperativity is attributed to alterations in noncovalent interactions between amino acids in adjacent subunits. These alterations, which are the result of subtle changes in the porphyrin ring structure of the heme upon oxygen binding, result in a conformational shift in the entire hemoglobin tetramer.

Hemoglobin also binds the excess hydrogen ions and carbon dioxide produced by respiring tissues.

H^+ can bind to a number of amino acid residues in the hemoglobin molecule, which stabilizes the deoxy form of hemoglobin. CO_2 can bind to amino-terminal residues to form carbaminohemoglobin; the binding of CO_2 forms additional ion pairs, which stabilize the deoxy form of hemoglobin. Binding of H^+ and CO_2 promotes the release of oxygen from hemoglobin to the tissues where it is needed. In the lungs, H^+ and CO_2 are released, increasing the affinity of hemoglobin for oxygen.

2,3-Bisphosphoglycerate (BPG) also regulates oxygen binding by hemoglobin.

A single BPG molecule binds to hemoglobin in the central cavity between the four subunits. Binding is due to interactions between negatively charged groups on BPG and positively charged amino acid residues that line the central cavity. BPG stabilizes the deoxy form of hemoglobin.

Sickle-cell hemoglobin (HbS) illustrates the importance of protein conformation to protein function.

The hemoglobin molecules of individuals with **sickle cell anemia** have different amino acid sequences than hemoglobin from normal individuals. A hydrophobic residue on the outer surface of the β subunit promotes clumping of hemoglobin molecules, producing abnormally shaped erythrocytes with decreased life span.

The vertebrate immune system protects the organism by utilizing incredibly specific interactions between white blood cells and their associated proteins, and nonself molecules.

T lymphocytes produce receptor proteins and B lymphocytes produce immunoglobulins. Recognition and binding by these proteins allow the organism to determine self from nonself. When a virus, bacterium, or other invader is detected, an immune response is mounted that involves the coordinated control of proliferation of specific B and T cells and their proteins for attack against the invading foreign proteins.

Immunoglobulins are the proteins produced by B cells.
Their shape, flexibility, and their combination of conserved and variable regions make them capable of recognizing an almost limitless variety of antigens.

The principles of protein flexibility and protein-protein interaction through noncovalent interactions are illustrated at an extreme level by the cyclic conformational changes that cause muscles to contract.
The proteins myosin and actin are organized into thick and thin filaments, which are themselves arranged in sarcomeres, the repeating units of myofibrils. Many myofibrils make up a muscle fiber; skeletal muscle consists of bundles of muscle fibers. The sliding of the myosin thick filaments along the actin thin filaments, coordinated by nerve impulses, produces the contraction of the muscle. Many other accessory proteins are involved in the structure and temporal control of these assemblies.

What to Review

Answering the following questions and reviewing the relevant concepts, which you have already studied, should make this chapter more understandable.

- Altering a single amino acid in a protein's sequence can have enormous consequences to the secondary, tertiary, and quaternary structure, and to the binding of ligands to proteins. Review the properties of the different R groups of amino acids (Fig. 3–5).
- Know the significance of variable and invariant residues (see section 3.4).
- Ionic interactions are critical to the cooperativity of oxygen binding in hemoglobin. How strong are ionic interactions compared to hydrogen bonds? To covalent bonds?
- Which reagents and/or techniques of protein purification denature proteins? Which leave them in their native conformation (pp. 85–92)?

Topics for Discussion

Answering each of the following questions, especially in the context of a study group discussion, should help you understand the important points of this chapter.

5.1 Reversible Binding of a Protein to a Ligand: Oxygen-Binding Proteins

Oxygen Can Bind to a Heme Prosthetic Group

1. Why is it important that iron is incorporated into the heme group, rather than free in the cell?

2. What is the shape of the heme group?

Myoglobin Has a Single Binding Site for Oxygen

3. How is the heme group attached to and positioned within the myoglobin molecule (Hint: look at Fig. 5–3)?

4. This section describes two ways to describe individual amino acids in the myoglobin molecule. Using Figure 5–7, what are the two ways to describe the conserved valine at position 10 in Hbα?

Protein-Ligand Interactions Can Be Described Quantitatively

Making a real effort to understand the quantitative relationships described in this section will help you to understand critical concepts in enzyme kinetics (Chapter 6).

5. In oxygen-binding curves for myoglobin, why is the term pO_2 used rather than $[O_2]$?

6. What does the term [PL] mean in terms of specific molecules?

7. Why are the total number of oxygen-binding sites in a solution of myoglobin equal to [PL]+[P]?

8. If you have a solution containing 1×10^6 molecules of myoglobin and $\theta = .75$, how many myoglobin molecules have oxygen bound?

Protein Structure Affects How Ligands Bind

For the following questions, go to www.rcsb.org. The following questions can be answered by examining the 3-D molecular model for myoglobin (1 MBO). Try **Select → Hetero → Ligand**, then **Display → Ball & Stick**, then **Color → CPK**.

- Can you identify the heme prosthetic group and helices that are closest to the heme?
- Which amino acid residues in these helices are closest to the heme group?
- What forces hold the heme prosthetic group in place?
- Where in the tertiary structure of the myoglobin protein is the oxygen-binding site located?

9. Given that free heme binds CO 20,000 times better than it binds O_2, why don't we all succumb to carbon monoxide poisoning?

Hemoglobin Transports Oxygen in Blood

10. Why is it important that erythrocytes are *small* cells?

Hemoglobin Subunits Are Structurally Similar to Myoglobin

For the following questions, go to www.rcsb.org. The following questions can be answered by examing the 3-D molecular models for deoxyhemoglobin (2HHB) and oxyhemoglobin (1gzx). Try **Select → Hetero → Ligand**, then **Display → Ball & Stick →** then **Color → CPK**.

- Identify the heme prosthetic groups and the two α helices that are closest to the heme. Are they the same ones you identified for myoglobin?
- Compare the tertiary structure of myoglobin and β subunit of hemoglobin. Where are the 3-D structures most similar and where do they differ?

11. Could a myoglobin molecule substitute for one of the subunits in the hemoglobin tetramer? Why or why not?

12. Why do you think the histidine residues E7 and F8 are invariant between myoglobin and the α and β chains of hemoglobin?

13. What weak bonds or interactions, *other than* ion pairs, are important to hemoglobin's quaternary contacts?

Hemoglobin Undergoes a Structural Change on Binding Oxygen

14. How does oxygen binding cause rearrangement of ion pairs and which amino acids are involved?

Hemoglobin Binds Oxygen Cooperatively

15. How does a sigmoid binding curve illustrate the presence of a low-affinity state and a high-affinity state?

16. Why does hemoglobin have a lower affinity for the first oxygen molecule it binds than it does for binding additional oxygen molecules? One way to think of this is by analogy to a block of four postage stamps, arranged in a **two-by-two array:**

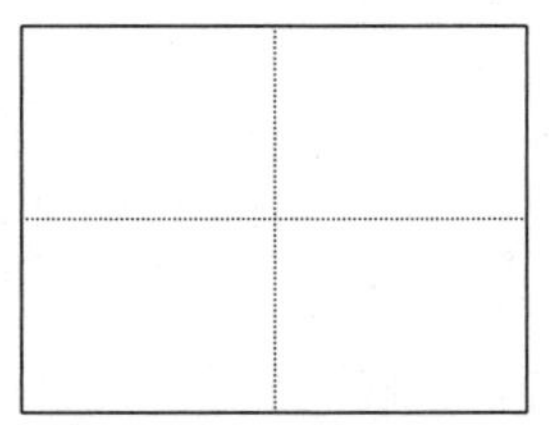

How many perforations (bonds) is it necessary to tear to free the first stamp from the other three? How many to free the second?

17. Oxygen acts as what type of modulator for hemoglobin?

Cooperative Ligand Binding Can Be Described Quantitatively

18. Under what circumstances is the Hill coefficient (n_H) of a multisubunit protein equal to 1? When is n_H greater than 1? When would it be less than 1?

Box 5–1 Carbon Monoxide: A Stealthy Killer

19. Use the data in Figure 2 to compare the percent of O_2 released from COHb to tissues to that from normal Hb.

Two Models Suggest Mechanisms for Cooperative Binding

20. Is the concerted model or the sequential model more consistent with current knowledge of hemoglobin's T → R transition?

Hemoglobin Also Transports H^+ and CO_2

21. The equation $HHb^+ + O_2 \leftrightarrow HbO_2 + H^+$ indicates that both O_2 and H^+ can bind to hemoglobin. Do these two ligands bind at the same site? Do they both cause the same changes in conformation?

22. Would a decrease in pH promote the "T" or the "R" conformation of hemoglobin?

23. Carbon dioxide also binds to hemoglobin. Do CO_2 and O_2 bind at the same site? Do they both cause the same changes in conformation?

24. How are the binding of H^+, CO_2, and O_2 interdependent?

Oxygen Binding to Hemoglobin Is Regulated by 2,3-Bisphosphoglycerate

Explore hemoglobin with 2,3-BPG bound (1B86) at www.rcsb.org.

25. The equation $HbBPG + O_2 \leftrightarrow HbO_2 + BPG$ indicates that oxygen and BPG can bind to hemoglobin. Do these two ligands bind at the same site? Do both cause the same changes in conformation?

26. How would the oxygen-binding curve of fetal hemoglobin compare to that of maternal hemoglobin? Does this make physiological sense?

Sickle-Cell Anemia Is a Molecular Disease of Hemoglobin

27. What protein purification techniques might be useful for separating HbS from HbA?

28. Would it be better for patients with sickle-cell disease to live at sea level or at a high altitude?

5.2 Complementary Interactions between Proteins and Ligands: The Immune System and Immunoglobulins

The Immune Response Features a Specialized Array of Cells and Proteins

29. What are the defense responsibilities of the humoral immune system? Of the cellular immune system?

30. Which type of white blood cell produces antibodies? What are the analogous recognition proteins produced by the cells of the cellular immune system?

31. Would a single amino acid be an effective antigen? A dipeptide? Hemoglobin?

32. What is the role of T_H cells in the protection of the organism under viral attack?

33. What is the role of B cells in the defensive strategy?

Antibodies Have Two Identical Antigen-Binding Sites

34. What type of bonds hold the portions of the heavy and light chains together in the Fab?

35. Antigen binds between the *variable* domains of the heavy and light chains. Why does this make sense for the specificity of the antigen/antibody interaction?

36. How are basophils and mast cells conscripted into the army of defensive cells of the immune response?

Antibodies Bind Tightly and Specifically to Antigen

37. Which amino acid residues are negatively charged? If these were found in an epitope of a particular antigen, which amino acid residues in the corresponding antibody would interact with them? Take a moment to review the categories of amino acid side chains.

38. How does induced fit increase the strength of the interaction between antibody and antigen?

The Antibody-Antigen Interaction Is the Basis for a Variety of Important Analytical Procedures

39. What type of chromatography is shown in Figure 3–18c?

40. The techniques illustrated in Figure 5–28 that use antibodies to detect proteins all require that the proteins be in native conformation. What reagents or conditions can therefore *not* be used in these procedures?

5.3 Protein Interactions Modulated by Chemical Energy: Actin, Myosin, and Molecular Motors

41. What are the important protein-protein interactions among motor proteins? (This is the major theme of this chapter.)

The Major Proteins of Muscle Are Myosin and Actin

42. Do thick filaments show directionality? What makes up the different "ends" of the thick filament?

43. What is the role of ATP in actin filament assembly?

Additional Proteins Organize the Thin and Thick Filaments into Ordered Structures

44. The Z disk anchors which type of filament? The M disk anchors which type?

45. The size hierarchy of the architectural levels of the muscle is a useful thing to commit to memory because it helps in understanding the interactions between all the proteins involved in contraction. Order the components, starting with the smallest (actin and myosin) and continuing up to the largest level—a muscle itself.

46. Calculate the (approximate) molecular weight of titin.

Myosin Thick Filaments Slide along Actin Thin Filaments

47. In terms of the protein-protein interactions that make muscle contraction possible, why is it critical to prevent overextension of the sarcomere?

48. What types of interactions are involved in the binding of the myosin head to actin? Are there any *covalent bonds* made or broken in the cycle of reactions that moves the thick filament relative to the thin filament?

49. What is the ligand that regulates contraction in skeletal muscles and what is the basis for this regulatory effect?

Discussion Questions for Study Groups

- Use Figure 5–4 to estimate the percentage of myoglobin that has oxygen bound at a pO_2 of 4 kPa, the partial pressure of O_2 normally found in muscle cells. What does this indicate about the oxygen-binding role of myoglobin in tissues?
- How do the effects of pH, CO_2, and O_2 contribute to the shape of the oxygen-binding curve for hemoglobin, and why do these effects make physiological sense?
- From the amino acid sequences given in Figure 5–7, what portions of hemoglobin and myoglobin contain invariant residues? Does the location of invariant residues make sense now that you know how the hemoglobin protein functions?

SELF-TEST

Do You Know the Terms?

ACROSS

4. _______ occurs when the binding of one ligand increases or decreases the binding of additional ligands.

7. The _______ immune system protects against bacterial infections.

9. This protein can exist in a globular or filamentous form; hydrolysis of ATP is necessary to convert one to the other.

10. Composed of many sarcomeres, many of these make up a muscle fiber.

13. The covalent binding of CO_2 to the amino termini of hemoglobin subunits favors the _______ form.

14. This protein has a hyperbolic O_2 binding curve and no quaternary structure; it serves as an O_2 "reservoir" in muscle cells.

15. The metabolic intermediate 2,3-_______ binds to hemoglobin molecules with a stoichiometry of 1:1 and promotes the release of O_2.

16. Red blood cell.

19. A prosthetic group containing iron.

21. Immunoglobulin.

22. Types of white blood cells; T and B cells.

23. A helper T cell can signal nearby lymphocytes by secretion of a signal protein called a(n) _______.

DOWN

1. The production of lactic acid in muscle tissue contributes to the _______ effect, which explains the link between lactate production and an increased release of O_2 from hemoglobin.

2. The iron in this prosthetic group can bind either CO or O_2 at its sixth coordination position.

3. A molecule reversibly bound by a protein.

5. Antibodies are produced by the immune system as part of a defense against invasion by a foreign particle known as a(n) _______.

6. Red blood cells transport CO_2 produced by respiring tissues in two forms: as bicarbonate ions and as _______.

7. Oxygen transport protein that binds O_2 with a stoichiometry of 4 O_2:1 molecule transport protein.

8. Individual molecules of _______ aggregate to form thick filaments.

11. All vertebrates have an immune system capable of distinguishing _______ from invader.

12. Cleavage of an IgG with the protease papain separates the basal fragment from the "branches," called _______.

17. Small molecules covalently attached to large proteins in the laboratory in order to elicit an immune response.

18. A particular molecular structure within an antigen that binds an individual antibody.

19. Heme groups are covalently bound to globin through the _______ histidine residue.

20. Allosteric proteins such as hemoglobin and IgG exhibit changes in their 3-D structure, a process known as _______ fit.

Do You Know the Facts?

1. Indicate whether each of the following statements about the heme-binding site in myoglobin is true or false. If you think a statement is false, explain why.

(a) The proximal histidine covalently binds iron.

(b) The distal histidine covalently binds oxygen.

(c) The distal histidine binds iron.

(d) Free heme binds CO with the Fe, C, and O atoms in a linear array.

(e) The iron in heme binds the oxygen atom of CO.

(f) A molecule of carbon monoxide can bind to myoglobin at the same time as a molecule of oxygen.

2. Which of the following statements is/are true concerning the structures of myoglobin and hemoglobin?

(a) The tertiary structure of myoglobin is similar to that of a subunit of hemoglobin.
(b) The quaternary structure of myoglobin is similar to that of a subunit of hemoglobin.
(c) Myoglobin could substitute for one of the subunits in hemoglobin in erythrocytes.
(d) Myoglobin contains one binding site for oxygen per molecule.
(e) Myoglobin contains one binding site for oxygen per heme.
(f) Hemoglobin contains one binding site for oxygen per molecule.
(g) Hemoglobin contains one binding site for oxygen per heme.

3. Which of the following statements is/are true of *both* hemoglobin and myoglobin?

(a) Acidic conditions lower the affinity for oxygen.
(b) The iron atom of the heme prosthetic group is bound at five of its six coordination sites to nitrogen atoms.
(c) The Hill coefficient is equal to the number of subunits in the molecule.
(d) O_2 binding occurs in the cleft where the polypeptides come into contact.
(e) None of the above is true of both molecules.

4. Which of the following explain(s) how the reaction of carbon dioxide with water helps contribute to the Bohr effect? Refer to the figure below.

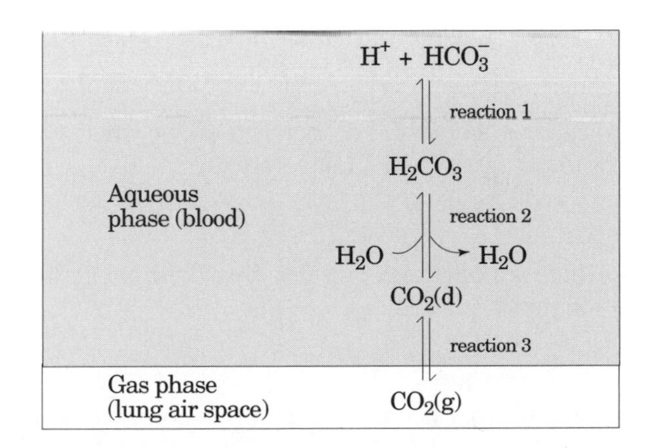

(a) The H^+ generated in this reaction decreases the pH of the blood, stabilizing the deoxy form of hemoglobin.
(b) The H_2CO_3 generated in this reaction stabilizes the deoxy form of hemoglobin.

(c) The H^+ generated by this reaction results in a lower affinity of hemoglobin for O_2, causing more O_2 to be released to the surrounding tissue.
(d) The H^+ forms a carbamate group with the amino termini of the four Hb chains.
(e) The increased concentration of HCO_3^- decreases the pH of the blood, stabilizing the deoxy form of hemoglobin.
(f) The CO_2 (dissolved) can form carbaminohemoglobin, which promotes the release of O_2.
(g) The H^+ generated interferes with the ionic interactions between the subunits, increasing the affinity for O_2.

5. High-altitude adaptation is a complex physiological process that involves an increase in both the number of hemoglobin molecules per erythrocyte and the total number of erythrocytes. It normally requires several weeks to complete. However, even after one day at high altitude, there is a significant degree of adaptation. This effect results from a rapid increase in the erythrocyte BPG concentration. **(a)** If oxygen-binding curves for both high-altitude-adapted hemoglobin and normal, unadapted hemoglobin were plotted together, would the curve for high-altitude-adapted hemoglobin be to the left of, to the right of, or the same as the curve for unadapted hemoglobin? **(b)** Is the O_2-binding affinity of high-altitude-adapted hemoglobin higher, lower, or the same as that of unadapted hemoglobin?

6. What is the relationship among θ (theta), K_d, and K_a?

7. What types of interactions are responsible for the close association of the nonidentical subunits (α and β) in the quaternary structure of hemoglobin?

8. Where does 2,3-bisphosphoglycerate interact with hemoglobin? What bonds or interactions are responsible for this interaction?

9. Which of the following describe the humoral immune system? The cellular immune system?
(a) It destroys host cells infected with viruses.
(b) It destroys some parasites.
(c) It responds to bacterial infections.
(d) It responds to extracellular viruses.
(e) The recognition proteins are immunoglobulins.
(f) The recognition proteins are cell surface receptors.
(g) The major cell type is B.
(h) The major cell types are T_C and T_H.

10. **(a)** Differentiate between antigen, epitope, and hapten. **(b)** Suppose a molecule of hemoglobin were bound by several different antibodies: one at each BPG binding site, one at each amino terminus of the α subunits, and one at each carboxyl terminus of the β subunits. How many total antibody molecules would be bound? How many antigens, epitopes, and haptens are involved in this scenario?

11. What is the role of T_H cells in the protection of the organism under viral attack? Why is the depletion of these cells (as in HIV infection) so devastating to the immune system?

12. Generate a hierarchical list for the components in muscles. Start with the individual actin and myosin molecules, and work up in complexity to full muscle tissue.

13. Place the following states or processes that occur during muscle contraction in their proper order. Use any step as a starting point.
(a) "Power stroke"; the thick filament moves with respect to the thin filament.
(b) ATP is hydrolyzed; the heads of myosin are displaced along the thin filament.

(c) ATP binds to the myosin head; reduces the affinity of the myosin head for actin.
(d) The myosin head rebinds to actin closer to the Z disk.

14. Consider the following events associated with the regulation of the contraction of skeletal muscle. Place the following events in the order in which they occur.
(a) A nerve impulse leads to the release of Ca^{2+} from the sarcoplasmic reticulum.
(b) Tropomyosin changes position on the thin filament.
(c) The formation of the complex between troponin and Ca^{2+} causes the tropomyosin-troponin complex to undergo a conformational change.
(d) The heads of myosin bind to actin.
(e) Ca^{2+} binds to troponin.

Applying What You Know

1. Though carbon monoxide is hindered in its binding to hemoglobin by the distal histidine, hemoglobin's affinity for CO is nonetheless about 200 times that for hemoglobin compared to that for O_2. Like oxygen, carbon monoxide shows **positive cooperativity** in binding to the heme sites of hemoglobin. A molecule of carbon monoxide and a molecule of oxygen can bind to the hemoglobin tetramer at the same time, though at different heme-binding sites. Use this information to analyze the following clinical situation.

Two people are present in a hospital emergency room. John's blood hemoglobin concentration is one-half the normal value (only one-half the normal number of hemoglobin molecules are present). David has been poisoned with carbon monoxide by car exhaust, such that the number of oxygen-binding sites available to bind oxygen is only one-half the normal value (because, on average, every hemoglobin molecule has two carbon monoxides bound to two of its four hemes). Assuming that John and David are otherwise similar, which patient is in more serious danger? Explain why.

2. Why is hemoglobin better suited for its job as the oxygen-carrying protein in the bloodstream than myoglobin? (Include a discussion of oxygen pressure in the lungs and peripheral tissues, and the O_2-binding curves of the two proteins.)

3. How do carbon dioxide levels both *indirectly* and *directly* affect the ability of hemoglobin to bind oxygen?

4. Antibodies against an injected protein can be generated in a laboratory animal. The antibodies can then be purified, and these antibodies themselves can be used as antigens to generate new antibodies in a different lab animal. For example, antibodies can be raised in a rabbit against the muscle protein α-actinin. A goat can then be injected with the rabbit immunoglobulin molecules to generate goat anti-rabbit immunoglobulins. The goat antibodies can then be tagged with a fluorescent molecule that will be visible under UV light. The rabbit anti-α-actinin antibodies will react with α-actinin, and the goat anti-rabbit antibodies will bind to the bound rabbit antibodies. The complex is then visualized under UV light. This technique is called indirect immunofluorescence. What is the advantage of using such a technique over *directly* tagging the rabbit antibodies themselves with a fluorescent tag in order to visualize the presence of α-actinin?

Biochemistry on the Internet

1. (a) In Chapter 4 you were presented with the case of a sick child with an abnormal form of hemoglobin named Hb Catonsville. This hemoglobin contains an additional amino acid, a glutamate, inserted between Pro^{37} and Thr^{38} in the α subunit. Where is this additional glutamate located in the three-dimensional structure of hemoglobin and how might this alteration in Hbα affect the function of the resulting hemoglobin $\alpha_2\beta_2$ tetramer?

There are a number of ways to answer these questions. Using protein structural information stored at the Protein Data Bank (www.rcsb.org/pdb/), you can look at the structure of normal hemoglobin and determine where the added glutamate would be located. Using your knowledge of how the hemoglobin molecule behaves, you can predict whether or not this mutation is likely to affect protein function. One structure for human hemoglobin that you can start with has the PDB ID number 1A3N.

- Enter "1A3N" in the search window and click **Site Search.**

- Click on the **Display Molecule** link and select either the Rasmol or Swiss-PDB viewer (you may have to download a plugin for your browser in order to view the file).
- Once you have a molecule to look at, right click on the molecule (or click on the MDL logo in the bottom right corner) to open a pop-up menu.
- In the pop-up menu click **Select → Residue → Pro** to select all the proline residues in the structure. (Proline is found just before the inserted glutamate in the mutant Hb.)
- Open the pop-up menu again, but this time click **Display → Spacefill → van der Waals Radii** to highlight the selected residues.
- Repeat these last two steps for Thr (the residue on the other side of the inserted glutamate). When you are done, click on the highlighted residues to locate Pro^{37} and Thr^{38}, or select **Options → Labels**.

(b) Based on your understanding of oxygen binding in hemoglobin, can you predict how the altered amino acid sequence of the α chain mutant in Hb Cantonsville will affect the function of the resulting hemoglobin $\alpha_2\beta_2$ tetramer.

2. Another Hb variant found in humans is the result of a deletion of eight amino acids in the α chain. It is named Hb J-Briska after the town in southeast Algeria where it was first discovered. The positions and identities of the amino acid residues in a normal α chain and in the Hb-J Briska variant are shown below:

Normal α Chain

1 VLSPADKTNV	11 KAAWGKVGAH	21 AGEYGAEALE	31 RMFLSFPTTK
41 TYFPHFDLSH	51 G*SAQVKGHG*K	61 KVADALTNAV	71 AHVDDMPNAL
81 SALSDLHAHK	91 LRVDPVNFKL	101 LSHCLLVTLA	111 AHLPAEFTPA
121 VHASLDKFLA	131 SVSTVLTSKY	141 R	

J-Briska α Chain

1 VLSPADKTNV	11 KAAWGKVGAH	21 AGEYGAEALE	31 RMFLSFPTTK
41 TYFPHFDLSH	51 GKKVADALTN	61 AVAHVDDMPN	71 ALSALSDLHA
81 HKLRVDPVNF	91 KLLSHCLLVT	101 LAAHLPAEFT	111 PAVHASLDKF
121 LASVSTVLTS	131 KYR		

(a) The residues that are deleted in the mutant α chain are shown in italics in the normal α-chain sequence. Using this sequence information see if you can determine where in the three-dimensional structure of the protein this deletion occurred.

(b) Based on your understanding of oxygen binding in hemoglobin, can you predict how the altered amino acid sequence of the Hb-J Briska α chain mutant will affect the function of the resulting hemoglobin $\alpha_2\beta_2$ tetramer?

ANSWERS

Do You Know the Terms?

ACROSS

4. cooperativity
7. humoral
9. actin
10. myofibrils
13. deoxy
14. myoglobin
15. bisphosphoglycerate
16. erythrocyte
20. porphyrin
21. antibody
22. lymphocytes
23. interleukin

DOWN

1. Bohr
2. heme
3. ligand
5. antigen
6. carbaminohemoglobin
7. hemoglobin
8. myosin
11. self
12. Fab
17. hapten
18. epitope
19. proximal
20. induced

Do You Know the Facts?

1. **(a)** T; **(b)** F; the distal histidine does bind oxygen, but it is a hydrogen bond, not a covalent one. **(c)** F; His^{93} (the proximal histidine) binds iron; **(d)** T; **(e)** F; it binds the carbon atom of CO. The details of the chemistry here are very important to remember. In carbon monoxide, carbon is *triple-bonded* to O. Carbon can make four bonds to other atoms; oxygen cannot. **(f)** F; there is only one binding site, which can be occupied by either O_2 or CO, but not both at the same time.
2. a, d, e, g
3. b
4. a, c, f
5. **(a)** right; **(b)** lower
6. K_a is the association constant of the equilibrium expression $P+L \leftrightarrow PL$, where P is a protein, L is a ligand, and PL represents protein bound to ligand.

$$K_a = \frac{[PL]}{[P][L]}$$

K_d, the dissociation constant, is the reciprocal of K_a.

$$K_d = \frac{[P][L]}{[PL]}$$

θ is the fraction of ligand-binding sites on the protein that are occupied by ligand.

$$\theta = \frac{[PL]}{[PL] + [P]}$$

The [L] at which half of the available ligand-binding sites are occupied (θ=0.5) corresponds to $1/K_a$.

7. Hydrophobic interactions, hydrogen bonds, and a few ion pairs. *All* of these are important, even though ionic interactions are given the most attention.
8. A single BPG molecule binds to hemoglobin in the central cavity formed by the four subunits. Negatively charged groups on BPG interact with positively charged amino acid residues lining the central cavity. BPG binding stabilizes the deoxy form of hemoglobin.
9. Humoral: (c), (d), (e), (g)

   Cellular: (a), (b), (f), (h)
10. **(a)** Antigen: any molecule or pathogen capable of eliciting an immune response. A large antigen may have several epitopes. Epitope: a particular molecular structure within the antigen that binds an individual antibody or T-cell receptor. Hapten: a small molecule (generally under M_r 5,000) that can be attached to larger molecules in order to elicit an immune response.

    (b) Five total antibodies bound, of three different binding specificities: two (identical) at the amino termini, two at the carboxy-termini, one at the single BPG site. Hb is one antigen; it has three epitopes; haptens are not involved.
11. Specific T_H-cell receptors recognize and bind to the viral peptides complexed to Class II MHC proteins displayed on the surface of macrophages and B lymphocytes. T_H cells have another receptor, CD4, that enhances the binding interaction. The two types of receptor binding activate the T_H cells. Activated T_H1 cells secrete interleukin-2, which stimulates the reproduction of nearby T_C and T_H cells with receptors for that particular interleukin. Activated T_H2 cells secrete interleukin-4, stimulating the reproduction of appropriate B cells that are nearby.

The T_H cells therefore act to signal the nearby cells, which will then actively defend the host. However, the T_H cells do not directly destroy the invader; they are helpers. Without their signals to cause proliferation of the other immune system cells and proteins, the entire immune system is incapacitated.

12. Individual myosin molecules aggregate into thick filaments. Individual actin (G-actin) molecules aggregate into F-actin, which, with troponin and tropomyosin, make up the thin filaments. Nebulin is also part of the thin filament.

Thick and thin filaments align and overlap in sarcomeres, which are the individual unit of contraction. Sarcomeres are bounded by Z disks, which are made up of the proteins α-actinin, desmin, and vimentin. The Z disks are the anchors for the two sets of thin filaments found in each sarcomere. The M disk, which is seen in the center of each sarcomere, anchors the thick filaments with the help of paramyosin, C-protein, and M-protein. Titins connect the thick filaments to the Z disk.

Many sarcomeres in linear array make up the myofibril. Many (~1,000) myofibrils in lateral alignment make up one muscle fiber. Long parallel bundles of muscle fibers (large, single, multinucleated cells) make up a muscle.

13. (c), (b), (d), (a)

14. (a), (e), (c), (b), (d)

Applying What You Know

1. Remember that both the number and affinity of the binding sites must be considered. Although John has half the normal number of hemoglobin molecules available, O_2 can bind to all four sites, some of which will be low-affinity sites, readily releasing oxygen in the peripheral tissues. David is in more serious danger. Although he and John have the same number of available binding sites for O_2, David's are all high-affinity sites because CO already occupies the low-affinity sites. David's bound O_2 will not be given up readily in the tissues.

2. An efficient O_2-carrying protein in the bloodstream must be able to bind oxygen efficiently in the lungs, where the pO_2 is about 13.3 kPa, and release oxygen in the peripheral tissues, where the pO_2 is about 4 kPa. Myoglobin, with its hyperbolic binding curve, is ill-suited to this function. Mb binds O_2 with high affinity, but will not release it in the tissues because of that same high affinity. Hemoglobin is a much better candidate because it undergoes a transition from a low-affinity state to a high-affinity state as more O_2 molecules are bound, as evidenced by its sigmoid O_2-binding curve. The transition from low- to high-affinity occurs cooperatively through the interaction of multiple subunits. The first O_2 that binds to (deoxy)hemoglobin binds weakly because it must bind to a heme in the T state. Its binding leads to changes in conformation that are communicated to adjacent subunits, making it easier for additional molecules of O_2 to bind. A single-subunit protein, like myoglobin, cannot produce a sigmoid binding curve.

3. Carbon dioxide levels indirectly affect the ability of hemoglobin to bind oxygen: Respiring tissues produce a large amount of CO_2, which diffuses into erythrocytes. There, CO_2 is converted to H_2CO_3 by the enzyme, carbonic anhydrase. H_2CO_3 dissociates into H^+ and HCO_3; the net result is a lowering of pH and a decrease in the affinity of hemoglobin for oxygen. Carbon dioxide levels directly affect the ability of hemoglobin to bind oxygen: CO_2 can bind to amino terminal residues to form carbamino-hemoglobin; the binding of CO_2 also forms additional salt bridges, which stabilize the deoxy form of hemoglobin. Binding of CO_2 promotes the release of oxygen in the tissues where it is needed. In the lungs, CO_2 is released, increasing the affinity of hemoglobin for oxygen.

4. This approach increases the sensitivity of the method because several fluorescently labeled goat antibodies can bind to each rabbit immunoglobulin; it is a large enough protein to have several epitopes.

Biochemistry on the Internet

1. **(a)** Pro^{37} and Thr^{38} are located on the surface of Hb α2 at a point where it comes into contact with Hb β1. In Hb-Cantonsville, therefore, the inserted glutamate must lie between these two residues at the αβ interface.

(b) The presence of a glutamate residue between Pro^{37} and Thr^{38} inserts a negative charge into a normally uncharged region. It would be reasonable to expect that this charge would affect the interactions between the two α/β dimers in the Hb tetramer. Changes in how the two dimers interact would, in turn, be expected to affect the function of the hemoglobin tetramer, specifically the cooperative nature of oxygen binding. In fact, oxygen-binding studies performed on this hemoglobin variant show a high affinity for oxygen and reduced cooperativity as well as a reduced Bohr effect (see Moo-Penn et al., Hb Cantonsville (Glutamic acid inserted between Pro-37(C2)α and Thr-38(C3)α). *The Journal of Biological Chemistry* (1989) Vol. 264: pp. 21,454–21,457).

2. **(a)** In order to determine how much of an effect the deleted amino acids might have on protein function, we need to see where they normally are located. Using structural information to display a 3-D model of human hemoglobin, you can locate the amino acid residues in positions 52–59 in the α subunit. You can click on portions of the 3-D structure to see the name and position of the corresponding residue, or select **Options** → **Labels**.

You can also locate specific residues that are deleted in the mutant Hb (Eg. Q^{54}) by clicking on **Select** → **Residue** → **Gln** and then **Display** → **Spacefill** → **van der Waals Radii** to display the glutamine residues. Because there is only one Gln in the α subunit, this will allow you to quickly identify the region you are looking for.

Select → **Protein** → **Backbone** and then **Display** → **Ribbons** allows you to identify the α helix that contains the deleted amino acids. This helix is adjacent to one face of the heme moiety and includes a key histidine residue (H^{58}) that restricts access to the oxygen-binding site on the heme and may explain why carbon monoxide binds less well to hemoglobin than it does to free heme. Note: you can display just one of the α subunits using the commands **Select** → **Chain** → **A** and then **Select** → **Hide** → **Unselected.**

(b) The absence of the positively charged distal His (H^{58}) in H-j Briska would be expected to reduce the steric hindrance that normally precludes the linear binding of CO to heme, increasing the affinity of Hb for CO. Removing this His residue could also affect the ability of oxygen to access the heme pocket and therefore would affect oxygen binding. In fact, the oxygen affinity of this variant is greater than that of normal hemoglobin, and the cooperativity of oxygen binding is decreased. However, the authors note that no abnormal red blood cell parameters were found in the patients with this Hb variant and that the distal histidine may not be crucial to the function of Hb because histidine is not present in the Hb of some animal species such as the opossum.

(See Wajcman et al., Haemoglobin J-Briska: a new mildly unstable α 1 gene variant with a deletion of eight residues (α 50–57, α 51–58 or α 52–59) including the distal histidine. *British Journal of Haematology,* (1998) Vol. 100: pp. 401–406.)

chapter

6 Enzymes

STEP-BY-STEP GUIDE

Major Concepts

Enzymes are biological catalysts.

Most enzymes are proteins. Many have multiple subunits and require additional components—cofactors and/or coenzymes—for activity. Enzymes are formally named according to the class of reaction they catalyze, but usually have a common or trivial name as well. Enzymes bind the substrate within an **active site,** forming an enzyme-substrate (ES) complex. The transition from reactant to product is limited by an energetic barrier. Enzymes increase the rate of a reaction by lowering this **activation energy.** The binding energy, derived from numerous weak interactions between enzyme and substrate, is the source of this rate increase. The binding energy also contributes to an enzyme's specificity because it allows an enzyme to discriminate between a substrate and a competing molecule. The **equilibrium** of a reaction is dependent on the difference in the free energy of the ground states of the reactants and the products. The equilibrium position is not influenced by the presence of an enzyme.

Understanding the kinetics of enzyme-catalyzed reactions can be very helpful for grasping the mechanisms of catalysis.

Through a series of simplifying assumptions and definitions, the initial reaction equation $\mathbf{E + S \rightleftharpoons ES \rightarrow E + P}$ can be manipulated to give the **Michaelis-Menten equation:**

$$V_0 = \frac{V_{max}\,[S]}{K_m\,[S]}$$

This equation describes the relationship among the initial velocity (V_0) of an enzymatic reaction, the substrate concentration ([S]), and the maximal velocity (V_{max}), via the Michaelis is constant K_m. K_m and V_{max} are measurable and are characteristic for each enzyme and each substrate. The turnover number, k_{cat}, is another useful parameter. It describes the number of substrate molecules that can be converted to product when the enzyme is fully saturated and functioning at its maximal velocity. K_m, V_{max}, and k_{cat}/K_m are all useful parameters for comparing the activities of different enzymes. Enzyme activity can be altered by environmental pH and by the presence of several distinct types of reversible inhibitors: competitive, uncompetitive, and mixed.

Although complete elucidation of enzymatic reaction mechanisms is not currently possible for many enzymes, certain enzymes can be used to illustrate key concepts.

Chymotrypsin provides an example of transition-state stabilization and general acid-base catalysis and covalent catalysis; hexokinase illustrates induced fit; enolase illustrates one type of metal ion catalysis, which occurs in two steps; the lysozyme reaction proceeds through covalent catalysis and general acid catalysis, and also provides an illustrative example of the use of the scientific method.

Regulatory enzymes control the overall rates of sequences of reactions.

Regulatory enzymes are either activated or inhibited by two major mechanisms: reversible, noncovalent allosteric interactions (e.g., the binding of an end product to the regulatory enzyme in a multienzyme pathway), or reversible, covalent modifications (e.g., addition of phosphoryl groups). Allosteric enzymes do not conform to Michaelis-Menten kinetics. Irreversible modification of enzymes also occurs, as in zymogen activation.

What to Review

Answering the following questions and reviewing the relevant concepts, which you have already studied, should make this chapter more understandable.

- You must be familiar with the ways energy is utilized and transferred in biological systems if you are to understand how enzymes facilitate reactions (review Chapter 1, pp. 20–26).
- The local pH and degree of ionization influence the types of interactions in which enzymes and their substrate molecules can participate. For this reason, you should remind yourself what the pK_a of an amino acid indicates (Table 3–1).
- Chymotrypsin is used as a model for enzymatic function in this chapter. Look up the proteolytic specificity of chymotrypsin (Table 3–7).
- A few amino acids have particularly critical roles in regulating the conformation of an enzyme and the types of noncovalent associations it can form. Review the structures and biochemical behaviors of three of these key amino acids: histidine, aspartate, and serine (Fig. 3–5).
- Antibodies can be engineered to function as enzymes. Review the structure and functions of antibodies (pp. 121–123).
- Hemoglobin exhibits cooperativity in the binding of oxygen by each of its subunits (pp. 160–162, Fig. 5-14). How is this cooperativity similar to allosteric effects observed for regulatory enzymes?

Topics for Discussion

Answering each of the following questions, especially in the context of a study group discussion, should help you understand the important points of this chapter.

6.1 An Introduction to Enzymes

Most Enzymes Are Proteins

1. What components besides amino acid residues are sometimes necessary for an enzyme to be active?

2. What are the functions of coenzymes? Note that many are derivatives of vitamins.

Enzymes Are Classified by the Reactions They Catalyze

3. By international agreement, every enzyme is classified according to the type of reaction it catalyzes. Refer back to an organic chemistry textbook, or page forward to the metabolism chapters (Chapters 13–23) in this biochemistry textbook, for examples.

6.2 How Enzymes Work

4. Why are catalysts necessary for reactions in living systems to proceed at a useful rate?

Enzymes Affect Reaction Rates, Not Equilibria

5. What is the difference between a transition state and a reaction intermediate?

6. What affects the reaction equilibrium between S and P? What affects the reaction rate of the conversion from S to P? Which aspect of a reaction can an enzyme alter?

Reaction Rates and Equilibria Have Precise Thermodynamic Definitions

7. What does a large, positive K'_{eq} (e.g., $K'_{eq} = 1{,}000$) for a reaction mean in terms of the final relative concentrations of product and reactants? What does a very small K'_{eq} (e.g., $K'_{eq} = 0.001$) mean?

8. What does K'_{eq} mean in terms of the standard free-energy change of the reaction? How does it relate to the speed at which equilibrium is reached?

9. In qualitative terms, what is the relationship between the rate constant K and the activation energy for an enzymatic reaction?

A Few Principles Explain the Catalytic Power and Specificity of Enzymes

10. What is the specific source of energy for lowering the activation energy barriers in enzyme-catalyzed reactions?

Weak Interactions between Enzyme and Substrate Are Optimized in the Transition State

11. Why is it important for an enzyme to be complementary to the reaction transition state rather than to the substrate?

12. What is one reason that some enzymes are very *large* molecules?

Binding Energy Contributes to Reaction Specificity and Catalysis

13. How does binding energy contribute to the high degree of specificity shown by enzymes?

14. Describe four physical and thermodynamic barriers to reaction, and explain how enzymatic catalysis overcomes them.

15. Explain the difference between the "lock and key" hypothesis and the "induced fit" mechanism.

Specific Catalytic Groups Contribute to Catalysis

16. When is general (as opposed to specific) acid-base catalysis observed?

17. Why must the covalent bond formed between enzyme and substrate in covalent catalysis be transient?

18. In what ways can metal ions participate in catalysis?

6.3 Enzyme Kinetics as an Approach to Understanding Mechanism

Substrate Concentration Affects the Rate of Enzyme-Catalyzed Reactions

19. What assumption concerning substrate concentration is made in the discussion of enzyme kinetics? Why is this important?

20. Explain the effect of saturating levels of substrate on enzyme-catalyzed reactions.

The Relationship between Substrate Concentration and Reaction Rate Can Be Expressed Quantitatively

21. In common terminology, what do the terms [S], V_0, V_{max}, E, ES, E_t, k_1, k_{-1}, and k_2 mean?

22. Why is there no k_{-2} in the equation describing the reaction from E + S to E + P?

23. What is the steady-state assumption?

24. What does the Michaelis-Menten equation describe?

25. When enzyme concentration is held constant, what is the relationship between V_0 and [S] at low [S]? At high [S]? How does the Michaelis-Menten equation illustrate these relationships mathematically?

26. What are the two definitions of K_m? What are its units?

Kinetic Parameters Are Used to Compare Enzyme Activities

27. Under what conditions does K_m (in this case called K_d) represent a measure of affinity of the enzyme for the substrate?

28. What does a high k_{cat} value mean? What does a high K_m value mean?

29. Why is k_{cat}/K_m, the specificity constant, the most useful parameter for discussing catalytic efficiency?

Box 6–1 Transformations of the Michaelis-Menten Equation: The Double-Reciprocal Plot

30. What makes transformation of the Michaelis-Menten equation to the Lineweaver-Burk equation useful? (Understanding the different ways to plot experimental data on enzyme activity is critical to grasping the material in this chapter. Working through the short-answer questions in the Self-Test section will help you to answer this question and to understand V_0 vs. [S] and double-reciprocal plots.)

Many Enzymes Catalyze Reactions with Two or More Substrates

31. Will increasing the concentration of S_2 in Figure 6-14b increase or decrease the overall V_{max} of the reaction?

Pre–Steady State Kinetics Can Provide Evidence for Specific Reaction Steps

32. What kinds of information can be obtained by the study of pre–steady state kinetics that cannot be obtained by steady state kinetics alone?

Enzymes Are Subject to Reversible or Irreversible Inhibition

33. Why are inhibitors necessary as enzymatic control mechanisms in biological systems?

34. Be sure you can explain how the different types of enzyme inhibition affect the apparent K_m and V_{max}.

Box 6–2 Kinetic Tests for Determining Inhibition Mechanisms

35. How would it be possible to distinguish experimentally between a competitive enzyme inhibitor and a mixed inhibitor?

36. Would you expect to find irreversible inhibitors as normal enzymatic control mechanisms in biological systems?

Enzyme Activity Depends on pH

37. Explain how alterations in the surrounding pH can affect enzyme activity.

6.4 Examples of Enzymatic Reactions

38. What types of experiments would be necessary to provide the information on enzyme mechanisms listed in numbered statements 1–5 of the text (p. 205)?

The Chymotrypsin Mechanism Involves Acylation and Deacylation of a Ser Residue

39. How do the serine, histidine, and aspartate residues in chymotrypsin act together to stabilize the transition state during catalysis?

40. How does Figure 6–19 show that the deacylation phase is rate-limiting? (Note that the x-axis of this graph is *not* [S].)

Box 6–3 Evidence for Enzyme–Transition State Complementarity

41. What techniques are used to study enzyme–transition state complementarity?

Hexokinase Undergoes Induced Fit on Substrate Binding

42. What is the basis for the specificity of hexokinase for its substrate?

The Enolase Reaction Mechanism Requires Metal Ions

43. Which residues are important in the mechanism of enolase?

Lysozyme Uses Two Successive Nucleophilic Displacement Reactions

44. How does the substrate for lysozyme become aligned in the active site?

An Understanding of Enzyme Mechanism Drives Important Advances in Medicine

45. HIV protease inhibitors bind to the substrate-binding site of their target enzyme. Why is this interaction so much stronger than the enzyme–substrate interaction?

6.5 Regulatory Enzymes

46. Why is a regulatory enzyme very often the first enzyme in a multiple reaction sequence?

47. What is the most significant difference in how the two major classes of regulatory enzymes are controlled? How are these enzymes similar?

Allosteric Enzymes Undergo Conformational Changes in Response to Modulator Binding

48. How are allosteric enzymes different from other enzymes?

In Many Pathways, Regulated Steps are Catalyzed by Allosteric Enzymes

49. What are the general characteristics of feedback inhibition, as illustrated by the regulation of the conversion of L-threonine to L-isoleucine?

The Kinetic Properties of Allosteric Enzymes Diverge from Michaelis-Menten Behavior

50. Why is the term $K_{0.5}$ (or $[S]_{0.5}$), rather than K_m, used to describe the substrate concentration that produces a half-maximal velocity in an allosteric enzyme-catalyzed reaction?

Some Enzymes Are Regulated by Reversible Covalent Modification

51. What functional groups other than $—PO_4^{2-}$ can be added to or removed from enzymes to turn them on or off?

Phosphoryl Groups Affect the Structure and Catalytic Activity of Enzymes

52. In what ways can phosphoryl groups interact with other groups on an enzyme? How can this affect catalysis?

53. Is glycogen phosphorylase made *more* or *less* active by the addition of phosphoryl groups? (Note: this may be different in other enzymes that are modified in this manner.)

Multiple Phosphorylations Allow Exquisite Regulatory Control

54. Table 6–10 lists myosin light chain kinase as one that recognizes a consensus sequence. What is the function of myosin light chain kinase? (Refer back to Chapter 5.)

55. Note that phosphorylation of glycogen synthase *inactivates* the enzyme; this is the reverse of the effect that phosphorylation has on glycogen phosphorylase. (There is a very good reason for this: see Chapter 15.)

Some Enzymes and Other Proteins Are Regulated by Proteolytic Cleavage of an Enzyme Precursor

56. Compare enzymes regulated by allosteric modulation, by reversible covalent modification, and by proteolytic cleavage with respect to how they are activated and inactivated.

Some Regulatory Enzymes Use Several Regulatory Mechanisms

57. Why is control of catalysis just as important as the fact of catalysis itself?

Discussion Questions for Study Groups

- Make sure to take advantage of the Living Graph opportunity for Equation 6–9. Go to bcs.whfreeman.com/lehninger; a chance to manipulate different variables in the Michaelis-Menten equation will aid greatly in understanding kinetics.
- Return to the Living Graph for Equations 6–28, 6–29, and 6–30, and compare your results with the summary information in Table 6–9.
- Refer to Figure 6–20. Can you estimate the approximate within-protein pKa values for the R-group of His^{57} and the α-amino group of Ile^{16} from this data?
- How are allosteric modulators different from uncompetitive and mixed inhibitors? Discuss in terms of the kinetics effects (graph them!).
- Why is it important in terms of regulatory control that a dephosphorylation reaction is not simply the reverse of the phosphorylation reaction?

SELF-TEST

Enzymes and the study of enzyme reaction rates (enzyme kinetics) are among the most difficult areas of biochemistry for students to assimilate. Consequently, more problems have been included in this chapter's Self-Test. If you work through these problems carefully, your understanding of this material will be greatly enhanced. However, these problems will be beneficial only if you work through them to completion without looking at the answers.

Do You Know the Terms?

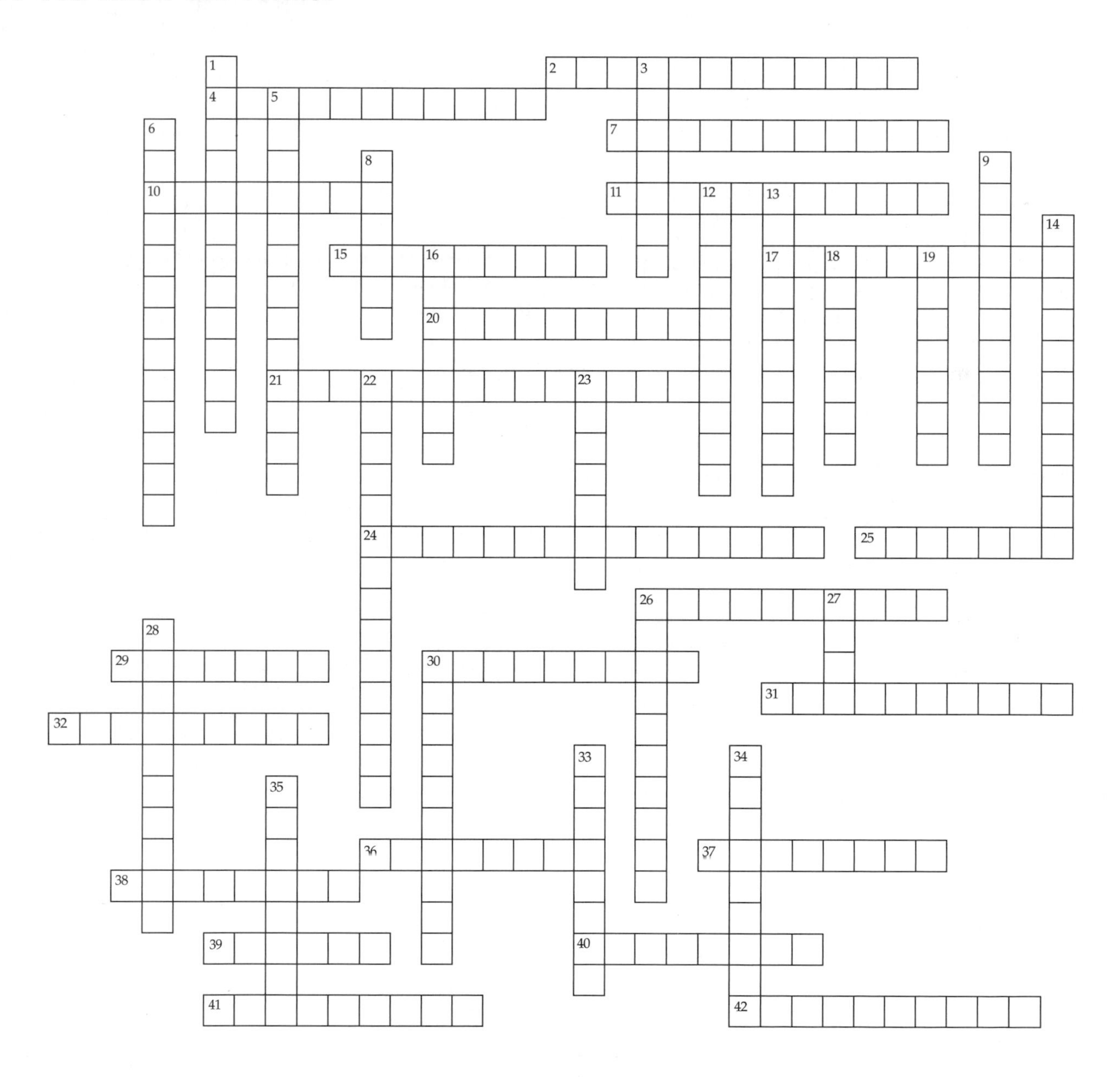

ACROSS

2. The slowest reaction in a sequence is the _______-_______ step.

4. State of a system in which no further net change is occurring.

7. The assumption that the rate of formation of ES is exactly equal to the rate of breakdown of ES is called the _______ _______ assumption. (2 words)

10. k_{cat} is known as the _______ number. At saturating substrate concentrations, $k_{cat} = V_{max}/[E_t]$.

11. Type of inhibitor that alters the K_m of an enzyme without altering V_{max}.

15. Molecule that binds to the active site of an enzyme.

17. Relatively small portion of an enzyme that is involved in substrate binding. (2 words)

20. Describes changes in the conformation of an enzyme upon substrate binding. (2 words)
21. A _______-_______ analog binds more tightly to the active site than does the substrate molecule.
24. _______-_______ kinetics describes the enzymatic activity of an idealized enzyme.
25. Trypsin is to trypsinogen as active enzyme is to _______.
26. Complete enzyme complex including all the protein subunits and prosthetic groups.
29. Type of energy derived from enzyme-substrate interactions that lowers the activation energy for a reaction.
30. An enzyme without its prosthetic group.
31. Hemoglobin without the heme, for example.
32. Common reocurring sequences often recognized by specific protein kinases are known as _______ sequences.
36. Organic cofactor required for certain enzymes to be active.
37. An enzyme that acquires a phosphate from ATP, releases the ATP, binds a second substrate and transfers the phosphate group to the second substrate uses a _______-_______ mechanism.
38. Type of inhibition in which increasing concentrations of a reaction pathway's product decrease the activity of an allosteric enzyme early in that pathway.
39. A specific enzyme that contains Ni^{2+} as a cofactor; it enhances the rate of its reaction by 10^{14}; Sumner's first enzyme crystallized.
40. A molecule essential to the functioning of an enzyme, but not part of the enzyme protein itself (e.g., a divalent cation).
41. Modulator that increases the overall rate of an enzyme-catalyzed reaction.
42. Enzymes that catalyze the rate-limiting step in a reaction sequence usually are _______ enzymes.

DOWN

1. An enzyme whose activity is regulated by a modulator other than its substrate.
3. Binding of a substrate molecule to an enzyme promotes catalysis by reducing the relative motions of the participants, and so reducing _______.
5. Inhibitor that binds only to the ES complex, and therefore cannot bind to the substrate-binding site.
6. EP and ES are examples of reaction _______.
8. Energetic state of a substrate or product molecule in its most energetically stable form.
9. Enzymes catalyze biological reactions by lowering the _______ energy needed for the reaction to proceed.
12. Some enzymes have a covalently linked _______ group essential to its activity.
13. The top of the energy hill is the _______ state.
14. _______ of a substrate occurs when hydrogen bonds between a substrate molecule and water are replaced by noncovalent interactions between the substrate molecule and an enzyme.
16. Inhibitors that irreversibly bind to an enzyme are known as _______ inactivators.
18. ES_1S_2 is a _______ complex.
19. "Ferments" according to Louis Pasteur; the names of many end in "ase."
22. A specific, rare type of mixed inhibitor that alters V_{max} without affecting K_m.
23. Some allosteric enzymes show _______ kinetic behavior, which reflects cooperativity.
26. Allosteric enzyme whose substrate is also a modulator of activity.
27. When the K'_{eq} = one, $\Delta G'^0$ = _______.
28. _______-_______ reactions involve a single substrate molecule, and the rate depends only on [S].
30. Class of regulatory enzymes that change their conformation when bound to a modulator.
33. The study of reaction rates in biological systems is referred to as enzyme _______.
34. Agent that reduces the overall rate of an enzyme-catalyzed reaction.
35. Regulation of enzyme activity by the reversible binding of a phosphoryl group is an example of regulation by _______ modification.

Do You Know the Facts?

1. In some enzymes, components other than amino acid residues are necessary for activity. These components are called ______________________________.
In such enzymes, the complete, active form is called a(n) ____________________ and the enzyme without its additional component is called a(n) ____________________.

2. Are enzymes always proteins, consisting only of amino acids? Be sure to explain your answer.

3. Why are enzymes necessary for the catalysis of reactions in living systems?

4. Why must the covalent bond formed between enzyme and substrate in covalent catalysis be transient?

5. Why are regulatory enzymes very often the first enzyme in a multireaction sequence?

6. How are allosteric enzymes different from other enzymes?

7. What is the difference between reversible covalent modification and proteolytic cleavage?

8. Label the double-reciprocal plot below with the letters corresponding to items (a)–(g).
   (a) A typical Michaelis-Menten enzyme in the absence of inhibitors
   (b) Enzyme activity in the presence of a noncompetitive inhibitor
   (c) Enzyme activity in the presence of a competitive inhibitor
   (d) 1/[S]
   (e) $1/V_0$
   (f) $-1/K_m$
   (g) $1/V_{max}$

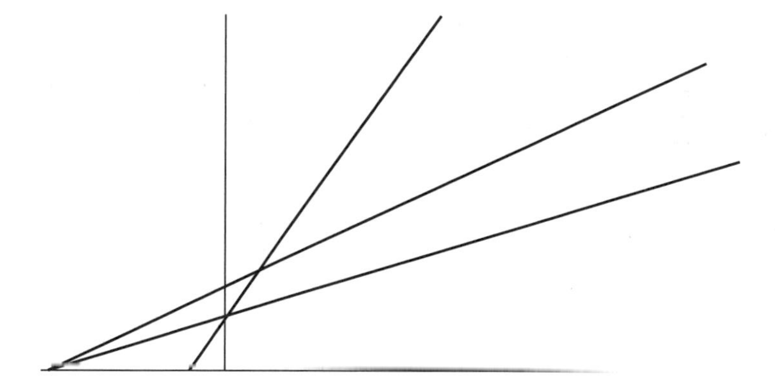

For the situations described in questions 9–12, indicate whether K_m, V_{max}, both parameters, or neither parameter would be altered. You can assume that [S] is in excess.

9. In the presence of a mixed inhibitor:

10. After [S] has been doubled:

11. In the presence of a competitive inhibitor:

12. After the [E] has been doubled:

13. An enzyme facilitates chemical reactions by:
A. increasing the free-energy difference between reactants and products.
B. decreasing the free-energy difference between reactants and products.
C. lowering the activation energy of the reaction.
D. raising the activation energy of the reaction.
E. none of the above.

14. Which of the following is not a cellular mechanism for regulating enzyme activity?
A. Binding of regulatory peptides via disulfide bonds
B. Proteolysis
C. Covalent modification
D. Induced changes in conformation

15. Using site-directed mutagenesis, you have created a mutant form of chymotrypsin that has alanine substituted for the usual serine at position 195. Which of the following effects would you expect to observe?
A. No effect or a slight increase in affinity for substrate coupled to a complete loss of enzyme activity
B. A decrease in the affinity for substrate coupled to a decrease in enzyme activity
C. An increase in the rate of peptide-bond cleavage due to an increase in the rate of acid-base catalysis
D. An increase in the rate of peptide-bond cleavage due to an increase in the rate of covalent catalysis
E. A complete loss of enzyme activity due to the inability to bind substrate

In questions 16–19, calculate the indicated parameters using the information below.
You wish to characterize a new enzyme that you have isolated. You set up a series of test tubes, each containing 0.1 μM enzyme and various substrate concentrations. You measure the activity of the enzyme in each tube and plot the enzyme activity (V_0) vs. [S]. The plot is shown below.

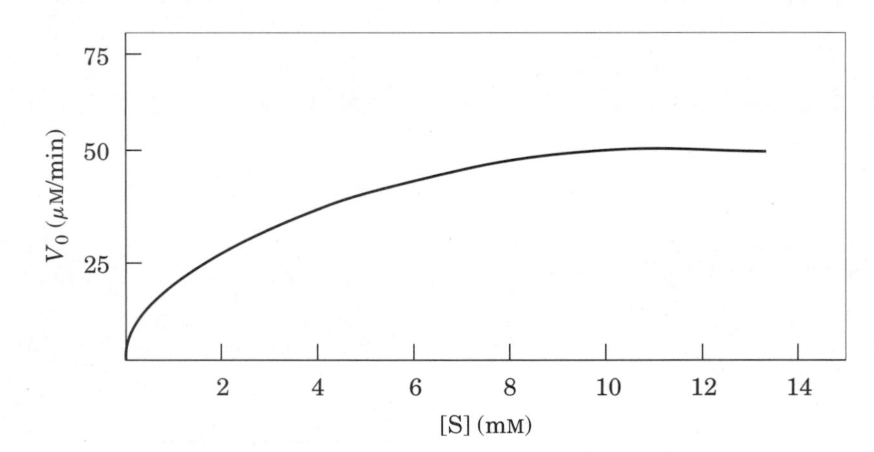

16. What is V_{max} for your enzyme with this substrate?
A. 10 μM/min
B. 30 μM/min
C. 0.8 μM/min
D. 0.05 mM/min
E. 50 mM/min

17. What is the K_m of your enzyme for this substrate?
A. 1.5 μM
B. 10 μM
C. 0.5 mM
D. 1.0 mM
E. 1.5 mM

18. What is the k_{cat} (turnover number) for your enzyme?
A. 5×10^2/s
B. 1.0/min
C. 5×10^2/min
D. 5×10^3/min
E. 1×10^2/s

19. Why does your plot eventually plateau?
A. The enzyme becomes inhibited at high substrate concentrations.
B. The enzyme affinity for substrate changes.
C. The enzyme is being degraded by proteases.
D. The active site is saturated with substrate.
E. The substrate is being degraded.

20. For the following enzyme-catalyzed reaction, why is the term k_{-2} not included in the Michaelis-Menten equation?

$$E + S \underset{k_{-1}}{\overset{k_1}{\rightleftharpoons}} ES \overset{k_2}{\rightleftharpoons} E + P$$

A. This reaction never occurs.
B. It simplifies the math.
C. The Michaelis-Menten equation only describes initial reaction rates.
D. It is part of the term K_m.
E. The Michaelis-Menten equation only describes the formation of the ES complex.

21. Indicate whether each of the following is true or false regarding Michaelis-Menten kinetics.

(a) K_m has the dimensions of rate per second.
(b) $K_m = k_2 + k_{-1}/k_1$.
(c) K_m is equal to the substrate concentration that will bring the reaction to maximal velocity.
(d) At very high substrate concentration, the velocity of the reaction is independent of substrate concentration.
(e) The concentration of the enzyme-substrate complex stays constant throughout the reaction.
(f) Enzyme concentration is much lower than substrate concentration.
(g) The velocity of the reaction is equal to the (mathematical) product of (concentration of enzyme-substrate complex) and (the rate constant), or $V_0 = [ES] \times k_2$.

22. Indicate whether each of the following statements about enzymes is true or false.

(a) To be effective, they must be present at the same concentration as their substrate.
(b) They increase the equilibrium constant for the reaction, thus favoring product formation.
(c) They increase the rate at which substrate is converted to product.
(d) They ensure that all substrate is converted to product.
(e) They ensure that the product is more thermodynamically stable than the substrate.
(f) They lower the activation energy for conversion of substrate to product.
(g) They are consumed in the reactions that they catalyze.

Applying What You Know

1. In your studies of the enzyme ribonuclease A, you obtain activity data for a wild-type enzyme and a mutant ribonuclease A. The two enzymes differ at one amino acid position in the protein. From the activity data, you calculate the following kinetic parameters:

	Maximum velocity	K_m
Wild type	100 μmol/min	10 mM
Mutant	1 μmol/min	0.1 mM

(a) Which enzyme has a higher affinity for substrate? (Assume a two-step reaction, with k_2 the rate-limiting step.)

(b) What is the initial velocity of the reaction catalyzed by the wild-type enzyme at a substrate concentration of 10 mM?

(c) Which enzyme shifts K_{eq}, the equilibrium constant, more in the direction of product?

2. The sensitivity of individuals of certain ethnic backgrounds to alcoholic beverages has a biochemical basis. In such individuals, much less ethanol is required to produce the vasodilation that results in facial flushing than is required to achieve the same effect in those without such sensitivity. This physiological effect arises from the acetaldehyde generated by liver **alcohol dehydrogenase (AD),** which catalyzes the following reaction:

$$\underset{\text{Ethanol}}{CH_3CH_2OH} + NAD^+ \underset{AD}{\rightleftharpoons} \underset{\text{Acetaldehyde}}{CH_3CHO} + H^+ + NADH$$

In sensitive individuals, it is the next step that confers sensitivity. Sensitive individuals have a different isozyme of the enzyme aldehyde dehydrogenase (ADD), which converts acetaldehyde to acetate.

$$\underset{\text{Acetaldehyde}}{CH_3CHO} + NAD^+ \underset{ADD}{\rightleftharpoons} \underset{\text{Acetate}}{CH_3COO^-} + H^+ + NADH$$

This isozyme allows a higher concentration of acetaldehyde to accumulate in the blood after alcohol consumption.

(a) Which ADD isozyme has a higher K_m for acetaldehyde?

(b) You are investigating the effects of several agents on the activity of alcohol dehydrogenase.

The enzyme activity data are shown in the table below. (Hint: If you put these data into a spreadsheet it will be much easier to work with. You can use the spreadsheet tools to do curve fitting and to perform a linear regression analysis.)

[Alcohol] (mM)	AD activity (V_0, mM/min)	AD activity + agent A (V_0, mM/min)	AD activity + agent B (V_0, mM/min)
0.1	14	2	5
0.5	50	7	8
1.0	65	10	30
2.0	72	12	45
4.0	80	14	62
8.0	85	15	75
32.0	90	16	90

Construct a **[substrate] vs. activity plot** and a **double-reciprocal plot** for this enzyme. Be sure to label all axes.

Determine the V_{max} and K_m for AD from the graphs in each type of plot.

Using the same plots, graph the data for AD activity in the presence of agent A and agent B. What information does this provide about the effects of these compounds on the activity of alcohol dehydrogenase?

ANSWERS

Do You Know the Terms?

ACROSS

2. rate-limiting
4. equilibrium
7. steady state
10. turnover
11. competitive
15. substrate
17. active site
20. induced fit
21. transition-state
24. Michaelis-Menten
25. zymogen
26. holoenzyme
29. binding
30. apoenzyme
31. apoprotein
32. consensus
36. coenzyme
37. Ping-Pong
38. feedback
39. urease
40. cofactor
41. activator
42. regulatory

DOWN

1. heterotropic
3. entropy
5. uncompetitive
6. intermediates
8. ground
9. activation
12. prosthetic
13. transition
14. desolvation
16. suicide
18. ternary
19. enzymes
22. noncompetitive
23. sigmoid
26. homotropic
27. zero
28. first-order
30. allosteric
33. kinetics
34. inhibitor
35. covalent

Do You Know the Facts?

1. Cofactors, coenzymes, and, if tightly bound, prosthetic groups; holoenzyme; apoenzyme
2. Enzymes are usually proteins, but some nucleic acids (RNA) show enzymatic activity. Many enzymes, though not all, also have non-amino acid portions, called cofactors or coenzymes, that are necessary for catalytic activity.
3. At the relatively mild temperature and pH of cells, most biomolecules are quite stable. Many common reactions in biochemistry do not occur at an apprc ciable rate. Enzymes circumvent these problems by providing a specific environment (the active site) within which a given reaction will occur at a significantly increased rate.
4. The enzyme must be regenerated in free form (the covalent bond to the substrate/product must be undone) to be able to catalyze the reaction again. Enzymes are catalysts and must be recycled many times.
5. In cell metabolism, groups of enzymes work together in sequential pathways to carry out a given metabolic process. Catalyzing even the first few reactions of a pathway leading to an unneeded product is a waste of energy and materials. Therefore, the first enzyme in the pathway is often the ideal place to turn on or off the whole pathway.
6. Allosteric enzymes are generally larger and more complex than simple enzymes. Most have two or more polypeptide subunits. In addition to active sites (catalytic sites), allosteric enzymes have at least one regulatory (allosteric) site that binds a modulator(s). Allosteric enzymes do not follow Michaelis-Menten kinetics.
7. Reversible covalent modification is just that—reversible. Phosphorylation, methylation, etc. are reversible modifications used to regulate the activity of enzymes and other proteins. Proteolytic cleavage is also a covalent modification of an enzyme, but it is irreversible. This process is also referred to as zymogen or proprotein activation.

8.

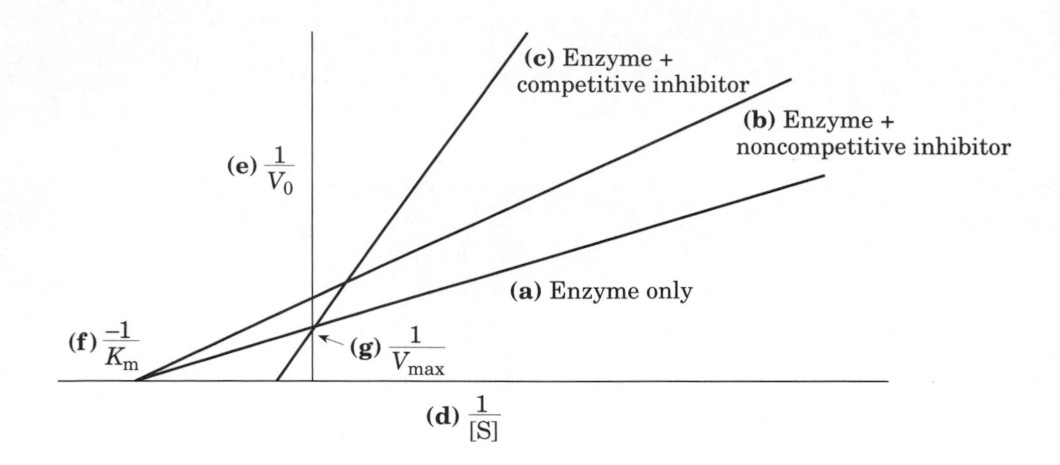

9. V_{max} and K_m both (usually). In the rare case of $\alpha = \alpha'$, which is called noncompetitive inhibition, K_m is not affected.

10. Neither

11. K_m

12. V_{max}

13. C

14. A

15. A

16. D

17. E

18. C $k_{cat} = V_{max}/E_t$ (the total enzyme present is 0.1 μM)
= (50 μM/min)/0.1 μM
= 500 (molecules of substrate transformed per molecule of enzyme)/min

19. D

20. C

21. **(a)** F; **(b)** T; **(c)** F; **(d)** T; **(e)** T; **(f)** T; **(g)** T

22. **(a)** F; **(b)** F; **(c)** T; **(d)** F; **(e)** F; **(f)** T **(g)** F

Applying What You Know

1. (a) The mutant enzyme binds substrate more strongly than does the wild-type enzyme because the amount of substrate that results in a half-maximal velocity is much lower for the mutant enzyme (0.1 mM) than for the wild-type enzyme (10 mM).

(b) The velocity of the enzyme can be calculated for any substrate concentration if you know the values of K_m and V_{max}.

$$V_0 = \frac{V_{max}\,[S]}{K_m + [S]}$$

$$= 100\,\frac{\mu\text{mol/min} \times 10\text{mM}}{10\text{ mM} + 10\text{ mM}}$$

$$= 50\mu\text{mol/min}$$

Be sure to use the same units when substituting values (i.e., either mM or μM).

(c) Neither enzyme alters the equilibrium position for this reaction; they only alter the rate at which that equilibrium position is reached.

2. (a) The K_m for the ADD isozyme in alcohol sensitive individuals is higher than that in nonsensitive individuals.

(b) See graph at right. From the V_0 vs. [S] plot, one can estimate the V_{max} for AD to be 90 mM/min. V_{max} in the presence of agent A can be estimated to be approximately 16 mM/min. V_{max} in the presence of agent B is unclear because it is not obvious that the curve has reached a plateau. K_m values can be obtained by estimated [S] at 1/2 V_{max}.

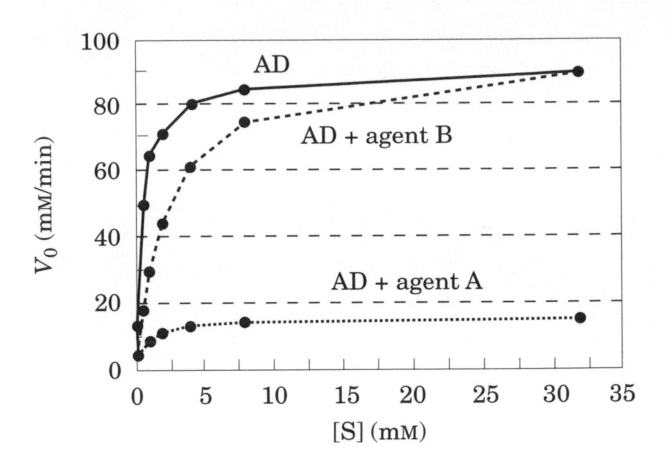

	AD Alone	AD + Agent A	AD + Agent B
V_{max} (1/b)	90 mM/min	16 mM/min	?
K_m ($m \times V_{max}$)	<1 mM	approx. 1 mM	?

The V_{max} can be calculated accurately for AD in all three samples. To do this, calculate the values for 1/[S] and $1/V_0$ for the enzyme and enzyme plus each inhibitor. Graph $1/V_0$ vs. 1/[S], and perform a linear regression on each set of data. This analysis will fit the data to a straight line using the formula ($y = mx + b$). The value for b will be the y-intercept, which is equal to $1/V_{max}$. The constant m is the slope of the line, which is equal to K_m/V_{max}. The value for R^2 tells you how well the data fit the formula; with a perfect fit, R^2 would be 1.0. Such analysis yields the following information:

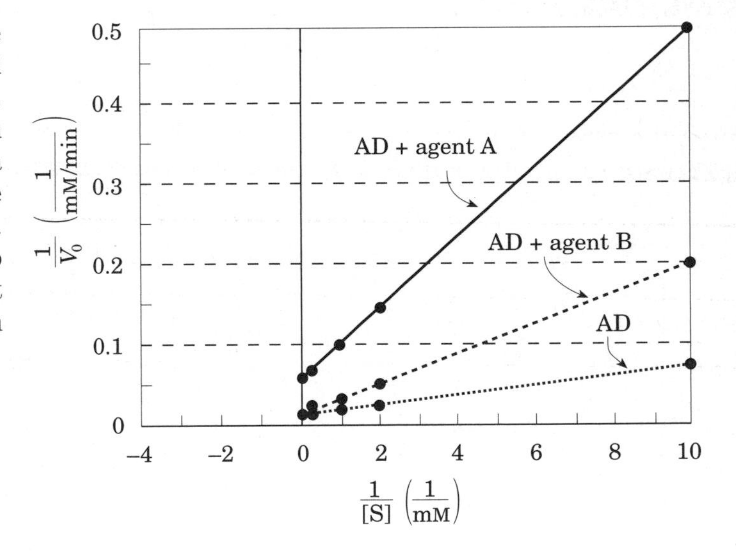

	AD Alone	AD + Agent A	AD + Agent B
M	0.006	0.044	0.019
B	0.010	0.060	0.013
V_{max} (1/b)	98 mM/min	18 mM/min	78 mM/min
K_m ($m \times V_{max}$)	0.59 mM	0.78 mM	1.5 mM
R^2	1.0	1.0	1.0

An examination of the values in the previous table provides the following insights as to the actions of agents A and B:

Agent A decreased the V_{max} of AD by 82%, while increasing K_m 34%. Agent A acts primarily as a mixed inhibitor of AD.

Agent B, on the other hand, does not affect the V_{max} as significantly as Agent A. (The maximal velocity decreased by 20%.) However, the K_m increased by more than 100% in the presence of agent B. This suggests that agent B may be a competitive inhibitor of AD or a mixed inhibitor. Further analysis is needed to determine which type of inhibitor it is.

chapter

7

Carbohydrates and Glycobiology

STEP-BY-STEP GUIDE

Major Concepts

Monosaccharides are the monomeric subunits from which di-, oligo-, and polysaccharides are formed.

They are classified according to the *position* of the carbonyl group (aldose or ketose), the total *number* of carbon atoms in the chain (triose to heptose), the *isomeric form* (L or D, **epimers, anomers**), whether or not they occur in *cyclic* form, and whether or not they have another **chemical substituent.**

Polysaccharides vary in monomeric composition, type of glycosidic bond connecting the monosaccharide units, chain length and degree of branching, and biological function.

Starch and **glycogen** are **homopolysaccharides** of glucose that function as fuel stores in plants and animals, respectively. Both starch and glycogen are chains of glucose monomers connected by linear linkages, (α**1→4**) glycosidic bonds, but they have different degrees of branching through (α**1→6**) glycosidic bonds. Branching allows faster degradation of a polysaccharide to individual monosaccharide units, which can then be used to generate biologically useful energy. **Cellulose** is also a homopolymer of glucose, but its (β**1→4**) linkages result in rigid extended polymers that are used by cells for support rather than as energy storage molecules. More complex **heteropolysaccharides,** composed of more than one type of monosaccharide, can be covalently cross-linked by peptides or proteins to form **peptidoglycans** or **proteoglycans.** These carbohydrate-protein aggregates function to provide structural support (bacterial cell wall), connectivity (extracellular matrix), and lubrication (synovial fluid).

Glycoconjugates include proteoglycans, glycoproteins, and glycolipids. These hybrid molecules have carbohydrate portions that alter or augment the functional properties of their protein and lipid components.

Proteoglycans are important components of the extracellular matrix of multicellular animals. They are composed primarily (by mass) of **glycosaminoglycan** chains that are connected to a protein and are important for structure and resilience. **Glycoproteins** contain smaller, more diverse oligosaccharides that are linked to amino acid residues of a protein via ***O***- or ***N***-glycosidic bonds; these information-rich carbohydrate portions commonly act as recognition sites for enzymes or receptors, or for targeting a protein to a specific cellular location. Membrane **glycolipids** such as the gangliosides, and bacterial **lipopolysaccharides,** are lipids that are covalently bound to oligosaccharide moieties.

The diversity of oligosaccharide structure found in glycoconjugates makes these molecules informational molecules, actually surpassing nucleic acids in the density of information carried.

Lectins, oligosaccharide-binding proteins, are central in cell-cell recognition and signaling processes. Lectin-carbohydrate recognition and binding are highly specific, and, like other biochemical contacts, are mediated by multiple weak, noncovalent interactions.

Chemical analysis of carbohydrates makes use of some of the same techniques used for protein analysis.

Full determination of the structure of complex heteropolysaccharides involves many steps not only because there are a variety of monosaccharide subunits, but also because there are a variety of linkages possible. Mass spectrometry and NMR spectroscopy are two important procedures used to characterize carbohydrates.

What to Review

Answering the following questions and reviewing the relevant concepts, which you have already studied, should make this chapter more understandable.

- Review the structure of aldehydes, ketones, and hydroxyl groups (Fig. 1–15); all of these groups are found in carbohydrates.
- Simple sugars exist in a variety of isomeric forms. Review what is meant by enantiomers (Fig. 1–19) and the various conformations of biological molecules (Figs. 1–18 through 1–23).
- Many sugars can be oxidized and are termed reducing sugars. Be sure you understand the overall concept of oxidation-reduction reactions.
- Review the various analytical techniques used in the study of amino acids and proteins (Figs. 1–8, 3–16 through 3–22). Which ones can be applied to the study of carbohydrates?

Topics for Discussion

Answering each of the following questions, especially in the context of a study group discussion, should help you understand the important points of this chapter.

7.1 Monosaccharides and Disaccharides

1. Why are monosaccharides appropriately named "*carbo hydrates*"?

The Two Families of Monosaccharides Are Aldoses and Ketoses

2. Why are monosaccharides soluble in water but not in nonpolar solvents?

3. What are the naming conventions for monosaccharides?

Monosaccharides Have Asymmetric Centers

4. The word "chiral" is derived from the Greek word meaning "hand." How does this relate to the structure of carbohydrates?

5. How many D isomers would an aldopentose have? How many total isomers?

6. What is a possible explanation for the observation that most of the hexoses found in living organisms are D isomers? (Hint: see Chapter 6.)

7. Is galactose an epimer of mannose?

The Common Monosaccharides Have Cyclic Structures

8. In an aqueous solution of D-glucose, why will there always be a very small amount of the linear form of the monosaccharide?

9. Does the discussion on conformations of cyclic forms of monosaccharides suggest an explanation for why the more common anomer of D-fructofuranose is the β anomer? (Hint: see Fig. 7–8.)

10. How many D isomers would a cyclized aldopentose have? How many total isomers?

Organisms Contain a Variety of Hexose Derivatives

11. What are some of the biologically more important monosaccharide derivatives?

12. Why is phosphorylation of sugars beneficial to a cell?

Monosaccharides Are Reducing Agents

13. What is reduced and what is oxidized in the reaction between a monosaccharide and a ferric ion?

Disaccharides Contain a Glycosidic Bond

14. Be able to draw and give the complete name of the disaccharides commonly known as **maltose, lactose,** and **sucrose.** Which of these is/are reducing sugars?

7.2 Polysaccharides

15. What are the structural and functional differences between homopolysaccharides and heteropolysaccharides?

16. Why are *homo*polysaccharides not useful as informational molecules?

Some Homopolysaccharides Are Stored Forms of Fuels

17. Suppose you had three polysaccharides, amylose, amylopectin, and glycogen, each with the same number of monosaccharide subunits. Which would be degraded the fastest, assuming that the enzymes acting to degrade the polysaccharides all worked at the same rate?

Some Homopolysaccharides Serve Structural Roles

18. What are the similarities and differences between cellulose and chitin?

Steric Factors and Hydrogen Bonding Influence Homopolysaccharide Folding

19. What are the most important noncovalent bonds or interactions in cellulose?

Bacterial and Algal Cell Walls Contain Structural Heteropolysaccharides

20. How does lysozyme act as a first defense against bacterial infection? How does penicillin combat bacterial infections?

21. What chemical property of agarose contributes to its ability to trap large amounts of water?

Glycosaminoglycans Are Heteropolysaccharides of the Extracellular Matrix

22. What chemical property of glycosaminoglycans contributes to their role as lubricants?

23. How does heparin work to inhibit blood coagulation?

7.3 Glycoconjugates: Proteoglycans, Glycoproteins, and Glycolipids

24. One of the distinctions between proteoglycans and glycoproteins is entirely relative. The second half of the name *usually* indicates the predominant species. Proteo*glycans* are primarily glycans (or polysaccharides), whereas glyco*proteins* usually contain more protein (by weight).

Proteoglycans Are Glycosaminoglycan-Containing Macromolecules of the Cell Surface and Extracellular Matrix

25. How does the R-group of Ser make it a good amino acid for connecting to a carbohydrate moiety?

26. If a proteoglycan aggregate is compared to an evergreen tree, with hyaluronate as the "trunk," what compound would be the "needles" on this tree? What role do the core proteins play?

27. What is the relationship among glycosaminoglycans, proteoglycans, fibrous proteins, integrins, and the extracellular matrix?

Glycoproteins Have Covalently Attached Oligosaccharides

28. How are the functions of glycoproteins different from those of glycosaminoglycans?

29. What types of linkages connect oligosaccharides to proteins?

30. What kinds of biologically important information are encoded by the oligosaccharide portions of glycoproteins?

31. How does the addition of carbohydrate moieties alter the chemistry of glycoproteins and glycolipids?

Glycolipids and Lipopolysaccharides Are Membrane Components

32. How do glycolipids and lipopolysaccharides differ?

7.4 Carbohydrates as Informational Molecules: The Sugar Code

33. How can so much distinguishing information be packed into an oligosaccharide of comparatively few monosaccharide units?

Lectins Are Proteins That Read the Sugar Code and Mediate Many Biological Processes

34. Why is it important that there are signals for removal and destruction of "old" cells and hormones?

35. Why do people of blood type O tend to have gastric ulcers more often than do people of blood type A or B?

Lectin-Carbohydrate Interactions Are Highly Specific and Often Polyvalent

36. A quick review of the pK_a of the histidine R-group will help in understanding the mannose 6-phosphate receptor-mannose 6-phosphate interaction.

7.5 Working with Carbohydrates

37. What makes analysis of carbohydrates so much more difficult than analysis of proteins?

38. How can the specificity of lectins and glycosidases be used to identify and analyze carbohydrates?

Discussion Questions for Study Groups

- How can such a simple difference—the difference between α and β glycosidic bonds—produce differences in function and three-dimensional structure as drastic as those seen between amylose and cellulose?
- S-domains of altered heparin sulfate bind to extracellular proteins and signaling molecules in several ways. Discuss the different noncovalent interactions possible for binding on the basis of the chemical groups present in these molecules.

SELF-TEST

Do You Know the Terms?

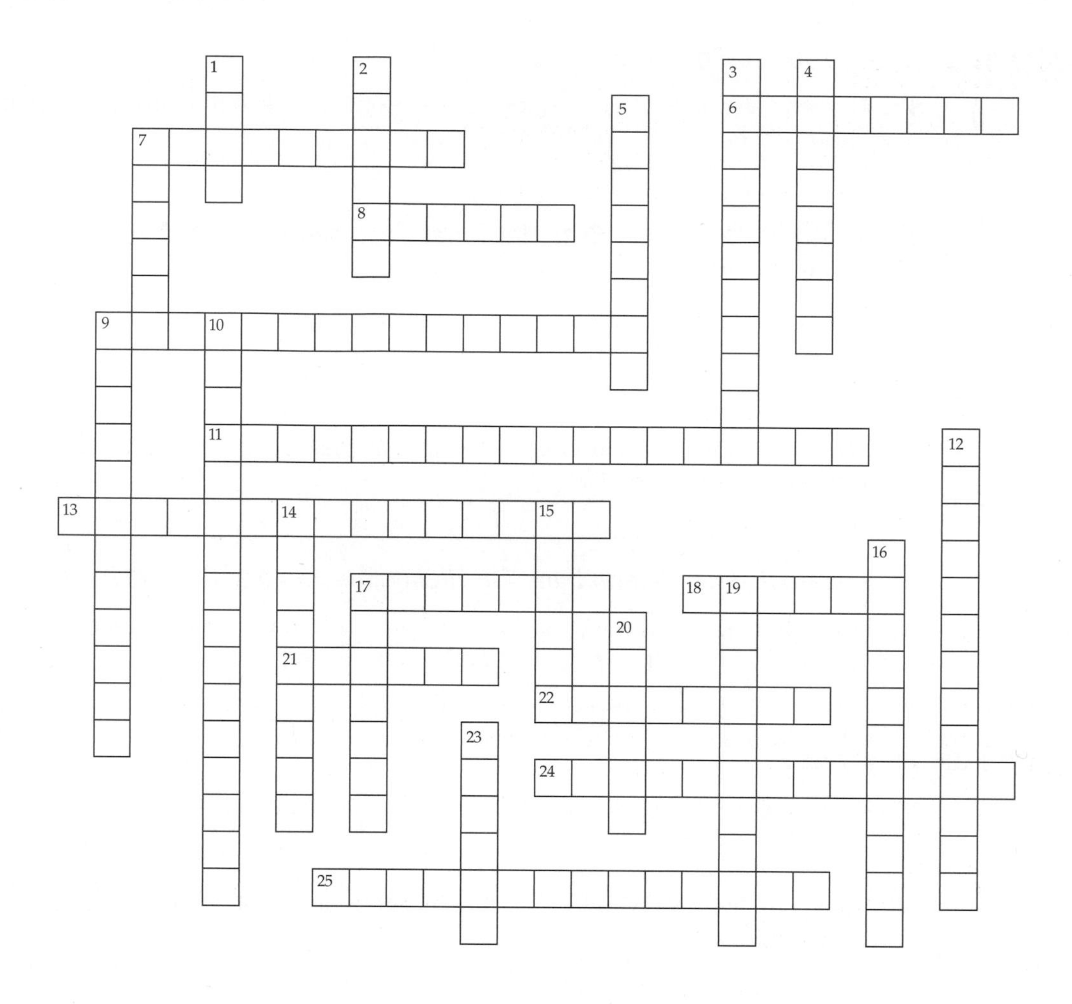

ACROSS

6. A homopolysaccharide of glucose; it is highly branched and found exclusively in animal cells.

7. Formed by cyclization of a ketose sugar.

8. A homopolysaccharide of glucose units connected by (α1→4) glycosidic bonds; found exclusively in plants.

9. Simple sugars.

11. Heteropolysaccharides such as hyaluronate.

13. Glycogen and cellulose, with thousands of simple sugar subunits, are examples.

17. Oxidation of the carbonyl carbon of sugars (except glucose) results in the formation of _______ acids.

18. A compound with an asymmetrical atom allowing formation of mirror-image isomers has one or more _______ centers.

21. β-D-glucuronate is an example of a _______ acid.

22. End of a polysaccharide chain that is not involved in a glycosidic bond and has a free anomeric carbon.

24. Lactose and sucrose are examples.

25. Polysaccharides cross-linked by peptides; found in bacterial cell walls.

DOWN

1. A _______ polysaccharide is a polymer of repeating monosaccharides.

2. A sugar with a carbonyl group at C-2 (or any position other than C-1).

3. Carbohydrate moieties are attached to glycoproteins through either *N*- or _______-_______ bonds.

4. Six-membered ring form of sugars.

5. Five-membered ring form of sugars.

7. A polysaccharide containing more than one type of sugar is a _______ polysaccharide; an example is chondroitin sulfate.

9. Process that interconverts isomers of pyranoses.

10. Lectins are proteins that bind to specific _______.

12. Glycoconjugates containing protein and oligosaccharide portions: for example, glycophorin A.

14. A homopolysaccharide of glucose units connected by (β1→4) glycosidic bonds; it is found exclusively in plants.

15. An isomer that differs at only one of two or more chiral centers.

16. Gangliosides, for example.
17. The α and β forms of a pyranose, for example.
19. In the formation of pyranoses, linkage between the aldehyde on C-1 and the alcohol on C-5.
20. A sugar with the carbonyl group at C-1.
23. Animal tissues have an extracellular ________ composed of glycoconjugates and fibrous proteins.

Do You Know the Facts?

1. Which of the following is *not* a characteristic of carbohydrates in cells?
A. They serve as energy stores in plants and animals.
B. They are major structural components of plant tissues.
C. They act as binding sites for proteins.
D. They are organic catalysts.
E. They play a role in cell-cell recognition.

2. Which of the following contributes to the structural rigidity of cellulose?
A. Adjacent glucose polymers are stabilized by hydrogen bonding.
B. Glucose residues are joined by (α1→4) linkages.
C. Cellulose is a highly branched molecule.
D. The conformation of the glucose polymer is a coiled structure.
E. Adjacent polymers are covalently linked by short peptides.

3. Which of the following is an epimer of glucose?
A. Talose
B. Idose
C. Gulose
D. Altrose
E. Allose

4. Why are sugars usually found as phosphorylated derivatives in cells?
A. Phosphorylated sugars are important in regulating cellular pH.
B. Unphosphorylated sugars can be transported across cell membranes.
C. Unphosphorylated sugars are rapidly degraded by cellular enzymes.
D. Phosphorylated sugars encode genetic information.
E. None of the above is a correct explanation.

5. Which of the following disaccharides could be extended to form a cellulose polymer?
A. Sucrose
B. Trehalose
C. Maltose
D. Lactose
E. None of the above

6. Which of the following is a heteropolysaccharide? What is its function?
A. Glycogen
B. Hyaluronate
C. Starch
D. Cellulose
E. Chitin

7. **(a)** Draw the pyranose form of β-D-glucose.

(b) How many isomers of glucopyranose are possible?

(c) Describe this monosaccharide using at least four chemical terms.

8. **(a)** Draw the furanose form of α-D-fructofuranose.

(b) How many isomers of fructofuranose are possible?

(c) Describe this monosaccharide using at least four chemical terms.

9. In addition to simple sugars, what other compounds have you encountered in the text that exist as enantiomers?

10. Cellulose is a homopolysaccharide of glucose units. Why can't this molecule be used by most organisms to generate energy?

11. Draw straight-chain (Fischer projection) and cyclic (Haworth perspective) representations of the following molecules. Identify your Fischer projections as showing the L or the D form, and identify your Haworth perspectives as the α or β form.

(a) Glyceraldehyde

(b) Dihydroxyacetone

(c) Ribose

(d) Galactose

12. Draw the Haworth perspective formulas for disaccharides containing two glucose molecules:

(a) in an (α1→4) linkage.

(b) in a (β1→4) linkage.

(c) in an (α1→6) linkage.

Applying What You Know

1. Why are many membrane proteins glycoproteins?

2. Describe the relationship between glycosaminoglycans and a proteoglycans aggregate. How is collagen involved in this arrangement?

3. What structural differences between glycogen and cellulose explain their functional differences?

4. Amylopectin is a homopolysaccharide containing 300–6,000 glucose residues. Because a single glucose residue is approximately 0.001 μm long, the largest amylopectin molecules would be 6 μm long, if all the residues were arranged in a single chain. Plant cells store amylopectin in granules whose diameters are 3–100 μm. How do plant cells maximize the storage capacity of these (relatively) small structures? What other factor(s) probably have influenced the way cells store glucose molecules?

5. The enzyme hyaluronidase is found as a component of some snake and insect toxins. How does it increase a toxin's effectiveness?

6. Consider the following experimental observations: (1) In many glycoproteins, the oligosaccharide portion attaches to the protein portion at sequences that form β bends. (2) The three-dimensional protein architecture of glycoproteins is *not* affected by removal of the associated oligosaccharide(s). What do these observations tell you about the spatial placement of the oligosaccharides in glycoproteins?

7. The bacterium *Helicobacter pylori* adheres to the surface of the stomach by interactions between bacterial membrane lectins and specific oligosaccharides of membrane glycoproteins of gastric epithelial cells. One of the binding sites recognized by *H. pylori* is the oligosaccharide called Le^b. Chemically synthesized analogs of the Le^b oligosaccharide can be administered orally to inhibit bacterial attachment. Refer to Figure 5–4a; draw a ligand binding curve to illustrate the *H. pylori* lectin-Le^b oligosaccharide interaction, with and without the presence of Le^b analog. What type of interaction is this similar to? (Hint: see Chapter 6.)

8. The starch present in barley consists of polymers of glucose that are linked by two types of bonds: (α1→4) and (α1→6) linkages. The enzymes found in malt (the dried, germinated barley that is rich in hydrolytic enzymes) are used to hydrolyze some of these bonds in starch to produce free glucose, which can then be fermented to ethanol and CO_2. There are other methods for making fermented beverages; for example, "spit beer" is made by a person chewing corn briefly and then spitting it into a container. The action of salivary amylase is similar to that of the barley enzymes. Neither the barley nor salivary enzymes can degrade starch completely; the pieces of starch left unhydrolyzed are highly branched and are referred to as limit dextrins. In "light" beer production, however, a mold enzyme is used to split all the limit dextrins to individual glucose molecules. What bond(s) must this accessory mold enzyme be able to cleave?

ANSWERS

Do You Know the Terms?

Across

6. glycogen
7. hemiketal
8. starch
9. monosaccharides
11. glycosaminoglycans
13. polysaccharides
17. aldonic
18. chiral
21. uronic
22. reducing
24. disaccharides
25. peptidoglycans

Down

1. homo
2. ketose
3. *O*-glycosidic
4. pyranose
5. furanose
7. hetero
9. mutarotation
10. oligosaccharides
12. glycoproteins
14. cellulose
15. epimer
16. glycolipids
17. anomers
19. hemiacetal
20. aldose
23. matrix

Do You Know the Facts?

1. D
2. A
3. E
4. B
5. E
6. B; Hyaluronate plays an important structural and lubricating role in the extracellular matrix of animal tissues.
7. (a)

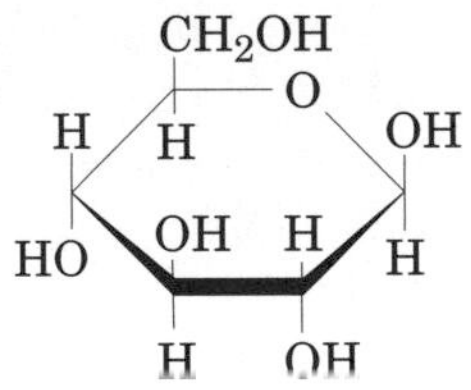

β-D-glucose

(b) 32 isomers (chiral carbons, $n = 5$; $2^n = 32$)
(c) hexose, aldose, pyranose, reducing sugar

8. (a)

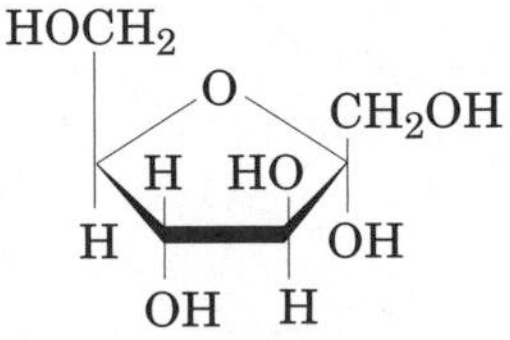

α-D-fructose

(b) 16 isomers (chiral carbons, $n = 4$; $2^n = 16$)
(c) hexose, ketose, furanose, reducing sugar

9. Amino acids other than glycine also have a chiral center—the α carbon.

10. The (β1→4) glycosidic bonds of cellulose can be hydrolyzed only by cellulase, which is produced by a few microorganisms, such as some bacteria and *Trichonympha*.

11. The different Fischer projection formulas are shown in Figure 9–3. The Haworth perspective formulas are shown below.

(a) Not found as a ring structure
(b) Not found as a ring structure
(c)

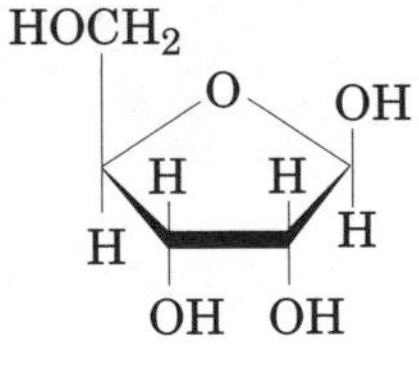

β-D-ribofuranose

(d)

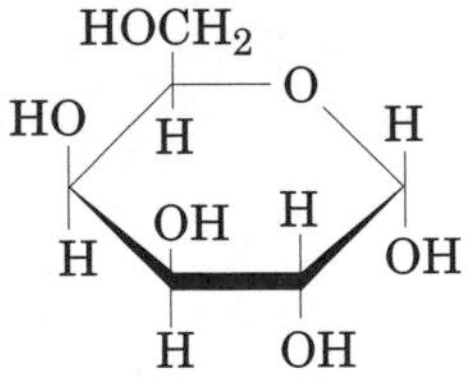

α-D-galactopyranose

12. (a) (α1→4) linkage

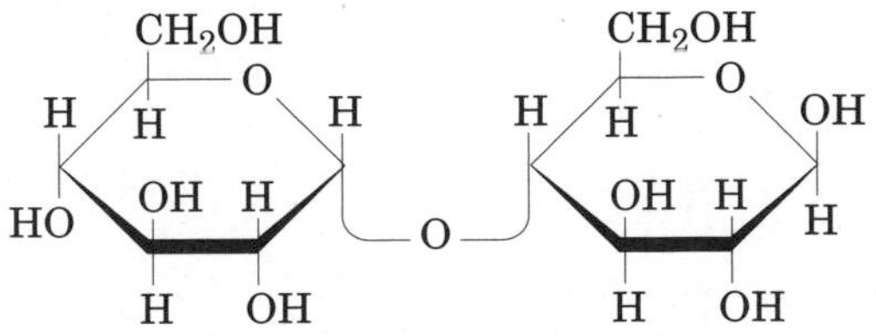

(b) (β1→4) linkage

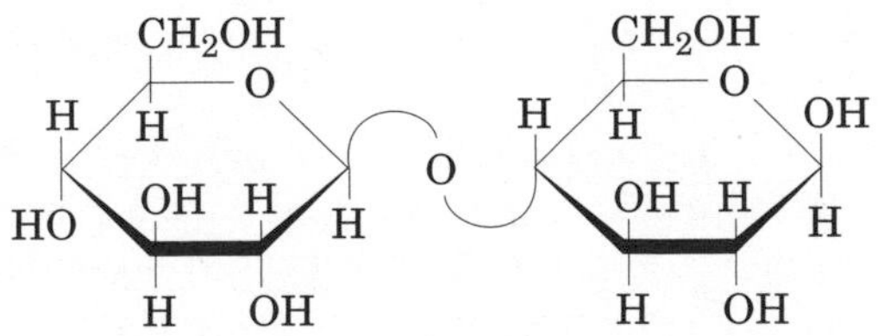

(c) (α1→6) linkage

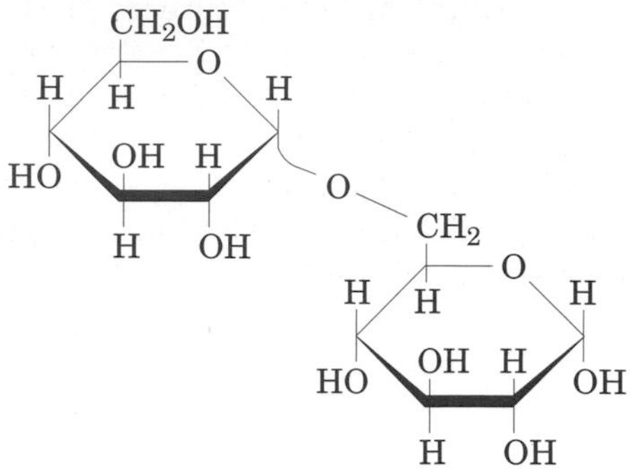

Applying What You Know

1. Many plasma membrane proteins are glycoproteins, with the oligosaccharide portion invariably located on the external surface of the membrane. Oligosaccharide moieties on glycoproteins are not monotonous, but rich in structural information. They act as biological markers, for example as "life-clocks" for individual proteins or cells, signaling whether they should be allowed to circulate or should be destroyed.

2. Glycosaminoglycans are heteropolysaccharides, a family of linear polymers composed of repeating disaccharide units. One of the two monosaccharides in the repeating disaccharide always has an amino group of some sort attached, hence glycos*amino*glycans. In a typical proteoglycan aggregate, a very long strand of a certain type of glycosaminoglycan (hyaluronate), which can be thought of as the central trunk of a tree, has numerous "branches" (aggrecan core proteins) attached. To these core protein branches are attached many smaller glycosaminoglycan molecules of other types (such as keratan sulfate). In a typical proteoglycan aggregate in human cartilage, each hyaluronate trunk has about 100 core protein branches, and each core protein branch has many smaller glycosaminoglycan branches. Interwoven with these enormous aggregates are fibrous proteins, such as collagen, which give strength and resilience to the matrix.

3. Both polysaccharides are homopolysaccharides, meaning they contain only a single type of monomeric unit; in this case, glucose. Cellulose is a linear, unbranched homopolysaccharide, whereas glycogen is branched. Both glycogen and cellulose have variable numbers of subunits, but both are very large molecules.

The critical difference between the two is that glucose residues in cellulose have a β configuration, whereas glucose residues in glycogen are linked in the α configuration. Glycogen is a polymer of (α1→4)-linked subunits of glucose, with (α1→6)-linked branches every 8–12 residues. The most favorable three-dimensional conformation for glycogen is a tightly coiled helical structure stabilized by hydrogen bonds. Cellulose is a polysaccharide of (β1→4) glycosidic bonds. Each glucose unit is turned 180° relative to its neighbors, yielding a straight, extended chain. Multiple hydrogen bonds form between neighboring chains on all sides.

These structural differences correspond to the particular functions of the polysaccharides. Glycogen's role is energy storage. When glucose is needed, its branched structure allows for rapid degradation (from many branch ends) by the cell. Cellulose has a structural role; it gives rigidity and strength to plant cell walls. Extensive hydrogen bonding between linear molecules of cellulose forms sheets, and further hydrogen bonding between stacked sheets makes for very tough, rigid masses of material.

4. Glucose storage molecules are branched, with branches occurring every 24–30 residues in amylopectin and every 8–12 residues in glycogen. The most stable conformation for the adjacent and relatively rigid residues is a curved chain. The extensively branched molecules, therefore, pack more residues into a tighter space.

Other factors probably influenced the evolution of very large, branched molecules for storage. The enzymes that act on these polymers to mobilize glucose for metabolism act only on their nonreducing ends. With extensive branching, more such ends are available for enzymatic attack than in the same quantity of glucose stored as a linear polymer. In effect, branched polymers increase the substrate concentration for these enzymes. Another significant consideration is the effect of the same number of *individual* glucose molecules on cell osmolarity. The increase in intracellular osmolarity due to the presence of hundreds of thousands of glucose molecules probably would be sufficient to rupture the cell.

5. Hyaluronidase degrades hyaluronate, an important structural glycosaminoglycan in animal tissues. It hydrolyzes the (β1→4) linkage between the repeating disaccharide units. This allows invasion of the tissue by the other components of the toxins such as those that act to disrupt cell membranes.

6. These two pieces of evidence, along with the highly hydrophilic character of oligosaccharides, indicate that the oligosaccharide portions of glycoproteins extend from the surface of the proteins, rather than being involved in the internal structure.

7. The presence of the analog of the Le^b oligosaccharide alters binding of the Le^b oligosaccharide to the *H. pylori* lectin in a manner similar to a competitive inhibitor of an enzyme. The analog competes with the Le^b oligosaccharide for the binding site of the lectin, much as a competitive inhibitor competes with substrate for the active site of an enzyme.

8. Barley and salivary enzymes can hydrolyze only the (α1→4) linkages; not only can they not hydrolyze the (α1→6) linkages at the branch points, usually the (α1→4) linkages just adjacent to the (α1→6) linkages are not cleaved. These short polymers remain in the final beer product, where they account for ~22% of the total starch (and therefore calories) in regular beer.

The mold enzyme used in light beer production can hydrolyze both the (α1→4) and (α1→6) linkages. Because it is added before the fermentation process, it hydrolyzes the starch completely, making all the glucose available to the yeast for fermentation. (This leads to higher alcohol content of the beer, which is then diluted, making light beer taste less than robust.)

chapter

8 Nucleotides and Nucleic Acids

STEP-BY-STEP GUIDE

Major Concepts

Nucleotides have a variety of roles in cells.

They serve as carriers of chemical energy, as components of enzyme cofactors, and as molecular messengers. Most important, they are the monomeric subunits of the polymers DNA and RNA. These nucleic acids carry the genetic information of every organism.

Nucleic acids are polymers of nucleotide residues.

Nucleotides contain three characteristic components: a nitrogenous base (purine or pyrimidine), a pentose (ribose or deoxyribose), and a phosphate group. The nucleotides in RNA contain ribose; in DNA, they contain 2′-deoxyribose. The pyrimidine bases in RNA are uracil and cytosine, while in DNA they are thymine and cytosine. The purines are adenine and guanine in RNA and DNA. A **nucleoside** is a nitrogenous base covalently bound in an *N*-β-glycosidic linkage to the 1′ carbon of a pentose sugar. A **nucleotide** is a nucleoside with at least one phosphate group attached to the 5′ carbon of the sugar. Successive nucleotide residues within a polymer are linked through their phosphate groups in phosphodiester linkages between the 5′ hydroxyl of one pentose and the 3′ hydroxyl of the next. Just as individual amino acid residues influence the overall structure of the proteins they make up, the chemical properties of the individual nucleotide residues influence the structure and behavior of the nucleic acids they form.

Nucleic acids have levels of structure, similar to proteins.

Polymers of DNA or RNA exhibit polarity in that the two ends of the molecule are different. The 3′ end has no nucleotide attached at the 3′ carbon of the pentose, whereas the 5′ end has no nucleotide attached through the 5′ phosphate. Two polymers (strands) of DNA are associated in an **antiparallel** arrangement; the orientation of one strand (with respect to its 3′ and 5′ ends) is opposite to that of the other strand. In the Watson-Crick DNA structure, also called B-form DNA, the strands are stabilized in a right-handed double-helix conformation that is held together by hydrogen bonds and base-stacking interactions. The ability of nucleotides in one strand to form *specific* base pairs with nucleotides in the other (adenine pairs with thymine; cytosine pairs with guanine) is the result of hydrogen bonds. Consequently, each strand in a double helix **complements** the other with respect to nucleotide sequence. The overall content of the different bases can affect the form of the DNA. A-DNA is relatively shorter and thicker than B-DNA; Z-DNA is left-handed. The occurrence of structural variants may have biological significance.

RNA is transcribed as a complementary, single-stranded molecule using a DNA molecule as a template.

Three species of RNA in cells are **messenger RNA** (mRNA), **ribosomal RNA** (rRNA), and **transfer RNA** (tRNA). These different types of RNA have various secondary structures, which are related to their base sequence and, in part, determine their functions.

An understanding of the chemistry of nucleic acids facilitates understanding of their functions.

Disruption of the noncovalent hydrogen bonds and hydrophobic interactions between bases in the DNA double helix causes it to denature or "melt" into separate strands. Laboratory measurements of the temperature at which a specific DNA denatures can yield information on its base composition (G≡C pairs form more hydrogen bonds than do A=T pairs and therefore lead to higher melting temperatures). Denaturation of double-helical DNA is the first step in **hybridization** experiments that form the basis of many techniques used in molecular genetics. The sequence of a strand of DNA can be determined using automated procedures, and DNA polymers containing specific sequences can be chemically synthesized.

DNA is susceptible to chemical changes that cause mutations.

Deamination reactions, hydrolysis of *N*-β-glycosidic bonds, irradiation-induced formation of pyrimidine dimers, alkylating reactions, and oxidative damage all can cause mutations in DNA. Not all alterations to the DNA are deleterious; enzymatic methylation of certain bases is common and affects the structure and function of DNA.

Nucleotides serve important cellular functions beyond their role as information molecules.

Nucleoside triphosphates, most importantly ATP, transfer energy in the cell. Many cofactors, including coenzyme A, contain nucleotides, as do **second messengers** such as cAMP.

What to Review

Answering the following questions and reviewing the relevant concepts, which you have already studied, should make this chapter more understandable.

- In a nucleotide, the pentose sugar is linked to the purine or pyrimidine base through an *N*-β-glycosidic bond. Review glycosidic bonds and remind yourself of other biologically important molecules that have glycosidic bonds (Figs. 7–11 and 7–29).
- Nucleotides contain a carbohydrate residue in the furanose form. Review the structure and nomenclature of furanoses. Draw a furanose sugar, including the numbering of its carbon skeleton (Fig. 7–7).
- Like proteins, nucleic acids exhibit several levels of structural organization. Compare the levels of protein structure (Chapters 3 and 4) to the levels of nucleic acid structure discussed in this chapter. What bonds and/or interactions are important to each class of biomolecule?
- The nucleotide ATP affects the reaction *equilibrium* and *rate* of many chemical reactions. Be sure you understand the distinction (pp. 186–188).

Topics for Discussion

Answering each of the following questions, especially in the context of a study group discussion, should help you understand the important points of this chapter.

8.1 Some Basics

1. What are the functions of DNA and of the various types of RNA?

2. What is the functional definition of a gene?

Nucleotides and Nucleic Acids Have Characteristic Bases and Pentoses

3. What are the general constituents of nucleosides and nucleotides, and what types of bonds are involved in each of these structures?

4. Which bases are commonly found in DNA? Which bases are found in RNA?

5. What minor bases are occasionally found in DNA? In RNA?

6. What are the possible phosphorylation sites in nucleotides?

Phosphodiester Bonds Link Successive Nucleotides in Nucleic Acids

7. What components form the backbone of DNA and RNA molecules?

8. Is DNA soluble in water? Why or why not?

9. Why is RNA sensitive to alkaline hydrolysis, whereas DNA is not?

The Properties of Nucleotide Bases Affect the Three-Dimensional Structure of Nucleic Acids

10. How does pH affect the structure of the purine and pyrimidine bases?

11. Are the free bases soluble in water?

8.2 Nucleic Acid Structure

12. What are the hierarchical levels of nucleic acid structure? What are the analogous levels of structure in proteins?

DNA Is a Double Helix That Stores Genetic Information

13. Which of "Chargaff's rules" do you think was the most important clue leading to the model postulated by Watson and Crick?

14. Refer to Figure 8–14 and identify the three components that make up a nucleotide, the ribose sugar, the phosphate group, and the nitrogenous base. Which of these are responsible for joining the nucleotides together in a chain and which are involved in the association of two nucleotide strands in a double helix?

15. What type of noncovalent interaction is critical for the specificity of base pairing between nucleotides of complementary strands of DNA?

16. What types of noncovalent interactions contribute to the overall stability of the double helix?

17. Draw, as best you can in two dimensions, the structure of double-helical DNA without looking at your text! Be sure you know where each of the nucleotide components is placed in relationship to the others, and the bonds and interactions involved. How does this structure suggest a mechanism for the transmission of genetic information?

DNA Can Occur in Different Three-Dimensional Forms

18. What are the similarities and differences among B-, A-, and Z-DNA?

19. What is the possible biological significance of the alternate forms of DNA?

Certain DNA Sequences Adopt Unusual Structures

20. What is the difference between a palindrome and a mirror repeat?

21. Why can't mirror repeats form hairpin structures?

22. What chemical conditions promote the formation of triplex DNA strands? What nucleotide sequences promote triplex formation?

23. Why are these unusual DNA structures of interest?

Messenger RNAs Code for Polypeptide Chains

24. What is the minimum length of mRNA required to code for a polypeptide chain of 80 amino acid residues?

25. What is one function of the noncoding regions of mRNA?

Many RNAs Have More Complex Three-Dimensional Structures

26. How does base pairing differ in DNA–DNA interactions and RNA–RNA interactions?

27. What bonds and interactions are important in some RNA secondary structures?

28. What is it about RNA molecules that allows them to form more complex structures than do molecules of DNA?

8.3 Nucleic Acid Chemistry

Double-Helical DNA and RNA Can Be Denatured

29. Which bonds and interactions are affected during denaturation of DNA? How does this compare to protein denaturation?

30. What physical changes allow scientists to determine the melting points of DNA strands with different nucleotide compositions? How are these changes experimentally measured?

31. Remind yourself why guanine can form three hydrogen bonds with cytosine, whereas adenine–thymine pairs have only two.

Nucleic Acids from Different Species Can Form Hybrids

32. How does the degree of evolutionary closeness between two species affect the extent of hybridization between their nucleic acids?

Nucleotides and Nucleic Acids Undergo Nonenzymatic Transformations

33. Explain in evolutionary terms why it is important that DNA contains thymine rather than uracil.

34. Physicians caution against excessive sun tanning. Explain the molecular basis for their concern.

Some Bases of DNA Are Methylated

35. List some reasons for methylation of certain DNA bases and sequences.

The Sequences of Long DNA Strands Can Be Determined

36. What are the general principles involved in the two methods of sequencing DNA?

37. What is the role of the dideoxynucleoside triphosphate in the Sanger method?

38. What component is labeled in the sequencing reactions using the Sanger method?

39. Be able to "read" a sequencing gel.

The Chemical Synthesis of DNA Has Been Automated

40. The ability to synthesize DNA has proved to be an extremely useful tool for a wide variety of biochemical and biological studies. Think of a biochemical question that can be addressed using synthetic polynucleotides.

8.4 Other Functions of Nucleotides

Nucleotides Carry Chemical Energy in Cells

41. Theoretically, how do ATP and PP_i compare in their ability to affect reaction equilibria?

42. How do ATP and PP_i compare in their ability to affect reaction rates?

Adenine Nucleotides Are Components of Many Enzyme Cofactors

43. Explain the concept of evolutionary economy. How does it relate to the adenosine component of many cofactors?

Some Nucleotides Are Regulatory Molecules

44. Why is cAMP called a *second* messenger?

Discussion Questions for Study Groups

- Why is DNA so well-suited to the role of molecular repository of genetic information and why is RNA not generally used for this "information storage" function?
- What are the cellular roles of the various types of RNA and how are prokaryotic and eukaryotic mRNA molecules different?
- With improvements in automated DNA sequencing it will someday become possible to obtain anyone's complete DNA sequence. What are some possible benefits and problems associated with the ability to obtain your personal DNA sequence?

SELF-TEST

Do You Know the Terms?

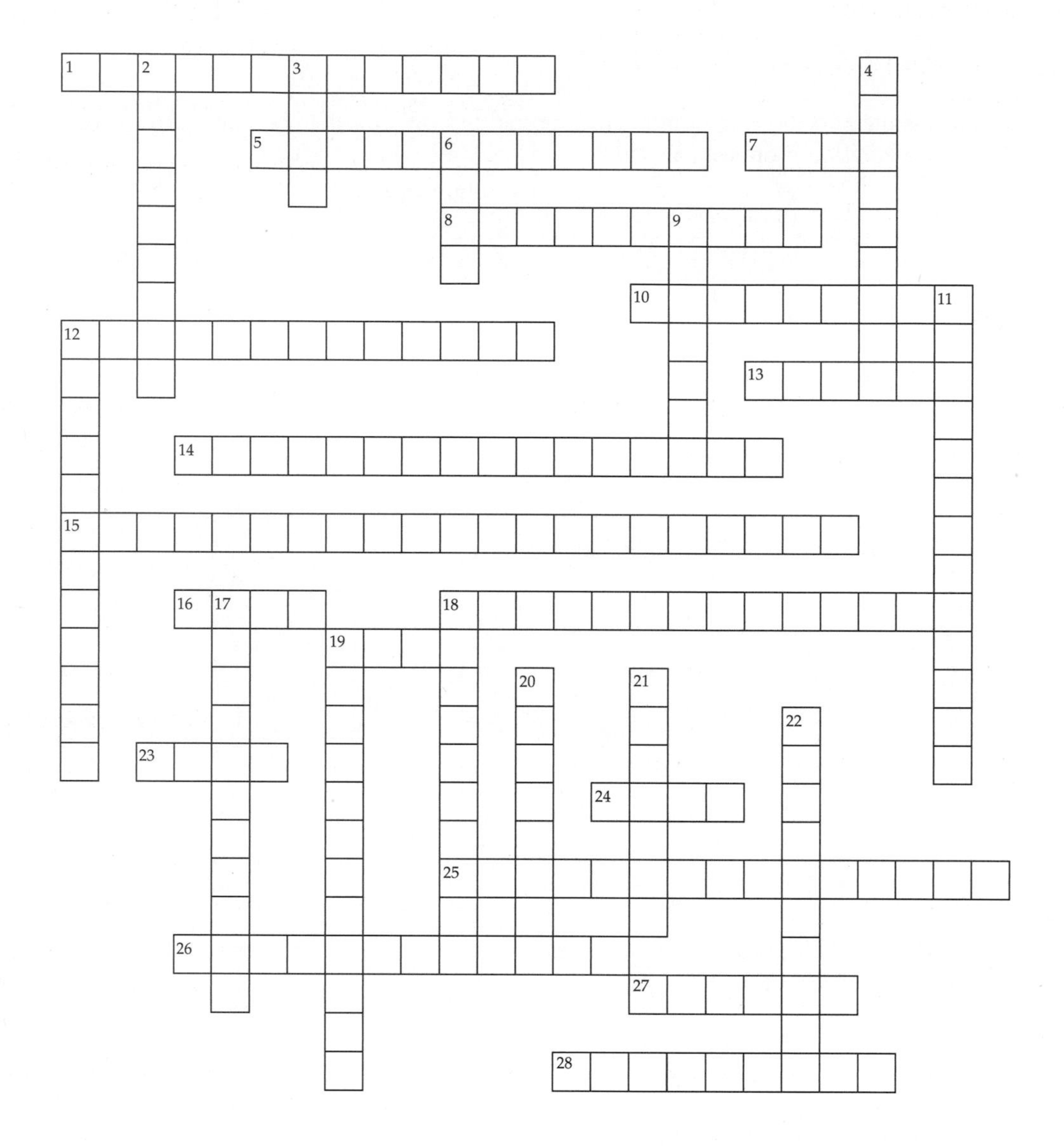

ACROSS

1. These two strands are _______ to each other.
   AATGCGGTCCTAT
   TTACGCCAGGATA
5. $3' \rightarrow 5'$
   $5' \leftarrow 3'$
7. A ribonucleic acid involved in protein synthesis; it binds amino acids.
8. Contains a phosphate group in an ester linkage to a ribose sugar and a nitrogenous base.
10. A common intracellular signaling molecule. (2 words)
12. Most eukaryotic mRNA codes for a single polypeptide and is _______.
13. What two complementary strands of DNA spontaneously do to form an intact duplex.
14. Thymidylate is a nucleotide found primarily in _______ acids.
15. A common protein domain found in proteins that bind ATP. (3 words)
16. A major structural component of the protein synthetic machinery of cells.
18. Covalent bonds that link the individual nucleotide residues in DNA and RNA.
19. Structure containing polypurine tracts and mirror repeats; forms a triple helix.
23. Left-handed double-helical structure.
24. Right-handed, Watson-Crick double helix.
25. Short polymers of nucleotides (50 or less), often used as complementary DNA "probes" for hybridization techniques.
26. AACCTTTTCCAA
    TTGGAAAAGGTT (2 words)
27. DNA duplex formed from DNA of different species.
28. Non–Watson-Crick, or _______ pairing; allows formation of triplex DNA strands.

DOWN

2. Hydrogen peroxide, a byproduct of aerobic metabolism, is a common cause of these in DNA.
3. Carries genetic information from DNA to the ribosomes.
4. Contains a purine attached to a phosphorylated ribose; the base was first isolated from bird manure.
6. Dehydrated, compact form of DNA.
9. A major pyrimidine; has a methyl group at C-5.
11. Describes mRNA that is translated into more than one protein.
12. Determined for a solution of DNA by measuring UV light absorption as a function of temperature. (2 words)
17. Uracil is a nitrogenous base found predominantly in _______ acids.
18. AACCAATTGGTT
TTGGTTAACCAA
19. The increase in UV light absorption when double-stranded DNA is denatured is referred to as the _______ effect.
20. A major pyrimidine; has an amino group at C-4.
21. Uracil attached through N-1 to ribose.
22. Purine or pyrimidine base covalently bound to furanose through an *N*-*β*-glycosidic bond.

Do You Know the Facts?

1. Draw the following structures.
(a) Adenylate **(b)** Deoxyguanylate **(c)** Deoxycytidylate

(d) Uridylate **(e)** Deoxythymidylate **(f)** Hypoxanthine

2. Compounds that contain a nitrogenous base, a pentose sugar, and a phosphate group are called **(a)** __________. Two purines found in DNA are **(b)** __________ and **(c)** __________. In DNA, the base pair **(d)** __________ - __________ is held together by three hydrogen bonds; the base pair **(e)** __________ - __________ has only two such bonds. In a solution of DNA, the purine and pyrimidine bases stack like coins because of their **(f)** hydro __________ nature. When the solution is heated, the base stacking is disrupted, and the optical absorbance (at 260 nm) of the solution increases; this is called the **(g)** __________.
3. Indicate whether the following statements about DNA are true or false.
_____ **(a)** A-form and B-form DNA are right-handed helices, but Z-form DNA is a left-handed helix found only in single-stranded DNA.
_____ **(b)** Palindromic sequences can potentially form cruciform structures.
_____ **(c)** The dideoxy method of sequencing DNA can be used on B-DNA, but not on Z-DNA.
_____ **(d)** Deoxyribose is bound to the nitrogenous base at C-1′.
_____ **(e)** A pyrimidine in one strand of DNA always hydrogen bonds with a purine in the opposite strand.
_____ **(f)** G≡C pairs share three glycosidic bonds.
_____ **(g)** All the monosaccharide units lack an —OH at C-2′.
_____ **(h)** The phosphodiester bonds that link adjacent nucleotides join the 3′ hydroxyl of one nucleotide to the 5′ hydroxyl of the next.
_____ **(i)** The two strands are aligned as parallel strands.

_____ **(j)** The phosphodiester bonds that link adjacent nucleotides are uncharged under physiological conditions.
_____ **(k)** B-form DNA predominates in aqueous solution; dehydration favors the A-form.
_____ **(l)** A-form DNA is shorter and has a larger diameter than the B-form DNA.
_____ **(m)** Nucleotide sequence has little or no effect on which form DNA takes.

4. ADP contains:
A. a furanose ring.
B. a ketose sugar.
C. two phosphoanhydride bonds.
D. a *β*-*O*-glycosidic bond.
E. a pyrimidine base.

5. Which of the following is true for DNA?
A. Phosphate groups project toward the middle of the double helix.
B. Deoxyribose units are connected by 3′,5′-phosphoanhydride bonds.
C. The 5′ ends of both strands are at the same end of the double helix.
D. G≡C pairs share three hydrogen bonds.
E. The ratio of A+T to G+C is constant for all naturally occurring DNA.

6. In the DNA sequencing by the Sanger (dideoxy) method:
A. the template strand of DNA is radioactive.
B. enzymes are used to cut the DNA into small pieces, which are then separated by electrophoresis.
C. ddTTP is added to each of four reaction mixtures prior to synthesis of complementary strands.
D. the role of ddATP is to occasionally terminate synthesis of DNA where dT occurs in the template strands.
E. the sequence is read from the top of the gel downward.

7. In living cells, various nucleotides and polynucleotides:
A. serve as structural components of enzyme cofactors.
B. alter the equilibria of chemical reactions.
C. carry metabolic energy.
D. serve as intracellular signaling molecules.
E. all of the above are true.

8. The compound that consists of ribose linked by an *N*-*β*-glycosidic bond to N-9 of adenine is:
A. a purine nucleotide.
B. a pyrimidine nucleotide.
C. adenosine.
D. AMP.
E. deoxyadenosine.

9. The phosphodiester bonds that link adjacent nucleotides in DNA:
A. are positively charged.
B. join the 3′ hydroxyl of one nucleotide to the 5′ hydroxyl of the next.
C. always link A with T and C with G.
D. are positively charged and always link A with T and C with G.
E. are positively charged and join the 3′ hydroxyl of one nucleotide to the 5′ hydroxyl of the next.

10. Nucleic acid samples have been isolated from three different organisms. The nucleic acids have the following base ratios (%):

	A	T	U	G	C	A + T/G + C	A + G/C + T
Sample 1	29	19	0	22	30	0.92	1
Sample 2	24	0	16	24	36	0.4	1.3
Sample 3	17	17	0	33	33	0.5	1

Which sample(s) are DNA? Which sample would you expect to have the highest t_m (melting point)?
A. 1 and 3; 3
B. 1, 2, and 3; 3
C. 1; 2
D. 1 and 3; 1
E. 2 and 3; 2

11. Two molecules of double-stranded DNA are the same length (1,000 base pairs), but differ in base composition. Molecule 1 contains 20% A+T; molecule 2 contains 60% A+T. Which molecule has a higher t_m (melting point)? How many C residues are there in the 60% A+T DNA molecule?
A. 2; 40
B. 1; 200
C. 2; 400
D. 1; 400
E. 2; 200

12. What crucial pieces of information from the Hershey-Chase experiment indicated that DNA is the genetic material?

13. Which of "Chargaff's rules" was the most important clue leading to the model postulated by Watson and Crick? Why?

14. Why does DNA contain thymine rather than uracil?

Applying What You Know

1. You wish to determine the relative degree of sequence similarity for DNA from different species of Galapagos finches (referred to as species A, B, and C). You make hybrid duplexes of the DNA between A and B, A and C, and B and C. You measure the increase in the absorption of UV light with increasing temperature for each of the duplex solutions and obtain the following data:

Hybrid Duplex	t_m
A + B	80 °C
A + C	87 °C
B + C	83 °C

(a) Which two species have a greater degree of similarity in their DNA sequences? Why?
(b) What factors should be taken into account when interpreting these data?

2. How would you use sequence information for a gene in *Drosophila* to identify and localize a similar gene from a mouse?

3. Why would reducing the concentration of salts (lowering the ionic strength) in a solution of double-stranded DNA lower the melting point of the DNA?

Biochemistry on the Internet

Nucleic acids are a functionally and structurally diverse group of molecules. Their secondary structures range from the familiar Watson-Crick double helix to the exceedingly complex ribosomal structures formed by rRNA. The structures of the various forms of nucleic acids provide insights into their different functions. You can view them first-hand using 3-D viewers and the Internet.

(a) First, look at the basic structure of a molecule of single-stranded DNA. Go to www.rcsb.org and type in the PDB ID code **116D** to find the file for Deoxyribonucleic Acid [DNA (5′-d(CpCpGpTpApCpGpTpApCpGpG)-3′)]. You can display this file as a molecular model using any of the listed viewers. Identify the various components of the sugar-phosphate backbone and the individual bases in this DNA strand. Which is the 5′ end of this single strand? Hint: See *Biochemistry on the Internet* questions in Chapters 3–6 of this guide or question 16 in the text for more information on viewing 3-D molecular models.

(b) To see how DNA forms a double-helical structure, find the file **1D92** and display the model using one of the viewers. This file contains a model of a synthetic double-stranded DNA molecule with the following sequence: (5′-d(GpGpGpGpCpTpCpC)-3′). How are these strands held together in a double-helical conformation? Two of the base pairs in this double helix are mismatched. Can you find them?

(c) To see the structure of a transfer RNA (tRNA) molecule, open the file **2TRA.** How many strands of RNA make up a tRNA molecule? Where are the 5′ and the 3′?

(d) File **1GIX** shows ribosomal RNA (rRNA) associated with messenger RNA (mRNA), tRNAs and a number of ribosomal proteins. See if you can identify the ribosome subunit and the three tRNA molecules. How many strands of RNA make up the ribosome subunit? Where is the piece of mRNA?

3-D VIEWING HINTS: To get a feel for the 3-D shape of each of the molecules above, display them in **Spacefill** format, which will allow you to visualize the surface of each molecule. You can also control other aspects of the model, such as the way molecules are colored. You can hold down shift and click and drag the molecule to zoom in or out.

ANSWERS

Do You Know the Terms?

ACROSS

1. complementary
5. antiparallel
7. tRNA
8. nucleotide
10. cyclicAMP
12. monocistronic
13. anneal
14. deoxyribonucleic
15. nucleotide-binding fold
16. rRNA
18. phosphodiester
19. H-DNA
23. Z-DNA
24. B-DNA
25. oligonucleotide
26. mirror repeat
27. hybrid
28. Hoogsteen

DOWN

2. mutations
3. mRNA
4. guanylate
6. A-DNA
9. thymine
11. polycistronic
12. melting point
17. ribonucleic
18. palindrome
19. hyperchromic
20. cytosine
21. uridine
22. nucleoside

Do You Know the Facts?

1. Refer to Figures 8–4 and 8–5 for the correct structures.
2. **(a)** nucleotides; **(b, c)** adenine, guanine; **(d)** G–C; **(e)** A–T; **(f)** phobic; **(g)** hyperchromic
3. **(a)** F; **(b)** T; **(c)** F; **(d)** T; **(e)** T; **(f)** F; **(g)** T; **(h)** T; **(i)** F; **(j)** F; **(k)** T; **(l)** T; **(m)** F
4. A
5. D
6. D
7. E
8. C
9. B
10. A
11. D; remember, 1,000 base pairs = 2,000 bases.
12. Two populations of bacteriophage T2 (bacterial virus) were radioactively labeled. The DNA of one batch was labeled with ^{32}P, and the *proteins* of a second batch were labeled with ^{35}S. The bacteriophage particles were then allowed to infect separate suspensions of *E. coli,* and the infected cells were collected and analyzed for the presence of radioactivity. Only the *E. coli* that were exposed to ^{32}P-labeled bacteriophage had associated radioactivity indicating that the ^{32}P labeled DNA had been transferred from the viral particles. ^{35}S, associated only with protein molecules, was not found in the infected *E. coli* cells. New viral particles were made in *both* suspensions of *E. coli,* but only the DNA-associated label was transferred to the infected bacteria; this strongly suggested that DNA was the source of the genetic information.
13. The rule stating that all DNA, regardless of the species, contains equal numbers of adenosine and thymidine residues (A=T) and that the number of guanosine residues equals the number of cytidine residues (G=C), suggested specific pairing between the bases. Watson and Crick's double-stranded structure, with A always H-bonding with T, and C always H-bonding with G, provides the molecular basis for the rule.
14. Spontaneous deamination of cytosine to uracil is relatively common—about 100 events per mammalian cell per day. Uracil is recognized as incorrect in DNA, and is removed by a repair system. If DNA normally contained uracil instead of thymine (as does RNA), recognition of these deamination mistakes would be more difficult. Unrepaired uracils would pair with adenine during replication, leading to permanent base-sequence changes (mutations).

Applying What You Know

1. **(a)** The melting point of DNA provides information concerning the degree of hydrogen bonding between two strands of DNA; the higher the melting point, the greater the extent of noncovalent bonding. Therefore, for the hybrid duplexes, a higher melting point indicates a greater degree of base pairing and a greater degree of sequence similarity between the two strands. The data suggest that species A and C share the highest degree of similarity in their DNA sequences.

   (b) Because G≡C pairs share three hydrogen bonds, whereas A=T pairs share only two, DNA hybrids having a high proportion of G≡C pairs will have a higher melting temperature than hybrids having more A=T pairs. Consequently, a hybrid with fewer matching bases that were all G≡C pairs could theoretically have a higher melting temperature than a hybrid with more matches all occurring between A and T.
2. Molecular biology techniques allow ready detection of genes with significant sequence similarity across species. These techniques rely on the ability of DNA strands to form hybrid duplexes.

   Step 1. Digest mouse genomic DNA into small fragments using specific enzymes, and separate fragments according to size by electrophoretic techniques.

   Step 2. Use sequence information from *Drosophila* to synthesize oligonucleotides with sequences complementary to the gene under consideration. These complementary oligonucleotides are tagged, with ^{32}P, for example, and used as "probes."

   Step 3. Expose the electrophoretically separated mouse DNA to the Drosophila "probe" and allow hybrids to form. Wash away unhybridized probe. Remaining probe is the result of the formation of hybrid duplexes and can be detected (e.g., using autoradiography).
3. The negative charges on the phosphate groups on each strand of DNA are shielded by positive ions in a high ionic strength solution. Lower ionic strength reduces the screening of the negative charges on the phosphate groups. The result is stronger charge-charge repulsion between the phosphate portions of the backbones of each strand, which favors strand separation.

Biochemistry on the Internet

(a) To identify the various structures, it may help to display the molecule as a ball-and-stick structure and to turn on CPK color in your viewer. (In the Jmol viewer menu choose **Select → Nucleic → All,** then **Render → Scheme → Ball and Stick.**) In this coloring scheme, the phosphates will be colored orange and their associated oxygens will be red, highlighting the backbone of this structure relative to the individual (and unpaired) bases. Notice that the deoxyribose (sugar) is oriented perpendicularly to the bases.

You can locate the 5′ end in this single strand of DNA in one of two ways: Either locate the end of the strand that has a free C5 on the ribose moiety or, given the nucleotide sequence in the PDB file, look for the end with a cytosine. The 3′ end is the one with a free C3 on the ribose or, alternatively, the end with a guanine.

(b) The strands in this DNA double helix are held together primarily by the numerous hydrogen bonds that form between adjacent bases. Display the molecule as a ball-and-stick structure in your viewer (see answer to part (a) above for details). In the Jmol viewer you can turn on the hydrogen bond display by selecting **Render → Hydrogen Bonds → Calculate.** The hydrogen bonds will be displayed as dashed lines.

The mismatched pairs are between G3 (A strand) and T14 (B strand) as well as T6 (A strand) and G11 (B strand). The mispairing is very evident with the hydrogen bond display turned on because the model shows no hydrogen bond formation between the mismatched bases.

(c) All transfer RNA molecules are composed of a single strand of RNA. The end that contains U1 is the 5′ end. Note the presence of a free phosphate group on C5 of the ribose portion of this nucleotide. C73 is at the 3′ end. Note the presence of a free C3 hydroxyl (represented by a red ball for the oxygen) on the ribose in this nucleotide. Also note the oxygen molecule on C2 of every ribose in this structure, which was not present on the ribose in the DNA models in (a) and (b); tRNA is a ribonucleic acid and not a *deoxy*ribonucleic acid after all.

(d) Ribosomal RNA (rRNA) provides the structural framework for the process of protein synthesis, which will be discussed in detail in Chapter 27. Most ribosomes are complexes composed of several different-sized pieces of RNA as well as numerous associated proteins. File 1GIX is a model of several of the components of ribosomes from *Thermus thermophilus,* a bacterium found in hot springs. This model contains the 30s ribosomal RNA subunit, as well as three molecules of transfer RNA and messenger RNA molecules (file 1GIY has the 50S ribosome subunit). To display the different components in the Jmol viewer, click **Select → Nucleic → All,** then **Render → Scheme → CPK Spacefill.** Chain A is the ribosomal RNA; chains B, C, and D are the transfer RNA molecules; and chain 1 is the mRNA. Chains E-X are ribosomal proteins; you can see the nucleic acid structures better if you click **Select → Protein → All** then **Render → Structures → Off** to hide the proteins. The relatively complex secondary structure seen in the rRNA and tRNA molecules is due to intramolecular base pairing. The association between the mRNA and tRNA is due to intermolecular base pairing.

chapter

9

DNA-Based Information Technologies

STEP-BY-STEP GUIDE

Major Concepts

The development of recombinant DNA technology has revolutionized the way we study cells and their biomolecules.

Cloning technology provides researchers with the ability to "cut and paste" selected pieces of DNA and to insert them into many different types of cells. The basic tools are the following: (1) the "scissors" are **sequence-specific endonucleases,** or **restriction endonucleases;** (2) the "paste" is the enzyme **DNA ligase;** and (3) the vehicles to move the DNA pieces around, called **vectors,** are self-replicating **DNAs** (**plasmids, bacteriophages,** and artificial chromosomes) that can insert themselves into host cells. The **recombinant DNA,** then, is the vector DNA with a small piece of foreign DNA that has been inserted with the aid of restriction endonucleases and DNA ligase.

A variety of techniques can be used to insert vectors containing foreign DNA into a host cell, which is often a strain of E. coli, *in a process called transformation.*

Once inside cells, the vector DNA replicates producing many copies of the foreign DNA along with the vector DNA. Not all cells are transformed by these procedures, however, and a process must be employed to **select** the cells that actually contain the recombinant DNA. The end result is a renewable source of a specific piece of DNA that can be replicated, transcribed, and translated in essentially unlimited quantities.

Genomics is the study of the DNA sequence of an organism.

Using cloning technology, the complete complement of **DNA** (genome) from a number of organisms has been cut into smaller, overlapping pieces and stored in "libraries." The DNA pieces in such libraries can then be sequenced and the location of the sequence on a chromosome can be determined. The result is that entire genomes have now been sequenced. This sequence information is being "mined" to discover new genes, to provide insights into genome organization and chromosome structure, to elucidate evolutionary relationships between organisms, and to develop new ways to treat diseases and much more.

Proteomics is a new field of investigation that attempts to ascribe a function to all the proteins encoded by the genes in a genome.

With a huge amount of sequence information available for a growing number of organisms, researchers are now faced with the larger challenge of determining the function of all the encoded proteins. Knowledge of protein structure is an important aspect of this area of research because a protein's function and its ability to interact with other cellular components is largely determined by its 3-D shape.

Recombinant DNA technology provides tools for use in research, medical therapies, and commercial production.

The ability to express cloned genes and obtain large quantities of the encoded protein is the basis for many of the applications of recombinant DNA technology (e.g., the production of erythropoietin). The ability to amplify specific DNA sequences using the polymerase chain reaction (PCR) makes it possible to detect as little as one copy of a DNA molecule (which is useful in the early detection of viral infections such as the human immunodeficiency virus). Site-directed mutagenesis techniques allow researchers to selectively alter genes, thereby manipulating protein functional domains and gaining insights into protein function (ultimately it may be possible to design proteins with specific functions "from scratch").

What to Review

Answering the following questions and reviewing the relevant concepts, which you have already studied, should make this chapter more understandable.

- Recombinant DNA technology makes use of many of the biochemical techniques already introduced in the text. A thorough understanding of the principles underlying these methods will make it easier for you to follow the discussion of recombinant DNA technology. These techniques include (but are not limited to):

  Gel electrophoresis (pp. 88–90)

  Chromatography techniques (pp. 85–87)

  Use of antibodies for isolation, detection, and localization of proteins (pp. 173–174)

  DNA sequencing techniques (pp. 293–294)
- Review the structure of nucleotides and polynucleotides from Chapter 8.
- The generation of recombinant DNA involves the joining of DNA pieces. Be sure you understand the structure of nucleotides and how they are joined to form polynucleotides. Be very sure you understand how the DNA strands in the double helix are oriented with respect to their 5′ and 3′ ends.
- Much of recombinant DNA technology relies upon the ability of DNA and RNA to form duplexes and hybrid duplexes. Be sure that you understand the molecular basis for base-pair formation between polynucleotide strands (pp. 288–289) and the concept of complementarity (pp. 279–280).

Topics for Discussion

Answering each of the following questions, especially in the context of a study group discussion, should help you understand the important points of this chapter.

9.1 DNA Cloning: The Basics

Use the information in the introduction to this section and draw your own diagrams or cartoons that illustrate each of the steps involved in cloning a piece of DNA; try to do this without copying from the text.

Restriction Endonucleases and DNA Ligase Yield Recombinant DNA

1. Endonucleases and ligases are normal constituents of some cells. What are the biological functions of these enzymes?

2. How do bacteria protect themselves from their own enzymes?

3. Why can restriction enzymes be thought of as molecular scissors? What bond in DNA do they cleave?

4. Why are restriction enzymes so crucial to the ability to clone DNA?

5. Why are Type II restriction endonucleases used for DNA cloning and not Types I and III?

6. How might the use of a restriction enzyme that produces sticky ends in a cloning protocol provide an advantage over the use of one that produces blunt ends? What makes "sticky ends" sticky?

7. Which restriction enzyme(s) would you use to cleave the polylinker region of the vector shown in Figure 9-3 if you wanted to insert the following "cut" pieces of DNA?

   a) GATCCTGCAGAAGCTTCCGGATCCCCGG
   GACGTCTTCGAAGGCCTAGGGGCCCTAG

   b) AATTCTGCAGAAGCTTCCGGATCCCCGG
   GACGTCTTCGAAGGCCTAGGGGCCTCGA

   Would the orientation of the inserted piece be the same in both cases?

Cloning Vectors Allow Amplification of Inserted DNA Segments

8. Why is it important that vectors be able to replicate autonomously?

   Given that plasmid DNA is self-replicating, why must it be inserted into a host cell?

9. After you read this section, use the diagram below and try to label the various elements of an *E. coli* plasmid without looking at the text. Label the following: *ori,* the origin of replication; *tet*^{R}, tetratcycline resistance gene; *amp*^{R}, the ampicillin resistance gene; the polylinker region.

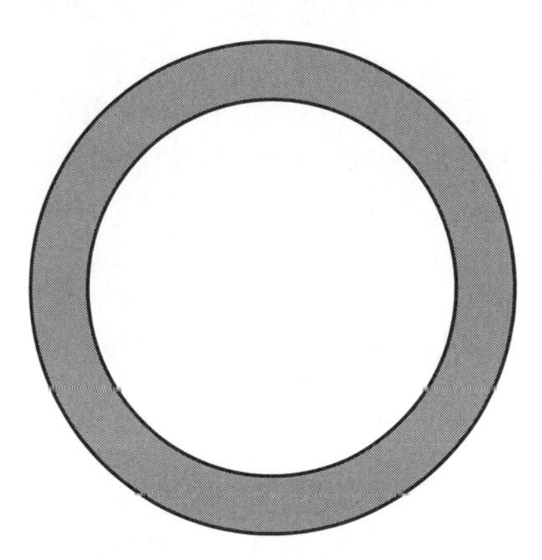

10. Why does the positioning of these elements in the vector matter? Where would you insert the foreign gene?

11. Why are two antibiotics used to screen for incorporation of a plasmid? Why do some of the transfected bacteria grow on plates containing ampicillin?

12. What is the advantage of using bacteriophages as vectors compared to plasmids?

13. In what ways does cloning in yeast differ from cloning in *E. coli*?

Specific DNA Sequences Are Detectable by Hybridization

14. What is the molecular basis for the use of labeled probes to identify clones containing specific DNA sequences?

15. What are the different ways that DNA probes can be generated?

Expression of Cloned Genes Produces Large Quantities of Protein

16. What elements do you need to add to the figure you generated in question 8 to create an expression vector?

Alterations in Cloned Genes Produce Modified Proteins

17. How might you use site-directed mutagenesis to study the function of human hemoglobin (see Chapter 5 in your text)?

18. What are the different strategies for generating specific mutations in cloned DNA sequences?

Terminal Tags Provide Binding Sites for Affinity Purification

19. *E. coli* expression vectors, e.g., Fig. 19–10, containing a terminal tag sequence, are commercially available. Indicate in your diagram from question 9 where these sequences must be located.

9.2 From Genes to Genomes

DNA Libraries Provide Specialized Catalogs of Genetic Information

20. How are the DNA pieces used to construct a genomic library obtained?

21. How are the DNA pieces used to construct a cDNA library generated?

22. What is absent from the cloned DNA in a cDNA library that is present in the DNA of a genomic library? (Hint: consider the source!)

The Polymerase Chain Reaction Amplifies Specific DNA Sequences

23. What characteristic of DNA duplexes is critical to the successful application of the PCR technique?

Genome Sequences Provide the Ultimate Genetic Libraries

24. How much of the human genome encodes functional proteins?

25. What are possible functions of the rest of the DNA?

26. In what ways can you envision information about the sequence of the human genome being used?

Box 9–1 A Potent Weapon in Forensic Medicine

27. In DNA fingerprinting, a DNA probe is used to label a few of the thousands of DNA "bands" that are generated when genomic DNA is digested with restriction enzymes. What type of "probe" is used in this procedure and how is it generated?

28. When using this technique to match DNA samples, what can one do to increase the odds that the DNA in two different samples comes from the same source?

9.3 From Genomes to Proteomes

29. What is the difference between genotype and phenotype?

Sequence or Structural Relationships Provide Information on Protein Function

30. What classes or types of genes would you expect to exhibit the most similarity between bacteria and humans? Between Drosophila and humans? Between rats and humans?

Cellular Expression Patterns Can Reveal the Cellular Function of a Gene

31. When genes are screened using DNA microarrays, what is the source of the probe? Is it the same type of probe used for DNA fingerprinting?

Detection of Protein-Protein Interactions Helps to Define Cellular and Molecular Function

32. What types of protein-protein interactions have you encountered so far in this text and what types of bonds are involved in these interactions? (See Chapters 4, 5, and 6.)

33. Can you come up with an example of a cellular function that could be defined by knowledge of a specific protein-protein interaction?

34. In a yeast two-hybrid screen, how do you know that you have found a set of proteins that interact?

9.4 Genome Alterations and New Products of Biotechnology

A Bacterial Plant Parasite Aids Cloning in Plants

35. In the transformation of plant cells, new genes are introduced using an engineered shuttle vector containing the gene to be inserted, a selectable marker, and two T-DNA repeats. Why is it also necessary to include an engineered Ti plasmid that no longer contains its T-DNA gene?

Manipulation of Animal Cell Genomes Provides Information on Chromosome Structure and Gene Expression

36. The Human Immunodeficiency Virus (HIV) is a retrovirus. In what way are retroviruses an important tool for use in gene therapy?

New Technologies Promise to Expedite the Discovery of New Pharmaceuticals

37. Information in this section should help you with your answer to question 33.

Box 9–2 The Human Genome and Human Gene Therapy

38. The therapeutic strategies discussed in this section involved gene transformation of bone marrow cells. Why would treatment of genetic diseases that involve other organs (e.g., cystic fibrosis is a genetic disease that affects lung function) require somewhat different strategies?

Recombinant DNA Technology Yields New Products and Challenges

39. Can you come up with additional genes/proteins (human or otherwise) that would be useful if cloned?

40. What are the benefits and potential hazards to the use of these genes or proteins?

Discussion Questions for Study Groups

- What are some of the practical problems that must be overcome in order for gene therapy to be successfuly applied to humans.
- What do you think are the major ethical hurdles to the practical use of recombinant DNA technologies?
- Do you think the potential benefits outweigh the potential problems?
- Can you think of a human gene that would be useful to clone?

SELF-TEST

Do You Know the Terms?

ACROSS

1. Vectors of choice for medium-size DNA fragments.
5. One approach to _______ - _______ mutagenesis is to use an oligonucleotide primer containing an altered base for the synthesis of a duplex DNA.
7. The new, improved *product* of two fused genes is a _______ protein.
8. Small, circular, extrachromosomal DNA molecule.
10. Viral _______ are often modified retroviruses.
12. A genomic _______ contains more information than its largest "homolog" at any university.
13. Synthetic DNA fragment with recognition sequences for several restriction endonucleases.
15. These cleave DNA at specific base sequences; the scalpels of molecular biology. (2 words)
16. In bacteria, event induced by a cold, calcium chloride bath followed by heat shock or a strong jolt of electricity.
18. Natural genetic engineer in plants. (2 words)
20. Synthetic DNA, complementary in sequence to an RNA template. (abbr.)
21. Describes ends of DNA fragments that have no overhang.
23. To _______ or not to _______ is no longer the question; everyone in biochemistry is doing it! (Hint: to make an identical copy of an organism, a cell, or a DNA segment.) (1 word)
24. Describes composite DNA molecules containing DNA from two or more species.

DOWN

2. A self-replicating DNA plasmid that contains an inducible promoter and a polylinker is called a(n) _______ vector.
3. Contains all the genetic information to make an organism.
4. Radioactive DNA fragment that can ferret out and bind to specific DNA sequences.
5. Self-replicating piece of DNA that is capable of surviving in both *E. coli* and *S.cerevisiae.* (2 words)
6. Shocking technique that makes cells transiently permeable to DNA.
9. It joins DNA strands; molecular "glue." (2 words)
10. _______ vectors are the equivalent of a molecular "syringe."
11. Differences in lengths and sequences of DNA fragments produced by random cutting of genomic DNA by restriction enzymes; vary from one individual to another. (abbr.)
13. Gene amplification technique that relies on the activity of a heat-stable polymerase (*Taq*I) isolated from hot-spring bacteria. (abbr.)
14. The fluorescent zebrafish in Figure 9-33 is an example of a _______ animal.
17. Restriction endonucleases that make staggered cuts produce _______. (2 words)
19. Following electrophoresis, proteins can be identified using antibodies in a Western blot, mRNA can be identified using DNA probes in a Northern blot, and DNA fragments can be identified using DNA probes in a _____ blot.
22. Used for cloning genomic DNA, these vectors are actually unstable when they contain inserts of less than 100,000 bp.
23. Although it's not the "micro" variety used in computers, this "DNA" version also contains an enormous amount of information.

Do You Know the Facts?

Questions 1–5 refer to the following illustration.

The illustration below shows a fragment of DNA that has been isolated from a genomic DNA library. The fragment contains a protein-coding region (indicated by the black box); the sites at which a number of restriction endonucleases cut (a restriction map) are indicated. Also shown is a plasmid that has been engineered to be used as an expression vector (the curved arrow indicates the direction of translation from the promoter); its restriction enzyme sites are indicated.

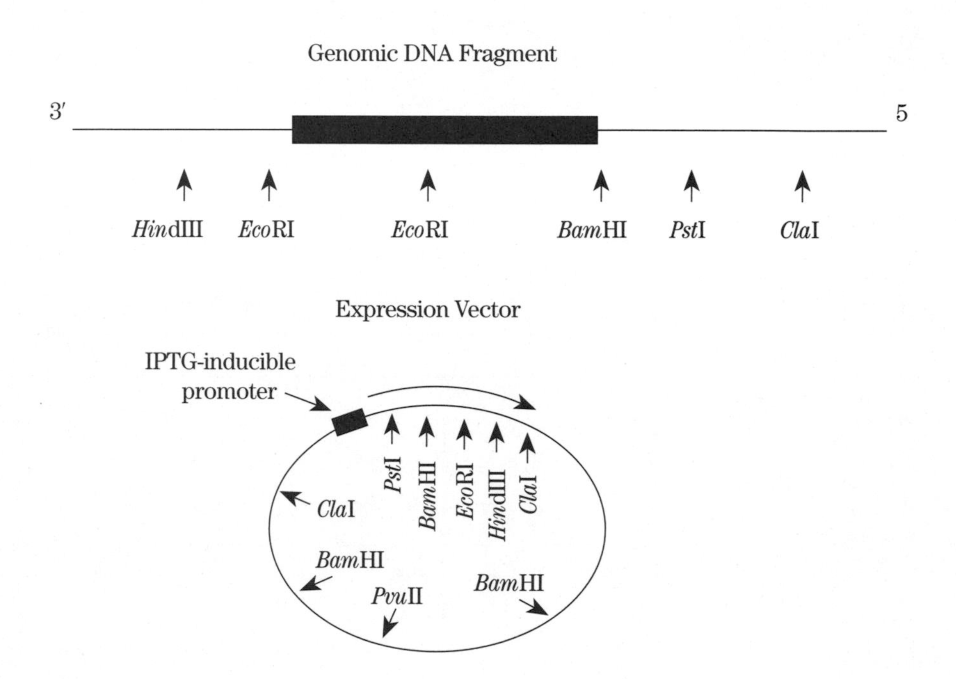

1. Name the region of the plasmid, near the promoter, where *Pst*I, *Bam*HI, *Eco*RI, *Hin*dIII, and *Cla*I have recognition sequences.
2. If you cut the vector with the restriction enzyme *Bam*HI, how many pieces will you get?
3. You wish to insert the coding region of the DNA fragment into the expression vector so that the gene will be expressed (i.e., produce the protein it codes for). Indicate where the gene fragment must be inserted into the vector and the orientation it must have.
4. If your goal is to obtain only a *portion* of the total protein for use in generating antibodies, which single enzyme could you use to cut the vector and the DNA fragment so that the vector could produce all or part of the protein? Would *all* vectors with inserts make protein?
5. If your goal is to produce a *complete, functional* protein, could you still use a single restriction enzyme? Which enzyme or combination of enzymes could you use to get enough of the gene fragment into the vector to ensure formation of the entire gene product?

Applying What You Know

1. You have received from a colleague a plasmid containing a piece of DNA that is expressed only in neuronal cells in *Drosophila.* You suspect that this gene codes for some type of signaling peptide.
   (a) What must you do to test your hypothesis? In other words, how do you get from a piece of DNA in a plasmid to the nucleotide sequence to a protein you can examine functionally?
   (b) How could you find out whether the gene is present in other species (e.g., a laboratory rat) and determine its pattern of expression?
   (c) Why would you or anyone care what proteins are present in *Drosophila* or in the lab rat?

2. How would you generate DNA probes for the following applications?
   (a) You have isolated an uncharacterized protein and you want to find out which cells and/or tissues actually express this protein.
   (b) You have isolated the mRNA for insulin from sheep pancreas. Your goal is to isolate a clone containing the insulin gene from a human cDNA library in order to produce authentic human insulin protein in *E. coli.*

ANSWERS

Do You Know the Terms?

ACROSS

1. bacteriophages
5. site-directed
7. fusion
8. plasmid
10. vectors
12. library
13. polylinker
15. restriction enzymes
16. transformation
18. ti plasmid
20. cDNA
21. blunt
23. clone
24. recombinant

DOWN

2. expression
3. genome
4. probe
5. shuttle vector
6. electroporation
9. DNA ligase
10. viral
11. RFLPs
13. PCR
14. transgenic
17. sticky ends
19. Southern
22. YACs
23. chip

Do You Know the Facts?

1. The polylinker region.
2. Three pieces.
3. The gene fragment must be inserted in the polylinker region, adjacent to and to the right of the promoter region. It must be oriented with its 5′ end to the left (at the beginning of the curved arrow) and its 3′ end to the right (at the arrowhead end): The DNA fragment will be flipped 180° relative to its orientation in the illustration.
4. You can use *Eco*RI to cut the vector and the DNA fragment; this will linearize the circular plasmid so that you can insert the fragment. You will also obtain a piece of the coding region that includes the genes for the carboxyl-terminal half of the protein. Because all the ends are cut with the same restriction enzyme, all are compatible. This means that the inserted DNA can be oriented in either the 5′→3′ or the 3′→5′ direction. However, only those plasmids with inserts oriented in the 5′→3′ direction will be able to translate the entire piece of DNA.
5. You could cut the vector and the DNA fragment with two restriction enzymes: *Pst*I and *Hin*dIII. When the fragment is ligated into the vector, its *Pst*I-cut 5′ end will be closest to the promoter (at the unique *Pst*I site in the polylinker) and its *Hin*dIII-cut 3′ end will be farthest from the promoter (inserting into the unique *Hin*dIII site in the polylinker). Note that you would not want to use *Bam*HI. It would insert the entire coding region into the vector in the correct orientation when used in concert with *Hin*dIII, but it also cuts at other sites within the vector.

Applying What You Know

1. **(a)** The first step is to determine the DNA sequence of the cloned gene. This can be done by purifying the plasmid vector DNA, excising the inserted DNA with restriction endonucleases, and using the dideoxy sequencing technique to sequence the excised fragment. If you obtained the DNA fragment from a genomic library, the sequence will likely include promoter regions, introns, exons, etc. Finding the initiation AUG sequence will provide you with the starting point of the open reading frame. If you obtained the DNA fragment from a cDNA library, determining the coding sequence will be more straightforward.

The second step is to deduce the amino acid sequence from the DNA sequence, which is now usually done with computer programs. This information can then be compared with the sequences in various databases to see whether any similar genes have been sequenced and characterized. (This capability often proves an invaluable tool for researchers; someone else may already have cloned a homologous gene, and this information may give clues about the function of the gene you are studying.)

Once you know the amino acid sequence, you can synthesize short segments of the protein for use in generating antibodies. These antibodies can localize the protein *in situ*, or they can be attached to column material for affinity-chromatography purification of the protein from homogenized *Drosophila*. An alternative approach would be to clone the DNA into a fusion protein vector and make the protein in *E. coli*.

Having obtained a functional protein, you are in a position to test whether it is a signaling peptide (neurotransmitter).

(b) You could look for homologous genes in other organisms using labeled DNA probes and tissue-specific cDNA libraries. If any of the clones in the libraries hybridizes to the probe, that clone most likely contains a gene similar to that from *Drosophila*. The source of the cDNA library will provide information about which tissues in a rat, for example, actually express the homologous gene.

(c) You should now be acutely aware that the same basic molecular mechanisms are used in most biological systems. Cellular and molecular mechanisms that operate in *Drosophila*—and those that operate in the rat—are likely to have counterparts in other organisms, including *Homo sapiens*. The homeobox genes that play important roles in the development of *Drosophila* embryos, for example, are closely related to the *Hox* genes that regulate development in vertebrates.

2. (a) To generate DNA probes given only an uncharacterized protein, you first have to establish the amino acid sequence of the protein so as to deduce the nucleotide sequence of its gene. It is then possible to generate a synthetic, **complementary** oligonucleotide containing sequences complementary to a portion of the deduced sequence. Ideally, the probe should be complementary to a region of the sequence that has minimal redundancy (see Fig. 9–10). You can label the oligonucleotide with a radioactive or fluorescent tag and perform *in situ* hybridization on fixed tissue slices to locate cells that contain mRNA that binds the probe. The results of this study will indicate which cells *might* contain the uncharacterized protein.

(b) Generating DNA probes when you have a supply of mRNA for a protein is relatively straightforward. Using reverse transcriptase and the mRNA as a template, you can make cDNA. This can be radioactively labeled and used to screen a genomic library. Once you have located a clone that contains the entire coding region, the foreign DNA fragment can be excised and cloned into an expression vector; it can then be used to produce quantities of the desired protein—in this case, insulin.

chapter

10

Lipids

STEP-BY-STEP GUIDE

Major Concepts

Biological lipids are chemically diverse and have a number of different functions in biological systems.

Lipids are poorly soluble in water. Consequently, they are well suited to serve as the major component of cell membranes, forming a barrier between the cell interior and the extracellular environment. The hydrocarbon chains of lipids are highly reduced, making them excellent compounds for energy storage. A few select lipids also serve very specific biological functions, for example, as intracellular messengers and redox cofactors.

Fatty acids are hydrocarbon chains, generally 4 to 36 carbons long.

Saturated fatty acids contain *no* double bonds; fatty acids *with* double bonds (usually cis) are **unsaturated.** The physical properties of fatty acids and the lipids that contain them are determined largely by the chain length and number of double bonds. Saturated fatty acids have full rotation around each carbon bond and are very flexible; they also have a relatively high melting point. Unsaturated fatty acids cannot rotate around their cis double bonds and are less flexible. Unsaturated fatty acids have lower melting points than saturated fatty acids. Fatty acids with long hydrocarbon chains tend to be less soluble in water than those with short chains.

Triacylglycerols are lipids that are used for energy storage by biological systems.

In mammals, these lipids are found primarily in the large storage vacuoles of adipocytes (fat cells). Triacylglycerols have a glycerol backbone to which three fatty acid chains are attached via ester linkages.

Membrane lipids include phospholipids, sphingolipids, and sterols (cholesterol in animals).

These lipids contain one or two hydrophobic hydrocarbon chains (tails) attached to a glycerol, sphingosine, or sterol backbone. A hydrophilic moiety (head) is also attached to that backbone. Therefore, unlike triacylglycerols, membrane lipids are **amphipathic** molecules, with both hydrophobic and hydrophilic regions. This property is essential to their ability to spontaneously form bilayer structures in aqueous solutions. Some membrane lipids, such as gangliosides, contain carbohydrate moieties as well. Many plant membrane lipids contain galactolipids and sulfolipids. Archaebacteria have unusual membrane lipids that help keep the membranes stable under the extreme conditions of their ecological niches.

Biologically active lipids can serve both intra- and intercellular signaling functions.

Hydrolysis of the membrane lipid phosphatidylinositol produces two **intracellular** signaling molecules, diacylglycerol and inositol triphosphate. *Localized* intercellular messengers derived from arachidonic acid are called **eicosanoids,** and include prostaglandins, thromboxanes, and leukotrienes. Hormones such as testosterone and estrogen are derived from cholesterol; these steroid hormones are **intercellular** messengers. Vitamins D, A, E, and K are fat soluble isoprenoids with essential metabolic roles. Quinones function as redox cofactors.

Lipid analysis is difficult because lipids are insoluble in water.

The extraction of lipids from tissues requires the use of organic solvents. Chromatographic techniques are employed to further separate charged lipids from neutral lipids. Mass spectrometry is often required for complete structure determination.

What to Review

Answering the following questions and reviewing the relevant concepts, which you have already studied, should make this chapter more understandable.

- Carboxyl, phosphoryl, and hydroxyl groups, and ether and ester bonds, all play roles in the formation, structure, and function of lipids. Remind yourself again of these functional groups of biomolecules (Fig. 1–15).
- The presence of cis double bonds in unsaturated fatty acids has important consequences for the properties of lipids. Cis and trans configurations are shown in Figure 1–18 to emphasize the structural differences between these isomers.
- In some circumstances, structural formulas are as useful as space-filling models, but the shapes of lipids are really well-illustrated only by the latter. See Figure 1–17 to remind yourself of the rules behind each model, and look at Figures 10–1, 10–2, and 10–14.
- Hydrophobic interactions and the behavior of amphipathic compounds in water are critical in explaining the behavior of lipids in biological systems. These were previewed in Chapter 2. What factor is most important in explaining the strength of the forces that hold nonpolar regions of micelles together?
- Lipid extraction, separation, and identification involve the use of organic solvents and chromatography techniques that should be familiar from a course in organic chemistry. If you do not remember these techniques, it would be wise to review an organic chemistry (laboratory) text.
- Glycolipids contain carbohydrate moieties: review the various oligosaccharides and linkages that are commonly found in glycolipids (Fig. 7–29).

Topics for Discussion

Answering each of the following questions, especially in the context of a study group discussion, should help you understand the important points of this chapter.

10.1 Storage Lipids

1. Note that the categorization of lipids is not as neat and clear-cut as for proteins; the defining feature of lipids is their insolubility in water.

Fatty Acids Are Hydrocarbon Derivatives

2. Draw linolenic acid in the structural (not space-filling) style seen in Figure 7–1a. Would it be solid or liquid at room temperature?

Triacylglycerols Are Fatty Acid Esters of Glycerol

3. Why is olive oil (among other fats) so insoluble in water?

Triacylglycerols Provide Stored Energy and Insulation

4. Why are triacylglycerols such a good storage fuel, whereas carbohydrates are better as quick sources of energy?

Box 10–1 Sperm Whales: Fatheads of the Deep

5. Why would solid spermaceti oil be more dense than liquid spermaceti oil? Contrast this property with the density change that occurs when going from solid water (ice) to liquid water.

Partial Hydrogenation of Cooking Oils Produces Trans Fatty Acids

6. Vegetable oil and solid shortening are both mixtures of triacylglycerols. Why is shortening a solid and vegetable oil a liquid at room temperature?

Waxes Serve as Energy Stores and Water Repellents

7. Sheep shearers often have extremely soft, young-looking skin on their hands. What aspect of their job might account for this?

10.2 Structural Lipids in Membranes

8. Although storage lipids are quite hydrophobic, membrane lipids are amphipathic and can have noncovalent interactions with polar molecules. For both types of lipids, function is dictated by structure. Why can't storage lipids function as membrane lipids?

Glycerophospholipids Are Derivatives of Phosphatidic Acid

9. What is the nonpolar portion of phosphatidylcholine? This portion of the lipid would face away from the aqueous surroundings.

Some Glycerophospholipids Have Ether-Linked Fatty Acids

10. What is the molecular basis for the suggestion that ether-linked fatty acids may confer resistance to phospholipases?

Chloroplasts Contain Galactolipids and Sulfolipids

11. Compare sulfolipids and phospholipids. How are they similar? How are they different?

Archae Contain Unique Membrane Lipids

12. How many carbon atoms serve to span the membrane lipid bilayer in eukaryotic cell membranes? Compare this to the number of carbons that span the archeal membrane in their glycerol tetraethers.

Sphingolipids Are Derivatives of Sphingosine

13. What is the polar portion of sphingomyelin? This portion of the lipid would face the aqueous surroundings.

14. Ceramides are structurally similar to diacylglycerols. Compare the attachment of head groups to the diacylglycerol unit in glycerophospholipids with their attachment to ceramide in sphingolipids.

Sphingolipids at Cell Surfaces Are Sites of Biological Recognition

15. What portions of glycosphingolipids confer the recognition function?

Phospholipids and Sphingolipids Are Degraded in Lysosomes

16. Why are lysosomes important to cellular metabolism?

Sterols Have Four Fused Carbon Rings

17. How are sterols similar to other membrane lipids? How are they different?

18. What is the structure of isoprene? Why is this structure biologically important?

Box 10–2 Abnormal Accumulations of Membrane Lipids: Some Inherited Human Diseases

19. Why must there be several phospholipases to degrade phospholipids?

20. What is the enzymatic defect in individuals affected with Tay-Sachs disease?

10.3 Lipids as Signals, Cofactors, and Pigments

Phosphatidylinositols and Sphingosine Derivatives Act as Intracellular Signals

21. How is the location of phosphatidylinositol critical to its functions?

Eicosanoids Carry Messages to Nearby Cells

22. How does ibuprofen act to reduce inflammation and pain?

Steroid Hormones Carry Messages between Tissues

23. How are steroid hormones able to be transported through the aqueous bloodstream?

24. How does prednisone act to reduce inflammation?

Vascular Plants Produce Thousands of Volatile Signals

25. Speculate on why lipids are often strong-smelling molecules.

Vitamins A and D Are Hormone Precursors

26. What are the symptoms of deficiencies of vitamins A and D?

Vitamins E and K and the Lipid Quinones Are Oxidation-Reduction Cofactors

27. What are the symptoms of deficiencies of vitamins E and K?

28. Quinones function as electron carriers *within* the lipid bilayer. What is it about their structures that make them so lipophilic?

Dolichols Activate Sugar Precursors for Biosynthesis

29. What structural aspect of dolichols makes them interact so strongly with the hydrophobic portions of membrane lipids?

10.4 Working with Lipids

Lipid Extraction Requires Organic Solvents

30. What would happen to proteins in a mixture of biomolecules if you were trying to extract lipids? What would happen to the lipids in the mixture as you attempted to extract proteins?

Adsorption Chromatography Separates Lipids of Different Polarity

31. To what protein purification technique is adsorption chromatography most similar?

Gas-Liquid Chromatography Resolves Mixtures of Volatile Lipid Derivatives

32. Does gas-liquid chromatography leave the lipid in biologically functioning form?

Specific Hydrolysis Aids in Determination of Lipid Structure

33. Why are specific enzymes so useful in the structural analysis of lipids?

Mass Spectrometry Reveals Complete Lipid Structure

34. How would mass spectrometry detect the difference in two isoprenoids with different numbers of isoprenoid units? (Hint: see the legend to Fig. 10–24.)

Lipidomics Seeks to Catalog All Lipids and Their Functions

35. Can you identify the general function(s) of each category of lipids listed in Table 10–3?

Discussion Questions for Study Groups

- How are steroid hormones, phosphatidylinositol, and eicosanoids similar and different in the locality of their effects? In their location in the cell? In their structure?

SELF-TEST

Do You Know the Terms?

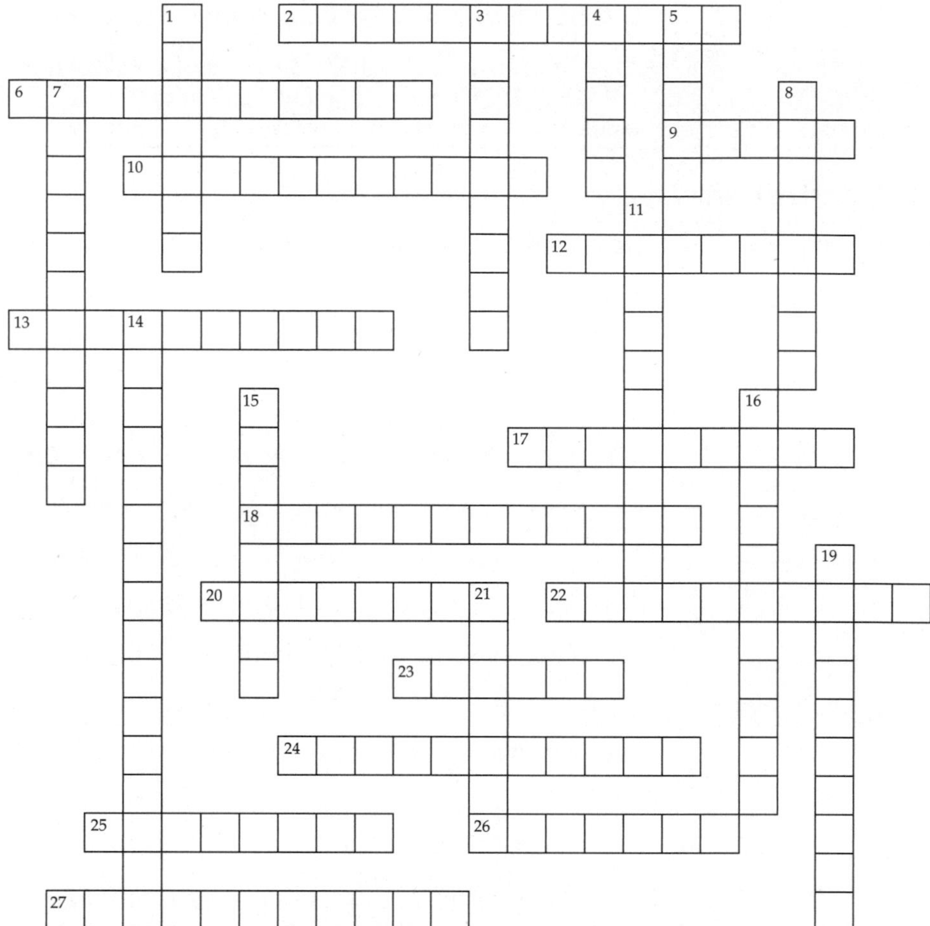

ACROSS

2. The possible head group of this class of lipids includes choline and sugars.
6. Sphingolipid with a very complex oligosaccharide head group.
9. Linkage joining the fatty acid tails of most glycerophospholipids to the glycerol backbone.
10. Describes the behavior of the lipid tail of a membrane lipid.
12. Intracellular organelle that compartmentalizes many degradative processes in cells.
13. Long hydrocarbon chains with carboxylic acid groups. (2 words)
17. Type of chromatography that separates lipids from a complex mixture on the basis of capillary action and differences in affinity for an immobile, polar matrix. (2 words)
18. Class of lipid whose hydrophilic moiety contains PO_4^{2-}.
20. H–C(H)(OH)–C(OH)(H)–C(H)(OH)–H
22. Type of chromatography that uses a polar, immobile substrate to selectively remove polar and charged lipids from a complex lipid mixture.
23. Vertebrate heart tissue is uniquely enriched in ether _______ such as plasmalogens.
24. Animal lipid having a rigid sterol nucleus.
25. Extremely hydrophobic isoprenoid compound that anchors sugars to cell membranes.
26. Phosphatidylinositols act as intracellular _______.
27. Describes fatty acids having one or more double bonds.

DOWN

1. The arrangement of lipids in membranes.
3. _______ - _______ chromatography uses differences in the ability to partition between an inert column matrix and an inert gas to separate lipids from a complex mixture. (2 words)
4. Cleavage of phosphatidylinositol bisphosphate by a lipase (phospholipase C) produces two _______ cellular messengers.
5. Steroid hormones are _______ cellular messengers because they carry messages between tissues.
7. Describes phospholipids, sphingolipids, and cholesterol, but not triacylglycerols.
8. Phosphatidic acid is to glycerophospholipids as _______ is to sphingolipids.
11. Ubiquinone and dolichols are both biologically active _______.
14. Lipids stored in adipocytes (fat cells).
15. Compounds in which electrons are shared equally between the atoms are _______ and hydrophobic; they can form few hydrogen bonds with water.
16. Describes the behavior of the head group of a membrane lipid.
19. Prostaglandins belong to this class of paracrine hormones derived from the fatty acid arachidonic acid.
21. Class of enzymes contained in adipocytes that are used for the mobilization of storage lipids.

Do You Know the Facts?

1. For each class of lipids on the left, put a check mark in the column under the components found in these lipids. If only *some* of the lipids in a category contain a component, write "some" in the box. Is the compound a membrane lipid? The first one has been done for you as an example.

	Fatty Acid	Glycerol	Sphingosine	Phosphate	Carbohydrate	Membrane Lipid?
Triacylglycerol	✔	✔				
Phospholipid						
Sphingolipid						
Ceramide						
Ganglioside						
Cholesterol						

2. What are the polar and nonpolar portions of each of the following classes of lipids?

(a) Triacylglycerol
(b) Glycerophospholipid
(c) Sphingolipid
(d) Ganglioside
(e) Cholesterol

3. The term amphipathic means:

A. branched, with at least two branch points.
B. having one region that is positively charged and one region that is negatively charged.
C. different on the inside and outside of the lipid bilayer.
D. having one region that is polar and one that is nonpolar.
E. having two different types of bonds.

4. Coconut oil contains only a very small amount of unsaturated fatty acids. How can it still have a low melting point?

A. It contains a lot of long-chain fatty acids.
B. It contains mostly short-chain fatty acids
C. It has only a few hydrogen bonds per fatty acid chain.
D. A and C are true.
E. B and C are true.

5. The polar head group of cholesterol is:

A. the alkyl side chain.
B. glycerol.
C. the steroid nucleus.
D. the hydroxyl group.
E. choline.

For questions 6–8, refer to the structures labeled (a)–(d).

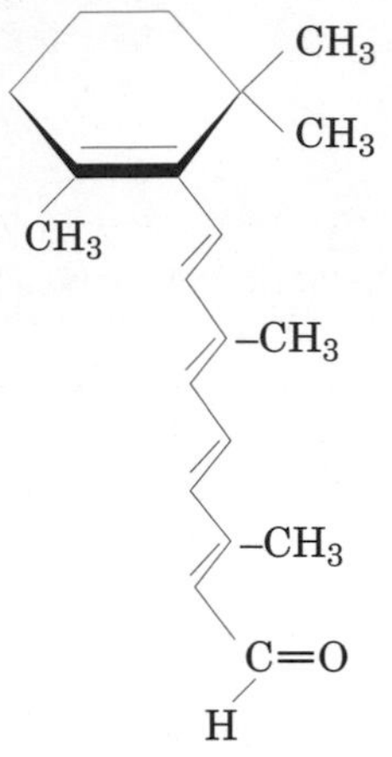

CH_3
CH_3
CH_3
$-CH_3$
$-CH_3$
C=O
H

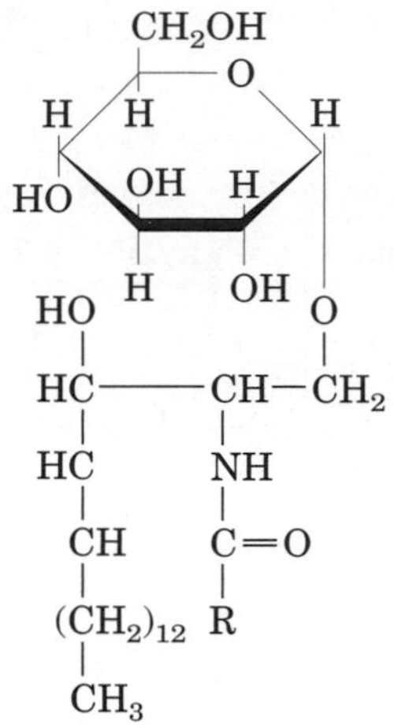

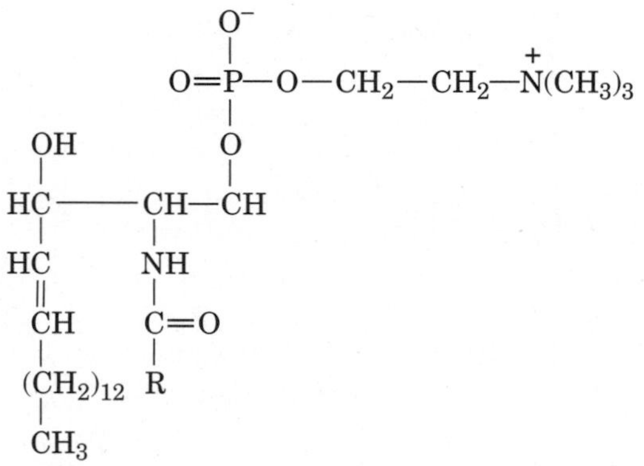

(a)

CH_2OH
O
H H H
HO OH H
H OH
HO O
HC—CH—CH_2
HC NH
CH C=O
$(CH_2)_{12}$ R
CH_3

(b)

O^-
$O{=}P{-}O{-}CH_2{-}CH_2{-}\overset{+}{N}(CH_3)_3$
O
OH
HC—CH—CH
HC NH
CH C=O
$(CH_2)_{12}$ R
CH_3

(c)

O^-
$O{=}P{-}O{-}CH_2{-}CH_2{-}CH_3$
H O
H_2C—C—CH_2
O O
O=C C=O
R_1 R_2

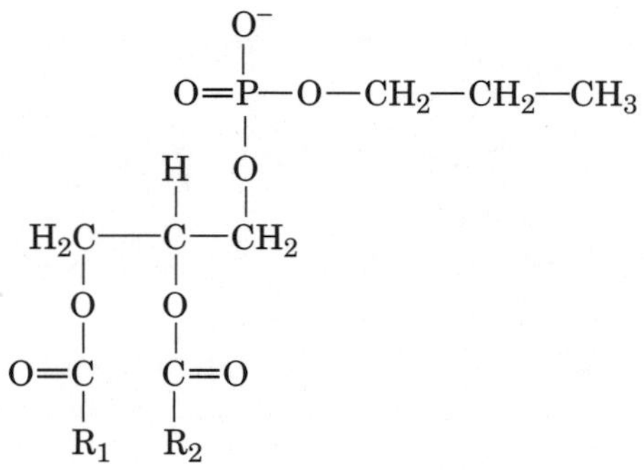

(d)

6. Which structure(s) contain(s) sphingosine?

7. Which structure(s) is a (are) glycolipid(s)?

8. Which structure(s) is a (are) glycerophospholipid(s)?

9. Match the following molecules to the descriptions:

(a) Testosterone and cortisol
(b) Phosphatidylinositol and its derivatives
(c) Eicosanoids (such as prostaglandins)
(d) Vitamin K and vitamin E

________ Fatty acid derivatives that act on the tissue in which they are produced

________ Isoprenoids that must be obtained from the diet

________ Intracellular messengers that are components of the plasma membrane

________ Steroid hormones that are produced in one tissue and carried through the bloodstream to target tissues

Applying What You Know

1. How are the chemical properties of storage lipids and membrane lipids different? How does this relate to their functional properties? (I.e., why wouldn't triacylglycerols make good membranes?)

2. Mixtures of lipids can be separated by adsorption column chromatography. A mixture containing the lipids listed below was applied to a silica gel column, and progressively more polar solvents were applied. What would be the order of elution of the individual lipids?

   ________ Triacontanylpalmitate, a component of beeswax
   ________ Lauric acid, a free fatty acid (12:0)
   ________ Cholesterol
   ________ Cholesterol palmitate, a cholesteryl ester (cholesterol esterified to a fatty acid at the —OH group)
   ________ Lignoceric acid, a free fatty acid (24:0)
   ________ Glucosylcerebroside, a sphingolipid with glucose as its "X" group

3. A naturally occurring fatty acid is systematically named 20:4 ($\Delta^{5,8,11,14}$) or cis-, cis-, cis-, cis- 5, 8, 11, 14-icosatetraenoic acid. What is the common name of this fatty acid? What is its shape? Of what class of compounds is this a precursor?

4. Naturally occurring unsaturated fatty acids are nearly all found in the cis configuration. However, trans fatty acids are found in some foods, most often as a byproduct of the saturation process (hydrogenation) used to harden natural oils in the manufacture of margarine. A small number of trans fatty acids are also found in foods derived from ruminant animals, produced by the action of microorganisms in the rumen.

   Oleic acid is a naturally occurring monounsaturated fatty acid, 18:1 (Δ^9). Elaidic acid is the trans form, also 18:1 (Δ^9), an unnatural isomer found in many margarine products. Would you expect a simple triaclyglycerol containing elaidic acid to have a higher or lower melting point relative to one containing oleic acid? Why?

5. Lipoproteins are lipid-protein complexes that transport lipids through the bloodstream from dietary and endogenous sources to peripheral tissues and to the liver. They include LDLs (the "bad" cholesterol"), HDLs (the "good" cholesterol), and others. They are essentially fat droplets, composed of triacylglycerols and cholesterol esters, enclosed by a monolayer of phospholipids, cholesterol, and protein. What is the functional significance of the monolayer of amphipathic lipids? How is it spatially arranged?

ANSWERS

Do You Know the Terms?

ACROSS

2. sphingolipid
6. ganglioside
9. ester
10. hydrophobic
12. lysosome
13. fatty acids
17. thin-layer
18. phospholipid
20. glycerol
22. adsorption
23. lipids
24. cholesterol
25. dolichol
26. signals
27. unsaturated

DOWN

1. bilayer
3. gas-liquid
4. intra
5. inter
7. amphipathic
8. ceramide
11. isoprenoids
14. triacylglycerols
15. nonpolar
16. hydrophilic
19. eicosanoid
21. lipases

Do You Know the Facts?

1.

	Fatty Acid	Glycerol	Sphingosine	Phosphate	Carbohydrate	Membrane Lipid?
Triacylglycerol	✔	✔				
Phospholipid	✔	✔ (some)	✔ (some)	✔	✔ (some)	✔
Sphingolipid	✔		✔	✔ (some)	✔ (some)	✔
Ceramide	✔		✔			✔
Ganglioside	✔		✔		✔	✔
Cholesterol	(has a short hydrocarbon side chain)					✔

2. **(a)** Fatty acid carboxylic acid groups are esterified to glycerol's —OH groups: the whole molecule is nonpolar.
   (b) Phosphate groups and attached head groups are polar; fatty acid tails are nonpolar.
   (c) The —OH groups on the alcohol sphingosine are polar; some sphingolipids have phosphate groups, some have sugar groups, both of which are polar; fatty acid tails are nonpolar.
   (d) The carbohydrate moiety is polar; fatty acid tails are nonpolar.
   (e) The —OH group is polar; the steroid nucleus and acyl chain are nonpolar.
3. D
4. B
5. D
6. b and c
7. b
8. d
9. c, d, b, a

Applying What You Know

1. Storage lipids include primarily fats and oils. These are highly reduced compounds, containing fatty acids that, when oxidized, provide a large amount of energy for the organism. Triacylglycerols are one type of storage lipid. These are made up of three fatty acids in ester linkage with a single glycerol molecule. Because the polar hydroxyls of glycerol and the polar carboxylates of the fatty acids are bound in ester linkages, triacylglycerols are nonpolar, hydrophobic molecules, essentially insoluble in water.

   Membrane lipids, on the other hand, are amphipathic. The orientation of their hydrophobic and hydrophilic regions directs their packing into membrane bilayers. The three general types of membrane lipids are glycerophospholipids, sphingolipids, and sterols. All have polar "heads" and nonpolar "tails," and all form bilayers, which constitute the central architectural feature of biological membranes. Hydrophobic triacylglycerols do not orient into such bilayers, having no hydrophilic portion to face toward the aqueous surroundings.
2. In order of first eluted to last eluted:
   - Cholesterol palmitate and triacontanylpalmitate (most nonpolar)
   - Cholesterol (still quite nonpolar because of the steroid nucleus, but with a polar —OH group)
   - Glucosylcerebroside (a neutral but polar lipid, due to its —OH group on the sphingosine backbone and the —OH groups of the glucose portion)
   - Lignoceric acid (a long, charged fatty acid)
   - Lauric acid (a charged fatty acid because of the $—COO^-$ group, but shorter than lignoceric acid so less nonpolar)
3. This is arachidonic acid. It is bent into a U-shape, due to the four cis double bonds, and is a precursor of the eicosanoids.
4. The triacyclglycerol containing the trans isomer of the fatty acid would likely have a higher melting point. In trans fatty acids, the acyl chains originate from the opposite side of the double bond, making the whole molecule relatively "straight." The cis isomer has the acyl chains originating on the same side of the double bond, giving the fatty acid a bend or "kink." The "straight" trans fatty acids will pack more tightly, increasing the interactions between them, and raising the melting point of the fat.
5. The monolayer of lipids are arranged so that their hydrophobic portions (the fatty acid tails of the phospholipids and the steroid nucleus of the cholesterol) are facing inward, in contact with the nonpolar center lipids. The hydrophilic portions (the polar heads of the phospholipids and the —OH group of the cholesterol) face outward in contact with the aqueous surroundings (the blood), increasing the solubility of the lipoproteins and therefore their transport in the bloodstream.

chapter

11 Biological Membranes and Transport

STEP-BY-STEP GUIDE

Major Concepts

Biological membranes define cells.

Biological membranes are composed of lipids and proteins. The lipid component, by weight percentage, is primarily phospholipids, but also includes varying amounts of sphingolipids and sterols. Membrane lipids are the primary *structural* component of biological membranes, but some serve as signaling molecules, as anchors for other membrane components, and probably in other roles, that have yet to be determined.

Membrane proteins are the primary *functional* entities of biological membranes. Proteins in membranes transport solutes, transduce extracellular signals, act as enzymes and enzyme activators, anchor other membrane components and intracellular structures, and provide intercellular contacts. The lipid and protein components of membranes vary from one cell to another in both their relative proportions and in the specific types of lipids and proteins.

Membrane structure is best described by the fluid mosaic model.

Membranes are organized as a lipid **bilayer;** proteins "float" within and on this fluid bilayer. Membranes are asymmetric with respect to the protein and lipid composition of the inner and outer layers. This asymmetry indicates that movement between the two faces of the bilayer must be restricted, although lateral movement within a single side of the bilayer is allowed and often is extremely rapid. The lipid bilayer is a thermodynamically stable structure that forms spontaneously, driven by hydrophobic interactions among the fatty acid tails of lipids and hydrophilic interactions among the polar head groups and surrounding water molecules.

Membrane proteins are associated with the lipid bilayer in a number of ways.

Peripheral membrane proteins are loosely associated with the membrane, usually through hydrogen bonds and ionic interactions. Some peripheral membrane proteins are attached through a covalent linkage to a membrane lipid "anchor." **Integral membrane proteins** span the lipid bilayer and contain regions that are relatively hydrophobic. These regions are buried in the core of the bilayer, where their association with the membrane is stabilized by hydrophobic interactions with membrane lipids. Integral proteins can contain one or several bilayer-spanning regions. These transmembrane portions are commonly α-helices or multistranded β-barrels, and can often be predicted from primary structure (amino acid sequence) data.

Membrane fluidity is critical for many biological processes.

Exocytosis, endocytosis, movement of many single-celled organisms, cell movement during development, cell division, and many other biological processes require a cell membrane that can change its shape. Membrane fluidity and the ability to reorganize spontaneously permit the fusion of biological membranes without disrupting the integrity of cells. Fusion is assisted by specific proteins called SNAREs.

The movement of many molecules across biological membranes is restricted.

The hydrophobic core of the lipid bilayer restricts the movement of polar or charged molecules. Water is an exception; it can diffuse slowly across the bilayer. However, water is also transported via **aquaporins** in cells when large fluxes of water are needed.

Transport can be categorized based on the need for energy, kinetic behavior, or the direction of movement. Metabolites and ions cannot cross the bilayer without specific **transporters,** which are specialized transmembrane proteins. Transporters are similar to enzymes in that they exhibit saturation and substrate specificity.

Passive transport moves solutes across membranes and down their respective concentration gradients, and therefore requires no direct input of energy. An example of passive transport is the movement of glucose into erythrocytes through the GluT1 transporter. **Active transport** moves solutes across membranes but against their respective electrochemical gradients, thus requiring the input of some form of energy. Primary active transport uses the energy of hydrolysis of ATP in a direct way. There are at least four types of ATPases in membranes. The electrochemical gradients generated by active transport often serve as potential energy sources to drive (via symport or antiport) the movement of other solutes against their concentration gradient. This process is **secondary active transport.**

Ion channels also provide routes for the transmembrane movement of specific ions. These do not show saturation behavior. Most ion channels are "gated" by either voltage or a particular ligand. The K^+ channel is an example of a membrane protein that forms a transmembrane pore or channel, which provides a route through the lipid bilayer. This channel is also an example of a multi-α-helical integral protein.

Porins are β-barrel integral proteins that transiently open to allow passage of specific solutes.

What to Review

Answering the following questions and reviewing the relevant concepts, which you have already studied, should make this chapter more understandable.

- Hydrogen bonds, van der Waals interactions, and hydrophobic interactions are extremely important forces in biological systems. Review how they apply to lipid (Chapter 2) and protein structure (Chapters 3 and 4).
- When describing cell membranes, a shorthand "cartoon" of membrane lipids is commonly used instead of drawing the complete structures. Be sure you understand what the components of that cartoon really represent (see Fig. 10–2).
- Lipids are the primary structural component of cell membrane, but not all lipids are found in cell membranes. Be sure you know what types of lipids are found in membranes, and know what properties they have in common (Fig. 10–7).
- Many proteins are associated with cell membranes. Two of the more common conformations of membrane-associated proteins are the α helix and β sheet. Review these secondary structures, and note how they contributes to the structural stability of proteins (Chapter 4, pp. 119–122).

Topics for Discussion

Answering each of the following questions, especially in the context of a study group discussion, should help you understand the important points of this chapter.

11.1 The Composition and Architecture of Membranes

Each Type of Membrane Has Characteristic Lipid and Proteins

1. What are the major classes of lipids found in biological membranes?

2. What do proteins contribute to the properties of biological membranes? How do the sugar moieties of membrane components contribute to the functions of membranes?

All Biological Membranes Share Some Fundamental Properties

3. What is the evidence that membranes really are organized in this bilayer fashion?

4. Biological membranes are described by the fluid mosaic model. Why are "fluid" and "mosaic" apt terms to describe membrane architecture?

A Lipid Bilayer Is the Basic Structural Element of Membranes

5. Under what conditions are each of the three types of amphipathic lipid aggregates favored?

6. What provides the energy that keeps membranes intact?

7. *Cell Map:* In the box that represents an enlargement of the lipid bilayer draw in the structures of several different types of membrane lipids. These lipids should be placed in the orientation that would allow them to form a stable bilayer arrangement. Indicate where the different types of noncovalent interactions occur that stabilize the orientation of these lipids in the membrane bilayer.

Three Types of Membrane Proteins Differ in Their Association with the Membrane

8. *Cell Map:* In the schematic diagram of the membrane bilayer at the top of the map, label the different components of cell membranes.

9. Compare the forces, interactions, or bonds that are most important in binding peripheral proteins and integral proteins to membranes.

10. Why are hydrophobic interactions NOT a factor in the association of peripheral proteins with membranes?

Many Membrane Proteins Span the Lipid Bilayer

11. Why is it important for determining the function of a membrane protein to know if it spans the bilayer or appears on only one face of the membrane?

12. What structural characteristics of glycophorin help maintain its asymmetric orientation? What prevents it from reorienting from one face of the bilayer to another by flip-flop diffusion?

Integral Proteins Are Held in the Membrane by Hydrophobic Interactions with Lipids

13. The figure below represents a cross section through a membrane and the seven circles represent a "top down" view of the seven α helices of bacteriorhodopsin. What classes of amino acid side chains would you expect to find projecting from the regions that interact with the membrane lipids, and what side chains would you expect to find projecting into the pore?

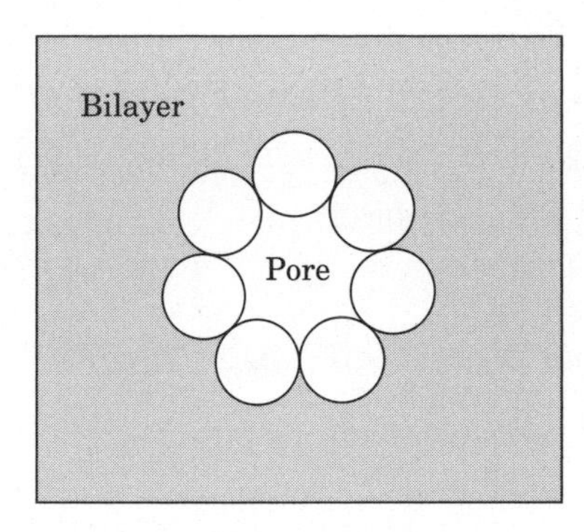

14. Why would a molecule of bacterorhodopisin (see Fig. 11–9) be insoluble in water? How, then, could it be solubilized for study?

The Topology of an Integral Membrane Protein Can Sometimes Be Predicted from Its Sequence

15. Note that hydropathy indices can be determined only if the amino acid sequence of the protein in question is already known.

16. What forces stabilize α helices formed from hydrophobic amino acid residues in transmembrane proteins? (Hint: review the section on α-helical structures in Chapter 4.)

17. Where would you predict you would find an amino acid residue with a hydropathy index of -2? What about an amino acid residue with an index of +2? What other information would you need to make a reasonable prediction?

18. What are some of the similarities and differences between α-helical and β-barrel transmembrane segments?

Covalently Attached Lipids Anchor Some Membrane Proteins

19. Besides attachment, what other functions may lipid anchors serve?

11.2 Membrane Dynamics

Acyl Groups in the Bilayer Interior Are Ordered to Varying Degrees

20. Which state would be favored by increasing the percentage of saturated fatty acids of a membrane? What other changes in lipid content of a membrane would produce the same effect?

Transbilayer Movement of Lipids Requires Catalysis

21. What forces and/or interactions would restrict lipid transbilayer diffusion ("flip-flop")?

Lipids and Proteins Diffuse Laterally in the Bilayer

22. Speculate on the functional reasons why some membrane proteins are anchored rather than free to diffuse laterally.

Sphingolipids and Cholesterol Cluster Together in Membrane Rafts

23. Why is it that sphingolipids and cholesterol preferentially cluster together and not other membrane lipids?

24. What is the significance of rafts in terms of the overall distribution of lipids and proteins in the membrane bilayer?

25. What is the role of calveolin in membrane structure?

Box 11–1 Atomic Force Microscopy to Visualize Membrane Proteins

26. In order to fully appreciate the atomic detail in these photos, you should compare the F_0 subunit of the chloroplast ATPase (Fig. 2c) with the cartoon of F_0 F_1 ATPase shown in Figure 11–39.

Membrane Curvature and Fusion Are Central to Many Biological Processes

27. What forces or interactions allow membranes to reorganize without loss of continuity?

28. When secretory vesicles fuse with the plasma membrane the SNARE and SNAP proteins bring the membranes together and induce membrane curvature. What else must occur before the membranes can fuse?

Integral Proteins of the Plasma Membrane Are Involved in Surface Adhesion, Signaling, and Other Cellular Processes

29. Why is it possible for integrins to have such a wide variety of specificities?

11.3 Solute Transport across Membranes

Passive Transport Is Facilitated by Membrane Proteins

30. What are the two components of the electrochemical gradient?

31. In what ways are membrane transporters and enzymes similar? In what ways are they different?

Transporters Can Be Grouped into Superfamilies Based on Their Structure

32. What generalities can be made concerning the protein structure of membrane carriers vs. membrane channels? Would these structural differences explain the differing transport properties?

The Glucose Transporter of Erythrocytes Mediates Passive Transport

33. What does the V_0 vs. $[S]_{out}$ plot for glucose entry into a cell tell you about the transport process itself?

34. What sets the upper limit on the concentration of glucose that can be transported into erythrocytes by GLUT1?

35. Does GLUT2 have a higher or lower affinity for glucose than GLUT1? What could be a physiological reason for this?

The Chloride-Bicarbonate Exchanger Catalyzes Electroneutral Cotransport of Anions across the Plasma Membrane

36. The passive transport process mediated by the chloride-bicarbonate exchanger is reversible. Why is this physiologically necessary?

Box 11–2 Defective Glucose and Water Transport in Two Forms of Diabetes

37. Some of the effects of diabetes include an inability to insert specific glucose and water transporters into the membranes of liver and kidney tubule cells. There are, however, other water and glucose transporters that are already present in the plasma membranes of these cells. Why are these transporters not sufficient to allow these cells to function normally? (Hint: What are the hallmarks of passive transport?)

Active Transport Results in Solute Movement against a Concentration or Electrochemical Gradient

38. Be sure you are clear on the definitions and characteristics of primary and secondary active transport for the rest of the chapter.

39. What information does $\Delta\psi$ supply?

P-type ATPases Undergo Phosphorylation during Their Catalytic Cycles

40. According to the description of how this class of transporter operates, what process actually converts (transduces) the chemical energy in ATP into the electrochemical energy of an ion gradient?

A P-Type ATPase Catalyzes Active Cotransport of Na^+ and K^+

41. Why doesn't intracellular K^+ bind to the two empty K^+ sites of the Na^+K^+ ATPase shown in part (1) of Figure 11–37?

42. What are some of the reasons why it is so important that the cytosolic concentrations of calcium be kept low?

43. What is the relationship between protein phosphorylation and ion binding in the SERCA pump and the Na^+K^+ ATPase?

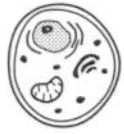

44. *Cell Map:* Use empty portions of the plasma membrane to draw in different transporter proteins and the solutes being transported. The molecular model for the Na^+K^+ ATPase has been provided. You should draw in the details (i.e., solutes and substrates bound and phosphorylation states) of the different conformations of this transporter. You can also diagram the various means by which glucose crosses cell membranes.

F-type ATPases Are Reversible, ATP-Driven Proton Pumps

45. What is the difference between an ATPase and ATP synthase?

46. In what ways are the F-type and V-type related?

ABC Transporters Use ATP to Drive Active Transport of a Wide Variety of Substrates

47. What features do all ABC transporters have in common?

Ion Gradients Provide the Energy for Secondary Active Transport

48. The mechanism by which membrane transporters work is based on changes in protein conformation. What is believed to be the basis for the protein conformation change underlying lactose transport?

49. How is the Na^+K^+ ATPase involved in transport of glucose and amino acids in intestinal epithelial cells?

50. How does the ionophore valinomycin facilitate the movement of potassium ions across membranes? What is the biological and medical significance of their function?

Box 11–3 A Defective Ion Channel in Cystic Fibrosis

51. Compare CTFR with the SERCA pump. What is the difference between an ion channel and an ion pump?

Aquaporins Form Hydrophilic Transmembrane Channels for the Passage of Water

52. Why is it so important that water molecules be able to cross the lipid bilayer?

53. Why is it so important that ONLY water molecules pass through the aquaporin channels?

Ion-Selective Channels Allow Rapid Movement of Ions across Membranes

54. How are ion channels different from ion transporters?

Ion-Channel Function Is Measured Electrically

55. In what other biochemical systems is signal amplification important?

56. Why is it functionally important that an ion channel remains open for only a very short time?

The Structure of a K^+ Channel Reveals the Basis for Its Specificity

57. What types of noncovalent interactions are critical to the ion specificity of the K^+ channel?

Gated Ion Channels Are Central in Neuronal Function

58. How are ligand-gated ion channels and voltage-gated ion channels similar? How are they different?

59. How does the nervous system rely on voltage-gated channels?

60. How does its molecular structure account for the channel's function?

Defective Ion Channels Can Have Severe Physiological Consequences

61. How can some toxins be so deadly at very low concentrations?

Discussion Questions for Study Groups

- Under what circumstances might the movement of a substance (e.g., Na^+) down its **concentration** gradient be energetically unfavorable?
- Experiments to establish membrane topography were described in the beginning of this chapter. In these experiments (p. 375), erythrocytes were treated with trypsin to preferentially degrade membrane proteins that are exposed to the extracellular compartment. How would you go on to experimentally determine whether any proteins had been affected by this treatment? (Hint: Consider the methods described in Chapter 3, Working with Proteins.)
- Ouabain is a potent poison obtained from seeds of plants found in Africa and Asia. Extracts from these seeds were applied to the tips of arrows allowing hunters to kill prey even if the prey were only slightly wounded. Ouabain works by inhibiting the Na^+/K^+ ATPase. What biological processes would be affected by this poison?

SELF-TEST

Do You Know the Terms?

ACROSS

2. A plot of _____ index vs. amino acid residue number in a protein predicts potential membrane-spanning α-helical regions of integral membrane proteins; such a plot is *not* useful for predicting β-barrel transmembrane segments
4. The structural organization of lipids in biological membranes.
8. One category of _____-_____ ATPases that are responsible for the production of ATP in mitochondria and chloroplasts; they are also known as ATP synthases.
9. _____ refers to the simultaneous transport of two solutes across a membrane in opposite directions.
13. _____ proteins are very firmly associated with the membrane via hydrophobic interactions with the fatty acid chains of membrane lipids.
16. The _____ potential takes into account the effects of the chemical concentration gradient and the electrical gradient.
20. An example of a(n) _____ -gated ion channel is the acetylcholine receptor.
21. The _____ family of integral proteins provides channels for rapid movement of water across plasma membranes.
23. Facilitated diffusion is also called _____ transport.
27. The transport of solutes against a concentration or electrochemical gradient that requires the input of energy is known as _____ transport.
29. Ion-selective _____ provide a route for the rapid movement of ions across membranes.
30. Simultaneous transport of two solutes across a membrane, in either the same or opposite directions.
31. A membrane protein in an intact erythrocyte that reacts with trypsin must have at least one domain exposed on the _____ face of the lipid bilayer.
32. _____; transport of a single solute across a membrane.

DOWN

1. An ion _____ is a source of potential energy that drives secondary transport processes in cells.
3. _____ interactions among lipid molecules in water drive the formation of micelles, bilayers, and liposomes.
5. The $Na^{+}K^{+}$ _____ is an example of a cotransporter that is critical to the function of all cells.
6. Type of diffusion that occurs down a concentration gradient.
7. Membrane component that can modulate membrane fluidity
10. The major class of membrane lipids, in terms of weight percent.
11. _____ - _____ ATPases are reversibly phosphorylated by ATP as part of the transport process.
12. _____ diffusion is mediated by an integral membrane protein that lowers the activation energy for transport; this process exhibits saturation kinetics.
14. Class of lipids containing covalently attached carbohydrates.
15. SNAREs are proteins required for membrane _____ in the process of exocytosis.
17. "Flip-flop" of lipids in membrane bilayers is also known as _____ diffusion; facilitated by flippases.
18. Describes the polar head groups of membrane lipids and peripheral membrane proteins.

19. The evocative name of the model describing the structure of biological membranes is the _____ mosaic model.

22. Proteins and phospholipids rarely exhibit this type of movement in membranes without an input of energy.

24. Face of the lipid bilayer where 2 K^+ ions are released by the Na^+K^+ ATPase.

25. Type of rapid diffusion exhibited by both lipids and proteins in membranes.

26. The transport of two solutes in the same direction across a lipid membrane.

28. _____ - _____ ATPases pump protons, regulating the pH of intracellular compartments.

Do You Know the Facts?

1. Would the following changes in the structure and/or composition of membrane lipids favor the liquid-ordered or liquid-disordered state?

(a) Unsaturated hydrocarbon chains are replaced by saturated hydrocarbon chains.

(b) The hydrocarbon chains are shortened.

(c) Cholesterol and sphingolipid content is increased

In questions 2–7, decide whether the statement is true or false, then explain your answer.

2. Proteins and lipids account for almost all of the mass of biological membranes.

3. Proteins and lipids are the only components of biological membranes.

4. The relative proportions of protein and lipid are the same in all biological membranes.

5. All the cellular membranes in a particular organism contain the same lipids and the same percentage of lipid to protein.

6. Membranes with different functions have different proteins.

7. Mitochondrial membranes commonly include covalently bound carbohydrate moieties.

8. Which of the following is *not* a membrane lipid? Why?

A. cholesterol

B. triacylglycerol

C. phosphatidylglycerol

D. cerebrosides

E. sphingomyelin

9. Which of the following is *not* a reason why the term "fluid" is appropriate to describe the fluid mosaic model of the biological membrane?

A. Many proteins found in membranes are not stationary, but undergo constant lateral motion.
B. The fatty acid chains of the membrane lipids are able to move by rotation about their carbon–carbon single bonds.
C. Individual lipid molecules can diffuse laterally in the plane of the bilayer.
D. Individual lipid molecules often flip-flop from one side of the bilayer to the other.
E. Cells undergo mitosis without leaking cytoplasmic components into the extracellular space.

10. Proteins A, B, and C are membrane proteins associated (either as peripheral or integral proteins) with the intact biological membranes of a culture of cells. After the cells containing these proteins are exposed to a severe drop in ionic strength of the cell culture medium, only protein A remains associated with the cells. One can conclude that:

A. B and C are peripheral proteins.
B. B and C are integral proteins.
C. A is a peripheral protein.
D. A is either an integral protein or a protein covalently attached to a membrane lipid.
E. both answers A and D apply.

11. Which of the following statements is true concerning integral membrane proteins?

A. Hydrophobic interactions anchor them within the membrane.
B. Ionic interactions and hydrogen bonds occur between the protein and the fatty acyl chains of the membrane lipids.
C. These proteins can be solubilized by a solution of high ionic strength.
D. Hydropathy plots can be used to determine the amino acid sequence of the protein.
E. All of the above.

12. Ruptured biological membranes are "self-sealing" due to all of the following *except:*

A. the amphipathic character of the lipids.
B. hydrophobic interactions between lipids.
C. hydrogen bonding between the head groups of the lipids and H_2O.
D. an increase in entropy of the system upon sealing.
E. covalent interactions among lipids.

13. The hydropathy plot below provides which of the following information?

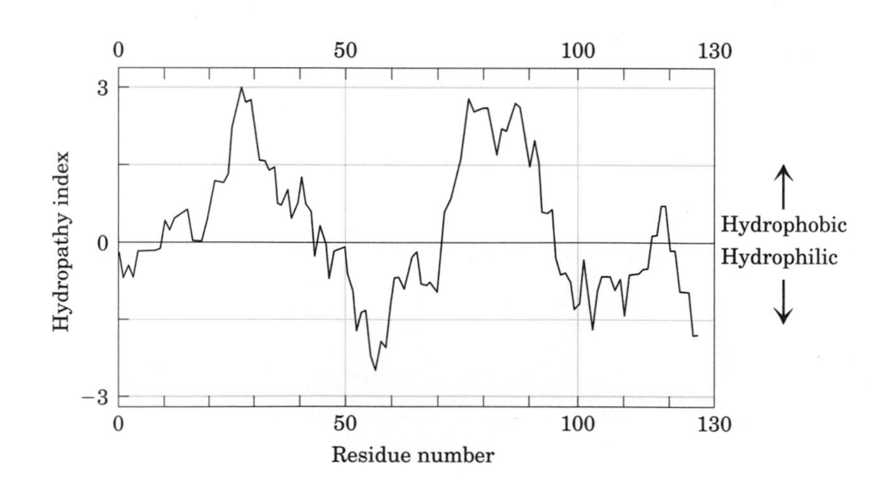

A. The protein being analyzed is a globular, peripheral membrane protein.
B. The protein being analyzed may be an integral membrane protein with two membrane-spanning domains.
C. The membrane being analyzed has a high degree of fluidity.
D. The membrane being analyzed has a high proportion of long-chain (as opposed to short-chain) fatty acids.
E. The protein being analyzed has extensive β-barrel domains.

14. The Na^+K^+ ATPase:

A. mediates active transport.
B. mediates cotransport of Na^+ and K^+.
C. is an integral membrane protein.
D. creates a transmembrane potential.
E. has all of the characteristics above.

15. You are investigating the function of the Na^+ K^+ ATPase in the cell membranes of cultured cells. You supply your cells with an analog of ATP (ATP γ-S), which binds to the protein but from which the terminal phosphate cannot be hydrolyzed due to the presence of the sulfate group. According to the model presented in your text, this drug would have which of the following effects on the transporter?

A. Na^+ would bind but would not be transported across the membrane.
B. Na^+ would bind and be transported across the membrane, but K^+ would not bind.
C. Na^+ would be transported, and K^+ would bind but would not be transported.
D. K^+ would bind and be transported but would not be released from its binding site.
E. The transporter would remain in the inactive state with no ions bound.

16. Arctic animals would be expected to have a higher cholesterol content in the cell membranes in their extremities because:

A. cholesterol increases hydrogen bonding between long-chain fatty acids in adjacent lipids, decreasing membrane fluidity at low temperatures.
B. increased noncovalent interactions between cholesterol's rigid, steroid ring structure and adjacent integral membrane proteins will increase membrane fluidity at low temperatures.
C. cholesterol's steroid nucleus prevents close packing of long-chain fatty acids in adjacent lipids, increasing membrane fluidity at low temperatures.
D. cholesterol's polar head group can form extensive interactions with water molecules, increasing membrane fluidity at low temperatures.
E. there are more van der Waals interactions between a molecule of cholesterol and a glycerophospholipid than can be formed between two glycerophospholipid molecules, increasing membrane fluidity at low temperatures.

17. Describe the lipid and protein content of the following membranes that are discussed in the text. How do these differences relate to their corresponding functions?

(a) The myelin sheath

(b) Vertebrate retina rod-cell

(c) Plasma membrane of an erythrocyte

18. What is the experimental evidence described in this chapter that supports the lipid bilayer as the basic structure of biological membranes?

19. Describe the effect of each of the treatments below on the various types of membrane proteins. Would the treatment release a peripheral protein from the membrane? An integral protein? A peripheral protein covalently attached to a lipid anchor?

(a) Changes in pH

(b) Changes in ionic strength

(c) Detergent

(d) Urea

(e) Phospholipase C

20. Describe the similarities and differences between transporter-mediated facilitated diffusion and transport via an ion-selective channel.

21. For each of the inhibitors listed below, describe first their *direct effect,* and then indicate the type of transport mechanism(s) that they block.

(a) Vanadate

(b) Cyanide

(c) Valinomycin

Applying What You Know

1. Two different species of bacteria have been isolated from very disparate environments: one, a hot spring with an average water temperature of ~40 °C, the other a glacial lake with an average water temperature of ~4 °C.

(a) Which of the two bacterial species would be expected to have more unsaturated fatty acids in its membrane lipids?

(b) Which would have longer-chain fatty acids?

(c) At 27 °C, which species would have a more fluid membrane?

2. You want to purify an integral membrane protein from the mitochondrial membranes of beef heart. (Because it is large and pumps constantly, heart muscle contains many energy-generating mitochondria per cell.) Compared with the precautions taken in cytosolic protein purification, what extra precautions must you take to purify the integral membrane protein?

3. In the *f*luorescence *r*ecovery *a*fter *p*hotobleaching **(FRAP)** experiment described in Figure 11–17, suppose that the chloride-bicarbonate exchange proteins in an erythrocyte had been fluorescently labeled. What would you predict would be your experimental results and why?

4. Ouabain is a specific inhibitor of the Na^+K^+ ATPase. You wish to determine how ouabain accomplishes this inhibition and perform a series of experiments; the results are presented below. What does each result tell you about the mechanism of inhibition of ouabain on the Na^+K^+ ATPase?

(a) Ouabain added to a solution containing intact cells inhibits movement of the ions across the plasma membrane (K^+ in and Na^+ out). Ouabain injected *inside* individual cells does *not* inhibit the movement of K^+ and Na^+ in those cells.

(b) Ouabain is added to a solution containing intact cells and allowed to effect full inhibition of the transporter. The cells are then collected, their membranes solubilized, and the membrane proteins are studied. All the molecules of Na^+K^+ ATPase from the ouabain-treated cells are found to be in the phosphorylated form (P-Enz_{II}).

(c) Based on the above results, which of the forms of the Na^+K^+ ATPase shown in Figure 11–37 represents the conformation that ouabain brings about?

5. Glucose uptake into adipocytes is mediated by the glucose transporter GLUT4. The majority of GLUT4 transporters are normally located in the membranes of intracellular vesicles but are translocated to the plasma membrane via membrane fusion in response to insulin; the resulting increase in the number of glucose transporters in the plasma membrane allows increase in uptake of glucose into the cells. Recently, a role for specific SNARE proteins has been described in the insulin-induced translocation of GLUT4 to the plasma membrane.

Before this mechanism was determined, other hypotheses existed to explain the behavior of GLUT4 uptake of glucose. Briefly describe an experiment that would investigate each of the following possible alternative explanations for the increase in glucose uptake by GLUT4 in response to insulin.

(a) Insulin increases the GLUT4 V_{max}, or decreases its K_m for glucose.

(b) Many molecules of GLUT4 are already present, but they are free in the cytoplasm and not located in the internal membranes.

(c) Membrane fusion is not a necessary part of the translocation process.

(d) SNAREs are not a necessary part of this process.

Biochemistry on the Internet

Nerve cells are able to transmit information due to their ability to generate electrical signals. These signals are the result of rapid changes in the electrical potential across the plasma membrane. If the change in the electrical potential (voltage change) is large enough, it creates a "domino effect" resulting in similar voltage changes that are transmitted along the entire length of the nerve cell. Such voltage changes are due to rapid changes in the distribution of charged molecules, primarily Na^+ and K^+ ions. Ion channels provide a favorable environment for the movement of ions through the uncharged and hydrophobic environment of the lipid bilayer.

The K^+ channel protein is an example of an integral membrane protein that facilitates the movement of ions through cell membranes. This channel is composed of four subunits, each of which has multiple membrane-spanning segments. A portion of this channel's structure has been determined and is depicted in the model with the PDB ID 1BL8. Try to answer the following questions based on the information provided by this model.

(a) How many subunits of the potassium channel are represented in this model?

(b) How would you expect this portion of the K^+ channel protein to be oriented in a plasma membrane; that is, which portions would face the aqueous environment and which portions would be associated with the membrane lipids?

(c) This ion channel is relatively selective for potassium ions. What portion(s) of this structure would you predict would contribute to its ability to allow the passage of K^+ while excluding other ions?

(d) What factors are important in preventing this structure from exhibiting "flip-flop" diffusion within the plasma membrane and in ensuring that the openings of the channel remain exposed to the aqueous environment?

ANSWERS

Do You Know the Terms?

ACROSS

2. hydropathy
4. bilayer
8. F type
9. antiport
13. integral
16. electrochemical
20. ligand
21. aquaporin
23. passive
27. active
29. channels
30. cotransport
31. outer
32. uniport

DOWN

1. gradient
3. hydrophobic
5. ATPase
6. simple
7. cholesterol
10. phospholipid
11. P type
12. facilitated
14. glycoplipid
15. fusion
17. transbilayer
18. hydrophilic
19. fluid
22. flip-flop
24. inner
25. lateral
26. symport
28. V type

Do You Know the Facts?

1. (a) liquid-ordered; (b) liquid-disordered; (c) liquid-ordered

2. T; see Table 11–1.

3. F; carbohydrates are also present, usually as glycoproteins and glycolipids.

4. F; proportions differ by organism and function; see Table 11–1.

5. F; content and proportion differ according to function, organ, role in cell, and location.

6. T; the proteins determine the function to a large degree.

7. F; plasma membranes contain glycoproteins on their outer face, but intracellular membranes rarely do.

8. B; triacylglycerols are not amphipathic and so do not aggregate into bilayers.

9. D

10. E

11. A

12. E

13. B

14. E

15. A

16. C

17. (a) The myelin sheath consists primarily of lipids; it functions as a passive insulator wrapped around the axons of myelinated neurons, and lipids are excellent insulators (see Ch.10, p. 346).

(b) In vertebrate retina rod-cell outer segments, a very large percentage (>90%) of the membrane proteins consists of just one single protein, the light-absorbing protein rhodopsin. The function of these cells is to detect light.

(c) The plasma membrane of an erythrocyte has many prominent proteins, not one specific one. This membrane is not specialized, but has many functions.

18. (a) In water, amphipathic lipids spontaneously form bilayers that are stabilized by hydrophobic interactions.

(b) Distribution of electron density, as shown by x-ray diffraction, is as expected for a bilayer structure (see Fig. 11–1).

(c) Laboratory-made liposomes behave similarly in terms of their impermeability to polar solutes.

(d) Some proteins are very difficult to extract from membrane lipids, suggesting they are embedded structures.

(e) Lateral diffusion of lipids in membranes is very rapid; transbilayer diffusion requires enzymatic catalysis.

19. The treatments described would release the membrane proteins that are: **(a)** peripheral; **(b)** peripheral; **(c)** integral and peripheral with anchor; **(d)** peripheral; **(e)** peripheral with anchor. Peripheral proteins associate with the membrane through electrostatic interactions and hydrogen bonds. They can therefore be released by changes in pH or ionic strength (which disrupt weak, noncovalent interactions), or by addition of urea (which breaks hydrogen bonds). Some peripheral proteins are attached to the membrane by a covalent lipid "anchor" as well; these can be released by the use of an appropriate enzyme. Integral proteins are more firmly attached, through hydrophobic interactions with the lipid bilayer. These can be released from the membrane only with detergents, organic solvents, or denaturants. Detergents form micelles with the individual proteins, keeping them in native conformation.

20. *Similarities:* Both types of transport require a protein that spans the bilayer (probably several times) and forms a transmembrane channel that provides a path through which a *specific* substrate can cross the membrane. Both allow flow of substrate only *down* its concentration gradient.

Differences: Transporters are saturatable; the rate of transport reaches a V_{max} at high substrate concentrations. Ion channels are not saturatable. While transporters speed up the rate at which their specific substrate can cross the membrane, the rate of flux through ion-selective channels can be orders of magnitude greater than for transporters. Transporters are permanently "open" to their substrate, as long as the concentration gradient exists. Ion channels are transient gates; they open or close in response to a specific signal.

21. **(a)** Vanadate is an analog of phosphate; it interferes with reversible phosphorylation, inhibiting P-type ATPases. Effects would include the loss of transmembrane electrochemical gradients and increases in intracellular Ca^{2+} levels.

(b) Cyanide (CN^-) blocks energy-yielding oxidation reactions that produce a H^+ (proton) gradient. It inhibits F-type ATPases (ATP synthases) and secondary active transport processes powered by proton gradients.

(c) Valinomycin binds K^+ in a hydrophilic central cavity, while presenting a hydrophobic exterior. This allows diffusion of the complex across the lipid bilayer of membranes, dissipating K^+ ion gradients. Valinomycin disrupts secondary active transport systems.

Applying What You Know

1. **(a)** The glacial lake bacterial species would have more unsaturated fatty acids in its membrane.

(b) The hot spring species would contain more long-chain fatty acids in its membranes.

(c) At 27 °C, the glacial lake species would most likely have a more fluid membrane.

2. Detergent must be included in all buffers, chromatographic columns, etc., or the integral protein will precipitate out as an insoluble aggregate.

3. Based on the information in this chapter and assuming that the chloride–bicarbonate exchange protein were the only protein labeled, one would predict that that there would be NO recovery of fluorescence following photobleaching. This is because the chloride-bicarbonate exchanger is an integral membrane protein that is tethered to the internal cytoskeletal protein spectrin and is not free to move in the membrane bilayer.

4. **(a)** Ouabain must bind specifically to a portion of the Na^+K^+ ATPase that is on the *outside* face of the plasma membrane.

(b) Ouabain binds to and "locks" the Na^+K^+ ATPase in its phosphorylated state.

(c) P-Enz_{II} with 2 K^+ bound and no Na^+ bound (the phosphorylated form has a high affinity for K^+ and a low affinity for Na^+), facing outside.

5. **(a)** Perform a series of [S] vs. V_0 trials (see Chapter 6) using a *specific number* (or amount of purified protein) of purified GLUT4 molecules in the presence and absence of insulin; determine the K_m, V_{max}, and K_{cat} (turnover number) values.

(b) Break apart adipocytes, separate soluble proteins from membrane-bound proteins, and determine the locale of the GLUT4 protein (e.g., using an antibody to GLUT4). This strategy could also be used to quantify movement of GLUT4 transport proteins from intracellular membranes to the plasma membranes by looking at insulin-induced vs. noninsulin-induced cells.

(c) Lower the temperature of the system and determine if the movement of GLUT4 to the plasma membrane slows down (glucose transport slows). Membrane fusion, like all processes that are based on the fluidity of the membrane, will slow down or halt at low temperatures. (This would not be the only explanation, though; other cellular processes are affected by temperature as well.)

(d) Generate anti-SNARE antibodies (see Chapter 5); use them as tools to block interaction of SNAREs in insulin-induced adipocytes, and determine if the rate of glucose transport slows down.

Biochemistry on the Internet

(a) Although the initial display format will vary depending on the viewer you are using, you should first display the structure in a **Ribbon** or **Backbone** format and then **"Color"** by **"Chain."** You will see four differently colored protein backbones displayed. Only a portion of the four subunits in this complex is shown.

(b) You can try to predict which portion of the polypeptide chain would be exposed to the aqueous environment and which would be associated with lipids based on the properties of the individual amino acid residues. If you click on each end of an α helix in these subunits, it is clear that they contain enough amino acid residues to span the lipid bilayer (24–26 residues). Select the hydrophobic amino acids **(Select → Protein → Hydrophobic)** and **Display** (or **Render**) them in a **Ball and Stick** format. Then select the charged amino acids **(Select → Protein → Charged)** and **Display** them in **Spacefill** format. The hydrophobic amino acid residues are primarily located in the transmembrane α-helical segments of the structure. The charged (and therefore hydrophilic) residues are clustered in two regions located on each end of the α-helical segments. These charged residues form ring-shaped structures that would be expected to associate with the aqueous environment on both sides of the lipid bilayer. Display the polar amino acid residues in **Spacefill** format. The polar residues (which are also capable of forming favorable interactions with water molecules) show a similar distribution to that of the charged residues.

Knowing the structure's orientation within a membrane, rotate the molecule so that you are viewing it from one of the ends that are in contact with the aqueous environment. It should be clear from this view how the protein complex allows the movement of molecules from one side of the membrane to the other. If your view is from the cytosolic or extracellular face, you should be able to see the central pore and several K^+ ions.

(c) Redisplay the entire molecule in a backbone format. Then **Select** the **Protein → Acidic** residues. **Display** (or **Render**) these residues in a **Spacefill** format. Because these residues are negatively charged, they would be expected to selectively interact with positively charged ions such as K^+. A total of four residues per subunit should be highlighted. These residues are located exclusively at each portion of the channel that is in contact with the aqueous environment. The presence of negative charges on the larger opening facilitates the attraction of positively charged K^+. The narrowing of the other opening combined with the presence of additional negative charges ensures that only molecules of a specific size, as well as charge, can pass through the opening.

In order to verify your assumptions, click **Select → Hetero** and then **Display → Spacefill** format. You will see three K^+ and an O_2 atom located in the region of the negatively charged residues.

(d) It should be clear from the above discussions that the presence of a significant number of hydrophilic residues at each end of the cylindrical channel will ensure that the opening of the channel is not buried in the lipid bilayer and remains associated with the aqueous environment. The hydrophobic residues ensure that the structure remains embedded in the lipid bilayer. Because these interactions are thermodynamically favorable, no energy needs to be expended to keep the molecule bound and oriented properly. In addition, models of the complete channel complex (e.g., PDB ID 1F6G) show an extensive cytoplasmic domain that would clearly prevent these proteins from experiencing "flip-flop" diffusion in the membrane.

chapter

12

Biosignaling

STEP-BY-STEP GUIDE

Major Concepts

In order to regulate and integrate their metabolic processes, cells have signaling mechanisms that relay information about the external environment to the interior of the cell.

Cells and their metabolic pathways do not exist in a vacuum. To function efficiently, cells must be able to coordinate their intracellular metabolism with the prevailing extracellular environment. In order to respond to information about their environment, cells must possess some kind of "sensor" or **receptor** that can interact with extracellular molecular signals. The production of a specific receptor in a cell, coupled with a fairly high affinity (low K_d) of that receptor for a given molecular signal, makes it possible for cells to be selective about the signals to which they respond.

Some of the extracellular signaling molecules to which cells can respond are lipid soluble, which allows them to pass unhampered through the lipid bilayer to the intracellular compartment. Receptors for lipid-soluble signals are found in the cytoplasmic and nuclear compartments of many cells. Some signaling molecules are hydrophilic; consequently, cells must have receptors for these signaling molecules on their plasma membrane. This latter signaling system requires a transducer element, which is "turned on" by the binding of the external signal, and which relays information by affecting the cellular machinery. In general, all signaling systems must be able to **amplify** the external signal, **integrate** information from a variety of sources, and turn themselves off when appropriate.

Cells use four basic mechanisms to transmit signals: (1) ion channels, (2) receptor enzymes, (3) G protein–linked receptors, and (4) steroid hormone receptors.

Ion channels allow the movement of ions across the plasma membrane in excitable cells.

Ion channels are integral membrane proteins that span the lipid bilayer several times, thus forming pores that allow charged molecules (e.g., Na^+, K^+, Ca^{2+}) to cross the membrane. The energy for the transmembrane movement through channels comes from the ion gradients established by the Na^+/K^+ ATPase. Channels are usually gated; that is, they can be opened and closed in response to specific signals. These signals include the binding of hormones to cell membrane receptors or changes in the electrical potential (voltage) across the cell membrane. The influx or efflux of ions through these pores can have a variety of effects on cellular metabolism. The nicotinic acetylcholine receptor that mediates muscle contraction in response to neural stimulation is an example of a receptor channel.

Overview of Neuronal Function

To understand signaling activity in neurons, it is helpful to categorize neuronal function into four distinct regions:

The **input region** is usually made up of cellular projections called dendrites. Dendrites contain **ligand-gated channels,** which respond to extracellular signals (e.g., neurotransmitter molecules). Binding of signaling molecules causes a conformational change that opens the associated channel. The movement of ions through channels in the plasma membrane results in a change in the **membrane potential (V_m),** which is passively transmitted by diffusion of ions away from the site where they entered the cell.

At the **integrating region,** called the spike-initiating zone, all of the changes in V_m caused by the opening of ligand-gated channels in the input region are added together. If the summed potentials are above a "set point," called the threshold potential, then an **action potential (AP)** is produced.

An AP is the rapid shift in V_m from a resting potential (the inside of the cell is negatively charged with respect to the outside) to a more positive intracellular potential. An AP is generated primarily by the activity of two **voltage-dependent ion channels,** the Na^+ and K^+ voltage-dependent channels. The initial, rapid and large depolarization of an AP is the result of the opening of **voltage-dependent Na^+ channels.** When these channels open, Na^+ ions enter the cell and flow down their concentration gradient. The V_m in the region of the open channels will move toward the equilibrium potential for Na^+ (which can be calculated using equation 12–1). The depolarization that occurs during an action potential ultimately causes the voltage-dependent Na^+ channels to close. **Voltage-dependent K^+ channels** open more slowly than do the Na^+ channels and therefore are not fully functional until after the initial Na^+-induced depolarization has occurred. Because the intracellular $[K^+]$ is higher than the extracellular $[K^+]$, this ion will exit the cell. Thus, the V_m will move toward the equilibrium potential for K^+, and the cell will be repolarized to the original resting potential.

The **conducting region** of a neuron is known as the axon and contains many voltage-dependent Na^+ and K^+ channels along its length. The sequential opening of these channels results in an AP that is conducted along the entire length of an axon.

The **output region** of a neuron is usually the nerve terminus. Here, the plasma membrane contains its own contingent of voltage-dependent channels; however, these channels are primarily permeable to Ca^{2+}. The Ca^{2+} ions that enter the terminal initiate a series of events that lead to the fusion of intracellular membrane-bound vesicles with the cell membrane. This fusion causes the release of thousands of neurotransmitter molecules into the extracellular space, called the **synapse,** that occurs between adjacent neurons.

Receptor enzymes have an extracellular receptor for the signal and a catalytic domain on the portion facing the cytosol.

In receptor enzymes, the binding of hormone to the extracellular domain often activates the enzymatic activity on the inside of the cell. The specific enzymatic activity that is turned on may vary, but often involves that of protein kinases. **Kinases** affect cellular metabolism by **phosphorylating** different target proteins such as metabolic enzymes. The insulin receptor is an example of a receptor enzyme in which a protein kinase activity is activated in response to insulin binding.

G protein–linked receptors use a GTP-binding protein to relay information to the interior of cells.

In some cases, the message carried by extracellular signals is transmitted to the interior of a cell through the generation of a second signaling molecule inside the cell. These molecules are referred to as **second messengers.** This type of signaling involves the binding of a hormone to an extracellular receptor, which activates a G protein. The G protein, in turn, activates an intracellular target protein (e.g., an enzyme or ion channel). It is this target protein that is responsible for generating the second messenger.

The G proteins can activate three major second messenger systems.

Cyclic nucleotides: Activated **G proteins** can affect the activity of a class of membrane-associated enzymes known collectively as cyclases. Two examples are adenylyl cyclase and guanylyl cyclase, which produce the second messengers **adenosine 3′,5′-cyclic nucleotide monophosphate (cAMP)** and **guanosine 3′,5′-cyclic nucleotide monophosphate (cGMP),** respectively. Many of the effects of cAMP and cGMP are exerted through their activation of specific cyclic nucleotide-dependent protein kinases. These activated kinases then affect cell metabolism by phosphorylating target proteins within the cell. There is also a cytoplasmic form of guanylyl cyclase that is activated by nitric oxide (NO) that can easily diffuse across the plasma membrane to enter the target cell.

Phosphatidylinositol bisphosphate metabolites: Hormonal activation of a different class of G proteins activates membrane-associated **phospholipase C (PLC).** This enzyme can cleave the membrane lipid **phosphatidylinositol bisphosphate (PIP_2)** producing two distinct intracellular messengers. One is **diacyl glycerol (DAG),** which is lipid soluble and therefore must exert its effects in or near the lipid bilayer. Typically, DAG (along with Ca^{2+}) activates the enzyme protein kinase C (PKC), which, like other protein kinases, exerts its effects on cells by phosphorylating target proteins. The second is **inositol triphosphate (IP_3),** which is water soluble and exerts its effects through binding to its own intracellular receptor on the surface of the endoplasmic reticulum membrane (ER) or sarcoplasmic reticulum membrane (SR). Binding to these intracellular receptors opens channels in the ER or SR membrane that allow Ca^{2+} that is stored in these structures to diffuse into the cytosol.

Calcium: In addition to the IP_3-mediated release of stored Ca^{2+} described above, there are other routes for getting this second messenger into cells. Normally, the level of Ca^{2+} inside cells is very low ($<1\mu M$). Extracellular hormones as well as changes in V_m open Ca^{2+} channels in the plasma membrane, resulting in an increase in the intracellular $[Ca^{2+}]$ as this ion passively flows down its concentration gradient into the cell. Ca^{2+} affects cell function in a number of ways, for example, by binding directly to enzymes and activating or inhibiting them or by binding to structural proteins and affecting their function. Alternatively, Ca^{2+} can also bind to its own intracellular receptor, **calmodulin.** This Ca^{2+}-protein complex then binds to cellular proteins, affecting their function.

Steroid hormones do not require intracellular messengers because they are lipid soluble and can diffuse across the membrane.

Because **steroid hormones** are relatively hydrophobic they usually must associate with a water-soluble molecule in order to be transported in the blood. When they enter the cell cytoplasm they must bind to specific hormone receptors in the nucleus before they can influence cellular function. Binding to these receptors causes conformational changes that allow the complex to associate with specific regions of DNA and influence gene expression.

Phosphorylation of cellular components is a molecular mechanism used by all cells to regulate cellular processes.

Kinases play a ubiquitous role in the regulation of cellular metabolism. Their influence is due to their ability to phosphorylate cellular components, thereby altering the conformation and thus the function of these substrates. The complicated pathways through which kinases are activated and through which they exert their effects provide multiple points where signal amplification as well as **integration** can occur. The specificity of kinase activity results primarily from the **localization** of the kinases and/or their substrates to specific regions in the cell.

What to Review

Answering the following questions and reviewing the relevant concepts, which you have already studied, should make this chapter more understandable.

- Review the topography of integral membrane proteins (Figs. 11–8, 11–9, and 11–12).
- The Na^+/K^+ ATPase is critical to the function of all cells and especially for neuronal signaling. Review the model for the Na^+/K^+ ATPase and be sure you understand why this pump is electrogenic (Fig. 11–37).
- Be sure you understand what is meant by membrane potential (Fig. 11–38).
- Hormonal signaling is critical to the regulation of metabolic pathways that you will learn about in coming chapters. The role of second messengers in signaling by peptide hormones is crucial. Two second messengers used by cells are derived from membrane lipids. Review the general structure of glycerophospholipids (Fig. 10–9) and note the sites of cleavage by lipases (Fig. 10–16).
- Many lipid-soluble hormones are derived from cholesterol. Review its structure and be sure you understand why it is lipid soluble. (Fig. 10–17).
- Protein phosphorylation plays a critical role in the regulation of enzyme activity by extracellular hormones and their intracellular messengers. Review which of the amino acids have available —OH groups that can be phosphorylated (Fig. 3–5).

Topics for Discussion

Answering each of the following questions, especially in the context of a study group discussion, should help you understand the important points of this chapter.

12.1 General Features of Signal Transduction

1. This section discusses six general mechanisms that cells use to relay extracellular signals to the interior of the cell. What features do each of these schemes have in common; what aspects are different?

2. How do hormones regulate metabolism only in the cells of specific target organs and not in cells of adjacent nontarget organs?

3. *Internet Question:* To see a model of a ligand bound to a receptor go to the Protein Data Bank and retrieve the file with PDB ID 1A22. This is a model of growth hormone bound to the extracellular portion of its receptor. What forces allow the ligand to associate with its binding site on the extracellular portion of the receptor? (Hint: try displaying the receptor chain in a Spacefill format and the growth hormone in a Ribbon structure. Then select charged amino acid residues and display them in a Ball and Stick format.)

4. *Cell Map:* As you are learning about the various signal transduction mechanisms, draw the components that make them work onto your cell map. (Hint: you may want to draw these schemes on a sheet of semitransparent paper that you use as an "overlay." This way, you will still have plenty of room to draw in the metabolic pathways you will be learning about in later chapters. Because these signaling mechanisms ultimately influence metabolic processes, you can use the overlay later to see how and where these interactions might occur.)

Box 12–1 Scatchard Analysis Quantifies the Receptor-Ligand Interaction

5. What Michaelis-Menten constant is analogous to the K_a for a receptor? B_{max}?

12.2 G Protein–Coupled Receptors and Second Messengers

6. *Cell Map:* As you read this section, construct your own molecular diagram showing how a G protein becomes activated in response to binding of a hormone to a G protein–coupled receptor.

The β-Adrenergic Receptor System Acts through the Second Messenger cAMP

7. In what ways can the effects of epinephrine, acting through the β-adrenergic receptor system, be turned off?

8. How does the G_s protein function as a transducer?

Box 12–2 G Proteins: Binary Switches in Health and Disease

9. What molecular activity is responsible for the "built-in timer" function of G proteins? Can the timer be altered?

Several Mechanisms Cause Termination of the β-Adrenergic Response

10. *Cell Map:* Expand your diagram from Question 6 to show how the activated G protein turns on the production of cAMP; include the effect of cAMP on its cellular target. Indicate the points at which the signal can be turned off.

The β-Adrenergic Receptor Is Desensitized by Phosphorylation and by Association with Arrestin.

11. In your model of β-adrenergic signaling (see Questions 6 and 10), where does desensitization occur? What aspect of desensitization cannot be illustrated in that model?

Cyclic AMP Acts as a Second Messenger for Many Regulatory Molecules

12. At which points in the molecular cascade (between receptor binding and kinase activation) can the cAMP second messenger system be regulated?

13. A large number of neurotransmitters and hormones exert their effects on cells by increasing cellular levels of cAMP. How is it possible for each signaling compound to produce a distinct cellular effect when they all use the same second messenger? (Hint: because of their small size we tend to think of cells as homogeneous "bags of enzymes," but at the microscopic levels they have a great deal of structure and organization.)

Diacylglycerol, Inositol Triphosphate, and Ca^{2+} Have Related Roles as Second Messengers

14. *Cell Map:* Both cleavage products of the membrane phospholipid phosphotidylinositol 4,5-bisphosphate (PIP_2) are important intracellular-signaling compounds. These products are diacylgylcerol (DAG) and phosphatidylinositol 3,4,5-triphosphate (IP_3). Using your cell map, draw a molecule of PIP_2. To the right of this membrane lipid draw the products of its cleavage by PLC. To which portion of the original membrane lipid do each of these signaling compounds correspond and where would you expect to find them in the cell?

15. What are the molecular consequences to the cell of increasing the [DAG] in a cell? IP_3?

Box 12–3 FRET: Biochemistry Visualized in a Living Cell

16. Why does loss of the FRET signal indicate a localized increase in cAMP levels?

17. How might this system be used to study the Ca^{2+}-dependent activation of CaM kinase?

Calcium Is a Second Messenger That May Be Localized in Space and Time

18. What is calmodulin and in what ways is its function similar to that of G proteins in signal transduction?

19. In what ways is the action of Ca^{2+} similar to that of cAMP in mediating signal transduction?

20. In what ways is ATP critical to the function of this second messenger system?

12.3 Receptor Tyrosine Kinases

Stimulation of the Insulin Receptor Initiates a Cascade of Protein Phosphorylation Reactions

21. *Cell Map:* Draw a diagram that illustrates the general mechanism whereby insulin exerts its effects on cell metabolism. Note that this intracellular signaling pathway will interact with the glucose transporters, enzymes involved in glycogen and lipid metabolism, as well as factors affecting gene expression.

The Membrane Phospholipid PIP_3 Functions at a Branch in Insulin Signaling

22. What cellular component links the insulin signaling pathway to PKB activation?

23. How does activation of PKB by insulin signaling result in an increased synthesis of glycogen?

The JAK-STAT Signaling System Also Involves Tyrosine Kinase Activity

24. What is the difference between the tyrosine kinase in the JAK-STAT signaling system and the tyrosine kinase in the insulin signaling system?

Cross Talk among Signaling Systems Is Common and Complex

25. The importance of cross talk in regulating cell metabolism will become much more apparant in Part II of the text.

12.4 Receptor Guanylyl Cyclases, cGMP, and Protein Kinase G

26. What is "cyclic" about cyclic GMP?

27. How is it that cGMP can generate different responses in different tissues?

28. Why is it possible for the guanylyl cyclase activated by NO to be located in the cytosol and not in the plasma membrane?

29. How is the intracellular signaling pathway mediated by the β-adrenergic receptor (section 12.2) different from that of ANF? In what ways are these two pathways similar?

12.5 Multivalent Adaptor Proteins and Membrane Rafts

Protein Modules Bind Phosphorylated Tyr, Ser, or Thr in Partner Proteins

30. "Combinatorial" mechanisms are common ways to control biological processes; what does "combinatorial control" mean in the context of this section?

31. What other class of proteins have you encountered that form "multivalent interactions"? What functions did these interactions serve?

Membrane Rafts and Caveolae Segregate Signaling Proteins

32. Segregation of raft components can be reversible. How might this be accomplished?

12.6 Gated Ion Channels

Ion Channels Underlie Electrical Signaling in Excitable Cells

33. How does the movement of ions through ion channels differ from the movement of ions across membranes achieved by the Na^+/K^+ cotransporter?

34. Why do neurons use ion channels instead of a transporter to generate the membrane voltage changes that are the basis for neuronal signaling?

35. What are the gradients that produce the "potential" in the electrochemical potential?

36. How does the general structure of a receptor channel differ from receptor enzymes?

37. What will happen to the V_m of a neuron when a neurotransmitter opens a channel that is permeable to more than one ionic species; for example, Na^+ *and* Ca^{2+}? (Hint: can you use equation 12–1 to predict the V_m when the membrane is permeable to more than one ionic species? What additional information might you need to know about the ionic permeabilities?)

38. How would V_m be affected by the opening of channels that are *equally* permeable to both Na^+ *and* Cl^-? (Hint: use Equation 12–1 to determine the separate equilibrium potentials and then factor in relative permeability.)

39. In what ways is the binding of acetylcholine to the nicotinic acetylcholine receptor similar to the binding of O_2 to hemoglobin?

Voltage-Gated Ion Channels Produce Neuronal Action Potentials

40. How might ion channel proteins "sense" voltage changes across the plasma membrane?

41. *Internet Question:* What classes of amino acid residues are likely to be involved in the sensing process? You can view models of different voltage-dependent channels (e.g., PDB ID 1J95, a voltage-dependent K^+ channel) to see if they support your hypothesis.

42. How would the membrane potential, V_m, and the production of action potentials be affected by the *simultaneous* opening of voltage-dependent Na^+ channels and voltage-dependent K^+ channels?

The Acetylcholine Receptor Is a Ligand-Gated Ion Channel

43. *Internet Question:* You can view the structure of the acetylcholine receptor channel using the PDB file 1OED. This model displays the five transmembrane subunits. Display the model in a ribbon format and rotate it until you can view the ion channel. See if you can locate the two α subunits to which acetylcholine binds.

Neurons Have Receptor Channels That Respond to Different Neurotransmitters

44. *Internet Question:* The interior of ion channels often includes a region referred to as a "selectivity filter." How is it that ion channels can "select" the type of ions that pass through them? What classes of amino acid residues are likely to be involved in this "selection" Process? Use the model you generated for Question 9 or 14 to see if you can confirm your hypothesis. (Hint: display the model as a ribbon format and then use the **"select"** and **"display"** commands to highlight different classes of amino acid residues.)

12.7 Integrins: Bidirectional Cell Adhesion Receptors

45. The integrin receptors described in this section and the receptor tyrosine kinases described in section 12.3 are both membrane receptors with a single transmembrane domain. Discuss the similarities and differences in the ways their extracellular and intracellular domains function to alter cellular physiology?
(Use Figs. 12-17 and 12-28 to focus your discussion.)

12.8 Regulation of Transcription by Steroid Hormones

46. Why do steroid hormones not require transducing elements, such as G proteins, to exert their effects on cellular metabolism?

47. Why must the drug tamoxifen, used in the treatment of breast cancer, be lipid soluble?

12.9 Signaling in Microorganisms and Plants

Bacterial Signaling Entails Phosphorylation in a Two-Component System

48. How does the two-component system described here compare to the insulin signaling pathway?

Signaling Systems of Plants Have Some of the Same Components Used by Microbes and Mammals

49. Which signaling systems used by all animal cells appear to be absent in plants?

Plants Detect Ethylene through a Two-Component System and a MAPK Cascade

50. What additional information from this section can you add to your response to Question 38 above?

51. Why do bananas and apples kept in a closed paper bag ripen faster than the same fruit left out in an open basket?

Receptorlike Protein Kinases Transduce Signals from Peptides and Brassinosteroids

52. What is the source of the signal that binds to receptorlike kinases in *Arabidopsis,* and what is the final effect of activation of this system?

12.10 Sensory Transduction in Vision, Olfaction, and Gustation

The Visual System Uses Classic GPCR Mechanisms

53. What happens to V_m when a cell is hyperpolarized, and how does a decrease in cGMP levels produce this effect?

54. In most neurons, it is the depolarization of membrane potential that leads to an influx of Ca^{2+} and the release of neurotransmitter by the excited neuron. Try to devise a cellular strategy to explain how hyperpolarization in rod cells leads to activation of the visual pathway. (Hint: neuronal signaling in higher organisms is mediated by a *series* of neurons, which relay information to higher brain centers. The signals released by these neurons can have excitatory as well as inhibitory effects on the cells that follow.)

55. Why are retinol and opsin needed in order to transduce light energy into something that can affect cellular processes?

Excited Rhodopsin Acts through the G Protein Transducin to Reduce the cGMP Concentration

56. How does the activation of transducin *decrease* cGMP levels?

57. How does this signaling system compare to that of epinephrine on cAMP levels? In what ways are the two systems similar? In what ways do they differ?

58. At which points in the transduction of light energy to a neuronal signal is the energy of the original light stimulus amplified?

59. Predict the effect of a cholera-like toxin on the visual system if it could interact with $G_{T\alpha}$ in the same way that it interacts with $G_{s\alpha}$. What disease symptoms would you expect to see in a patient exposed to this hypothetical visual system toxin?

The Visual Signal Is Quickly Terminated

60. What are the different cellular components that contribute to the rapid termination of the visual signal? (Hint: it may be helpful to construct a cellular model when discussing this question.)

61. The protein recoverin plays a role that is analogous to which other intracellular signaling protein discussed in this chapter?

62. Why do Ca^{2+} levels *decrease* during stimulation of the rod cell by light, and what role does Ca^{2+} play in desensitization to light?

Cone Cells Specialize in Color Vision

63. What is meant by the phrase "slightly different environments" in this section?

Vertebrate Olfaction and Gustation Use Mechanisms Similar to the Visual System

64. List all the different forms of G proteins that have been discussed in this chapter.

65. In what ways are the actions of the different G proteins similar? In what ways are they different?

66. What is the cellular location of each of the different forms of G protein and what are the immediate cellular targets of each form?

67. Why does the closing of K^+ channels cause depolarization in neurons?

GPCRs of the Sensory Systems Share Several Features with GPCRs of Hormone Signaling Systems

68. The relatively complex G protein–coupled serpentine receptor system appears to have been evolutionarily "chosen" as the basis for so many different transmembrane signaling pathways. What are the likely reasons that cells don't use much simpler pathways for signaling?

12.11 Regulation of the Cell Cycle by Protein Kinases

The Cell Cycle Has Four Stages

69. How would the daughter cells be affected by a genetic mutation that results in the shortening of the G2 stage of the cell cycle?

Levels of Cyclin-Dependent Protein Kinases Oscillate

70. The cyclin-dependent protein kinase, CDK, is regulated in a manner that is exactly the opposite of the cAMP-dependent protein kinase, PKA. Explain.

Regulation of CDKs by Phosphorylation

71. What are the two molecular effects of CDK phosphorylation that result in the inhibition of catalytic activity?

72. Under what cellular circumstances would it be important to be able to inhibit this enzyme's activity?

Controlled Degradation of Cyclin

73. What would happen to the length of the cell cycle in a cell with a mutation in the gene for DBRP phosphatase? (Assume the mutation renders the gene product inactive.)

Regulated Synthesis of CDKs and Cyclins

74. Progression through the cell cycle depends not only upon the increased synthesis of CDKs but also on the increased synthesis of which other key proteins?

Inhibition of CDKs

75. How might the direct binding of a protein to CDK inhibit its activity?

CDKs Regulate Cell Division by Phosphorylating Critical Proteins

76. You have already seen how phosphorylation of enzymes can affect cell metabolism by increasing or decreasing enzyme activity. What are some of the other functional consequences of protein phosphorylation that are discussed in this chapter?

12.12 Oncogenes, Tumor Suppressor Genes, and Programmed Cell Death

Oncogenes Are Mutant Forms of the Genes for Proteins That Regulate the Cell Cycle

77. Many of the chemotherapy agents used to treat cancer work by destroying all rapidly dividing cells in an individual. Why is this a good strategy for controlling cancer, and why don't these drugs destroy all the other cells in an individual with cancer?

78. Suggest a reason why so many of the known oncogenes code for proteins involved in transmembrane signaling? Why would you expect alterations in signaling to result in cancer?

Defects in Certain Genes Remove Normal Restraints on Cell Division

79. In molecular terms, why are pRb, p53, and p21 all tumor suppressor genes?

Box 12–5 Development of Protein Kinase Inhibitors for Cancer Treatment

80. What aspects of enzyme kinetics are involved in the different approaches to drug development discussed in this section?

Apoptosis Is Programmed Cell Suicide

81. Why must cell death be programmed?

82. Do you think the role of proteolytic cleavage (vs. reversible protein phosphorylation) in the activation of protease 8 is significant?

Discussion Questions for Study Groups

- Would you expect a single neurotransmitter or hormone (e.g., epinephrine) to simultaneously activate protein kinases and protein phosphatases in the same cell?
- How does the existence of such proteins as AKAP help to explain why relatively few intracellular messengers (cAMP, cGMP, Ca^{2+}, IP_3, DAG) are needed to mediate the effects of a wide variety of extracellular signaling molecules?
- Protein dephosphorylation by phosphatases is a potent regulatory mechanism. What characteristics must phosphatases possess if they are to play regulatory roles in cells?
- The human visual system is sensitive enough to be able to detect a single photon of light. What are the various biochemical mechanisms that allow relatively weak extracellular signals to trigger a significant cellular response?

SELF-TEST

Do You Know the Terms?

ACROSS

1. This protein mediates many of the actions of the intracellular messenger Ca^{2+}.
2. The G protein that transduces light signals into an electrical signal in the vertebrate rod.
7. This is necessary for the cellular reception of all biologically relevant signals.
9. Activation of signaling pathways often results in the phosphorylation of specific cellular proteins; signaling in this type of pathway can be reversed through the actions of this class of enzyme.
11. Ion ______________ provide a thermodynamically favorable route for the movements of ions across the lipid bilayer.
12. The direction in which Na^{+} ions move through voltage-dependent ion channels is determined by the ______________ potential, which is a combination of the chemical and charge gradients that exist across the cell membrane.
16. The insulin receptor can also be classified as this type of enzyme.
17. "Regulates the Cell Cycle" describes the normal function of this class of mutant gene.
18. One way to reduce a cell's response to an extracellular signal is to reduce the ability of the cell's receptors to respond to the signal; a process referred to as ______________.
19. In neurons, a(n) ______________ is initiated by a localized redistribution of ions across the cell membrane, resulting in a shift of V_m to more positive values. (2 words)
22. GTP-binding transducer molecule. (2 words)
23. The channels that permit entry of Na^{+} and initiates an action potential is an example of a(n) -gated channel.
24. When phosphorylated, the β-adrenergic receptor binds this protein, arresting the flow of information through this signaling pathway. (2 words)
25. Hormonal signals in the circulatory system of vertebrates are usually found in relatively low concentrations. These weak extracellular signals are usually amplified inside the cell as a result of an enzyme ______________.

DOWN

1. The activation of Gsα by epinephrine results in the production of this second messenger.
3. The β-adrenergic receptor is one example of this class of integral membrane protein receptor that contains multiple membrane-spanning regions.
4. The receptor-enzyme complex that synthesizes this second messenger molecule can be found either in the cytosol or associated with the membranes of cells. (abbr.)
5. Extracellular signals that are lipid soluble bind to intracellular receptors. The resulting complex then influences gene expression by binding to specific regulatory sequences in DNA called ______________. (abbr.)
6. G protein–mediated activation of phospholipase C results in the production of two intracellular second messengers, IP_3 and ______________.(abbr.)
8. The R subunit is to PKA as ______________ is to CDK.
10. When V_m changes to a more negative value a cell is said to be ______________.
13. The process of converting a light stimulus into an electrical signal in neurons or a hormonal signal into altered cellular metabolism in hepatocytes are examples of signal ______________.
14. A protein kinase that regulates the β-adrenergic receptor. (abbr.)
15. IP_3, cAMP, cGMP, and Ca^{2+} are all examples of ______________ messengers.
20. Ducks do not form distinct claws during embryogenesis, as do most other birds; this is because the cells present between the forming toe bones do not experience programmed cell death or ______________.
21. The cell membranes of most neurons do not actually touch; usually there is a small space between them called a(n) _____ through which neurotransmitters must diffuse.

Do You Know the Facts?

1. All the membrane proteins known as G proteins:

A. hydrolyze GTP.
B. have a subunit that activates adenylate cyclase.
C. are gated ion channels.
D. are homotrimeric proteins.
E. mediate the effects of insulin on cells.

2. GTP plays all of the following roles in the cAMP-mediated second messenger pathway *except:*

A. it associates with the β subunit of the G-protein complex.
B. hydrolysis of GTP drives the adenylate cyclase catalyzed reaction.
C. its binding causes dissociation of the G protein.
D. its effect is terminated by the GTPase activity of the G protein.
E. it transduces information from a hormone-receptor complex to an intracellular system that produces an allosteric effector.

3. Second messengers, such as cAMP, are important to cell function for all of the following reasons *except* they:

A. provide information about the extracellular environment.
B. permit a large intracellular response to relatively weak extracellular signals.
C. provide a site for integrating the metabolic requirements of cells.
D. reversibly regulate enzymatic activity in cells.
E. are a source of energy for cells.

4. Which of the following is *not* a signal transducer?

A. G_i
B. β-adrenergic receptor
C. Rhodopsin
D. Ubiquitin
E. PLC

5. cAMP acts as an intracellular second messenger in neurons. The intracellular levels of this molecule can be increased through which of the following routes?

A. By opening voltage-gated channels
B. Ligand binding to G_s-coupled receptors
C. Binding of neurotransmitters to ligand-gated channels
D. Increasing the activity of the G_i α subunit G proteins
E. By hyperpolarizing the neuron

6. Which statement correctly describes steroid hormones?

A. Their intracellular actions are mediated by integral membrane proteins.
B. Their effects on cells require a water-soluble intracellular signal.
C. Their effects are mediated by binding to a water-soluble receptor protein.
D. They are synthesized directly from amino acid precursors.
E. Their effects usually involve the activation of other intracellular enzymes.

7. Which of the following correctly describes the role of phospholipase C in mediating the cellular effects of hormones?

A. It directly activates a protein kinase.
B. It degrades triacylglycerols to fatty acids and glycerol.
C. It indirectly increases intracellular calcium levels.
D. It directly activates adenylate cyclase.
E. It indirectly increases cAMP levels.

8. Which of the following statements does *not* describe the phosphoinositide signaling pathway?

A. Inositol triphosphate activates protein kinase C.
B. One effect of activation of this pathway is to increase intracellular [Ca^{2+}].
C. The pathway requires GTP.
D. Protein kinase C is inactive in its soluble form.
E. Phospholipase C activation produces two intracellular messengers.

9. G_p-mediated activation of phospholipase C would be turned off by increasing:

A. the concentration of α_p subunits.
B. the concentration of α_s subunits.
C. hydrolysis of GTP.
D. the concentration of β subunits.
E. the concentration of γ subunits.

10. A specific extracellular neurotransmitter or hormone (e.g., epinephrine) may have differing effects on different target tissues because:

A. different types of receptors are activated in different tissues.
B. channels may be opened in one tissue and closed in another in response to the same signaling molecule.
C. different second messenger pathways can be activated in different tissues.
D. the target proteins of the activated intracellular messenger pathway may differ.
E. All of the above are true.

11. Some neurons in the human body have cellular projections (axons) that extend up to a meter in length, for example, the nerves that run from your spinal cord to your toes. These neurons are able to send signals to the muscles in your toes because:

A. release of neurotransmitter in the spinal cord depolarizes toe neurons so strongly that they can act as an electrical cable and can passively transmit the signal for several meters.
B. small depolarizations by input from spinal cord neurons trigger a "domino effect" in the voltage-dependent Na^+ channels, allowing the depolarizing signal to proceed undiminished along long axon lengths.
C. Diffusion of Ca^{2+} stored in the endoplasmic reticulum in the cell body down the axon to the cell terminal causes release of neurotransmitter at the synapse.
D. Second messengers, produced as a result of signaling from spinal cord neurons, can diffuse to effector enzymes in nerve terminals, which produce neurotransmitter.
E. B and C.

12. Neurons transmit electrical signals due to the presence of voltage-dependent Na^+ channels in the axonal membrane. Opening these channels causes V_m to depolarize to the Na^+ equilibrium potential. The V_m does not remain depolarized but returns very rapidly to its original resting potential so that the neuron is ready to generate another signal. This repolarizing phase of the action potential is most likely due to which of the following?

A. Voltage-dependent K^+ channels open allowing K^+ to exit the neuron flowing passively down their electrochemical gradient.
B. The voltage-dependent Na^+ channels eventually close.
C. Ca^{2+} must be actively pumped out of the cell.
D. Cl^- flows out of the cell down its concentration gradient.
E. A and B.

13. Which of the following is *true* for a light-stimulated rod cell?

A. Na^+/Ca^{2+} cation channels are closed.
B. The intracellular levels of cGMP are increased.
C. PDE is tightly bound to an inhibitory subunit.
D. Tα has GDP bound.
E. The membrane potential is depolarized.

14. Which of the following is *not* an example of signal amplification?

A. The activation of PKA by cAMP in adipose cells
B. The phosphorylation of IRS-1 by the insulin receptor in liver cells
C. The conversion of cGMP to 5′-GMP by PDE in rod cells
D. The activation of Raf kinase by Ras in hepatocytes
E. A and D

Applying What You Know

1. Sildenafil (Viagra), vardenafil, and tadalafil are available worldwide for the treatment of erectile dysfunction (ED). However, individuals with heart disease who use organic nitrates for the treatment of angina and also take one of these agents for ED may experience a disastrous, and sometimes fatal, drop in blood pressure. Can you explain why this might happen?

2. Discuss the specific molecular effects that you would expect as a result of the following:
   (a) Overexpression of GEFs on signaling by the protein hormone glucagon in hepatocytes.

   (b) Overexpression of GAPs on signaling through the β-adrenergic pathway in myocytes.

   (c) Overexpression of β-adrenergic receptor kinase on signaling through the β-adrenergic pathway in myocytes.

   (d) *Bordetella pertussis* infection on cAMP levels in lung epithelial cells.

   (e) A mutation that decreases the GTPase activity of G_α on cAMP levels in lung epithelial cells.
   (**Hint:** Drawing a cellular diagram of what is going on in each instance may help you answer these questions.)

3. Normally, synaptic input to a neuron occurs at the dendritic or "input" region of the cell, which is distinct from the axon or "output" region. Depolarization of the cell membrane generally occurs unidirectionally, starting with neurotransmitter-induced depolarization at the input end, followed by the production of an action potential at the spike initiating zone, which is then transmitted along the full length of the axon to the cell terminus. What would you predict would be the effect of artificially depolarizing an axon in the *middle* of its length?

4. The ability of a squid to avoid becoming a calamari appetizer depends upon the rapid transmission of action potentials by its giant neuron that initiates its escape reflex. The initiation and transmission of an action potential depends upon the opening of voltage-dependent Na^+-selective ion channels in the axon of the neuron. When ion channels open, ions flow in a direction dictated by the electrochemical gradient until a new equilibrium is established.
   (a) Given the ion concentrations shown in Table 12–2, what will be the membrane potential (V_m) for a squid axon when these Na^+-selective channels are open? (Assume these are the only channels that are open and T = 18 °C.)

(b) What was the membrane potential before an action potential was initiated? (Hint: at rest, neuron cell membranes are primarily permeable to K^+.)

5. Olfactory stimuli, acting through serpentine receptors and G proteins, trigger cellular responses either through an increase in the second messenger [cAMP] or an increase in the second messenger [Ca^{2+}].

(a) Outline the steps involved from receptor activation to the production of the second messenger for both cAMP and Ca^{2+}.

(b) In many second messenger pathways, the "goal" is to activate protein kinases, which affect cellular metabolism by phosphorylating cellular proteins (e.g., enzymes). What protein kinase(s) are activated in each of the pathways described above?

(c) If you separated activated olfactory cells into membrane and cytosolic fractions, where would you expect to find the labeled kinases?

6. Peptide hormones and other extracellular signaling molecules are used by cells to regulate growth and development. For example, an immortal cell line cultured from an adrenal tumor, called PC12 cells, can be induced to differentiate into neurons in response to the extracellular application of peptide nerve growth factor (NGF). Bradykinin, a peptide hormone, which can act as a vasodilator, also induces differentiation in this cell line. It has been proposed that the effects of both of these signaling molecules are mediated by the intracellular second messenger, cAMP.

In order to test this hypothesis, you decide to measure the levels of cAMP in these cells following the extracellular application of NGF and/or bradykinin. In addition, you examine the effects of these agents on cells that have been injected with $\beta\gamma$ subunits of G proteins and obtain the following data:

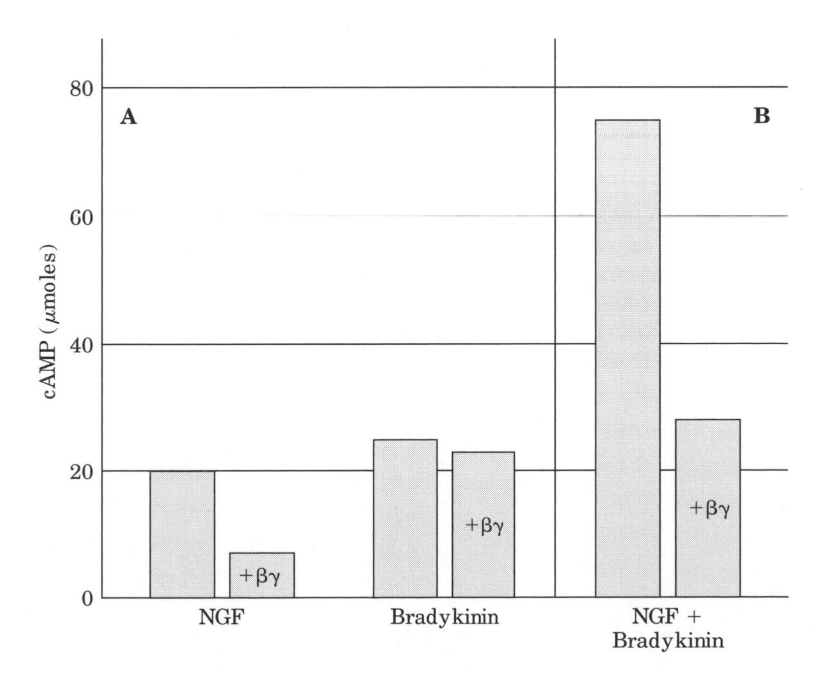

(a) At the molecular level, what would be the effect of adding excess $G_{\beta\gamma}$ on the cAMP pathway?

(b) Based on the results in part A of this figure, would you conclude that both bradykinin and NGF activate adenylate cyclase through a G protein? Explain your reasoning!

(c) What do the combined effects of NGF and bradykinin suggest about this system?

(d) Suggest a molecular model to explain the effects of these two hormones on cAMP levels in these cells.

ANSWERS

Do You Know the Terms?

ACROSS

1. calmodulin
2. transducin
7. receptor
9. phosphatase
11. channels
12. electrochemical
16. kinase
17. oncogene
18. desensitization
19. action potential
22. G protein
23. voltage
24. β arrestin
25. cascade

DOWN

1. cAMP
3. serpentine
4. cGMP
5. HREs
6. DAG
8. cyclin
10. hyperpolarized
13. transduction
14. βARK
15. second
20. apoptosis
21. synapse

Do You Know the Facts?

1. A
2. B
3. E
4. D
5. B
6. C
7. C
8. A
9. C
10. E
11. B
12. E
13. A
14. E

Applying What You Know

1. Sildenafil (Viagra), vardenafil, and tadalafil all inhibit cGMP phosphodiesterase, the enzyme that converts the second messenger cGMP to its inactive form, 5' GMP. These inhibitors selectively affect phosphodiesterase 5 (PDE5), a subclass of phosphodiesterase that is found in vascular smooth muscle cells. Inhibition of cGMP degradation by PDE5 results in an elevation in cellular cGMP levels, smooth muscle relaxation, and local vasodilation. Organic nitrates used to treat patients with heart disease release NO, which activates the cytoplasmic form of guanylyl cyclase causing an increase in cellular cGMP levels. The combined effects of both an increase in cGMP synthesis and a decrease in cGMP degradation may result in excessive vasodilation, resulting in a dramatic drop in blood pressure and a heart attack.

2. **(a)** GEFs are guanosine nucleotide exchange factors that stimulate the replacement of GDP by GTP on G proteins. GEF proteins include hormone receptors such as those that bind glucagon. Interactions between GEFs and G proteins result in binding of GTP by the G protein and its dissociation into the G_α and $G_{\beta\gamma}$ subunits. The now active GTP-G_α can in turn activate downstream targets such as adenylyl cyclase.

   (b) GAPs (GTPase activator proteins) and RGSs (regulators of G protein signaling) bind to G proteins resulting in an increase in the inherent ATPase activity of the G protein. It is the conversion of GTP to GDP that switches the G protein from an "active" to an "inactive" state. Thus, overexpression of these modulatory proteins would be expected to speed up the rate at which G proteins are turned off, resulting in a reduction in the amount of cAMP that would normally be produced in response to β-adrenergic signaling.

(c) The β-adrenergic receptor kinase (or βARK) is an enzyme that binds to the $Gs_{\beta\gamma}$ portion of Gs proteins. Consequently, binding only occurs when G proteins have been activated. This binding event localizes the kinase to the cell membrane where it can phosphorylate the β-adrenergic receptor. Once phosphorylated, the receptor then binds β-arrestin, which prevents further interactions between the receptor and G proteins terminating β-adrenergic receptor signaling. Binding of β-arrestin also results in the sequestration and removal of receptors from the plasma membrane further reducing the effect of subsequent β-adrenergic signaling.

(d) *Bordetella pertussis* bacteria produce a toxin (pertussis toxin), a portion of this protein catalyzes the ADP ribosylation of the α subunit of G_i preventing GDP-GTP exchange. With GDP more or less permanently bound, G_i can no longer inactivate adenylyl cyclase resulting in an increase in the normal cAMP production in affected cells. (Inactivation of an inactivator usually results in activation.)

(e) Mutations that decrease the inherent GTPase activity of the stimulatory G protein, G_s, would result in an increase in the proportion of $G_{s\alpha}$ bound to GTP. This active form of Gs_α stimulates the activity of adenylyl cyclase, resulting in an increase in cAMP production.

3. During an AP, the voltage-dependent Na^+ channels open in advance of the voltage-dependent K^+ channels, and they close in response to the depolarization of the membrane potential that they produce. This closing is followed by a "refractory period" during which the channels cannot reopen. Because most APs proceed from the cell body toward the terminus, this refractory period ensures that the depolarizing signal moves in only one direction. When an axon is artificially depolarized in its middle, the channels on either side of the stimulation site are *not* in a refractory state. Therefore, a signal can be generated that is propagated in *both* directions from the point of stimulation.

4. (a) Assuming that the neuron membrane is permeable to Na^+ only during the depolarizing phase of an action potential (a valid assumption because the permeability to Na^+ is usually *much* greater than for any other ions), then the membrane potential (V_m) for the squid axon will approach the equilibrium potential for Na^+. This can be calculated from the Equation 12–1 using the ionic concentrations given in Table 12–2.

(Eq. 12–1) $\Delta G = RT \ln (C_{in}/C_{out}) + Z\mathcal{J}V_m$

At equilibrium, $\Delta G = 0$

$$-Z\mathcal{J}V_m = RT \ln (C_{out}/C_{in})$$

$$V_m = -\frac{RT}{Z\mathcal{J}} \ln \left(\frac{C_{in}}{C_{out}}\right)$$

$$= -(.025) \ln \left(\frac{50}{440}\right) = +.055 \text{ V}$$

Thus, the V_m will approach +55 mV when the voltage-dependent Na^+ channels are open.

(b) Assuming that the neuron membrane is permeable to K^+ only at the resting membrane potential, the V_m for the squid axon will approach the equilibrium potential for K^+. This can be calculated using the ionic concentrations given in Table 12–2.

$$\Delta G = RT \ln \left(\frac{K^+_{in}}{K^+_{out}}\right) + Z\mathcal{J}V_m$$

$$-Z\mathcal{J}V_m = RT \ln \left(\frac{400}{20}\right)$$

$$V_m = -(.025)\ (+2.996) = -.075 \text{ V}$$

Thus, the V_m will approach −75 mV when only K^+-selective channels are open.

5. (a) cAMP: Receptor activation results in the activation of G_s protein, the displacement of GDP for GTP, dissociation of G_s protein into $G_{s\alpha}$ and $G_{s\beta\gamma}$, activation of adenylate cyclase, and the production of cAMP from ATP.

Ca^{2+}: Receptor activation results in the activation of G_p protein, the displacement of GDP for GTP, dissociation of G_p protein into $G_{p\alpha}$ and $G_{p\beta\gamma}$, activation of phospholipase C, and the production of the second messengers diacylglycerol (DAG) and inositol triphosphate (IP_3) from phosphatidylinositol bisphosphate (PIP_2). IP_3 then binds to receptors located on the membrane of the endoplasmic reticulum (sarcoplasmic reticulum in muscle cells), where it opens Ca^{2+} channels and releases stored Ca^{2+}.

(b) cAMP pathway: cAMP-dependent protein kinase

Ca^{2+} pathway: Ca^{2+} plus the other product of PLC activation, DAG, cooperate in the activation of protein kinase C. It is also possible for Ca^{2+} to bind to the ubiquitous cellular protein calmodulin, which can then activate the multifunctional Ca^{2+}/calmodulin-dependent protein kinase, CaM kinase.

(c) cAMP pathway: cAMP (like ATP) is a water soluble molecule. PKA is also soluble, so this activated kinase would be expected to copurify with the cytosolic components of a cell. Ca^{2+} pathway: Ca^{2+}, calmodulin, and CaM kinase are all water-soluble compounds, so this activated kinase would also be expected to copurify with the cytosolic components of a cell. However, DAG is a lipid soluble molecule and cannot leave the lipid bilayer. Consequently, the activated form of PKC should be associated with the cell membrane and should copurify with the membane components of these cells.

6. (a) The $\beta\gamma$ subunits normally limit the effects of $G\alpha$ by binding to the α subunit in its GDP bound form and preventing the exchange of GDP for GTP. This "turns off" the effects of the α subunit.

(b) No; inhibiting the effect of $G_{s\alpha}$ by adding excess $\beta\gamma$ subunits has no effect on the ability of bradykinin to increase cAMP levels. This indicates that bradykinin does not affect adenylate cyclase through the direct activation of a G protein. NGF's effects are inhibited by the injection of $\beta\gamma$ subunits indicating that the effects of this hormone *are* mediated by α_s.

(c) The combined effects of NGF and bradykinin are more than additive, suggesting a synergistic effect between the two signaling molecules. Physiologically, this means that the two extracellular signals, when present at the same time, can convey different information than they can individually.

(d) There are a number of possibilities: Bradykinin could enhance the effects of NGF on adenylate cyclase by inhibiting phosphodiesterase, which hydrolyzes cAMP and terminates the intracellular signal. This might occur if bradykinin could reduce the intracellular levels of free calcium (e.g., by closing a Ca^{2+} channel), which is a known activator of phosphodiesterase. Another possibility is that bradykinin could activate adenylate cyclase through a non-G-protein–related pathway. Specifically, it could activate the cyclase directly by increasing the levels of calcium in the cell. This would turn on the calcium-sensitive form of adenylate cyclase. Synergistic effects of several extracellular signaling molecules on a calcium-sensitive form of cyclase has been proposed as the molecular basis for cellular "learning" in *Aplysia*. Finally, bradykinin could activate a non-cAMP dependent protein kinase (such as the calcium/calmodulin-dependent protein kinase) that phosphorylates components of the NGF signaling pathway.

chapter

13 Bioenergetics and Biochemical Reaction Types

STEP-BY-STEP GUIDE

Major Concepts

Basic principles of thermodynamics govern the transfer of energy in biological systems.

The first law of thermodynamics states that all energy is conserved; although the form of the energy may change, the total amount in the universe remains constant. The second law of thermodynamics states that the total disorder (entropy) of the universe increases.

A thermodynamic system can be a single, simple chemical reaction or an entire organism.

The transfer of energy in a single reaction or in an organism can be described by three thermodynamic quantities: (Gibbs) **free energy, G; enthalpy, H; and entropy, S.** These quantities are related by the equation

$$\Delta G = \Delta H - (T\,\Delta S)$$

where T is the absolute temperature (in degrees Kelvin, or K). ΔG describes the change in free energy that occurs in a chemical reaction.

Reactions proceed spontaneously only if ΔG is negative, meaning energy is released by the reaction.

If ΔG is positive, the reaction requires an input of energy. Reactions are at equilibrium if $\Delta G = 0$.

The standard free-energy change, $\Delta G'^\circ$ is a physical constant and can be calculated from the equilibrium constant, K'_{eq}: $\Delta G'^\circ = -RT \ln K'_{eq}$.

For a system at equilibrium, the rates of the forward and reverse reactions are equal; no further net change occurs in the system. For the reaction

$$A + B \rightleftharpoons C + D$$

at equilibrium

$$\frac{[C][D]}{[A][B]} = K'_{eq}$$

The *actual* free-energy change (ΔG) of a specific reaction depends on the concentrations of products and reactants, temperature, and pH, but is not affected by the reaction pathway or by enzyme catalysis. Under *standard* conditions (25 °C, 1 M, 1 atm, pH = 7), $\Delta G'^\circ$ can be measured. The *actual* ΔG for a reaction can be related to the free-energy change $\Delta G'^\circ$ that occurs under *standard* conditions by substituting the actual concentrations of reactants and products:

$$\Delta G = \Delta G^{\circ\prime} + RT \ln \frac{[C][D]}{[A][B]}$$

Because at equilibrium $\Delta G = 0$, the equation reduces to

$$\mathbf{\Delta G'^\circ = -RT \ln K'_{eq}}$$

The $\Delta G'^\circ$ values for sequential reactions are additive so that individual reactions with an unfavorable (positive) $\Delta G'^\circ$ can proceed if they are coupled to reactions with a favorable (negative) $\Delta G'^\circ$. (K'_{eq}s are multiplicative.) Biological systems use the free energy of hydrolysis of "high-energy compounds" to drive energetically unfavorable reactions. Also, in biological systems, a displacement from equilibrium can produce a favorable ΔG, *even if $\Delta G'^\circ$ is positive,* by giving the term RT ln (products/reactants) a larger value than $\Delta G'^\circ$.

Most of the reactions in living cells fall into one of five general categories: (1) reactions that make or break carbon–carbon bonds; (2) internal rearrangements, isomerizations, and eliminations; (3) free-radical reactions; (4) group transfers; and (5) oxidation-reductions.

ATP is a common source of free energy for biological systems.

The hydrolysis of ATP to ADP and P_i (inorganic phosphate) has a $\Delta G'^\circ$ of −30.5 kJ/mol (the actual ΔG may vary depending on the microenvironment). Although the hydrolysis of phosphoryl, pyrophosphoryl, or adenylyl groups from ATP provides energy, without some mechanism for **coupling** this hydrolysis to other reactions the energy would be released as heat and would not be available to drive other reactions in cells. The coupling mechanism involves the transfer of these groups from ATP to an enzyme or to its substrate. This is called a **group transfer reaction.** ATP has a high **phosphoryl group transfer potential.**

Compounds other than ATP have high group transfer potentials and are capable of driving biological reactions.

GTP and some other phosphorylated compounds, such as phosphorylated sugars, have a large, negative $\Delta G'^{\circ}$ of hydrolysis that can be used to drive other reactions. In addition, the hydrolysis of thioesters, such as acetyl-CoA, has a large, negative $\Delta G'^{\circ}$. Free energy is released by the hydrolysis of these compounds because the products are much more stable than the initial reactants. There are a number of reasons why this is so: electrostatic repulsion (and resulting bond strain) is relieved by hydrolysis, and the products of hydrolysis may be stabilized by ionization, by formation of a resonance hybrid, by isomerization (tautomerization), and/or by solvation. Despite the availability of other "high-energy" compounds, ATP is the one most widely used by cells. ATP occupies a middle ground in terms of its energy-carrying capacity and thus can serve as a link between energy-producing catabolic and energy-requiring anabolic reactions in cells.

Phosphorylation of certain compounds effectively "primes" them for participation in biological reactions.

Glucose is primed for entry into the glycolytic pathway by the addition of a phosphoryl group to form glucose 6-phosphate. Fatty acids used in the synthesis of lipids and amino acids used in the synthesis of proteins are first primed by hydrolysis of ATP and the covalent addition of AMP to the amino acid or fatty acid molecule. Cleavage of PP_i (pyrophosphate) from activated nucleoside triphosphates is required for assembly of the nucleotide units in RNA and DNA synthesis.

ATP has other roles in cells.

It provides energy for active transport and for muscle contraction.

Electron flow in oxidation-reduction reactions can be used to do "biological work."

Oxidation-reduction reactions involve two components: an electron donor (reducing agent) and an electron acceptor (oxidizing agent). Therefore, two simultaneous reactions (half-reactions) occur in any oxidation-reduction process. Electrons can be transferred in several ways: (1) by direct transfer, as when a metal ion is involved in a reaction (e.g., $Fe^{2+} + Cu^{2+} \rightleftharpoons Fe^{3+} + Cu^{+}$); (2) as a hydrogen atom (as opposed to a hydrogen ion); (3) as a hydride ion, H^-; and (4) when oxygen is covalently attached to an organic compound. **Standard reduction potential,** E'°, describes the affinity of any chemical species for electrons *relative* to the affinity of hydrogen ions (H^+) for electrons (at pH 7). The higher the value of E'°, the greater the affinity for electrons.

Electrons are transferred by a variety of cofactors that act as carriers.

Nucleotide carriers include NAD^+, $NADP^+$, FMN, and FAD. NAD^+ and $NADP^+$ are water-soluble, recyclable cofactors that accept a hydride ion to become reduced (to NADH and NADPH). NAD^+ generally functions to accept hydride ions during catabolic reactions, and NADP (as NADPH) usually functions to transfer hydride ions for anabolic reductions. FAD and FMN can accept either one or two electrons in the form of hydrogen atoms. These two cofactors tend to be tightly bound to their enzymes.

What to Review

Answering the following questions and reviewing the relevant concepts, which you have already studied, should make this chapter more understandable.

- The following terms and reaction classes are in the Introduction to Part II of the text: **autotroph, heterotroph, catabolism, anabolism.** These terms and reactions should all be familiar from previous biology classes; a complete understanding of their meaning is critical for the material in this chapter.
- Be sure you understand what an equilibrium constant represents (pp. 24–25).
- Remind yourself what enzymes can and cannot do with respect to reaction rates, equilibria, and other reaction parameters (p. 25).

Topics for Discussion

Answering each of the following questions, especially in the context of a study group discussion, should help you understand the important points of this chapter.

13.1 Bioenergetics and Thermodynamics

Biological Energy Transformations Obey the Laws of Thermodynamics

1. Explain how defining the "system" and "surroundings" allows living organisms to operate within the second law of thermodynamics.

2. What is the definition of exergonic? Of endergonic? Gibbs free energy, and entropy, and enthalpy? Make sure all the terms in this section are thoroughly understood.

Cells Require Sources of Free Energy

3. From where do cells acquire their necessary free energy? Why can't cells use heat as a free-energy source?

Standard Free-Energy Change Is Directly Related to the Equilibrium Constant

4. What exactly does "at equilibrium" mean in terms of (a) the *rate* of both the forward and reverse reactions and (b) the *concentrations* of the reactants and products?

5. What reaction conditions are used to measure the standard free-energy change, $\Delta G'^{\circ}$?

6. If, at equilibrium, the concentration of products is greater than the concentration of reactants, is $\Delta G'^{\circ}$ positive or negative? What can you say about the value of K'_{eq}?

Actual Free-Energy Changes Depend on Reactant and Product Concentrations

7. What effect does the presence of an enzyme have on the $\Delta G'^{\circ}$ of the reaction it catalyzes?

8. Under what circumstances can ΔG be negative if $\Delta G'^{\circ}$ is positive? Could cells use this strategy to drive thermodynamically unfavorable reactions?

9. Does the value of $\Delta G'^{\circ}$ (or ΔG) tell you anything about (a) the rate at which a reaction occurs or (b) the pathway by which the final product is formed?

Standard Free-Energy Changes Are Additive

10. How can the coupling of a thermodynamically unfavorable reaction to a thermodynamically favorable reaction increase the K'_{eq} of the overall equation?

11. Explain why relatively small changes in $\Delta G'^{\circ}$ correspond to large changes in K'_{eq}.

13.2 Chemical Logic and Common Biochemical Reactions

12. Why might potentially relevant reactions fail to occur in living systems?

13. When will a carbon atom act as a nucleophile; when as an electrophile?

14. List the five general categories of reactions that commonly occur in living cells, and include at least one example for each. Leave room on your list to add reactions as you encounter them in the following chapters.

Biochemical and Chemical Equations Are Not Identical

15. When is it important to use chemical rather than biochemical equations?

13.3 Phosphoryl Group Transfers and ATP

The Free-Energy Change for ATP Hydrolysis Is Large and Negative

16. What physical and chemical factors contribute to the free-energy change of ATP hydrolysis?

17. How do ΔG, $\Delta G'^{\circ}$, and ΔG_p differ?

Other Phosphorylated Compounds and Thioesters Also Have Large Free Energies of Hydrolysis

18. What are the main reasons for the high $\Delta G'^{\circ}$ values for hydrolysis of phosphoenolpyruvate, 1,3-bisphosphoglycerate, phosphocreatine, acetylCoA, and other similar compounds?

ATP Provides Energy by Group Transfers, Not by Simple Hydrolysis

19. Why is the "single arrow" representation of the conversion of ATP to ADP and P_i deceiving?

20. The reactions that produce conformational changes linked to ATP (or GTP) hydrolysis are different from other ATP-driven reactions. Explain.

21. What is the relative position of ATP in the hierarchy of compounds with phosphoryl group transfer potentials?

22. What is the difference between *thermodynamic* stability (or instability) and *kinetic* stability?

ATP Donates Phosphoryl, Pyrophosphoryl, and Adenylyl Groups

23. Why are adenylylation reactions so thermodynamically favorable?

Assembly of Informational Macromolecules Requires Energy

24. What are the similarities and differences between the assembly of DNA and RNA from their component nucleotides and the activation of fatty acids and amino acids?

Box 13–1 Firefly Flashes: Glowing Reports of ATP

25. Speculate on what use could be made of a sensitive assay for ATP concentration in a biological sample.

ATP Energizes Active Transport and Muscle Contraction

26. What is the *specific* role of ATP in the transport of solutes across membranes (i.e., what does the transfer of a phosphoryl group from ATP to the pump protein accomplish)?

Transphosphorylations between Nucleotides Occur in All Cell Types

27. What are the cellular roles of nucleoside diphosphate kinase, adenylate kinase, and creatine kinase? Why are they so important?

Inorganic Polyphosphate Is a Potential Phosphoryl Group Donor

28. Why is polyP an interesting molecule to those who study cellular evolution?

13.4 Biological Oxidation-Reduction Reactions

29. What are the different sources of electrons (i.e., the different electron donating compounds) that provide energy for the work done by biological organisms?

The Flow of Electrons Can Do Biological Work

30. Why are cell membranes critical to the generation of a proton-motive force in cells?

Oxidation-Reductions Can Be Described as Half-Reactions

31. In the equation $Fe^{2+} + Cu^{2+} \rightleftharpoons Fe^{3+} + Cu^{+}$, which of the iron species is more oxidized? Which of the copper species is more reduced?

Biological Oxidations Often Involve Dehydrogenation

32. What are the different oxidation states of carbon?

33. Besides the transfer of electrons in the form of hydrogen atoms, in what other ways does electron transfer occur?

Reduction Potentials Measure Affinity for Electrons

34. How is the hydrogen electrode used as *the* reference half-cell?

35. Which has the higher (more positive) reduction potential, NADH or cytochrome *b* (Fe^{3+})? In which direction will electrons flow in a system that contains these two compounds?

Standard Reduction Potentials Can Be Used to Calculate Free-Energy Change

36. Why is it important to have a universal standard for measuring reduction potentials?

Cellular Oxidation of Glucose to Carbon Dioxide Requires Specialized Electron Carriers

37. Calculate approximately how many molecules of ATP could be synthesized from the complete oxidation of glucose ($\Delta G'^{\circ} = -2840$ kJ/mol). Revisit this estimate later.

A Few Types of Coenzymes and Proteins Serve as Universal Electron Carriers

38. What is the significance of the various categories of electron carriers (water-soluble vs. lipid-soluble; mobile vs. bound; associated with peripheral vs. integral membrane proteins) in terms of an electron carrier's functions?

NADH and NADPH Act with Dehydrogenases as Soluble Electron Carriers

39. How do the normal concentration ratios of NAD^+/NADH and $NADP^+$/NADPH in cells reflect the different metabolic roles of these electron carriers?

Dietary Deficiency of Niacin, the Vitamin Form of NADH and NADPH, Causes Pellagra

40. Why is it important that vegetarians eat a variety of vegetables and vegetable-based foods?

Flavin Nucleotides Are Tightly Bound in Flavoproteins

41. Why are flavoproteins, as a class, involved in a greater variety of reactions than are NAD-linked enzymes?

Discussion Questions for Study Groups

- Return to some of the "Discussion Questions for Study Groups" in the Study Guide from the Enzyme chapter (Chapter 6), and reevaluate your answers based on your new knowledge of bioenergetic principles.

SELF-TEST

Do You Know the Terms?

ACROSS

1. Describes the NAD^+/NADH pair. (2 words)
4. ΔG_p; also called the compound's ____________ potential.
6. The transfer of phosphoryl, pyrophosphoryl, or adenylyl groups from ____________ couples the energy of breakdown to endergonic transformations.
7. A category of the dehydrogenases, this class of enzymes catalyzes reactions involving transfer of a hydride ion to NAD^+ or $NADP^+$.
9. Another way of describing energy-consuming biosynthetic reactions.
11. ____________ potential: the ability of ATP to donate a phosphate group, or of acetyl-CoA to donate an acetyl group. (2 words)
12. Term for a single electron participating in an oxidation-reduction reaction, regardless of the form in which the electron is transferred. (2 words)
14. Standard ____________: the affinity of a compound for electrons, relative to a hydrogen electrode. (2 words)
16. ____________ free energy is the amount of free energy released in a cell and available to do work. (Compare to 17 down.)
18. Class of enzymes containing a cofactor derived from the vitamin riboflavin, including FAD and FMN.
20. This is minimal in a completed jigsaw puzzle.
21. ____________ trophs use ingested nutrients as carbon source for metabolic processes.
22. ____________ trophs use CO_2 as carbon source for metabolic processes.
23. General class of enzymes involved in oxidation-reduction reactions.

DOWN

1. Describes reactions involved in the degradation of ingested nutrients.
2. Source of reducing equivalents used primarily in anabolic pathways.
3. Type of bond in acetyl-CoA that has a large, negative $\Delta G'^\circ$ of hydrolysis.
5. Electron transfer in cells occurs as coupled ____________ - ____________ reactions.
6. Cells do not rely solely on de novo synthesis to resupply depleted ATP stores; for example, this enzyme catalyzes the generation of ATP and AMP from 2 ADP. (2 words)
8. NADH is the electron ____________ in the reaction:

$$\text{Acetaldehyde} + \text{NADH} \longrightarrow \text{ethanol} + NAD^+.$$

10. Value of this quantity is negative in chemical reactions that release heat.
13. $NAD^+ + H^+ + 2e^- \rightleftharpoons NADH$ and $2O_2 + 2H^+ + 2e^- \rightleftharpoons H_2O$ are examples of ____________ - ____________.
15. State of a reaction when $\Delta G'^\circ = 0$.
16. Oxygen is the electron ____________ in the reaction:

$$\text{Glucose} + 6O_2 \longrightarrow 6CO_2 + 6H_2O.$$

17. Describes the available free energy of a chemical reaction occurring under defined conditions of 1 M reactant and substrate concentrations, pH 7, and 25 °C.
19. Electron carrier used primarily in catabolic pathways.

Do You Know the Facts?

1. Indicate whether each of the following statements about ATP is true or false.

_________ **(a)** It contains a β-*N*-glycosyl linkage.

_________ **(b)** It contains a furanose ring.

_________ **(c)** It is the highest-energy compound in cells.

_________ **(d)** It is used as a long-term storage form of energy in cells.

_________ **(e)** Its free energy of hydrolysis can be used to drive other reactions.

_________ **(f)** It is synthesized from ADP and P_i in an exergonic reaction.

_________ **(g)** It contains three phosphoanhydride bonds.

_________ **(h)** Repulsion between the negatively charged phosphoryl groups is reduced when ATP is hydrolyzed.

_________ **(i)** Its sugar moiety is glucose.

_________ **(j)** Its phosphoryl group transfer potential is higher than that of phosphoenolpyruvate.

2. Which of the following BEST describes the relationship between $\Delta G'^{\circ}$ and the rate of a reaction?

A. $\Delta G'^{\circ}$ is linearly proportional to the rate.
B. $\Delta G'^{\circ}$ is inversely proportional to the rate.
C. If $\Delta G'^{\circ}$ is positive, the reaction is spontaneous in the forward direction.
D. If $\Delta G'^{\circ}$ is negative, the reaction is at equilibrium.
E. $\Delta G'^{\circ}$ provides no information about the rate.

3. ATP is intermediate in the hierarchy of phosphorylated compounds with high standard free energies of hydrolysis (see Table 13–6). Why is it advantageous for cells to use ATP as their primary energy-carrying molecule?

A. ATP can be regenerated by coupling with a reaction that releases more free energy than does ATP hydrolysis.
B. ATP is thermodynamically unstable but kinetically stable.
C. Reactions that release more free energy than ATP hydrolysis do not occur in living cells.
D. The phosphoryl groups of other high-energy compounds cannot be removed.
E. The phosphoryl group transfer potential of ATP is higher than that of any other phosphorylated compound.

4. Which of the following does NOT contribute to the high phosphoryl group transfer potential of ATP? ADP and P_i have _____________ than does ATP.

A. More resonance stabilization
B. Less electrostatic repulsion
C. A greater degree of ionization
D. A greater ability to isomerize
E. A greater degree of hydration

In questions 5–8, match each electron carrier with the correct statement.

5. NAD^+:

A. accepts 2 electrons and 2 hydrogen ions.
B. accepts 2 electrons and 1 hydrogen ion.
C. accepts 1 electron and 1 hydrogen ion.
D. transfers electrons in reductive biosynthesis.
E. in its oxidized form is NADH.

6. $FADH_2$:

A. accepts 2 electrons and 2 hydrogen ions.
B. accepts 2 electrons and 1 hydrogen ion.
C. accepts 1 electron and 1 hydrogen ion.
D. transfers electrons in reductive biosynthesis.
E. in its oxidized form is FAD.

7. NADH:
A. accepts 2 electrons and 2 hydrogen ions.
B. accepts 2 electrons and 1 hydrogen ion.
C. accepts 1 electron and 1 hydrogen ion.
D. can be used as a source of electrons for ATP synthesis.
E. transfers electrons in reductive biosynthesis.

8. NADPH:
A. accepts 2 electrons and 2 hydrogen ions.
B. accepts 2 electrons and 1 hydrogen ion.
C. accepts 1 electron and 1 hydrogen ion.
D. transfers electrons in reductive biosynthesis.
E. in its oxidized form is NAD^+.

9. The bacterium *Pseudomonas saccharophilia* contains sucrose phosphorylase, an enzyme that catalyzes the phosphorolytic cleavage of sucrose:

$$\text{Sucrose} + P_i \longrightarrow \text{glucose 1-phosphate} + \text{fructose}$$

Given the data below, what is the standard free-energy change for the phosphorolysis of sucrose?

$$H_2O + \text{sucrose} \longrightarrow \text{glucose} + \text{fructose} \qquad \Delta G'^\circ = -29 \text{ kJ/mol}$$

$$H_2O + \text{glucose 1-phosphate} \longrightarrow \text{glucose} + P_i \qquad \Delta G'^\circ = -21 \text{ kJ/mol}$$

A. -8 kJ/mol
B. 8 kJ/mol
C. -50 kJ/mol
D. 50 kJ/mol
E. Cannot be determined from the data given.

10. Given the $\Delta G'^\circ$ values for the reactions below, what is the $\Delta G'^\circ$ for the hydrolysis of acetyl-CoA?

Reaction	$\Delta G'^\circ$ (kJ/mol)
Oxaloacetate + acetyl-CoA + H_2O → citrate + CoASH + H^+	−32.2
Malate → fumarate + H_2O	3.8
Oxaloacetate + acetate → citrate	−0.8
Malate + NAD^+ → oxaloacetate + NADH + H^+	29.7
Oxaloacetate + $2H^+$ + $2e^-$ → malate^{2-}	−32.04

A. -2.5 kJ/mol
B. -31.4 kJ/mol
C. 61.9 kJ/mol
D. 31.4 kJ/mol
E. -61.9 kJ/mol

11. Consider the general reaction A → B, where $\Delta G'^\circ = -60$ kJ/mol. Initially, 10 mM of A and 0 mM of B are present. After 24 hours, analysis reveals the presence of 2 mM of B and 8 mM of A. What can you conclude from this result?
A. A and B have reached equilibrium concentrations.
B. Formation of B is thermodynamically unfavorable.
C. The result described is impossible, given the $\Delta G'^\circ$ of the reaction.
D. Formation of B is kinetically slow; equilibrium has not been reached at 24 hours.
E. An enzyme has shifted the equilibrium toward formation of A.

12. What does "at equilibrium" mean in terms of (a) the rates of forward and reverse reactions and (b) the concentrations of reactants and products?

13. (a) Define oxidation and reduction. Can an oxidation occur without a simultaneous reduction? Why or why not?

(b) What is a conjugate redox pair and how does it relate to standard reduction potential?

Applying What You Know

1. In glycolysis, the enzyme phosphofructokinase I catalyzes the reaction

Fructose 6-phosphate + ATP $\longrightarrow$ fructose 1,6-bisphosphate + ADP

Given the data below, calculate the equilibrium constant for this reaction.

$R = 8.315$ J/ mol · K

$T = 25$ °C

ATP $\longrightarrow$ ADP + P_i $\quad \Delta G'^\circ = -30.5$ kJ/mol

Fructose 1,6-bisphosphate $\longrightarrow$ fructose 6-phosphate + P_i $\quad \Delta G'^\circ = -16.0$ kJ/mol

2. (a) Does the oxidation of glucose, represented by the equation

$C_6H_{12}O_6 + 6O_2 \longrightarrow 6CO_2 + 6H_2O$,

involve an increase or decrease in entropy? Why?

(b) Using the relationship $\Delta G = \Delta H - T\,\Delta S$, determine whether ΔG for the above reaction is positive or negative.

3. Consider the reaction

Glucose 1-phosphate $\rightleftharpoons$ glucose 6-phosphate

If, at equilibrium, the concentration of glucose 6-phosphate is greater than the concentration of glucose 1-phosphate, is the $\Delta G'^\circ$ of the reaction positive or negative? Explain.

4. The standard free-energy change for the reaction ADP + $P_i \rightarrow$ ATP is +30.5 kJ/mol. If all of the reactants are *initially* present at equal concentrations of 1 M, what can you say about the relative concentrations of ADP, P_i, and ATP *at equilibrium* (assume standard conditions of pH and temperature)?

ANSWERS

Do You Know the Terms?

ACROSS

1. conjugate redox
4. phosphorylation
6. ATP
7. oxidoreductases
9. anabolic
11. group transfer
12. reducing equivalent
14. reduction potential
16. actual
18. flavoproteins
20. entropy
21. hetero
22. auto
23. dehydrogenases

DOWN

1. catabolic
2. NADPH
3. thioester
5. oxidation reduction
6. adenylate kinase
8. donor
10. enthalpy
13. half-reactions
15. equilibrium
16. acceptor
17. standard
19. NADH

Do You Know the Facts?

1. **(a)** T; **(b)** T; **(c)** F; **(d)** F; **(e)** T; **(f)** F; **(g)** F; **(h)** T; **(i)** F; **(j)** F
2. E
3. A
4. D
5. B
6. E
7. D
8. D
9. A
10. B
11. D
12. **(a)** At equilibrium, the *rates* of the forward and reverse reactions are exactly equal and no further net change occurs in the system.

    (b) The concentrations of reactants and products *at equilibrium* define the equilibrium constant, K'_{eq}. In the general reaction $aA + bB \rightleftharpoons cC + dD$, where a, b, c, and d are the number of molecules of A, B, C, and D participating, the equilibrium constant is given by

    $$K'_{eq} = \frac{[C]^c\,[D]^d}{[A]^a\,[B]^b}$$

    where [A], [B], [C], and [D] are the molar concentrations of the reactants at the point of equilibrium.
13. **(a)** Oxidation is the loss of electrons; reduction is the gain of electrons. Free electrons are unstable and do not exist in biological systems. Whenever an electron is released in the oxidation of one chemical species, an electron must be accepted by another species, which is thus reduced. (Remember from beginning chemistry: "LEO says GER!" **L**oss of **E**lectrons is **O**xidation; **G**ain of **E**lectrons is **R**eduction!)

    (b) In a redox reaction, the general equation is

    $$\text{electron donor} \rightleftharpoons e^- + \text{electron acceptor}$$

    The electron donor and the electron acceptor together constitute a conjugate redox pair. Reduction potentials are a measure of affinity for electrons, and standard reduction potentials allow for the calculation of free-energy change. When two conjugate redox pairs are together in solution, electron transfer from the electron donor of one pair to the electron acceptor of the other pair may occur spontaneously. The tendency of such a reaction to occur depends on the relative affinity of the electron acceptor of each redox pair for electrons. The standard reduction potential, E'°, is a measure of this affinity and can be determined by experiment.

Applying What You Know

1. The reaction catalyzed by phosphofructokinase is the sum of two reactions, for each of which $\Delta G'^\circ$ is given (for the first reaction below, note that the sign changes for $\Delta G'^\circ$):

    Fructose 6-phosphate + P_i →
    fructose 1,6-bisphosphate $\Delta G'^\circ = 16.0$ kJ/mol
    ATP → ADP + P_i $\Delta G'^\circ = -30.5$ kJ/mol
    Sum: Fructose 6-phosphate + ATP →
    fructose 1,6-bisphosphate + ADP
    $\Delta G'^\circ = -14.5$ kJ/mol

    Using $\Delta G'^\circ = -RT \ln K'_{eq}$ and solving for K'_{eq}:

    $$\ln K'_{eq} = -\Delta G'^\circ/RT$$
    $$= -(-14.5 \text{ kJ/mol})/[(8.315 \text{ J/mol} \cdot \text{K})(298 \text{ K})]$$
    $$= 5.8518$$
    $$K'_{eq} = 348$$
2. **(a)** Whenever a reaction proceeds so that there is an increase in the number of molecules (in the case of glucose oxidation, for every 7 reactant molecules there are 12 product molecules), and/or when a solid (such as glucose) is converted into liquid or gaseous products (water and carbon dioxide), which have more freedom than a solid to move and fill space, there is an increase in molecular disorder and thus an *increase in entropy.*

    (b) For the glucose oxidation reaction, ΔS is positive (because entropy increases); ΔH is negative (assuming this is a combustion reaction that gives off heat to the surroundings); and T is positive (because the units are K). Given that $\Delta G = \Delta H - T\,\Delta S$, the ΔG for this reaction must be negative.
3. If the concentration of products is greater than concentration of reactants at equilibrium, then $K'_{eq} > 1$. Given that $\Delta G'^\circ = -RT \ln K'_{eq}$, if $K'_{eq} > 1$, then $\ln K'_{eq}$ is positive and $\Delta G'^\circ$ must be negative.

    Another way of looking at this is in terms of the final ratio of products and reactants. If the concentration of products exceeds the concentration of reactants, the reaction proceeds in the forward direction and thus $\Delta G'^\circ$ must be negative.
4. A positive value of $\Delta G'^\circ$ means that the product(s) of the reaction contain more free energy than the reactants. The reaction will therefore tend to go in the reverse direction if we start with 1.0 M concentrations of all components. At equilibrium the concentrations of the reactants ADP and P_i will be greater than 1 M and the concentration of the product ATP will be less than 1 M. Keep in mind that if the concentrations of ADP and P_i increase, the concentration of ATP must correspondingly decrease because ATP → ADP + P_i.

Glycolysis, Gluconeogenesis, and the Pentose Phosphate Pathway

chapter

14

STEP-BY-STEP GUIDE

Major Concepts

Glycolysis is a series of reactions in which one molecule of glucose is degraded to two molecules of pyruvate, yielding biologically useful energy in the form of ATP and NADH.

Glycolysis can be subdivided into a preparatory phase (five reactions) requiring an input of energy and a payoff phase (five reactions) with a net production of energy in the form of ATP and NADH. In the **preparatory phase** two molecules of ATP are used to phosphorylate and thus "activate" each molecule of glucose. The phosphorylated hexose is then cleaved to form two three-carbon molecules. In the **payoff phase** energy is transferred from the three-carbon intermediates to ADP by substrate-level phosphorylation. Two molecules of ATP are formed for each three-carbon molecule passing through this phase; the three-carbon intermediate is ultimately converted to **pyruvate.** NADH is also formed during this phase. The NADH is either oxidized by the reduction of pyruvate to regenerate NAD^+ or is used to generate additional ATP via the electron transfer chain. The net equation for glycolysis is:

$$\text{Glucose} + 2\text{NAD}^+ + 2\text{ADP} + 2\text{P}_i \longrightarrow 2\text{ pyruvate} + 2\text{NADH} + 2\text{H}^+ + 2\text{ATP} + 2\text{H}_2\text{O}$$

Carbohydrates other than glucose also feed into the glycolytic pathway.

The energy-storage macromolecules **glycogen** and **starch** can be degraded to a phosphorylated form of glucose, which can enter glycolysis. Other monosaccharides, and disaccharides that have been hydrolyzed to monosaccharides, can enter glycolysis if they are first converted to one of the phosphorylated glycolytic intermediates.

Pyruvate has three possible fates after glycolysis.

Under aerobic conditions, pyruvate is oxidized to acetate, which can enter the citric acid cycle and thus be further oxidized to CO_2 and H_2O. Under anaerobic conditions, pyruvate is reduced by NADH to either **lactate** or **ethanol** and CO_2, depending on the organism. These reactions function to regenerate NAD^+ so that glycolysis can continue.

Gluconeogenesis is the ubiquitous pathway for synthesis of glucose from noncarbohydrate precursors.

Biosynthesis of glucose is essential in mammals because some tissues, including the brain, use glucose as their sole or primary fuel source. Gluconeogenesis is not simply the reversal of glycolysis, although the two pathways do share seven enzymes. Three irreversible reactions of glycolysis must be bypassed, or circumvented, in gluconeogenesis: these three steps of gluconeogenesis are conversion of pyruvate to phosphoenolpyruvate, conversion of fructose 1,6-bisphosphate to fructose 6-phosphate by hydrolysis of the C-1 phosphate, and conversion of glucose 6-phosphate to free glucose. The enzyme catalyzing this third reaction is found in the liver, where most gluconeogenesis occurs, but not in the muscle or brain. Net synthesis of glucose can occur from starting points in the citric acid cycle and from the breakdown of some amino acids, but no net conversion of fatty acids to glucose can take place in mammals. The formation of glucose is an energy-requiring process, and gluconeogenesis and glycolysis are reciprocally regulated to avoid futile cycling in cells.

There are alternative pathways for the oxidation of glucose.

Glucose breakdown via the **pentose phosphate pathway** produces NADPH and ribose 5-phosphate; the ratios produced vary with tissue type and metabolic demands.

What to Review

Answering the following questions and reviewing the relevant concepts, which you have already studied, should make this chapter more understandable.

- Most organisms use **glucose** as their primary fuel molecule. Review its structure and be sure you remember the difference between an aldose and a ketose (p. 237).
- The principles of bioenergetics (Chapter 13), especially with respect to ATP use by cells, will be important to your understanding of this chapter as well as of the following metabolism chapters.
- The GLUT family of transporters control glucose uptake into cells. Review their mechanism of action.

Topics for Discussion

Answering each of the following questions, especially in the context of a study group discussion, should help you understand the important points of this chapter.

14.1 Glycolysis

1. Why does the glycolytic pathway occupy a central position in cell metabolism?

An Overview: Glycolysis Has Two Phases

2. What is accomplished in the preparatory phase of glycolysis? What is accomplished in the payoff phase of glycolysis?

3. How is the phosphorylation of glyceraldehyde 3-phosphate in the payoff phase of glycolysis different from the phosphorylation of glucose in the preparatory phase?

Fates of Pyruvate

4. Describe three catabolic fates of pyruvate.

ATP and NADH Formation Coupled to Glycolysis

5. What is the efficiency of recovery, in the form of ATP, of the energy released by glycolysis under *standard* conditions?

6. Why is the efficiency of energy recovery higher under the conditions that exist in living cells?

Energy Remaining in Pyruvate

7. How can the biologically available energy remaining in pyruvate be extracted?

Importance of Phosphorylated Intermediates

8. What are the three functions of the phosphoryl groups of phosphorylated intermediates?

The Preparatory Phase of Glycolysis Requires ATP

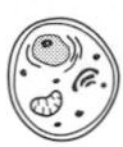

9. *Cell Map:* You should know the structures and names of the intermediates and the names of the enzymes and cofactors involved in each step of the preparatory phase (and payoff phase). These should be entered in the spaces provided on your Cell Map.

① Phosphorylation of Glucose

10. Why is the $\Delta G'^{\circ}$ for this reaction negative?

11. What is the general role of a *kinase*?

12. The $\Delta G'^{\circ}$ for hydrolysis of ATP is −30.5 kJ/mol. How much of this energy is conserved in the phosphorylation of glucose by hexokinase?

② Conversion of Glucose 6-Phosphate to Fructose 6-Phosphate

13. The conversion of glucose 6-phosphate to fructose 6-phosphate is readily reversible: What cellular conditions would bias the reaction in the direction of fructose 6-phosphate formation?

14. Why is this isomerization critical to the subsequent steps?

③ Phosphorylation of Fructose 6-Phosphate to Fructose 1,6-Bisphosphate

15. Why is phosphofructokinase-1 (PFK-1) a good candidate for a regulatory enzyme?

16. What conditions activate PFK-1? What conditions inactivate PFK-1?

17. If PFK-1 is inhibited, in which direction will the reaction catalyzed by phosphohexose isomerase proceed?

④ Cleavage of Fructose 1,6-Bisphosphate

18. The aldolase reaction has a $\Delta G'^{\circ}$ of +23.8 kJ/mol. What mechanism do cells use to pull this endergonic reaction in the direction of cleavage?

⑤ Interconversion of the Triose Phosphates

19. If C-3 of glucose is radioactively labeled and the glucose is allowed to go through the preparatory phase of glycolysis, which carbon in glyceraldehyde 3-phosphate will carry the radioactive label? Will *all* glyceraldehyde 3-phosphate molecules be labeled?

The Payoff Phase of Glycolysis Yields ATP and NADH

⑥ Oxidation of Glyceraldehyde 3-Phosphate to 1,3-Bisphosphoglycerate

20. What is being oxidized in this reaction?

21. What is being reduced in this reaction?

⑦ *Phosphoryl Transfer from 1,3-Bisphosphoglycerate to ADP*

22. How does step ⑦ "pull" step ⑥ forward?

23. Why is this process called *substrate-level* phosphorylation?

24. Where in the cell is this ATP produced? (It is important to keep track of the cellular locations of metabolic processes.)

⑧ *Conversion of 3-Phosphoglycerate to 2-Phosphoglycerate*

25. What is the essential coenzyme in this reaction, and what is its contribution to the formation of 2-phosphoglycerate?

⑨ *Dehydration of 2-Phosphoglycerate to Phosphoenolpyruvate*

26. What is the significance of this dehydration reaction to the payoff phase of glycolysis?

⑩ *Transfer of the Phosphoryl Group from Phosphoenolpyruvate to ADP*

27. Why does this reaction have a large, negative $\Delta G'^{\circ}$?

28. Which of the payoff reactions is/are irreversible in the cell?

The Overall Balance Sheet Shows a Net Gain of ATP

29. Use the overall equation of glycolysis under aerobic conditions to remind yourself of the pathways of carbon, phosphoryl groups, and electrons.

Glycolysis Is under Tight Regulation

30. What is the "Pasteur effect"?

Glucose Uptake Is Deficient in Type I Diabetes Mellitus

31. To what fuel(s) do tissues turn when glucose uptake is deficient?

Box 14–1 High Rate of Glycolysis in Tumors Suggests Targets for Chemotheraphy and Facilitates Diagnosis

32. Why do many cancer cells rely on anaerobic glycolysis?

14.2 Feeder Pathways for Glycolysis

Dietary Polysaccharides and Disaccharides Undergo Hydrolysis to Monosaccharides

33. This is a good time to review the different types of glycosidic bonds.

Endogenous Glycogen and Starch Are Degraded by Phosphorolysis

34. Does the degradation of glycogen (or starch) to a form of glucose that can enter the glycolytic pathway require an input of energy? Explain.

Other Monosaccharides Enter the Glycolytic Pathway at Several Points

35. Is it more energetically expensive to feed monosaccharides other than glucose into the glycolytic pathway? Explain.

14.3 Fates of Pyruvate under Anaerobic Conditions: Fermentation

36. Why is the cell's ability to regenerate NAD^+ critical to glycolysis? Which glycolytic enzyme requires NAD^+?

Pyruvate Is the Terminal Electron Acceptor in Lactic Acid Fermentation

37. What is the most important consequence to the cell of the reduction of pyruvate to lactate by lactate dehydrogenase?

38. What is the net yield of ATP per glucose molecule by this pathway?

Ethanol Is the Reduced Product in Ethanol Fermentation

39. What is the most important consequence to the cell of the reduction of pyruvate to ethanol by pyruvate decarboxylase and alcohol dehydrogenase?

40. Do all organisms perform alcohol fermentation and lactic acid fermentation?

41. What is the net yield of ATP per glucose molecule by this pathway?

Box 14–2 Athletes, Alligators, and Coelacanths: Glycolysis at Limiting Concentrations of Oxygen

42. How does the Cori cycle relate to the deep breaths taken by an athlete as she recovers from a sprint?

Thiamine Pyrophosphate Carries "Active Aldehyde" Groups

43. What structural aspect of TPP allows it to act as an "electron sink"?

44. In general, what is the role of TPP in biological reactions?

Box 14–3 Ethanol Fermentations: Brewing Beer and Producing Biofuels

45. Why are both aerobic and anaerobic stages of yeast growth allowed in beer brewing, even though the aerobic stage produces no ethanol?

Fermentations Are Used to Produce Some Common Foods and Industrial Chemicals

46. Why must industrial fermentations be carried out under conditions that exclude all but the desired microorganisms?

14.4 Gluconeogenesis

47. Why is the biosynthesis of glucose so important in mammals?

48. How can both glycolysis and gluconeogenesis be irreversible processes in the cell?

Conversion of Pyruvate to Phosphoenolpyruvate Requires Two Exergonic Reactions

49. What is the energetic cost of converting pyruvate to PEP?

50. Some of the reactions for the conversion of pyruvate to PEP occur in the mitochondrial matrix, but the end product is made in the cytosol. Besides the transfer of oxaloacetate to the cytosol, what other important molecule is made available for gluconeogenesis by this path?

51. What cellular factor allows the shorter pathway from lactate to PEP to occur?

Conversion of Fructose 1,6 Bisphosphate to Fructose 6-Phosphate is the Second Bypass

52. What drives this reaction forward?

Conversion of Glucose 6-Phosphate to Glucose Is the Third Bypass

53. Why can't gluconeogenesis occur in muscle and brain cells?

Gluconeogenesis Is Energetically Expensive, but Essential

54. Compare the net equations of glycolysis and gluconeogenesis. Why is this high cost of synthesis necessary?

Citric Acid Cycle Intermediates and Some Amino Acids Are Glucogenic

55. Compare the structures of the amino acids alanine and glutamine and their entry points for glucose synthesis, pyruvate and α-ketoglutarate.

Mammals Cannot Convert Fatty Acids to Glucose

56. Why are some amino acids, and fatty acids, *not* glucogenic?

Glycolysis and Gluconeogenesis Are Regulated Reciprocally

57. Why is reciprocal regulation of opposing pathways a "cool" idea?

58. *Cell Map:* On your map, draw in the gluconeogenic bypasses next to the reactions of glycolysis. We suggest using a different color to highlight these reactions.

14.5 Pentose Phosphate Pathway of Glucose Oxidation

59. In what tissues is the pentose phosphate pathway most active?

60. What are the products of this pathway and what roles do they play in cellular metabolism?

Box 14–4 Why Pythagoras Wouldn't Eat Falafel: Glucose 6-Phosphate Dehydrogenase Deficiency

61. In what specific reaction(s) is NADPH crucial for protection of cellular structures?

The Oxidative Phase Produces Pentose Phosphates and NADPH

62. Which glucose-carbon is lost as CO_2?

The Nonoxidative Phase Recycles Pentose Phosphates to Glucose 6-Phosphate

63. What are the roles of transaldolase and transketolase?

Wernicke-Korsakoff Syndrome Is Exacerbated by a Defect in Transketolase

64. What kinds of interactions might be critical in holding a coenzyme in place in its enzyme?

Glucose 6-Phosphate Is Partitioned between Glycolysis and the Pentose Phosphate Pathway

65. What controls the activity of G6PD?

Discussion Questions for Study Groups

- Where are the processes of glycolysis, gluconeogenesis, and the pentose phosphate pathway located in the cell? How are they connected?

SELF-TEST

Do You Know the Terms?

ACROSS

2. A carboxylic acid anhydride with a very high standard free energy of hydrolysis. (2 words)

6. Glucose to lactate, lactate to glucose. (2 words)

8. The ____ pathway produces NADPH, the source of reducing equivalents for biosynthetic processes, and ribose 5-phosphate, an essential precursor for nucleotide synthesis. (2 words)

11. Enzyme regulated by allosteric mechanisms and by covalent phosphorylation/dephosphorylation. Differentially regulated in the liver and muscles.

13. Hexokinase and glucokinase are examples.

15. Describes amino acids that can be degraded to acetyl-CoA.

16. Process in which lactate or pyruvate is used to form new molecules of glucose.

17. The hexokinase reaction is the first step in glycolysis but is *not* the ____ step because the product, glucose 6-phosphate, can enter either the glycolytic or the pentose phosphate pathway.

19. AMP is a(n)_____ regulator of phosphorylase *b*.

20. Describes amino acids that can be degraded to one of the citric acid cycle intermediates.

21. Balance achieved in the rate of formation and rate of utilization of, for example, glucose 6-phosphate achieved by the feedback inhibition of hexokinase.

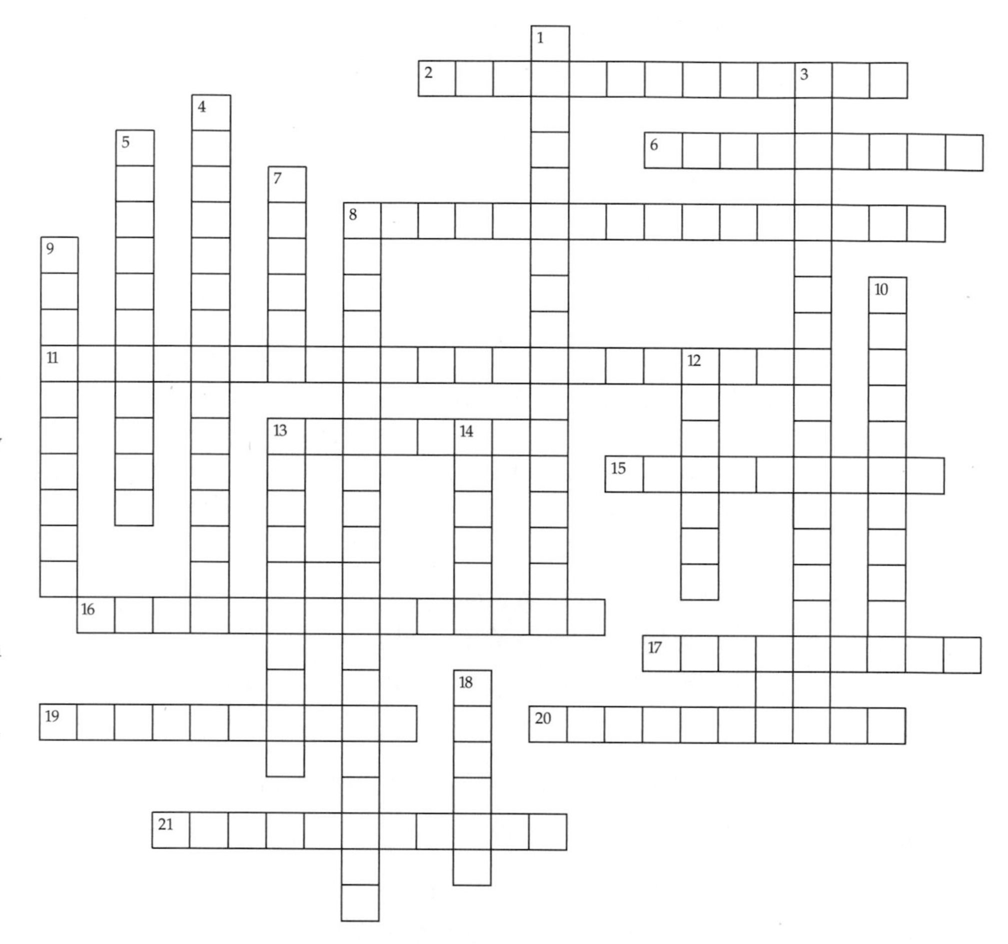

DOWN

1. Initial product of phosphorolysis of glycogen by phosphorylase.

3. Process whereby NAD^+ needed for glycolysis is regenerated by reduction of acetaldehyde. (2 words)

4. Process in which lactate or pyruvate is used to form new molecules of glucose.

5. In the synthesis of hexoses from nonhexose precursors, the first step toward rephosphorylation of pyruvate to phosphoenolpyruvate is catalyzed by pyruvate______.

7. Enzyme that transfers a phosphoryl group between two compounds.

8. Key rate-limiting glycolytic enzyme.

9. Physiological state indicated by excess lactate in the blood. (2 words)

10. The toxic ingredient of fava beans.

12. Its formation in very active muscle regenerates NAD^+ for use in glycolysis.

13. Class of enzymes including that which catalyzes the rearrangement of dihydroxyacetone phosphate to glyceraldehyde 3-phosphate.

14. Type of enzyme that transfers a functional group from one position to another in the same molecule; for example, the transfer of a phosphoryl group from C-3 to C-2 of phosphoglycerate.

18. Cells "trap" glucose molecules by phosphorylating them in an irreversible reaction catalyzed by hexokinase. Liver cells release free glucose into the blood by reversing this process in a ____ reaction catalyzed by glucose 6-phosphatase.

Do You Know the Facts?

For questions 1–10, identify the enzyme that catalyzes each of the ten glycolytic reactions.

GLUCOSE

1. ↓

Glucose 6-phosphate

2. ↓

Fructose 6-phosphate

3. ↓

Fructose 1,6-bisphosphate

4. ↓

Glyceraldehyde 3-phosphate + dihydroxyacetone phosphate

5. ↓

Glyceraldehyde 3-phosphate

6. ↓

1,3-Bisphosphoglycerate

7. ↓

3-Phosphoglycerate

8. ↓

2-Phosphoglycerate

9. ↓

Phosphoenolpyruvate

10. ↓

PYRUVATE

In questions 11–15, refer to the numbered reactions in the previous diagram.

11. Which require(s) an input of energy in the form of ATP?

12. Which involve(s) substrate-level phosphorylation?

13. Which reduce(s) NAD^+?

14. Which is/are irreversible under intracellular conditions?

15. Which is the committed step?

16. In human tissues, ALL of the pathways that pyruvate can take:

A. are aerobic.
B. eventually reduce NAD^+ to NADH.
C. eventually reoxidize NADH to NAD^+.
D. provide equal amounts of ATP to the cell.
E. lower the pH of the cell.

17. Indicate whether each of the following statements about the pentose phosphate pathway is true or false.

_____ **(a)** It generates NADH for reductive biosyntheses.
_____ **(b)** The reactions occur in the cytosol.
_____ **(c)** It is more active in muscle cells than in fat-storage cells.
_____ **(d)** It interconverts trioses, tetroses, pentoses, hexoses, and heptoses.
_____ **(e)** Through this pathway, excess ribose 5-phosphate can be completely converted into glycolytic intermediates.

18. Indicate whether each of the following statements about gluconeogenesis is true or false.

_____ **(a)** It occurs completely in the mitochondrial matrix.
_____ **(b)** Pyruvate carboxylase, catalyzing an anaplerotic reaction, is the first regulatory enzyme in the pathway.
_____ **(c)** Precursors of hexoses include lactate, pyruvate, glycerol, and alanine.

_____ **(d)** The three bypass reactions have ΔG values near zero, whereas other gluconeogenic reactions have large, negative ΔG values.

_____ **(e)** Some reactions occur in the mitochondrial matrix and some in the cytoplasm.

19. Which of the following correctly matches the *glycolytic reaction* with the *gluconeogenic enzyme* used in the corresponding bypass reaction?

A. Glucose → glucose 6-phosphate; glucose 6-phosphatase
B. Fructose 6-phosphate → fructose 1,6-bisphosphate; phosphofructokinase-2
C. Fructose 1,6-bisphosphate → dihydroxyacetone phosphate + glyceraldehyde 3-phosphate; glyceraldehyde 3-phosphate dehydrogenase
D. 2-Phosphoglycerate → phosphoenolpyruvate; phosphoglycerate kinase
E. Phosphoenolpyruvate → pyruvate; pyruvate kinase

20. Explain how both glycolysis and gluconeogenesis can be irreversible processes in the cell.

Applying What You Know

1. The glycolytic reaction:

Glyceraldehyde 3-phosphate + P_i ⟶ 1,3-bisphosphoglycerate

is endergonic ($\Delta G'^{\circ} = 6.3$ kJ/mol). In cells, the reaction is driven in the direction of 1,3-bisphosphoglycerate formation by the next reaction in the glycolytic pathway:

1,3-Bisphosphoglycerate + ADP ⟶ 3-phosphoglycerate + ATP

How are these two reactions connected? Which component in the equation $\Delta G = \Delta G'^{\circ} + RT \ln K'_{eq}$ is being affected?

2. Pyruvate and ATP are the end products of glycolysis. In active muscle cells, pyruvate is converted to lactate. Lactate is transported in the blood to the liver where it is recycled by gluconeogenesis to glucose, which is transported back to muscle for additional ATP production. Why don't active muscle cells export pyruvate, which can also be converted to glucose via gluconeogenesis?

3. If C-1 of glucose is radioactively labeled and then enters glycolysis, which carbon of glyceraldehyde 3-phosphate will be labeled? What fraction of the molecules will carry the radioactive label?

4. The compound 2,3-bisphosphoglycerate (2,3-BPG) acts as a coenzyme in the glycolytic reaction catalyzed by phosphoglycerate mutase. Though in most cells 2,3-BPG is present only in trace amounts—enough to act in its role as coenzyme—it is present in relatively high concentration in erythrocytes, where it acts as a regulator of the affinity of hemoglobin for oxygen (see Chapter 5, pp. 166–167). Because erythrocytes synthesize and degrade 2,3-BPG via a detour from the glycolytic pathway, the rate of glycolysis and therefore the rate of generation of glycolytic intermediates has an impact on the concentration of 2,3-BPG. It follows that defects in the glycolytic pathway in erythrocytes can affect the ability of hemoglobin to carry oxygen.

(a) How would the concentration of 2,3-BPG, and therefore the affinity of hemoglobin for oxygen, be affected in erythrocytes with a deficiency of hexokinase?

(b) How would a pyruvate kinase deficiency affect hemoglobin's affinity for oxygen?

5. Many bacterial species can carry out the reactions of gluconeogenesis. Would you expect to find the gluconeogenic enzyme glucose 6-phosphatase in such bacteria? Why or why not?

ANSWERS

Do You Know the Terms?

ACROSS

2. acyl phosphate
6. Cori cycle
8. pentose phosphate
11. glycogen phosphorylase
13. isozymes
15. ketogenic
16. substrate-level
17. committed
19. allosteric
20. glucogenic
21. homeostasis

DOWN

1. glucose 1-phosphate
3. alcohol fermentation
4. gluconeogenesis
5. carboxylase
7. kinase
8. phosphofructokinase-I
9. oxygen debt
10. divicine
12. lactate
13. isomerases
14. mutase
18. bypass

Do You Know the Facts?

1. hexokinase
2. phosphohexose isomerase
3. phosphofructokinase-1
4. aldolase
5. triose phosphate isomerase
6. glyceraldehyde 3-phosphate dehydrogenase
7. phosphoglycerate kinase
8. phosphoglycerate mutase
9. enolase
10. pyruvate kinase
11. 1 and 3
12. 7 and 10
13. 6
14. 1, 3, and 10
15. 3
16. C
17. **(a)** F; **(b)** T; **(c)** F; **(d)** T; **(e)** T
18. **(a)** F; **(b)** T; **(c)** T; **(d)** F; **(e)** T
19. A
20. Seven of the ten enzymatic steps of gluconeogenesis are the reverse of glycolytic reactions. Three glycolytic reactions are essentially irreversible in cells and cannot be used in gluconeogenesis: the conversion of glucose to glucose 6-phosphate by hexokinase; the phosphorylation of fructose 6-phosphate to fructose 1,6-bisphosphate by phosphofructokinase; and the conversion of phosphoenolpyruvate to pyruvate by pyruvate kinase. These three reactions are characterized by a large, negative free-energy change, whereas the other reactions of glycolysis (also used in gluconeogenesis) have ΔG values near zero. The three irreversible glycolytic steps are bypassed by a separate set of enzymes in gluconeogenesis; these steps also have large, negative free-energy changes, but in the direction of glucose synthesis rather than glucose breakdown. Gluconeogenesis and glycolysis are independently regulated through controls exerted at the enzymatic steps that are not common to both pathways.

Applying What You Know

1. Because of the rapid removal of 1,3-bisphosphoglycerate by the phosphoglycerate kinase reaction, the preceding glycolytic reaction catalyzed by glyceraldehyde 3-phosphate dehydrogenase

$$\text{Glyceraldehyde 3-phosphate} + P_i \longrightarrow \text{1,3-bisphosphoglycerate} \qquad \Delta G'^{\circ} = 6.3 \text{ kJ/mol}$$

has a *negative* ΔG. Under cellular conditions, the free-energy change can have a negative value, even though $\Delta G'^{\circ}$ is positive, when the value for $RT \ln \frac{[\text{products}]}{[\text{reactants}]}$ is large and negative. Because of the rapid removal of the product 1,3-bisphosphoglycerate, the glyceraldehyde 3-phosphate dehydrogenase reaction has a $\frac{[\text{products}]}{[\text{reactants}]}$ ratio of less than one and the value for $RT \ln \frac{[\text{products}]}{[\text{reactants}]}$ is negative.

2. For glycolysis to proceed, cells must have a constant supply of NAD^+, that is, a way of regenerating NAD^+ from NADH. In active muscle cells, pyruvate is the oxidizing agent used to oxidize NADH. If the cells were to export pyruvate, the continued production of ATP by glycolysis would not be possible.

3. The C-3 of glyceraldehyde 3-phosphate will be labeled. This carbon carries a phosphoryl group, and was *either* C-1 or C-6 in the six-carbon molecule fructose 1,6-bisphosphate. Thus, *one-half* of the glyceraldehyde 3-phosphate molecules will be labeled because half were derived from C-1 of glucose (which was radioactively labeled) and half from C-6 (unlabeled).

4. (a) When hexokinase activity is lowered, the concentrations of glycolytic intermediates decrease, and correspondingly the concentration of 2,3-BPG drops. This results in a higher affinity of hemoglobin for oxygen; the oxygen-binding curve is shifted to the left.

(b) If pyruvate kinase activity is decreased, the intermediates of glycolysis build up, increasing the 2,3-BPG concentration and its binding to hemoglobin, which decreases the affinity for oxygen; the oxygen-binding curve is shifted to the right.

5. Bacteria, as single-celled organisms, have no need to export free glucose to other organs or cells. Formation of free glucose by hydrolysis of glucose 6-phosphate would essentially waste the energy of a phosphoanhydride bond. It is more likely that glucose 6-phosphate is converted to glucose 1-phosphate for incorporation into a storage polysaccharide or some other compound.

Principles of Metabolic Regulation

chapter

15

STEP-BY-STEP GUIDE

Major Concepts

Metabolic regulation is one of the most remarkable features of a living cell.

In metabolically active cells, concentrations of key intermediates, such as ATP and NADH, are kept at the steady state. Changes in enzyme activities that maintain these metabolites compensate for changes in the cell's environment. Enzyme activities can be altered by an increase or decrease in actual enzyme molecules available, via the rate of synthesis or degradation; by allosteric or covalent alteration of existing enzyme molecules; or by spatial separation of the enzyme from its substrate. Allosteric adjustments are usually the fastest, while synthesis of new enzyme molecules represents the slowest adjustment. Metabolic control analysis has provided a powerful tool for determining which enzymes actually exert the most control over the overall flux of metabolites through a particular biochemical pathway. Three experimentally determined measures are used in metabolic control analysis: C, the flux control coefficient; ε, the elasticity coefficient; and R, the response coefficient, which is a function of C and ε. Metabolic control analysis allows an engineering-style approach to understanding the control and regulation of metabolite fluxes in cells.

Glycolysis and gluconeogenesis are tightly and reciprocally regulated.

The mechanisms used in the regulation of glycolysis and gluconeogenesis provide models for understanding the control of many other metabolic pathways. The regulatory mechanisms include allosteric, covalent, and hormonal regulation. Three glycolytic and two gluconeogenic enzymes are subject to allosteric regulation. In the liver of animals this regulation is largely mediated by **fructose 2,6-bisphosphate,** an allosteric effector of the glycolytic enzyme **phosphofructokinase-1** and the gluconeogenic enzyme **fructose 1,6 bisphosphatase.** The concentration of this effector is decreased by the hormones epinephrine and glucagon and increased by the action of insulin. Other allosteric effectors of glycolytic and gluconeogenic enzymes include ATP, AMP, citrate, acetyl CoA, and glucose 6-phosphate.

Glycogen metabolism in animals occurs largely in the liver and muscle cells.

The synthesis of glycogen takes place under conditions of excess glucose availability in animals. The process uses **UDP-glucose** (a **sugar nucleotide**); these sugar derivatives have properties that make them especially suitable for biosynthetic reactions. The enzyme **glycogen synthase** catalyzes the addition of glucose units to the growing glycogen polymer, a branching enzyme adds ($\alpha 1 \rightarrow 6$) linkages at branch points, and an unusual protein called **glycogenin** serves as primer and catalyst.

The breakdown of glycogen occurs through phosphorolytic cleavage at the nonreducing ends of the glycogen polymer, producing glucose 1-phosphate, which is then converted to glucose 6-phosphate; this product can enter glycolysis or be converted to free blood glucose according to tissue type.

The synthesis and breakdown of glycogen is regulated by the hormones glucagon and epinephrine, which stimulate glycogen breakdown, and by insulin, which stimulates glycogen synthesis. The regulatory mechanisms include the reciprocal inactivation and activation of glycogen synthase and glycogen phosphorylase by phosphorylation (by specific kinases) and dephosphorylation (by specific phosphatases). Insulin also exerts an effect by triggering the movement of glucose transporters. In addition, faster control is exerted on glycogen phosphorylase and synthase by allosteric effectors. The end results of all this regulation are somewhat different in liver and muscle tissue, reflecting the different roles of these tissues in the body: the liver acts to keep blood glucose levels relatively even, and muscles need to be able to contract and move the animal to safety.

What to Review

Answering the following questions and reviewing the relevant concepts, which you have already studied, should make this chapter more understandable.

- Because this chapter deals with regulation of biosynthetic pathways, it would be useful to review the section opener of Part II (pp. 485–488), which introduces the basic concepts and discusses the coordination of metabolism.
- What does the K_m of an enzyme mean? If an enzyme has two substrates, does just knowing the K_m of each substrate tell you at what rate the enzyme catalyzes each reaction (pp. 196–200)?
- Review what is meant by **reaction equilibria** and the relationships among $\boldsymbol{\Delta G}$, $\boldsymbol{\Delta G'^{\circ}}$, and $\boldsymbol{K'_{eq}}$ (pp. 491–494). These concepts are important to an understanding of the regulation of metabolic pathways.
- Review enzyme cascades and the mechanism of action of insulin, glucagon, and epinephrine (Chapter 12).

Topics for Discussion

Answering each of the following questions, especially in the context of a study group discussion, should help you understand the important points of this chapter.

15.1 Regulation of Metabolic Pathways

Cells and Organisms Maintain a Dynamic Steady State

1. What is the difference between "at equilibrium" and "in a dynamic steady state"?

Both the Amount and the Catalytic Activity of an Enzyme Can Be Regulated

2. Which of the following mechanisms for enzyme regulation works the fastest? Which takes the longest to exert its regulatory effect?

a. Allosteric regulation
b. Hormonal action
c. Synthesis/degradation of enzymes

3. How do allostenic effectors convert enzyme kinetics? (This should be a review from Chapter 5.)

Reactions Far from Equilibrium in Cells Are Common Points of Regulation

4. Make sure you understand the difference between Q and K'_{eq}.

5. Why would allowing some cellular reactions to reach equilibrium be such a dangerous state for the cell?

Adenine Nucleotides Play Special Roles in Metabolic Regulation

6. Why is AMP a more sensitive indicator of the cell's energetic state than ATP?

7. What is the reaction catalyzed by adenylate kinase?

8. What is the reaction catalyzed by AMP-dependent protein kinase?

9. What regulatory mechanisms (which demonstrate good cellular economic sense!) have evolved to keep fuel and energy needs in alignment?

15.2 Analysis of Metabolic Control

The Contribution of Each Enzyme to Flux through a Pathway Is Experimentally Measurable

10. How would the experiment described in Figure 15–7 differ from a similar experiment done in intact cells?

The Control Coefficient Quantifies the Effect of a Change in Enzyme Activity on Metabolite Flux through a Pathway

11. Define C. What factors contribute to its value?

Box 15–1 Metabolic Control Analysis: Quantitative Aspects

12. For an enzyme that has typical Michaelis-Menten kinetics, what are the ε values at very low [S]? at very high [S]?

The Elasticity Coefficient Is Related to an Enzyme's Responsiveness to Changes in Metabolite or Regulator Concentrations

13. Define ε. Compare it to C. How can the value of ε exceed 1.0?

The Response Coefficient Expresses the Effect of an Outside Controller on Flux through a Pathway

14. Define R. Compare it to ε and C. What is P? How are they all related?

Metabolic Control Analysis Has Been Applied to Carbohydrate Metabolism, with Surprising Results

15. What has metabolic control analysis determined to be the real role of PFK-1 regulation?

16. Use Figure 15–10 to review the mechanisms of insulin action on glycogen synthesis in muscle cells. What effects of insulin actually increase glucose flux through the pathway?

Metabolic Control Analysis Suggests a General Method for Increasing Flux through a Pathway

17. Henrik Kacser predicted that increase of flux through one pathway can be accomplished without altering fluxes through other pathways by increasing the concentrations of every enzyme in that initial pathway. This may seem logical and intuitive now, but for years the idea of one or a few rate-determining enzymes was the preferred explanation.

15.3 Coordinated Regulation of Glycolysis and Gluconeogenesis

18. Carbohydrate catabolism is a highly regulated process: the rate must be responsive to cellular demands for the ATP produced by glucose oxidation *and* to cellular demands for biosynthetic precursors. What categories of biomolecules require carbohydrates for their synthesis and/or function?

19. Why does it make sense to the overall economy of the cell that essentially irreversible (highly exergonic) reactions are sites of metabolic regulation?

20. What are the three essentially irreversible reactions of glycolysis that must be circumvented in gluconeogenesis?

Hexokinase Isozymes of Muscle and Liver Are Affected Differently by Their Product, Glucose 6-Phosphate

21. Which of the two isozymes, hexokinase II or hexokinase IV (glucokinase), reaches $V_{max}/2$ at lower substrate concentrations? Why does this make sense physiologically?

22. How does fructose 6-phosphate exert control over hexokinase IV activity?

Box 15–2 Isozymes: Different Proteins That Catalyze the Same Reaction

23. List four factors for the differential distribution of isozyme forms in cells and organs.

Hexokinase IV (Glucokinase) and Glucose 6-Phosphatase Are Transcriptionally Regulated

24. How quickly (or slowly) will the concentrations of these enzymes change?

Phosphofructokinase-1 and Fructose 1,6-bisphosphatase Are Reciprocally Regulated

25. How do ATP and ADP/AMP interact in the control of PFK-1?

Fructose 2,6-Bisphosphate Is a Potent Allosteric Regulator of PFK-1 and FBPase-1

26. How is the level of fructose 2,6-bisphosphate controlled?

Xylulose 5-Phosphate Is a Key Regulator of Carbohydrate and Fat Metabolism

27. How does xylulose 5-phosphate act to connect the intake of high carbohydrate foods to an increase in fatty acid synthesis?

The Glycolytic Enzyme Pyruvate Kinase Is Allosterically Inhibited by ATP

28. How is pyruvate kinase differentially controlled by hormones in liver and muscle?

The Gluconeogenic Conversioin of Pyruvate to Phosphoenol Pyruvate Is Under Multiple Types of Regulation

29. What are the control points for reciprocal regulation of gluconeogenesis and glycolysis?

Transcriptional Regulation of Glycolysis and Gluconeogenesis Changes the Number of Enzyme Molecules

30. Describe the "reciprocal" transcriptional regulation of insulin on the enzymes of glycolysis and gluconeogenesis.

31. This is indeed a complicated story; Figure 15-41 will help to pull all the concepts together, and it is useful to look ahead at this figure now.

15.4 The Metabolism of Glycogen in Animals

32. Where is glycogen stored in animals? What are the various roles of this storage polysaccharide in these tissues?

33. The terms for the breakdown and synthesis of glucose and glycogen may appear too similar at first to keep straight! But the breakdown processes are *lytic* (glyco*lysis*, glycogeno*lysis*), while the anabolic processes *generate* new molecules (gluconeo*genesis*, glyco*genesis*).

Glycogen Breakdown Is Catalyzed by Glycogen Phosphorylase

34. Does the degradation of glycogen to a form of glucose that can enter the glycolytic pathway require an input of energy? Explain.

35. What kind of bond is cleaved by the debranching enzyme but not by the phosphorylase?

Glucose 1-Phosphate Can Enter Glycolysis or, in Liver, Replenish Blood Glucose

36. In the liver, how is the process of producing free glucose for export segregated from the process of glycolysis?

The Sugar Nucleotide UDP-Glucose Donates Glucose for Glycogen Synthesis

37. What properties of sugar nucleotides make them particularly suitable for biosynthetic reactions?

38. What drives the UDP-glucose pyrophosphorylase reaction forward?

Box 15–4 Carl and Gerty Cori: Pioneers in Glycogen Metabolism and Disease

39. Use Table 1 in Box 15–1 to test yourself: WHY would a defective glycogen synthase lead to low blood glucose?

Glycogenin Primes the Initial Sugar Residues in Glycogen

40. Glycogenin remains buried in the glycogen particle it helped to build. How is this fate different from the fate of most enzymes?

41. What are the necessary components (substrates, enzymes, and cofactors) for building glycogen, beginning with the primer glycogenin?

15.5 Coordinated Regulation of Glycogen Synthesis and Breakdown

Glycogen Phosphorylase Is Regulated Allosterically and Hormonally

42. What does a phosphorylase do? What does a kinase do?

43. What is the hormone that triggers glycogen breakdown in the muscle? In the liver?

44. How do ATP and AMP affect the activity of muscle phosphorylase?

45. How does glucose affect the activity of liver phosphorylase?

Glycogen Synthase Is Also Regulated by Phosphorylation and Dephosphorylation

46. How is liver glycogen synthase a glucose 6-phosphate sensor?

47. Compare the mechanisms of regulation of glycogen synthase and glycogen phosphorylase. What are the roles of glucagon, epinephrine, and insulin in these mechanisms? Which are allosteric controls and which are covalent?

Glycogen Synthase Kinase 3 Mediates Some of the Actions of Insulin

48. What are the advantages to a cell of having a hierarchical control strategy (e.g., the necessity of priming phosphorylation reactions)?

Phosphoprotein Phosphatase 1 Is Central to Glycogen Metabolism

49. Why is it critical that the activity of PP1 itself is regulated?

Allosteric and Hormonal Signals Coordinate Carbohydrate Metabolism Globally

50. Why do liver cells have GLUT2 glucose transporters, and muscle and adipose tissue cells have GLUT4 glucose transporters?

51. In order to keep track of all that is occurring in this summary of the interacting levels of control of carbohydrate metabolism in two different tissue types, keep in mind the specific roles of the liver and muscles in the body: the liver is generous, the muscles are selfish.

Carbohydrate and Lipid Metabolism Are Integrated by Hormonal and Allosteric Mechanisms

52. What are the rapid (in milliseconds) versus less rapid (seconds to minutes) regulatory strategies for carbohydrate metabolism in muscle? Which is/are hormonally induced and which is/are allosterically induced?

SELF-TEST

Do You Know the Terms?

ACROSS

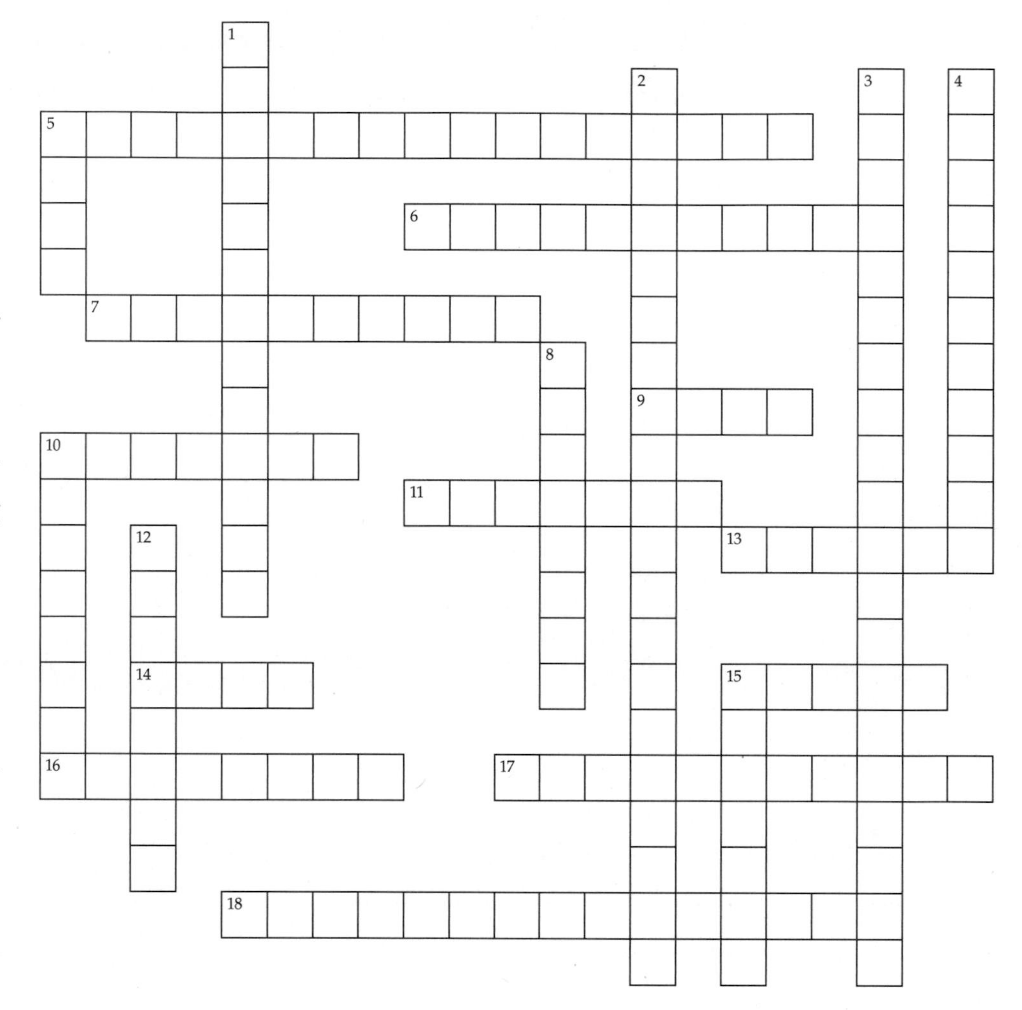

5. Initial product of phosphorolysis of glycogen by phosphorylase.
6. Balance achieved in the rate of formation and rate of utilization of, for example, glucose 6-phosphate achieved by the feedback inhibition of muscle hexokinase.
7. AMP is a(n) _____ regulator of muscle phosphorylase *b*.
9. *C*, the _____ control coefficient.
10. Peptide hormone that regulates the metabolism of glucose and triacylglycerols.
11. Increased concentrations of intracellular free _____ cause muscle contraction and breakdown of stored glycogen.
13. Enzyme that transfers a phosphoryl group between two compounds.
14. Activation of this enzyme in the liver indirectly activates phosphofructokinase-1. (abbr.)
15. Product of the third step in glycolysis; must be dephosphorylated in gluconeogenesis. (abbr.)
16. Glycogen _____ *a* catalyzes the formation of ($\alpha 1 \rightarrow 4$) glycosidic bonds.
17. Hormone that stimulates phosphorylation of phosphorylase *b* kinase, glycogen synthase, and glycogen phosphorylase; its receptors are found primarily on muscle cells.
18. Pathway by which the glycerol backbone of triacylglycerols can be made into fuel for brain cells; occurs primarily in liver cells.

DOWN

1. Glycogen _____ *a* catalyzes the cleavage of ($\alpha 1 \rightarrow 4$) glycosidic bonds.
2. Third glycolytic enzyme: regulated by several effectors.
3. Activated by phosphorylation, this enzyme relays information about the binding of epinephrine to the enzymes responsible for glycogen breakdown. (3 words)
4. General name for an enzyme that reverses the effects of a kinase.
5. The most important regulatory kinase in the inactivation of glycogen synthase. (abbr.)
8. Hormone that stimulates phosphorylation of phosphorylase *b* kinase, glycogen synthase, and glycogen phosphorylase; its receptors are located mainly on liver cells.
10. Hexokinase II and hexokinase IV (glucokinase) are examples.
12. *R*, the _____ coefficient.
15. Activation of this enzyme in the liver results in decreased (fructose 2,6-bisphosphate). (abbr.)

Do You Know the Facts?

In questions 1–3, indicate the effect (increase, decrease, or no effect) of each situation on the rate of glycolysis. If there is a regulatory effect, identify the enzyme(s) involved and the regulatory mechanism(s) employed.

1. Increased concentration of glucose 6-phosphate.

2. Increased concentration of fructose 2,6-bisphosphate.

3. Increased concentration of ATP.

In questions 4–6, indicate the (immediate) effect (increase, decrease, no change) of each situation on the rate of gluconeogenesis.

4. Decreased concentration of acetyl-CoA.

5. Phosphorylation of phosphofructokinase-2.

6. Increase in [AMP]/[ATP] ratio.

7. Which statement is *not* true of phosphofructokinase-1?

A. It is inhibited by fructose 2,6-bisphosphate.
B. It is activated by AMP.
C. It is inhibited by citrate.
D. It is inhibited by ATP.
E. ATP increases its $K_{0.5}$ for fructose 6-phosphate.

8. Which of the following is true of the control of gluconeogenesis?

A. Glucagon stimulates adenylyl cyclase, causing the formation of cAMP.
B. Cyclic AMP stimulates the phosphorylation and thus increases the activity of FBPase-2.
C. FBPase-2 activity lowers the level of fructose 2,6-bisphosphate, thus increasing the rate of gluconeogenesis.
D. It is reciprocally linked to the control of glycolysis.
E. All of the above are true.

9. Indicate whether each of the following descriptions of glycogen and its synthesis and breakdown is true or false.

_____ **(a)** Glycogen synthase is regulated by hormone-dependent phosphorylation and is active in its phosphorylated form.

_____ **(b)** Glycogen phosphorylase and glycogen synthase are usually active simultaneously.

_____ **(c)** Glycogen synthase and glycogen phosphorylase are reciprocally regulated.

_____ **(d)** Phosphoprotein phosphatase I removes phosphate groups from Ser residues on glycogen synthase *b*, converting it to the active form.

_____ **(e)** The primer protein glycogenin is required for initiation of glycogen synthesis.

10. Which of the following has the (immediate) effect of increasing the rate of glycogen breakdown

A. Increased concentration of cAMP
B. Increase in the [AMP]/[ATP] ratio
C. Increased secretion of glucagon
D. A and C
E. A, B, and C

11. Which of the following has the (immediate) effect of increasing the rate of glycogen synthesis?
   A. Increased concentration of cAMP
   B. Increase in the [AMP]/[ATP] ratio
   C. Increased secretion of insulin
   D. A and C
   E. A, B, and C

12. Describe the reciprocal regulation of glycogen phosphorylase *a* and *b* and glycogen synthase *a* and *b*.

13. Sugar nucleotides are the substrates for polymerization of monosaccharides into di- and polysaccharides. What properties of sugar nucleotides make them especially suitable for biosynthetic reactions?

14. Explain how glycolysis and gluconeogenesis can be irreversible processes in the cell.

Applying What You Know

1. Many bacterial species can carry out the reactions of gluconeogenesis. Would you expect to find the gluconeogenic enzyme glucose 6-phosphatase in such bacteria? Why or why not?

2. Explain fully the regulatory role of fructose 2,6-bisphosphate in gluconeogenesis and glycolysis, and the mechanism by which it affects fructose 1,6-bisphosphatase-1 and phosphofructokinase-1. How is the concentration of fructose 2,6-bisphosphate regulated?

ANSWERS

Do You Know the Terms?

ACROSS

5. glucose 1-phosphate
6. homeostasis
7. allosteric
9. flux
10. insulin
11. calcium
13. kinase
14. PFK-2
15. F1,6-BP
16. synthase
17. epinephrine
18. gluconeogenesis

DOWN

1. phosphorylase
2. phosphofructokinase-1
3. phosphorylase *b* kinase
4. phosphatase
5. GSK 3
8. glucagon
10. isozymes
12. response
15. FBPase-2

Do You Know the Facts?

1. Decrease, by product inhibition of hexokinase
2. Increase, by allosteric activation of phosphofructokinase-1
3. Decrease, by allosteric inhibition of phosphofructokinase-1 and pyruvate kinase
4. Decrease
5. Increase
6. Decrease
7. A
8. E
9. **(a)** F; **(b)** F; **(c)** T; **(d)** T; **(e)** T
10. E
11. C
12. Glycogen phosphorylase is the enzyme that cleaves glycogen into individual glucose residues (glucose 1-phosphate). It exists in two forms: glycogen phosphorylase *b* (a relatively inactive form) and glycogen phosphorylase *a* (a phosphorylated and fully active, glycogen-degrading form). Only the *b* form of this enzyme can be *allosterically* regulated; the allosteric *activator* is AMP and the *inhibitor* is ATP, making this enzyme responsive to cellular energy levels. Glycogen phosphorylase is also covalently modified in response to hormonal stimulation. Binding of epinephrine to receptors on muscle cells or

glucagon binding to receptors on hepatocytes results in the production of cAMP, which activates an enzyme cascade. This enzyme cascade ultimately results in the activation of phosphorylase *b kinase* and the phosphorylation of glycogen phosphorylase *b*. This is the covalent modification that converts glycogen phosphorylase to its fully active *a* form. In addition, both hormones inactivate phosphorylase *a phosphatase,* which reverses the effects of covalent modification by removing the phosphate group.

Glycogen synthase is the enzyme that synthesizes glycogen. This enzyme also exists in two forms: glycogen synthase *b* (the inactive form) and glycogen synthase *a* (the active, glycogen-synthesizing form). In contrast to glycogen phosphorylase, the active form of glycogen synthase is *not* phosphorylated, whereas the inactive, *b* form *is* phosphorylated. The inactivation of glycogen synthase *a* is the result of phosphorylation by a cAMP-dependent protein kinase. Thus, the same hormones that activate glycogen phosphorylase by increasing cAMP levels *also* inactivate glycogen synthase. Due to the opposing effects of phosphorylation on glycogen phosphorylase and glycogen synthase, the anabolic and catabolic pathways can be reciprocally regulated by the same hormonal input.

13. The properties of sugar nucleotides include the following: (a) Their formation splits one high-energy bond, releasing PP_i, which is then hydrolyzed by inorganic pyrophosphatase. The large, negative free-energy change makes the synthetic reaction thermodynamically favorable. (b) They offer many potential groups for noncovalent interactions with enzymes. (c) The nucleotidyl group is an excellent leaving group. (d) "Tagging" some hexoses with nucleotidyl groups specifies the use of those hexoses for a specific cellular purpose.

14. Seven of the ten enzymatic steps of gluconeogenesis are the reverse of glycolytic reactions. Three glycolytic reactions are essentially irreversible in cells and cannot be used in gluconeogenesis: the conversion of glucose to glucose 6-phosphate by hexokinase, the phosphorylation of fructose 6-phosphate to fructose 1,6-bisphosphate by phosphofructokinase, and the conversion of phosphoenolpyruvate to pyruvate by pyruvate kinase. These three reactions are characterized by a large, negative free-energy change, whereas the other reactions of glycolysis (also used in gluconeogenesis) have ΔG values near zero. The three irreversible glycolytic steps are bypassed by a separate set of enzymes in gluconeogenesis; these steps also have large, negative free-energy changes, but in the direction of glucose synthesis rather than glucose breakdown. Gluconeogenesis and glycolysis are independently regulated through controls exerted at the enzymatic steps that are not common to both pathways.

Applying What You Know

1. Bacteria, as single-celled organisms, have no need to export free glucose to other organs or cells. Formation of free glucose by hydrolysis of glucose 6-phosphate would essentially waste the energy of a phosphoanhydride bond. It is more likely that glucose 6-phosphate is converted to glucose 1-phosphate for incorporation into a storage polysaccharide or some other compound.

2. Fructose 2,6-bisphosphate (F 2,6-BP; not to be confused with fructose 1,6-bisphosphate) is *not* an intermediate in gluconeogenesis or glycolysis, but an allosteric effector that acts as a mediator for the hormonal regulation of these processes. High [F 2,6-BP] stimulates glycolysis by activating PFK-1 and inhibiting FBPase-1, which slows gluconeogenesis. Low [F 2,6-BP] stimulates gluconeogenesis.

The concentration of F 2,6-BP is controlled by the relative rates of its formation and breakdown: by the activity of *phosphofructokinase-2* (PFK-2), which catalyzes the formation of F 2,6-BP by phosphorylation of fructose 6-phosphate; and by *fructose 2,6-bisphosphatase* (FBPase-2), which catalyzes the dephosphorylation of F 2,6-BP. These two enzymes are part of a single protein. The balance of these two activities in the liver, and therefore the level of F 2,6-BP, is regulated by glucagon. Glucagon stimulates adenylyl cyclase, causing the formation of cAMP. Cyclic AMP stimulates a kinase, which transfers a phosphate group from ATP to the PFK-2/FBPase-2 protein. Phosphorylation leads to increased FBPase-2 activity, lowering [F 2,6-BP] and thus inhibiting glycolysis and stimulating gluconeogenesis. Glucagon therefore inhibits glycolysis, increasing gluconeogenesis and enabling the liver to replenish blood glucose levels.

chapter

16 The Citric Acid Cycle

STEP-BY-STEP GUIDE

Major Concepts

Respiration is the complete oxidation of organic fuels to CO_2 and H_2O.

Cellular respiration occurs in three major stages. The first stage oxidizes fuel molecules to the two-carbon molecule **acetyl-CoA.** In the second stage (the citric acid cycle), acetyl-CoA is oxidized to CO_2 and the electron carriers NAD^+ and FAD are reduced. In the third stage (oxidative phosphorylation, discussed in Ch. 19), electrons from the oxidation of fuel molecules are transferred to O_2; ATP is formed as a result of this electron transfer process.

The first stages of cellular respiration include glycolysis and the production of acetate.

Pyruvate, the product of glycolysis, is converted into acetyl-CoA and CO_2 by the pyruvate dehydrogenase complex. This large enzyme complex consisting of three different enzymes is regulated by allosteric mechanisms and by covalent modification. This is a good example of substrate channeling by an enzyme.

The citric acid cycle is a series of eight chemical transformations.

For each turn of the cycle, two carbons enter as an acetyl group and two carbons leave as molecules of CO_2. **Oxaloacetate** combines with acetyl-CoA to form **citrate,** which "carries" the carbons from acetyl-CoA into the cycle. Oxaloacetate is regenerated at the end of the cycle. The citric acid cycle produces energy for cells in the form of nucleotide triphosphates (ATP or GTP) and reduced electron carriers $FADH_2$ and NADH. Citric acid cycle intermediates also provide precursors for the synthesis of a number of biomolecules. Intermediates "lost" from the cycle in this way are replenished through a variety of processes called anaplerotic reactions.

The citric acid cycle, as befits a reaction series with diverse functions, is under tight regulation.

Regulation occurs at the conversion of pyruvate to acetyl-CoA and at three points in the cycle itself: the entry of acetyl-CoA into the cycle and both of the oxidative decarboxylation steps. The primary enzymes involved in regulation are the pyruvate dehydrogenase complex, citrate synthase, isocitrate dehydrogenase, and α-ketoglutarate dehydrogenase.

The glyoxylate cycle is a variation of the citric acid cycle that occurs in some microorganisms and all plants.

This pathway (which occurs in addition to the citric acid cycle, the two being coordinately regulated) results in the net formation of oxaloacetate from two molecules of acetate. The glyoxylate cycle makes it possible to convert the carbons of acetyl-CoA into glucose molecules, an anaplerotic pathway not present in vertebrate systems.

What to Review

Answering the following questions and reviewing the relevant concepts, which you have already studied, should make this chapter more understandable.

- Review the roles of **FAD** and **NAD$^+$** (pp. 516–519) as electron carriers. Reduced forms of these nucleotides are generated in the citric acid cycle and play a critical role in the oxidative phosphorylation of ADP (Chapter 19).
- **TPP** is an important cofactor used by fermentative and citric acid cycle enzymes. Review its role in ethanol fermentation (pp. 547–550).
- Review what is meant by **reaction equilibria** and the relationships among ΔG, $\Delta G'^{\circ}$, and K'_{eq} (pp. 491–494). These concepts are important to an understanding of the regulation of metabolic pathways such as the citric acid cycle.

Topics for Discussion

Answering each of the following questions, especially in the context of a study group discussion, should help you understand the important points of this chapter.

16.1 Production of Acetyl-CoA (Activated Acetate)

Pyruvate Is Oxidized to Acetyl-CoA and CO_2

1. The oxidative decarboxylation of pyruvate is a highly exergonic reaction. How is the energy released by this reaction conserved?

The Pyruvate Dehydrogenase Complex Requires Five Coenzymes

2. Make sure you can recognize the structure of coenzyme A; pay attention to the components of the molecule and to its general function. Why is the thioester bond important to the function of coenzyme A?

3. What are the two possible roles of lipoate in enzymatic reactions?

The Pyruvate Dehydrogenase Complex Consists of Three Distinct Enzymes

4. Be sure you understand how the different enzyme subunits of pyruvate dehydrogenase and their cofactors act together to catalyze the decarboxylation of pyruvate.

In Substrate Channeling, Intermediates Never Leave the Enzyme Surface

5. What is the result of the first three reactions of the overall pyruvate dehydrogenase complex reaction? What is the source of energy for this set of reactions?

6. What is the result of steps ④ and ⑤ in this process?

7. Why is it advantageous to have three different enzymatic activities clustered into a single enzyme complex?

8. Why is thiamine deficiency a serious condition?

16.2 Reactions of the Citric Acid Cycle

9. What are the roles of the citric acid cycle?

10. Where in eukaryotes do these reactions take place? Where do they occur in prokaryotes?

The Citric Acid Cycle Has Eight Steps

11. *Cell Map:* Draw in the structures and names of the intermediates and the names of the enzymes and cofactors for each step of the citric acid cycle on your cell map at this point.

① *Formation of Citrate*

12. The concentration of oxaloacetate in the cell is normally quite low. What factor in the citrate synthase reaction prevents this from hindering the operation of the cycle?

② *Formation of Isocitrate via cis-Aconitate*

13. The $\Delta G'^{\circ}$ of the aconitase reaction is 13.3 kJ/mol. What drives the aconitase reaction forward?

14. What is the general function of the iron-sulfur center in aconitase?

③ *Oxidation of Isocitrate to α-Ketoglutarate and CO_2*

15. Why do eukaryotic cells need two isozymes of isocitrate ehydrogenase?

Box 16–1 Moonlighting Enzymes: Proteins with More Than One Job

16. How does IRP 1 differ from "normal" aconitase?

④ *Oxidation of α-Ketoglutarate to Succinyl-CoA and CO_2*

17. How is the energy of oxidation of α-ketoglutarate conserved?

18. What are the similarities between the α-ketoglutarate dehydrogenase complex and the pyruvate dehydrogenase complex? What are the differences?

⑤ *Conversion of Succinyl-CoA to Succinate*

19. What is the source of the energy used to drive the substrate-level phosphorylation of GDP?

20. What are the differences between substrate-level phosphorylation and respiration-linked phosphorylation?

21. Given that the $\Delta G'^{\circ}$ of the nucleoside diphosphate kinase reaction is 0 kJ/mol, what factor would encourage the reaction to proceed in the direction of ATP formation?

Box 16–2 Synthases and Synthetases; Ligases and Lyases; Kinases, Phosphatases, and Phosphorylases: Yes, the Names Are Confusing!

22. You can more easily learn and remember the functions of specific enzymes if you learn the names and functions of the general classes to which they belong. In your text, find an enzyme for each of the enzyme classes listed in the title above, and write the reaction it catalyzes.

⑥ Oxidation of Succinate to Fumarate

23. Explain why it is significant that the electron carrier FAD is covalently bound to succinate dehydrogenase, whereas NAD^+ is in a soluble, unconjugated form.

⑦ Hydration of Fumarate to Malate

24. This reaction is readily reversible, with a $\Delta G'^{\circ}$ of -3.8 kJ/mol. Why do you think it proceeds in the direction of malate formation in vivo?

⑧ Oxidation of Malate to Oxaloacetate

25. The $\Delta G'^{\circ}$ of the malate dehydrogenase reaction is 29.7 kJ/mol. What drives this reaction forward?

Box 16–3 Citrate: A Symmetric Molecule That Reacts Asymmetrically

26. What is necessary in an enzyme's active site in order for a symmetric molecule to react asymmetrically with the enzyme?

27. What is the net equation for one turn of the citric acid cycle?

28. What is the net energy yield per molecule of glucose for the combined reactions of glycolysis, the pyruvate dehydrogenase reaction, and the citric acid cycle?

The Energy of Oxidations in the Cycle Is Efficiently Conserved

29. Which reaction(s) of the citric acid cycle store(s) the energy derived from oxidations as NADH? Which store(s) it as $FADH_2$?

Why Is the Oxidation of Acetate So Complicated?

30. Why is a process as complex as the citric acid cycle actually an economical way for cells to do their metabolic business?

Citric Acid Cycle Components Are Important Biosynthetic Intermediates

31. *Cell Map:* Because the citric acid cycle is the "hub" of cellular metabolism, it is important that you have a firm grasp of all the reactions and intermediates involved in this pathway. Add in arrows from the CAC intermediates out to their various cellular products on your cell map.

Anaplerotic Reactions Replenish Citric Acid Cycle Intermediates

32. Under what cellular circumstances is pyruvate carboxylase activity stimulated?

Box 16–4 Citrate Synthase, Soda Pop, and the World Food Supply

33. How does citrate act to immobilize Al^{3+}?

Biotin in Pyruvate Carboxylase Carries CO_2 Groups

34. What is the actual source of CO_2 groups for biotin?

35. What structural component of biotin is directly involved in carboxylation reactions?

16.3 Regulation of the Citric Acid Cycle

Production of Acetyl-CoA by the Pyruvate Dehydrogenase Complex Is Regulated by Allosteric and Covalent Mechanisms

36. What do all the allosteric *activators* of the pyruvate dehydrogenase complex signal about the cell's energy state? What do the allosteric *inhibitors* signal?

37. How does covalent modification regulate the pyruvate dehydrogenase complex?

The Citric Acid Cycle Is Regulated at Its Three Exergonic Steps

38. Why are citrate synthase, isocitrate dehydrogenase, and α-ketoglutarate dehydrogenase good candidates for regulatory enzymes?

39. What is the role of calcium ions in regulation of the citric acid cycle?

Substrate Channeling through Multienzyme Complexes May Occur in the Citric Acid Cycle

40. What are the advantages to the cell of substrate channeling?

Some Mutations in Enzymes of the Citric Acid Cycle Lead to Cancer

41. Based on your current knowledge, list all the problems that would follow from an accumulation of fumarate?

16.4 The Glyoxylate Cycle

The Glyoxylate Cycle Produces Four-Carbon Compounds from Acetate

42. What is the net equation of the glyoxylate cycle?

43. What is the advantage to plants of using fatty acids as energy storage molecules in seeds, rather than complex carbohydrates such as starch?

44. In what plant tissue types are glyoxysomes abundant?

The Citric Acid and Glyoxylate Cycles Are Coordinately Regulated

45. Why is the coordinated regulation of isocitrate lyase and isocitrate dehydrogenase of advantage to the organism?

46. Do the intermediates of the citric acid cycle and glycolysis exert their regulatory effects directly or indirectly on these enzymes?

Discussion Questions for Study Groups

- Does the covalent regulation of the pyruvate dehydrogenase complex and the citric acid cycle enzymes provide any information about the cell's energy state? How about the allosteric regulation?

SELF-TEST

Do You Know the Terms?

ACROSS

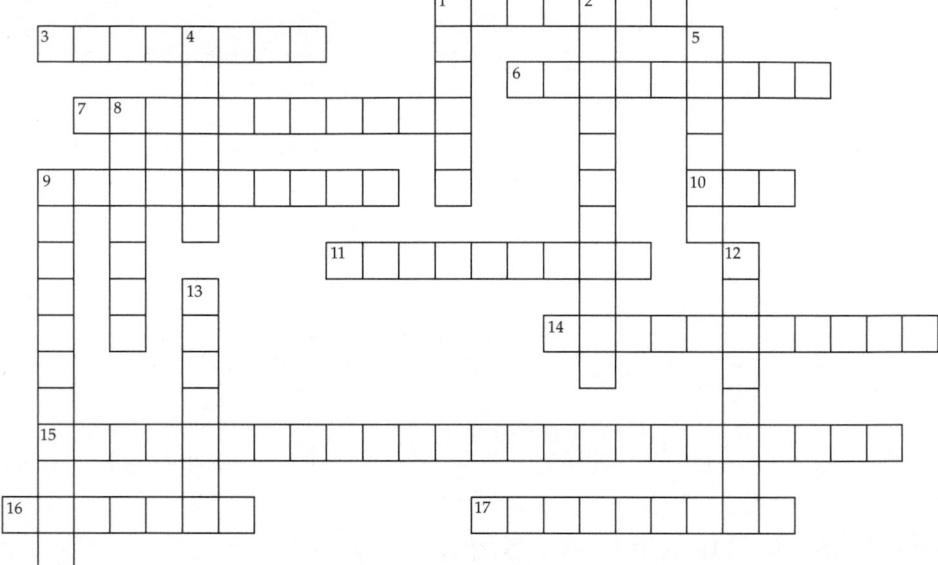

1. Enzymes, including synthetases, that catalyze condensation reactions requiring an input of energy.

3. Disease resulting from a dietary deficiency of the vitamin thiamin.

6. Enzymes that catalyze condensation reactions that do not require a nucleotide triphosphate for an energy source, such as: oxaloacetate + acetyl-CoA + $H_2O \rightarrow$ citrate + CoA-SH.

7. Organelles containing enzymes of fatty acid degradation and the glyoxylate cycle.

9. Describes pathways used in anabolism and catabolism; for example, the citric acid cycle, in which oxaloacetate is an intermediate in the degradation and the synthesis of glucose.

10. Cofactor involved in the decarboxylation of pyruvate and the initial binding of the resulting acetyl group to pyruvate dehydrogenase. (abbr.)

11. Describes symmetric molecules that bind to a substrate-binding site in only one of two possible orientations.

14. Process of passing electrons from fuel molecules to O_2.

15. General type of reaction catalyzed by isocitrate dehydrogenase and by α-ketoglutarate dehydrogenase complex. (2 words)

16. Enzymes that catalyze group-transfer reactions involving a phosphoryl group.

17. Acyl groups are linked to coenzyme A through a _____ bond and are thus activated for group transfer.

DOWN

1. Enzymes that catalyze cleavage reactions involving double bonds and electron rearrangements.

2. Enzymes that catalyze reactions involving a nucleoside triphosphate; for example:

Succinyl-CoA + GDP + $P_i \rightleftharpoons$
succinate + GTP + Co A-SH

4. Consuming too much eggnog or too many raw eggs may cause a deficiency in the vitamin _____, which is required as a cofactor for pyruvate carboxylase.

5. Compound formed by the condensation of glyoxylate and acetyl-CoA.

8. Pyruvate dehydrogenase cofactor involved in the transfer of acyl groups and of electrons in the form of hydrogen.

9. Describes reactions that produce citric acid cycle intermediates.

12. End product of glycolysis.

13. Citric acid cycle intermediate that inhibits phosphofructokinase-1.

Do You Know the Facts?

In questions 1–10, fill in the names of the missing intermediates in the diagram below.

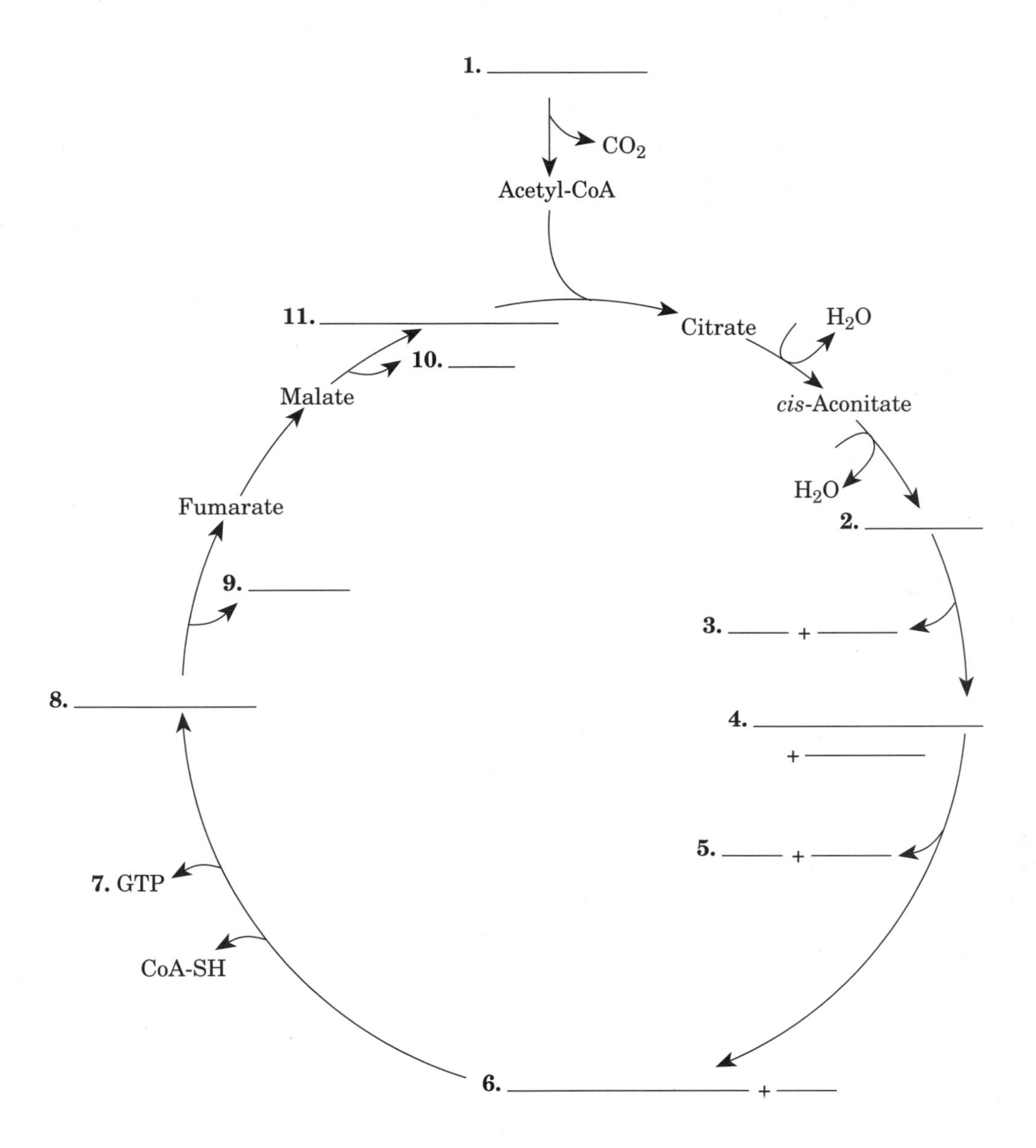

11. Indicate whether each of the following metabolic effects would or would not occur (yes/no) in cells lacking the phosphoprotein phosphatase of the pyruvate dehydrogenase complex.

_____ **(a)** Increased [lactate]
_____ **(b)** Increased pH
_____ **(c)** Increased [citrate]
_____ **(d)** Decreased [NADH] and [ATP]

12. Which of the following correctly describes the citric acid cycle?

A. Oxygen is required to regenerate electron acceptors.
B. Citrate is dehydrated and hydrated by the same enzyme, aconitase.
C. Ten high-energy phosphate bonds are eventually formed as a result of one round of the cycle.
D. Succinate dehydrogenase links the citric acid cycle to oxidative phosphorylation.
E. All of the above are true.

13. In order to examine the citric acid cycle, you have obtained a *pure preparation of isolated,* ***intact*** *mitochondria.* You add some succinyl-CoA to the suspension of mitochondria. How many moles of ATP would you expect to be generated in one turn of the citric acid cycle from each mole of succinyl-CoA added to the test tube?

A. 3
B. 4
C. 5
D. 5.5
E. No ATP would form under these conditions.

14. You have discovered a compound that inhibits fumarase. How many moles of ATP would you expect to be generated from one mole of acetyl-CoA in the presence of this inhibitor?

A. 5
B. 6
C. 6.5
D. 7.5
E. No ATP would form under these conditions.

15. All of the following enzymes are linked to the reduction of NADH *except:*

A. isocitrate dehydrogenase.
B. lactate dehydrogenase.
C. succinate dehydrogenase.
D. pyruvate dehydrogenase.
E. α-ketoglutarate dehydrogenase.

In questions 16–22, match the descriptions to the appropriate enzyme(s) from the list below. (There can be more than one answer to a question.)

(a) Citrate synthase
(b) Isocitrate dehydrogenase
(c) Pyruvate dehydrogenase
(d) Succinate dehydrogenase
(e) Succinyl-CoA synthetase
(f) α-Ketoglutarate dehydrogenase

_____ **16.** Allosterically activated by calcium.
_____ **17.** Catalyzes the committed step in the citric acid cycle.
_____ **18.** The only membrane-bound enzyme in the citric acid cycle.
_____ **19.** Catalyzes the substrate-level phosphorylation of ADP or GDP.
_____ **20.** Regulated by reversible phosphorylation.
_____ **21.** Inhibited by NADH.
_____ **22.** Activated by AMP or ADP.

In questions 23–26, indicate the effect (increase, decrease, no change) of each situation on the overall rate of the citric acid cycle. If there is a regulatory effect, identify the enzyme(s) involved and the type of regulatory mechanism(s) employed.

23. Increased concentration of citrate.

24. Increased concentration of ATP.

25. Increased concentration of glucose 6-phosphate.

26. Increased concentration of succinyl-CoA.

27. Indicate whether each of the following statements about the glyoxylate cycle is true or false.

_____ **(a)** It allows the products of fatty acid oxidation to be converted, eventually, to glucose.
_____ **(b)** It provides intermediates for the citric acid cycle during periods when amino acids are being synthesized.
_____ **(c)** It depletes cellular stores of oxaloacetate.
_____ **(d)** It occurs instead of the citric acid cycle in plants.

28. Although molecular oxygen does not participate directly in any of the reactions in the citric acid cycle, the cycle operates only when oxygen is present. This is because O_2:

A. is necessary as an activator for several enzymatic dehydrogenations (oxygenations) in the cycle.
B. is necessary for producing water, which is crucial for all cellular processes.
C. accepts electrons from the electron transfer chain, allowing reoxidation of NADH to NAD^+.
D. removes toxic byproducts of the citric acid cycle.
E. has all of the above functions.

Applying What You Know

1. The flow of metabolites into the glycolytic pathway and the citric acid cycle is reduced during periods of DNA replication. How might cells coordinate DNA synthesis and glucose metabolism?

2. You and your fellow students find it difficult to believe that the oxidation of a two-carbon acyl compound requires a pathway as complex as the citric acid cycle. In order to verify the metabolic pathways you have been learning about, you conduct an experiment using a sample of glucose radioactively labeled with ^{14}C at C-1. If the information you have learned is correct, in which glycolytic or citric acid cycle intermediate, and on which carbon, should you find the label:

(a) if O_2 were unavailable?

(b) in the presence of malonate?

(c) after one round of the citric acid cycle?

(d) after three rounds of the citric acid cycle?

3. You wish to determine the metabolic fate of oxaloacetate in cells. In an experiment using oxaloacetate radioactively labeled with ^{14}C at C-1, you find that much of the label ends up in CO_2. Can you conclude from this result that the overall cellular levels of oxaloacetate in the sample are being depleted?

4. Calculate the amount of metabolic energy produced from the following carbon sources under the specified cellular conditions. Which of these yields the most energy and which yields the least?

(a) 1 mol of glyceraldehyde 3-phosphate; O_2 is available.

(b) 1 mol of succinate and 1 mol of pyruvate; O_2 is available.

(c) 8 mol of glucose; anaerobic conditions.

(d) 1 mol of acetyl-CoA; O_2 is available.

ANSWERS

Do You Know the Terms?

ACROSS

1. ligases
3. beriberi
6. synthases
7. glyoxysomes
9. amphibolic
10. TPP
11. prochiral
14. respiration
15. oxidative decarboxylation
16. kinases
17. thioester

DOWN

1. lyases
2. synthetases
4. biotin
5. malate
8. lipoate
9. anaplerotic
12. pyruvate
13. citrate

Do You Know the Facts?

1. pyruvate
2. isocitrate
3. CO_2 + NADH
4. α-ketoglutarate + CoA-SH
5. CO_2 + NADH
6. succinyl-CoA + GDP
7. succinate + GTP
8. $FADH_2$
9. NADH
10. oxaloacetate
11. **(a)** Yes; **(b)** No; **(c)** No; **(d)** Yes
12. E
13. E; succinyl-CoA could not enter the *intact* mitochondrion
14. D
15. C
16. (b), (c), and (f)
17. (a)
18. (d)
19. (e)
20. (c) and (b)
21. (a), (c), and (f)
22. (a), (b), and (c)
23. Decrease. Citrate synthase is inhibited by a negative feedback mechanism through product accumulation. The amount of acetyl-CoA available for entry into the cycle will decrease due to allosteric inhibition of PFK-1.
24. Decrease. ATP inhibits citrate synthase and isocitrate dehydrogenase through product accumulation. ATP also allosterically inhibits PFK-1, pyruvate kinase, and pyruvate dehydrogenase.
25. No effect.
26. Decrease. α-Ketoglutarate dehydrogenase is inhibited by product accumulation.
27. **(a)** T; **(b)** T; **(c)** F; **(d)** F
28. C

Applying What You Know

1. DNA synthesis syphons off oxaloacetate for the production of the pyrimidine bases thymine and cytosine. As levels of oxaloacetate decrease, acetyl-CoA accumulates, decreasing the activity of pyruvate dehydrogenase. The net result is a slowing of the overall rate of the citric acid cycle. The pentose phosphate pathway, an alternative route of glucose catabolism, reroutes glucose 6-phosphate from glycolysis to generate ribose 5-phosphate, slowing glycolysis. This pentose is necessary for the biosynthesis of nucleotides, which are needed for DNA synthesis.

2. **(a)** Lactate at C-3, or ethanol in the methyl group, depending on the organism.

   (b) Succinate at either C-2 or C-3.

   (c) Oxaloacetate in the carbonyl carbon and at C-3.

   (d) CO_2 and oxaloacetate, at all carbons. Remember that fumarate is a symmetric molecule; in hydration of the double bond, a hydroxyl group is added to either C-2 or C-3 with equal probability.

3. You cannot conclude that oxaloacetate levels are being depleted. Although the oxaloacetate carbons (including the ^{14}C label) are lost as CO_2 in the citric acid cycle, new carbons are introduced when acetyl-CoA condenses with oxaloacetate. A molecule of oxaloacetate is regenerated after each turn of the citric acid cycle, but this contains two new carbon atoms obtained from the incoming acetyl-CoA.

4. **(a)** 1 mol of glyceraldehyde 3-phosphate produces 16–17 mol of ATP under aerobic conditions (depending on the shuttle used to transfer electrons into the mitochondrion from the NADH formed during glycolysis).

   (b) 1 mol of succinate and 1 mol of pyruvate produce 16.5 mol of ATP under aerobic conditions.

   (c) 8 mol of glucose produce 16 mol of ATP under anaerobic conditions.

   (d) 1 mol of acetyl-CoA produces 10 mol of ATP under aerobic conditions.

   Thus, under the conditions specified, (a) and (b) yield the most energy, (d) the least. However, on a *per mole* basis, (c) produces the least energy.

chapter

17 Fatty Acid Catabolism

STEP-BY-STEP GUIDE

Major Concepts

The insolubility of triacylglycerols poses a problem for the absorption of dietary lipids and for the mobilization of triacylglycerols stored in adipocytes.

Absorption of dietary fats is dependent on the presence of amphipathic bile salts, which disrupt fat globules into small micelles. The triacylglycerols in the micelles are then cleaved by **lipases** into mono-acyl- and diacylglycerols, free fatty acids, and glycerol, which can be absorbed by the intestinal epithelial cells. These components are then reassembled into triacylglycerols and transported in the blood by complexing to lipid-binding proteins, **apolipoproteins,** to form **chylomicrons,** which can form favorable associations with the aqueous environment.

This process of triacylglycerol cleavage and passage of the components into cells is repeated in the capillaries of muscle and adipose tissue. In muscle cells, fatty acids are oxidized for energy; in adipocytes, they are reesterified to triacylglycerols for storage. Stored triacylglycerols are subsequently mobilized when hormonal cues signal a need for metabolic energy. The fatty acids liberated from adipocytes are carried in the bloodstream by serum albumin.

Fatty acids are oxidized in mitochondria.

Mitochondria have two distinct membrane layers: an outer membrane that faces the cytosol and an inner membrane that is thrown into numerous folds called **cristae** (see Fig. 19–1, p. 708). The space between the outer and inner membranes is called the **intermembrane space,** and the interior of the mitochondrion is called the **matrix.** Fatty acids are oxidized in the matrix.

Fatty acids must be transported across the inner mitochondrial membrane.

The fatty acids are first activated by enzymes in the outer mitochondrial membrane, the **acyl–CoA synthetases.** These enzymes catalyze the formation of a thioester linkage between the fatty acid and coenzyme A (CoA-SH) to form a **fatty acyl–CoA,** a reaction coupled to the cleavage of ATP to AMP and PP_i. Passage of the fatty acyl-CoA across the inner membrane requires the assistance of a carrier compound, **carnitine.** The fatty acyl–carnitine ester, formed by the action of **carnitine acyltransferase I,** is carried into the matrix by facilitated diffusion via the **acyl-carnitine/carnitine transporter.** This entry process is rate limiting and commits the fatty acid to oxidation in the mitochondrial matrix.

Oxidation of fatty acids in the mitochondria has three stages.

The first stage is **β oxidation.** This involves oxidation of the fatty acid at the β carbon (C-3), followed by a hydration, an oxidation, and a thiolysis, a series of reactions resulting in the removal of C-1 and C-2 as a molecule of acetyl-CoA. One round of β oxidation (removal of one two-carbon unit) produces one molecule each of $FADH_2$ and NADH. In the second stage of fatty acid oxidation, the acetyl-CoA is further oxidized in the citric acid cycle, producing three NADH, one $FADH_2$, and one GTP. In the third stage, the NADH and $FADH_2$ formed during stages 1 and 2 pass their electrons to O_2 via the electron transfer chain. Counting each $FADH_2$ as equivalent to 1.5 ATP, and each NADH as equivalent to 2.5 ATP; the yield of ATP produced from all three stages of oxidation for an even-numbered, saturated fatty acid containing n carbons is $14(n/2)-4$. To account for the two ATP equivalents used to activate the fatty acid to fatty acyl-CoA, two more ATPs must be subtracted; therefore, the net equation is $14(n/2)-6$.

Oxidation of unsaturated and odd-chain fatty acids requires additional reactions.
For monounsaturated fatty acids, an isomerization reaction changes the cis double bond to a trans; for polyunsaturated fatty acids, an additional reduction reaction is required. These reactions produce a saturated fatty acid that can enter β oxidation. Odd-chain fatty acids proceed through the same pathway as do the even-numbered chains until the final round of β oxidation, which produces acetyl-CoA and a three-carbon compound, propionyl-CoA. Propionyl-CoA is converted to succinyl-CoA in three steps: a carboxylation, a group-transfer reaction, and an isomerization requiring coenzyme B_{12}. Succinyl-CoA is then oxidized in the citric acid cycle.

Fatty acid oxidation is highly regulated.
The carnitine acyltransferase I reaction is inhibited by malonyl-CoA, the first intermediate in the *synthesis* of fatty acids. In addition, two of the β-oxidative enzymes are inhibited when the overall energy levels of the system are high. On a longer time scale, the PPAR family of transcription factors affect genes essential for fatty acid oxidation.

Peroxisomes and glyoxysomes are organelles that carry out some oxidative breakdown of fatty acids.
The oxidation of fatty acids in peroxisomes differs from that in mitochondria. The initial reaction produces a double bond between the α and β carbons, with electrons transferred directly to O_2, producing H_2O_2. No ATP is produced. The H_2O_2 is cleaved by catalase to H_2O and O_2. In glyoxysomes, β-oxidation allows conversion of stored lipids into carbohydrates.

Omega oxidation is another pathway for fatty acid breakdown.
This pathway, which occurs in the endoplasmic reticulum, produces dicarboxylic acids such as succinate. Genetic defects in the enzyme medium-chain acyl-CoA dehydrogenase (MCAD) can produce a serious disease characterized by the inability to oxidize medium-chain fatty acids via β oxidation. It is indicated by high urinary concentrations of medium-chain dicarboxylic acids produced via omega oxidation.

In mammals, an alternative pathway for acetyl-CoA produces ketone bodies.
The key to the fate of acetyl-CoA is the concentration of oxaloacetate, the intermediate that is required for entry of acetyl-CoA into the citric acid cycle and is used in the formation of glucose (gluconeogenesis). When the demand for glucose is high, oxaloacetate levels are reduced. Under these conditions, acetyl-CoA does not enter the citric acid cycle; it is converted to the ketone bodies **acetoacetate, D-β-hydroxybutyrate,** and **acetone** in the liver. These molecules can be transported in the blood to tissues such as muscle and brain, where they are converted back to acetyl-CoA and oxidized via the citric acid cycle.

What to Review

Answering the following questions and reviewing the relevant concepts, which you have already studied, should make this chapter more understandable.

- Review the basic structure of a triacylglycerol (Fig. 10–3).
- Where do lipases cleave triacylglycerols (Fig. 10–15)? Review the properties of fatty acids in aqueous solution (Fig. 2–7).
- The oxidation of fatty acids produces **acetyl-CoA,** which can be further metabolized to produce additional energy for cells. Review how acetyl-CoA enters the citric acid cycle (Fig. 16–7, p. 621). Review the **glyoxylate cycle.** What is the role of the glyoxylate cycle in plant seeds (Fig. 16–22, p. 639)?
- A number of steps in the oxidation of fatty acids result in the production of NADH and $FADH_2$, as well as compounds that can be further metabolized in the citric acid cycle. Be sure you remember how much ATP is generated from reoxidation of NADH and $FADH_2$ (p. 630, Table 16–1). And be sure you remember which steps in the citric acid cycle produce reduced electron carriers and high-energy compounds (Fig. 16–13, p. 630).
- Fatty acids must be transported across mitochondrial membranes in order to be metabolized. Make sure you understand the characteristics of the type of membrane transport called **facilitated diffusion** (pp. 389–395).

Topics for Discussion

Answering each of the following questions, especially in the context of a study group discussion, should help you understand the important points of this chapter.

17.1 Digestion, Mobilization, and Transport of Fats

Dietary Fats Are Absorbed in the Small Intestine

1. How do the chemical properties of lipids affect the way they must be mobilized in biological systems?

2. How are the different types of lipids transported in the blood *to* adipose tissue?

Hormones Trigger Mobilization of Stored Triacylglycerols

3. How are lipids transported in the blood *from* adipose tissue?

4. Why does it make sense that low blood glucose triggers mobilization of fat stores?

Fatty Acids Are Activated and Transported into Mitochondria

5. *Cell Map:* On your cell map, find the acyl-carnitine/carnitine transporter. Fill in the enzymes that activate fatty acids for transport, that attach fatty acyl groups to carnitine, and that reattach acyl groups to CoA in the matrix.

6. What is the energetic "cost" of the formation of a fatty acyl-CoA?

7. What are the roles of the two pools of coenzyme A? How does the acyl-carnitine/carnitine transporter maintain the separation of these pools?

17.2 Oxidation of Fatty Acids

The β Oxidation of Saturated Fatty Acids Has Four Basic Steps

8. *Cell Map:* To the right of the citric acid cycle, enter onto your cell map the four steps of β oxidation. Where do the acetyl-CoA, NADH, and $FADH_2$ go from here?

9. What *types* of reactions make up the four basic steps of β oxidation?

10. What is the overall effect of the first three reactions of β oxidation on the bond between the α and β carbons?

11. At what steps are reduced electron carriers generated in the β oxidation of fatty acids?

12. To which enzyme complexes do the reduced electron carriers pass their electrons?

The Four β-Oxidation Steps Are Repeated to Yield Acetyl-CoA and ATP

13. How many ATPs are generated by the reoxidation of each type of electron carrier?

14. How many ATPs are generated from this first stage (i.e., β oxidation) of the overall process of fatty acid oxidation?

Acetyl-CoA Can Be Further Oxidized in the Citric Acid Cycle

15. Why are *two* ATP equivalents subtracted in the calculation of the net yield of fatty acid oxidation?

Box 17–1 Fat Bears Carry out β Oxidation in Their Sleep

16. By what reaction is water generated during hibernation? How is blood glucose produced?

Oxidation of Unsaturated Fatty Acids Requires Two Additional Reactions

17. What structural property of mono- and polyunsaturated fatty acids prevents oxidation of these compounds by the β-oxidation pathway? What additional *types* of reactions are needed?

18. Is there any difference in the amount of ATP formed by saturated and unsaturated fatty acid oxidation?

Complete Oxidation of Odd-Number Fatty Acids Requires Three Extra Reactions

19. Is there any difference in the amount of ATP formed by even-chain and odd-chain fatty acid oxidation?

20. What two cofactors are necessary in the extra reactions required for odd-chain fatty acid oxidation?

Fatty Acid Oxidation Is Tightly Regulated

21. The processes of fatty acid oxidation and synthesis take place in separate cellular compartments. What molecule acts as the regulatory signal to prevent the catabolic and anabolic processes from occurring simultaneously?

Transcription Factors Turn on the Synthesis of Proteins for Lipid Catabolism

22. In what tissue(s) is PPARα activated in the transition from fetal to neonatal metabolism? During endurance training?

Genetic Defects in Fatty Acyl-CoA Dehydrogenases Cause Serious Disease

23. Why do individuals with a mutation in the MCAD gene have high urinary levels of medium-chain dicarboxylic acids?

Box 17–2 Coenzyme B_{12}: A Radical Solution to a Perplexing Problem

24. How does coenzyme B_{12} illustrate the necessity for trace elements in our diet?

Peroxisomes Also Carry Out β Oxidation

25. Is there any difference in the amount of ATP formed by fatty acid oxidation in a mitochondrion and in a peroxisome?

Plant Peroxisomes and Glyoxysomes Use Acetyl-CoA from β Oxidation as a Biosynthetic Precursor

26. What is the overall role of plant glyoxysomes? Which metabolite connecting β oxidation to the glyoxylate cycle makes this role possible?

The β-Oxidation Enzymes of Different Organelles Have Diverged during Evolution

27. What is a possible advantage of a multifunctional protein with several enzymatic activities?

The ω Oxidation of Fatty Acids Occurs in the Endoplasmic Reticulum

28. What are the oxidation steps that occur in omega oxidation?

Phytanic Acid Undergoes α Oxidation in Peroxisomes

29. Will α oxidation followed by β oxidation of phytanic acid generate more or less ATP than β oxidation of a straight-chain fatty acid of equal length?

17.3 Ketone Bodies

Ketone Bodies, Formed in the Liver, Are Exported to Other Organs as Fuel

30. Where in the cell, and from what metabolite, are ketone bodies made?

31. How and where are ketone bodies used?

Ketone Bodies Are Overproduced in Diabetes and during Starvation

32. How does formation of ketone bodies allow continuation of fatty acid oxidation in the liver?

33. What do starvation and untreated diabetes have in common as a trigger to induce increased production of ketone bodies?

Discussion Questions for Study Groups

- At this stage in your study of biochemistry, you should begin to see connections: using your Cell Map, discuss the interconnections between carbohydrate catabolism and fat catabolism.

SELF-TEST

Do You Know the Terms?

ACROSS

1. Association of hydrophobic lipids with these molecules permits lipid transport in the blood.
4. Organelles in which hydrogen peroxide (H_2O_2) is produced in the first step of β oxidation.
6. Enzyme catalyzing the second step in β oxidation: removal of a double bond by the addition of water. (2 words)
8. Enzyme catalyzing activation of the fatty acid in which ATP is converted to AMP+PP_i. (2 words)
10. Hormone-sensitive ________ lipase links hormonal signaling to mobilization of stored energy in adipose tissue.
13. Enzyme that transfers the fatty acyl group from Co-A to a carrier molecule. (2 words)
15. Alternative name for acyl-CoA acyltransferase, enzyme catalyzing the fourth reaction in β oxidation, removing a two-carbon unit and transferring remainder of the fatty acid to new CoA-SH.
16. In animal cells, fatty acid synthesis is carried out in the
19. Site of β oxidation of fatty acids, citric acid cycle, and electron transfer chain is the mitochondrial ________.
20. Intermediate in synthesis of long-chain fatty acids that inhibits carnitine acyltransferase I, thus regulating fatty acid oxidation at the fatty acid transport step.
21. In the third step of β oxidation, electrons are transferred to NAD^+; NADH then donates its electrons to ________ in the electron-transfer chain in a reaction catalyzed by β-hydroxyacyl-CoA dehydrogenase.
22. First step in β oxidation transfers electrons from fatty acyl-CoA to the FAD prosthetic group of acyl-CoA ________.
23. When oxaloacetate concentration is low, these compounds are formed from excess acetyl-CoA produced by β oxidation of fatty acids. (2 words)
24. The final three-carbon compound generated by oxidation of odd-chain fatty acids is converted from methylmalonyl-CoA to succinyl-CoA by methylmalonyl-CoA ________.
25. ________ oxidation of fatty acids in the endoplasmic reticulum produces dicarboxylic acids such as succinate.

DOWN

2. Before it can be oxidized, the final three-carbon compound generated by oxidation of odd-chain fatty acids is carboxylated by this biotin-containing enzyme. (2 words)
3. ________ lipase: extracellular enzyme in muscle and fat tissues that hydrolyzes triacylglycerols in blood.
5. Oxidation of unsaturated fatty acid produces a fatty acid with a cis double bond that can undergo further β oxidation only after a cis to trans conversion by enoyl-CoA ________.
7. General name for enzyme that hydrolyzes triacylglycerols.
9. Carrier compound required for transport of fatty acids across the inner mitochondrial membrane.
11. Fatty acids, in the form of acyl-carnitine esters, enter the mitochondrial matrix and are transferred to intramitochondrial ________ by carnitine acyltransferase II.

12. Enzyme that catalyzes the production of H_2O and O_2 from the H_2O_2 produced by extramitochondrial β oxidation of fatty acids.

14. The rate-limiting step in fatty acid oxidation is the transport of fatty acids from the ________ space into the mitochondrial matrix.

17. Glycerol ________ converts the glycerol backbone of triacylglycerols to glyceraldehyde 3-phosphate, which can enter the glycolytic pathway.

18. Pathway for conversion of fatty acids to acetyl-CoA.

21. Structures formed by folding of the inner mitochondrial membrane.

Do You Know the Facts?

1. Which of the following describes fatty acid transport into the mitochondrial matrix?

A. It is the rate-limiting step in fatty acid oxidation.

B. It is regulated by [malonyl-CoA].

C. The cytosolic and matrix pools of CoA are distinct and separate.

D. Once fatty acyl groups have entered the matrix, they are committed to oxidation to acetyl-CoA.

E. All of the above are true.

2. Which of the following is true of β oxidation of fatty acids?

A. In a single round, one molecule of $FADH_2$ and one molecule of NADPH are produced.
B. It is the same for both saturated and unsaturated fatty acids.
C. Lipoprotein lipase catalyzes the first step.
D. Fatty acids are oxidized at C-3 to remove a two-carbon unit.
E. It occurs in the intermembrane space of mitochondria.

3. Place the following steps in lipid oxidation in their proper sequence. (Note: this is *not* a complete list of steps.)

(a) Thiolysis

(b) Reaction of fatty acyl-CoA with carnitine

(c) Oxidation requiring NAD^+

(d) Hydrolysis of triacylglycerol by lipase

(e) Activation of fatty acid by joining to CoA

(f) Hydration

In questions 4–8, match the role in fatty acid oxidation and/or mobilization to the appropriate component from the list below.

(a) Bile salt

(b) Serum albumin

(c) ApoC-II

(d) Apolipoprotein

(e) Carnitine

______ **4.** Acts as a "carrier" of fatty acids across the inner mitochondrial membrane.
______ **5.** Acts as a biological detergent, disrupting fat globules into small mixed micelles.
______ **6.** Binds and transports triacylglycerols, phospholipids, and cholesterol between organs.
______ **7.** Activates lipoprotein lipase, which cleaves triacylglycerols into their components.
______ **8.** Binds some fatty acids mobilized from adipocytes and transports them in the blood to heart and skeletal muscle.

9. (a) At which steps in the β oxidation of saturated fatty acids are reduced electron carriers generated?

(b) For each two-carbon increase in the length of a saturated fatty acid chain, how many additional moles of ATP can be formed upon complete oxidation of one mole of the fatty acid to CO_2 and H_2O?

10. What is the net result of the additional reactions required for the β oxidation of unsaturated fatty acids?

11. How do the functions for which plants use the products of triacylglycerol oxidation differ from those in mammals?

12. How does the production of ketone bodies allow continued oxidation of fatty acids? How does this affect the cytosolic coenzyme A concentration?

Applying What You Know

1. Compare the net ATP production in the following situations:
   (a) Complete oxidation of palmitate (16-C, saturated fatty acid).

   (b) Complete oxidation of palmitoleic acid (16-C, monounsaturated fatty acid with a cis double bond between C-9 and C-10).

   (c) Partial oxidation of palmitate to the ketone body acetoacetate.

   (d) Oxidation of palmitate in a liver peroxisome.

ANSWERS

Do You Know the Terms?

ACROSS

1. apolipoproteins
4. peroxisomes
6. enoyl-CoA hydratase
8. acyl-CoA synthetase
10. triacylglycerol
13. carnitine acyltransferase I
15. thiolase
16. cytoplasm
19. matrix
20. malonyl-CoA
21. Complex I
22. dehydrogenase
23. ketone bodies
24. mutase
25. omega

DOWN

2. propionyl-CoA carboxylase
3. lipoprotein
5. isomerase
7. lipase
9. carnitine
11. coenzyme A
12. catalase
14. intermembrane
17. kinase
18. β oxidation
21. cristae

Do You Know the Facts?

1. E

2. D

3. (d), (e), (b), (f), (c), (a)

4. (e)

5. (a)

6. (d)

7. (c)

8. (b)

9. (a) $FADH_2$ is formed in the first step of β oxidation, the oxidation reaction catalyzed by acetyl-CoA dehydrogenase. NADH is formed in the oxidation reaction catalyzed by β-hydroxyacyl-CoA dehydrogenase. Passage of electrons from the $FADH_2$ and NADH through the electron transfer chain produces ATP.

(b) For each extra two-carbon unit oxidized, 14 more ATP molecules are formed. The two oxidations of the β-oxidation pathway produce 1 $FADH_2$ and 1 NADH, which yield 1.5 and 2.5 ATP, respectively, by oxidative phosphorylation. The extra acetyl-CoA, when oxidized via the citric acid cycle, yields another 10 ATP equivalents: 3 NADH, 1 $FADH_2$, and 1 ATP (or GTP).

10. The two additional reactions, an isomerization and a reduction reaction, produce a saturated fatty acid, which can pass through the β-oxidation pathway.

11. *Plants* do not use the oxidation of triacylglycerols to generate metabolic energy in any significant amount; plant mitochondria lack the appropriate enzymes for this process. Triacylglycerols can be degraded to fatty acids, which are activated to their CoA derivatives, then oxidized in peroxisomes and glyoxysomes (found in germinating seeds). This oxidation produces acetyl-CoA, which is routed through the glyoxylate cycle and into gluconeogenesis. The glucose so formed is the precursor of a variety of metabolic intermediates. *Mammals* use triacylglycerol breakdown and fatty acid oxidation to generate large amounts of ATP; this takes place in the mitochondria. Mammals cannot use acetyl-CoA as a precursor of glucose; they lack crucial enzymes of the glyoxylate cycle.

12. The generation of ketone bodies (which contain no CoA moiety) from acetyl-CoA frees up mitochondrial coenzyme A for continued fatty acid oxidation. The separation of the coenzyme A pools of the cytosol and the mitochondrial matrix is maintained by the acyl-carnitine/carnitine transport system. Ketone body production therefore does not affect the cell's cytosolic CoA concentration.

Applying What You Know

1. (a) Complete oxidation of palmitate involves 7 rounds of β oxidation, producing:

- 7 $FADH_2$ and 7 NADH
- 8 acetyl-CoA, which are oxidized in the citric acid cycle to produce 24 NADH, 8 $FADH_2$, and *8 ATP* (from GTP)

Thus, 15 $FADH_2$ and 31 NADH are produced. Electron transfer and oxidative phosphorylation produce:

- 1.5 ATP per $FADH_2$ = *22.5 ATP*
- 2.5 ATP per NADH = *77.5 ATP*

Thus, a total of 22.5 + 77.5 + 8 = *108 ATP* is produced. From this, we must *subtract 2 ATP* because the activation of palmitate to palmitoyl-CoA required 2 ATP equivalents.

- Net energy produced = **106 ATP**

(b) Complete oxidation of palmitoleic acid involves:

- 3 rounds of β oxidation, producing 3 $FADH_2$ and 3 NADH
- 1 round with a "miss" of formation of $FADH_2$ during the acyl-CoA dehydrogenase step; the cis double bond must be converted to trans in order for it to be acted on by enoyl-CoA hydratase. Enoyl-CoA isomerase catalyzes this conversion and allows the rest of this round of β oxidation, producing 1 NADH.
- 3 more rounds of β oxidation, producing 3 $FADH_2$ and 3 NADH
- Formation of 8 acetyl-CoA, which (as above) produces 24 NADH, 8 $FADH_2$, and *8 ATP*

Thus, 14 $FADH_2$ and 31 NADH are produced, which give rise to:

- *21 ATP* and *77.5 ATP*

Thus, a total of 21 + 77.5 + 8 = *106.5 ATP* is produced. Again, subtracting *2 ATP:*

- Net energy produced = **104.5 ATP**

(c) Partial oxidation of palmitate to acetoacetate involves 7 rounds of β oxidation, producing:

- 7 $FADH_2$ and 7 NADH
- 8 acetyl-CoA, which combine to form 4 acetoacetate

Thus, 7 $FADH_2$ and 7 NADH are produced, which give rise to:

- *10.5 ATP* and *17.5 ATP*

Thus, the total is 10.5 + 17.5 = *28 ATP.* Again, subtracting *2 ATP:*

- Net energy produced = **26 ATP**

(d) Oxidation of palmitate in a liver peroxisome involves 7 rounds of β oxidation, producing:

- $FADH_2$, which does not pass its electrons through the respiratory chain to O_2, so *no ATP* is derived from this step; instead, the electrons are passed directly to O_2 to form H_2O_2.
- 7 NADH, which cannot be reoxidized within the peroxisome; NADH is exported to the cytosol, from where it eventually passes its reducing equivalents into mitochondria.
- 8 acetyl-CoA, which are exported to the cytosol as acetate because liver peroxisomes do not contain the enzymes of the citric acid cycle.

Thus, *within the peroxisome itself,* **no ATP** is produced. Assuming no energetic cost for transporting NADH and acetyl-CoA out of the peroxisome and into the mitochondrial matrix, the 7 NADH and 8 acetyl-CoA could eventually produce **95.5 ATP.**

chapter

18 Amino Acid Oxidation and the Production of Urea

STEP-BY-STEP GUIDE

Major Concepts

Dietary proteins are the primary source of biologically useful nitrogen in animals.

Ingested proteins are hydrolyzed to individual amino acids by a series of enzymatic reactions beginning in the stomach and continuing in the small intestine. The enzymes involved in proteolysis are usually secreted into the digestive tract as inactive precursors called **zymogens,** which are then activated by other enzymes; this protects the secretory cells from the action of the proteolytic enzymes. Free amino acids are transported across the intestinal epithelium and carried in the circulatory system to the liver.

The general scheme for the further metabolism of these amino acids involves the transfer of the amino group to α-ketoglutarate, forming glutamate plus an α-keto acid.

These reactions are catalyzed by **aminotransferases,** or **transaminases.** All aminotransferases have a common prosthetic group, **pyridoxal phosphate (PLP),** which acts as a carrier of the amino group and as an electron sink stabilizing the transamination reaction.

The glutamate produced is transported to liver mitochondria and deaminated by glutamate dehydrogenase.

Glutamate dehydrogenase, present only in the mitochondrial matrix, is a complex allosteric enzyme using NAD^+ or $NADP^+$ as the electron acceptor. The enzyme is positively modulated by high [ADP] in the cell, negatively modulated by high [GTP], and catalyzes the deamination of glutamate to α-ketoglutarate and NH_4^+.

Glutamine and alanine transport ammonia formed in other tissues to the liver.

Ammonia produced in extrahepatic tissues can be transported in the blood to the liver in the form of glutamine. This amino acid converted to glutamate and NH_4^+ in liver mitochondria. In active muscle cells the excess pyruvate formed from glycolysis is combined with NH_4^+ to form alanine. This nontoxic carrier of ammonia is then transported in the circulatory system to the liver, where it is reconverted to pyruvate and ammonia. Through gluconeogenesis, pyruvate is a source of additional glucose molecules that can be transported back to and used as a fuel source by muscles.

Nitrogen is excreted as ammonia, urea, or uric acid.

Amino groups resulting from the oxidative degradation of amino acids cannot be released as ammonia because this is highly toxic to cells. Therefore, amino groups are either reused or excreted. Aquatic organisms, including bacteria, that have access to unlimited quantities of water can excrete nitrogen as ammonia because it is diluted in the aqueous surroundings. Terrestrial organisms for which conserving water is especially important excrete nitrogen in the form of **uric acid,** a semisolid. Other terrestrial organisms excrete nitrogen in the form of **urea,** which is less toxic than ammonia but nevertheless can be tolerated only at low concentrations in most biological systems.

Urea is formed from ammonia in a series of reactions called the urea cycle.

The urea cycle begins in the mitochondrial matrix. NH_4^+ is first combined with HCO_3^- to yield **carbamoyl phosphate.** This reaction requires hydrolysis of two molecules of ATP and is essentially irreversible. In the urea cycle itself, the carbamoyl moiety of carbamoyl phosphate is passed to **ornithine,** forming **citrulline,** which passes to the cytosol. Here, citrulline acquires an amino group from **aspartate** to yield **argininosuccinate,** in a reaction requiring ATP. The argininosuccinate is cleaved to **arginine** and **fumarate.**

The fumarate can be shunted to the citric acid cycle (thus linking the two cycles) where it is converted to oxaloacetate; this can be aminated to regenerate aspartate. The arginine is cleaved to form **urea** and ornithine, which reenters the mitochondrial matrix. **(Note that aspartate and ornithine are not consumed in the urea cycle but act as carriers of ammonia in the form of amino groups. The final product is one molecule of urea, which contains two nitrogen atoms and a single oxidized carbon atom.)**

Urea cycle activity is regulated.

On a long-term basis, urea cycle activity can be altered by regulating the synthesis of the urea cycle enzymes. Short-term regulation involves allosteric activation of the first enzyme in the pathway (carbamoyl phosphate synthetase I) by *N*-acetylglutamate.

Deaminated amino acids produce carbon skeletons that can be funneled into the citric acid cycle.

Each of the 20 standard amino acids produces a different carbon skeleton upon deamination. However, these compounds ultimately form only six major products: **pyruvate**, **acetyl-CoA, α-ketoglutarate, succinyl-CoA, fumarate,** or **oxaloacetate.** Although the specific reaction pathways for the degradation of each amino acid differ, they have two types of reactions in common: **transaminations** by enzymes requiring PLP and **one-carbon transfer reactions.** The one-carbon transfers involve one of several cofactors: **biotin,** which transfers CO_2; **tetrahydrofolate,** which usually transfers —HC=O, —HCOH, or sometimes $—CH_3$ groups (covalently linked to N-5 or N-10 of tetrahydrofolate); and ***S*-adenosylmethionine,** which transfers methyl groups, $—CH_3$.

Some amino acids are ketogenic, some are glucogenic, and some are both.

Ketogenic amino acids are degraded to acetoacetyl-CoA and/or acetyl-CoA, which can be converted to ketone bodies. Glucogenic amino acids are degraded to compounds that can be converted into glucose via gluconeogenesis.

What to Review

Answering the following questions and reviewing the relevant concepts, which you have already studied, should make this chapter more understandable.

- Be sure to know the structures of amino acids (p. 75). Be able to see which carbons of their skeletons will be part of citric acid cycle intermediates.
- The carbon skeletons generated by deamination of amino acids can be funneled into the citric acid cycle for energy production. A review of the structures of the intermediates in the citric acid cycle (Fig. 16–7) will help you learn the specific points in the cycle where these carbon skeletons enter.
- What are **ketone bodies** and what is their function (pp. 666–668)? Recall that ketone bodies can be used as an alternative fuel source by muscle, kidney, and brain tissue. Ketone bodies are also produced by the catabolism of amino acids.
- Once again, review the structure of mitochondria (see Fig. 19–1 for the best view). Some of the urea cycle reactions occur in the cytosol and some in the mitochondrial matrix.

Topics for Discussion

Answering each of the following questions, especially in the context of a study group discussion, should help you understand the important points of this chapter.

18.1 Metabolic Fates of Amino Groups

1. What are the fates of proteins degraded in plants? In carnivorous animals? In herbivorous animals?

Dietary Protein Is Enzymatically Degraded to Amino Acids

2. What is the role of the low pH of gastric juice in degrading dietary proteins?

3. Why is it important that proteases, which cleave proteins to their individual amino acid components, are produced and secreted as zymogens?

4. Why is a *series* of enzymes required for degradation of dietary proteins to free amino acids?

Pyridoxal Phosphate Participates in the Transfer of α-Amino Groups to α-Ketoglutarate

5. What are the common characteristics of many aminotransferases?

6. What are the two substrates, and in what order do they react, in the Ping-Pong reaction catalyzed by an aminotransferase?

Glutamate Releases Its Amino Group as Ammonia in the Liver

7. How is the glutamate dehydrogenase reaction an "intersection" of carbon and nitrogen metabolism?

Glutamine Transports Ammonia in the Bloodstream

8. What are the different roles of glutamine and glutamate in amino group metabolism and transport?

Box 18–1 Assays for Tissue Damage

9. Why would you *not* expect to observe measurable quantities of aminotransferases in the blood in the absence of tissue damage?

Alanine Transports Ammonia from Skeletal Muscles to the Liver

10. Why is the use of alanine as a transporter of ammonia from muscle to liver a "kill-two-birds-with-one-stone" solution?

Ammonia Is Toxic to Animals

11. Why would a depletion of cellular ATP be so dangerous in brain cells?

18.2 Nitrogen Excretion and the Urea Cycle

Urea Is Produced from Ammonia in Five Enzymatic Steps

12. Why is the reaction catalyzed by carbamoyl phosphate synthetase I also a "kill-two-birds-with-one-stone" situation?

13. The role of ornithine in the urea cycle is analogous to the role of oxaloacetate in the citric acid cycle. To what citric acid cycle intermediate is citrulline analogous?

14. Which four metabolites involved in the urea cycle must traverse the mitochondrial membrane?

15. *Cell Map:* Add to your cell map the structures and enzymes of the urea cycle below and to the right of the citric acid cycle, near the mitochondrial inner membrane.

The Citric Acid and Urea Cycles Can Be Linked

16. Aspartate is a participant in the urea cycle. From where is its amino group derived?

17. *Cell Map:* Add to your cell map the interconnections between the citric acid and urea cycles as indicated in Figure 18–11.

The Activity of the Urea Cycle Is Regulated at Two Levels

18. What are the short- and long-term strategies for coping with changes in the demand for urea cycle activity?

Pathway Interconnections Reduce the Energetic Cost of Urea Synthesis

19. The urea cycle produces one molecule of urea at the expense at least 1.5 ATPs overall. Why is this a worthwhile, and necessary, investment?

Genetic Defects in the Urea Cycle Can Be Life-Threatening

20. How does dietary administration of aromatic acids remove ammonia from the bloodstream?

18.3 Pathways of Amino Acid Degradation

21. How significant is the contribution of amino acid catabolism to overall energy production in humans?

22. How does the degradation of amino acids feed into the central catabolic processes of the cell?

Some Amino Acids Are Converted to Glucose, Others to Ketone Bodies

23. Why are ketogenic amino acids especially important under conditions of starvation or untreated diabetes?

Several Enzyme Cofactors Play Important Roles in Amino Acid Catabolism

24. Which cofactors involved in amino acid degradation carry which one-carbon groups?

Six Amino Acids Are Degraded to Pyruvate

25. Which reaction pathway(s) will convert the carbon skeleton of alanine to a ketone body? Which will convert alanine to glucose?

Seven Amino Acids Are Degraded to Acetyl-CoA

26. List the amino acids degraded to acetyl-CoA. Note that several degradative pathways share certain reactions, enzymes, and coenzymes.

Phenylalanine Catabolism Is Genetically Defective in Some People

27. Persons with phenylketonuria sometimes exhibit very light coloration of the skin, eyes, and some internal organs where melanin is normally accumulated. Suggest a reason for this.

Five Amino Acids Are Converted to α-Ketoglutarate

28. Which five amino acids are degraded to the citric acid cycle intermediate α-ketoglutarate? Note that they do so through the initial production of glutamate.

Four Amino Acids Are Converted to Succinyl-CoA

29. What are these four amino acids? Note once again the use of certain reactions and enzymes for more than one pathway. Some reactions occurring in the oxidation of odd-chain fatty acids are also used in the degradation of some amino acids to succinyl-CoA.

Box 18–2 Scientific Sleuths Solve a Murder Mystery

30. Why are very limited amounts of isoleucine and valine *not* fatal to individuals with methylmalonic acidemia (MMA)?

Branched-Chain Amino Acids Are Not Degraded in the Liver

31. Where in the body are branched-chain aminotransferases found?

Asparagine and Aspartate Are Degraded to Oxaloacetate

32. What are the overall final products of amino acid degradation?

Discussion Questions for Study Groups

- What types of organisms in what types of habitats excrete nitrogen as ammonia? As uric acid? As urea? Why is it energetically more favorable (if possible) for an organism to excrete nitrogen as ammonia than as urea?

SELF-TEST

Do You Know the Terms?

ACROSS

3. Compound regulating activity of the urea cycle in the short term.

5. Prosthetic group of all aminotransferases; acts as an electron sink. (2 words)

8. Cofactor used in methyl group transfers; reduced form of the vitamin folate.

9. Enzymes that remove α-amino groups from amino acids.

12. Describes amino acids that can be converted to intermediates in gluconeogenesis.

13. Synthesis of a molecule of urea requires four high-energy phosphate groups; two of these come from a single molecule of ATP in the form of _____, which is hydrolyzed to yield two P_i.

15. Cofactor used in methyl group transfers; its synthesis from ATP and methionine releases a *triphosphate.*

17. In the liver glutamate undergoes _____ deamination, catalyzed by glutamate dehydrogenase.

18. The formation of citrulline, the first intermediate in the urea cycle, is catalyzed by _____ transcarbamoylase.

19. Intermediate directly linking the urea and citric acid cycles.

20. The enzyme ______ dehydrogenase, present only in the mitochondrial matrix, can use either NAD^+ or $NADP^+$ as acceptor of reducing equivalents.

22. Transaminases are also referred to as _____ transferases.

24. Keto acid that accepts an amino group to form aspartate, which eventually donates the amino group to urea.

26. Argininosuccinate is cleaved to arginine and fumarate by argininosuccinate ______.

28. Pepsinogen is to pepsin as a ______ is to an active enzyme.

29. Describes organisms with access to (relatively) unlimited supplies of water that use an energy-saving method of nitrogen excretion.

30. In most animals, excess nitrogen produced by amino acid breakdown is transported in the form of this amino acid; contains *two* amino groups.

31. Describes organisms that use $CO(NH_2)_2$ as nitrogenous waste product.

DOWN

1. Cellular location of carbamoyl phosphate synthetase II.

2. Electron acceptor not generally used by enzymes in catabolism but *can* be used by an enzyme involved in the urea cycle.

4. Intermediate that carries NH_4^+ generated in liver mitochondria into the urea cycle. (2 words)

6. Produced, directly or indirectly, by breakdown of the carbon skeletons of seven amino acids.

7. Carbamoyl phosphate synthetase I is located in the ______ of mitochondria.

10. Urea cycle intermediate containing four nitrogen-containing groups, one contributed by aspartate and three by citrulline.
11. Ornithine is to oxaloacetate as citrulline is to ______.
14. Mitochondrial enzyme that "releases" transported amino group from glutamine in the liver.
16. Cofactor used in one-carbon transfers that carries carbon groups in their most oxidized form, CO_2.
21. Describes organisms that use uric acid as nitrogenous waste product, relying on its insolubility to minimize its toxicity.
22. Urea contains two amino groups: one derived from deamination of glutamate in mitochondria; the other from deamination of ______, also generated in mitochondria.
23. Describes amino acids that are degraded to acetyl-CoA or acetoacetyl-CoA.
25. Enzyme that catalyzes regeneration of ornithine and formation of urea.
27. In mitochondria, this molecule condenses with bicarbonate to form an "activated" compound that can enter the urea cycle.

Do You Know the Facts?

1. Indicate whether each of the following is or is not (yes/no) common to the degradation of *all* amino acids.
______ **(a)** Separation of the amino group(s) from the carbon skeleton
______ **(b)** Use of the amino groups for synthesis of new amino acids or other nitrogenous products
______ **(c)** Excretion of excess amino groups in a form appropriate to the organism and its environment
______ **(d)** Passage of the carbon skeletons to the gluconeogenic pathway
______ **(e)** Conversion to α-ketoglutarate
______ **(f)** No associated net gain of energy for the cell
______ **(g)** In mammals, a process occurring mainly in the liver
______ **(h)** Usually, transfer of amino groups to pyruvate or glutamate for transport to the liver

2. Which of the following is a characteristic of many aminotransferase reactions?
A. They have a large, negative $\Delta G'^{\circ}$.
B. The amino group is transferred to an α-keto acid (such as α-ketoglutarate) to form the corresponding amino acid.
C. The amino group is transferred from an ammonia molecule.
D. They are catalyzed by the same enzyme.
E. They require the cofactor *S*-adenosylmethionine.

3. In some respects the urea and citric acid cycles are analogous processes. Ornithine and citrulline have roles that are similar to those of oxaloacetate and what other citric acid cycle intermediate?
A. Acetyl-CoA
B. Citrate
C. CO_2
D. Malate
E. Ammonia

4. Which compound can serve as a direct acceptor of an additional amino group derived from amino acid catabolism?
A. Glutamine
B. Asparagine
C. α-Ketoglutarate
D. Fumarate
E. Glycerol

5. The carbon skeleton produced by deamination of alanine enters the citric acid cycle as:
A. malate.
B. oxaloacetate.
C. fumarate.
D. succinate.
E. acetyl-CoA.

6. *Not* taking into account the NADH generated in the malate dehydrogenase reaction, how many high-energy phosphate bonds are used to form a molecule of urea, starting from ammonia and HCO_3^-?
A. 1
B. 2
C. 3
D. 4
E. 0

7. The citric acid cycle and the urea cycle overlap to form what has sometimes been called the "Krebs bicycle." Which of the following statements is relevant to the interactions between these two metabolic cycles?
   A. Oxaloacetate is converted to aspartate.
   B. Aspartate combines with citrulline to produce argininosuccinate in the cytosol.
   C. Argininosuccinate is cleaved to fumarate and arginine.
   D. Fumarate is a citric acid cycle intermediate.
   E. All of the above are true.
8. Early in its life, a tadpole lives in an aqueous environment and excretes much of its excess nitrogen as ammonia. Once it matures into an adult, the frog spends more time on dry land and becomes ureotelic. Which of the following enzyme activities would be most likely to increase drastically in the adult frog?
   A. Carbamoyl phosphate synthetase I
   B. Glutamine synthetase
   C. Glutaminase
   D. α-Ketoglutarate dehydrogenase
   E. Carboxypeptidase
9. The reactions of the urea cycle occur in two different cellular compartments. Which urea cycle intermediate(s) must be transported across the inner mitochondrial membrane?
   A. Argininosuccinate
   B. Citrulline
   C. Ornithine
   D. A and C
   E. B and C
10. Which cofactor involved in amino acid degradation is correctly matched to the one-carbon group it transfers?
   A. Biotin: CHO
   B. Tetrahydrofolate: CHOH
   C. *S*-adenosylmethionine: CH_2OH
   D. Pyridoxal phosphate: CH_3
   E. Pepsinogen: CO_2
11. Degradation of amino acids yields compounds that are common intermediates in the major metabolic pathways. Explain the distinction between ketogenic and glucogenic amino acids in terms of their metabolic fates. Give three examples each of ketogenic and glucogenic amino acids.
12. Describe the role and reactions of the glucose-alanine cycle. What cofactor is required for this and all other transamination reactions?

Applying What You Know

1. Suppose you are stranded on a desert island and have nothing but protein to eat. Outline *briefly* how your body will use this source of food for its energy needs.
2. Define "zymogen" and describe the role of one zymogen in protein digestion. Why is it important that proteases, which cleave proteins to their individual amino acid components, are produced as zymogens?
3. Why does a mammal go to the energetic expense of making urea from ammonia rather than simply excreting ammonia, as do bacteria?
4. Many Europeans enjoy eating "sweetbreads," which are organ meats, including pancreas tissue. The pancreas is the source of the polypeptide hormone insulin, which is important in regulating carbohydrate metabolism. Why don't such gourmets have problems with their carbohydrate metabolism?

ANSWERS

Do You Know the Terms?

ACROSS

3. *N*-acetylglutamate
5. pyridoxal phosphate
8. tetrahydrofolate
9. transaminases
12. glucogenic
13. pyrophosphate
15. *S*-adenosylmethionine
17. oxidative
18. ornithine
19. fumarate
20. glutamate
22. amino
24. oxaloacetate
26. lyase
28. zymogen
29. ammonotelic
30. glutamine
31. ureotelic

DOWN

1. cytosol
2. NADP
4. carbamoyl phosphate
6. acetyl-CoA
7. matrix
10. argininosuccinate
11. citrate
14. glutaminase
16. biotin
21. uricotelic
22. aspartate
23. ketogenic
25. arginase
27. ammonia

Do You Know the Facts?

1. **(a)** Y; **(b)** Y; **(c)** Y; **(d)** N; **(e)** N; **(f)** N; **(g)** Y; **(h)** Y
2. B
3. B
4. C
5. E
6. D
7. E
8. A
9. E
10. B
11. *Ketogenic* amino acids are catabolized to yield acetyl-CoA or acetoacetyl-CoA, the precursors of ketone bodies. Examples are Trp, Phe, Tyr, Ile, Leu, Thr, and Lys. *Glucogenic* amino acids are catabolized to intermediates that can serve as substrates for gluconeogenesis: that is, pyruvate and/or any of the four- or five-carbon intermediates of the citric acid cycle. Examples are Ala, Asp, Asn, Cys, Arg, Gly, Ser, Trp, Phe, Tyr, Ile, Met, Thr, Val, Gln, His, and Pro.
12. Alanine is the nontoxic form in which ammonia from amino acid catabolism in muscle is transported to the liver. Alanine is formed in muscle by transamination of pyruvate, produced from the breakdown of glucose in glycolysis; glutamate is the amino group donor. In the liver, alanine is reconverted to pyruvate by transamination; its amino group is eventually converted to urea. The pyruvate is converted to glucose by gluconeogenesis then exported back to the muscle. The necessary cofactor for the transaminations is PLP.

Applying What You Know

1. On the desert island your body will: (a) Break down ingested proteins to their constituent amino acids. (b) Remove or transaminate amino groups. (c) Synthesize needed proteins and other compounds that incorporate nitrogen. (d) Use excess amino acid carbon skeletons for energy (they enter the citric acid cycle at various points), *or* for gluconeogenesis (for export of glucose to specific tissues that require glucose as fuel), *or* for production of ketone bodies (also for export to specific tissues.)
2. A zymogen is an inactive form of an enzyme, which can be activated by proteolytic cleavage. The pancreatic enzymes chymotrypsinogen, trypsinogen, and procarboxypeptidases A and B are inactive (zymogen) forms of proteases, which are activated by proteolytic cleavage after their release into the small intestine. Formation and secretion of proteases as zymogens protects the secretory cells from being digested by the enzymes they produce and secrete.
3. Because bacteria are unicellular organisms they release their waste nitrogen, in the form of ammonia,

directly into the surrounding medium. The excreted ammonia is immediately diluted to nontoxic levels (1 g of nitrogen, in the form of ammonia, requires 300–500 mL of H_2O to dilute it to nontoxic levels). The ammonia produced by amino acid catabolism in mammals cannot be sufficiently diluted in the tissues and the blood to avoid accumulation to toxic levels. Dissolved urea is much less toxic than ammonia (about 50 mL of H_2O is required to dilute 1 g of nitrogen, in the form of urea, to nontoxic levels).

4. The biological activity of insulin (a polypeptide) is destroyed by the low pH in the stomach and by proteases in the small intestine. Even if insulin were not degraded to its constituent amino acids in the intestine, it could not enter the bloodstream; the epithelial cells that line the intestine transport (absorb) free amino acids, not whole polypeptides or proteins.

chapter

19 Oxidative Phosphorylation and Photophosphorylation

STEP-BY-STEP GUIDE

Major Concepts

Oxidative phosphorylation is the enzymatic synthesis of ATP coupled to the transfer of electrons to molecular oxygen.

This process, occurring in the inner mitochondrial membrane of eukaryotes and in the plasma membrane of prokaryotes, involves the transfer of electrons from NADH and $FADH_2$ to O_2 via a series of protein complexes and coenzymes. The energy for ATP production is ultimately driven by this electron flow.

The mitochondrial respiratory chain is an ordered array of electron carriers arranged in complexes.

Electrons flow through this chain in a series of steps, from carriers of relatively low, to relatively high, E'°. The flow is spontaneous, but the serial transfer from one carrier to another ensures that the electron transfer occurs in a controlled and energetically useful fashion.

Complex I is a transmembrane protein complex of the inner membrane.

The complex is positioned so that it interacts with NADH produced in reactions catalyzed by enzymes located in the matrix. Reducing equivalents are transferred from NADH, first to the **FMN** prosthetic group and then to the **Fe-S** centers of Complex **I.** The complex then passes its electrons to **ubiquinone (Q).** Complex I functions as a **proton pump,** moving protons from the matrix to the intermembrane space.

Complex II, or succinate dehydrogenase, transfers electrons from succinate to the complex's covalently bound FAD and then to its Fe-S centers.

Electrons that first enter the electron transfer chain at Complex II are passed to Q. However, Complex II does not act as a proton pump, and no protons are pumped into the intermembrane space. Other proteins that transfer electrons to Q include **ETFP,** which transfers electrons from the $FADH_2$ generated by fatty acid oxidation in the matrix to Q. The production of dihydroxyacetone phosphate from glycerol 3-phosphate in the cytosol catalyzed by **glycerol 3-phosphate dehydrogenase** also results in the transfer of electrons to Q. Because NADH cannot cross the inner mitochondrial membrane, this latter reaction pathway shuttles reducing equivalents generated in the cytosol to the inner mitochondrial membrane, where they can be used to generate ATP. (The malate-aspartate shuttle can also be used to transport reducing equivalents from cytosolic NADH into the matrix.)

Complex III transfers electrons from QH_2 to cytochrome c.

The protons generated when QH_2 is reoxidized to Q via the Q-cycle are released into the intermembrane space, further elevating the $[H^+]$ of the intermembrane space relative to the mitochondrial matrix.

Cytochrome c *is a mobile protein that "shuttles" electrons, in this case carrying them from Complex III to Complex IV.*

Complex IV contains **cytochromes *a* and *a*$_3$,** which accept electrons from cytochrome *c* and transfer them to oxygen, forming H_2O. In the process, additional protons are pumped from the matrix to the intermembrane space.

The chemiosmotic model explains in molecular terms how the proton gradient generated by the flow of electrons through the respiratory chain drives the synthesis of ATP.

A chemical (ΔpH) and an electrical ($\Delta\psi$) component make up the **proton-motive force.** ATP is synthesized from ADP and P_i by **ATP synthase** in the inner mitochondrial membrane. This enzyme is made up of two complexes: $\mathbf{F_o}$ is a membrane-spanning protein complex, and $\mathbf{F_1}$ is located on the matrix side of the inner membrane. Because the inner membrane is impermeable to $\mathbf{H^+}$**,** the only path for protons to reenter the matrix is through the pore formed by the F_o subunit of ATP synthase. ADP is phosphorylated by the enzymatic activity of the F_1 complex, driven by the proton-motive force.

The proton-motive force also drives other cell activities.
The transfer of electrons to O_2 and concomitant pumping of H^+ into the intermembrane space accounts for most of the ATP produced in aerobic organisms. The proton gradient that is established also provides the energy for the active transport of some molecules and ions across membranes, for the generation of heat in certain tissues, and for the rotary motion of bacterial flagella.

In photosynthesis, light energy produces electron flow that is coupled to the phosphorylation of ADP to ATP.
The critical difference between oxidative phosphorylation and photophosphorylation is that in the former process, the energy needed to drive the phosphorylation of ADP is provided by the input of a good electron donor (NADH or $FADH_2$), whereas in the latter process the energy provided by light *creates* a good electron donor (in higher plants, P680* and P700*).

The photosystems of plants contain pigment molecules that absorb photons, then release the absorbed energy in a controlled fashion.
Plant photosystems are located in the inner membranes of chloroplasts. These inner membranes, like the inner membrane of mitochondria, are relatively impermeable and form saclike structures called **thylakoids.** Chloroplasts also have an outer membrane that is permeable to small molecules and ions. In **photosystem II, chlorophyll** and **accessory pigments** absorb light energy and transfer this energy to the **reaction center, P680.** The excited reaction center (P680*) loses an electron, which is transferred through a series of electron carriers. The electrons lost by P680* are replaced by electrons from H_2O, resulting in the production of O_2. In **photosystem I,** the excited **P700*** reaction center also donates an electron which is passed through a series of carriers, ultimately reducing $NADP^+$ to NADPH. The electron lost by P700* is replaced by an electron from the soluble protein **plastocyanin.**

Electrons from photosystem II are transferred to photosystem I by the cytochrome b_6f complex.
The pathway of electron flow from photosystem II to photosystem I is diagramed by the **Z scheme.** In this cooperative process, H^+ ions are pumped across the **thylakoid** membrane into the lumen of the thylakoid. The proton gradient generated is analogous to that generated across the inner mitochondrial membrane in oxidative phosphorylation but is oriented in the "opposite" direction. The ATP synthase of thylakoids is very similar to the complex in mitochondria.

ATP can be generated by photosystem I without production of NADPH or O_2 in a process called cyclic photophosphorylation.
In a "detour" from the noncyclic route from ferredoxin to NADPH, electrons loop back to the cytochrome b_6f complex. Protons are pumped across the thykaloid membrane by the cytochrome b_6f complex, generating a proton gradient that can be used to synthesize ATP.

Photosynthesis can occur with alternative hydrogen donors.
Hydrogen sulfide (H_2S) or reduced organic compounds (depending on the organism) can be substituted for H_2O in the general equation for photosynthesis:

$$2\,H_2D + CO_2 \xrightarrow{\text{light}} (CH_2O) + H_2O + 2D$$

What to Review

Answering the following questions and reviewing the relevant concepts, which you have already studied, should make this chapter more understandable.

- What is indicated by a low or high standard reduction potential, E'° (pp. 514–516)? If you had a mixture consisting only of ferredoxin, cytochrome *c*, and oxygen, which would be the final electron acceptor?
- This chapter deals with the fate of the reducing equivalents carried by NADH and $FADH_2$. In which reaction(s) of the citric acid cycle (Chapter 16) and the β oxidation of fatty acids (Chapter 17) are NADH and $FADH_2$ produced?
- ATP synthases are members of a group of proteins known as **F-type ATPases.** What are the characteristics of these enzymes (Figs. 11–39, 11–40)?
- The glycolytic reaction catalyzed by glyceraldehyde 3-phosphate dehydrogenase provides a model of an oxidation coupled to a phosphorylation. The

product of this reaction led researchers to look for a similar chemical intermediate in mitochondrial oxidation coupled to phosphorylation. What are the reactant(s) and product(s) of the glyceraldehyde 3-phosphate dehydrogenase reaction (pp. 535–536)?

- **NADH** is a carrier of reducing equivalents. Review its structure in order to understand why the NADH molecule cannot be transported across the inner mitochondrial membrane (p. 517).

Topics for Discussion

Answering each of the following questions, especially in the context of a study group discussion, should help you understand the important points of this chapter.

1. What are the three fundamental similarities between oxidative phosphorylation and photophosphorylation? Keeping these similarities in mind (and noting further ones) as you read this chapter will help make overall sense of many of the details.

OXIDATIVE PHOSPHORYLATION

19.1 Electron-Transfer Reactions in Mitochondria

2. Why does it make physiological sense that heart mitochondria have more sets of electron transfer components than liver mitochondria?

3. How does the structure of the mitochondrion provide for physical separation of metabolic processes in the cell?

Electrons Are Funneled to Universal Electron Acceptors

4. Categorize the electron carriers NAD^+, $NADP^+$, FMN, and FAD according to the number of hydrogen atoms and/or electrons they carry and the type of association with dehydrogenases.

5. What is the cellular role of NADH? Of NADPH?

Electrons Pass through a Series of Membrane-Bound Carriers

6. As you proceed through this section, categorize each electron carrier in the respiratory chain according to which of the four types of electron transfer it accomplishes.

7. Which chemical characteristics of ubiquinone make it uniquely useful in the electron transfer chain?

8. What are the similarities among cytochromes, iron-sulfur proteins, and flavoproteins as electron carriers? What are the differences?

9. What proteins have you already encountered that contain heme prosthetic groups?

10. How was the order of electron carriers in the mitochondrial inner membrane determined?

Electron Carriers Function in Multienzyme Complexes

11. Categorize each of the mitochondrial electron-carrier complexes in terms of the overall reaction it catalyzes and whether or not it acts as a proton pump. Use the subscripts "P" and "N" to keep track of the location of the protons.

12. By which route do electrons from the β oxidation of fatty acids enter the respiratory chain? What is the route from glycerol 3-phosphate oxidation?

13. What is the net equation (and the net effect) of the Q cycle?

Mitochondrial Complexes May Associate in Respirasomes

14. Based on your biochemical knowledge, what kinds of interactions would you guess act to hold together the complexes in a respirasome?

The Energy of Electron Transfer Is Efficiently Conserved in a Proton Gradient

15. Why is the *actual* free-energy change of the $NADH + H^+ + \frac{1}{2}O_2 \longrightarrow H_2O + NAD^+$ reaction not the same as the standard free-energy change?

16. What are the two components of the proton-motive force in mitochondria?

Reactive Oxygen Species Are Generated during Oxidative Phosphorylation

17. What do the enzymes superoxide dismutase and glutathione peroxidase do? Why are they so critical to maintenance of cellular health?

Plant Mitochondria Have Alternative Mechanisms for Oxidizing NADH

18. What happens to the energy in NADH in plant mitochondria when NAD^+ must be regenerated?

19. *Cell Map:* Find in the mitochondrion the components of the electron transfer chain. Show where electrons from NADH and $FADH_2$ enter; indicate where protons are pumped.

Box 19–1 Hot, Stinking Plants and Alternative Respiratory Pathways

20. What are the three routes for entry of electrons into the respiratory pathway that result in the generation of heat rather than ATP?

19.2 ATP Synthesis

21. What is the mechanism of uncoupler action? What would occur if DNP were added to a cyanide-treated system as described in Figure 19–20 (a)?

22. How are both components of the proton-motive force induced in the experimental setup shown in Figure 19–22?

ATP Synthase Has Two Functional Domains, F_0 and F_1

23. Why are both F_1 and F_o necessary for ATP synthesis?

24. How would the protein purification protocols for F_1 and F_0 differ?

ATP Is Stabilized Relative to ADP on the Surface of F_1

25. The free energy change for ATP synthesis with purified F_1 acting as the catalyst is close to zero. From where does the energy come to drive the equilibrium of the reaction toward the formation of product?

The Proton Gradient Drives the Release of ATP from the Enzyme Surface

26. Where is the "hill" of activation energy in the ATP synthase reaction? (Hint: draw a reaction coordinate diagram of the reaction process.)

Each β subunit of ATP Synthase Can Assume Three Different Conformations

27. Which subunit(s) of the ATP synthase complex contain the catalytic site(s) for ATP synthesis?

Rotational Catalysis Is Key to the Binding-Change Mechanism for ATP Synthesis

28. What are three conformations that the β subunits can assume?

29. Which subunit of F_0 is postulated to act like a drive shaft?

30. Which other allosteric protein that you have encountered shows alterations in one ligand-binding site depending on the ligand-binding state of other protein subunits?

Chemiosmotic Coupling Allows Nonintegral Stoichiometries of O_2 Consumption and ATP Synthesis

31. Why is it so technically difficult to measure P/O ratios? To measure proton fluxes?

32. How many protons are pumped out per NADH? Per succinate? How many protons must flow in to produce one ATP? Finally, after all these metabolism chapters, the math becomes clear for the P/O ratios of 2.5 with NADH as the electron donor and 1.5 for succinate!

33. *Cell Map:* Find in the mitochondrion an ATP synthase; indicate the direction of H^+ movement.

The Proton-Motive Force Energizes Active Transport

34. Which component of the proton-motive force is dissipated by the adenine nucleotide translocase? Which component is dissipated by the phosphate translocase?

35. *Cell Map:* Find the two translocases to your map; indicate what enters and exits.

Shuttle Systems Indirectly Convey Cytosolic NADH into Mitochondria for Oxidation

36. What characteristics of NADH itself and of the NADH dehydrogenase make it necessary to have shuttle systems for transporting cytosolic-reducing equivalents into the matrix?

37. What are the similarities and differences between the two shuttle systems in terms of tissue location, number of ATPs made, and involvement of membrane transport proteins?

38. *Cell Map:* Indicate on the two shuttle systems provided on the map how reducing equivalents from cytosolic NADH are transported into the respiratory chain.

19.3 Regulation of Oxidative Phosphorylation

Oxidative Phosphorylation Is Regulated by Cellular Energy Needs

39. Why is the cellular concentration of ATP maintained at such a steady level? Why doesn't the cell "store" ATP just as it stores polysaccharides and lipids?

An Inhibitory Protein Prevents ATP Hydrolysis during Hypoxia

40. How is IF_1 regulated by pH?

Hypoxia Leads to ROS Production and Several Adaptive Responses

41. How do all of the changes mediated by HIF-1 work together to reduce ROS formation?

ATP-Producing Pathways Are Coordinately Regulated

42. Figure 19–33 may be one of the most useful in the textbook. Make sure all the accelerations and inhibitions of the various metabolic steps by ADP, ATP, NAD^+, and NADH make sense to you.

19.4 Mitochondria in Thermogenesis, Steroid Synthesis, and Apoptosis

Uncoupled Mitochondria in Brown Adipose Tissue Produce Heat

43. Do you think brown adipose tissue mitochondria have sufficient molecules of thermogenin so that no ATP is formed by oxidative phosphorylation in these specialized adipose cells? Why or why not?

Mitochondrial P-450 Oxygenases Catalyze Steroid Hydroxylations

44. How is the synthesis of steroid hormones different from other biosynthetic reactions you have studied?

Mitochondria Are Central to the Initiation of Apoptosis

45. What are the two roles of cytochrome *c* in the eukaryotic cell?

19.5 Mitochondrial Genes: Their Origin and the Effects of Mutations

Mitochondria Evolved from Endosymbiotic Bacteria

46. During operation of bacterial electron transfer chains, into what space are protons pumped? Does this create a problem in terms of "retrieval" of these protons for energy generation?

Mutations in Mitochondrial DNA Accumulate throughout the Life of the Organism

47. Why is the mitochondrial genome particularly susceptible to accumulation of defects?

Some Mutations in Mitochondrial Genomes Cause Disease

48. Why are neurons especially vulnerable to defects in their ATP-generating systems?

Diabetes Can Result from Defects in the Mitochondria of Pancreatic β Cells

49. What is the mechanistic connection between [ATP] and release of insulin from β cells?

PHOTOSYNTHESIS: HARVESTING LIGHT ENERGY

50. What is the overall equation for photosynthesis in vascular plants?

19.6 General Features of Photophosphorylation

51. Which processes of photosynthesis occur only in the light? Which processes do not require light?

Photosynthesis in Plants Takes Place in Chloroplasts

52. Do plants have mitochondria? Explain.

Light Drives Electron Flow in Chloroplasts

53. What is the biological electron acceptor in chloroplasts that corresponds to the acceptor "A" in the Hill equation?

54. What is the ultimate electron donor and the ultimate electron acceptor in chloroplasts? In mitochondria? In each organelle, is energy necessary as an input to the process, or is it produced as a consequence?

19.7 Light Absorption

55. Why do you think that infrared and ultraviolet light are not useful for photosynthesis?

Chlorophylls Absorb Light Energy for Photosynthesis

56. What structural property of the photopigments accounts for their ability to absorb light?

57. Why are plants green?

Accessory Pigments Extend the Range of Light Absorption

58. What advantage would a plant gain by having several different types of light-absorbing pigments?

Chlorophyll Funnels the Absorbed Energy to Reaction Centers by Exciton Transfer

59. What is the series of reactions that occurs to bring about electric charge separation in the reaction-center chlorophyll? It is complicated, but very important to understand.

19.8 The Central Photochemical Event: Light-Driven Electron Flow

Bacteria Have One of Two Types of Single Photochemical Reaction Center

60. What is the end result of the electron transfer process through the Type II reaction center?

61. What additional product (over the Type II reaction center) does the Type I reaction center produce?

Kinetic and Thermodynamic Factors Prevent the Dissipation of Energy by Internal Conversion

62. How are diffusion and random collisions minimized in reaction centers?

In Plants, Two Reaction Centers Act in Tandem

63. What is the connecting protein between photosystems II and I?

64. What does the "Z" shape of the "Z scheme" actually represent?

65. How many photons are required to oxidize a molecule of water?

66. Which component of photosystem II finally receives the four electrons abstracted from water?

Antenna Chlorophylls Are Tightly Integrated with Electron Carriers

67. Why must the components of the PS1 be so closely aligned?

The Cytochrome b_6f Complex Links Photosystems II and I

68. What are the similarities and differences between the cytochrome b_6f complex of chloroplasts and Complex III of mitochondria?

Cyclic Electron Flow between PSI and the Cytochrome b_6f Complex Increases the Production of ATP Relative to NADPH

69. Under what cellular circumstances does cyclic electron flow rather than noncyclic electron flow occur?

State Transitions Change the Distribution of LHCII between the Two Photosystems

70. Where in the thylakoid membrane is PSII located? Where are PSI and ATP synthase? Where is the cytochrome b_6f complex?

Water Is Split by the Oxygen-Evolving Complex

71. What is the stoichiometry of number of photons absorbed to number of electrons "boosted" in each photosystem?

19.9 ATP Synthesis by Photophosphorylation

A Proton Gradient Couples Electron Flow and Phosphorylation

72. In order for this coupling to occur, why must the thylakoid membrane, like the inner mitochondrial membrane, be impermeable to protons?

The Approximate Stoichiometry of Photophosphorylation Has Been Established

73. What component of the proton motive force is most important in chloroplasts? Why?

74. What is the overall equation for noncyclic photophosphorylation?

The ATP Synthase of Chloroplasts Is Like That of Mitochondria

75. On which side of the thylakoid membrane is the CF_1 protein complex? How is this different from and similar to the placement of the mitochondrial F_1?

76. *Cell Map:* In the chloroplast, place the components of the proton and electron circuits of thylakoids. (The most useful figure for this is 19–63.)

19.10 The Evolution of Oxygenic Photosynthesis

Chloroplasts Evolved from Ancient Endosymbiotic Bacteria

77. What characteristics of chloroplasts link them to their probable photosynthetic prokaryotic ancestors?

78. Can photosynthetic bacteria produce ATP? NADPH? O_2? Explain.

79. What are the "double roles" played by cytochrome c_6 and cytochrome b_6f?

In Halobacterium, *a Single Protein Absorbs Light and Pumps Protons to Drive ATP Synthesis*

80. How is the phototransducing machinery of *H. salinarum* dissimilar to that of cyanobacteria and plants?

SELF-TEST

Do You Know the Terms?

ACROSS

2. An electrochemical _____-_____ force is used to drive the synthesis of ATP in oxidative phosphorylation and photosynthesis.
5. Passes an electron to an Fe-S protein in PSI.
6. Compound that supplies electrons to replace those "lost" by photo-system II in noncyclic photophosphorylation.
8. Electrons are transferred from ferredoxin to $NADP^+$ by the enzyme ferredoxin-$NADP^+$ _____.
10. The catalytic portion of Complex I of the respiratory chain. (2 words)
13. _____ molecules: pigments that absorb light of wavelengths other than those absorbed by chlorophylls *a* and *b*, thus "grabbing" stray light energy that would otherwise be missed.
15. Photosynthetic pigments are located in the _____ membrane of chloroplasts; analogous to the cristae of mitochondria.
17. Involved in mitochondrial electron transfers, this compound relies, in part, on its mobility within membranes for its function.
18. Only enzyme of the citric acid cycle that is membrane-bound; directly links the cycle to the electron transfer chain. (2 words)
19. The _____-_____ shuttle translocates reducing equivalents from cytosolic NADH into the mitochondrion, at no cost in terms of ATP.
24. Iron-sulfur protein that can transfer its electron *either* to the cytochrome b_6f complex and then P700 *or* to a flavoprotein and then $NADP^+$.
25. Mobile electron transfer protein that links photosystem II and photosystem I.
26. The _____ model explains how a proton gradient is used to drive ATP synthesis.
27. Diagram of the interaction between photosystem I and photosystem II.
28. The _____ complex interacts with cytochrome *c;* a site of proton pumping.
29. Respiratory complex that transfers electrons to molecular oxygen. (2 words)

DOWN

1. The _____ complex interacts with the quinone PQ_B; a site of proton pumping.
2. Electron transfer compound receiving its electrons directly from $Pheo^-$.
3. Photosystem _____ is responsible for the production of NADPH. (word)
4. Adenine nucleotide _____ is required for cells to use the ATP generated inside mitochondria by oxidative phosphorylation.
5. Without this *transmembrane* transport system, ATP synthase would quickly run out of substrate. (2 words)
7. P680* and P700* are excellent _____ in photosystems II and I, respectively. (2 words)

9. Causes the transport of ions across a membrane, down their concentration gradient; can increase the intracellular levels of calcium, for example.
11. Net production of carbohydrate occurs during the carbon-fixation, or _____ reactions of photosynthesis.
12. ATP and NADPH are produced in the _____-dependent reactions of photosynthesis.
14. Photosystem _____ is involved only in noncyclic photophosphorylation. (word)
16. Enzyme complex containing F_1 and F_o.
20. Location in chloroplast where carbon fixation takes place.
21. Substance involved in electron transfer in photophosphorylation; receives an electron from P680*.
22. Proteins present in mitochondria, chloroplasts, and bacterial membranes; contain iron-containing prosthetic groups.
23. Prosthetic group involved in electron transfers; bound to proteins through Cys residues. (2 words)

Do You Know The Facts?

1. List (or construct a diagram showing) the sequence of electron transfer complexes and mobile electron carriers that are associated with the inner mitochondrial membrane. (Be sure to include the original electron donor and the ultimate electron acceptor.) Which has the highest, and which the lowest, E'°?

2. Indicate whether each of the following statements about the mitochondrial electron transfer chain and oxidative phosphorylation is true or false.

_____ **(a)** NADH dehydrogenase complex, cytochrome bc_1 complex, and cytochrome oxidase are transmembrane proteins.
_____ **(b)** Synthesized ATP must be transported into the intermembrane space before it can enter the cytosol.
_____ **(c)** Cytochrome *c* and the F_1 subunit of ATPase are peripheral membrane proteins.
_____ **(d)** Complexes I, II, III, and IV all are proton pumps.
_____ **(e)** Ubiquinone is a hydrophilic molecule.
_____ **(f)** Ubiquinone and the F_o subunit of ATP synthase are peripheral membrane proteins.
_____ **(g)** The final electron acceptor is H_2O.

3. Which of the following experimental observations would *not* support the chemiosmotic model of oxidative phosphorylation?

A. If mitochondrial membranes are ruptured, oxidative phosphorylation cannot occur.
B. Raising the pH of the fluid in the intermembrane space results in ATP synthesis in the matrix.
C. Transfer of electrons through the respiratory chain results in formation of a proton gradient across the inner mitochondrial membrane.
D. The orientation of the enzyme complexes of the electron transfer chain results in a unidirectional flow of H^+.
E. Radioactively labeled inorganic phosphate is incorporated into cytosolic ATP only in the presence of an H^+ gradient across the inner mitochondrial membrane.

4. The proton-motive force generated by the electron transfer chain:

A. includes a pH-gradient component.
B. includes an electrical-potential-gradient component.
C. is used for active transport processes.
D. is used to synthesize ATP.
E. has all of the above characteristics.

5. Where do the electrons from NADH and $FADH_2$, respectively, enter the electron transfer chain? Why must they enter at different sites? What effect does this have on the number of ATPs produced from reoxidation of each carrier?

6. Why do various cell types differ in the maximum number of ATPs they can produce per molecule of glucose aerobically oxidized?

7. Of all the components of the "Z scheme," which has the lowest $E'°$ (i.e., is the best reducing agent)?

A. P700
B. P700*
C. O_2
D. H_2O
E. NADPH

8. Which of the following statements about photosystem II is correct?

A. It is located in the inner mitochondrial membrane.
B. It contains the electron carrier with the most negative $E'°$ in the entire photosynthetic system.
C. It contains an Mn-containing complex that splits water.
D. Its final electron acceptor is NADPH.
E. H_2O is the only electron donor capable of regenerating P680 from P680*.

9. Which of the following statements about cyclic photophosphorylation and noncyclic photophosphorylation is correct?

A. Cyclic photophosphorylation involves only photosystem II and produces only ATP; noncyclic photophosphorylation involves photosystems I and II and produces only ATP.
B. Both pathways liberate oxygen.
C. Both pathways involve photosystems I and II.
D. Cyclic photophosphorylation reduces $NADP^+$ and liberates oxygen; noncyclic photophosphorylation reduces $NADP^+$ but does not liberate oxygen.
E. Noncyclic photophosphorylation reduces $NADP^+$ and liberates oxygen; cyclic photophosphorylation produces ATP but does not liberate oxygen.

10. For each of the following statements, indicate whether it is true of only chloroplasts, only mitochondria, or both.

	Chloroplasts	Mitochondria	Both
(a) One source of electrons is NADH.	____	____	____
(b) Electron transfer leads to establishment of a proton gradient.	____	____	____
(c) The organelle requires a system of intact membranes to generate ATP.	____	____	____
(d) The organelle contains cytochromes and flavins in its electron transfer chain.	____	____	____
(e) The final electron acceptor is $NADP^+$.	____	____	____

11. Some photosynthetic prokaryotes use H_2S, hydrogen sulfide, instead of water as their photosynthetic hydrogen donor. How does this *change* the ultimate products of photosynthesis?

A. Carbohydrate (CH_2O) is not produced.
B. H_2O is not produced.
C. Oxygen is not produced.
D. ATP is not produced.
E. The products do not change.

Applying What You Know

1. Oxidative phosphorylation and photophosphorylation resemble each other in certain respects. Describe the ways in which the two processes are similar, then list the significant differences.

2. Artificial, but functional, electron transfer systems can be made in the lab by building artificial membrane-bound vesicles. This is done by combining detergent-solubilized, purified respiratory complexes and membrane lipids. When the mixture is dialyzed to remove the detergent, liposomes spontaneously form that contain the protein complexes integrated into the "membrane." The central cavity of the liposome can be made to contain certain molecules in aqueous solution; the surrounding medium can also be manipulated.

   Using this protocol, you create the following electron transfer systems in liposomes, containing the listed set of components (not necessarily in their functioning order) along with the specified initial electron donors. Place the components in their correct functional sequence and indicate the final electron acceptor in each case.

   (a) NADH as initial electron donor; Q and Complexes I, III, and IV in the liposomes; oxygen is present.

   (b) NADH as initial electron donor; Complexes I, II, and IV in the liposomes; oxygen is present.

   (c) Succinate as initial electron donor; Q, cytochrome *c*, and Complexes II, III, and IV in the liposomes; oxygen is present.

3. **(a)** If the orientation of the mitochondrial ATP synthase were reversed so that the F_1 unit were on the opposite side of the inner mitochondrial membrane, and assuming that nothing else is changed in the cell, what would be the consequences to the cell? Explain.

   (b) What would be the consequences if just the F_o complex were flipped within its membranous surroundings, so that its normally matrix-facing surface faced the intermembrane space? Assume the F_1 complex is still on the matrix side.

4. You have been given three different species of bacteria by your research advisor. You are told that one species lives by fermentation and is a facultative anaerobe; one lives by oxidation of glucose via the citric acid cycle and oxidative phosphorylation; and one lives by photosynthesis. All are capable of consuming glucose. You manage, somehow, to mix up the three bacterial cultures. Without resorting to any biochemical tests (which would require help from a grouchy graduate student), how can you use *different laboratory growth conditions* to determine which culture is which?

ANSWERS

Do You Know the Terms?

ACROSS

2. proton-motive
5. phylloquinone
6. water
8. oxidoreductase
10. NADH dehydrogenase
13. antenna
15. thylakoid
17. ubiquinone
18. succinate dehydrogenase
19. malate-aspartate
24. ferredoxin
25. plastocyanin
26. chemiosmotic
27. Z scheme
28. cytochrome bc_1
29. cytochrome oxidase

DOWN

1. cytochrome b_6f
2. plastoquinone
3. one
4. translocase
5. phosphate translocase
7. electron donors
9. ionophore
11. dark
12. light
14. two
16. ATP synthase
20. stroma
21. pheophytin
22. cytochromes
23. Fe-S center

Do You Know the Facts?

1. The sequence of complexes and mobile carriers is as follows:

(a) Complex I, NADH dehydrogenase: transfers electrons from NADH (lowest E'°) to Q.
(b) Ubiquinone (Q)
or
(a) Complex II, succinate dehydrogenase: transfers electrons from $FADH_2$ to Q.
(b) Ubiquinone
then
(c) Complex III, cytochrome bc_1 complex: transfers electrons from Q to cytochrome *c*.
(d) Complex IV, cytochrome oxidase: transfers electrons from cytochrome *c* to O_2 (highest E'°).

2. **(a)** T; **(b)** T; **(c)** T; **(d)** F; **(e)** F; **(f)** F; **(g)** F

3. B

4. E

5. Electrons from NADH enter at the NADH dehydrogenase complex, Complex I. Electrons from the $FADH_2$ of FAD-linked enzymes, which can be derived from several different reactions, enter through Complex II (succinate dehydrogenase), glycerol 3-phosphate dehydrogenase, or ETFP. All of these enzymes transfer electrons to Q. The E'° of $FADH_2$ in each of these enzymes is higher than that of NADH, so electrons must enter at a "later" point in the respiratory chain, missing the first proton pumping site, Complex I. Thus, whereas electrons derived from NADH yield 2.5 ATP, those from $FADH_2$ yield only 1.5 ATP.

6. The maximum number of ATPs depends on the shuttle system. NADH from glycolysis must be reoxidized in the mitochondrial matrix, but it cannot traverse the inner mitochondrial membrane as NADH. Some types of cells use the malate-aspartate shuttle, which results in the production of 2.5 ATP per NADH. Other cells use the glycerol 3-phosphate shuttle, which yields only 1.5 ATP per NADH. This results in two fewer ATPs produced per glucose molecule.

7. B

8. C

9. E

10. **(a)** mitochondria; **(b)** both; **(c)** both; **(d)** both; **(e)** chloroplasts

11. C

Applying What You Know

1. Similarities between oxidative phosphorylation and photophosphorylation include:

- A sequential chain of membrane-bound electron carriers.
- Cytochromes and flavins in the electron transfer chains.
- Electron transfer leading to establishment of a proton gradient.
- A system of intact membranes to separate protons "inside" and "outside."
- An ATP synthase as a coupling factor.

- The F_1 portion of the ATP synthase located on the more alkaline side of the membrane, so the direction of flow of protons is down their concentration gradient.

Differences include:

- The organelle in which the process occurs: mitochondrion vs. chloroplast.
- The initial source of electrons: NADH and $FADH_2$ vs. H_2O (or H_2S, or organic hydrogen donors such as lactate).
- Source of energy for the electron donors: fuel molecules vs. photons.
- Final electron acceptor: O_2 vs. $NADP^+$.
- Sidedness of the F_1 component of the ATP synthase: "inside" facing the matrix vs. "outside" facing the stroma.

2. (a) Electrons will pass from NADH to Complex I to Q to Complex III, the final electron acceptor, because there is no cytochrome *c* to transfer the electrons to Complex IV.

(b) Complex I is the first and final electron acceptor because no Q is present.

(c) Succinate donates electrons to the FAD of Complex II. Electrons are then transferred to Q, then to Complex III, then to cytochrome *c*, then to Complex IV, which finally passes the electrons to O_2.

3. (a) Protons could not flow down their concentration gradient into the matrix because they can pass through the inner mitochondrial membrane only via the *normally* oriented ATP synthase, first through the F_o channel then through F_1 (which catalyzes the formation of ATP from ADP and P_i). Here, the F_1 units would be on the opposite side of the inner membrane. Furthermore, ADP + P_i are at a high concentration in the matrix, not in the intermembrane space, and would not be readily available to the F_1 subunit to make ATP. Thus, the cell would be unable to generate ATP by oxidative phosphorylation. Furthermore, as electron transfer continued, the protons would continue to increase in concentration in the intermembrane space; a limit would be reached, and electron transfer, proton pumping, and oxidation of electron carriers such as NADH would cease. Substrate-level phosphorylation in glycolysis would also eventually stop because the cell would be unable to regenerate NAD^+. Having lost its ability to form ATP, the cell would die.

(b) Although F_o acts as a proton channel or pore through the membrane, it is not symmetrical, or equal, on each side of the membrane. F_1 is held to the matrix side of the F_o unit by specific interactions. Because most biomolecules have sidedness or direction, the same interactions could not occur on the surface that normally faces the intermembrane space. In addition, the ability of protons to enter the proton channel is likely to be influenced by the nature of the amino acid residues present in the portion of the channel that faces the intermembrane space. Therefore, flipping the F_o complex would, most likely, inhibit the ability of protons to flow through the pore.

4. You should transfer three samples of each culture into separate test tubes. For each culture, put one test tube in each of the following growth conditions: dark with no oxygen; dark with oxygen present; and light with no oxygen. Wait a few days, and see what grows and what dies. A chart, like the one below, will help organize the information:

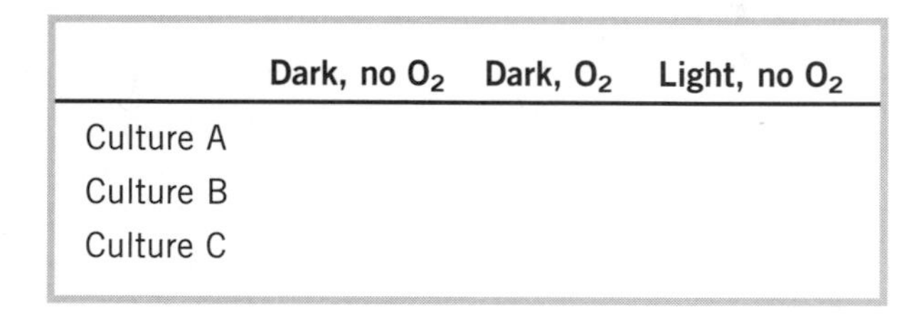

	Dark, no O_2	Dark, O_2	Light, no O_2
Culture A			
Culture B			
Culture C			

The culture that grows under all three conditions is the fermenter, the facultative anaerobe. The culture that grows only in light is the photosynthesizer. The culture that grows only in the presence of oxygen is the third species, which depends on aerobic glucose oxidation.

chapter

20 Carbohydrate Biosynthesis in Plants and Bacteria

STEP-BY-STEP GUIDE

Major Concepts

Photosynthetic organisms synthesize carbohydrates from carbon dioxide and water.

Carbon assimilation occurs in three stages. In the first stage, CO_2 condenses with ribulose 1,5-bisphosphate, a reaction catalyzed by **ribulose 1,5-bisphosphate carboxylase/oxygenase,** or **rubisco.** This enzyme, the key enzyme in the formation of organic biomolecules from inorganic carbon (CO_2), is regulated by substrate availability, cofactor concentration, covalent modification, and pH. The six-carbon intermediate formed in the fixation reaction is broken down to two molecules of 3-phosphoglycerate. The second stage of carbon assimilation, the conversion of 3-phosphoglycerate to glyceraldehyde 3-phosphate, uses the ATP and NADPH synthesized in the light reactions. Glyceraldehyde 3-phosphate has several fates in the plant cell, including entry into glycolysis or conversion into hexoses. The third stage regenerates ribulose 1,5-bisphosphate via a pathway that is essentially the reversal of the oxidative pentose phosphate pathway. **The net reaction of the carbon fixation pathway (the Calvin cycle) is synthesis of one molecule of glyceraldehyde 3-phosphate from three CO_2 molecules, at a cost of six NADPH and nine ATP.**

Carbon assimilation requires a P-triose phosphate antiport system in the chloroplast inner membrane.

This system exchanges cytosolic P_i for stromal dihydroxyacetone phosphate, allowing continued carbon fixation; it effectively shuttles ATP and reducing equivalents as well.

Regulation of carbohydrate metabolism in plants is complicated.

The major enzyme of carbon fixation, rubisco, is under tight control as noted above. Specific enzymes of the Calvin cycle are also subject to regulation by light (mediated by changes in pH), alterations in cofactor concentrations, thioredoxin, and covalent modification. Gluconeogenesis and glycolysis are also regulated. Light plays a regulatory role in that the triose phosphates produced by photosynthesis inhibit phosphofructokinase-2 activity; this lowers the levels of fructose 2,6-bisphosphate, resulting in a decrease in glycolysis and an increase in gluconeogenesis.

Besides its carboxylase activity, rubisco also has an oxygenase function that catalyzes the condensation of O_2 with ribulose-1,5-bisphosphate.

This reaction is followed by a series of carbon "salvage" reactions, which, along with the oxygenase activity, consume O_2 and produce CO_2 in an energy-expending process known as **photorespiration. C_4 plants,** so called because they initially fix CO_2 into a four-carbon compound, have decreased levels of photorespiration. By increasing the local concentration of CO_2 in certain specialized cells, C_4 plants are able to outcompete **C_3 plants** under hot and bright environmental conditions.

What to Review

Answering the following questions and reviewing the relevant concepts, which you have already studied, should make this chapter more understandable.

- Rubisco, the key enzyme of the carbon fixation reactions in photosynthetic organisms, has two substrates, oxygen and carbon dioxide, each with its own K_m value. What does the K_m of an enzyme mean? If an enzyme has two substrates, does just knowing the K_m of each substrate tell you at what rate the enzyme catalyzes each reaction (pp. 196–200)?

- In plant cells, the P_i-triose phosphate antiporter in the chloroplast inner membrane provides a conduit for the products of photosynthesis to enter the cytosol. What are the characteristics of antiport systems (p. 395)?
- **Sucrose** is the principal form in which sugar is transported from leaves to the rest of the plant (p. 244); **glycogen** and **starch** are the storage polysaccharides of animals and plants, respectively (pp. 244–246). What are the monosaccharide units of these compounds? What type(s) of bonds connect the monomers?
- The **glyoxylate cycle** allows plants and some bacteria to synthesize carbohydrates from acetyl-CoA, an ability that vertebrate animals lack. What is the net reaction of the glyoxylate cycle? What enzymes are present in this pathway that are not present in the citric acid cycle (pp. 638–640)?

Topics for Discussion

Answering each of the following questions, especially in the context of a study group discussion, should help you understand the important points of this chapter.

20.1 Photosynthetic Carbohydrate Synthesis

1. Does net CO_2 "fixation" occur in heterotrophic organisms in the pyruvate carboxylase, acetyl-CoA carboxylase, or carbamoyl phosphate synthetase I reactions?

Plastids Are Organelles Unique to Plant Cells and Algae

2. Differentiate between amyloplasts, protoplastids, and chloroplasts.

Carbon Dioxide Assimilation Occurs in Three Stages

3. What is the net reaction, and the actual stoichiometry, for each stage of the Calvin cycle?

Stage 1: Fixation of CO_2 into 3-Phosphoglycerate

4. Where in the ribulose 1,5-bisphosphate molecule is the CO_2 that is being "fixed" added?

5. Where is rubisco located in plants? (Although "rubisco" is a nice short name, the full name gives all the information about the two activities of this enzyme.)

6. What is the advantage of rubisco activity being decreased in the dark by the "nocturnal inhibitor"?

Stage 2: Conversion of 3-Phosphoglycerate to Glyceraldehyde 3-Phosphate

7. Could glycolysis occur in the chloroplast stroma?

8. What are the various roles of glyceraldehyde 3-phosphate in plant cells?

Stage 3: Regeneration of Ribulose 1,5-Bisphosphate from Triose Phosphates

9. What is the general role of the prosthetic group thiamin pyrophosphate?

10. Compare the final products of the "reductive" pentose phosphate pathway with those of the oxidative pentose phosphate pathway.

Synthesis of Each Triose Phosphate from CO_2 Requires Six NADPH and Nine ATP

11. *Cell Map:* What is the net reaction of the Calvin cycle? What is its stoichiometry? Add the (abbreviated) version of the Calvin cycle in Figure 20–14 to the chloroplast on your cell map; connect the inputs of ATP and NADPH from photophosphorylation.

12. What is the source of energy for this process?

13. Why is the term "carbon assimilation reactions" much more accurate than "dark reactions"?

14. What prevents animals from being able to assimilate CO_2?

A Transport System Exports Triose Phosphates from the Chloroplast and Imports Phosphate

15. Why is the chloroplast membrane (or any membrane) impermeable to phosphorylated compounds?

16. *Cell Map:* What are the functions of the P_i–triose phosphate antiport system in the inner chloroplast membrane? Add this transporter to the chloroplast membrane on your map.

Four Enzymes of the Calvin Cycle Are Indirectly Activated by Light

17. How are stromal environmental conditions influenced by the presence of light? What enzymes are affected by these conditions?

18. Describe the mechanism of action of thioredoxin.

20.2 Photorespiration and the C_4 and CAM Pathways

Photorespiration Results from Rubisco's Oxygenase Activity

19. What is the net result of photorespiration? How is it similar to and different from mitochondrial respiration?

20. Suppose you were conducting an experiment on rubisco in intact plant cells, with environmental conditions of 9 μm CO_2 and 350 μm O_2 (and a temperature under 28 °C). What more would you need to know in order to determine the proportion of rubisco activity involved in carbon assimilation and the proportion involved in oxygen "fixation"?

The Salvage of Phosphoglycolate Is Costly

21. What is the net reaction of the glycine decarboxylase complex?

In C_4 Plants CO_2 Fixation and Rubisco Activity Are Spatially Separated

22. Which of the following mechanisms is used by C_4 plants to increase the efficiency of the carbon-fixing activity of rubisco in bundle-sheath cells: increasing the local concentration of CO_2, decreasing the local concentration of O_2, decreasing the K_m for CO_2, or increasing the K_m for O_2?

23. Which uses more energy, the fixation of CO_2 in C_4 plants or in C_3 plants?

In CAM Plants, CO_2 Capture and Rubisco Activity Are Temporally Separated

24. Into what four-carbon compound is CO_2 initially fixed?

20.3 Biosynthesis of Starch and Sucrose

ADP-Glucose Is the Substrate for Starch Synthesis in Plant Plastids and for Glycogen Synthesis in Bacteria

25. What are the similarities and differences between glycogen and starch synthesis?

UDP-Glucose Is the Substrate for Sucrose Synthesis in the Cytosol of Leaf Cells

26. What makes sucrose a good "transport" form of carbon?

Conversion of Triose Phosphates to Sucrose and Starch Is Tightly Regulated

27. What proportion of the triose phosphates produced during carbon fixation is consumed in starch synthesis? In sucrose synthesis? How does this proportion relate to the P_i-triose phosphate antiport system?

28. How is the concentration of fructose 2,6-bisphosphate controlled?

29. How is the activity of sucrose 6-phosphate synthase controlled?

20.4 Synthesis of Cell Wall Polysaccharides: Plant Cellulose and Bacterial Peptidoglycan

Cellulose Is Synthesized by Supramolecular Structures in the Plasma Membrane

30. Though a simple homopolysaccharide of glucose like starch and glycogen, the necessity of exportation of cellulose greatly complicates its synthesis. List the intermediates, enzymes, and steps necessary for this process, and compare them to the machinery needed for glycogen synthesis.

Lipid-Linked Oligosaccharides Are Precursors for Bacterial Cell Wall Synthesis

31. Note that the construction of bacterial cell walls involves the participation of nucleotides, amino acids, heteropolysaccharides, and isoprenoids—all of biochemistry in one macromolecular construct!

20.5 Integration of Carbohydrate Metabolism in the Plant Cell

Gluconeogenesis Converts Fats and Proteins to Glucose in Germinating Seeds

32. Why are seeds so high in protein and fat?

33. How does the glycerol from triacylglycerols enter gluconeogenesis?

34. How does the compartmentation provided by glyoxysomes and mitochondria aid in gluconeogenesis in plant seeds?

Pools of Common Intermediates Link Pathways in Different Organelles

35. Use Figure 20–35 as a self-test: What enzymes are responsible for the conversion reactions between the hexose phosphates, the pentose phosphates, the triose phosphates?

36. What transport mechanisms are responsible for the movement of the shared intermediates between organelles in the plant cell?

SELF-TEST

Do You Know the Terms?

ACROSS

1. Plants are able to synthesize sugars from the product of fatty acid breakdown by the combined actions of the _____ cycle and gluconeogenesis.
5. _____ plants, such as cactus and pineapple, avoid losing water vapor with their variation on carbon dioxide fixation.
6. Organelle that encloses most biosynthetic activities in plants.
8. Each triose phosphate synthesized from CO_2 costs 6 _____ and 9 ATP.
10. Carbon fixation and _____ are competitive pathways in plants.
12. Ribulose 1,5-bisphosphate carboxylase/oxygenase.
14. Glycogen is the glucose storage molecule in animals; _____ is the glucose storage molecule in plants.
15. Carbon assimilation is also known as the _____ cycle.
16. Starch _____ inserts glucosyl residues onto the reducing end of the growing starch chain.

DOWN

1. The _____ pathway converts two molecules of 2-phosphoglycolate to one molecule of serine and one of carbon dioxide.
2. Phosphoenolpyruvate _____ catalyzes the fixation of CO_2 into oxaloacetate in C_4 plants.
3. Sugar _____ : activated forms of sugars such as ADP-glucose.
4. Molecule that couples light energy to enzyme activation.
7. An organism able to convert inorganic carbon dioxide into organic compounds.
9. Synthesis of this bacterial cell wall polysaccharide involves lipid-linked oligosaccharides.
11. A disaccharide with no reducing end; the transport form of carbon in plants.
13. Synthesis of this polysaccharide occurs in rosettes in the plasma membrane.

Do You Know the Facts?

1. Given that plants can produce ATP and NADPH from photophosphorylation, for what purpose(s) do they reduce CO_2 to glucose?

A. At night, plants need the ATP produced by glycolysis and the citric acid cycle in the dark.
B. Plants need glucose to produce starch and cellulose.
C. Plants need glucose as a precursor of components of nucleic acids, lipids, and proteins.
D. B and C
E. A, B, and C

2. Carbon fixation involves a condensation reaction between CO_2 and:

A. 3-phosphoglycerate.
B. phosphoglycolate.
C. ribulose 1,5-bisphosphate.
D. fructose 6-phosphate.
E. ribose 5-phosphate.

3. Which equation correctly summarizes the carbon fixation reactions of photosynthesis?

A. $6H_2O + 6CO_2 \longrightarrow 6O_2 + C_6H_{12}O_6$
B. $6CO_2 + 12NADPH + 18ATP + 12H^+ \longrightarrow 12NADP^+ + 18ADP + 18P_i + C_6H_{12}O_6$
C. $6H_2O + 6ADP + 6P_i + 6NADP \longrightarrow 6O_2 + 6ATP + NADPH + 6H^+$
D. $2H_2O \longrightarrow O_2 + 4H^+ + 4e^-$
E. $2C_3O_3H_5 + CO_2 \longrightarrow 2C_3O_3H_3 + H_2O + CH_2O$

4. Ribulose 1,5-bisphosphate carboxylase/oxygenase is arguably the most important enzyme on earth because nearly all life is dependent, ultimately, on its action. The reactions catalyzed by this enzyme are influenced by:

A. pH.
B. substrate concentration.
C. Mg^{2+} concentration.
D. temperature.
E. all of the above.

5. Under hot, dry conditions, temperate grass species (C_3 grasses) are outcompeted by other grasses, notably crabgrass (C_4 plants). Why?

A. C_4 plants do not use as much ATP to fix CO_2 to hexose as do C_3 plants.
B. C_4 plants use CO_2 more efficiently under hot conditions because they increase $[CO_2]$ in bundle-sheath cells.
C. The K_m of the oxygenase reaction of rubisco is higher in C_3 plants.
D. C_4 plants are able to bypass the Calvin cycle and so save energy.
E. All of the above are true.

6. Sugar cane is a C_4 plant. Which of the following is true for these (often) tropical plants?

A. The affinity of rubisco for oxygen decreases with increasing temperature.
B. "C_4" refers to the four-carbon compounds by which CO_2 is shuttled from mesophyll cells to bundle-sheath cells.
C. The net result of the C_4 strategy is to lower the K_m of rubisco for CO_2.
D. C_4 plants can outcompete C_3 plants in hot environments because they have a different enzyme for carbon fixation.
E. All of the above are true.

7. What is the net reaction of the Calvin cycle? Include energy inputs and electron carriers.

8. What prevents animals from being able to assimilate CO_2?

9. What makes sucrose a good "transport" form of carbon?

Applying What You Know

1. Why is it important, and logical, that some enzymes of the Calvin cycle are regulated by the presence or absence of light?

2. Why are plant seeds such a good source of fat and protein for animals that eat them?

3. Compare C_3,C_4, and CAM plants in terms of their first step in initial CO_2 assimilation, and whether or not they release CO_2 in the light (their apparent photorespiration).

ANSWERS

Do You Know the Terms?

ACROSS

1. glyoxylate
5. CAM
6. plastid
8. NADPH
10. photorespiration
12. rubisco
14. starch
15. Calvin
16. synthase

DOWN

1. glycolate
2. carboxylase
3. nucleotides
4. thioredoxin
7. autotroph
9. peptidoglycan
11. sucrose
13. cellulose

Do You Know the Facts?

1. E
2. C
3. B
4. E
5. B
6. B
7. The net reaction is

$$9ATP + 3CO_2 + 6NADPH + H^+ \longrightarrow 1G\ 3\text{-}P + 9ADP + 8P_i + 6NADP^+$$

where G 3-P represents glyceraldehyde 3-phosphate. This is not given in the text as such. Make sure you understand that the "lost" P_i is part of glyceraldehyde 3-phosphate, and that 3 CO_2 are necessary to make one triose molecule.

8. All the reactions of the Calvin cycle except those catalyzed by rubisco, sedoheptulose 1,7-bisphosphatase, and ribulose 5-phosphate kinase also take place in animal tissues. For lack of these three enzymes (and the abundant ATP and NADPH provided by the light reactions of photosynthesis), animals cannot carry out net conversion of CO_2 into glucose.

9. In the course of evolution, sucrose may have been selected as the transport form of carbon because of its unusual linkage between the anomeric C-1 of glucose and the anomeric C-2 of fructose. This bond is not hydrolyzed by amylases or other common carbohydrate-cleaving enzymes, and the unavailability of the anomeric carbons prevents sucrose from reacting nonenzymatically (as does glucose) with amino acids and proteins.

Applying What You Know

1. Plants get energy (ATP) and reducing power (NADPH) from the light reactions of photosynthesis, both of which are necessary for CO_2 fixation. If the plant lacks sufficient cellular energy (ATP) and reducing power (NADPH), it is of no advantage for the Calvin cycle to be running. Plants have developed mechanisms to signal when light reactions are occurring, such as making the stromal compartment alkaline so that enzymes can work at their optimal pH.

2. Many plants store lipids and proteins in their seeds, to be used as sources of energy and biosynthetic precursors during germination, before photosynthetic mechanisms can supply both. Plants can convert acetyl-CoA derived from fatty acid oxidation into glucose. Glucogenic amino acids derived from the breakdown of stored seed proteins also yield precursors for gluconeogenesis. Active gluconeogenesis must be able to occur in germinating seeds, in order to provide glucose for the synthesis of sucrose, polysaccharides, and the many metabolites derived from hexoses.

3. In C_3 plants the first step in carbon assimilation is the reaction of CO_2 with ribulose 1,5-bisphosphate, catalyzed by rubisco; these plants do show CO_2 release in the light. In C_4 plants the first step in carbon assimilation is the reaction of CO_2 (actually, HCO_3^-, which is favored relative to CO_2 in the aqueous equilibrium) with PEP, to make the four-carbon intermediate oxaloacetate. This reaction is catalyzed by PEP carboxylase. In CAM plants at night, CO_2 is fixed into oxaloacetate by PEP carboxylase as well. CAM and C_4 plants show relatively little CO_2 release (low apparent photorespiration).

chapter

21

Lipid Biosynthesis

STEP-BY-STEP GUIDE

Major Concepts

Biosynthesis of fatty acids does not proceed as simple reversal of fatty acid oxidation.

Malonyl-CoA, a three-carbon intermediate, is formed from acetyl-CoA and bicarbonate by the biotin-containing enzyme **acetyl-CoA carboxylase.** The assembly of fatty acids from these malonyl-CoA units and an initial acetyl group is catalyzed by a multienzyme complex, **fatty acid synthase.** A four-step sequence is repeated seven times to build palmitate, the 16-carbon saturated fatty acid that is the principal product of fatty acid synthesis in animals. The sequence includes: **condensation** of an activated acetyl group and malonyl group; **reduction** of the carbonyl group at C-3; **dehydration** at C-2 and C-3, yielding a double bond; and finally, **reduction** of the double bond. Further reactions after the fatty acid synthase series can include elongation and/or desaturation of the palmitate product. Some fatty acids are not made in mammals; these essential fatty acids must be obtained in the diet. The signaling lipids known as **eicosanoids** are formed from the essential fatty acid linoleate in response to hormonal or other stimuli.

These reactions are under tight control because the process is energetically expensive.

The first enzyme, acetyl-CoA carboxylase, is inhibited by palmitoyl-CoA, the final product of biosynthesis, and activated by citrate, which carries acetyl-CoA from the mitochondrial matrix into the cytosol where the synthetic reactions take place. This enzyme is also regulated by covalent alteration, which is under hormonal control.

Fatty acid synthesis and oxidation are coordinately regulated so that they do not occur simultaneously.

They take place in separate cellular locations (synthesis in the cytosol and breakdown in the mitochondrial matrix), and the breakdown process is inhibited by malonyl-CoA, the first intermediate in the synthetic process.

Synthesis of storage and membrane lipids from fatty acids is determined by the metabolic needs of the organism.

Triacylglycerols and glycerophospholipids are synthesized from common precursors, fatty acyl–CoA and glycerol 3-phosphate. Triacylglycerol synthesis is regulated by hormones; synthesis occurs when the fuel intake exceeds the energy needs of the organism. Membrane lipids are synthesized when the cell is growing rapidly. Glycerophospholipid and sphingolipid biosynthesis share some precursors and enzyme mechanisms. The systemic balance between triacylglycerol synthesis and mobilization is also regulated by hormones (glucagon and epinephrine) and as well by the process of glyceroneogenesis, itself controlled by glucocorticoid hormones that regulate the expression of the enzyme PEP carboxykinase.

Cholesterol is synthesized from acetyl-CoA.

This process occurs in four stages: the first stage is the condensation of three acetyl-CoA molecules to form **HMG-CoA.** This compound is reduced to mevalonate by **HMG-CoA reductase,** a reaction that is the committed step of the pathway. The second stage is the conversion of mevalonate to two **activated isoprene** compounds; this requires the input of phosphate groups from ATP. In the third stage, six isoprene units condense to form **squalene,** a 30-carbon compound. Finally, squalene is rearranged into the four-ring steroid nucleus. Cholesterol synthesis is controlled by transcriptional regulation of the gene, feedback inhibition, and hormonal signals that mediate covalent modifications, all of the rate-limiting enzyme HMG-CoA reductase.

Once synthesized, cholesterol has several fates.

It is a component of cell membranes; it is converted into bile salts for use in digestive processes; and it is the precursor of steroid hormones, which have diverse and crucial functions. These hormones are produced from cholesterol by alteration of its side chain and introduction of oxygen into the steroid ring system. In addition, intermediates in cholesterol biosynthesis have a wide variety of critical biological roles; these include vitamins, hormones, and electron carriers.

Cholesterol and other lipids are transported in the blood as lipoproteins, which have a variety of densities, compositions, and specific functions.

Chylomicrons carry exogenous (dietary) lipids from the small intestine to peripheral tissues and are the largest, but least dense, of the lipoproteins. **VLDLs** carry endogenous lipids (synthesized in the liver) to muscle and adipose tissue. **LDLs** are derived from VLDLs and are very rich in cholesterol and cholesteryl esters; high levels of LDLs have been correlated with **atherosclerosis. HDLs** scavenge cholesterol from chylomicrons and VLDLs, returning it to the liver, which is the only route for its ultimate excretion. There is a negative correlation between HDL levels and arterial disease.

Uptake of cholesterol by cells is accomplished by receptor-mediated endocytosis.

This uptake, as well as the biosynthesis and dietary intake of cholesterol, influences the concentration of this lipid in human blood.

What to Review

Answering the following questions and reviewing the relevant concepts, which you have already studied, should make this chapter more understandable.

- Aspirin and ibuprofen work to alleviate pain and/or fever by interfering in the synthesis of **prostaglandins** and **thromboxanes.** Review the functions of these eicosanoids (pp. 358–359).
- **Membrane fusion** processes are important in the intracellular transport of newly synthesized membrane lipids from the site of synthesis to the site of function, and in receptor-mediated endocytosis of LDLs (pp. 387–388).
- **Biotin** functions as a prosthetic group in the carboxylation reaction that produces malonyl-CoA from acetyl-CoA and bicarbonate. What other metabolic roles have already been discussed for this vitamin (Fig. 16–16; Fig. 18–16)?
- *Some* of the reactions of fatty acid synthesis are simple reversals of those in fatty acid oxidation. The control of fatty acid synthesis and oxidation must be coordinated to prevent futile cycling. Outline the sequence of reaction steps and the regulation of **fatty acid oxidation;** be sure to note *where* in the cell these reactions take place (Fig. 17–12, p. 661).
- The acyl carrier protein that is part of the fatty acid synthase complex carries acyl groups attached through a thioester linkage. Why are **thioesters** important in these carrying functions (pp. 505–506)?
- When a cell or organism has excess fuel, the excess is converted to triacylglycerols for storage. Why are triacylglycerols such good stores of energy (pp. 346–348)?

Topics for Discussion

Answering each of the following questions, especially in the context of a study group discussion, should help you understand the important points of this chapter.

21.1 Biosynthesis of Fatty Acids and Eicosanoids

Malonyl-CoA Is Formed from Acetyl-CoA and Bicarbonate

1. What are the three functional regions of the acetyl-CoA carboxylase enzyme?

Fatty Acid Synthesis Proceeds in a Repeating Reaction Sequence

2. From this preview, how are fatty acid oxidation and biosynthesis similar and different?

The Mammalian Fatty Acid Synthase Has Multiple Active Sites

3. What is the role of the acyl carrier protein?

Fatty Acid Synthase Receives the Acetyl and Malonyl Groups

4. What reactions must take place before the condensation reaction occurs?

5. *Cell Map:* Fatty acid synthesis takes place in the cytosol and starts with acetyl-CoA. Indicate the subsequent steps in the pathway by drawing the appropriate structures on the schematic diagram of the enzyme complex. Next to each reaction, indicate the reaction type (e.g., condensation, reduction, etc.).

Step ① Condensation

6. What is the source of energy that drives the condensation reaction? To which reaction in gluconeogenesis is this step similar?

7. Is any CO_2 actually "fixed" in this process?

8. What is the product of the condensation reaction?

Step ② Reduction of the Carbonyl Group

9. What is the electron donor in the reduction of both the carbonyl group and the double bond?

Step ③ Dehydration

10. From where are the elements of water removed in the dehydration step?

Step ④ Reduction of the Double Bond

11. What is the final product after one pass through the fatty acid synthase complex?

The Fatty Acid Synthase Reactions Are Repeated to Form Palmitate

12. How many carbons are in the acyl chain after one pass through the fatty acid synthase complex? After two passes? After seven passes?

13. For which processes are ATP and NADPH required?

Fatty Acid Synthesis Occurs in the Cytosol of Many Organisms but in the Chloroplasts of Plants

14. Why is the location of fatty acid synthesis (and other anabolic processes) in the cytosol "logical" for animal cells but not for plant cells?

Acetate Is Shuttled Out of Mitochondria as Citrate

15. What are the sources of acetyl-CoA for fatty acid synthesis?

16. In addition to transporting acetyl-CoA out of mitochondria, the shuttle systems described in this section "feed" a number of metabolic pathways in the cytosol and the mitochondrial matrix. How many alternative pathways can you list for the various intermediates shown in Figure 21–10?

17. *Cell Map:* Fatty acid synthesis is fueled by intermediates from the citric acid cycle. Show how the citrate, malate-α-ketoglutarate, and pyruvate shuttle systems are involved in providing necessary components.

Fatty Acid Biosynthesis Is Tightly Regulated

18. What are the allosteric and covalent mechanisms controlling the rate-limiting step of fatty acid synthesis in vertebrates? Which hormones are involved?

19. What factors activate acetyl-CoA carboxylase in plants?

20. Why don't bacteria store triacylglycerols as energy reserves?

21. How is the simultaneous functioning of fatty acid oxidation and synthesis avoided?

Long-Chain Saturated Fatty Acids Are Synthesized from Palmitate

22. Where in the cell does elongation of fatty acids occur?

23. What is the acyl carrier in the elongation of palmitate?

Desaturation of Fatty Acids Requires a Mixed-Function Oxidase

24. What type of reaction is the fatty acyl-CoA desaturase reaction?

25. Linoleate and α-linolenate are considered "essential fatty acids." Why?

Box 21–1 Mixed-Function Oxidases, Oxygenases, and Cytochrome P-450

26. What is accomplished by the hydroxylation of substances, such as drugs and xenobiotics, by cytochrome P-450?

Eicosanoids Are Formed from 20-Carbon Polyunsaturated Fatty Acids

27. How do aspirin and ibuprofen alleviate pain and/or fever, and possibly prevent stroke?

28. Why does aspirin use cause stomach irritation? How do the new NSAIDs avoid this side effect?

21.2 Biosynthesis of Triacylglycerols

Triacylglycerols and Glycerophospholipids Are Synthesized from the Same Precursors

29. What are the precursors of triacylglycerols?

Triacylglycerol Biosynthesis in Animals Is Regulated by Hormones

30. Why do people with severe diabetes mellitus lose weight?

31. Between which two tissues does the systemic triacylglycerol cycle occur?

Adipose Tissue Generates Glycerol 3-Phosphate by Glyceroneogenesis

32. What enzyme common to gluconeogenesis and glyceroneogenesis is reciprocally controlled in the liver and adipose tissue?

Thiazolidinediones Treat Type 2 Diabetes by Increasing Glyceroneogenesis

33. The PPAR family of transcription factors was first introduced in Chapter 17. This new information about PPAR should fit into your growing understanding of the integration of carbohydrate and lipid metabolism; more to come in Chapter 23!

21.3 Biosynthesis of Membrane Phospholipids

34. What steps are required for assembly of phospholipids? Where in the cell does this assembly occur?

Cells Have Two Strategies for Attaching Phospholipid Head Groups

35. What does cytidine diphosphate contribute to the glycerophospholipid product in both strategies of head-group attachment?

Phospholipid Synthesis in E. coli *Employs CDP-Diacylglycerol*

36. When in the synthetic process does head-group modification occur?

Eukaryotes Synthesize Anionic Phospholipids from CDP-Diacylglycerol

37. Remind yourself of the general function of kinases.

Eukaryotic Pathways to Phosphatidylserine, Phosphatidylethanolamine, and Phosphatidylcholine Are Interrelated

38. Does it matter exactly what types of phospholipids are in a membrane?

Plasmalogen Synthesis Requires Formation of an Ether-Linked Fatty Alcohol

39. What type of enzyme is responsible for the formation of the double bond in plasmalogens?

Sphingolipid and Glycerophospholipid Synthesis Share Precursors and Some Mechanisms

40. How do the sugar moieties of cerebrosides and gangliosides enter these molecules during their biosynthesis?

Polar Lipids Are Targeted to Specific Cellular Membranes

41. Why must membrane lipids be specially transported from their site of synthesis?

21.4 Biosynthesis of Cholesterol, Steroids, and Isoprenoids

42. What are the functions of cholesterol in humans?

Cholesterol Is Made from Acetyl-CoA in Four Stages

Stage ① Synthesis of Mevalonate from Acetate

43. Which is the committed step of cholesterol synthesis? What enzyme catalyzes this step?

Stage ② Conversion of Mevalonate to Two Activated Isoprenes

44. How are the two activated isoprene species structurally related?

Stage ③ Condensation of Six Activated Isoprene Units to Form Squalene

45. What drives these reactions? Are they irreversible?

Stage ④ Conversion of Squalene to the Four-Ring Steroid Nucleus

46. Note the participation once again of a mixed-function oxidase.

Cholesterol Has Several Fates

47. For what reason(s) is cholesterol converted into cholesteryl esters?

Cholesterol and Other Lipids Are Carried on Plasma Lipoproteins

48. Why is a *mono*layer of phospholipids found on the surface of lipoproteins, rather than the *bi*layer occurring in membranes?

49. Which are the most dense lipoproteins? The least dense? How does this relate to their lipid:protein ratios? (It may help to remember that in chicken soup, the fat floats on the surface but the meat [mostly protein] sinks.)

50. Categorize the various lipoproteins according to: site of origin; lipids transported within them; apolipoproteins contained; lipid:protein ratio; function; and site of degradation or uptake.

Box 21–2 ApoE Alleles Predict Incidence of Alzheimer's Disease

51. What is one known function of apoE2 and apoE3 that is not shared by apoE4?

Cholesteryl Esters Enter Cells by Receptor-Mediated Endocytosis

52. What part of the LDL particle is recognized by the LDL receptor?

Cholesterol Biosynthesis Is Regulated at Several Levels

53. What are the allosteric, hormonal, and covalent strategies for regulation of cholesterol biosynthesis?

54. Based on the function of HDLs, why would you expect a negative correlation between HDL levels and arterial disease?

55. Why is it important that the half-life of the amino-terminal domain of SREBP is short?

Box 21–3 The Lipid Hypothesis and the Development of Statins

56. At what stage in the process of cholesterol synthesis do lovastatin and compactin exert their inhibition?

Steroid Hormones Are Formed by Side-Chain Cleavage and Oxidation of Cholesterol

57. What are the general functions of mineralocorticoids, glucocorticoids, and the sex hormones?

Intermediates in Cholesterol Biosynthesis Have Many Alternative Fates

58. Be aware of how the use of a single starting material to make many different end products is an energy-saving strategy in cells.

59. What does prenylation of a protein accomplish?

Discussion Questions for Study Groups

- Compare and contrast gluconeogenesis and glyceroneogenesis. Are there any between-organ cycles for glucose or precursors of glucose that are similar to the triacylglycerol cycle?

SELF-TEST

Do You Know the Terms?

ACROSS

1. Class of hormones derived from cholesterol that affect the ability of the kidney to reabsorb Na^+ and Cl^- from blood.
4. Cytochrome in membranes of the endoplasmic reticulum (*not* in mitochondrial membranes); involved in hydroxylations that render substances more water soluble and excretable.
5. Acetylsalicylate acts through inhibition of enzymes that synthesize this class of compounds.
7. General name for enzyme that oxidizes compounds by direct incorporation of oxygen atoms.
10. The rate-limiting step of fatty acid biosynthesis is catalyzed by acetyl-CoA _____.
11. Protein containing phosphopantetheine, similar to coenzyme A; carries fatty acyl groups attached by a thioester bond. (abbr.)
13. General name for enzyme that oxidizes substrates by transferring electrons to oxygen, producing H_2O or H_2O_2.
15. Enzyme responsible for the cholesterol and lipid scavenging ability of high-density lipoproteins. (abbr.)
17. Regulated by phosphorylation triggered by glucagon and ephinephrine. (2 words)
18. Generated in adipose tissue by glyceroneogenesis; used for triacylglycerol synthesis.
22. Competitive inhibitor of key enzyme in cholesterol synthesis; shows promise for treatment of hypercholesterolemia.
24. Receptor-mediated _____: mechanism of uptake of lipoproteins by peripheral (nonhepatic) tissues.
26. Class of hormones derived from cholesterol that increase the synthesis of PEP carboxykinase, fructose 1,6-bisphosphate, glucose 6-phosphate, and glycogen synthase.
27. Lipoproteins synthesized in the liver that transport primarily triacylglycerols. (abbr.)
28. Complex of multifunctional proteins involved in lipid synthesis; channels intermediates between its different active sites. (3 words)
29. _____ reductase catalyzes the committed step in cholesterol biosynthesis. (abbr.)

DOWN

2. Compound important in synthesis of regulatory lipids; formed only from essential fatty acids in the diet.
3. Precursor of cholesterol; formed from acetyl-CoA.
4. Isoprenyl derivatives, which appear to serve as signals targeting proteins to cell membranes, are attached to proteins through a _____ reaction.
5. Another name for diacylglycerol 3-phosphate.
6. Process that increases efficiency of multiple enzyme-catalyzed reactions. (2 words)
8. Estrogen, testosterone, and cholesterol are examples of this class of enzymes.
10. Attachment of this compound in final step of glycerophospholipid synthesis forms an activated compound to which the polar head group is attached. (abbr.)
11. Enzyme that converts cholesterol to cholesteryl esters in the liver. (abbr.)

12. Primary carrier of dietary triacylglycerols in the blood; synthesized in epithelial cells of the small intestine.
14. Compounds derived from arachidonic acid; include several classes of "local" hormones released by most body cells that cause smooth muscle contraction, vasoconstriction, localized tissue swelling, and platelet aggregation.
16. Lipoproteins rich in cholesterol and cholesteryl esters; synthesized in the liver.
19. Macromolecular complexes that associate with lipids, making them water soluble for transport in the blood.
20. Lipoproteins that transport lipids and cholesterol to the liver, where they can be excreted.
21. Product of the committed step in cholesterol biosynthesis.
23. Compound formed by condensation of six isoprene units.
25. Cofactor for key enzymes of fatty acid synthesis and gluconeogenesis; involved in one-carbon transfers.

Do You Know the Facts?

1. A sample of malonyl-CoA synthesized from radioactive (^{14}C-labeled) HCO_3^- and unlabeled acetyl-CoA is used in fatty acid synthesis. In which carbon(s) will the final fatty acid be labeled? (Recall that the carboxyl carbon is C-1.)
A. Every carbon
B. Every odd-numbered carbon
C. Every even-numbered carbon
D. Only the carbon farthest from C-1
E. No part of the molecule will be labeled.

2. Fatty acid synthase of mammalian cells:
A. uses a four-step sequence to lengthen a growing fatty acyl chain by one carbon.
B. contains ACP, which carries acyl groups attached through thioester linkages.
C. requires chemical energy in just one form: the reducing power of NADPH.
D. is activated by glucagon and epinephrine.
E. is activated by palmitoyl-CoA, the principal product of fatty acid synthesis.

3. List the steps of fatty acid synthesis in their correct order.
(a) The double bond is reduced to form butyryl-ACP.
(b) Condensation of the acetyl and malonyl groups produces an acetoacetyl group bound to ACP.
(c) The elements of water are removed to yield a double bond.
(d) The carbonyl group of acetoacetyl-ACP is reduced.
(e) The fatty acid synthase complex is charged with the correct acyl groups.

4. The synthesis of palmitate requires:
A. 8 acetyl-CoA.
B. 14 NADH.
C. 7 ATP.
D. A and C.
E. A, B, and C.

5. In mammalian cells, fatty acid synthesis occurs in the cytosol. The beginning substrate for this series of reactions is acetyl-CoA, which is formed in the mitochondrial matrix. How does the acetyl-CoA move from matrix to cytosol?
A. Acetyl-CoA reacts with oxaloacetate and leaves the matrix as citrate via the citrate transporter.
B. Acetyl-CoA combines with bicarbonate and is transported out of the matrix as pyruvate.
C. There is a specific transport protein for acetyl-CoA.
D. Acetyl-CoA is a nonpolar molecule and can diffuse across all membranes.
E. Inner and outer mitochondrial membranes are freely permeable to acetyl-CoA due to the presence of transmembrane pores or channels.

6. How are fatty acid oxidation and synthesis controlled so that futile cycling does not occur?
A. They occur in different cellular compartments.
B. They employ different electron carriers.
C. The product of the first oxidation reaction inhibits the rate-limiting step of biosynthesis.
D. A and B
E. A, B, and C

7. Which of the following reactions is catalyzed by a mixed-function oxidase?

A. $AH + BH_2 + O_2 \longrightarrow A{-}OH + B + H_2O$
B. $RH + O_2 + 2NADPH \longrightarrow R{-}OH + H_2O + 2NADP^+$
C. $R^1{-}O{-}CH_2{-}CH_2{-}R^2 + NADH + H^+ + O_2 \longrightarrow R^1{-}O{-}CH{=}{=}CH{-}R^2 + 2H_2O + NAD^+$
D. A and B
E. A, B, and C

8. Which of the following is *not* true of cholesterol synthesis?

A. Reduction of HMG-CoA to mevalonate catalyzed by HMG-CoA reductase is the committed step.
B. Acetyl-CoA is the ultimate source of all 27 carbon atoms of cholesterol.
C. It is inhibited by elevated levels of intracellular cholesterol.
D. It is hormonally inactivated by glucagon and activated by insulin.
E. It occurs in the mitochondrial matrix.

9. Cholesterol and other lipids are transported in the human body by lipoproteins. Indicate whether each of the following statements about these molecules is true or false.

_____ **(a)** The more lipid in a particular lipoprotein, the denser is the lipoprotein.
_____ **(b)** Lipoproteins "solubilize" hydrophobic lipids by surrounding them with a monolayer of amphipathic lipids.
_____ **(c)** Endogenous (internally synthesized) triacylglycerols follow the same route in the body as exogenous triacylglycerols (obtained from the diet).
_____ **(d)** Cholesterol and cholesteryl esters are transported in lipoproteins.
_____ **(e)** Chylomicrons are the largest but least dense lipoproteins, containing a high proportion of triacylglycerols.
_____ **(f)** Chylomicrons carry fatty acids obtained in the diet to tissues for use or storage.
_____ **(g)** Very low-density lipoproteins perform the same role as chylomicrons, but with endogenous lipids rather than dietary lipids.
_____ **(h)** As triacylglycerols are removed from VLDLs, the VLDLs become chylomicrons.
_____ **(i)** VLDLs, LDLs, and HDLs are progressively denser versions of the same initial molecule.
_____ **(j)** Low-density lipoproteins are very high in cholesterol and cholesteryl esters.
_____ **(k)** High-density lipoproteins carry the enzyme LCAT, which removes cholesterol from chylomicrons and VLDLs, and transport cholesterol back to the liver.
_____ **(l)** Some of the apolipoproteins of lipoproteins act as enzymes for degradation of the lipids once they reach their target tissue.
_____ **(m)** The liver is the only route through which cholesterol leaves the body (through bile salts lost in digestive processes).

10. Plasmalogens are:

A. membrane lipids.
B. synthesized from fatty acyl-CoA.
C. synthesized from the glycolytic intermediate dihydroxyacetone phosphate.
D. A and B.
E. A, B, and C.

Applying What You Know

1. Which two hormones affect the activity of acetyl-CoA carboxylase in fatty acid biosynthesis? Why does this make physiological sense?

2. What are the two possible fates of newly synthesized fatty acids in vertebrates? Under what conditions is one fate more likely than the other?

3. Why do untreated diabetics lose weight?

4. Glucagon has been shown to reduce the activity of HMG-CoA reductase. Why does this make physiological sense?

5. High levels of cholesterol in the blood have been positively correlated with the incidence of atherosclerosis. Recently, the LDL:HDL ratio has been shown to be a better indicator. Explain why this ratio might be a predictor of coronary artery obstruction.

6. List the ways in which fatty acid oxidation and fatty acid synthesis differ. (Suggestions: What are the electron carriers? What are the "activating groups"? Where in the cell do the reactions take place? What group is added or taken away?)

ANSWERS

Do You Know the Terms?

ACROSS

1. mineralocorticoids
4. P-450
5. prostaglandins
7. oxygenase
10. carboxylase
11. ACP
13. oxidase
15. LCAT
17. acetyl-CoA carboxylase
18. glycerol 3-phosphate
22. lovastatin
24. endocytosis
26. glucocorticoids
27. VLDLs
28. fatty acid synthase
29. HMG-CoA

DOWN

2. arachidonate
3. isoprene
4. prenylation
5. phosphatidate
6. substrate channeling
8. sterols
10. CDP
11. ACAT
12. chylomicrons
14. eicosanoids
16. LDLs
19. lipoproteins
20. HDLs
21. mevalonate
23. squalene
25. biotin

Do You Know the Facts?

1. E
2. B
3. (e), (b), (d), (c), (a)
4. D
5. A
6. D
7. E
8. E
9. **(a)** F; **(b)** T; **(c)** F; **(d)** T; **(e)** T; **(f)** T; **(g)** T; **(h)** F; **(i)** F; **(j)** T; **(k)** T; **(l)** T; **(m)** T
10. E

Applying What You Know

1. Acetyl-CoA carboxylase, which forms malonyl-CoA from acetyl-CoA and HCO_3^-, is inactivated by glucagon and epinephrine (through enzyme phosphorylation, which causes dissociation of the enzyme into its inactive monomeric subunits). Glucagon and epinephrine indicate the body's need for blood glucose. It makes sense that fatty acid synthesis does not proceed when an immediately available source of energy (i.e., glucose) is needed, rather than a storage form of energy (i.e., fatty acids).

2. The two possible fates of fatty acids are storage as metabolic energy or incorporation into phospholipids destined for membranes. During rapid cell growth, membrane phospholipid synthesis is the more likely fate. Cells not undergoing growth, and having a surplus of energy, will store the fatty acids (e.g., as triacylglycerols in humans).

3. The rate of triacylglycerol biosynthesis is affected by the action of several hormones, one of which is insulin. Insulin promotes the conversion of carbohydrates to triacylglycerols. People with severe diabetes, due to failure of insulin secretion or action, not only are unable to use glucose properly but also fail to synthesize fatty acids from carbohydrates or amino acids. Untreated diabetics have increased rates of fat oxidation and ketone body formation and, as a consequence, lose weight.

4. The presence of glucagon is a signal that blood glucose is low and must be diverted to those organs that require it to generate energy. The need for catabolically generated energy in general suppresses biosynthetic reactions; for example, the inhibition of HMG-CoA reductase suppresses cholesterol biosynthesis.

5. LDLs shuttle cholesterol from the liver to other sites in the body (including the coronary arteries), and can be viewed as "bad" cholesterol. They contribute to the high blood cholesterol levels associated with atherosclerotic plaque formation. HDLs, on the other hand, scavenge cholesterol from the various lipoproteins in the bloodstream, removing it from LDLs and VLDLs and transporting it back to the liver. The liver is the only route for excretion of cholesterol (via bile salts lost in digestive processes). A high LDL:HDL ratio indicates a high level of circulating cholesterol; a low LDL:HDL ratio, even if overall cholesterol in the body is high, indicates that HDL levels are sufficient to keep the blood cholesterol concentration low.

6.

	Fatty acid synthesis	**Fatty acid oxidation**
Electron carrier:	NADPH	NADH
Acyl groups bound to:	ACP	CoA
Cellular location:	Cytosol (in animal cells)	Mitochondrial matrix
Process involves:	Addition of malonyl groups	Removal of acetyl groups

chapter

22 Biosynthesis of Amino Acids, Nucleotides, and Related Molecules

STEP-BY-STEP GUIDE

Major Concepts

Prokaryotes cycle nitrogen into and out of biologically available forms.

Nitrogen fixation is carried out only by prokaryotes, symbiotic as well as free-living, which convert atmospheric nitrogen (N_2) to ammonia through activity of the enzymes of the **nitrogenase complex.** The NH_4^+ produced can be oxidized to nitrite and nitrate by the bacterial process known as **nitrification.** Plants and microorganisms incorporate ammonia into amino acids, and these amino acids are used by animals as a nitrogen source. When organisms die, it is again prokaryotes that degrade the proteins to ammonia and, ultimately, to N_2.

α-Ketoglutarate and glutamate act as the entry point for nitrogen in biosynthetic reactions.

The incorporation of ammonia into these biomolecules is catalyzed by glutamine synthetase and glutamate synthase. As in amino acid degradation, glutamate and glutamine are the primary transport forms of NH_4^+ for biosynthetic reactions. **Glutamine synthetase** is a key regulatory enzyme in nitrogen metabolism. This enzyme is regulated through allosteric inhibition by the end products of glutamine metabolism, and through covalent modification: Removal of an AMP (deadenylylation) activates and addition of AMP (adenylylation) decreases its activity. Glutamine sythetase is under indirect control imposed by regulation of the enzyme **adenylyl transferase,** which catalyzes this adenylylation and deadenylylation. The activity of adenylyl transferase is modulated by a regulatory protein that is, itself, covalently modified in response to various metabolites and end products. In bacteria, transfer of an amino group from glutamine to α-ketoglutarate to form two molecules of glutamate is catalyzed by **glutamate synthase.** The net effect of these two enzymatic reactions is the synthesis of one glutamate molecule from ammonia and α-ketoglutarate.

Amino acids are synthesized from intermediates in glycolysis, the citric acid cycle, and the pentose phosphate pathway.

Ten amino acids are derived from the citric acid cycle intermediates **α-ketoglutarate** and **oxaloacetate.** Nine amino acids are derived from the glycolytic intermediates **3-phosphoglycerate, phosphoenolpyruvate,** and **pyruvate.** Glucose 6-phosphate can be diverted from glycolysis into the pentose phosphate pathway to produce **ribose 5-phosphate,** a precursor of histidine. PLP is an important cofactor in many of these pathways, and NADPH is the primary source of reducing power. Several of the amino acid synthetic pathways have a common intermediate, **phosphoribosyl pyrophosphate (PRPP),** which is also of importance in nucleotide biosynthesis.

Amino acid biosynthesis is regulated by a variety of feedback inhibition strategies.

In **concerted inhibition,** many modulators act on the enzyme catalyzing the first, usually irreversible, reaction in a sequence, and the effect of these modulators is more than additive. **Enzyme multiplicity** often occurs when there are several possible fates for the metabolic intermediates in a pathway. In this case, different isozymes are independently controlled by differing sets of modulators; consequently, the entire reaction pathway is not shut down by one biosynthetic product when other products of the same pathway are needed. In **sequential feedback inhibition** a metabolite inhibits its own formation at several different points in a pathway; these inhibitory effects *overlap* because they usually inhibit several of the isozymes and affect different enzymatic steps in the pathway.

Amino acids are the precursors of many extremely important biomolecules, in addition to proteins.

Porphyrins, such as heme, and **bilirubin,** the degradation product of heme, are derived from glycine and succinyl-CoA. **Phosphocreatine** and **glutathione** have roles in energy generation and redox reactions, respectively. Such plant products as **lignin, tannins,** and **auxin** are derived from aromatic amino acids. Neurotransmitters are often primary or secondary amines derived from amino acids; these include **dopamine, norepinephrine, epinephrine,** and **GABA.** The vasodilator **histamine** is formed from histidine, and the polyamines **spermine** and **spermidine** used in DNA packaging are also derived from amino acids.

Nucleotides are synthesized de novo and through salvage pathways.

In **de novo purine synthesis,** AMP and GMP are synthesized from PRPP, glutamine, glycine, N^{10}-formyl H_4 folate, aspartate, and fumarate. The purine ring structure is built upon 5-phospho-β-D-ribosylamine one or a few moieties at a time. **Inosinate** (IMP) is the intermediate common to AMP and GMP formation. IMP, AMP, and GMP all regulate the activity of key enzymes in the pathway by feedback inhibition. In **de novo pyrimidine synthesis,** CMP and UMP are synthesized from aspartate, PRPP, and carbamoyl phosphate. The pyrimidine ring structure, orotate, is constructed first and then attached to PRPP to form **orotidylate.** The pyrimidines are synthesized from this intermediate. The carbamoyl phosphate component of the pyrimidine ring is produced in the cytosol by a different enzyme from that which synthesizes carbamoyl phosphate in the mitochondrion for the urea cycle. As for purine biosynthesis, this pathway is regulated by feedback inhibition.

Nucleotide monophosphates are phosphorylated by phosphate transfer from ATP.

AMP is converted to ADP by **adenylate kinase.** ADP is then converted to ATP by oxidative phosphorylation or by glycolytic enzymes. ATP is used to convert the other nucleotide monophosphates to triphosphates by the action of **nucleoside monophosphate kinases** and **nucleoside diphosphate kinase.**

Ribonucleotides are reduced to deoxyribonucleotides by ribonucleotide reductase.

The reaction catalyzed by this enzyme involves the generation of a free radical. The activity and substrate specificity of the reductase are regulated. Activity is regulated by the binding of either ATP (activator) or dATP (inactivator). Substrate specificity is regulated by the binding of ATP, dATP, dTTP, or dGTP (acting as effector molecules), which induce different conformations in the enzyme. These controls all ensure a balanced pool of precursors for DNA synthesis.

Purines are degraded to uric acid and pyrimidines to urea.

Uric acid is excreted by a few organisms (notably birds), but in most organisms it is further degraded to urea or ammonia. Metabolic degradation of nucleotides produces free bases that can be recycled in **salvage pathways.** AMP is synthesized from adenine and PRPP by the action of **adenosine phosphoribosyltransferase.** GMP is synthesized from guanine and hypoxanthine by **hypoxanthine-guanine phosphoribosyltransferase.** Pyrimidines are not salvaged in significant amounts in mammals.

Because cancer cells grow more rapidly than normal cells and therefore have a greater need for nucleotides, they are vulnerable to disruptions in nucleotide synthetic processes.

Pharmaceuticals that take advantage of this fact include **azaserine, fluorouracil, methotrexate,** and **aminopterin.**

What to Review

Answering the following questions and reviewing the relevant concepts, which you have already studied, should make this chapter more understandable.

- **Aminotransferases** are important in the transfer of α-amino groups in amino acid biosynthesis; review their role in amino acid oxidation, including the importance of their cofactor **pyridoxal phosphate** (PLP) in the transfer of amino groups (pp. 677–679).
- The enzyme glutamine synthetase is central to ammonia assimilation in bacteria. Glutamine also plays an important role in mammals, that of ammonia transporter. Why must NH_4^+ be converted into glutamine for transport in the bloodstream (pp. 680–682)?
- Other cofactors involved in the reactions discussed in this chapter include **tetrahydrofolate** and ***S*-adenosylmethionine.** What other reactions require these cofactors (pp. 689–691)?
- Amino acids are built from precursors that are derived from the central carbohydrate metabolic pathways. Review the structures of these precursors from glycolysis (Chapter 14), the citric acid cycle (Chapter 16), and the pentose phosphate pathway (Chapter 14).
- **Mechanism-based inactivators,** or **suicide inhibitors,** are coming into use as pharmaceutical agents for attacking parasites or processes that use certain pathways and intermediates at a greater rate than do the host's normal processes. How do they work? Why are they so powerful and specific in their inhibition (p. 204)?

Topics for Discussion

Answering each of the following questions, especially in the context of a study group discussion, should help you understand the important points of this chapter.

22.1 Overview of Nitrogen Metabolism

The Nitrogen Cycle Maintains a Pool of Biologically Available Nitrogen

1. What are the forms of nitrogen in the nitrogen cycle, from the most oxidized to the most reduced form?

Nitrogen Is Fixed by Enzymes of the Nitrogenase Complex

2. What organisms are involved in nitrogen fixation? From where do they get the energy to fix nitrogen? From where do they get the reducing power?

3. How is the role of ATP in the process of nitrogen fixation unusual?

4. How do the various nitrogen-fixing organisms cope with the "oxygen problem" during nitrogen fixation?

Box 22–1 Unusual Lifestyles of the Obscure but Abundant

5. Why are the aeration costs associated with nitrification lower when anammox bacteria are used in the waste treatment process? Review Figure 22–1, as well as the information in this Box, for help in answering this question.

Ammonia Is Incorporated into Biomolecules through Glutamate and Glutamine

6. Why is the nitrogen required for amino acid and nucleotide synthesis usually supplied by glutamate or glutamine?

Glutamine Synthetase Is a Primary Regulatory Point in Nitrogen Metabolism

7. Why is the activity of glutamine synthetase so tightly regulated? What are the various levels of regulation of this enzyme?

Several Classes of Reactions Play Special Roles in the Biosynthesis of Amino Acids and Nucleotides

8. What is the main difference between glutamine amidotransferases and glutaminase?

22.2 Biosynthesis of Amino Acids

9. Note that the terms "essential" and "nonessential" as applied to amino acids refer to the ability or inability of *mammals* to synthesize these compounds.

10. What are the six primary precursor molecules from which all 20 standard amino acids are derived?

11. From what is PRPP synthesized?

α-Ketoglutarate Gives Rise to Glutamate, Glutamine, Proline, and Arginine

12. What is the other product of the argininosuccinase reaction that produces arginine?

Serine, Glycine, and Cysteine Are Derived from 3-Phosphoglycerate

13. What is the role of PLP in the biosynthesis of these amino acids?

Three Nonessential and Six Essential Amino Acids Are Synthesized from Oxaloacetate and Pyruvate

14. Which amino acids are synthesized by simple transamination of carbohydrate metabolites?

15. What contributes the reducing power for these biosynthetic reactions?

Chorismate Is a Key Intermediate in the Synthesis of Tryptophan, Phenylalanine, and Tyrosine

16. Why is tyrosine not considered an essential amino acid?

Histidine Biosynthesis Uses Precursors of Purine Biosynthesis

17. How is the role of ATP in this pathway unusual?

Amino Acid Biosynthesis Is under Allosteric Regulation

18. What three types of regulation are used in coordinating amino acid biosynthesis? Under what conditions is each type of regulation more useful than the other two?

22.3 Molecules Derived from Amino Acids

Glycine Is a Precursor of Porphyrins

19. What structural characteristic of porphyrin-type molecules causes them to react strongly to ultraviolet light?

Box 22–2 On Kings and Vampires

20. Why do you think that α amino β ketoadipate does not accumulate in individuals with acute intermittent porphyria?

Heme Is the Source of Bile Pigments

21. Generate a "color wheel" for the heme degradation products to help you remember the various chemical changes. The names of some of the products should also be helpful (bili<u>verd</u>in is green)?

Amino Acids Are Precursors of Creatine and Glutathione

22. What are the functions of phosphocreatine and glutathione?

D-Amino Acids Are Found Primarily in Bacteria

23. What is the enzyme target of the antibacterial agent cycloserine?

Aromatic Amino Acids Are Precursors of Many Plant Substances

24. What are the functions of the plant compounds lignin and auxin?

Biological Amines Are Products of Amino Acid Decarboxylation

25. What do all the amino acid decarboxylases have in common?

Box 22–3 Curing African Sleeping Sickness with a Biochemical Trojan Horse

26. Why does DFMO cause no harm to human patients, yet kill the trypanosome parasites?

Arginine Is the Precursor for Biological Synthesis of Nitric Oxide

27. What category of hormones does nitric oxide represent?

28. What intracellular signal turns on the production of nitric oxide?

22.4 Biosynthesis and Degradation of Nucleotides

29. What are the roles of nucleotides? Do these functions provide clues as to why their synthesis includes both de novo and salvage pathways?

De Novo Purine Nucleotide Synthesis Begins with PRPP

30. Which molecules contribute the nitrogen atoms in purine synthesis? What is the source of energy? From where are the carbons derived?

Purine Nucleotide Biosynthesis Is Regulated by Feedback Inhibition

31. Sequential and concerted feedback inhibition regulate purine biosynthesis. Is enzyme multiplicity a control mechanism here as well?

Pyrimidine Nucleotides Are Made from Aspartate, PRPP, and Carbamoyl Phosphate

32. What are the general similarities and the differences in the de novo synthetic schemes of purines and pyrimidines?

Pyrimidine Nucleotide Biosynthesis Is Regulated by Feedback Inhibition

33. Why is it logical that ATP prevents the inhibition of aspartate transcarbamoylase by CTP?

Nucleoside Monophosphates Are Converted to Nucleoside Triphosphates

34. Why is the relative nonspecificity of nucleoside monophosphate kinases and of nucleoside diphosphate kinase so important in cells?

Ribonucleotides Are the Precursors of Deoxyribonucleotides

35. What are the sources of reducing power for the reduction of the D-ribose moiety of ribonucleotides?

36. How many active sites are there in a functional dimer of ribonucleotide reductase? How many regulatory sites?

37. How do ATP, low [dATP], and high [dATP] each affect the activity of ribonucleotide reductase?

38. Why is it crucial that deoxyribonucleotide concentrations are balanced within the cell?

Thymidylate Is Derived from dCDP and dUMP

39. Why must the concentration of dUTP be kept low?

Degradation of Purines and Pyrimidines Produces Uric Acid and Urea, Respectively

40. In humans, is the excretion of uric acid a large fraction of the total excreted nitrogen?

Purine and Pyrimidine Bases Are Recycled by Salvage Pathways

41. Calculate the input of energy necessary to synthesize nucleotide bases; this will give you an idea about the need for salvage pathways for these bases.

Excess Uric Acid Causes Gout

42. How are gout and Lesch-Nyhan syndrome similar, and how are they different?

Many Chemotherapeutic Agents Target Enzymes in the Nucleotide Biosynthetic Pathways

43. What is the mechanism for azaserine's inhibition of nucleotide biosynthetic pathways?

44. Which enzymes are inactivated by fluorouracil and methotrexate?

45. *Cell Map:* We recommend that you don't include all the amino acid and nucleotide synthesis and degradation pathways on your map because it would become incomprehensible. Instead, indicate the glycolytic, pentose phosphate pathway; amino acid breakdown; and/or citric acid cycle intermediates that are used to generate the various amino acids and nucleotides.

Discussion Questions for Study Groups

- What structural feature(s) of bilirubin make it useful as an antioxidant? What other biomolecules have you studied that can act as antioxidants?

SELF-TEST

Do You Know the Terms?

ACROSS

1. Cys, Ser, and Gly are all derivatives of this common intermediate.
6. Type of inhibition in which each enzyme of an enzymatic series is inhibited by the product of its reaction. (2 words)
7. Series of reactions by which nitrogen in the atmosphere and in the proteins of organisms is circulated, salvaged, and reused. (2 words)
9. Product of the committed step in pyrimidine biosynthesis.
11. Common metabolic intermediate in the synthesis of the purines, AMP, GMP, and IMP.
13. Enzyme with 12 subunits that are regulated allosterically and covalently; produces an intermediate for at least six separate metabolic pathways. (2 words)
16. Process in which electrons from oxidative phosphorylation are transferred to NO_3^-.
18. Regulatory strategy known as enzyme ________ uses different isozymes, which can be regulated independently, to catalyze the same reaction.
20. Class of enzymes catalyzing synthesis of di- and triphosphates from nucleoside monophosphates.
21. Recycling is something biological systems have been doing for millennia; an example is the ________ pathways for purine and pyrimidine bases.
22. Type of enzyme involved in biosynthesis of most amino acids.
23. ________ transferase "senses" [ATP] in cells and activates glutamine synthetase by deadenylylation when [ATP] is high.
24. Common precursor for the biosynthesis of Ile, Val, Leu, and Ala.
25. ________ inhibition: synergistic (more than additive) effects of feedback, covalent, and allosteric inhibition of an enzyme.

DOWN

2. Important intermediate in biosynthesis of all nucleotides; generated from glucose via the pentose phosphate pathway. (abbr.)
3. In pyrimidine biosynthesis, compound formed from attachment of the ribose 5-phosphate moiety to orotate.
4. Important reducing agent when present as a tripeptide; when oxidized, consists of two tripeptides linked by a disulfide bond.
5. *Fixation* of atmospheric nitrogen is a complex process that ________ N_2 to NH_3 or NH_4^+.
8. A number of important signaling molecules are derived from the decarboxylation of amino acids; for example, this amino acid is decarboxylated to yield dopamine, norepinephrine, and epinephrine.
9. Supramolecular structure containing dinitrogenase reductase and dinitrogenase. (2 words)
10. More efficient nitrogen carrier than glutamate.
12. In mammals, most amino acids are derived from a few common precursor molecules using relatively simple enzymatic pathways; the remainder are ________ amino acids that cannot be synthesized endogenously.
14. Process of converting "fixed" nitrogen to NO_2^- and NO_3^-.
15. Aspartate can be used to generate these five amino acids. (one-letter abbrs.)
17. The bacterial enzyme ________ synthase incorporates reduced nitrogen into proteins and nucleotides.
19. Phe, Trp, and Tyr are all derived from this common precursor.

Do You Know the Facts?

1. Which of the following correctly describes the nitrogen cycle?

A. Fixation of atmospheric nitrogen by nitrogen-fixing bacteria yields bioavailable nitrate.
B. Nitrate is reduced to ammonia in a process known as denitrification.
C. Biological nitrogen fixation is carried out by a complex of proteins called the nitrogenase complex.
D. A and B
E. A, B, and C

2. Which of the following is *not* true of nitrogen fixation by the nitrogenase complex?

A. This enzyme complex is inactivated by exposure to oxygen.
B. ATP binding causes a change in the conformation of the reductase moiety.
C. The primary role of ATP is to drive nitrogen fixation through the hydrolysis of PP_i.
D. Nitrogen fixation occurs only in prokaryotes.
E. The final electron acceptor in this process is N_2.

3. Which of the following is the immediate product of nitrogen fixation (i.e., the product of the reaction catalyzed by the nitrogenase complex)?

A. N_2
B. NO_3^-
C. NH_4^+
D. Urea
E. Amino acids

4. Which cofactor is essential to all transamination reactions?

A. PRPP
B. CoA
C. ATP
D. PLP
E. adoMet (*S*-adenosylmethionine)

5. Synthesis of the amino acid glutamine, catalyzed by glutamine synthetase, has been studied extensively in bacteria. Indicate whether each of the following statements about glutamine synthetase and/or the reaction it catalyzes is true or false (in bacteria, unless otherwise noted).

_______ **(a)** The enzyme catalyzes the reaction: glutamate + NH_4^+ + ATP → glutamine + ADP + P_i + H^+.
_______ **(b)** In mammals, this is the main pathway for converting toxin-free ammonia into nontoxic glutamine for transport in the blood.
_______ **(c)** Each of the enzyme's 12 subunits is subject to allosteric regulation.
_______ **(d)** The enzyme is covalently inactivated by the addition of a phosphate group.
_______ **(e)** The enzyme is activated by the covalent attachment of AMP.
_______ **(f)** The complex of adenylyl transferase and P_{II}-UMP catalyzes deadenylylation of the enzyme.
_______ **(g)** ATP is a reactant in the reaction.
_______ **(h)** ATP is a regulator in the reaction.
_______ **(i)** Uridylylation of the enzyme is stimulated by α-ketoglutarate.
_______ **(j)** High concentrations of glutamine itself, and of end products of glutamine metabolism, decrease the enzyme's activity.
_______ **(k)** The enzyme is covalently activated by uridylylation.

6. Which of the following does *not* provide a carbon skeleton for the synthesis of amino acids?

A. succinate
B. α-ketoglutarate
C. pyruvate
D. oxaloacetate
E. ribose 5-phosphate

7. Shown below are the structures of three metabolic intermediates that can be used in amino acid synthesis. Which is correctly paired with its amino acid end product(s)?

#1

COO^-
CH_2
$HO-C-COO^-$
CH_2
COO^-

#2

COO^-
$C=O$
CH_2
COO^-

#3

O S-CoA
C
CH_2
CH_2
COO^-

A. #1; serine, glycine, cysteine
B. #2; alanine, valine, leucine
C. #3; glutamate, glutamine, proline
D. #1; histidine
E. #2; methionine, threonine, lysine

8. What does the term "essential" mean in terms of amino acids in the human diet?

A. Necessary for all protein synthesis
B. Only available in animal protein
C. Cannot be synthesized by humans
D. Cannot be coded for by DNA
E. Cannot be degraded in the liver

9. Which of the following is *not* a physiological role of nucleotides?

A. Allosteric regulators
B. Intermediates for biosynthetic processes
C. Components of many proteins
D. Components of the coenzymes NAD, FAD, and CoA
E. Intracellular signaling molecules

10. How is the type of feedback inhibition known as enzyme multiplicity correctly described?

A. Different isozymes in a pathway are independently controlled by different modulators; this prevents one biosynthetic product from shutting down key steps in a pathway when other products of the same pathway are needed.
B. A metabolite inhibits its own formation at several different points with multiple, overlapping inhibition in the synthetic pathway.
C. Metabolite synthesis is inhibited at one enzymatic point by more than one end product of its own metabolism; effects of the different inhibitors on the enzyme are more than additive, and all the inhibitors together more or less shut down the enzyme.
D. Enzyme multiplicity refers to an enzyme cascade that results in the covalent modification of a key regulatory enzyme.
E. All of the above are true.

11. Consider the following series of enzymatic reactions involved in the synthesis of two newly discovered bacterial amino acids:

Amino acid 1, vivekine, is formed by oxygenation of a compound called previvekine in a reaction catalyzed by previvekine oxygenase; this enzyme is inhibited by high concentrations of the product vivekine.

Amino acid 2, avanic acid, is formed by carboxylation of vivekine, catalyzed by vivekine carboxylase; this enzyme is inhibited by high concentrations of avanic acid, which also inhibits previvekine oxygenase.

Previvekine is synthesized from a compound called proctorol in a reaction catalyzed by proctorol reductase; this enzyme is inhibited by vivekine and by avanic acid; when both of these products are present, the inhibitory effect is more than additive.

Regulation of this reaction sequence is an example of:

A. sequential feedback inhibition.
B. enzyme multiplicity.
C. concerted feedback inhibition.
D. A and C.
E. A, B, and C.

12. Indicate (yes/no) whether each of the following describes a way in which orotate is used in pyrimidine biosynthesis.

_________ **(a)** It condenses with PRPP to form orotidylate and then UMP.
_________ **(b)** It contributes a portion of its carbon atoms to the formation of PRPP.
_________ **(c)** It is constructed on PRPP through 10 separate reactions.
_________ **(d)** It contributes a portion of its carbon atoms to the formation of the pyrimidine base via tetrahydrofolate.
_________ **(e)** An amino group is transferred to the C-1 of orotate to form CTP.
_________ **(f)** It is formed from cytosolic carbamoyl phosphate.
_________ **(g)** It is formed from mitochondrial carbamoyl phosphate.

13. Indicate (yes/no) whether each of the following is a way in which inosinate is used in purine biosynthesis.

_________ **(a)** It condenses with PRPP to form IMP.
_________ **(b)** An amino group is transferred to inosinate to form AMP.
_________ **(c)** It is deaminated to form GMP.
_________ **(d)** It contributes a portion of its carbon atoms to the formation of the purine base via tetrahydrofolate.
_________ **(e)** It is a precursor of PRPP.
_________ **(f)** It is constructed on PRPP through 10 separate reactions.
_________ **(g)** It is formed from cytosolic carbamoyl phosphate.

14. In nucleotide metabolism, all of the following are true *except:*

A. The committed step in purine biosynthesis is the transfer of an amino group to PRPP.
B. Purine and pyrimidine biosynthesis are regulated by end-product inhibition.
C. Nucleotides can be synthesized in a single reaction via salvage pathways.
D. De novo pyrimidine synthesis begins with a molecule of PRPP.
E. Orotidylate is the common precursor in the biosynthesis of pyrimidines, and inosinate is the common precursor in the biosynthesis of the purines ATP and GTP.

15. Which of the following describes the activity and regulation of ribonucleotide reductase and/or its importance to the cell?

A. Both its activity and its substrate specificity are regulated by the binding of effector molecules.
B. ATP increases the overall activity of the enzyme.
C. Control of the enzyme's activity ensures a balanced pool of precursors for DNA synthesis.
D. Balanced pools of deoxyribonucleotides are necessary in DNA synthesis, given the complementary base-pairing of nucleotides in double-stranded DNA.
E. All of the above are true.

Applying What You Know

1. The hereditary condition called orotic aciduria has symptoms of retarded growth and severe anemia, and high levels of orotate (orotic acid) are excreted. In this condition, either one or both of the enzymes orotate phosphoribosyltransferase and orotidylate carboxylase are absent. When patients with this disease are fed uridine or cytidine, there is a significant improvement in the anemia and a decrease in the production of orotate. How can these effects be explained?

ANSWERS

Do You Know the Terms?

ACROSS

1. 3-phosphoglycerate
6. sequential feedback
7. nitrogen cycle
9. *N*-carbamoylaspartate
11. inosinate
13. glutamine synthetase
16. denitrification
18. multiplicity
20. kinases
21. salvage
22. aminotransferase
23. adenylyl
24. pyruvate
25. concerted

DOWN

2. PRPP
3. orotidylate
4. glutathione
5. reduces
8. tyrosine
9. nitrogenase complex
10. glutamine
12. essential
14. nitrification
15. KIMNT (or a variation of this)
17. glutamate
19. chorismate

Do You Know the Facts?

1. C
2. C
3. C
4. D
5. **(a)** T; **(b)** T; **(c)** T; **(d)** F; **(e)** F; **(f)** T; **(g)** T; **(h)** T; **(i)** F; **(j)** T; **(k)** F
6. A
7. E
8. C
9. C
10. A
11. D
12. **(a)** Y; **(b)** N; **(c)** N; **(d)** N; **(e)** Y; **(f)** Y; **(g)** N
13. **(a)** N; **(b)** Y; **(c)** N; **(d)** N; **(e)** N; **(f)** Y; **(g)** N
14. D
15. E

Applying What You Know

1. Orotate phosphoribosyltransferase catalyzes the conversion of orotate to orotidylate and orotidylate carboxylase converts orotidylate to uridylate (UMP), a precursor in the biosynthesis of both UTP and CTP. Uridine (UMP) and cytidine (CMP) circumvent these enzymatic steps and can be phosphorylated to provide the necessary nucleotides, UTP and CTP, for DNA synthesis in rapidly growing red blood cells. This would improve the anemia. A logical explanation for the decrease in orotate excreted might seem to be that the CTP produced could act as a feedback inhibitor of the enzyme aspartate transcarbamoylase, so that orotate would not be synthesized in high concentrations. In *animals,* however, ATCase is not the regulatory enzyme, as it is in bacteria. Pyrimidine synthesis is controlled at the level of carbamoyl phosphate synthetase II, which is inhibited by UTP, and this is the reason for the decrease in orotic acid excretion in the treated patients.

chapter

23 Hormonal Regulation and Integration of Mammalian Metabolism

STEP-BY-STEP GUIDE

Major Concepts

Mammalian tissues are specialized for different metabolic functions.

Although the anabolic and catabolic pathways that are critical to the function of organisms occur within individual cells, many cells are specialized for particular functions. Cells may associate into tissues that specialize in different aspects of cellular metabolism.

The liver monitors and maintains the blood levels of key metabolites, specifically sugars, amino acids, and fatty acids.

Glucose 6-phosphate has a central role in five major metabolic pathways: (1) Only liver cells contain the enzyme that can dephosphorylate glucose 6-phosphate to glucose, which can be released into the blood for use by other tissues; alternatively, (2) glucose 6-phosphate is converted into glycogen for storage. (3) In a pinch, glucose 6-phosphate can be oxidized to produce ATP, although liver cells prefer to use fatty acids. (4) Excess glucose 6-phosphate is used to make acetyl-CoA for the production of triacylglycerols and lipids. (5) Glucose 6-phosphate can be shunted into the pentose phosphate pathway, producing NADPH and/or ribose 5-phosphate. **Amino acids** can be (1) used as synthetic precursors for proteins, nucleotides, and hormones in the liver itself, or (2) exported in the blood to other tissues; alternatively, (3) they are degraded to acetyl-CoA and citric acid cycle intermediates. (4) Alanine can also be deaminated to pyruvate and converted to glucose for export in the blood. **Fatty acids** from lipids provide (1) a major source of acetyl-CoA for energy production in liver, (2) a source of ketone bodies for energy production in extrahepatic tissues, (3) precursors for liver lipid synthesis, and (4) precursors for cholesterol biosynthesis. They can also be exported (5) to adipose tissue for storage or (6) for use as an energy source by muscle.

Adipose tissue stores and releases fatty acids.

This tissue is specialized for the uptake of lipids from the blood and for lipid storage in large fat droplets within cells. Under the appropriate conditions, stored triacylglycerols can be hydrolyzed by lipases and their components released into the blood.

Muscle tissue contains contractile proteins that use ATP to do mechanical work.

Glucose, fatty acids, and ketone bodies can all be converted to acetyl-CoA, which is completely oxidized to CO_2 via the citric acid cycle. Glucose from stored glycogen can be rapidly converted to lactic acid, producing ATP by anaerobic fermentation. The high phosphate transfer potential of stored phosphocreatine is also used to generate ATP from ADP.

Brain tissue is a specialized signal transduction organ.

It receives information from body systems as well as from external sources, integrates this information, and then releases hormones that regulate the metabolic demands of different tissues. Brain cells rely primarily upon glucose metabolism to meet the high energy demands of nerve cells.

Blood cells connect the various body tissues by relaying information from the brain (via hormones) and by transporting metabolites and oxygen.

Signals from the brain cause cells of the neuroendocrine system to release hormones into the circulatory system. The circulatory system carries hormones to peripheral tissues where they regulate metabolic pathways.

The neuroendocrine system has a hierarchical structure.

The **hypothalamus** is the interface between the informational input from the brain and the hormonal output of the pituitary. The hypothalamus secretes releasing factors into a local network of blood vessels; these factors bind to receptors on the **anterior pituitary** causing them to release hormones into the general circulation. The **posterior pituitary** contains the endings of hypothalamic neurons, which release hormones directly into the circulatory system.

Hormone molecules are often grouped into the following major categories: peptide hormones, amine hormones, steroid hormones, thyroid hormones, and eicosanoids.

Water-soluble hormones such as peptide hormones and catecholamines interact with specific target cells via membrane-bound receptors. Lipid-soluble hormones, such as the steroid and thyroid hormones as well as vitamin D hormone and nitric oxide, are all lipid soluble and therefore do not require membrane-bound receptors to influence metabolic processes; these hormones interact with intracellular receptors.

Metabolic processes are tightly regulated by the concerted actions of hormones.

For example, the amine hormone **epinephrine** causes a number of physiological and metabolic effects that result in increased mobilization of glucose (i.e., it promotes glycogen breakdown) and a decrease in enzyme activity in fuel storage pathways. **Glucagon,** a peptide hormone, also increases blood glucose levels. Like epinephrine, it promotes the breakdown of glycogen but it also increases gluconeogenesis in the liver. The net effect is that more glucose is exported to the blood. The peptide hormone insulin opposes the effects of epinephrine and glucagon. **Insulin** is released in response to elevated levels of blood glucose and stimulates glucose uptake and storage by a variety of cell types. The connection between high blood glucose and insulin secretion is made in cells of the pancreas where glucose uptake stimulates ATP production via glycolysis. This ATP acts as an intracellular signal to close K^+ channels, depolarizing the cells and causing insulin secretion. **Diabetes mellitus,** a disease caused by a deficiency in the secretion or action of insulin, causes numerous metabolic changes that can be life-threatening, illustrating the importance of balanced control of carbohydrate metabolism.

Regulation of body mass involves multiple signaling pathways.

Control of food consumption represents regulation at the behavioral level. Leptin, a peptide hormone, is a feedback signal produced by adipose cells that alters feeding behavior by acting on the brain to reduce food intake. Insulin, in addition to its effects on enzyme activity, also acts on the brain to decrease food consumption. Conversely, neuropeptide Y (NPY) is a signal released during starvation that stimulates feeding behavior.

Regulation of food intake involves a complex cascade of regulatory events.

Leptin, insulin, ghrelin, and $PYY_{3\text{-}36}$ are key players in the hormonal feedback loops designed to maintain body weight. Leptin, insulin, and $PYY_{3\text{-}36}$ all inhibit appetite and/or stimulate metabolism and weight loss. For example, leptin causes an increase in the production of the mitochondrial uncoupling protein in adipocytes leading to thermogenesis and the expenditure of energy. Ghrelin opposes the actions of these hormones, causing a stimulation of appetite resulting in weight gain. The actions of these hormones are primarily exerted through neurons located in the hypothalamus.

What to Review

Answering the following questions and reviewing the relevant concepts, which you have already studied, should make this chapter more understandable.

- All of the following metabolic pathways must be integrated and regulated in cells in order to keep the supplies of metabolites balanced with cellular demand. You should now be familiar with all of these pathways; an overview or summary of each can be found in the figure indicated.

  Glycolysis (Fig. 14–2)

  Fates of pyruvate (Fig. 14–3)

  Pentose phosphate pathway, an alternative to glycolysis (Fig. 14–20)

  Citric acid cycle (Fig. 16–7)

  Glyoxylate cycle, a variation of the citric acid cycle in some organisms (Fig. 16–20)

  Fatty acid oxidation (Fig. 17–8)

  Amino acid oxidation (Fig. 18–1)

  Urea cycle (Fig. 18–10)

  Oxidative phosphorylation:

  Electron transfer chain (Fig. 19–15)

  ATP synthesis and the chemiosmotic model (Figs. 19–17 and 19–24)

  Shuttle systems for cytosolic NADH generated by glycolysis:

  Malate-aspartate shuttle (Fig. 19–27)

  Glycerol 3-phosphate shuttle (Fig. 19–28)

  Photophosphorylation (Fig. 19–49)

  Gluconeogenesis, compared with the glycolytic pathway (Fig. 14–16)

  Carbon fixation, overview (Fig. 20–4)

  Fatty acid biosynthesis (Figs. 21–2 and 21–4)

  Triacylglycerol synthesis (Figs. 21–17 and 21–18)

Nitrogen fixation (Fig. 22–2)
Amino acid synthesis (Fig. 22–9)
Nucleotide synthesis:
  Purines (Fig. 22–35)
  Pyrimidines (Fig. 22–36)
Nucleotide degradation:
  Purines (Fig. 22–45)
  Pyrimidines (Fig. 22–46)

- You should now have a greater appreciation of Figures 16–1 and 16–15. These figures show how a number of key metabolic pathways are interconnected.
- Several **molecular shuttles** are used to transport metabolic intermediates between tissues. Review the glucose-alanine cycle (Fig. 18–9) and the Cori cycle (Box 14–2). Note how these shuttles contribute to the specialization of tissue functions.
- Information carried by hormones is critical to the regulation of metabolic pathways. This information is often relayed to the interior of cells by intracellular messengers; two intracellular messengers are derived from membrane lipids. Review the structure of **membrane lipids** (Fig. 10–7) and note the sites of cleavage by lipases (Fig. 10–16)
- Protein phosphorylation plays a critical role in the regulation of enzyme activity by hormones and their intracellular messengers. Review which of the amino acids have an available —OH group that can be phosphorylated (Fig. 3–5).

Topics for Discussion

Answering each of the following questions, especially in the context of a study group discussion, should help you understand the important points of this chapter.

This chapter is about the integration of biochemical pathways. This can become pretty complicated, so if you have been using your cell map to keep track of the various biochemical pathways, you can use it as a guide for understanding this chapter. If you have not yet started recording pathway information on your map, you can use it now to review what you have learned and to visually organize the information in a cellular context.

23.1 Hormones: Diverse Structures for Diverse Functions

The Detection and Purification of Hormones Requires a Bioassay

1. How did scientists measure the effects of insulin?

2. What features of the radioimmunoassay technique make it so sensitive?

Hormones Act through Specific High-Affinity Cellular Receptors

3. How is it possible that hormones found in tiny concentrations in the blood can have such profound effects (e.g., "flight or fight" response) on specific organs or organ systems?

4. Draw a model or "cartoon" that illustrates how each type of receptor works and then list the different signaling molecules that use each type of receptor under the corresponding model.

Hormones Are Chemically Diverse

5. Which classes of hormones require cell membrane localized receptors, and which can bind to intracellular receptors?

6. How is nitric oxide (NO) released from a cell? Would you expect this signaling molecule to act in endocrine, paracrine, or autocrine fashion(s)?

Hormone Release is Regulated by a Hierarchy of Neuronal and Hormonal Signals

7. In what ways can the hypothalamus be viewed as the coordinating center of the endocrine system?

8. Why is it significant that coordination of the endocrine system takes place in a region of the brain rather than in endocrine tissue?

23.2 Tissue-Specific Metabolism: The Division of Labor

The Liver Processes and Distributes Nutrients

9. Which pathways for energy production in the liver use glucose 6-phosphate?

10. What are the possible fates of acetyl-CoA produced by the liver?

11. What tissue specializations make the liver such a critical organ in the regulation of blood glucose levels?

Adipose Tissues Store and Supply Fatty Acids

12. What types of lipids are stored in adipocytes?

13. Under what conditions are these lipids "mobilized" (i.e., hydrolyzed to release fatty acids that can be oxidized, producing acetyl-CoA and ultimately ATP)?

14. How are the enzymes (lipases) that hydrolyze lipids activated?

Brown Adipose Tissue is Thermogenic

15. What are the different roles of the chemiosmotic proton gradient in BAT and WAT?

Muscle Uses ATP for Mechanical Work

16. Which compounds are used as energy sources in muscle tissue and under what conditions?

17. How do muscle cells use the energy in ATP to do mechanical work?

18. What adaptations do muscles employ to ensure adequate supplies of energy during periods of intense activity?

The Brain Uses Energy for Transmission of Electrical Impulses

19. Which compounds are used as energy sources by the brain?

20. Why does the brain consume as much as 20% of the O_2 required by a resting human (i.e., what is the major energy-consuming metabolic process in brain cells)?

Blood Carries Oxygen, Metabolites, and Hormones

21. Name some of the metabolic intermediates you would expect to find in the blood.

23.3 Hormonal Regulation of Fuel Metabolism

Insulin Counters High Blood Glucose

22. How does the binding of insulin to tyrosine kinase receptors on hepatocytes affect metabolism of those cells? (Hint: see Chapter 12.)

23. Why does the insulin-stimulated increase in the uptake and conversion of blood glucose to glucose 6-phosphate make sense physiologically?

24. Based on what you have learned about the effects of epinephrine on glycogen phosphorylase, predict how insulin regulates this enzyme's activity.

Pancreatic β Cells Secrete Insulin in Response to Changes in Blood Glucose

25. What is the biochemical connection between increased glucose metabolism and increased insulin release?

26. Use Fig. 23–28 to explain why individuals with mutations in the ATP-gated K^+ channel that result in a constantly open channel require insulin therapy.

Glucagon Counters Low Blood Glucose

27. Draw a molecular model that explains the effects of glucagon on liver cell metabolism.

28. At which point(s) is it possible for the actions of epinephrine and glucagon to be integrated (i.e., where do these two systems interact in the above model)?

29. What are the other targets of glucagon that complement its effects on liver cells that raise blood glucose levels?

During Fasting and Starvation, Metabolism Shifts to Provide Fuel for the Brain

30. Why does the citric acid cycle in hepatocytes slow and eventually stop during fasting, even though acetyl-CoA (from lipid oxidation) is available?

31. What happens to the acetyl-CoA that becomes available from stored fats during starvation?

Epinephrine Signals Impending Activity

32. How is it possible for a single hormone, such as epinephrine, to affect the activity of different enzymes in three separate tissue types (liver, muscle, and adipose tissue)?

33. What is the *molecular mechanism,* initiated by epinephrine, that results in a stimulation of glycogen breakdown in liver, a stimulation of glycolysis in muscle, and an increase in fat mobilization in adipose tissue?

34. How does epinephrine *coordinate* enzyme activity in the gluconeogenic and glycogen breakdown pathways?

Cortisol Signals Stress, Including Low Blood Glucose

35. Cortisol is derived from cholesterol; how does this hormone exert its effects on cell metabolism?

36. What enzyme activities does it influence, and how do these effects compare to those of epinephrine and glucagon?

Diabetes Mellitus Arises from Defects in Insulin Production or Action

37. Diabetics have increased levels of acetyl-CoA due to the increased oxidation of lipids, yet this acetyl-CoA is used to produce acetoacetyl-CoA and β-hydroxybutyrate (ketone bodies) rather than energy via the citric acid cycle. Explain why this is so.

23.4 Obesity and the Regulation of Body Mass

Adipose Tissue Has Important Endocrine Functions

38. Leptin is a signaling molecule made by fat cells. Would you expect an individual with excess fat to have higher or lower levels of leptin than a leaner individual?

39. What is the cellular role of leptin in weight regulation?

Leptin Stimulates Production of Anorexigenic Peptide Hormones

40. Would you expect leptin's effects on the neurons in the brain that stimulate eating (the NPY-releasing neurons) to be positive or negative?

41. What would you predict leptin's effects would be on the neurons that suppress eating behavior (the α-MSH releasing neurons)?

Leptin Triggers a Signaling Cascade That Regulates Gene Expression

42. What other signaling systems require receptor dimerization and tyrosine phosphorylation?

The Leptin System May Have Evolved to Regulate the Starvation Response

43. Why is it thought that leptin plays a role in starvation responses and not primarily in preventing obesity? (Hint: Is obesity generally a threat to survival in animal populations other than human?)

Insulin Acts in the Arcuate Nucleus to Regulate Eating and Energy Conservation

44. What would be the effect on cell metabolism of leptin activation of PI3 kinase (a "downstream event" triggered by insulin)?

Adiponectin Acts through AMPK to Increase Insulin Sensitivity

45. Assuming that adiponectin exerts its effects via AMPK, propose a mechanism whereby an increase in AMPK activity would enhance the effects of insulin. (Hint: think of "downstream events.")

Diet Regulates the Expression of Genes Central to Maintaining Body Mass

46. What are the signaling molecules that activate PPARs?

47. Where in Figure 23–35 would PPARδ be expected to exert its effects?

Short-Term Eating Behavior Is Influenced by Ghrelin and PYY_{3-36}

48. Figure 23–39 shows a dramatic cycling in the plasma levels of ghrelin; where would you expect to see peaks in PYY_{3-36} levels?

Discussion Questions for Study Groups

- The only difference between the effects of a neurotransmitter and the effects of a hormone lies in the timing of the response. Do you agree with this statement? Why or why not?
- Can you make any generalizations concerning the cellular targets of the different classes of hormones? How does the time course for the physiological effects of hormones correlate with their molecular/cellular effects?
- What are the advantages of having a neuroendocrine system with such a complex cellular hierarchy?

SELF-TEST

DO YOU KNOW THE TERMS?

ACROSS

2. The inability to regulate blood glucose levels causes _____, which is the production of copious amounts of urine containing large amounts of glucose.

3. General name for an enzyme that catalyzes protein phosphorylation.

6. Carbon backbones from this class of amino acids can be used by the liver for gluconeogenesis.

7. A peptide hormone that decreases the rate of gluconeogenesis and glycogen breakdown in hepatocytes.

9. Oxytocin and vasopressin belong to this class of hormone.

10. Home pregnancy tests are based on this technique, which uses an enzyme-linked immunosorbent to detect small quantities of the hormone HCG present in urine. (abbr.)

11. Overproduction of ketone bodies from metabolized fatty acids occurs in fasting individuals and is characterized by the odor of acetone on their breath; this condition is called ______.

12. Prostaglandins are examples of local regulators that act in a ______ fashion; they are secreted by immune system cells and diffuse to neighboring target cells where they exert their effects.

14. This second messenger is produced by a cyclase that is located in the cytosol and not in the plasma membrane. (abbr.)

15. Derived from membrane phospholipids, one member of this class of hormones is secreted by cells of the placenta causing the nearby muscle of the uterus to contract, inducing labor.

19. It is believed that chronic overeating leads to a loss of responsiveness to insulin, producing a form of diabetes known as ______. (abbr.)

20. Multiple hormonal "messages" can be carved out of this.

22. One way to reverse the regulatory effects of a protein kinase is through the activation of a protein _____.

24. The leptin signal is mediated by signal transducers and activators of transcription, also known as ______ (abbr.), that move to the nucleus, bind to specific DNA sequences, and stimulate the expression of specific genes.

26. An essential component in the signaling pathway for all hormones and neurotransmitters.

27. A structure in the brain that releases hormones from neurons as well as endocrine cells.

28. A common metabolic disorder that can be treated by the exogenous replacement of insulin (abbr.).

DOWN

1. For this class of hormones, the secretory cell is also the target.

2. Low K_m isozyme of hexokinase that serves as a glucose "sensor" for the liver.

4. Uncontrolled diabetes can overwhelm the capacity of the bicarbonate buffering system of blood causing a condition called ______.

5. Endocrine cells in this portion of the pituitary secrete hormones in response to blood-borne signals from the hypothalamus.

6. Although this hormone is secreted when blood glucose levels (and system energy levels) are low, one of its primary effects is to inhibit glycolysis in hepatocytes.

8. A sensitive assay technique that relies on competition between unlabeled and radioactively labeled compounds for binding sites on antibody molecules is the ______ assay.

10. Neurons secrete neurotransmitters; ______ cells secrete hormones.
13. One of the main targets for hormones regulating metabolism and the storage site for triacylglycerols is ______ tissue.
16. Measurements of urine output from kidneys provide a ______ for the identification of hormones that can affect water balance such as the antidiuretic hormone.
17. (Clues 17 & 18) ______ secrete . . .
18. . . . ______, a peptide hormone that acts on the hypothalamus to curb feeding in mice.
21. This second messenger mediates the effects of epinephrine and/or glucagon on cells.
23. ______ hormones are derived from cholesterol and are lipid soluble; they do not require intracellular, second messengers.
25. Derived from the precursor protein thyroglobulin, ______ hormones stimulate energy-yielding metabolism by activating the expression of genes encoding key catabolic enzymes.

Do You Know the Facts?

1. Which compound links glycolysis, nucleotide synthesis, and glycogen synthesis?
A. Acetyl-CoA
B. Oxaloacetate
C. Citrate
D. Glucose 6-phosphate
E. Glycerol 3-phosphate

2. Anabolic and catabolic pathways for a given biomolecule almost never involve identical sequences of reactions because:
A. enzyme-catalyzed reactions are usually at equilibrium.
B. catabolic and anabolic pathways do not take place in the same cell.
C. they would be difficult to regulate and futile cycling would most likely occur.
D. the $\Delta G'^{\circ}$ for the reactions always favors the anabolic pathway.
E. it would be too easy for students of biochemistry to learn them.

3. Ingested fatty acids are degraded to acetyl-CoA, which is an allosteric activator of pyruvate carboxylase. What is the product of this enzyme reaction, and what is its effect on metabolism?
A. Oxaloacetate; increased rate of gluconeogenesis and the citric acid cycle
B. Oxaloacetate; decreased rate of fatty acid synthesis
C. Lactate; increased rate of glycolysis and gluconeogenesis
D. Acetyl-CoA; increased rate of the citric acid cycle
E. Acetoacetyl-CoA; decreased rate of triacylglycerol synthesis

4. Acetyl-CoA is a common metabolic intermediate in a number of metabolic pathways; this intermediate enables mammals to convert:
A. proteins to lipids.
B. sugars to triacylglycerol.
C. lipids to glycogen.
D. A and B only.
E. A, B, and C.

5. Rank the following energy sources according to the order in which they would be used by heart muscle tissue under fasting conditions.
(a) Cellular proteins
(b) Muscle glycogen
(c) Liver glycogen
(d) Triacylglycerol in adipose tissue
(e) Blood glucose
(f) ATP from other body tissues
(g) Blood acetoacetate

6. Which pair correctly matches the enzyme with its allosteric activator?
A. Hexokinase; ATP
B. Phosphofructokinase-1; AMP
C. Pyruvate kinase; ATP
D. Pyruvate dehydrogenase; NADH
E. Pyruvate carboxylase; ADP

7. How can glycolysis *and* gluconeogenesis be irreversible processes in cells?
A. They occur in separate cellular compartments.
B. Key reactions in each pathway are characterized by large and negative free-energy changes.
C. All the enzymes in the pathways are different, so the reaction pathways don't overlap.
D. A and B
E. A, B, and C

8. A cell that is deficient in pyruvate dehydrogenase phosphatase would exhibit all of the following metabolic effects except:
A. an increase in lactate production.
B. a decrease in pH.
C. an increase in the levels of citrate.
D. a decrease in the levels of NADH and ATP.
E. an increase in glycolytic activity.

9. Mature erythrocytes (red blood cells) are *full* of hemoglobin and have *no* organelles. They metabolize glucose at a high rate, generating lactate, which is transported to the liver for use in gluconeogenesis. Why do erythrocytes metabolize glucose to lactate in order to generate energy?
A. They have no citric acid cycle.
B. Oxygen is not available for the aerobic oxidation of glucose.
C. Anaerobic oxidation of pyruvate regenerates the NAD^+ needed for glycolysis to continue.
D. A and C
E. B and C

10. Predict the effect of a duplication of the gene for phosphorylase kinase in mutant liver cells relative to normal liver cells that contain only a single copy of this gene.
A. Glycogen synthase activity would be higher in mutant cells.
B. The rate of glycolysis would be increased in mutant liver cells.
C. The ratio of unphosphorylated/phosphorylated glycogen phosphorylase will be higher in mutant cells.
D. The activity of glycogen phosphorylase will be increased in mutant cells.
E. All of the above would be seen in mutant cells.

11. Which correctly describes the receptor-mediated effects of epinephrine in glucose metabolism? The hormone:
A. activates phosphorylase kinase.
B. stimulates the activity of glycogen synthase.
C. activates phosphodiesterase.
D. activates phospholipase C.
E. does all of the above.

12. Indicate whether each of the following is true or false.
______ **(a)** Insulin increases the intracellular concentration of glucose in hepatocytes.
______ **(b)** Glucagon is secreted in response to low blood glucose levels.
______ **(c)** Insulin increases the capacity of the liver to synthesize glycogen.
______ **(d)** Insulin has effects opposite to those of epinephrine and glucagon.
______ **(e)** Insulin, epinephrine, and glucagon all activate adenylate cyclase.
______ **(f)** Insulin activates phospholipase C.
______ **(g)** Glucagon increases triacylglycerol synthesis.
______ **(h)** Glycogen phosphorylase catalyzes glycogen breakdown in the absence of glucagon.

13. Which statement correctly describes steroid hormones?
A. Their intracellular actions are mediated by integral membrane proteins.
B. Their effects on cells require a water-soluble intracellular signal.
C. Their effects are mediated by binding to a water-soluble receptor protein.
D. They are synthesized directly from amino acid precursors.
E. Their effects usually involve the activation of other intracellular enzymes.

14. Of the following, the most *efficient* way to turn off glycogen degradation is to:
A. decrease the activity of phosphorylase kinase.
B. decrease the activity of phosphodiesterases.
C. increase the activity of phosphatases.
D. increase the activity of glycogen synthase.
E. decrease the intracellular levels of cAMP.

15. Untreated diabetics experience hyperglycemia, which is an excess of glucose in their blood. Any glucose present in the blood normally gets reabsorbed by transporters present in the nephrons of the kidney. In diabetics not all blood glucose is reabsorbed and excreted in the urine (hence, the name *diabetes mellitus,* meaning "sweet urine"). What is the most likely molecular explanation for this effect?
A. The transporters are inhibited by the presence of high levels of glucose.
B. The binding affinity of these transporters is increased in diabetics.
C. The V_{max} of the transporter is reached at lower blood glucose levels.
D. The binding capacity of the transporters has been exceeded.
E. C and D

16. Chromium is often taken as a dietary supplement and possible appetite suppressant. Data suggests that chromium functions as part of a complex that facilitates the binding of insulin to its receptor, thus enhancing the effect of insulin on cellular metabolism. Which of the following would be expected to occur in someone taking chromium supplements who had previously been deficient in this compound?
A. Enhanced activity of glucokinase, enhanced uptake of dietary fatty acids, and subsequent glycogen and triacylglycerol synthesis.
B. Increased rates of glycogen phosphorylase leading to an increase in the breakdown of stored glycogen.
C. Increased activity of lipases in adipocytes results in the breakdown of stored triacylglycerols.
D. Increased storage of glucose as glycogen in muscle cells.
E. Increased activity of fructose 1,6-bisphosphatase resulting in an increase in the rate of glycolysis in hepatocytes.

17. The *porcine stress syndrome,* where pigs die suddenly on their way to the slaughter house, is due to a genetic defect in metabolism. The symptoms are a dramatic rise in body temperature and both metabolic and respiratory acidosis, which occur in affected pigs that are exposed to high-stress situations. Uncontrolled futile cycling, in which ATP hydrolysis is greatly accelerated, is thought to be the basis for the rise in body temperature. Which of the following pairs of enzymes, assuming both are fully activated, would produce a futile cycle?
A. Glucose 6-phosphatase: fructose 1,6-bisphosphatase
B. Phosphofructokinase-1: phosphofructokinase-2
C. Glycogen phosphorylase kinase: glycogen synthase
D. Pyruvate carboxylase: phosphoenolpyruvate carboxykinase
E. Glycogen phosphorylase: hexokinase

Applying What You Know

1. Explain, in terms of enzyme kinetics, why glucokinase is responsible for channeling glucose into the glycogen synthesis pathway in the liver, and why hexokinase (an isozyme of glucokinase) regulates the entry of glucose into the glycolytic pathway in muscle.

2. Two patients exhibiting problems with their glucose metabolism had their blood tested for oxygen-binding ability. The oxygen-binding curves for the hemoglobin of each patient are shown below, along with the data for hemoglobin from a "normal" individual. Both patients were later tested for deficiencies in several glycolytic enzymes. Patient A was shown to be deficient in hexokinase, and patient B deficient in pyruvate kinase.

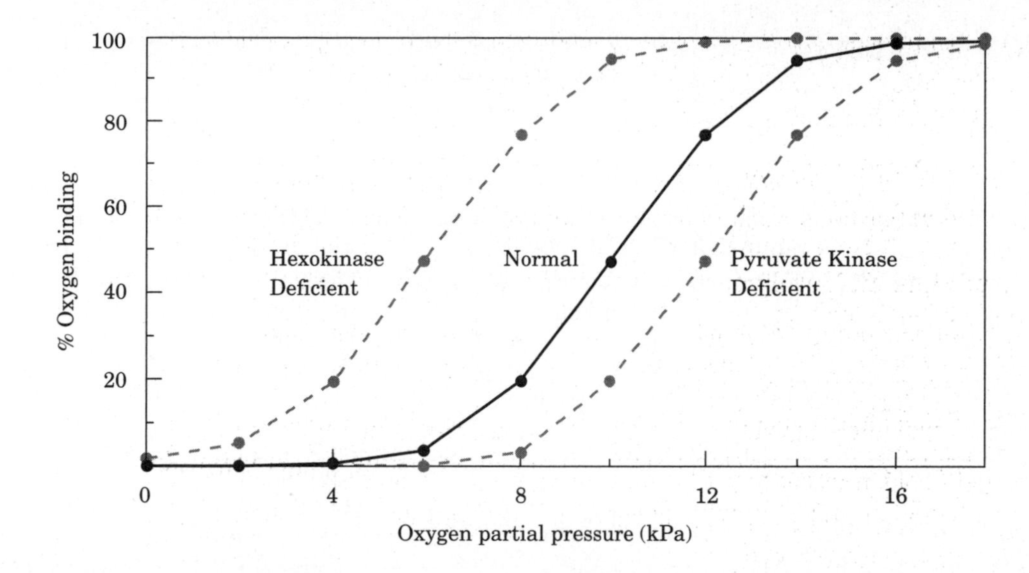

Why did the attending physician order the blood test? Why was the oxygen-binding test the first test to be done? Explain why the binding curves for the two patients are shifted relative to that for the normal individual.

3. In diabetes, **untreated** individuals have been said to "starve amidst plenty." Describe the actual metabolic situation that this phrase describes with respect to carbohydrate metabolism.
4. Untreated diabetics are often mistakenly thought to be inebriated due to the excessive production of ketone bodies resulting from the oxidation of fatty acids. The normal *inhibitory* effects of insulin on the oxidation of fatty acids are absent in diabetics. However, *positive* regulation of fatty acid oxidation also contributes to the increased levels of acetoacetate in diabetics. Describe the hormonal regulation of fatty acid oxidation and explain why untreated diabetics have ketones on their breath.

ANSWERS

Do You Know the Terms?

ACROSS

2. glucosuria
3. kinase
6. glucogenic
7. insulin
9. peptide
10. ELISA
11. ketosis
12. paracrine
14. cGMP
15. eicosanoids
19. NIDDM
20. prohormone
22. phosphatase
24. STATS
26. receptor
27. pituitary
28. IDDM

DOWN

1. autocrine
2. glucokinase
4. acidosis
5. anterior
6. glucagon
8. radioimmuno
10. endocrine
13. adipose
16. bioassay
17. adipocytes
18. leptin
21. cAMP
23. steroid
25. thyroid

Do You Know the Facts?

1. D
2. C
3. A
4. D
5. (e), (b), (c), (d), (g), (a). Note that (f) will *not* contribute as an energy source to heart muscle; ATP does not leave individual cells in large amounts and so is not a "transportable" energy source.
6. B
7. B
8. C
9. D
10. D
11. A
12. **(a)** T; **(b)** T; **(c)** T; **(d)** T; **(e)** F; **(f)** F; **(g)** F; **(h)** T
13. C
14. E
15. D
16. A
17. C

Applying What You Know

1. *Hexokinase* phosphorylates glucose to glucose 6-phosphate. This enzyme binds glucose with a relatively high affinity: it is half-saturated with substrate at [glucose] = 0.1 mM (about 50-fold less than the normal [glucose] in blood). At normal blood glucose levels, hexokinase binds glucose and "traps" it inside cells. In the liver, *glucokinase* (or hexokinase D) is half-saturated with substrate at [glucose] = 10 mM, 100-fold greater than the half-saturation level for hexokinase, but only twice the [glucose] in normal blood. Consequently, glucokinase is only slightly active at normal blood glucose levels, and makes a significant contribution to the glucose level in hepatocytes only when blood [glucose] is high.

Insulin is also released when blood [glucose] is elevated, resulting in the activation of glycogen synthase and inactivation of glycogen phosphorylase. The "trapped" glucose is therefore funneled into the glycogen synthesis pathway. This does not happen in muscle cells because they lack glucokinase and their energy demands are quite high, so that most of the available glucose is funneled into the glycolytic pathway.

2. As you now know, carbohydrate metabolism involves many interconnected anabolic and catabolic pathways. A defect in just one of the enzymes in any of these pathways can have profound effects on the way an individual metabolizes carbohydrates. It would be cumbersome (and expensive) to test *each* enzyme in all of these pathways for possible defects. However, there are ways to narrow the field to a small subset of enzymes. For example, defects in any of the glycolytic enzymes would be expected to affect the levels of 1,3-bisphosphoglycerate. This glycolytic intermediate is converted into 2,3-bisphosphoglycerate (BPG) by phosphoglycerate mutase. Because BPG has a profound effect on O_2 binding by hemoglobin, examining the O_2 binding of hemoglobin is a quick way to screen for defects in glycolytic enzymes. BPG binds in the center of the hemoglobin tetramer and stabilizes the deoxy form.

The patient with a leftward shift in the O_2-binding curve (patient A) has hemoglobin that is half-saturated at lower O_2 partial pressures, indicating a higher affinity for O_2. Reduced levels of BPG (reflecting reduced levels of 1,3-bisphosphoglycerate) could account for this shift, suggesting a defect in one of the glycolytic enzymes that catalyzes a reaction *before* the production of 1,3-bisphosphoglycerate. Conversely, increased levels of BPG (reflecting increased levels of 1,3-bisphosphoglycerate) would shift the binding curve to the right (patient B) because more hemoglobin would be stabilized in the deoxy form. A rightward shift in the curve suggests a defect in a glycolytic enzyme that catalyzes a reaction *after* the production of 1,3-bisphosphoglycerate.

3. Diabetics are unable to either produce or respond to the hormone insulin. Without insulin, liver cells cannot increase the rate of glucose uptake and cannot store glucose as glycogen due to the fact that glycogen synthase cannot be activated. This means that glucose in the blood following a meal cannot be effectively taken up and stored by cells. In addition, glycogen phosphorylase cannot be inactivated and the glucose that is already stored in liver cells as glycogen will be released and exported into the blood. The combined effects result in unnaturally high blood glucose levels that cells are unable to utilize.

4. In addition to the effects on carbohydrate metabolism mentioned above, insulin also antagonizes the effects of epinephrine and glucagon on lipid metabolism. Epinephrine- or glucagon-induced increases in [cAMP] result in the phosphorylation (and activation) of adipocyte lipases that hydrolyze lipids and release free fatty acids. In addition, increased intracellular [cAMP] also result in the inhibition of acetyl-CoA carboxylase. The consequent decrease in [malonyl-CoA] results in a concomitant activation of carnitine acyltransferase I, which increases the rate of fatty acid β oxidation producing acetyl CoA and ultimately ketone bodies. Normally ketone bodies are produced for use by the brain and muscle under conditions of low blood glucose. It is the large quantity of ketone bodies in the blood that produce an alcoholic smell on the breath of untreated diabetics (and, by the way, on the breath of individuals who are fasting or dieting).

chapter

24

Genes and Chromosomes

STEP-BY-STEP GUIDE

Major Concepts

All living organisms contain genetic material that directs the processes of life.

Prokaryotic and eukaryotic cells contain DNA as their genetic material; viral genomes are either DNA or RNA. Viral DNA molecules are small because the virus itself does not have to direct functions that the host cell provides. Bacteria carry much more DNA, in a single, circular **chromosome.** They can also contain small extrachromosomal circular molecules of DNA called **plasmids,** which often code for specialized enzymes such as those involved in the degradation of antibiotics and unusual food sources. Eukaryotic cells have more DNA than prokaryotic cells. The DNA is arranged into chromosomes, with the amount of DNA per chromosome and the number of chromosomes varying with species. Mitochondria and chloroplasts of eukaryotic cells also contain small amounts of functioning, nonchromosomal DNA.

A gene is currently defined as all the DNA needed to code for a polypeptide chain or for an RNA that has structural or catalytic function.

Genes coding for proteins include the **introns,** which are nontranslated interruptions of the coding sequence for a polypeptide, and transcriptional control regions. Some genes code for RNA molecules that are not translated into a protein product, the RNA found in ribosomes is one example. There are a number of other identifiable sequences on chromosomes. These include **regulatory sequences,** which function as signals to turn on and off DNA-dependent processes; **highly repetitive sequences,** noncoding DNA that is associated with **centromeres** and **telomeres;** and **moderately repetitive DNA,** which is scattered throughout the chromosome and has no known function.

Supercoiling of the DNA molecule aids in compaction but, paradoxically, makes the DNA accessible for replication and transcription processes.

Most DNA in cells is **underwound,** which promotes strand separation but "strains" the DNA molecule. DNA supercoiling reduces this strain and leaves the DNA in its more accessible, underwound state; this is because supercoiling is energetically more favorable than actual strand separation, which requires the breaking of hydrogen bonds between base pairs. Because the process of supercoiling is critical to DNA replication and transcription, it is actively controlled by cells, primarily by topoisomerases. The extent of underwinding (and therefore supercoiling) in closed circular DNA is described by a set of parameters including: **linking number,** which specifies the number of helical turns in the DNA molecule; **specific linking difference,** a measure of the degree of underwinding; **twist,** which describes the spatial relationship of neighboring base pairs; and **writhe,** a measure of the coiling of the helical axis. Identical DNA molecules that differ only in their topology (e.g., linking number or linking difference) are **topoisomers.**

Eukaryotic chromosomes are made of chromatin, which includes not only DNA but also significant amounts of protein and small amounts of RNA.

Important chromatin proteins are **histones,** which are small, basic proteins. The strict conservation of the amino acid sequences of histones across species is an indication of their critical functions. Histones act as central cores for the wrapping of the DNA double helix. Such a wrapped kernel makes up a **nucleosome,** which is considered the first level of DNA packaging. Further compression of the DNA is accomplished by formation of **30 nm fibers,** wrapping of the nucleosomes into a structure that resembles the stalk of a brussel sprout plant. Still tighter packaging is accomplished with loops of 30 nm fibers radiating out from a nuclear scaffold. This elaborate compacting strategy is necessary in order to fit the enormous length of DNA into the cell, while still maintaining sufficient organization to allow unwinding, and accessibility, of specific portions of the DNA.

What to Review

Answering the following questions and reviewing the relevant concepts, which you have already studied, should make this chapter more understandable.

- Much of the discussion in this chapter assumes a very clear knowledge of the **structure of DNA** as described in Chapter 8. Review the components that make up DNA, the covalent bonds and noncovalent interactions that stabilize its structure, and the chemical behavior of the molecule as a whole.
- Also review **cruciform structures** and **Z-DNA** conformations of DNA that are affected by DNA supercoiling (Chapter 8).
- There are many references to the differences between **prokaryotes** and **eukaryotes** in this chapter. Make sure you understand the critical differences between the two. Where do **viruses** fit into this categorization?
- **Mutations** are alterations in the DNA sequence that result in a difference in the amino acid sequence of a protein, often leading to decreased or no function. Recall the effect of a *single* amino acid alteration on the functioning of hemoglobin in sickle-cell anemia (pp. 168–169).

Topics for Discussion

Answering each of the following questions, especially in the context of a study group discussion, should help you understand the important points of this chapter.

24.1 Chromosomal Elements

Genes Are Segments of DNA That Code for Polypeptide Chains and RNAs

1. How did the "classical" definition of a gene differ from the "one gene–one polypeptide" definition? What is the current concept of a "gene"? (There will be more on the concept of the "whole" gene in subsequent chapters.)

2. Do regulatory sequences encode a polypeptide product?

3. How large a polypeptide could be produced from an "average" gene, 1,050 base pairs long?

DNA Molecules Are Much Longer Than the Cellular or Viral Packages That Contain Them

4. Eukaryotic cells are larger than prokaryotic cells. Using the average length of the DNA from a human somatic cell provided in this section and assuming a minimum cell size of 10μm, calculate the ratio of DNA length to cell length.

5. Will the amount of compaction needed to fit the DNA into a human somatic cell be more or less than that required for an *E. coli* cell?

6. Where is DNA located in prokaryotic cells? In cells from eukaryotes?

Eukaryotic Genes and Chromosomes Are Very Complex

7. What are the relationships between an exon, an intron, a nontranslated insert, and the protein product of the gene?

8. Where in the eukaryotic chromosome are highly repetitive segments located and in what cellular functions are these sequences involved?

24.2 DNA Supercoiling

9. What are two important functions of the DNA folding mechanisms?

10. Is DNA supercoiling a random process?

Most Cellular DNA Is Underwound

11. What is a significant consequence of underwound DNA *in vivo*?

DNA Underwinding Is Defined by Topological Linking Number

12. Why is linking number a topological property?

13. What would be the specific linking difference for a DNA molecule of 420 base pairs from which eight turns were removed?

14. What is the difference between positive and negative supercoiling?

15. Why are twist and writhe *not* topological properties?

Topoisomerases Catalyze Changes in the Linking Number of DNA

16. What are the similarities and differences between the type I and type II topoisomerases? How do they affect linking number?

Box 24–1 Curing Disease by Inhibiting Topoisomerases

17. What is the molecular basis for the anticancer properties of topoisomerase inhibitors?

DNA Compaction Requires a Special Form of Supercoiling

18. Long lengths of fishing line can be "compacted" onto a fishing reel; which type of supercoiling does this represent?

24.3 The Structure of Chromosomes

Chromatin Consits of DNA and Proteins

19. During which phase of eukarytoic cell cycle is the chromosomal material the most condensed? Why?

20. Electron micrographs of DNA have the appearance of beads on a string; what do the "beads" represent?

Histones Are Small, Basic Proteins

21. DNA is a negatively charged molecule. Does this offer a clue as to why histones have such high percentages of basic amino acids?

22. Are histones the only types of protein associated with DNA in chromatin?

Nucleosomes Are the Fundamental Organizational Units of Chromatin

23. Which histone proteins are part of the nucleosome core?

24. What factors influence positioning of nucleosomes on DNA?

Box 24–2 Epigenetics, Nucleosome Structure, and Histone Variants

25. What are the ways information can be inherited epigenetically?

Nucleosomes Are Packed into Successively Higher-Order Structures

26. Why would regions of DNA that were actively being transcribed need to be less ordered?

27. What is the nucleosome scaffold made of?

Condensed Chromosome Structures Are Maintained by SMC Proteins

28. SMC proteins have an ATPase activity; what does this suggest about the condensation of chromosomes in cells?

29. What is the major distinction between the roles of cohesions and condensins?

Bacterial DNA Is Also Highly Organized

30. What are the reasons that bacterial DNA is less condensed, in general, than eukaryotic DNA?

Discussion Questions for Study Groups

- Why is it that only the "minimum" size of a gene can be estimated from the size of its polypeptide product? (Try to draw a figure that illustrates the relationship between a "typical" eukaryotic gene and its polypeptide product.)
- To visualize how nucleosomes aid in compaction of DNA, try winding a thick, closed circular rubber band around a "nucleosome." Use a thick rubber band, folded so that it is flat, to represent the DNA double helix and a pencil to represent the nucleosome core. Hold one end of the rubber band against the pencil with your thumb and use the other hand to wind the doubled rubber band around the pencil. What happens to the unwound portion of the rubber band? (A compensatory supercoil should form.)
- How might histone H1 interact with the structure illustrated in Figure 24–26. (Try drawing a molecule of H1 on the model.)

SELF-TEST

Do You Know the Terms?

ACROSS

1. State of DNA double helices that is thought to promote transcription by facilitating strand separation.
5. DNA plus two each of histone 2A, 2B, 3, and 4.
7. Supercoiled DNA structure assumed in most cells (due to the association of proteins), known as _______ supercoiling.
10. Higher-order structure of DNA, referred to as a 30 nm fiber, requires the association of the protein _______.
11. Structural units made up of DNA and associated proteins; contain heritable genetic information.
13. Nontranslated regions of DNA in the genes of most eukaryotes.
15. DNA supercoiling is the result of _______ strain on the molecule.
17. Describes sequences of DNA usually associated with centromeres and telomeres. (2 words)
18. The linking number for underwound DNA is _______ than that for relaxed DNA.
20. Describes plasmid and bacterial DNA, which lack telomeres.
21. Again, describes DNA molecules in plasmids and bacteria.
22. *In vivo* state of most double-stranded DNA.
24. Circular, extrachromosomal DNA that can replicate independently of genomic DNA.
26. Regions of DNA that influence *where* gene transcription starts and ends and the *rate* of transcription; called _______ sequences.
27. Proteins that contain a relative abundance of the amino acids H, R, and K.
28. Coding regions of DNA that usually specify amino acid sequences for only a portion of the final protein.

DOWN

2. Organisms that package most of their genetic material in a membrane-bounded nucleus.
3. Region of a bacterial cell that contains its genetic information.
4. All of the information needed to make an entire organism.
6. Property of an organism that is the product of the information contained in the genome.
8. Treating chromatin with DNase degrades _______ DNA.
9. Its genetic information is encoded in either DNA or RNA.
11. Structure essential for the separation of chromosomes during cell division.
12. Extrachromosomal DNA in eukaryotes includes _______ DNA.
14. Enzymes that alter the linking number for a given molecule of DNA; especially important in the process of DNA replication.
16. Formed by repetitive DNA sequences; help to stabilize linear DNA.
19. Organisms that lack cellular organelles; for example, bacteria.
20. The stuff (in nondividing cells) of which chromosomes are made.
23. DNA that is not supercoiled and not underwound is said to be in a _______ state.
25. Most DNA exists as a _______ of polynucleotide strands.

Do You Know the Facts?

1. DNA supercoiling:

A. results in compaction of the DNA structure.
B. makes the DNA molecule relaxed.
C. occurs only in viruses.
D. occurs in only one direction (i.e., is only positive).
E. usually results from overwinding.

2. Human chromosomes are extremely large, complex structures. Which of the following statements correctly describes the organization of human chromosomes?

A. All the genetic information of the cell is encoded in the nuclear, chromosomal DNA.
B. Genes for histones and ribosomal RNAs are structural genes.
C. Most of the chromosomal DNA codes for proteins.
D. A and B.
E. A, B, and C.

3. Plasmids are:

A. part of the bacterial chromosome.
B. found only in mitochondria.
C. composed of RNA and protein.
D. closed circular DNA molecules.
E. found only in viruses.

4. There are several levels of protein architecture, and the same is true for chromosomal DNA. Indicate whether each of the following statements about the structure and packaging of chromosomal DNA is true or false.

_____ **(a)** Histones are basic proteins that make up about half the mass of chromatin.
_____ **(b)** Histones grip the DNA double helix like a fist, with the DNA in the center.
_____ **(c)** The nucleosome structure acts to condense DNA by decreasing its length.
_____ **(d)** Formation of the nucleosome structure is the only condensing step in chromosomal DNA.
_____ **(e)** DNA constitutes approximately 90% of the total mass of chromatin; the rest is protein and a small amount of RNA.
_____ **(f)** The amino acid sequences of histones vary widely from species to species.
_____ **(g)** Between the nucleosomes are protease-sensitive regions of DNA.

5. In a certain organism, the gene for the glycolytic enzyme hexokinase has 21,000 bases. The molecular weight of hexokinase is approximately 110,000. Is this organism a prokaryote or eukaryote? How do you know? (Average molecular weight of an amino acid $\approx$110; of a base $\approx$450.)

Applying What You Know

1. Prokaryotic topoisomerase II, also called DNA gyrase, is specifically inhibited by an antibiotic called novobiocin, which binds to the β subunit of the enzyme. This antibiotic inhibits bacterial DNA replication. What does DNA gyrase do and suggest why a compound that inhibits this enzyme would be an effective antibiotic.

Biochemistry on the Internet

Approximately 2 meters of DNA must be packaged into the nucleus of most eukaryotic cells. In order to achieve this level of compaction, it is necessary to organize the DNA strands into higher-order structures. The structure known as the nucleosome represents the first level of DNA compaction and includes a protein-based core around which double-helical DNA is wound. The nucleosome structure achieves about a sevenfold compaction, reducing the 2 meters of DNA to about 30 cm in length. The protein core of the nucleosome is made up of two copies of each of the four histone proteins: H2A, H2B, H3, and H4. A model of a nucleosome can be obtained from the Protein Data Bank (www.rcsb.org/pdb) using the PDB ID code 1AOI.

(a) Proteins in the nucleosome core are arranged as histone dimers. According to this molecular model, which of the histones listed above form dimers with each other?

(b) The DNA-protein interactions that stabilize the nucleosome structure are noncovalent. Looking at the molecular model of the nucleosome, can you discern pattern in the way the DNA duplex interacts with the histone proteins in this complex? How might these interactions contribute to the overall structure and function of the nucleosome?

ANSWERS

Do You Know the Terms?

ACROSS

1. underwound
5. nucleosome
7. solenoidal
10. H1
11. chromosomes
13. introns
15. structural
17. highly repetitive
18. less
20. circular
21. closed
22. supercoiled
24. plasmid
26. regulatory
27. histones
28. exons

DOWN

2. eukaryotes
3. nucleoid
4. genome
6. phenotype
8. linker
9. virus
11. centromere
12. mitochondrial
14. topoisomerases
16. telomeres
19. prokaryotes
20. chromatin
23. relaxed
25. duplex

Do You Know the Facts?

1. A
2. B
3. D
4. **(a)** T; **(b)** F; **(c)** T; **(d)** F; **(e)** F; **(f)** F; **(g)** F
5. The gene is ~7 times as long as it needs to be to code for the ~1,000 amino acids that make up this protein. The presence of untranslated regions of DNA within the coding sequence indicates that this gene must contain introns and that the organism, therefore, is a eukaryote.

Applying What You Know

1. Topoisomerase II or DNA gyrase in bacteria decrease the linking number by cleaving the DNA duplex and another section of the bacteria's circular DNA duplex to pass through the break (look at Fig. 24–15, DNA gyrase activity would convert the structure in part b to that seen in part a). Inhibition of DNA gyrase by novobiocin prevents the addition of negative supercoils necessary for strand separation during replication. Look at Figure 24–12, which illustrates what happens when DNA strands separate during replication without concomitant gyrase induced underwinding; clearly, inhibition of DNA gyrase would inhibit DNA replication.

Biochemistry on the Internet

(a) According to the molecular description of this complex, chains A and E represent the two copies of histone H3, chains B and F are histone H4, chains C and G are histone H2A, and chains D and H are histone H2B. One way to visualize the different chains in this model is to display them in **Ribbon** format and then change the color setting to **Chain.** With these settings, each chain of the nucleosome is given a different color. By clicking on the various chains, it is possible to identify each one. Chains A-B and E-F form two of the four dimers in the nucleosome core, indicating that H3 and H4 can dimerize with each other. Chains C-D and G-H form the remaining two dimers, indicating that H2A and H2B can also form dimers.

(b) You can visualize the points of contact between the DNA and protein core by using the viewer menu to **Select** the **Protein** portion of the complex and then displaying it in **Spacefill** format.

Visualized in this format, the points of contact with the DNA double helix become more obvious. Careful examination of the points of interaction shows that almost all involve the minor groove of the DNA double helix. In addition, most of the sites of contact occur on the inner face of the helix, suggesting that it is the DNA backbone that is involved. In fact, approximately 50% of the noncovalent interactions between DNA and the histone core in a nucleosome are hydrogen bond interactions between the protein main chain and the phosphodiester backbone atoms in the DNA. This means that the formation of nucleosomes on a DNA strand is not completely dictated by the specific nucleotide sequence in that stretch of DNA.

There are also several sites at which the histone tails appear to penetrate the DNA supercoil. These tail regions thus represent additional sites at which the nucleosome core can bind DNA. In fact, the H3 tail has been shown to interact not only with DNA but also with the H3 tail from adjacent nucelosomes. In addition, the histone H4 tail has been shown to interact with the H2A-H2B dimer in adjacent nucleosomes. These observations suggest a role for histone tails in both the stabilization of the nucleosome structure as well as the formation of higher-order structures in chromatin.

chapter

25

DNA Metabolism

STEP-BY-STEP GUIDE

Major Concepts

DNA replication adheres to certain precepts in every cellular organism.

This process is semiconservative. It begins at a specific origin, or many specific origins, and proceeds bidirectionally in a 5′→3′ direction only. It is continuous on one strand and discontinuous on the other strand. The polymerization reaction requires a primer to begin synthesis and a template to direct the addition of each monomer.

Replication is an extremely complicated process and requires more than 20 enzymes and proteins.

The fidelity of transfer of the genetic message is the most important consideration, and, unlike most other biochemical reactions, the energetic expense is *not* a significant factor. Replication proceeds in stages: **initiation, elongation,** and **termination.** Each stage has many specific enzymes, though some enzymes are necessary for more than one stage.

Central to all stages of replication are DNA polymerases.

DNA polymerases catalyze extremely accurate template-driven polymerization of deoxyribonucleotides into polynucleotides in the 5′→3′ direction. Additional, and extremely important, functions of some DNA polymerases (or polymerase subunits) include **exonuclease activity,** in either the 3′→5′ or 5′→3′ direction, and **proofreading.** Polymerases are also categorized by a property called **processivity,** which refers to the number of nucleotides that the polymerase can add to the polymer before it dissociates from the DNA.

Because DNA synthesis can occur only in the 5′→3′ direction, it proceeds continuously in one direction in the leading strand and in pieces, or discontinuously, in the lagging strand.

These small pieces are called **Okazaki fragments.** Synthesis of the lagging strand requires more enzymes and proteins than does leading-strand synthesis. Not surprisingly, replication in eukaryotes is more complex than in prokaryotes, and not as well understood.

Mutations are permanent changes in the base sequence of DNA; all cells have mechanisms to detect and repair these before they can be replicated and transmitted to the next generation.

DNA damage on one strand is detectable and repairable because the complementary strand contains the correct directions for restoration. Cells have multiple repair systems. **Mismatch repair** finds and replaces incorrectly matched (i.e., wrongly base-paired) bases after replication. Discrimination between parental and daughter strands is achieved through methylation of the parental strand. **Base-excision repair** recognizes lesions such as those produced by deamination of purines and pyrimidines; it first removes the base itself, then the rest of the nucleotide, and finally refills the gap using DNA polymerase I. Large distortions in the helical structure of DNA require the mechanism of **nucleotide-excision repair,** which uses a special enzyme, **ABC excinuclease.** This enzyme is unusual in that it makes two cuts, not one, in the DNA in order to remove the damaged portion. **Direct repair** does not involve extraction of a base or nucleotide and is thus limited to those types of damage that can be fixed by removing or reversing the damage—usually a methyl group or extra bonds between pyrimidines. In extreme cases of extensive damage to the DNA, an **error-prone repair** pathway is induced. This system allows replication even without a complete template, which increases mutation rates. Though risky, such a strategy is preferable to an inability to replicate at all.

DNA recombination processes maintain genetic diversity, repair DNA damage, and regulate gene expression.

There are different types of recombination, varying in their need for homology between DNA sequences that are recombined and in when the recombination occurs. **Homologous genetic recombination** occurs between pieces of the chromosome that are similar, but not identical, leading to small but often significant differences in the recombinants. This type of recombination is important during the crossing over of chromosomes in meiosis. It is also critical to DNA repair when there is a double-strand break or a lesion in a single strand; this is called **recombinational repair.** A model to explain the physical positioning of the recombination structures is the **Holliday intermediate,** a crossover structure that illustrates how different strands interact during the process. **Site-specific recombination** occurs only at unique DNA sequences and is responsible for a variety of functions, the most notable being the regulation of gene expression. An example of this type of recombination is the process that generates a virtually unlimited number of immunoglobulin proteins (needed to recognize all of the potential antigens that might invade a system) from a relatively limited number of immunoglobulin genes. Recombination by **transposons** is unusual in that neither homology nor specific sites are required; transposition, which must be tightly regulated, allows segments of DNA to hop from one position in the chromosome to another, in mostly random fashion. Each type of recombination requires specialized enzymes.

What to Review

Answering the following questions and reviewing the relevant concepts, which you have already studied, should make this chapter more understandable.

- Why are areas of DNA that have high concentrations of A=T base pairs easier to denature (p. 288)?
- How do noncovalent bonds contribute to the energetics (favorable or unfavorable) of a reaction (pp. 188–191)?
- Review what the **5′ end** and **3′ end** of a DNA strand are; this information is essential in understanding replication (and transcription) processes (pp. 274–275).
- Review the unusual tautomeric forms of nucleotide bases that can cause incorrect base pairing; this incorrect pairing can cause mutations (p. 276).
- NAD^+ can donate an AMP moiety to DNA ligase during catalysis. Review the structure of **NAD** to reassure yourself that this is, in fact, possible (Fig. 8–38).
- How do **topoisomerases** relieve the topological stress created by strand separation (pp. 954–957)?
- Repair of several different types of damage that can occur to DNA are discussed in this chapter; review the causes and effects of DNA damage (pp. 289–292).

Topics for Discussion

Answering each of the following questions, especially in the context of a study group discussion, should help you understand the important points of this chapter.

A Word about Terminology

1. *umuC* and *umuD* and MutL are involved in the repair of damaged DNA (details on this later in the chapter); which of these are genes and which are the protein products?

25.1 DNA Replication

DNA Replication Follows a Set of Fundamental Rules:

- **DNA Replication Is Semiconservative**

2. If replication were *conservative,* how many different DNA bands would appear in a cesium chloride density gradient after 100 rounds of replication?

- **Replication Begins at an Origin and Usually Proceeds Bidirectionally**

3. Why does addition of radioactive thymidine to the growth medium generate the labeled DNA molecules shown in the autoradiograms in Figure 25–3? Make sure that you really understand these experiments.

- **DNA Synthesis Proceeds in a 5′→3′ Direction and Is Semidiscontinuous**

4. In what direction are leading strands synthesized relative to the direction of movement of the replication fork? Is the synthesis continuous? In what direction is the lagging strand synthesized relative to the direction of movement of the replication fork? Is the synthesis continuous?

DNA Is Degraded by Nucleases

5. What is the primary difference between an endonuclease and an exonuclease?

6. As you read through this chapter, it may be useful to make a list of the different roles played by endo- and exonucleases in DNA metabolism.

DNA Is Synthesized by DNA Polymerases

7. What is the general reaction equation of DNA polymerases?

8. What other polymerization reaction have you already learned about that requires a primer for synthesis?

9. Could it be said that a DNA polymerase with high processivity is using substrate channeling?

10. Why are relatively weak, noncovalent interactions of more importance than strong covalent bonds to the energetics of polymerization?

11. Which noncovalent interactions play the most significant roles in the process of DNA polymerization, and what are those roles?

Replication Is Very Accurate

12. List the two strategies for maintaining the very high degree of fidelity of DNA replication during polymerization itself.

E. coli *Has at Least Five DNA Polymerases*

13. Following the discovery of DNA polymerase I, what experimental evidence suggested that another enzyme is the primary enzyme of replication?

14. What properties of DNA polymerase III make it a more appropriate enzyme than DNA polymerase I for the primary role in *E. coli* DNA replication?

15. Why is the action of DNA polymerase I called "nick translation"?

16. How is $5' \rightarrow 3'$ exonuclease activity different from $3' \rightarrow 5'$ proofreading activity?

DNA Replication Requires Many Enzymes and Protein Factors

17. Luckily for students of biochemistry, most of the major enzymes involved in replication have names that simply describe their function. List the enzymes involved and their functions.

Replication of the E. coli *Chromosome Proceeds in Stages*

- **Initiation**

18. Why is it significant that the replication origin contains many A=T base pairs?

19. What category of enzyme is the DnaB protein? What is its role in the initiation of DNA replication?

20. Why is it important that initiation of DNA replication is carefully regulated? What is the proposed role of DnaA in this process?

- **Elongation**

21. What extra proteins are necessary for lagging-strand synthesis that are not needed in leading-strand synthesis?

22. In the synthesis of Okazaki fragments, in which direction does the primosome travel relative to movement of the replication fork? In which direction does DNA polymerase III travel?

23. What is the role of DNA polymerase I in this process?

24. What is the nature of the bond formed by DNA ligase? What entities does the bond link?

- **Termination**

25. What terminates the bidirectional replication process?

26. Why is the activity of a topoisomerase required before cell division can occur?

27. What must occur after chromosome replication is complete, before an *E. Coli* cell can divide?

Replication in Eukaryotic Cells Is Both Similar and More Complex

28. How do eukaryotic cells (probably) make up for the relatively slow movement of the replication fork?

29. PCNA of eukaryotic cells is analogous to which protein in prokaryotes? RPA of eukaryotic cells is analogous to which protein in prokaryotes?

Viral DNA Polymerases Provide Targets for Antiviral Therapy

30. Where else are nucleotide triphosphastes lacking 3′ hydroxyls put to good use?

25.2 DNA Repair

Mutations Are Linked to Cancer

31. Define substitution, insertion, deletion, and silent mutations.

32. Does the Ames test directly measure the carcinogenicity of compounds?

33. Why don't all cells eventually accumulate enough mutations to become cancerous?

All Cells Have Multiple DNA Repair Systems

34. What property of DNA molecules makes it possible for DNA damage to be detected and correctly repaired?

- **Mismatch Repair**

35. Why is there a time limit during which mismatch repair must occur?

36. How does the MutH protein mark the correct strand for repair?

- **Base-Excision Repair**

37. Why does this type of DNA repair mechanism suggest a reason for the presence of thymine in DNA?

38. Why do you think the presence of uracil instead of thymine in RNA does not pose a problem for cells?

- **Nucleotide-Excision Repair**

39. In what way is the ABC excinuclease unusual?

- **Direct Repair**

40. How is direct repair different from the other methods of DNA damage repair?

41. Why must the methyltransferase repair occur *before* replication?

The Interaction of Replication Forks with DNA Damage Can Lead to Error-Prone Translesion DNA Synthesis

42. When is error-prone repair the preferable option? What are the other options?

Box 25–1 DNA Repair and Cancer

43. Defects in both the nucleotide-excision repair system and DNA polymerase η cause xeroderma pigmentosa. Why are defects in DNA polymerase η more mutagenic?

25.3 DNA Recombination

44. Why is DNA recombination desirable?

Homologous Genetic Recombination Has Several Functions

45. During which stage of meiosis does crossing over at chiasmata occur?

46. Name the three functions of homologous genetic recombination.

47. What types of damage require recombinational repair?

48. Which enzymes does branch migration require?

49. Homologous means "comparable" or "much the same." Does homologous recombination alter the base sequence of the DNA strand on which it occurs?

Recombination during Meiosis Is Initiated with Double-Strand Breaks

50. Why are the 3′ ends of nicked DNA used to initiate genetic exchange in homologous recombination?

Recombination Requires a Host of Enzymes and Other Proteins

51. What are the functions of the *chi* sequences?

52. How does the RecA protein interact with DNA?

53. What causes Holliday structures to form, and how are the chromosomes involved in this structure "untangled"?

All Aspects of DNA Metabolism Come Together to Repair Stalled Replication Forks

54. How does homologous recombination provide a new complementary strand from which the damaged strand can be repaired?

Site-Specific Recombination Results in Precise DNA Rearrangements

55. What are the components of a site-specific recombination system?

56. How are the various outcomes of inversion, deletion, or insertion determined?

57. Does the integration of λ phage into a bacterial chromosome disrupt the bacterial genome's functioning? Explain.

Complete Chromosome Replication Can Require Site-Specific Recombination

58. Why may DNA repair by homologous recombination also require site-specific recombination?

Transposable Genetic Elements Move from One Location to Another

59. How is transposition different from the other classes of recombination discussed?

60. What would be the consequences of unregulated transposition?

Immunoglobulin Genes Are Assembled by Recombination

61. Note that assembly of diverse antibodies is made possible by recombination processes (at the DNA level), posttranscriptional processes (at the mRNA level), and protein folding and assembly (at the protein level).

62. In what cells does the antibody generating recombination process take place?

63. Why is genetic recombination essential to the function of the immune system?

Discussion Questions for Study Groups

- What is the basis for the increased processivity seen for DNA polymerase III? Why is this feature important?
- Try to construct a diagram that includes a double-stranded DNA and all the components of DNA polymerase III, then add the components needed to form a replisome.
- Whenever possible, cells appear to use the fewest possible enzymes and enzyme pathways to accomplish their goals (e.g., consider the overlap in enzymes catalyzing reactions in glycolysis and gluconeogenesis). Why, then, do cells go to the energetic expense of using multiple enzyme systems to accomplish the single task of DNA repair?

SELF-TEST

Do You Know the Terms?

ACROSS

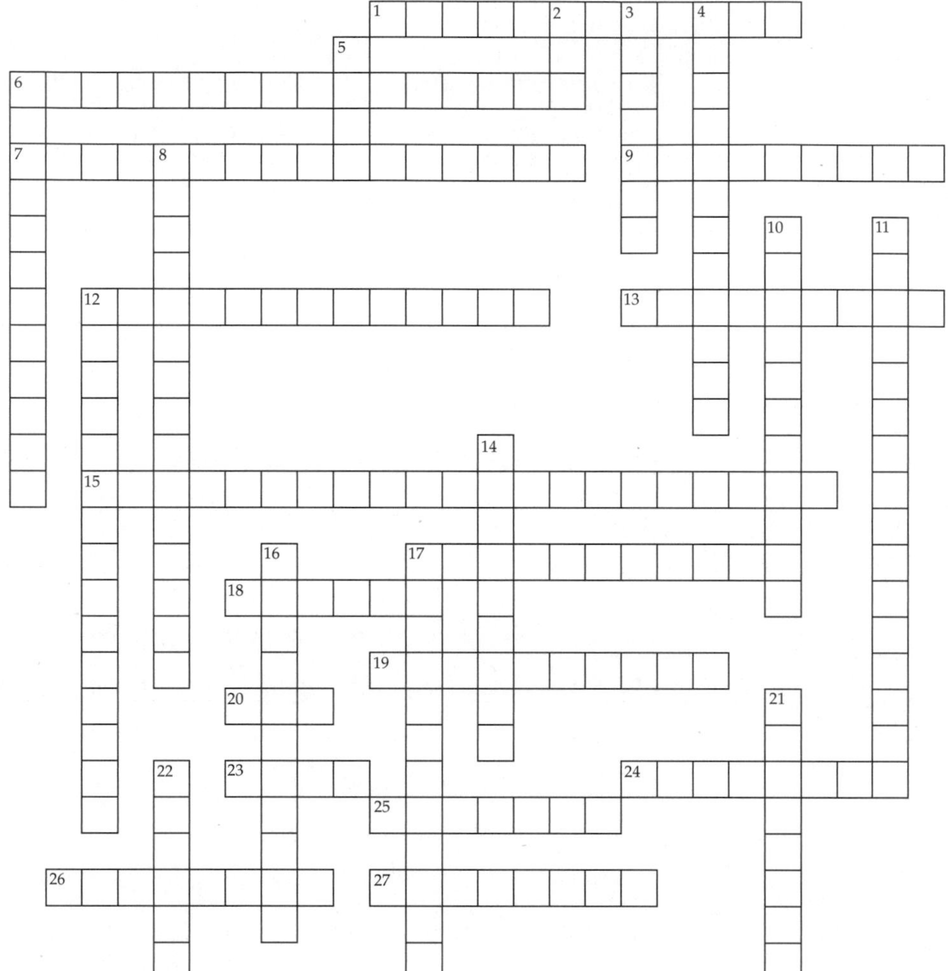

1. Enzymes used in base-excision repair to create abasic sites.

6. DNA lesions repaired directly or by ABC excinuclease in nucleotide-excision repair. (2 words)

7. Small pieces of DNA synthesized in the direction opposite to the direction of movement of the replication fork. (2 words)

9. Type of mutation involving infiltration of an extra base pair into the DNA sequence.

12. It would take twice as long for an *E. coli* chromosome to be duplicated if replication were not _______.

13. Traveling protein machine that helps in lagging-strand synthesis.

15. Crossover structures in homologous genetic recombination. (2 words)

17. *E. coli* has (at least) three types: I, II and II; eukaryotes have three types: α, δ, ε; all types catalyze $5' \rightarrow 3'$ synthesis of DNA.

18. Like glycogen synthesis, DNA polymerization requires this to begin.

19. Similar but not exactly alike; also describes a type of genetic recombination.

20. DNA-binding protein necessary in initiation and elongation steps of replication. (abbr.)

23. The replication _______ is a moving opening that leads the replication process.

24. Parental strand that provides guidance for synthesis of new DNA, using the Watson-Crick base-pairing rules.

25. Describes the DNA strand made in a discontinuous fashion.

26. The Ames test is a simple test for these.

27. Type of repair that discriminates between the parental strand and the newly synthesized strand by recognizing methylation of the template strand.

DOWN

2. In bacteria, when DNA damage is extensive, the _______ response kicks in and initiates error-prone repair.

3. Describes the DNA strand that is continuously synthesized.

4. _______ - _______ recombination is important in regulation of expression of certain genes; uses a recombinase.

5. Nucleoprotein filament that assembles cooperatively on single-stranded DNA.

6. Number of nucleotides added before dissociation of a polymerase from DNA; a measure of "hold."

8. Enzymes that act after formation of abasic sites.

10. Enzyme that catalyzes $3' \rightarrow 5'$ error correction or $5' \rightarrow 3'$ removal of RNA primers.

11. Describes DNA replication in which the newly synthesized DNA duplex has one newly made strand and one strand from the parental duplex.

12. Process occurring during homologous genetic recombination in which the extent of base-pairing between a template strand and each of two complementary strands is altered. (2 words)

14. All the necessary DNA replication enzymes and proteins, in one neat package; has not yet been isolated as such.

16. "Jumping genes."

17. Mechanism that helps to ensure the integrity of DNA; occurs during polymerization.

21. Enzyme that unwinds DNA.

22. Enzyme that catalyzes formation of a phosphodiester bond between a 3′-hydroxyl at the end of one DNA strand and a 5′-phosphate at the end of another.

Do You Know the Facts?

1. One of the most demanding and remarkable of all biological processes is the replication of DNA. More than 20 proteins are involved in this complex process. Some of the more important proteins involved in DNA replication in *E. coli* are listed in the table below. Identify the stage of replication each is involved in and match it with its function from the accompanying list.

Protein	Stage	Function
DnaG Primase		
DNA pol I		
SSB		
DnaA protein		
DnaB protein		
DNA gyrase		
DNA ligase		
DNA pol III		
Tus		
Dam methylase		

(a) Erases primer and fills gaps
(b) Synthesizes DNA
(c) Stabilizes single-stranded regions
(d) Helicase that unwinds the double helix
(e) Opens duplex at specific sites
(f) Introducc3 ncgativc 3upcrcoil3
(g) Synthesizes RNA primers
(h) Protein that binds to a 20 base pair *Ter* sequence
(i) Selectively methylates DNA at the *ori*C region
(j) Joins the ends of DNA

2. In the laboratory, some *E. coli* cells have been grown with transcription inhibitors so that only replication could occur. You break open the cells; gently purify the DNA, separating it from the rest of the cell contents; then treat the intact DNA with radioactively labeled antibodies to DNA ligase and DNA polymerase I. Where do you expect to find the highest concentration of these enzyme antibodies?

A. On the leading strand, in one direction from the point of origin
B. On the leading strands, in both directions from the point of origin
C. On the lagging strand, in one direction from the point of origin
D. On the lagging strands, in both directions from the point of origin
E. On all parts of the DNA molecule equally

3. Although DNA replication has very high fidelity, mutations do occur. Which of the following types of single base-pair mutations would be most likely to be a lethal mutation?

A. Substitution
B. Insertion
C. Deletion
D. Silent
E. B and C

4. You have been sitting in the sun a lot lately, though you know full well that the UV in sunlight can cause the formation of pyrimidine dimers (covalent links between adjacent pyrimidine residues) in the DNA of your skin cells. Luckily, skin cells have repair mechanisms that will most likely fix the lesions formed by UV light. Which of the following describes enzymes or mechanisms used in such repair processes?

A. ABC excinuclease recognizes the lesion and cuts out the damaged DNA.
B. RNA polymerase III fills in the damaged area.
C. DNA glycosylases remove the affected bases.
D. The MutH protein marks the strand for repair.
E. Methyltransferase transfers a methyl group from DNA to its Cys residue.

5. General recombination *and* transposon-type recombination:

A. occur between homologous regions on chromosomes.
B. are important in the repair of damaged DNA.
C. involve RecA protein.
D. generate genetic diversity.
E. have nothing in common.

6. Explain how each of the following facets of DNA structure or metabolism has evolved to aid in maintaining the integrity of the genetic message—that is, to prevent alterations in the DNA sequence through repair systems or "quality control."

(a) Thymine instead of uracil in the DNA strand

(b) Watson-Crick base pairing

(c) Methylation of the parental strand

7. The immune system is capable of generating antibody molecules that can bind to (and thereby inactivate) an unlimited number of foreign substances. Discuss the mechanisms used by the immune system to generate diversity in antibody binding sites.

Applying What You Know

1. What kind of results would Meselson and Stahl have seen after one generation on the new ^{14}N medium if DNA ligase were suddenly inactivated in the *E. coli* cells?

2. In eukaryotic DNA replication, DNA polymerase α has a primase activity and relatively low processivity, and DNA polymerase δ has a $3' \rightarrow 5'$ proofreading activity, lacks a primase activity, and has a very high processivity when complexed with PCNA. Both enzymes replicate DNA by extending a primer in the $5' \rightarrow 3'$ direction under the direction of a single-stranded DNA template. Why is it likely that DNA polymerase δ acts as the leading-strand synthesizing enzyme and DNA polymerase α as the lagging-strand synthesizing enzyme?

3. The kind of recombination displayed by transposons (sometimes called RecA-independent recombination) has been regarded as having far greater evolutionary significance than does general (RecA-dependent) recombination. Why would this be so?
4. Why should mutagenesis of *E. coli* using UV radiation as the mutagen be done in a laboratory without windows?

ANSWERS

Do You Know the Terms?

ACROSS

1. glycosylases
6. pyrimidine dimers
7. Okazaki fragments
9. insertion
12. bidirectional
13. primosome
15. Holliday intermediates
17. polymerases
18. primer
19. homologous
20. SSB
23. fork
24. template
25. lagging
26. mutagens
27. mismatch

DOWN

2. SOS
3. leading
4. site-specific
5. RecA
6. processivity
8. AP endonucleases
10. exonuclease
11. semiconservative
12. branch migration
14. replisome
16. transposons
17. proofreading
21. helicase
22. ligase

Do You Know the Facts?

1.

Protein	Stage	Function
DnaG Primase	Elongation	(g)
DNA pol I	Elongation	(a) and (b)
SSB	Initiation	(c)
DnaA protein	Initiation	(e)
DnaB protein	Initiation	(d)
DNA gyrase	Initiation, elongation	(f)
DNA ligase	Elongation, termination(?)	(j)
DNA pol III	Elongation	(b)
Tus	Termination	(h)
Dam methylase	Initiation	(i)

2. D
3. E
4. A
5. D
6. (a) Under typical cellular conditions, deamination of cytosine in DNA will occur in ~1 of every 10^7 cytosines in 24 hours. The product of cytosine deamination—uracil—is recognized as foreign in DNA and is removed by the base-excision repair system. If DNA normally contained uracil rather than thymine, distinguishing between the "correct" uracils and those formed from cytosine deamination would be difficult.

(b) In synthesis of a daughter strand of DNA, the template DNA strand provides the instruction through formation of the proper hydrogen bonds between A and T and between C and G. These base-pairing rules are the basis for the fidelity of replication.

(c) Methylation of the parental (template) strand distinguishes between the old and the new. Because it is the newly made strand that may have mistakes made during replication, the mismatch repair system has to be able to recognize which strand is new. Methylation also serves to differentiate "self" from "foreign" DNA. For example, restriction enzymes are produced by bacteria to degrade invading DNA; methylation of the bacterial DNA protects it from these enzymes (see p. 296).

7. Recombination mechanisms at the DNA level bring together different combinations of V and J segments in the light-chain genes of different B lymphocytes. Imperfect DNA splicing during the rearrangement of antibody genes increases the already numerous versions of each coding region. The final joining of the V-J combinations to the C region is carried out by an RNA-splicing reaction after transcription; because there are a number of different C genes, this further increases the number of possible combinations. Finally, mutation of the selected V sequences occurs at an unusually high rate during B-cell differentiation, once again increasing the number of different mRNA molecules (and ultimately polypeptides) that can be produced from a limited set of immunoglobulin genes.

Applying What You Know

1. DNA ligase is required to join together the Okazaki fragments of the lagging strand and to link the ends of the leading strand at the end of replication. Without this enzyme activity, there would be no neat duplexes, each with one ^{15}N strand and one ^{14}N strand. Instead, at least briefly, there would be full-length, circular ^{15}N strands (the parental strands); a number of long fragments representing the noncircularized ^{14}N leading strands; and many small ^{14}N Okazaki fragments. The single-stranded pieces would be unlikely to exist for very long; nucleases present in many cells recognize single-stranded DNA as foreign and degrade it.

2. The leading-strand replicase (replication enzyme) needs high processivity but only occasionally needs a primer. The lagging-strand replicase requires frequent priming, but processivity is not as important because the lagging strand is synthesized in short pieces. The two eukaryotic DNA polymerases have the appropriate properties for their distinct functions: the δ form constructing the leading strand and the α form the lagging strand.

3. Because general recombination can occur only between homologous segments of chromosomes, the resulting recombinant genome will not contain drastic alterations in the encoded information and the alterations that do occur are unlikely to be lethal. Transposon recombination allows for insertions, deletions, and duplications of genes. These rearrangements, though often lethal, occasionally produce a new combination that provides some evolutionary advantage.

4. UV light induces mutations by causing the formation of pyrimidine dimers, especially between two adjacent (on the same strand) thymines. This distorts the DNA helix and interferes with proper replication, leading to an increased number of mutations in the daughter cells that survive. Many cells are so badly damaged that they are unable to replicate at all. The usual repair mechanism is excision of the damaged area of DNA followed by insertion of new bases to replace the excised portion; this is the nucleotide-excision repair system. Visible light (in this case, sunshine coming in through windows) activates another repair enzyme, a photolyase. The photolyase of *E. coli* binds to the thymine dimers and is activated on exposure to visible light, splitting the dimer and thus repairing it before replication. This decreases the number of mutations induced by the UV light. If the goal is to produce *E. coli* mutants, sunshine should be excluded from the lab.

chapter

26

RNA Metabolism

STEP-BY-STEP GUIDE

Major Concepts

RNA synthesis, or transcription, is a template-dependent process.

The enzymes of transcription, the **DNA-dependent RNA polymerases,** add ribonucleotide units to the 3′ end of the growing RNA chain by following the instructions contained in (i.e., by base pairing with) one strand of a DNA duplex, the **template strand.** The added ribonucleotides adhere to the base-pairing rules except for the addition of U instead of T. The RNA therefore has a sequence identical to that of the **nontemplate strand** of the DNA, except for the substitution of a U for every T in the DNA. Polymerization occurs only in the $5' \rightarrow 3'$ direction, as does DNA synthesis. No primer is required for transcription.

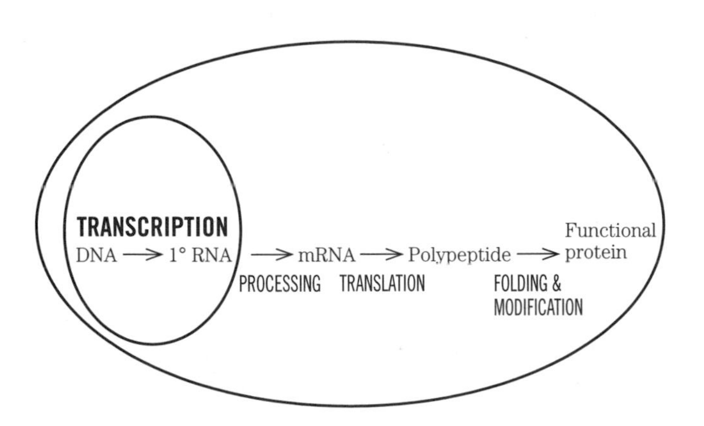

RNA synthesis is initiated at specific DNA sequences called promoters.

The RNA polymerase of *E. coli* has a specific subunit, the σ subunit, that directs the enzyme to the correct promoter. Many promoters from different bacterial species have been found to contain similarities in their sequences at positions −35 and −10 base pairs "upstream" from the start site of transcription. It is presumably these **consensus sequences** that are recognized by the σ subunit. The regulation of this step of gene transcription is probably the most critical regulatory point in the process of gene expression; some of the mechanisms employed by cells to regulate transcription will be discussed in Chapter 28.

Elongation of RNA transcripts requires the unwinding of DNA.

Once RNA polymerase is properly positioned, a portion of the DNA is unwound by the enzyme forming the **transcription bubble.** The unwinding of the DNA duplex generates supercoils in front of the transcribing enzyme, which can be relaxed by topoisomerase. The newly polymerized RNA is found as a short RNA-DNA hybrid within the transcription bubble.

Termination of transcription is signaled by specific DNA sequences.

One class of termination signal relies on the aid of a protein called **rho** (ρ) and the other class does not. In both types of termination, the formation of an RNA hairpin structure helps in dissociation of the RNA transcript from the DNA template.

Eukaryotic transcription has three types of RNA polymerases.

RNA polymerase I synthesizes preribosomal RNA, which contains the precursors for the 5.8S, 18S, and 28S **ribosomal RNAs (rRNAs).** The major role of RNA polymerase II is to synthesize **messenger RNAs (mRNAs),** which carry the genetic message from the chromosome to the ribosomes. RNA polymerase III synthesizes **transfer RNAs (tRNAs),** which act as adapters between mRNA and amino acids. Eukaryotic promoters are more variable than are those of prokaryotes; eukaryotes also need **transcription factors** to modulate the binding of RNA polymerases to the promoter sequences.

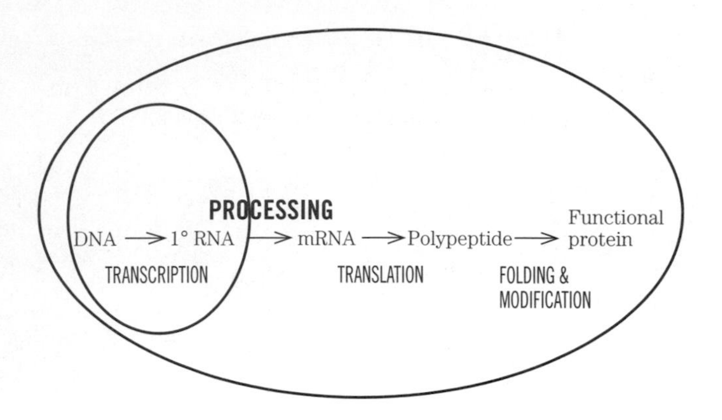

Once transcribed, the RNA in eukaryotes is extensively processed.

A variety of alterations to the primary transcripts occur in the nucleus. One process involves removal of noncoding regions of the RNA called **introns** and the careful **splicing** of the remaining **exons** to form a contiguous sequence. There are four classes of introns, and their cutting and splicing reactions vary in their requirement for high-energy cofactors and for specialized RNA-protein complexes. In some cases, the same primary transcript can be spliced in different ways (alternative splicing) to produce a variety of different mRNAs and thus different protein products. One of the biggest surprises in biochemistry came from the study of splicing in introns: in some cases, protein enzymes are not used in catalysis and the catalytic component is the RNA itself. Studies of these **ribozymes** have shown them to have many of the properties of protein enzymes. Other modifications of the primary transcript include addition of a **7-methylguanosine cap** on the 5′ end and a **poly(A) tail** to the 3′ end. Both of these steps are thought to protect the mRNA from enzymatic attack by nucleases in the cytoplasm, to which they are exported after the processing reactions are completed. This is not long-term protection, however; mRNAs must be degraded after their protein products are no longer needed. Additional modifications of RNA molecules include base methylation, deamination, and reduction.

DNA and RNA can be synthesized from an RNA template.

Retroviruses use RNA as their genetic material and can "reverse transcribe" RNA into DNA. They contain the enzyme **reverse transcriptase.** This enzyme has some properties similar to DNA and RNA polymerases, but its job is to catalyze the synthesis of a DNA duplex that is complementary to the viral RNA. The DNA so formed is then integrated into the host chromosome, where it can hide for many cell generations, eventually reemerging to generate new virus particles. All reverse transcriptases, but in particular that of **HIV,** lack any proofreading capability and so have a higher error rate. This means a faster rate of evolution of new viral strains, making treatment of the host's virus-related disease particularly difficult. Some retroviruses contain cancer-causing genes called **oncogenes.** These sequences are derived from normal cellular genes that often encode proteins involved in cell growth and development.

The enzyme **telomerase,** which prevents the progressive shortening of linear eukaryotic chromosomes, acts as a specialized reverse transcriptase in that it uses an internal RNA template for the synthesis of a DNA segment. Some viruses have RNA-dependent RNA polymerases, or **replicases,** which act only on their own RNA. The structural and functional complexity of RNA has led to speculation that the "RNA world" was important in the transition from prebiotic chemistry to life.

What to Review

Answering the following questions and reviewing the relevant concepts, which you have already studied, should make this chapter more understandable.

- The term **consensus sequence** is important in this chapter in relation to promoters. It means "a DNA or amino acid sequence consisting of residues that occur most commonly at each position within a set of similar sequences." These sequences were first mentioned in reference to proteins regulated by cAMP. Why do these proteins and DNA have such sequences (p. 225)?
- Transposon insertion in bacteria is similar to integration of the Ti plasmid in plants and retroviral DNA into its host cell DNA. Review the properties of these "hopping genes" (pp. 330–335).
- Cell-growth and cell-division factors, such as tyrosine kinases, growth factors and their receptors, and G proteins, can be altered so that their activity is uncontrolled; cell growth is then unfettered and cancer can result. Oncogenes are cancer-causing genes, some of which are derived from the normal proto-oncogenes for such growth factors. Review the mechanisms by which these factors wield such power (Chapters 12 and 23).
- The discussion about the "RNA world" is extended in this chapter. Review the possible scenario for the transition from the prebiotic RNA world to the biotic DNA world presented in Chapter 1 (Fig. 1–34).

Topics for Discussion

Answering each of the following questions, especially in the context of a study group discussion, should help you understand the important points of this chapter.

26.1 DNA-Dependent Synthesis of RNA

1. What are the different types of ribonucleic acids called and what are their different biological functions?

RNA Is Synthesized by RNA Polymerases

2. How is transcription similar to and different from replication?

3. What type of supercoil is generated in front of the moving RNA polymerase? Behind the polymerase? How is the strain from this supercoiling relieved?

RNA Synthesis Begins at Promoters

4. Why do you think there is a consensus sequence for bacterial promoters?

5. Note that the sequences of the promoters shown in Figure 26–5 are those of the *coding* strand. What would be the DNA sequence of the *template* strand for each of these promoters?

6. What is the function of the σ subunit of RNA polymerase? Biochemically, how can different σ subunits coordinate the expression of different sets of genes?

Box 26-1 RNA Polymerase Leaves Its Footprint on a Promoter

7. Does the "footprint" on the gel appear because of *more* cuts in the DNA, or *fewer*?

Transcription Is Regulated at Several Levels

8. Why does it make energetic sense for the cell to regulate transcription at its early stages?

Specific Sequences Signal Termination of RNA Synthesis

9. In ρ-independent termination, in which nucleic acid (DNA or RNA) is the termination sequence? Which nucleic acid forms the hairpin structure?

10. Why might a poly A=U hybrid be relatively unstable?

Eukaryotic Cells Have Three Kinds of Nuclear RNA Polymerases

11. List the functions of the three different eukaryotic RNA polymerases.

RNA Polymerase II Requires Many Other Protein Factors for Its Activity

It may help you learn the processes involved in transcription if you draw your own model of the factors that are crucial to each step and chart their functions. Be sure to compare your work with Table 26–2 and Figures 26–10 and 26–12.

- **Assembly of RNA Polymerase and Transcription Factors at a Promotor**

12. Which of the transcription factors act to unwind the DNA?

- **RNA Strand Initiation and Promoter Clearance**

13. Which of the transcription factors has a protein kinase activity and why is that significant?

- **Elongation, Termination, and Release**

14. Why could elongation factors also be named "suppression-of-arrest" factors?

- **Regulation of RNA Polymerase II Activity**

15. Why must the regulation of this enzyme be so complex?

- **Diverse Functions of TFIIH**

16. How do the roles of TFIIH explain the observation that DNA damage is repaired more efficiently in the template strand than the nontemplate strand?

DNA-Dependent RNA Polymerase Undergoes Selective Inhibition

17. How do intercalators inhibit transcription? Compare this with the mechanisms of inhibition by rifampicin and α-amanitin.

26.2 RNA Processing

18. In RNA processing, what is removed from primary transcripts?

19. Where in the eukaryotic cell do RNA processing reactions take place?

Eukaryotic mRNAs Are Capped at the 5′ End

20. What are the functions of the "cap"?

Both Introns and Exons Are Transcribed from DNA into RNA

21. What are the ways that introns and exons are different?

RNA Catalyzes the Splicing of Introns

22. What is unusual about the use of a guanosine nucleotide as a cofactor in group I intron splicing reactions?

23. The spliceosome complex represents a major investment by the cell, indicating that the splicing process is very important. Why must the intron splicing reactions be so exquisitely accurate?

Eukaryotic mRNAs Have a Distinctive 3′ End Structure

24. Attaching the poly(A) tail uses a lot of ATP. What is the (probable) function of the poly(A) tail that would make this energy outlay worthwhile to the cell?

25. What percentage of the length of the ovalbumin gene is actually reflected in the mature transcript?

A Gene Can Give Rise to Multiple Products by Differential RNA Processing

26. In what ways does the mRNA processing shown in Figure 26–20a differ from that shown in Figure 26–20b?

Ribosomal RNAs and tRNAs Also Undergo Processing

27. What further processing reactions of precursor transcripts are required to produce mature rRNAs and tRNAs?

Special-Function RNAs Undergo Several Types of Processing

28. What is the difference in terms of function between snRNAs and miRNAs?

RNA Enzymes Are the Catalysts of Some Events in RNA Metabolism

29. Why would RNA be a good substrate for ribozymes?

- **Enzymatic Properties of Group I Introns**

30. What can L-19 IVS do? Does it do this *in* the cell?

- **Characteristics of Other Ribozymes**

31. What are the similarities between ribozymes and protein enzymes? What are the differences?

Cellular mRNAs Are Degraded at Different Rates

32. Why is it essential to cell function that mRNAs are eventually degraded?

Polynucleotide Phosphorylase Makes Random RNA-like Polymers

33. How does polynucleotide phosphorylase differ from other enzymes that synthesize nucleic acids?

26.3 RNA-Dependent Synthesis of RNA and DNA

34. What is the "central dogma" concerning the flow of genetic information? What processes had to be added when the role of RNA templates was discovered?

Reverse Transcriptase Produces DNA from Viral RNA

35. What are the three typical retroviral genes? What protein product(s) does each code for?

36. How are reverse transcriptases similar to DNA and RNA polymerases? How are they different?

37. Why is the lack of a proofreading function in reverse transcriptases medically important?

Some Retroviruses Cause Cancer and AIDS

38. What is the relationship between oncogenes and proto-oncogenes?

39. What is the *mechanism* by which overexpression of the *src* gene in Rous sarcoma virus causes unregulated cell division? (Hint: see Chapter 13.)

40. How is the HIV reverse transcriptase different from other known reverse transcriptases? Why is this of concern?

Many Transposons, Retroviruses, and Introns May Have a Common Evolutionary Origin

41. Why are retrotransposons trapped within a single cell?

42. What is the evidence that introns originated as molecular "parasites"?

Box 26–2 Fighting AIDS with Inhibitors of Reverse Transcriptase

43. Why must AZT be given as the *unphosphorylated* nucleoside?

Telomerase Is a Specialized Reverse Transcriptase

44. What problem does telomerase solve?

45. What is the function of a T loop?

Some Viral RNAs Are Replicated by RNA-Dependent RNA Polymerase

46. Why aren't RNA replicases generally useful in recombinant DNA technology?

RNA Synthesis Offers Important Clues to Biochemical Evolution

47. What is the evidence that implies that adenine was the first nucleotide constituent?

Box 26–3 The Selex Method for Generating RNA Polymers with New Functions

48. Evolution is defined as the genetic change in a population over generations. It occurs when natural selection produces changes in the relative frequencies of alleles in a population's gene pool. What is the (*unnatural*) selection present in the SELEX technique?

Discussion Questions for Study Groups

- Is the template strand always the same DNA strand for different genes along the chromosome?
- Why is the nontemplate strand of DNA referred to as the coding strand?
- RNA polymerases lack a proofreading active site. Why is a mistake in an RNA molecule less of a catastrophe than a mistake in DNA?
- Why is DNA a better molecule than RNA for the purpose of long-term storage of genetic information? (Hint: see Chapter 8.)

SELF-TEST

Do You Know the Terms?

ACROSS

2. Technique that can pinpoint promoter locations.
5. Complex ____ produce two or more *different* mRNAs and polypeptides.
8. Proteins necessary for the activity of RNA Pol II are ____ factors.
9. "Accelerated evolution in a test tube."
11. DNA-directed enzymes that require a DNA template and ATP, GTP, UTP, and CTP. (2 words)
13. The newly synthesized RNA molecule is the ____ transcript.
14. As in film editing, cutting and ____ an RNA by one frame too few or too many can be devastating to the final product.
15. Molecule that carries the genetic message from the chromosome to the ribosomes. (abbr.)
16. Protein factor that signals termination. (word)
18. RNA-dependent DNA polymerase. (2 words)
19. DNA strand identical in sequence to the RNA transcript, except with T's instead of U's.
23. Term describing the form of RNA polymerase that contains the σ subunit.
24. Areas of DNA containing consensus sequences.
27. A poison; why you shouldn't eat strange mushrooms.

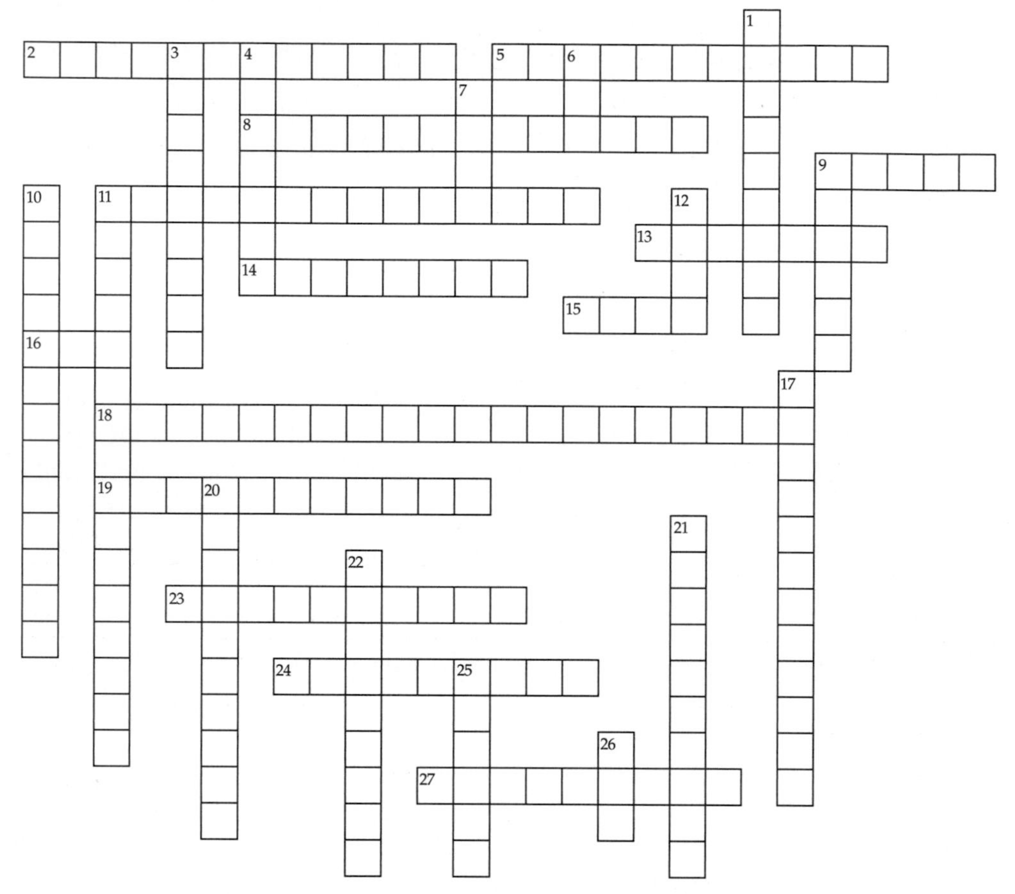

DOWN

1. Type of enzymes; RNase P, for example.
3. __(AAAAAAAAAAAAAAAAAAAA)$_4$__OH(3′), at least. (2 words)
4. These are lacking in histone genes but present in almost all other vertebrate genes.
6. Structural analog of deoxythymidine. (abbr.)
7. Adapter molecule between mRNA and amino acid. (abbr.)
9. Not Saturday morning cartoon characters, but complexes containing RNA and protein and important in group III splicing reactions. (abbr.)
10. Actinomycin D and acridine, for example.
11. "Defective viruses, trapped in one cell."
12. Nucleic acid serving as a component of ribosomes. (abbr.)
17. Molecular proof that genetic information can flow "backward."
20. Enzyme that prevents the slow nibbling away of the ends of chromosomes.
21. Positive ____ form ahead of the transcription bubble.
22. Good genes gone bad; for example, the *src* gene.
25. Consensus sequence of the −10 region in *E. coli.*
26. Virus containing a tragically inaccurate reverse transcriptase. (abbr.)

Do You Know the Facts?

1. The processes of DNA replication and transcription of RNA are similar in some respects and different in others. Indicate whether each of the following describes replication only, transcription only, or both.

	Replication	Transcription	Both
(a) A specific region of the DNA is recognized and bound by the polymerase.	______	______	______
(b) The direction of polymerization is 5′→3′.	______	______	______
(c) The direction of enzyme movement on the template strand is 3′→ 5′.	______	______	______
(d) The process includes its own 3′→ 5′ exonuclease proofreading mechanism.	______	______	______
(e) The mechanism of reaction is attack by the 3′-OH group of the pentose on the α-phosphorus of an incoming nucleoside triphosphate	______	______	______
(f) The process requires a primer.	______	______	______

2. Match each enzyme in the table with its function in eukaryotic RNA synthesis, from the list below. Indicate (with a check mark) which inhibitor(s) affect each enzyme.

		Inhibited by:		
Enzyme	**Principal Function**	**α-Amanitin**	**Rifampcin**	**Actinomycin D**
RNA polymerase I				
RNA polymerase II				
RNA polymerase III				

(a) Synthesizes mRNA.
(b) Synthesizes tRNAs.
(c) Synthesizes preribosomal RNA.

3. Consider the mRNA sequence: (5′) AAUGCAGCUUUAGCA (3′). The sequence of the *coding* strand of DNA is:
A. (5′) ACGATTTCGACGTAA (3′).
B. (3′) TTACGTCGAAATCGA (5′).
C. (5′) AATGCAGCTTTAGCA (3′).
D. (5′) AAUGCAGCUUUAGCA (3′).
E. (3′) AATGCAGCTTTAGCA (5′).

4. ________ is found *within* the transcription bubble.
A. Positive supercoils in the DNA
B. Negative supercoils in the DNA
C. An intact DNA duplex
D. A short DNA-RNA hybrid
E. An RNA primer

5. The σ^{70} subunit of the *E. coli* RNA polymerase:
A. acts as the catalytic site for polymerization.
B. recognizes promoters.
C. has a proofreading function.
D. increases the processivity of the enzyme.
E. recognizes termination signals.

6. In ρ-independent termination, the termination sequence is on the ____ and the hairpin structure is formed by the ____.

A. DNA; RNA
B. RNA; DNA
C. DNA; ρ protein
D. RNA; RNA
E. DNA; DNA-RNA hybrid

7. Indicate (yes/no) whether each of the following events can occur during the processing of eukaryotic mRNA transcripts.

________ **(a)** Attachment of a poly(A) tail to the 5′ end of the transcript.
________ **(b)** Methylation of all G residues.
________ **(c)** Excision of introns.
________ **(d)** Conversion of standard bases to modified bases such as inosine.
________ **(e)** Splicing together of exons.
________ **(f)** Differential cutting and splicing to produce two different proteins.

8. On Mars, it seems, the genetic material is double-stranded RNA rather than DNA. Martian RNA polymerase uses the double-stranded RNA as a template to make mRNA; it is an RNA-directed RNA polymerase. Assume all other aspects of transcription and translation and the functions and integrity of the enzymes involved are the same on Mars as on Earth (i.e., RNA has the same function, the same base pairs are complementary to each other, and chemistry in general is the same). Over time, the integrity of the genetic message in Martian RNA would: (Hint: check Ch. 8 as well!)

A. decline more rapidly than that of DNA because of the lack of editing function of RNA polymerase.
B. be comparable with that of DNA.
C. decline more rapidly than that of DNA because the deamination of cytosine to uracil in RNA would confuse the message.
D. be greater than that of DNA because the message would be more direct: RNA → protein, with one less step in which mistakes could happen.
E. A and C are true.

9. Indicate whether each of the following is true of ribozymes only, protein enzymes only, or both.

	Ribozymes	**Protein enzymes**	**Both**
(a) Base-pairing reactions can help align an RNA substrate for catalysis.	______	______	______
(b) They can be denatured.	______	______	______
(c) They are saturable.	______	______	______
(d) They are recycled for use in the same reaction several to many times.	______	______	______
(e) They are subject to inhibition by substrate analogs.	______	______	______

10. What are the various functions of TFIIH?

Applying What You Know

1. The antibiotic *cordycepin,* an adenosine analog that lacks a 3′-OH group, inhibits bacterial RNA synthesis. What is the mechanism of this inhibition?

2. Ethidium bromide is an intercalator. It is used as a fluorescent dye specific for nucleic acids and is often used to stain DNA in agarose gels. When handling this substance, laboratory personnel should always wear gloves. Why?

3. Suppose you wished to separate out all the *mature* (fully processed) mRNAs from a particular eukaryotic cell lysate. Suggest a way to partition the mature mRNAs from the primary transcripts.

4. The only mature eukaryotic mRNAs that generally lack poly(A) tails are histone mRNAs.

(a) What could you guess about these mRNAs, given what is known about the function of poly(A) tails?

(b) Histone genes are among the very few vertebrate genes that lack introns. Can you suggest a reason why histones do not have introns?

5. β-Thalassemia is a class of diseases in which the β-globin gene of hemoglobin is barely expressed. One form of the disease appears to result from a single nucleotide change. If this base change does *not* result in an alteration in the encoded amino acid sequence, what other reasons can you think of for its having such severe consequences for the protein product?

ANSWERS

Do You Know the Terms?

ACROSS

2. footprinting
5. transcripts
8. transcription
9. SELEX
11. RNA polymerases
13. primary
14. splicing
15. mRNA
16. rho
18. reverse transcriptase
19. nontemplate
23. holoenzyme
24. promoters
27. α-amanitin

DOWN

1. ribozymes
3. poly(A) tail
4. introns
6. AZT
7. tRNA
9. snRNPs
10. intercalators
11. retrotransposons
12. rRNA
17. retroviruses
20. telomerase
21. supercoils
22. oncogenes
25. TATAAT
26. HIV

Do You Know the Facts?

1. **(a), (b), (c),** and **(e)** both; **(d)** and **(f)** replication only.
2.

Enzyme	Principal Function	Inhibited by:		
		α-Amanitin	Rifampcin	Actinomycin D
RNA polymerase I	(c)			✓
RNA polymerase II	(a)	✓		✓
RNA polymerase III	(b)	✓ (at high conc.)		✓

Note that rifampicin would not inhibit any of these enzymes; it specifically reacts with the β subunit of *prokaryotic* RNA polymerases.

3. C
4. D
5. B
6. A
7. **(a)** No; **(b)** No; **(c)** Yes; **(d)** No; **(e)** Yes; **(f)** Yes
8. E
9. **(a)** ribozymes only; **(b)** both; **(c)** both; **(d)** in general, protein enzymes, but true for some ribozymes; **(e)** both
10. TFIIH:
    - acts as a necessary part of the closed complex.
    - acts as a helicase to unwind DNA.
    - has a kinase activity that phosphorylates RNA Pol II, causing a conformational change that initiates transcription.
    - interacts with damaged DNA and recruits the nucleotide-excision repair complex.

Applying What You Know

1. The addition of cordycepin to the 3′ end of a growing RNA strand, as will occur in 5′ → 3′ chain growth, prevents any further elongation because the incorporated antibiotic lacks a 3′-OH group and the next nucleoside triphosphate cannot be added.
2. Intercalating agents such as ethidium bromide inhibit RNA polymerases; actinomycin D and acridine are other examples. They tightly bind to duplex DNA and strongly inhibit transcription, presumably by interfering with the passage of the polymerases. Intercalators therefore cause inhibition of transcription—clearly to be avoided!
3. Affinity chromatography using poly(T) molecules as ligand would preferentially bind mature mRNAs by their poly(A) tails.
4. **(a)** Histone mRNAs have much shorter lifetimes than do other mRNAs: <30 min in the cytosol, compared with hours or days for other mRNAs. Although there are other factors involved in the rate of histone mRNA degradation, this is one piece of evidence that poly(A) tails provide RNAs with some protection (though not permanently) from degradation in the cytosol.

   (b) Given the importance of histones' function and their relatively conserved sequences across species, perhaps the possibility of incorrect splicing has proved too great a danger.
5. The nucleotide change could be at an intron-exon junction, altering the removal of introns and/or splicing of exons. Another possibility is a defect in the promoter of the gene, leading to inefficient transcription. Premature termination is another feasible cause, due to a change in the nucleotide sequence producing an early signal for termination. All of these single-nucleotide effects would reduce β-globin gene expression.

chapter

27

Protein Metabolism

STEP-BY-STEP GUIDE

Major Concepts

The amino acid sequence of proteins is determined by the sequence of nucleotides in coding regions of DNA.

Regions of DNA that code for proteins are transcribed into a complementary **messenger RNA** (mRNA) molecule. Each amino acid is coded for by a sequence of three nucleotide residues in the mRNA. The code is degenerate in that more than one triplet, or **codon,** can be used to specify a given amino acid. For most codons, the third base tends to pair more loosely with the corresponding base of its tRNA anticodons (they "wobble"), and this has some interesting consequences for the code. The **initiation** and **termination** points of the protein-coding region are also specified by triplet codons. In general, regions of DNA containing an initiation codon followed by at least 50 codons specifying amino acids usually produce functional proteins. Codons do not usually overlap (though there are exceptions in viral systems) and there is no "punctuation" between codons.

Protein synthesis occurs in the cytosol on ribosomes.

Ribosomes are cytoplasmic complexes made up of proteins and **ribosomal RNA** (rRNA). After leaving the nucleus, mRNA molecules become associated with ribosomes. The molecule of mRNA serves as a template and the ribosome provides the structure and catalytic activity for protein synthesis.

The mRNA codons are recognized by transfer RNA (tRNA) anticodons.

Transfer RNA molecules contain a single strand of RNA that is folded (due to intramolecular base pairing) into a cloverleaf secondary structure. Codon recognition is the result of base pairing between the nucleotides of the mRNA strand and a three-nucleotide region **(anticodon)** on one "arm" of the tRNA molecule. Because a given tRNA must recognize a *specific* mRNA codon and a specific amino acid, tRNA molecules can be thought of as adapters linking two dissimilar entities.

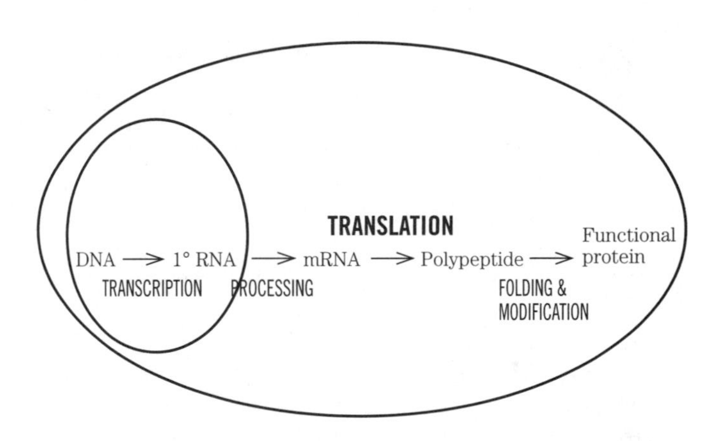

The 3′ end of the tRNA molecule is the point of attachment for its amino acid.

Amino acids are first activated by ATP and then attached to their tRNA to form an **aminoacyl-tRNA.** This reaction, driven by the hydrolysis of pyrophosphate and essentially irreversible, is catalyzed by **aminoacyl-tRNA synthetases.** Each enzyme is specific for the tRNA *and* the amino acid in order to ensure that the genetic information is faithfully translated into a specific amino acid sequence.

Polypeptide synthesis is initiated at the amino-terminal end of the protein and requires an AUG codon.

The AUG start codon, which codes for methionine, and a specific class of tRNA that binds *N*-formylmethionine (fMet) is required for prokaryotic **initiation.** Protein synthesis in bacteria begins with the formation of the 70S initiation complex, which contains the following: the 30S and 50S ribosomal subunits, the mRNA molecule, the fMet-tRNA, three proteins called **initiation factors,** GTP, and Mg^{2+}. The fMet-tRNA is bound to one of two tRNA-binding sites on the initiation complex, referred to as the **P** (for peptidyl) **site.**

Polypeptide elongation involves the formation of peptide bonds between amino acids on adjacent tRNAs.

Elongation requires the 70S initiation complex, the tRNA specified by the next codon in the mRNA sequence, three proteins called **elongation factors,** and GTP. The incoming tRNA is first bound to the **A** (for aminoacyl) **site** of the ribosome complex. Next, a peptide bond is formed between the initial methionine and the amino acid on the next tRNA; this reaction is catalyzed by the 23S rRNA of the 50S ribosomal subunit. The dipeptide is then translocated to the P site of the ribosome complex and the uncharged tRNA is released from the E (for exit) site in preparation for a new elongation cycle. Proofreading is limited at this stage to the codon-anticodon pairing; no check of the amino acid attached to the tRNA is carried out during elongation.

Termination of polypeptide synthesis requires one of three termination codons and one of three termination factors.

In **termination,** each of the three termination codons is recognized by one of the three **termination** or **release factors.** Binding of the release factor causes hydrolysis of the polypeptide chain from the P-site tRNA, release of these components from the P site, and dissociation of the ribosomal complex into its subunits. Protein synthesis is energetically very expensive, but understandably so; the cell must synthesize its molecular "machinery," which includes the enzymes, transport proteins, defense systems, regulatory proteins, and all other polypeptides that direct a cell's metabolism.

Some antibiotics are naturally occurring inhibitors of prokaryotic protein synthesis.

They are useful clinically and as tools for basic research into the processes of protein synthesis.

Polypeptide chains undergo posttranslational modification.

Protein folding and processing occur following protein synthesis. **Protein folding** is largely determined by noncovalent interactions between amino acids in the polypeptide chain. **Protein modification** includes (1) removal of the amino-terminal Met residue; (2) acetylation of the carboxyl terminus; (3) covalent modification of individual amino acids in the chain, including phosphorylation, glycosylation, methylation, addition of isoprenyl groups, and addition of prosthetic groups; (4) proteolytic cleavage of precursor proteins into smaller, active forms; and (5) formation of disulfide bonds.

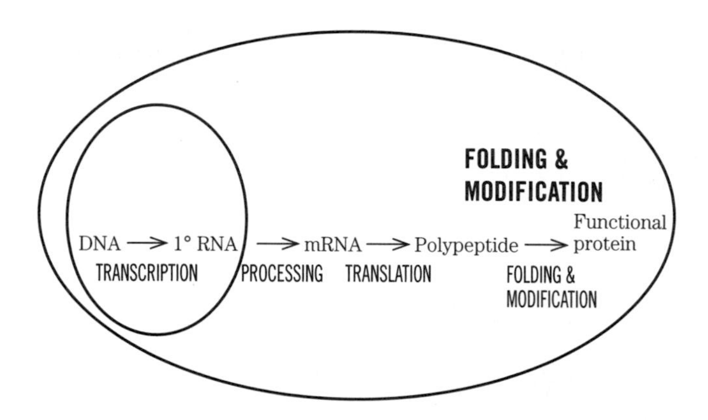

In addition, certain proteins are directed to specific regions in the cell by information contained in signal sequences.

These sequences determine whether a protein will be inserted into the membrane or translocated into the lumen of the endoplasmic reticulum (where certain of the posttranslational modification reactions occur in eukaryotes) for eventual secretion. Following translocation, signal sequences are proteolytically removed. In addition, there are nuclear localization signals, which are not removed, that direct some proteins to the nucleus.

The mechanisms triggering the degradation of proteins are poorly understood.

One possible signal may be the specific amino-terminal residue of a protein: certain amino-terminal amino acids appear to have a stabilizing effect (in terms of half-life of the protein), whereas others have a destabilizing effect. In eukaryotes, proteins are marked for destruction by the covalent attachment of one or more molecules of the protein **ubiquitin.** The mechanism of targeting by ubiquitin is unknown.

What to Review

Answering the following questions and reviewing the relevant concepts, which you have already studied, should make this chapter more understandable.

- Because proteins are composed of **amino acids** linked by **peptide bonds,** it will be helpful to review the properties of individual amino acids and the details of peptide bond structure (Chapter 3, especially Fig. 3–5 and pp. 82–83).

- The function of proteins is highly dependent on their structure, which is the result of protein folding and posttranslational modifications. Review the various factors that affect **protein folding** and review the **levels of protein structure** (Chapter 4).
- Protein folding and final structure are influenced by the presence of hydrophobic regions. Review the properties of soluble proteins compared with membrane-associated proteins (pp. 375–379).
- **Noncovalent interactions** between nucleic acids are critically important in the translation of mRNA sequences into proteins. Review the different types of interactions that affect base-pair formation (Fig. 8–14), the formation of hybrid duplexes (pp. 288–289), and the formation of secondary and tertiary structures by DNA and RNA molecules (pp. 281–282).

Topics for Discussion

Answering each of the following questions, especially in the context of a study group discussion, should help you understand the important points of this chapter.

27.1 The Genetic Code

1. What are the essential elements of the protein synthetic machinery in cells?

2. Why is DNA not *directly* involved in this process?

The Genetic Code Was Cracked Using Artificial mRNA Templates

3. What pieces of evidence indicated that the genetic code was based on "three-letter words"?

4. What is the difference between overlapping and nonoverlapping codes?

5. What is meant by "reading frame" and why is it so critical to the production of a functional protein?

6. What technological developments have occurred since the breaking of the genetic code that would have made this accomplishment relatively straightforward? (Hint: see Chapters 8 and 9.)

7. Why are termination codons also called "nonsense" codons? Why isn't the initiation codon a "nonsense" codon?

8. What is the relationship between an "open reading frame" and a gene?

9. At which position in the codon is the genetic code *usually* degenerate? What are the exceptions?

Wobble Allows Some tRNAs to Recognize More than One Codon

10. What are the cellular advantages of having a "wobble" nucleotide in codons?

Translational Frameshifting and RNA Editing Affect How the Code is Read

11. Most examples of functional proteins produced by frameshifting and "genes within genes" are found in viruses and bacteria. What are the probable reasons for this?

Box 27–1 Exceptions That Prove the Rule: Natural Variations in the Genetic Code

12. How are the changes in the mitochondrial code a kind of "genomic streamlining"?

27.2 Protein Synthesis

Protein Biosynthesis Takes Place in Five Stages

13. What are the energy requirements for each stage of protein synthesis?

The Ribosome Is a Complex Supramolecular Machine

14. How do RNA molecules contribute to the function of ribosomes? What are the contributions of the protein components of ribosomes?

Box 27–2 From an RNA World to a Protein World

15. What are some of the "negatives" of being an RNA enzyme?

Transfer RNAs Have Characteristic Structural Features

16. What forces are responsible for the three-dimensional structure of tRNA?

17. Why does the anticodon arm always contain seven unpaired nucleotides?

Stage 1: Aminoacyl-tRNA Synthetases Attach the Correct Amino Acids to Their tRNAs

18. How are amino acids correctly paired with (added to) the tRNA that contains the appropriate anticodon? What makes this process irreversible?

19. What is the nature of the bond that is formed?

- **Proofreading by Aminoacyl-tRNA Synthetases**

20. Which aminoacyl-tRNA synthetases are *most* likely to require a proofreading capability? Which aminoacyl-tRNA synthetases probably do *not* have a proofreading capability?

21. Why is a proofreading capability more important for the enzymes involved in DNA replication than for aminoacyl-tRNA synthetases?

- **Interaction between an Aminoacyl-tRNA Synthetase and a tRNA: A "Second Genetic Code"**

22. What constitutes a "codon" in the second genetic code?

Box 27–3 Natural and Unnatural Expansion of the Genetic Code

23. How do *E. coli* produce tRNA "charged" with selenocysteine in the absence of a selenocysteine-tRNA synthetase?

24. What additional elements are required to get selenocysteine incorporated into a protein?

Stage 2: A Specific Amino Acid Initiates Protein Synthesis

25. Which end of a protein is coded for by the 3′ end of an mRNA molecule? Which end is coded for by the 5′ end?

26. Why is an initiation codon critical to correct protein synthesis?

27. Suggest why there is only *one* initiation codon, but three termination codons.

28. Is fMet found in the proteins of vertebrate cells?

- **The Three Steps of Initiation**

29. What additional information can you now add to your answer to the question: How do RNA molecules contribute to the function of ribosomes?

30. What is the role of GTP in this process?

31. What is the maximum number of tRNA molecules that can bind to a ribosome?

32. *Cell Map:* If you still have room, draw an initiation complex using different colors to indicate the different species of RNA involved.

- **Initiation in Eukaryotic Cells**

33. How is the initial AUG distinguished from internal AUGs in eukaryotes?

Stage 3: Peptide Bonds Are Formed in the Elongation Stage

34. What additional elements are required for protein elongation, besides the 70S initiation complex?

- **Elongation Step 1: Binding of an Incoming Aminoacyl-tRNA**

35. What elements are required for binding tRNA to the A site?

- **Elongation Step 2: Peptide Bond Formation**

36. What additional information can you now add to your answer to the question: How do RNA molecules contribute to the function of ribosomes?

- **Elongation Step 3: Translocation**

37. What are the roles of ATP and GTP in the elongation process? How much energy (in terms of high-energy phosphate bonds) is expended in the formation of one peptide bond?

- **Proofreading on the Ribosome**

38. What is the molecular basis for proofreading during elongation? In what way does this molecular mechanism support the concept of a second genetic code?

Stage 4: Termination of Polypeptide Synthesis Requires a Special Signal

39. How do termination factors "recognize" the UAA, UAG, or UGA termination codons?

- **Energy Cost of Fidelity in Protein Synthesis**

40. At which point(s) is an energy input required in the process of translation?

- **Rapid Translation of a Single Message by Polysomes**

41. Why are polysomes especially important to the process of protein synthesis in bacteria?

Stage 5: Newly Synthesized Polypeptide Chains Undergo Folding and Processing

42. Does protein folding require an input of energy?

43. Which amino acid residues can be modified following translation?

44. Are all posttranslational modifications permanent?

45. Give some of the reasons why proteins require posttranslational modification.

Box 27–4 Induced Variation in the Genetic Code: Nonsense Suppression

46. How do "two wrongs make a right" in nonsense suppression?

Protein Synthesis Is Inhibited by Many Antibiotics and Toxins

47. Some antibiotics selectively block bacterial protein synthesis, which is the basis for their therapeutic effects. What are some of the differences (encountered in this chapter) between bacterial and eukaryotic protein synthesis that may account for the selectivity of these agents?

27.3 Protein Targeting and Degradation

48. Which properties of amino acids could contribute to the targeting function of signal sequences?

Posttranslational Modification of Many Eukaryotic Proteins Begins in the Endoplasmic Reticulum

49. Signal sequences on proteins destined for the endoplasmic reticulum (ER) usually contain a core of hydrophobic amino acids. What does this suggest about the signal recognition particle?

50. Proteins destined to be secreted by cells must somehow make their way into the lumen of the ER. Why is a peptide translocation complex required for the successful production of these secretory proteins? How do you think this complex functions? (Hint: consider what you have learned about the properties of proteins and cell membranes.)

Glycosylation Plays a Key Role in Protein Targeting

51. Most membrane proteins and secretory proteins are glycosylated in the lumen of the ER. What consequence does this have on the protein's topology?

52. Based on the information on hydrolase sorting, what appears to be the molecular basis for the use of oligosaccharides as targeting signals?

Signal Sequences for Nuclear Transport Are Not Cleaved

53. Why can the NLS be located in the interior of a protein?

54. *Cell Map:* Look at the various cellular components that you have included on your map. Which ones required signal sequences?

Bacteria Also Use Signal Sequences for Protein Targeting

55. Suggest why a protein in its final, three-dimensional form would no longer be "translocation-competent."

Cells Import Proteins by Receptor-Mediated Endocytosis

56. How do cells "select" the specific extracellular items that they want to import via clathrin-coated pits? Is this mechanism foolproof?

Protein Degradation Is Mediated by Specialized Systems in All Cells

57. Which different mechanisms do cells use to ensure that only certain proteins (e.g., old or defective ones) and not other cellular components are degraded?

58. What aspect of protein structure influences the life span of cellular proteins?

SELF-TEST

Do You Know the Terms?

ACROSS

2. Cellular complex that increases the speed of protein synthesis.
5. Sometimes contains the nucleotide inosinate; portion of tRNA responsible for decoding the mRNA message.
6. "Start" codons are distinguished from other, identical codons occurring later in the mRNA sequence by ______-______ interactions between the Shine-Dalgarno sequence in the mRNA and stretches of rRNA near the P site.
13. Hydrolysis of this nucleotide triphosphate is required for translocation of an mRNA molecule during the elongation phase of protein synthesis. (abbr.)
14. Polynucleotide that contains the bases adenine, guanine, cytosine, and uracil and contains the code for a polypeptide. (abbr.)
16. "Word" in genetic terms.
18. Region of the ribosome that binds all charged tRNAs except $tRNA^{fMet}$.
22. "Start" in genetic terms.
23. Polynucleotide that contains the bases adenine, guanine, cytosine, and uracil and forms covalent associations with an amino acid. (abbr.)
24. Complex of proteins and ribonucleotides that has structural and catalytic functions in protein synthesis.
25. Strong base pairing between the first two nucleotides in a codon and the complementary nucleotides in a tRNA anticodon contributes to the *specificity* of codon recognition; the weaker pairing, or ______, that occurs between the third pair of nucleotides influences the *speed* of protein synthesis.
26. Describe the incipient or initial protein formed on ribosomes.
28. Frame shifting is an example of ______ modification.
29. Covalent attachment of an amino acid to a specific tRNA requires energy in the form of ATP; the product of this reaction is commonly referred to as being ______.

DOWN

1. Proteins, such as the Na^+/K^+-ATPase, require a ______ sequence in the DNA code.
2. The "start" codon of an mRNA binds to ribosomes at the P, or ______, site.
3. "I do not like green eggs and ham. I do not like them Sam-I-Am" is analogous to an ______ reading frame, in genetic terms.
4. Stage of protein synthesis in which an aminoacyl-tRNA binds to AUG on the A site of a ribosome.
7. Recognition and binding of a specific tRNA by a specific aminoacyl-tRNA synthetase is the basis for the ______ genetic code.
8. Stage of protein synthesis in which an aminoacyl-tRNA binds to AUG on the P site of a ribosome.
9. Cytosolic protein involved in targeting proteins to the endoplasmic reticulum. (abbr.)
10. The peptide ______ complex serves as a docking protein as well as a transmembrane channel.
12. Posttranslational protein modification includes such processes as ______ and phosphorylation.
15. Scaffold:Building construction as ______: Protein synthesis. (abbr.)
17. Each codon specifies only one amino acid, meaning the genetic code is unambiguous; however, any one amino acid may have more than one codon, meaning the genetic code is ______.
19. Stage of protein synthesis in which the ester bond of peptidyl-tRNA is hydrated.
20. Protein occurring *ubiquitously* in cells; signals protein destruction.
21. Activated form of tRNA, known as ______-tRNA, that binds to the A site of ribosomes.
27. The process of protein synthesis requires a number of "accessory" proteins known as initiation, elongation, and releasing ______.

Do You Know the Facts?

1. Following the 5′ → 3′ convention of writing nucleotide sequences, indicate (yes/no) whether each of the following mRNA codons can be recognized by the tRNA anticodon ICG.

_____ **(a)** UGC
_____ **(b)** CGA
_____ **(c)** UGA
_____ **(d)** CGU
_____ **(e)** CGC

2. Indicate (yes/no) whether nascent insulin molecules are altered by each of the following posttranslational modifications.

_____ **(a)** Cleavage of a signal sequence
_____ **(b)** Carboxylation of Asp residues
_____ **(c)** Phosphorylation of Tyr residues
_____ **(d)** Proteolytic cleavage of internal sequences
_____ **(e)** Formation of disulfide cross-links

3. The enzymatic process of "charging" a molecule of tRNA is similar to other enzyme catalyzed reactions you have encountered in previous chapters. Which of the following correctly describes such similarities?

A. Formation of an acyl-adenylate intermediate coupled to the hydrolysis of pyrophosphate is similar to reactions involved in fatty acid activation.
B. Proofreading and correcting abilities that prevent incorporation of the wrong molecule are analogous to the action of DNA polymerases.
C. The aminoacyl-tRNA synthetase is a relatively nonspecific enzyme (i.e., it can activate many different amino acids) and in this respect is similar to ribonucleotide reductase.
D. A and B
E. B and C

4. A new compound, vivekine, was recently discovered by a clever graduate student. It was isolated from bacteria found in deep sea-dwelling organisms. Vivekine inhibits protein synthesis in eukaryotes: protein synthesis can initiate, but only dipeptides are formed and these remain bound to the ribosome. This toxin affects eukaryotic protein synthesis by blocking the:

A. binding of formylmethionyl-tRNA to ribosomes.
B. activity of elongation factors.
C. activation of amino acids.
D. recognition of stop signals.
E. formation of peptide bonds.

5. Which of the following describes the structure of transfer RNAs?

A. Their unusual bases are coded for by the DNA template.
B. Hydrophobic interactions between adjacent aromatic rings stabilize the three-dimensional structure.
C. The activated amino acid recognizes the codon.
D. All four "arms" have double-helical structures.
E. All of the above are true.

6. Indicate whether each of the following statements about prokaryotic translation is true or false.

_____ **(a)** An aminoacyl-tRNA synthetase catalyzes formation of an ester bond.
_____ **(b)** An mRNA molecule cannot be used to direct protein synthesis until it has been completely transcribed.
_____ **(c)** The positioning of fMet-tRNA on the A site defines the reading frame.
_____ **(d)** Incoming aminoacyl-tRNAs are first bound to the A site.
_____ **(e)** Formation of the 70S initiation complex requires an input of energy.
_____ **(f)** The carboxyl group of the amino acid on the aminoacyl-tRNA is transferred to the amino group of a peptidyl-tRNA.
_____ **(g)** Release factors cause the peptidyl transferase activity of the ribosome to use H_2O as a substrate.

7. Based on what you now know about DNA and RNA, suggest why cells don't use mRNA instead of DNA as the repository of genetic information, thus eliminating the need for a transcription step.

8. Which would you expect to be more critical to the production of a functional protein: the precise positioning of the initiation codon or the precise positioning of the termination codon?

Applying What You Know

1. Certain bacteria have suppressor tRNAs that suppress the effects of mutations in the DNA. In theory, in what *two* ways could suppressor tRNA differ from normal tRNA to account for this function?

2. One of the most clinically important missense mutations is that which causes sickle-cell disease: the alteration from a glutamate to a valine in the sixth position of the α chains of the hemoglobin tetramer. What would be the simplest explanation for how such a change occurred?

3. (a) You wish to set up a system for in vitro protein translation. What components would the system require? Assume you are working with a prokaryotic system.

(b) You repeat the experiment above using mRNA for insulin and partially purified eukaryotic components obtained from a *crude extract* (a common phrase used to denote the possible presence of undefined items in the mixture). You conclude that protein synthesis in eukaryotes must be different from that in prokaryotes because translation stops after a relatively short time. Examination of the components of your system reveals that a short (200 amino acid) peptide has been synthesized but is "stuck" on the ribosome, as is the mRNA molecule. Explain this result. What must you add to your in vitro system to permit complete protein synthesis?

ANSWERS

Do You Know the Terms?

ACROSS

2. polysome
5. anticodon
6. base-pair
11. reading frame
13. GTP
14. mRNA
16. codon
18. A site
22. AUG
23. tRNA
24. ribosome
25. wobble
26. nascent
28. translational
29. charged

DOWN

1. signal
2. peptidyl
3. open
4. elongation
7. second
8. initiation
9. SRP
10. translocation
12. glycosylation
15. rRNA
17. degenerate
19. termination
20. ubiquitin
21. aminoacyl
27. factors

Do You Know the Facts?

1. **(a)** No; **(b)** Yes; **(c)** No; **(d)** Yes; **(e)** Yes

2. **(a)** Yes; **(b)** No; **(c)** No; **(d)** Yes; **(e)** Yes

3. D

4. B

5. B

6. **(a)** T; **(b)** F; **(c)** F; **(d)** T; **(e)** T; (f) F; **(g)** T

7. Some organisms *do* use RNA for storing their genetic information (e.g., retroviruses, such as the virus that causes AIDS), but they do not sidestep the transcription requirement. DNA is used as the repository of genetic information by most organisms primarily because it is much more stable than RNA: DNA exists as a double-stranded molecule and lacks the reactive —OH group present on the ribose residues of RNA molecules. In addition, uracils formed by spontaneous deamination of cytosine cannot be recognized as "foreign" in RNA molecules and therefore cannot be "repaired," as they are in DNA. In fact, DNA is the only biomolecule for which repair systems exist—a critical property for a molecule with so much depending on the constancy of its sequences.

8. The precise positioning of the initiation codon is more critical to protein synthesis because the reading frame is determined by this position. If the reading frame is off by one or two codons the *entire* sequence of the protein will be altered; a misplaced start codon has a 66.6% chance of producing a garbled protein. The mispositioning of a termination codon, assuming that it remains somewhere near the end of the protein-coding sequence, means the loss (or gain) of some carboxyl-terminal amino acids. This would most likely have much less of an impact on protein function than a completely altered amino acid sequence.

Applying What You Know

1. Suppressor tRNAs could have complementary mutations in their anticodons that allow them to base-pair with the mutated mRNA codon, negating the effect of the DNA mutation. Such mutations have been observed (see Box 27–4). In theory, it is also possible that a mutation in a particular aminoacyl-tRNA synthetase could change its binding specificity for either the tRNA or the amino acid, such that an amino acid now pairs with an anticodon that recognizes the mutated mRNA codon. The problem with this mechanism, and the reason it is probably not found in vivo, is that *all* of the cell's tRNAs with the anticodon for the mutant code would be improperly aminoacylated, and thus all other proteins in the cell would also contain inappropriate amino acids.

2. The codon for glutamate is GAA or GAG. A single substitution of a U for an A gives GUA or GUG, both of which are codons for valine. Note that a change of the second A in GAA to a U would not be as calamitous; it would alter the Glu to an Asp, which is also a negatively charged amino acid.

3. **(a)** The process of prokaryotic protein synthesis in a test tube would require the following:

Ribosomes (bacterial variety, 50S and 30S subunits)
mRNA containing initiation and termination codons
All 20 of the amino acids

fMet-tRNA
At least 20 tRNAs that can bind the 20 amino acids
All the aminoacyl-tRNA synthetases
ATP, GTP, and Mg^{2+}
Initiation factors: IF-1, IF-2, and IF-3
Elongation factors: EF-Tu, EF-Ts, and EF-G
Termination (release) factors: RF_1, RF_2, and RF_3

(b) Synthesis stopped after about 200 amino acid residues because the polypeptide product of the insulin mRNA has a signal sequence to which a signal recognition particle (SRP) binds, halting translation. Synthesis will resume if you add vesicles composed of endoplasmic reticulum membrane on which the SRP-ribosomal complex can dock. Docking allows the SRP to detach and protein synthesis to resume.

chapter

28

Regulation of Gene Expression

STEP-BY-STEP GUIDE

Major Concepts

The regulation of gene expression determines which proteins are made by which cells and how much of each protein is produced.

The transcription of DNA to mRNA, catalyzed by RNA polymerases, is most often the point at which gene expression is regulated. Nevertheless, regulation can also occur at each of the other steps involved in transforming the genetic message into functional proteins: RNA processing, translation, protein (posttranslational) processing, protein targeting, and degradation. The structural characteristics of proteins and polynucleotides provide the molecular basis for the various regulatory strategies used by cells.

Protein "factors" play key roles in the regulation of transcription.

In prokaryotes these proteins include **specificity factors,** which regulate the ability of RNA polymerase to bind to specific promoter regions on the DNA; **repressors,** which block polymerase binding; and **activators,** which enhance polymerase binding. In eukaryotes, gene activators are called **transcription factors.** These proteins have amino acid domains that bind DNA (some can recognize specific DNA sequences) and/or bind to other DNA-binding proteins. The mechanisms used to regulate gene expression are, even for simple organisms, highly complex, reflecting the complex compositions of the intracellular and extracellular environments to which cells must adapt.

RNA molecules also play roles in regulating gene expression.

Non-coding RNAs (ncRNAs) have been shown to regulate both transcription and translation, primarily by binding to protein factors and modulating their activity. In addition, ncRNAs such as snRNA and snoRNA are involved in the processing of RNA transcripts, and complementary miRNAs "silence" mRNA transcripts by forming duplexes, which are then inactivated and/or degraded.

Regulation of gene expression can be either positive or negative and often involves the interactions of multiple regulatory factors.

The precise control of gene expression is ensured by the use of **combinations** of regulators. This use of multiple regulators is an example of cellular economy: a relatively small pool of factors can be used in different combinations to regulate the expression of a large number of genes. In general, it appears that gene expression is not regulated in an "all-or-none" fashion. The use of multiple regulators makes it possible to *partially* turn on or turn off expression of a gene, making cells more responsive to fluctuations in their environment.

Prokaryotes and eukaryotes differ in some of their regulatory strategies.

Prokaryotes have smaller cell sizes than do eukaryotes and correspondingly smaller genomes. They compensate by clustering the genes for related proteins together in units called **operons** and by producing **polycistronic mRNA** that can be translated into a number of different proteins. Examples of operons that have been closely studied are the *lac* and *trp* operons of *E. coli.* The larger genomes of **eukaryotes** are highly condensed in order to fit into the nucleus. Consequently, in regions containing the genes to be expressed the DNA structure must be altered for transcription to occur. The regulation of gene expression in eukaryotes is usually through **activation** of transcription (i.e., positive regulation), as opposed to repression (negative regulation). Because of the complexity and large size of eukaryotic genomes, the possibility of nonspecific binding of transcription factors and inappropriate gene expression is increased. As a result, eukaryotic regulatory mechanisms are more complex and employ larger numbers of regulatory factors than those of prokaryotes. Finally, the physical separation of the genomic information in the nucleus from the protein-synthesizing machinery in the cytoplasm provides more points for regulation of gene expression than are available in prokaryotes, in which transcription and translation are tightly coupled.

What to Review

Answering the following questions and reviewing the relevant concepts, which you have already studied, should make this chapter more understandable.

- Gene expression includes (but is not limited to) transcription and translation. Briefly review the steps involved in each of these processes (nicely summarized in Figs. 26–10 and 27–33).
- You have already been introduced to factors that alter cell function by affecting gene expression; these factors are the **steroid** and **thyroid hormones.** Review their structure and function (pp. 905–906, 908–909).
- The concept of **protein domains** is critical to an understanding of how proteins interact with other proteins and with DNA to regulate gene expression. Be sure that you are clear on the relationship between protein structure, protein function, and protein domains (Chapter 4).
- The use of **transcription terminators** is an important molecular mechanism in gene regulation schemes. Review the ways in which transcription can be terminated (Fig. 26–8).
- DNA and RNA structure are crucial factors in the regulation of gene expression. Review the structure of DNA presented in Chapter 24, especially the way in which DNA is condensed in the nucleus (pp. 961–962).

Topics for Discussion

Answering each of the following questions, especially in the context of a study group discussion, should help you understand the important points of this chapter.

1. Why is the regulation of gene expression critical to the function of all organisms?

2. Make a flowchart listing the steps from the DNA code to the production of a functional protein. Show the points at which gene expression can be regulated. You may want to locate where each of these steps occurs on your cell map.

28.1 Principles of Gene Regulation

3. What proteins have you encountered in your study of biochemistry that are likely to be coded for by housekeeping genes?

4. In addition to those cited in the text, what other proteins can you think of whose synthesis can be induced or repressed?

RNA Polymerase Binds to DNA at Promoters

5. What factors affect the binding of RNA polymerase to DNA?

6. How does RNA polymerase binding affect protein production?

7. Is the DNA binding by RNA polymerase an “all-or-none” process?

Transcription Initiation Is Regulated by Proteins That Bind to or Near Promoters

8. What *molecular mechanisms* might account for the effects of specificity factors? Repressors? Activators?

Many Bacterial Genes Are Clustered and Regulated in Operons

9. What are the different regions of DNA that make up an operon?

10. Why does it make biological sense for bacteria to have operons, whereas most higher eukaryotes do not?

The lac *Operon Is Subject to Negative Regulation*

11. What role does lactose play in the regulation of gene expression in *E. coli*?

12. How does the Lac repressor protein exert its effects?

Regulatory Proteins Have Discrete DNA-Binding Domains

13. What types of interactions can occur between DNA and proteins that would allow proteins to recognize specific nucleotide sequences?

14. Which region of a DNA molecule provides the greatest opportunity for such interactions to occur?

15. Why does a Gln residue only interact with adenine in DNA and not, for example, with guanine? (Hint: see Fig. 28–9.) Do other amino acid residues also interact with adenine?

Helix-Turn-Helix

16. The DNA-binding region of the helix-turn-helix motif is a relatively short stretch of 7 to 9 amino acids that interact with only a portion of the DNA recognition sequence. These structures are usually unstable. How do transcription factors that contain this motif function in vivo? (Hint: see Fig. 28–11 for an example.)

Zinc Finger

17. How is the structure of this type of DNA-binding motif stabilized?

18. Would you expect a protein with a single zinc finger to bind to DNA with more or less specificity than a protein with 37 zinc fingers?

Homeodomain

19. In what way is the homeodomain related to the helix-turn-helix motif?

Regulatory Proteins Also Have Protein-Protein Interaction Domains

20. Why might it be important for regulatory proteins to be able to interact with RNA polymerase?

Leucine Zipper

21. What forces are responsible for the dimerization of leucine-zipper transcription factors?

Basic Helix-Loop-Helix

22. What molecular mechanisms are used by helix-loop-helix proteins to permit dimerization and DNA binding?

23. What is the significance of short stretches of basic amino acid residues in these proteins?

Subunit Mixing in Eukaryotic Regulatory Proteins

24. In what ways can protein-protein interactions contribute to the *specificity* of protein-DNA interactions? (Hint: regulatory proteins bind to DNA as *homo*dimers as well as *hetero*dimers.)

25. With which other cellular components might transcription factors interact in order to regulate gene expression? (Hint: you can find clues to the answer of this question in the section on the *lac* operon.)

28.2 Regulation of Gene Expression in Bacteria

The *lac* operon provides important insights into the molecular strategies used by cells to regulate gene expression. Keep in mind as you work through this section that the information gained from studies of prokaryotes suggests things to look for in eukaryotes.

The lac *Operon Undergoes Positive Regulation*

26. Why do even relatively simple organisms, such as bacteria, use complex regulatory schemes to regulate gene expression?

27. How do bacteria respond to the presence of lactose (i.e., how do they transmit information about the presence of lactose to their DNA)?

28. How do bacteria respond to the presence of glucose?

29. Are these responses "all-or-none" phenomena?

30. Why is it useful for cells to have a few regulatory molecules, such as CRP and cAMP, control many different genes?

Many Genes for Amino Acid Biosynthetic Enzymes Are Regulated by Transcription Attenuation

31. In the case of the *trp* operon, how does DNA structure contribute to the regulation of gene expression?

32. The regulatory "leader protein" of the *trp* operon has *no* function in the cell other than as a "sensor" for tryptophan levels. What is its "sensory" mechanism?

33. Why is the tight coupling of transcription and translation in prokaryotes critical to the transcription attenuation mechanism?

34. In what ways is transcription attenuation similar to feedback inhibition of enzymes by their reaction product?

Induction of the SOS Response Requires Destruction of Repressor Proteins

35. How does the RecA protein act as a sensor of DNA damage in cells?

36. What is the molecular basis for RecA's signaling actions?

Synthesis of Ribosomal Proteins Is Coordinated with rRNA Synthesis

37. How do bacteria coordinate the rate of ribosomal protein synthesis with that of rRNA?

38. What intracellular signaling pathway is used to relay information about the levels of amino acids in bacterial cells?

39. How do lowered amino acid levels decrease translation in bacteria?

The Function of Some mRNAs Is Regulated by Small RNAs in Cis or in Trans

40. In what ways is TPP binding to mRNA similar to hormone-receptor interactions?

Some Genes Are Regulated by Genetic Recombination

41. How does genetic recombination cause a switch in the transcription of different flagellin genes in *Salmonella*?

42. How can this recombination event be so specific that it does not affect the expression of other genes necessary to the function of the organism?

43. Is this regulation an "all-or-none" response?

(Note: a similar strategy is used by the human immunodeficiency virus (HIV) to escape detection by the immune system. The development of a vaccine to protect against HIV has been greatly hindered by the virus's ability to continuously alter the conformation of proteins in its coat.)

28.3 Regulation of Gene Expression in Eukaryotes

44. Which major structural differences between eukaryotic and prokaryotic cells affect the way their gene expression is regulated?

45. Why does "positive regulation" of gene expression predominate in eukaryotes?

Transcriptionally Active Chromatin Is Structurally Distinct from Inactive Chromatin

46. What are "hypersensitive sites," and why is it likely that they correspond to areas of active gene transcription?

47. What is the effect of methylation on DNA structure?

Chromatin Is Remodeled by Acetylation and Nucleosomal Displacement/Repositioning

48. Why does histone acetylation promote gene transcription?

Many Eukaryotic Promoters Are Positively Regulated

49. In what ways is the initiation of gene transcription in cells analogous to opening a combination lock?

DNA-Binding Transactivators and Coactivators Facilitate Assembly of the General Transcription Factors

50. Where do each of the three classes of proteins bind and what is their role in gene activation?

51. Why does the regulation of eukaryotic genes usually involve multiple regulatory sites (i.e., promoter and enhancer regions) and multiple regulatory proteins (i.e., general and specific transcription factors)?

Transcription Activators

52. With what components of gene transcription do these activators interact?

Coactivator Protein Complexes

53. Use Figure 28–29 to help you visualize how all three classes of proteins work together to promote transcription.

TATA-Binding Protein

54. Recall how this protein works in the initiation of transcription (see Figs. 26–9 and 26–10).

Choreography of Transcriptional Activation

55. Do you think the process of transcriptional activation is more like an on-off switch or a rheostat?

Reversible Transcriptional Activation

56. What is the likely reason that inhibition of gene expression is rare in eukaryotes?

DNA-Binding Transactivators

57. Transactivators bind to enhancer elements that affect transcription of eukaryotic genes that are usually located 100 to 5,000 base pairs away from the gene they regulate, on either the 3′ *or* the 5′ side. Why is this possible? (Compare this with TATA boxes, which are located 25 to 30 base pairs "upstream," on the 5′ side of the gene.)

The Genes of Galactose Metabolism in Yeast Are Subject to Both Positive and Negative Regulation

58. Identify the positive and negative regulators in this system.

Transcription Activators Have a Modular Structure

59. Can you think of a mechanism that would account for the *specific* activation of a gene by factors that bind relatively *nonspecifically* to DNA domains, such as the glutamine-rich, proline-rich, or acidic activation domains?

Eukaryotic Gene Expression Can Be Regulated by Intercellular and Intracellular Signals

60. What is the transducer that couples a steroid hormone signal to alterations in gene expression?

61. Where does the steroid hormone-receptor complex fit into the general scheme shown in Figure 28–29?

Regulation Can Result from Phosphorylation of Nuclear Transcription Factors

62. Why do nonsteroid hormones require second messengers to affect transcription?

63. What is the primary difference that distinguishes the effects of nonsteroid hormones from steroid hormones on transcriptional regulation?

Many Eukaryotic mRNAs Are Subject to Translational Repression

64. Why is translational repression an important regulatory mechanism in eukaryotes but not in prokaryotes?

65. What are the molecular mechanisms used by eukaryotes to achieve translational repression?

Posttranscriptional Gene Silencing is Mediated by RNA Interference

66. What is the basis for the regulation of gene expression by small RNAs?

RNA-Mediated Regulation of Gene Expression Takes Many Forms in Eukaryotes

67. Answer Question 1 in "Do You Know the Facts" and then indicate where the different ncRNAs exert their effects.

Development is Controlled by Cascades of Regulatory Proteins

68. Suggest why *Drosophila* oocytes require a complement of *maternal* mRNA.

69. What are the different classes of regulatory genes involved in pattern formation during *Drosophila* embryogenesis and where do each originate?

Box 28–1 Of Fins, Wings, Beaks, and Things

70. Using what you have learned about the relationship between genes, protein structure, and protein function, evaluate the statement: "Very few mutations are required, and the needed mutations affect regulation."

Discussion Questions for Study Groups

- One of the effects of exercise is to increase muscle mass by increasing the amount of muscle protein. Discuss the possible ways that exercise might influence the expression of genes encoding muscle proteins, such as actin and myosin.
- The development of the anterior-posterior axis in *Drosophila* depends upon the establishment of mRNA and protein gradients in developing embryos. The anteriorly *localized bicoid* mRNA is involved in specifying the "head" end of the fly, and the posteriorly *localized nanos* mRNA specifies the "tail" end of the fly. Can you think of ways in which these gradients could also specify the developmental fates of cells in intermediate regions such as the thorax? How do you think the dorsal-ventral axis is established?

SELF-TEST

Do You Know the Terms?

ACROSS

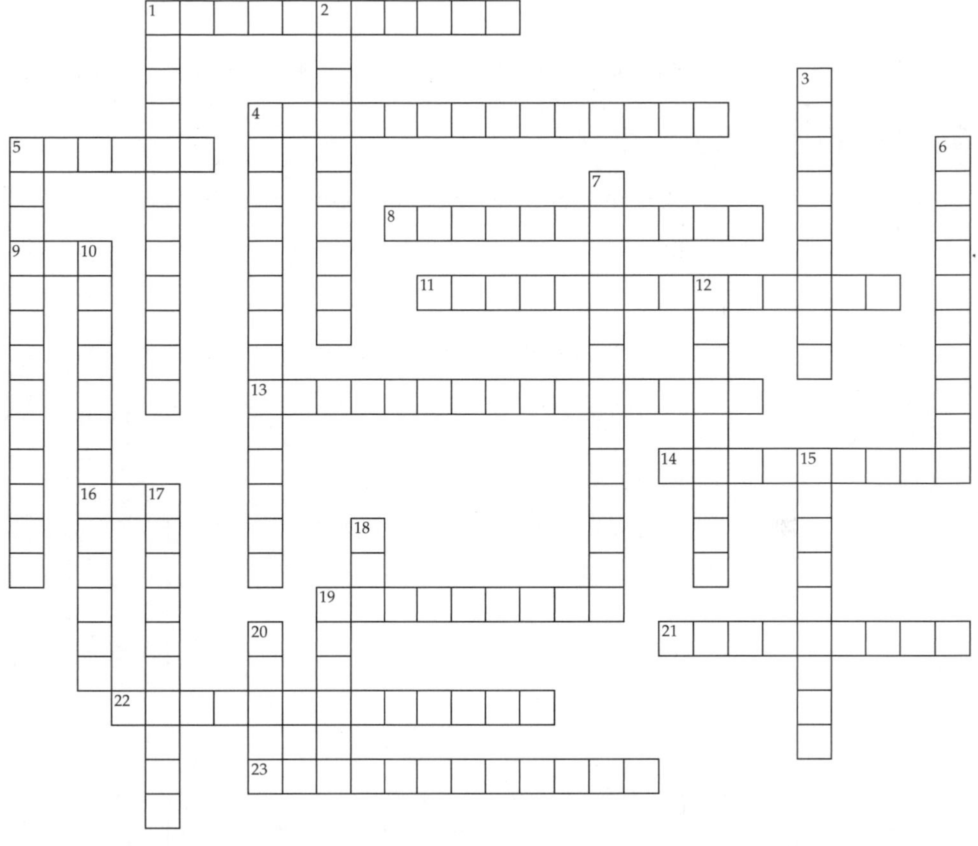

1. Region of transcription factor with a structure similar to that of the DNA-binding region of helix-turn-helix proteins.
4. Region of regulatory protein, the ______-______-______, contains two amphipathic α helices connected by extended amino acid chains of varying lengths.
5. Region of mRNA transcribed from the *trp* operon that "senses" the levels of tryptophan in *E. coli* cells.
8. Hairpin loops of nascent mRNA sometimes act as ______, abruptly terminating DNA transcription.
9. Transcription factor that regulates transcription of the *lac* operon genes and acts as a glucose sensor. (abbr.)
11. The ability to control yourself, for example.
13. Adenylate cyclase is to cAMP as ______ is to ppGpp. (2 words)
14. Molecule inactivated by binding of lactose to a regulatory protein, with consequent increase in expression of genes in the *lac* operon.
16. One of the gene products of the ______ operon is tryptophan synthetase.
19. Regions of DNA that usually include some or all of the promoter region.
21. Describes genes of the *lac* operon, whose gene products are needed only when [glucose] is low and [lactose] is high.
22. ______ factors regulate the formation of mRNA.
23. Describes transcription of *lac* genes in cells with mutations in the Lac repressor binding region.

DOWN

1. Class of genes used by all cells and essential to normal function; the genes for ribosomal proteins, for example.
2. In eukaryotes, a nucleic acid conformation that contributes to the interaction of enhancer and promoter elements. (2 words)
3. In the absence of glucose, CAP binds cAMP and undergoes a conformational change that allows it to bind to the *lac* promoter, resulting in the ______ of gene transcription; in the presence of glucose, cAMP levels fall and CAP is prevented from binding to DNA.
4. Describes sites in DNA that lack nucleosomes and tend to be found near transcription start sites; they are especially susceptible to the action of DNases.
5. Region of transcription factor; dimer held together by hydrophobic interactions between two α helices. (2 words)
6. Region of transcription factor; has one or more prominent loops of amino acids stabilized by interactions between cysteine (and sometimes histidine) and a zinc atom. (2 words)
7. These are transcribed into mRNA that is eventually packaged into unfertilized oocytes. (2 words)
10. Describes prokaryotic mRNA produced from gene clusters that are transcribed together.
12. During induction of the SOS response in a severely damaged cell, the RecA protein facilitates cleavage of specific repressor proteins that normally prevent viruses from switching between the ______ life cycle to the lytic, or replication competent, form.
15. UASs are to yeast as ______ are to higher eukaryotes.
17. The DNA to which the *lac* repressor binds contains the following inverted repeat, which is an almost perfect ______.

AATTGTGAGCGGATAACAATT
TTAACACTCGCCTATTGTTAA

18. Class of segmentation genes transcribed in *Drosophila* embryos following fertilization; often their translation is induced by transcription factors produced from maternal mRNA.

19. Region of DNA that contains the binding sites for promoters, repressors, and RNA polymerases, and the coding regions for two or more proteins.

20. Describes amino acids that make up short stretches of the DNA-binding domains of many transcription factors.

Do You Know the Facts?

1. A cartoon of a "typical" eukaryotic cell is shown below. Identify each of the numbered steps involved in gene expression.

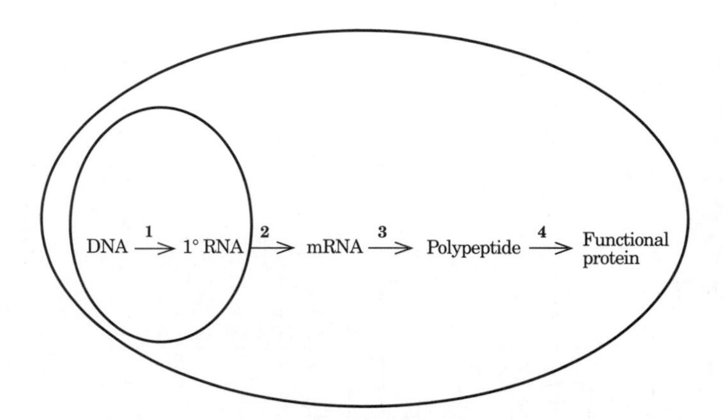

2. Provide a specific, molecular example of how regulation can occur at each step in question 1.

3. Indicate whether each statement describes transcription and/or translation in prokaryotes or in eukaryotes.

(a) Transcription and translation of the same gene occur almost simultaneously.
(b) Messenger RNAs undergo extensive processing.
(c) Transcription is both positively and negatively regulated.
(d) A single mRNA codes for several proteins.
(e) Transcription and translation occur in separate cellular compartments.
(f) The DNA code for mRNA is noncontiguous.
(g) Unwinding from histones is probably necessary for transcriptional activation of the DNA.

4. Indicate (yes/no) whether each type of bond or interaction contributes to the ability of transcription factors to recognize specific DNA sequences.

_____ **(a)** Hydrophobic interactions
_____ **(b)** Van der Waals interactions
_____ **(c)** Hydrogen bonds
_____ **(d)** Covalent bonds
_____ **(e)** Phosphoanhydride bonds

5. Which of the following statements about enhancer elements is correct?

A. They can be located on either side of the gene they activate.
B. They will *not* function unless located at some distance from the gene they control.
C. They bind transcription factors.
D. An enhancer element activating a specific gene is not functional in all cell types.
E. All of the above are true.

6. The *lac* operon is shown diagrammatically below. Match the elements of the *lac* operon (listed below the diagram) with the appropriate features or functions (more than one may apply per element).

DNA	P_I	I	O_3	P	O_1	Z	O_2	Y	A

_____ P_I
_____ I
_____ P
_____ O_1
_____ O_2
_____ O_3
_____ Z
_____ Y
_____ A
_____ Lac repressor

(a) Binding site for the experimental inducer IPTG.
(b) Contains a binding site for RNA polymerase.
(c) Encodes a protein that binds to allolactose or to DNA.
(d) Encodes the protein β-galactosidase.
(e) Binding site for the Lac repressor protein.
(f) Encodes the protein galactoside permease.
(g) Secondary Lac repressor binding site.
(h) The glucose sensor.
(i) Encodes the protein thiogalactoside transacetylase.

7. Indicate whether each of the following statements about the *lac* operon is true or false.

_____ **(a)** All seven genes are transcribed into mRNA, which is then translated into five different proteins.
_____ **(b)** The repressor binds to the structural genes, preventing their transcription.
_____ **(c)** Efficient binding of RNA polymerase to DNA occurs only in the presence of CRP and cAMP.
_____ **(d)** cAMP acts as an inducer by binding to the Lac repressor.
_____ **(e)** In the absence of lactose, the repressor binds the operator.
_____ **(f)** A mutation in the I gene such that no gene product is made leads to constitutive expression of the Z, Y, and A genes.

8. CRP has been found to contain a typical helix-turn-helix motif. Which of the following best describes the interactions between CRP and the *lac* operon?

A. Covalent bonds occur between the DNA helix and CRP helices.
B. Nonpolar interactions and hydrogen bonds contribute to the specificity of the CRP-DNA interactions.
C. The double helix of the operon DNA interacts with the double helix of the CRP DNA.
D. The base sequence of the DNA has no influence on the binding of CRP.
E. The specific amino acid sequence of CRP has no influence on DNA binding.

9. Prokaryotic *and* eukaryotic promoters:

A. contain AT-rich regions.
B. interact with transcription factors.
C. are influenced by base sequences thousands of base pairs away, which increase their activity.
D. are recognized by three different types of RNA polymerases, depending on location in the genome and type of transcript.
E. have all of the above properties.

Applying What You Know

1. (a) The *Homo sapiens* genome has approximately 3×10^9 base pairs; if an "average" protein contains 400 amino acids, what is the maximum number of proteins that can be encoded by the human genome?

(b) Analysis of the sequence of the human genome suggests that there are only 30–40,000 protein-encoding genes, and it is estimated that human cells make ten times this number of proteins due to differential splicing, etc. Using the information from part (a), what percentage of the total genome is actually used to encode proteins? How would you account (in molecular terms, of course) for the answers you obtained for (a) and (b)?

(c) The calculation made in part (b) above is an underestimation of the percentage of the genome used to encode proteins. Why?

2. The regulation of gene expression that occurs during the development of multicellular organisms is an intricately orchestrated series of "molecular switches." These molecular switches are responsible for turning on developmental pathways that result in the differentiation of a single, pluripotent egg cell into the thousands of different cell types that make up an organism. Many of these switches are *transcription factors* that can turn on whole groups of genes needed to produce, say, a head structure. The maternal gene *bicoid* codes for just such a transcription factor in *Drosophila.* The *bicoid* gene product is localized at one end of the unfertilized egg due to the presence of a 3′ untranslated region (UTR) in the maternal mRNA that "tethers" mRNA molecules to some intracellular structure in this future "head" end. If a mutation occurred in the *bicoid* gene that deleted all of the 3′ UTR, what would be the effect on the phenotype of the adult fly?

Biochemistry on the Internet

The study of "gene cascades" involved in *Drosophila* development have provided new insights into the roles of nucleotide binding proteins and gene regulation. The genetic mechanisms for establishment of the anterior-posterior axes in the early stages of embryological development have been determined and a number of nucleotide binding proteins have been implicated.

1. One gene that has been implicated in the establishment of the posterior system is a maternal factor, *nanos* (*nos*). The *nanos* mRNA is localized in the posterior region of the embryo. Its gene product is a putative zinc-finger protein. These types of nucleotide-binding proteins have a protruding "finger" of amino acids that consists of about 30 amino acid residues, four of which (either four Cys or two Cys and two His) coordinate a single Zn^{2+} atom, thereby stabilizing the structure.

(a) At what level of gene expression does this protein function, and how does its function contribute to the generation of a posterior axis?

(b) In the process of characterizing the functional domains of this protein, you make a mutant strain in which all the Cys residues in this protein are replaced by acidic amino acid residues. How would you expect this substitution to affect the function of this protein?

(c) How many zinc fingers is this protein likely to have? (Hint: you can use FlyBase, the information source for research on *Drosophila* located at http://flybase.bio.indiana.edu/.)

2. The *Drosophila* gene *engrailed* encodes a protein product that is involved in determining the anterior-posterior axis in the wing of the fly. The name is derived from the fact that the mutant fly, which lacks this gene product, has a wing margin with a scalloped or engrailed edge. In the normal fly, this edge is typical of the anterior portion of the wing, whereas the posterior region has a smoother edge. In the *engrailed* mutant, the posterior portion of the wing is replaced by a mirror image of the anterior wing structures, giving the wing an overall "engrailed" appearance.

(a) The *engrailed* gene product is a transcription factor. What kind of DNA-binding motif does it use to bind to DNA and alter transcription? (Hint: you can use FlyBase or try the SCOP database at http://scop.berkeley.edu and do a keyword search.)

(b) What is the amino acid sequence of the Engrailed protein? (Hint: continue using FlyBase and find the amino acid sequence for the *engrailed* protein product.)

(c) Use the Protein Explorer and PDB file 3HDD (a molecular model of the Engrailed protein binding to DNA) to answer the following questions:

1.) What is the secondary structure of the DNA-binding region of the *engrailed* protein?

2.) Can you locate the amino acid residues in the *engrailed* protein that are likely to form noncovalent interactions with the DNA? (Hint: focus on protein Chain A. Nonocovalent interactions generally occur with intermolecular distances between 3.5 and 4.5 Å. Select the charged residues in the protein and

1																													30
E	K	R	P	R	T	A	F	S	S	E	Q	L	A	R	L	K	R	E	F	N	E	N	R	Y	L	T	E	R	R

31																													60
R	Q	Q	L	S	S	E	L	G	L	N	E	A	Q	I	K	I	W	F	Q	N	K	R	A	K	I	K	K	S	T

display them as a space fill model.) The diagram of the amino acid sequence for the DNA-binding region of 3HDD may help you to keep track of the amino acids that are binding to the DNA.

3.) A general consensus amino acid sequence for the homeodomain DNA-binding motif has been described. It was obtained by comparing more than 300 homeobox proteins and is shown below. In this sequence, seven of the residues are absolutely conserved (marked with *), six are highly conserved (+), and the rest of the conserved positions (indicated by capital letters) have no more than nine different amino acid residues.

```
    +          *   *     +             +    + +** * * * +
xxxxRxxYxxxQLxxLExxFxxxxYLxxxxRxxLAxxLxLxxxQVKIWFQNRRxKxKxxx
```

Source: *Guidebook to the Homeobox Genes* (Duboule, D., ed.), Oxford University Press, 1994.

Which of these conserved residues correspond to those identified in part **(2)** above?

4.) What is the sequence of the portion of DNA that the *engrailed* protein interacts with?

5.) The consensus DNA-binding sequence for this protein is "TAATTA." Can you use the molecular model to predict the identity of the residues that play the biggest role in recognizing a specific DNA sequence? How could you test your hypothesis?

ANSWERS

Do You Know the Terms?

ACROSS

1. homeodomain
4. helix-turn-helix
5. leader
8. attenuators
9. CRP
11. autoregulation
13. stringent factor
14. repressor
16. *Trp*
19. operators
21. inducible
22. transcription
23. constitutive

DOWN

1. housekeeping
2. DNA looping
3. induction
4. hypersensitive
5. leucine zipper
6. zinc finger
7. maternal genes
10. polycistronic
12. lysogenic
15. enhancers
17. palindrome
18. gap
19. operon
20. basic

Do You Know the Facts?

1. The major steps linking the genetic information residing in DNA sequences to functional proteins and a functional organism are:

(1) *Transcription* of DNA into RNA molecules in the nucleus.

(2) *Posttranscriptional processing* of RNA and export to the cytosol.

(3) *Translation* of the mRNA transcript to a polypeptide.

(4) *Folding and posttranslational processing and modification* of the polypeptide.

2. The following are just a few examples of the types of regulation that can occur at each step in gene expression (this list is *not* exhaustive).

(1) Transcriptional control

- Induction by activators and transcription factors; for example, CRP in the regulation of *lac* operon transcription.
- Repression; for example, the Lac repressor protein.
- Regulation by a form of feedback inhibition; for example, the regulation of transcription of the *trp* operon by tryptophan.
- DNA structure can determine which genes are available for transcription; for example, DNA-associated proteins, such as histones, affect DNA packing.
- Genetic recombination; for example, used by *Salmonella* to regulate which of two genes for flagellin proteins are transcribed.

(2) RNA processing
- Splicing of introns by snRNAs and self-splicing mechanisms.
- Differential (or alternative) splicing of complex transcripts; for example, the production of calcitonin in the thyroid and calcitonin gene–related peptide in the brain from a common primary transcript.
- Base modification by methylation, deamination, or reduction in tRNAs.
- Addition of 5′ caps and 3′ poly(A) tails to mRNAs.
- Differential degradation of mRNAs; this type of regulation may be related to base modification.

(3) Translational control
- A variety of initiation, elongation, and release factors control when and how much of a gene product is made; the activity of these factors is often modulated by protein phosphorylation or the availability of GTP.
- Regulation of the number of ribosomes available for translation in bacteria.
- Translation repressor proteins regulate the synthesis of ribosomal proteins in bacteria.

(4) Protein processing
- Proteolytic processing; for example, formation of a functional insulin dimer from a large, nonfunctional precursor molecule.
- Glycosylation; often targets proteins to specific cellular locations.
- Isoprenylation; also helps in localizing proteins (anchors them in membranes).
- Amino acid modification; for example, reversible phosphorylation of Ser, Thr, or Tyr residues that activates or inactivates enzymes.
- Formation of disulfide bonds; for example, in insulin, three disulfide bonds must be formed between the correct pairs of Cys residues to produce a functional protein.

3. **(a), (c), (d),** prokaryotes; **(b), (e), (f), (g),** eukaryotes; **(c)** also occurs in eukaryotes, but is a relatively rare event.

4. **(a)** Y; **(b)** N; **(c)** Y; **(d)** N; **(e)** N

5. E

6. P_I **(b);** I **(c);** P **(b), (g);** O_1 **(e);** O_2 **(g);** O_3 **(g);** Z **(d);** Y **(f);** A **(i);** Lac repressor **(a)**

7. **(a)** F; **(b)** F; **(c)** T; **(d)** F; **(e)** T; **(f)** T

8. B

9. A

Applying What You Know

1. (a) If the average protein contains 400 amino acids, the minimum size for a gene is 1,200 base pairs. Dividing the total number of base pairs (3×10^9) by 1,200 gives a maximum of **2.5×10^6 proteins** that can be encoded by the human genome.

(b) In actuality, only about **1.6% of the genome** is used to encode proteins. This figure is obtained by determining the number of base pairs (bp) minimally needed to encode 40,000 protein-encoding genes (40,000 proteins $\times$ 1,200 bp per protein from part a above = 4.8×10^7 bp) and then dividing that number by the total number of base pairs in the genome (3×10^9 bp). The "extra" 98.4% is noncoding DNA containing a variety of types of sequences, including what some refer to as "junk" DNA. Also included in the extra DNA are regulatory regions (enhancers and promoters); spacer DNA; duplicated portions that are essentially redundant; pseudogenes; repetitious DNA; centromeric and telomeric regions; and other regions that remain to be identified. Much of this seemingly extraneous DNA is fodder for the evolutionary "mill."

(c) The calculations made above assume that the genes are composed only of exons; introns were not included in our approximation of the amount of DNA needed to encode a protein. The average size of a human gene has been estimated to be approximately 27,000 bp (based on the transcribed region only). Using this number the percent of the human geneome used to encode proteins inceases to 36%, which still leaves a large proportion (64%) of the geneome as noncoding sequences.

2. The absence of the 3′ UTR in the maternal mRNA for the *bicoid* transcription factor would prevent the mRNA from binding to internal structures in one end of the egg. Consequently, the maternal message would *diffuse further* into the egg than in an egg with the unmutated (wild-type) message. Recall that the *bicoid* transcription factor is responsible for turning on a cascade of genes, ultimately resulting in the formation of head (anterior) structures. The presence of the factor in new locations in the egg would cause these head structures to form in inappropriate (ectopic) positions. The resulting phenotype would be a *Drosophila* with a *larger than normal, or inappropriately positioned,* "head" region.

Biochemistry on the Internet

1. (a) This morphogen is a translational repressor that binds to mRNA primarily in the posterior region of the embryo. It acts by suppressing production of proteins involved in the formation of anterior structures.

(b) The *nanos* gene product is a putative zinc-finger protein. The zinc-finger structure is thought to bind to specific DNA sequences and regulate transcription. Substitution of other amino acid residues for Cys residues in this protein would prevent the formation of the "finger" structure and interfere with this protein's ability to interact with specific DNA sequences and regulate gene expression.

(c) At FlyBase, perform a search for the gene *nanos.* Click on "*nos*" in the search output and then look up either of the Protein Accession numbers presented to get the amino acid sequence of the *nanos* gene product. There are relatively few of the Cys residues needed to form a zinc finger. All six of the Cys residues are located in the carboxyl end of this protein. Probably only a single zinc finger can be generated by this protein.

2. (a) This protein is identified as a homeodomain-containing DNA-binding protein (there are two homeobox domains in this protein). It also contains a helix-turn-helix domain.

(b) Search results for "engrailed" in Flybase provide links to information on the engrailed gene (symbol en). You can get the amino acid sequence for the engrailed protein product by clicking on "Gene Products & Expression," "Polypeptide Data," and the protein name.

The complete sequence is shown below with one of the homeodomain regions indicated in bold.

MALEDRCSPQ	SAPSPITLQM	
QHLHHQQQQQ	QQQQQQMQHL	
HQLQQLQQLH	QQQLAAGVFH	60
HPAMAFDAAA	AAAAAAAAAA	
AHAHAAALQQ	RLSGSGSPAS	
CSTPASSTPL	TIKEEESDSV	120
IGDMSFHNQT	HTTNEEEEAE	
EDDDIDVDVD	DTSAGGRLPP	
PAHQQQSTAK	PSLAFSISNI	180
LSDRFGDVQK	PGKSIENQAS	
IFRPFEANRS	QTATPSAFTR	
VDLLEFSRQQ	QAAAAAATAA	240
MMLERANFLN	CFNPAAYPRI	
HEEIVQSRLR	RSAANAVIPP	
PMSSKMSDAN	PEKSALGSLC	300
KAVSQIG QPA	APTMTQPPLS	
SSASSLASPP	PASNASTISS	
TSSVATSSSS	SSSGCSSAAS	360
SLNSSPSSRL	GASGSGVNAS	
SPQPQPIPPP	SAVSRDSGME	
SSDDTRSETG	STTTEGGKNE	420
MWPAWVYCTR	YSDRPSSGPR	
YRRPKQPKDK	TND**EKRPRTA**	
FSSEQLARLK	**REFNENRYLT**	480
ERRRQQLSSE	**LGLNEAQIKI**	
WFQNKRAKIK	**KST**GSKNPLA	
LQLMAQGLYN	HTTVPLTKEE	540
EELEMRMNGQ	IP	552

Most proteins that contain a homeobox domain can be further classified into three subfamilies based on their sequence characteristics—engrailed, antennapedia, and paired. Proteins currently known to belong to the engrailed subfamily include two *Drosophila* proteins (Engrailed and Invected); two homologous proteins in silk moths; E30 and E60 proteins in honeybees; G-En protein in grasshoppers; En-1 and En-2 in mammals and birds; Eng-1, -2, -3 in zebrafish; SU-HB-en in sea urchins; Ht-En in leeches; and ceh-16 in *C. elegans.*

(c) 1. To examine the interactions between this homeodomain-containing transcription factor and DNA, use Jmol viewer or one of the other molecular viewers available at www.rcsb.org. Load the molecule of interest (3HDD). SELECT "Nucleic" and then RENDER or DISPLAY "Spacefill," and you should be able to distinguish the duplex DNA and the two associated Engrailed protein fragments. One of the alpha helices in each fragment is clearly in contact with the DNA. (Note: you can click on a structure and drag it to move the molecular structure, this improves 3D visualization. If you get stuck, you can always refresh the window and start over.)

2. To explore the interactions between the engrailed protein and DNA, display the DNA as a "Ball & Stick" model and the protein as a "Spacefill" model. By clicking on the atoms that come close to the DNA you should be able to identify the following amino acid residues that could make noncovalent interactions with the DNA: Arg^{3}, Arg^{5}, Thr^{6}, Phe^{8}, Tyr^{25}, Ile^{47}, Gln^{50}, Asn^{51}, Arg^{53}, Lys^{55}, Lys^{57}.

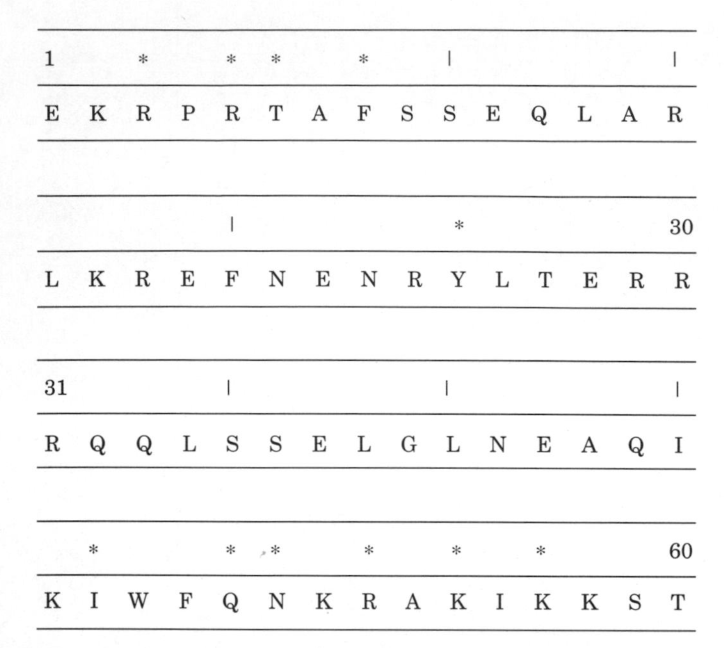

1		*		*	*		*		|					|
E	K	R	P	R	T	A	F	S	S	E	Q	L	A	R
				|					*					30
L	K	R	E	F	N	E	N	R	Y	L	T	E	R	R
31				|					|				|
R	Q	Q	L	S	S	E	L	G	L	N	E	A	Q	I
	*			*	*		*		*		*			60
K	I	W	F	Q	N	K	R	A	K	I	K	K	S	T

3. Most of the residues in the DNA-binding homeodomain of Engrailed that are capable of forming noncovalent interactions with the DNA correspond to conserved amino acids in the consensus sequence for this motif as indicated in bold below:

+ * * +
xxxx**R**xxYxxxQLxxLExxFxxxx**Y**LxxxxRx

\+ + +* * * * * +
xLAxxLxLxxxQVK**I**WF**QN**R**R**x**K**x**K**xxx

4. Interactions that permit binding of the Engrailed protein to **specific** DNA sequences will be those that occur between the DNA binding α helix of the homeodomain region and the nitrogenous bases themselves, and not those occurring with atoms in the phosphate backbone. To look at the contact sites on the intact protein, click on RESET VIEW and SELECT the engrailed homeodomain protein "Chain A." The new display will show the DNA contacts made on the protein. Repeat the above procedure to identify the nucleotides that are involved in the protein-DNA interactions. You should be able to identify the following nucleotides in the DNA Chain C: T211, A212, A213, T214; contacts involving T211 and A213 are between the protein and the nitrogenous bases.

5. One approach is to display the DNA/protein complex as a spacefill model. This provides a good visualization of how the Engrailed homeodomain "grips" the DNA helix. The 3D shape of the protein gives the impression of a "clamplike" structure. Although it is clear that a number of amino acids residues contact the DNA, two of the amino acid residues in particular seem to insert into the grooves of the helix. Clicking on this part of the protein identifies the residues as Arg^{5} and Asn^{51}. If these residues are crucial to DNA recognition, then experimentally changing the identity of these residues would alter the ability of this protein to recognize and bind to this specific DNA sequence.

Solutions Manual

chapter

1

The Foundations of Biochemistry

1. The Size of Cells and Their Components

(a) If you were to magnify a cell 10,000-fold (typical of the magnification achieved using an electron microscope), how big would it appear? Assume you are viewing a "typical" eukaryotic cell with a cellular diameter of 50 μm.

(b) If this cell were a muscle cell (myocyte), how many molecules of actin could it hold? (Assume the cell is spherical and no other cellular components are present; actin molecules are spherical, with a diameter of 3.6 nm. The volume of a sphere is 4/3 πr^3.)

(c) If this were a liver cell (hepatocyte) of the same dimensions, how many mitochondria could it hold? (Assume the cell is spherical; no other cellular components are present; and the mitochondria are spherical, with a diameter of 1.5 μm.)

(d) Glucose is the major energy-yielding nutrient for most cells. Assuming a cellular concentration of 1 mM, calculate how many molecules of glucose would be present in our hypothetical (and spherical) eukaryotic cell. (Avogadro's number, the number of molecules in 1 mol of a nonionized substance, is 6.02×10^{23}.)

(e) Hexokinase is an important enzyme in the metabolism of glucose. If the concentration of hexokinase in our eukaryotic cell is 20 μM, how many glucose molecules are present per hexokinase molecule?

Answer

(a) The magnified cell would have a diameter of 50×10^4 μm = 500×10^3 μm = 500 mm, or 20 inches—about the diameter of a large pizza.

(b) The radius of a globular actin molecule is 3.6 nm/2 = 1.8 nm; the volume of the molecule, in cubic meters, is $(4/3)(3.14)(1.8 \times 10^{-9}\ \text{m})^3 = 2.4 \times 10^{-26}\ \text{m}^3$.*

The number of actin molecules that could fit inside the cell is found by dividing the cell volume (radius = 25 μm) by the actin molecule volume. Cell volume = $(4/3)(3.14)(25 \times 10^{-6}\ \text{m})^3 = 6.5 \times 10^{-14}\ \text{m}^3$. Thus, the number of actin molecules in the hypothetical muscle cell is

$$(6.5 \times 10^{-14}\ \text{m}^3)/(2.4 \times 10^{-26}\ \text{m}^3) = 2.7 \times 10^{12} \text{ molecules}$$

or 2.7 trillion actin molecules.

**Significant figures:* In multiplication and division, the answer can be expressed with no more significant figures than the least precise value in the calculation. Because some of the data in these problems are derived from measured values, we must round off the calculated answer to reflect this. In this first example, the radius of the actin (1.8 nm) has two significant figures, so the answer (volume of actin = $2.4 \times 10^{-26}\ \text{m}^3$) can be expressed with no more than two significant figures. It will be standard practice in these expanded answers to round off answers to the proper number of significant figures.

(c) The radius of the spherical mitochondrion is 1.5 μm/2 = 0.75 μm, therefore the volume is $(4/3)(3.14)(0.75 \times 10^{-6}\text{ m})^3 = 1.8 \times 10^{-18}\text{ m}^3$. The number of mitochondria in the hypothetical liver cell is

$$(6.5 \times 10^{-14}\text{ m}^3)/(1.8 \times 10^{-18}\text{ m}^3) = 36 \times 10^3 \text{ mitochondria}$$

(d) The volume of the eukaryotic cell is $6.5 \times 10^{-14}\text{ m}^3$, which is $6.5 \times 10^{-8}\text{ cm}^3$ or 6.5×10^{-8} mL. One liter of a 1 mM solution of glucose has $(0.001\text{ mol/1000 mL})(6.02 \times 10^{23}\text{ molecules/mol}) = 6.02 \times 10^{17}$ molecules/mL. The number of glucose molecules in the cell is the product of the cell volume and glucose concentration:

$$(6.5 \times 10^{-8}\text{ mL})(6.02 \times 10^{17}\text{ molecules/mL}) = 3.9 \times 10^{10}\text{ molecules}$$

or 39 billion glucose molecules.

(e) The concentration ratio of glucose/hexokinase is 0.001 M/0.00002 M, or 50/1, meaning that each enzyme molecule would have about 50 molecules of glucose available as substrate.

2. Components of *E. coli* *E. coli* cells are rod-shaped, about 2 μm long and 0.8 μm in diameter. The volume of a cylinder is $\pi r^2 h$, where h is the height of the cylinder.

(a) If the average density of *E. coli* (mostly water) is 1.1×10^3 g/L, what is the mass of a single cell?

(b) *E. coli* has a protective cell envelope 10 nm thick. What percentage of the total volume of the bacterium does the cell envelope occupy?

(c) *E. coli* is capable of growing and multiplying rapidly because it contains some 15,000 spherical ribosomes (diameter 18 nm), which carry out protein synthesis. What percentage of the cell volume do the ribosomes occupy?

Answer

(a) The volume of a single *E. coli* cell can be calculated from $\pi r^2 h$ (radius = 0.4 μm):

$$3.14(4 \times 10^{-5}\text{ cm})^2(2 \times 10^{-4}\text{ cm}) = 1.0 \times 10^{-12}\text{ cm}^3 = 1 \times 10^{-15}\text{ m}^3 = 1 \times 10^{-15}\text{ L}$$

Density (g/L) multiplied by volume (L) gives the mass of a single cell:

$$(1.1 \times 10^3\text{ g/L})(1 \times 10^{-15}\text{ L}) = 1 \times 10^{-12}\text{ g}$$

or a mass of 1 pg.

(b) First, calculate the proportion of cell volume that does *not* include the cell envelope, that is, the cell volume *without* the envelope—with $r = 0.4\ \mu\text{m} - 0.01\ \mu\text{m}$; and $h = 2\ \mu\text{m} - 2(0.01\ \mu\text{m})$—divided by the total volume.

$$\text{Volume without envelope} = \pi(0.39\ \mu\text{m})^2(1.98\ \mu\text{m})$$

$$\text{Volume with envelope} = \pi(0.4\ \mu\text{m})^2(2\ \mu\text{m})$$

So the percentage of cell that does *not* include the envelope is

$$\frac{\pi(0.39\ \mu\text{m})^2(1.98\ \mu\text{m}) \times 100}{\pi(0.4\ \mu\text{m})^2(2\ \mu\text{m})} = 90\%$$

(Note that we had to calculate to one significant figure, rounding down the 94% to 90%, which here makes a large difference to the answer.) The cell envelope must account for 10% of the total volume of this bacterium.

(c) The volume of all the ribosomes (each ribosome of radius 9 nm) $= 15{,}000 \times (4/3)\pi(9 \times 10^{-3}\ \mu\text{m})^3$

The volume of the cell $= \pi(0.4\ \mu\text{m})^2(2\ \mu\text{m})$

So the percentage of cell volume occupied by the ribosomes is

$$\frac{15{,}000 \times (4/3)\pi(9 \times 10^{-3}\ \mu\text{m})^3 \times 100}{\pi(0.4\ \mu\text{m})^2(2\ \mu\text{m})} = 5\%$$

3. Genetic Information in *E. Coli* DNA The genetic information contained in DNA consists of a linear sequence of coding units, known as codons. Each codon is a specific sequence of three deoxyribonucleotides (three deoxyribonucleotide pairs in double-stranded DNA), and each codon codes for a single amino acid unit in a protein. The molecular weight of an *E. coli* DNA molecule is about 3.1×10^9 g/mol. The average molecular weight of a nucleotide pair is 660 g/mol, and each nucleotide pair contributes 0.34 nm to the length of DNA.

(a) Calculate the length of an *E. coli* DNA molecule. Compare the length of the DNA molecule with the cell dimensions (see Problem 2). How does the DNA molecule fit into the cell?

(b) Assume that the average protein in *E. coli* consists of a chain of 400 amino acids. What is the maximum number of proteins that can be coded by an *E. coli* DNA molecule?

Answer

(a) The number of nucleotide pairs in the DNA molecule is calculated by dividing the molecular weight of DNA by that of a single pair:

$$(3.1 \times 10^9 \text{ g/mol})/(0.66 \times 10^3 \text{ g/mol}) = 4.7 \times 10^6 \text{ pairs}$$

Multiplying the number of pairs by the length per pair gives

$$(4.7 \times 10^6 \text{ pairs})(0.34 \text{ nm/pair}) = 1.6 \times 10^6 \text{ nm} = 1.6 \text{ mm}$$

The length of the cell is 2 μm (from Problem 2), or 0.002 mm, which means the DNA is (1.6 mm)/(0.002 mm) = 800 times longer than the cell. The DNA must be tightly coiled to fit into the cell.

(b) Because the DNA molecule has 4.7×10^6 nucleotide pairs, as calculated in (a), it must have one-third this number of triplet codons:

$$(4.7 \times 10^6)/3 = 1.6 \times 10^6 \text{ codons}$$

If each protein has an average of 400 amino acids, each requiring one codon, the number of proteins that can be coded by *E. coli* DNA is

$$(1.6 \times 10^6 \text{ codons})(1 \text{ amino acid/codon})/(400 \text{ amino acids/protein}) = 4{,}000 \text{ proteins}$$

4. The High Rate of Bacterial Metabolism Bacterial cells have a much higher rate of metabolism than animal cells. Under ideal conditions some bacteria double in size and divide every 20 min, whereas most animal cells under rapid growth conditions require 24 hours. The high rate of bacterial metabolism requires a high ratio of surface area to cell volume.

(a) Why does surface-to-volume ratio affect the maximum rate of metabolism?

(b) Calculate the surface-to-volume ratio for the spherical bacterium *Neisseria gonorrhoeae* (diameter 0.5 μm), responsible for the disease gonorrhea. Compare it with the surface-to-volume ratio for a globular amoeba, a large eukaryotic cell (diameter 150 μm). The surface area of a sphere is $4\pi r^2$.

Answer

(a) Metabolic rate is limited by diffusion of fuels into the cell and waste products out of the cell. This diffusion in turn is limited by the surface area of the cell. As the ratio of surface area to volume decreases, the rate of diffusion cannot keep up with the rate of metabolism within the cell.

(b) For a sphere, surface area = $4\pi r^2$ and volume = $4/3\ \pi r^3$. The ratio of the two is the surface-to-volume ratio, S/V, which is $3/r$ or $6/D$, where D = diameter. Thus, rather than calculating S and V separately for each cell, we can rapidly calculate and compare S/V ratios for cells of different diameters.

$$S/V \text{ for } N.\ gonorrhoeae = 6/(0.5\ \mu\text{m}) = 12\ \mu\text{m}^{-1}$$

$$S/V \text{ for amoeba} = 6/(150\ \mu\text{m}) = 0.04\ \mu\text{m}^{-1}$$

$$\frac{S/V \text{ for bacterium}}{S/V \text{ for amoeba}} = \frac{12\mu\text{m}^{-1}}{0.04\ \mu\text{m}^{-1}} = 300$$

Thus, the surface-to-volume ratio is 300 times greater for the bacterium.

5. **Fast Axonal Transport** Neurons have long thin processes called axons, structures specialized for conducting signals throughout the organism's nervous system. Some axonal processes can be as long as 2 m—for example, the axons that originate in your spinal cord and terminate in the muscles of your toes. Small membrane-enclosed vesicles carrying materials essential to axonal function move along microtubules of the cytoskeleton, from the cell body to the tips of the axons. If the average velocity of a vesicle is 1 μm/s, how long does it take a vesicle to move from a cell body in the spinal cord to the axonal tip in the toes?

Answer Transport time equals distance traveled/velocity, or

$$(2 \times 10^6\ \mu\text{m})/(1\ \mu\text{m/s}) = 2 \times 10^6\ \text{s}$$

or about 23 days!

6. **Is Synthetic Vitamin C as Good as the Natural Vitamin?** A claim put forth by some purveyors of health foods is that vitamins obtained from natural sources are more healthful than those obtained by chemical synthesis. For example, pure L-ascorbic acid (vitamin C) extracted from rose hips is better than pure L-ascorbic acid manufactured in a chemical plant. Are the vitamins from the two sources different? Can the body distinguish a vitamin's source?

Answer The properties of the vitamin—like any other compound—are determined by its chemical structure. Because vitamin molecules from the two sources are structurally identical, their properties are identical, and no organism can distinguish between them. If different vitamin preparations contain different impurities, the biological effects of the *mixtures* may vary with the source. The ascorbic acid in such preparations, however, is identical.

7. **Identification of Functional Groups** Figures 1–15 and 1–16 show some common functional groups of biomolecules. Because the properties and biological activities of biomolecules are largely determined by their functional groups, it is important to be able to identify them. In each of the compounds below, circle and identify by name each functional group.

```
     H  H
 +   |  |
H3N—C——C—OH
     |  |
     H  H
```
Ethanolamine
(a)

```
    H
    |
H—C—OH
    |
H—C—OH
    |
H—C—OH
    |
    H
```
Glycerol
(b)

```
        O
        ||
  HO—P—O⁻
        |
 H      O
  \     |
   C==C—COO⁻
  /
 H
```
Phosphoenolpyruvate, an intermediate in glucose metabolism
(c)

```
        COO⁻
 +      |
H3N—C—H
        |
    H—C—OH
        |
        CH3
```
Threonine, an amino acid
(d)

```
  ⁻O    O
     \  //
       C
       |
       CH2
       |
       CH2
       |
       NH
       |
       C==O
       |
    H—C—OH
       |
H3C—C—CH3
       |
       CH2OH
```
Pantothenate, a vitamin
(e)

```
    H    O
     \  //
       C
       |   +
    H—C—NH3
       |
  HO—C—H
       |
    H—C—OH
       |
    H—C—OH
       |
       CH2OH
```
D-Glucosamine
(f)

Answer

(a) —NH_3^+ = amino; —OH = hydroxyl

(b) —OH = hydroxyl (three)

(c) —$P(OH)O_2^-$ = phosphoryl (in its ionized form); —COO^- = carboxyl

(d) —COO^- = carboxyl; —NH_3^+ = amino; —OH = hydroxyl; —CH_3 = methyl (two)

(e) —COO^- = carboxyl; —CO—NH— = amide; —OH = hydroxyl (two); —CH_3 = methyl (two)

(f) —CHO = aldehyde; —NH_3^+ = amino; —OH = hydroxyl (four)

8. Drug Activity and Stereochemistry The quantitative differences in biological activity between the two enantiomers of a compound are sometimes quite large. For example, the D isomer of the drug isoproterenol, used to treat mild asthma, is 50 to 80 times more effective as a bronchodilator than the L isomer. Identify the chiral center in isoproterenol. Why do the two enantiomers have such radically different bioactivity?

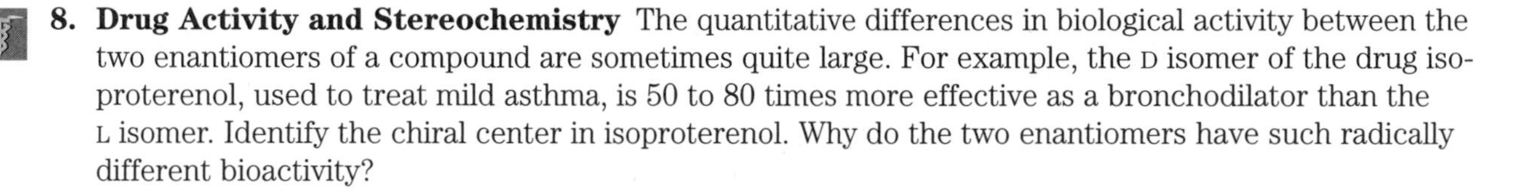

Isoproterenol

Answer A chiral center, or chiral carbon, is a carbon atom that is bonded to four different groups. A molecule with a single chiral center has two enantiomers, designated D and L (or in the RS system, *S* and *R*). In isoproterenol, only one carbon (asterisk) has four different groups around it; this is the chiral center:

The bioactivity of a drug is the result of interaction with a biological "receptor," a protein molecule with a binding site that is also chiral and stereospecific. The interaction of the D isomer of a drug with a chiral receptor site will differ from the interaction of the L isomer with that site.

9. Separating Biomolecules In studying a particular biomolecule (a protein, nucleic acid, carbohydrate, or lipid) in the laboratory, the biochemist first needs to separate it from other biomolecules in the sample—that is, to *purify* it. Specific purification techniques are described later in the text. However, by looking at the monomeric subunits of a biomolecule, you should have some ideas about the characteristics of the molecule that would allow you to separate it from other molecules. For example, how would you separate **(a)** amino acids from fatty acids and **(b)** nucleotides from glucose?

Answer

(a) Amino acids and fatty acids have carboxyl groups, whereas only the amino acids have amino groups. Thus, you could use a technique that separates molecules on the basis of the properties (charge or binding affinity) of amino groups. Fatty acids have long hydrocarbon chains and therefore are less soluble in water than amino acids. And finally, the sizes and shapes of these two types of molecules are quite different. Any one or more of these properties may provide ways to separate the two types of compounds.

(b) A nucleotide molecule has three components: a nitrogenous organic base, a five-carbon sugar, and phosphate. Glucose is a six-carbon sugar; it is smaller than a nucleotide. The size difference could be used to separate the molecules. Alternatively, you could use the nitrogenous bases and/or the phosphate groups characteristic of the nucleotides to separate them (based on differences in solubility, charge) from glucose.

10. Silicon-Based Life? Silicon is in the same group of the periodic table as carbon and, like carbon, can form up to four single bonds. Many science fiction stories have been based on the premise of silicon-based life. Is this realistic? What characteristics of silicon make it *less* well adapted than carbon as the central organizing element for life? To answer this question, consider what you have learned about carbon's bonding versatility, and refer to a beginning inorganic chemistry textbook for silicon's bonding properties.

Answer It is improbable that silicon could serve as the central organizing element for life under such conditions as those found on Earth for several reasons. Long chains of silicon atoms are not readily synthesized, and thus the polymeric macromolecules necessary for more complex functions would not readily form. Also, oxygen disrupts bonds between two silicon atoms, so silicon-based life-forms would be unstable in an oxygen-containing atmosphere. Once formed, the bonds between silicon and oxygen are extremely stable and difficult to break, which would prevent the breaking and making (degradation and synthesis) of biomolecules that is essential to the processes of living organisms.

11. Drug Action and Shape of Molecules Several years ago two drug companies marketed a drug under the trade names Dexedrine and Benzedrine. The structure of the drug is shown below.

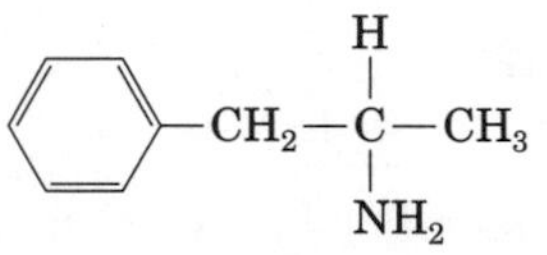

$$C_6H_5-CH_2-\underset{NH_2}{\overset{H}{C}}-CH_3$$

The physical properties (C, H, and N analysis, melting point, solubility, etc.) of Dexedrine and Benzedrine were identical. The recommended oral dosage of Dexedrine (which is still available) was 5 mg/day, but the recommended dosage of Benzedrine (no longer available) was twice that. Apparently, it required considerably more Benzedrine than Dexedrine to yield the same physiological response. Explain this apparent contradiction.

Answer Only one of the two enantiomers of the drug molecule (which has a chiral center) is physiologically active, for reasons described in the answer to Problem 3 (interaction with a stereospecific receptor site). Dexedrine, as manufactured, consists of the single enantiomer (D-amphetamine) recognized by the receptor site. Benzedrine was a racemic mixture (equal amounts of D and L isomers), so a much larger dose was required to obtain the same effect.

12. Components of Complex Biomolecules Figure 1–10 shows the major components of complex biomolecules. For each of the three important biomolecules below (shown in their ionized forms at physiological pH), identify the constituents.

(a) Guanosine triphosphate (GTP), an energy-rich nucleotide that serves as a precursor to RNA:

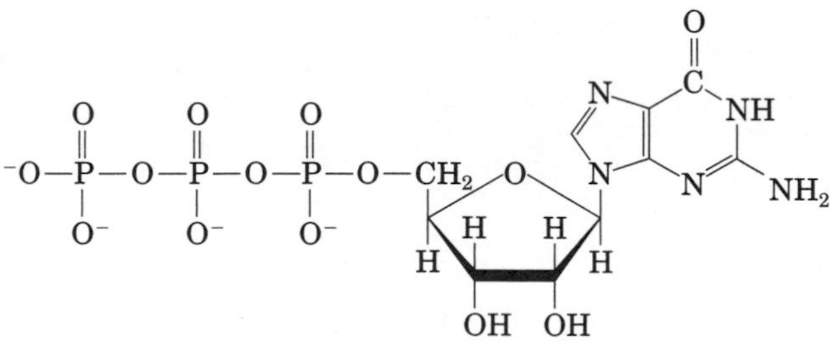

$$^{-}O-\underset{O^-}{\overset{O}{P}}-O-\underset{O^-}{\overset{O}{P}}-O-\underset{O^-}{\overset{O}{P}}-O-CH_2\text{(ribose: H, H, H, H; OH OH)}-\text{guanine (C=O, NH, N, N, }NH_2)$$

(b) Methionine enkephalin, the brain's own opiate:

```
                                                     (C6H5)
                                                       |
               H  O      H      H  H  O               CH2    H  H
               |  ||     |      |  |  ||               |     |  |
HO-(C6H4)-CH2--C--C--N---C--C---N--C--C---N---C---C---N---C--COO-
               |     |   |  ||     |      |   |   ||      |
              NH2    H   H  O      H      H   H   O      CH2
                                                          |
                                                         CH2
                                                          |
                                                          S
                                                          |
                                                         CH3
```

(c) Phosphatidylcholine, a component of many membranes:

```
      CH3               O-
      |                 |
CH3--+N--CH2--CH2--O----P--O--CH2                   H  H
      |                 ||    |                     |  |
      CH3               O     HC--O--C--(CH2)7--C==C--(CH2)7--CH3
                              |      ||
                              |      O
                              CH2--O--C--(CH2)14--CH3
                                      ||
                                      O
```

Answer

(a) Three phosphoric acid groups (linked by two anhydride bonds), esterified to an α-D-ribose (at the 5′ position), which is attached at C-1 to guanine.

(b) Tyrosine, two glycine, phenylalanine, and methionine residues, all linked by peptide bonds.

(c) Choline esterified to a phosphoric acid group, which is esterified to glycerol, which is esterified to two fatty acids, oleic acid and palmitic acid.

13. Determination of the Structure of a Biomolecule An unknown substance, X, was isolated from rabbit muscle. Its structure was determined from the following observations and experiments. Qualitative analysis showed that X was composed entirely of C, H, and O. A weighed sample of X was completely oxidized, and the H_2O and CO_2 produced were measured; this quantitative analysis revealed that X contained 40.00% C, 6.71% H, and 53.29% O by weight. The molecular mass of X, determined by mass spectrometry, was 90.00 u (atomic mass units; see Box 1–1). Infrared spectroscopy showed that X contained one double bond. X dissolved readily in water to give an acidic solution; the solution demonstrated optical activity when tested in a polarimeter.

(a) Determine the empirical and molecular formula of X.

(b) Draw the possible structures of X that fit the molecular formula and contain one double bond. Consider *only* linear or branched structures and disregard cyclic structures. Note that oxygen makes very poor bonds to itself.

(c) What is the structural significance of the observed optical activity? Which structures in **(b)** are consistent with the observation?

(d) What is the structural significance of the observation that a solution of X was acidic? Which structures in **(b)** are consistent with the observation?

(e) What is the structure of X? Is more than one structure consistent with all the data?

Answer

(a) From the C, H, and O analysis, and knowing the mass of X is 90.00 u, we can calculate the relative atomic proportions by dividing the weight percents by the atomic weights:

Atom	Relative atomic proportion	No. of atoms relative to O
C	(90.00 u)(40.00/100)/(12 u) = 3	3/3 = 1
H	(90.00 u)(6.71/100)/(1.008 u) = 6	6/3 = 2
O	(90.00 u)(53.29/100)/(16.0 u) = 3	3/3 = 1

Thus, the empirical formula is CH_2O, with a formula weight of 12 + 2 + 16 = 30. The molecular formula, based on X having a mass of 90.00 u, must be $C_3H_6O_3$.

(b) Twelve possible structures are shown below. Structures **1** through **5** can be eliminated because they are unstable enol isomers of the corresponding carbonyl derivatives. Structures **9**, **10**, and **12** can also be eliminated on the basis of their instability: they are hydrated carbonyl derivatives (vicinal diols).

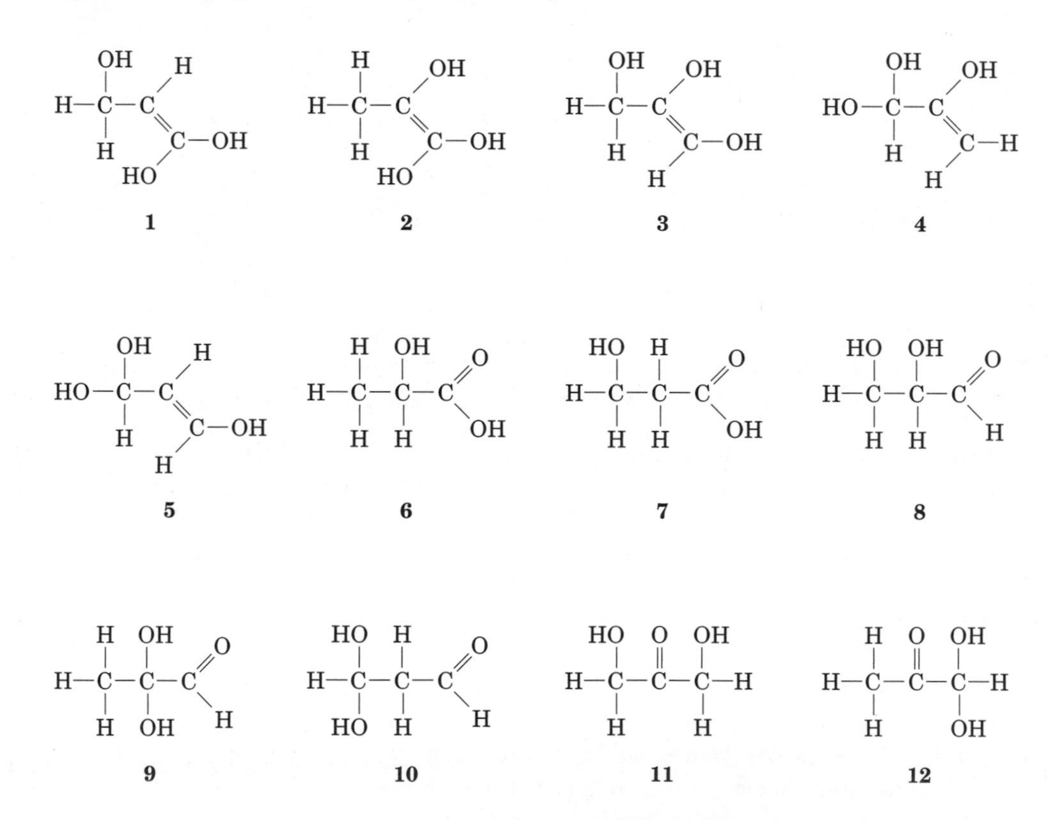

(c) Optical activity indicates the presence of a chiral center (a carbon atom surrounded by four different groups). Only structures **6** and **8** have chiral centers.

(d) Of structures **6** and **8**, only **6** contains an acidic group: a carboxyl group.

(e) Structure **6** is substance X. This compound exists in two enantiomeric forms that cannot be distinguished, even by measuring specific rotation. One could determine absolute stereochemistry by x-ray crystallography.

Data Analysis Problem

14. Sweet-Tasting Molecules Many compounds taste sweet to humans. Sweet taste results when a molecule binds to the sweet receptor, one type of taste receptor, on the surface of certain tongue cells. The stronger the binding, the lower the concentration required to saturate the receptor and the sweeter a given concentration of that substance tastes. The standard free-energy change, $\Delta G°$, of the binding reaction between a sweet molecule and a sweet receptor can be measured in kilojoules or kilocalories per mole.

Sweet taste can be quantified in units of "molar relative sweetness" (MRS), a measure that compares the sweetness of a substance to the sweetness of sucrose. For example, saccharin has an MRS of 161; this means that saccharin is 161 times sweeter than sucrose. In practical terms, this is measured by asking human subjects to compare the sweetness of solutions containing different concentrations of each compound. Sucrose and saccharin taste equally sweet when sucrose is at a concentration 161 times higher than that of saccharin.

(a) What is the relationship between MRS and the $\Delta G°$ of the binding reaction? Specifically, would a more negative $\Delta G°$ correspond to a higher or lower MRS? Explain your reasoning.

Shown below are the structures of 10 compounds, all of which taste sweet to humans. The MRS and $\Delta G°$ for binding to the sweet receptor are given for each substance.

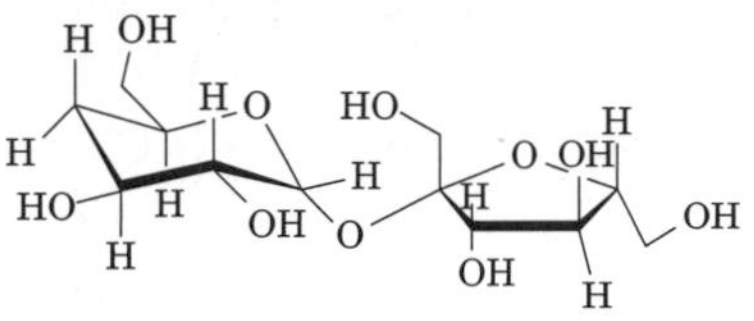

Deoxysucrose
MRS = 0.95
$\Delta G° = -6.67$ kcal/mol

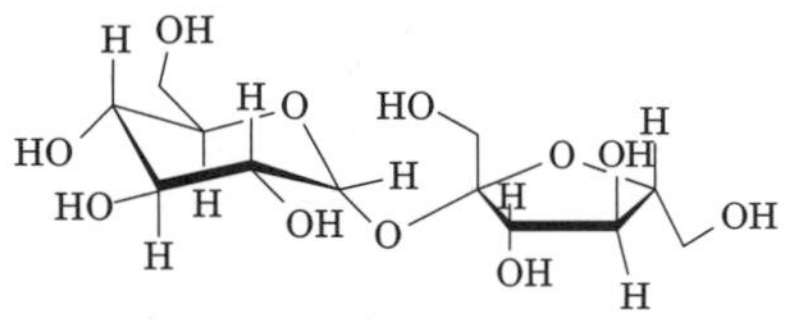

Sucrose
MRS = 1
$\Delta G° = -6.71$ kcal/mol

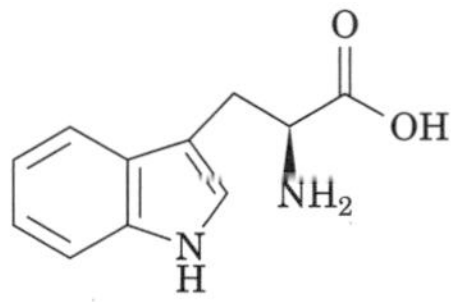

D-Tryptophan
MRS = 21
$\Delta G° = -8.5$ kcal/mol

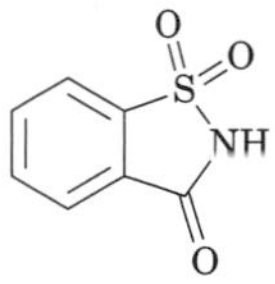

Saccharin
MRS = 161
$\Delta G° = -9.7$ kcal/mol

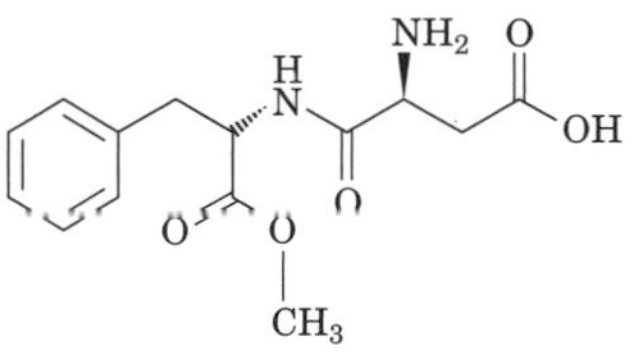

Aspartame
MRS = 172
$\Delta G° = -9.7$ kcal/mol

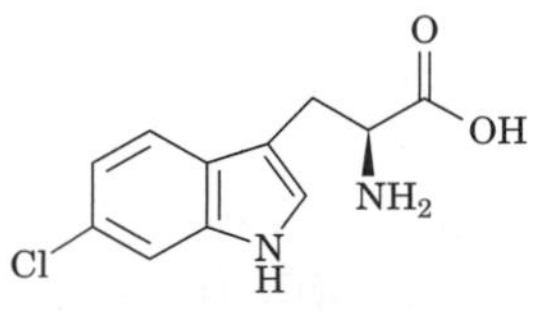

6-Chloro-D-tryptophan
MRS = 906
$\Delta G° = -10.7$ kcal/mol

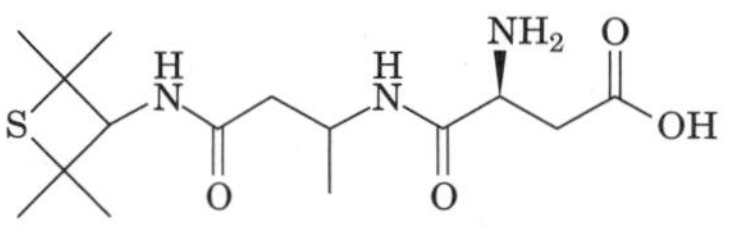

Alitame
MRS = 1,937
$\Delta G° = -11.1$ kcal/mol

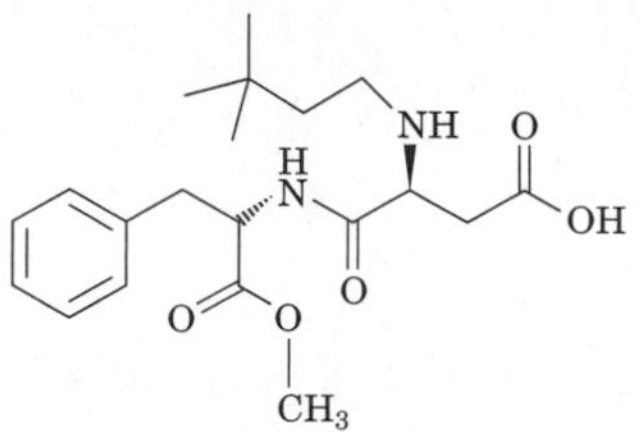

Neotame
MRS = 11,057
$\Delta G^\circ = -12.1$ kcal/mol

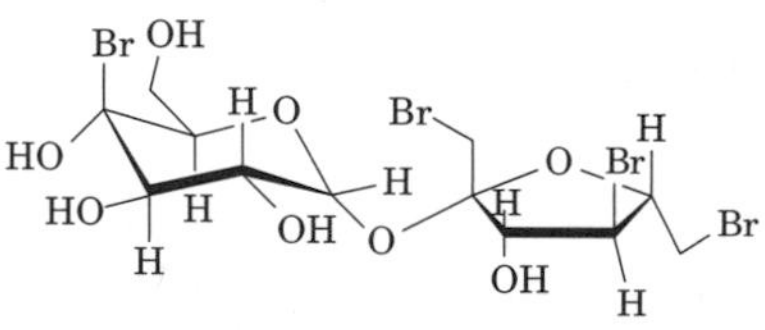

Tetrabromosucrose
MRS = 13,012
$\Delta G^\circ = -12.2$ kcal/mol

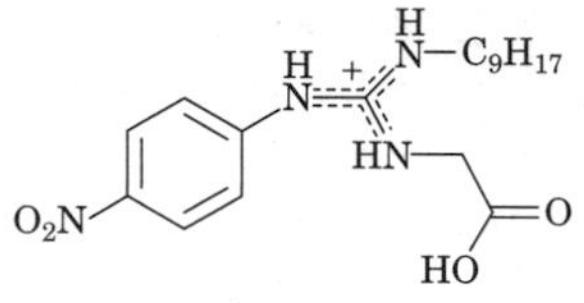

Sucronic acid
MRS = 200,000
$\Delta G^\circ = -13.8$ kcal/mol

Morini, Bassoli, and Temussi (2005) used computer-based methods (often referred to as "in silico" methods) to model the binding of sweet molecules to the sweet receptor.

(b) Why is it useful to have a computer model to predict the sweetness of molecules, instead of a human- or animal-based taste assay?

In earlier work, Schallenberger and Acree (1967) had suggested that all sweet molecules include an "AH-B" structural group, in which "A and B are electronegative atoms separated by a distance of greater than 2.5 Å [0.25 nm] but less than 4 Å [0.4 nm]. H is a hydrogen atom attached to one of the electronegative atoms by a covalent bond" (p. 481).

(c) Given that the length of a "typical" single bond is about 0.15 nm, identify the AH-B group(s) in each of the molecules shown above.

(d) Based on your findings from **(c),** give two objections to the statement that "molecules containing an AH-B structure will taste sweet."

(e) For two of the molecules shown above, the AH-B model *can* be used to explain the difference in MRS and ΔG°. Which two molecules are these, and how would you use them to support the AH-B model?

(f) Several of the molecules have closely related structures but very different MRS and ΔG° values. Give two such examples, and use these to argue that the AH-B model is unable to explain the observed differences in sweetness.

In their computer-modeling study, Morini and coauthors used the three-dimensional structure of the sweet receptor and a molecular dynamics modeling program called GRAMM to predict the ΔG° of binding of sweet molecules to the sweet receptor. First, they "trained" their model—that is, they refined the parameters so that the ΔG° values predicted by the model matched the known ΔG° values for one set of sweet molecules (the "training set"). They then "tested" the model by asking it to predict the ΔG° values for a new set of molecules (the "test set").

(g) Why did Morini and colleagues need to test their model against a different set of molecules from the set it was trained on?

(h) The researchers found that the predicted ΔG° values for the test set differed from the actual values by, on average, 1.3 kcal/mol. Using the values given with the structures above, estimate the resulting error in MRS values.

Answer

(a) A more negative $\Delta G°$ corresponds to a larger K_{eq} for the binding reaction, so the equilibrium is shifted more toward products and tighter binding—and thus greater sweetness and higher MRS.

(b) Animal-based sweetness assays are time-consuming. A computer program to predict sweetness, even if not always completely accurate, would allow chemists to design effective sweeteners much faster. Candidate molecules could then be tested in the conventional assay.

(c) The range 0.25 to 0.4 nm corresponds to about 1.5 to 2.5 single-bond lengths. The figure below can be used to construct an approximate ruler; any atoms in the gray rectangle are between 0.25 and 0.4 nm from the origin of the ruler.

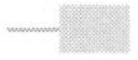

There are many possible AH-B groups in the molecules; a few are shown here.

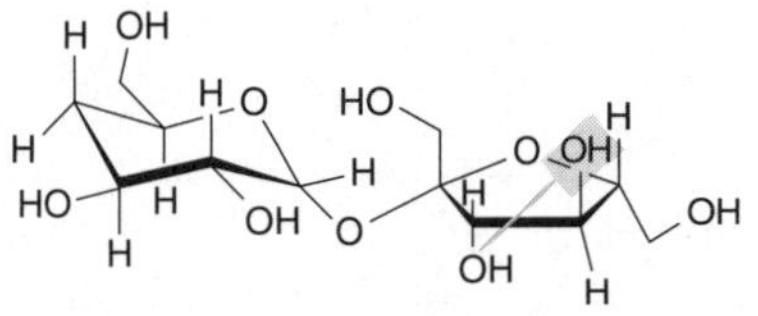

Deoxysucrose

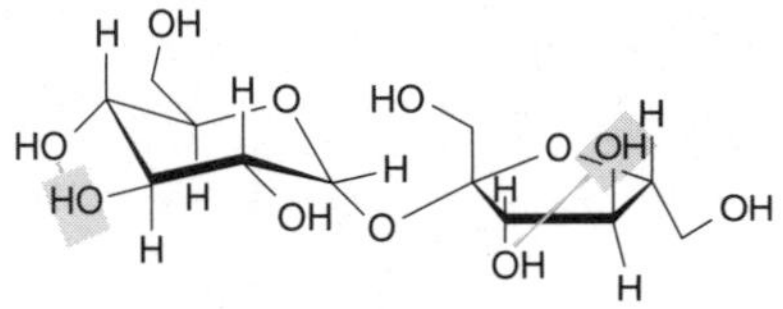

Sucrose

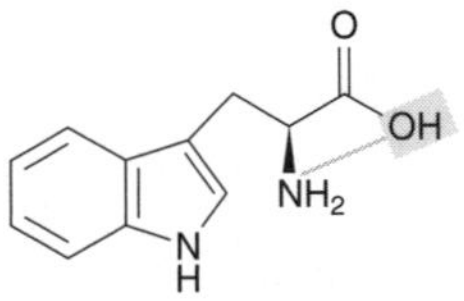

D-Tryptophan

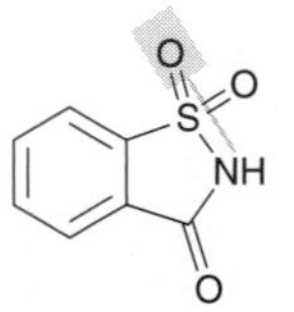

Saccharin

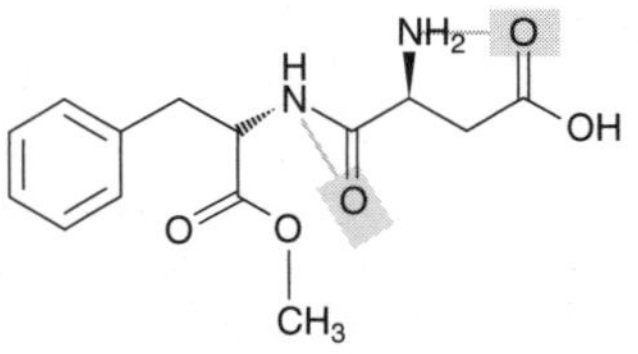

Aspartame

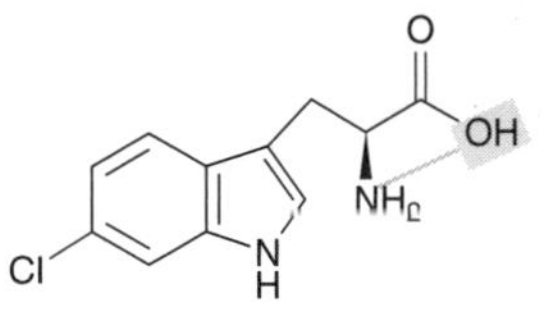

6-Chloro-D-tryptophan

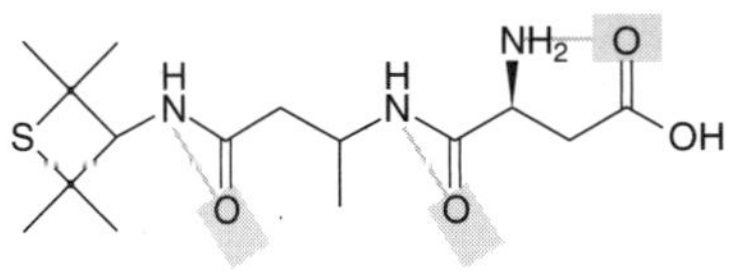

Alitame

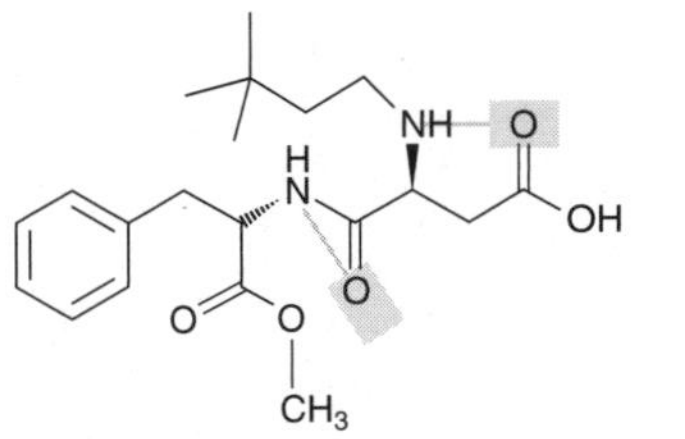

Neotame

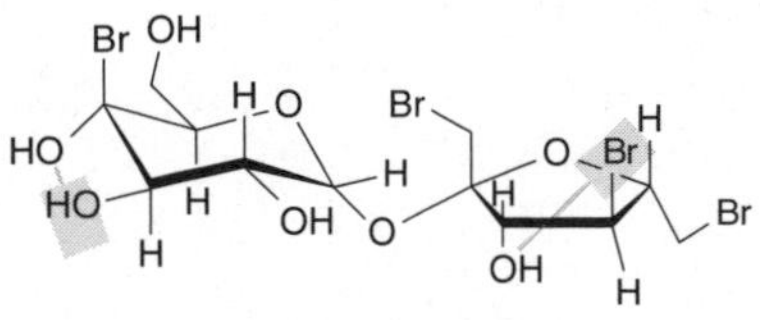

Tetrabromosucrose

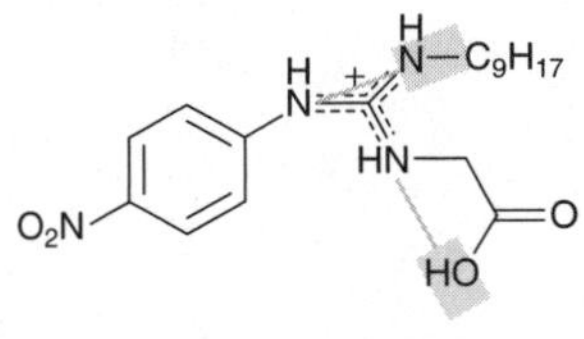

Sucronic acid

(d) First, each molecule has multiple AH-B groups, so it is difficult to know which is the important one. Second, because the AH-B motif is very simple, many nonsweet molecules will have this group.

(e) Sucrose and deoxysucrose. Deoxysucrose lacks one of the AH-B groups present in sucrose and has a slightly lower MRS than sucrose—as is expected if the AH-B groups are important for sweetness.

(f) There are many such examples; here are a few: (1) D-Tryptophan and 6-chloro-D-tryptophan have the same AH-B group but very different MRS values. (2) Aspartame and neotame have the same AH-B groups but very different MRS values. (3) Neotame has two AH-B groups and alitame has three, yet neotame is more than five times sweeter than alitame. (4) Bromine is less electronegative than oxygen and thus is expected to weaken an AH-B group, yet tetrabromosucrose is much sweeter than sucrose.

(g) Given enough "tweaking" of parameters, any model can be made to fit a defined dataset. Because the objective was to create a model to predict $\Delta G°$ for molecules not tested in vivo, the researchers needed to show that the model worked well for molecules it had not been trained on. The degree of inaccuracy with the test set could give researchers an idea of how the model would behave for novel molecules.

(h) MRS is related to K_{eq}, which is related exponentially to $\Delta G°$, so adding a constant amount to $\Delta G°$ multiplies the MRS by a constant amount. Based on the values given with the structures, a change in $\Delta G°$ of 1.3 kcal/mol corresponds to a 10-fold change in MRS.

References

Morini, G., Bassoli, A., & Temussi, P.A. (2005) From small sweeteners to sweet proteins: anatomy of the binding sites of the human T1R2_T1R3 receptor. *J. Med. Chem.* **48,** 5520–5529.

Schallenberger, R.S. & Acree, T.E. (1967) Molecular theory of sweet taste. *Nature* **216,** 480–482.

chapter

2

Water

1. **Solubility of Ethanol in Water** Explain why ethanol (CH_3CH_2OH) is more soluble in water than is ethane (CH_3CH_3).

Answer Ethanol is polar; ethane is not. The ethanol —OH group can hydrogen-bond with water.

2. **Calculation of pH from Hydrogen Ion Concentration** What is the pH of a solution that has an H^+ concentration of **(a)** 1.75×10^{-5} mol/L; **(b)** 6.50×10^{-10} mol/L; **(c)** 1.0×10^{-4} mol/L; **(d)** 1.50×10^{-5} mol/L?

Answer Using pH = $-\log [H^+]$:
(a) $-\log (1.75 \times 10^{-5}) = 4.76$; **(b)** $-\log (6.50 \times 10^{-10}) = 9.19$; **(c)** $-\log (1.0 \times 10^{-4}) = 4.0$;
(d) $-\log (1.50 \times 10^{-5}) = 4.82$.

3. **Calculation of Hydrogen Ion Concentration from pH** What is the H^+ concentration of a solution with pH of **(a)** 3.82; **(b)** 6.52; **(c)** 11.11?

Answer Using $[H^+] = 10^{-pH}$:
(a) $[H^+] = 10^{-3.82} = 1.51 \times 10^{-4}$ M; **(b)** $[H^+] = 10^{-6.52} = 3.02 \times 10^{-7}$ M;
(c) $[H^+] = 10^{-11.11} = 7.76 \times 10^{-12}$ M.

4. **Acidity of Gastric HCl** In a hospital laboratory, a 10.0 mL sample of gastric juice, obtained several hours after a meal, was titrated with 0.1 M NaOH to neutrality; 7.2 mL of NaOH was required. The patient's stomach contained no ingested food or drink, thus assume that no buffers were present. What was the pH of the gastric juice?

Answer Multiplying volume (L) by molar concentration (mol/L) gives the number of moles in that volume of solution. If x is the concentration of gastric HCl (mol/L),

$$(0.010 \text{ L})x = (0.0072 \text{ L})(0.1 \text{ mol/L})$$

$$x = 0.072 \text{ M gastric HCl}$$

Given that pH = $-\log [H^+]$ and that HCl is a strong acid,

$$\text{pH} = -\log (7.2 \times 10^{-2}) = 1.1$$

5. **Calculation of the pH of a Strong Acid or Base** **(a)** Write out the acid dissociation reaction for hydrochloric acid. **(b)** Calculate the pH of a solution of 5.0×10^{-4} M HCl. **(c)** Write out the acid dissociation reaction for sodium hydroxide. **(d)** Calculate the pH of a solution of 7.0×10^{-5} M NaOH.

Answer

(a) $HCl \rightleftharpoons H^+ + Cl^-$

(b) HCl is a strong acid and fully dissociates into H^+ and Cl^-. Thus, $[H^+] = [Cl^-] = [HCl]$.
$pH = -\log [H^+] = -\log (5.0 \times 10^{-4}\ \text{M}) = 3.3$ (two significant figures)

(c) $NaOH \rightleftharpoons Na^+ + OH^-$

(d) NaOH is a strong base; dissociation in aqueous solution is essentially complete, so $[Na^+] = [OH^-] = [NaOH]$.
$pH + pOH = 14$
$pOH = -\log [OH^-]$
$pH = 14 + \log [OH^-]$
$= 14 + \log (7.0 \times 10^{-5}) = 9.8$ (two significant figures)

6. **Calculation of pH from Concentration of Strong Acid** Calculate the pH of a solution prepared by diluting 3.0 mL of 2.5 M HCl to a final volume of 100 mL with H_2O.

Answer Because HCl is a strong acid, it dissociates completely to $H^+ + Cl^-$. Therefore, 3.0 mL $\times$ 2.5 M HCl = 7.5 meq of H^+. In 100 mL of solution, this is 0.075 M H^+.
$pH = -\log [H^+] = -\log (0.075) = -(-1.1) = 1.1$ (two significant figures)

7. **Measurement of Acetylcholine Levels by pH Changes** The concentration of acetylcholine (a neurotransmitter) in a sample can be determined from the pH changes that accompany its hydrolysis. When the sample is incubated with the enzyme acetylcholinesterase, acetylcholine is quantitatively converted into choline and acetic acid, which dissociates to yield acetate and a hydrogen ion:

$$CH_3{-}\overset{\overset{\displaystyle O}{\|}}{C}{-}O{-}CH_2{-}CH_2{-}{}^{+}\underset{\underset{\displaystyle CH_3}{|}}{\overset{\overset{\displaystyle CH_3}{|}}{N}}{-}CH_3 \xrightarrow{H_2O} HO{-}CH_2{-}CH_2{-}{}^{+}\underset{\underset{\displaystyle CH_3}{|}}{\overset{\overset{\displaystyle CH_3}{|}}{N}}{-}CH_3 + CH_3{-}\underset{\underset{\displaystyle O}{\|}}{C}{-}O^- + H^+$$

Acetylcholine Choline Acetate

In a typical analysis, 15 mL of an aqueous solution containing an unknown amount of acetylcholine had a pH of 7.65. When incubated with acetylcholinesterase, the pH of the solution decreased to 6.87. Assuming that there was no buffer in the assay mixture, determine the number of moles of acetylcholine in the 15 mL sample.

Answer Given that $pH = -\log [H^+]$, we can calculate $[H^+]$ at the beginning and at the end of the reaction:

At pH 7.65, $\log [H^+] = -7.65$ $\quad [H^+] = 10^{-7.65} = 2.24 \times 10^{-8}$ M

At pH 6.87, $\log [H^+] = -6.87$ $\quad [H^+] = 10^{-6.87} = 1.35 \times 10^{-7}$ M

The difference in $[H^+]$ is

$$(1.35 - 0.22) \times 10^{-7}\ \text{M} = 1.13 \times 10^{-7}\ \text{M}$$

For a volume of 15 mL, or 0.015 L, multiplying volume by molarity gives

$$(0.015\ \text{L})(1.13 \times 10^{-7}\ \text{mol/L}) = 1.7 \times 10^{-9}\ \text{mol of acetylcholine}$$

8. **Physical Meaning of $\mathbf{p}K_a$** Which of the following aqueous solutions has the lowest pH: 0.1 M HCl; 0.1 M acetic acid ($pK_a = 4.86$); 0.1 M formic acid ($pK_a = 3.75$)?

Answer A 0.1 M HCl solution has the lowest pH because HCl is a strong acid and dissociates completely to $H^+ + Cl^-$, yielding the highest $[H^+]$.

9. Simulated Vinegar One way to make vinegar (*not* the preferred way) is to prepare a solution of acetic acid, the sole acid component of vinegar, at the proper pH (see Fig. 2–14) and add appropriate flavoring agents. Acetic acid (M_r 60) is a liquid at 25 °C, with a density of 1.049 g/mL. Calculate the volume that must be added to distilled water to make 1 L of simulated vinegar (see Fig. 2–15).

Answer From Figure 2–15, the pK_a of acetic acid is 4.76. From Figure 2–14, the pH of vinegar is ~3; we will calculate for a solution of pH 3.0. Using the Henderson-Hasselbalch equation

$$\text{pH} = \text{p}K_a + \log \frac{[\text{A}^-]}{[\text{HA}]}$$

and the fact that dissociation of HA gives equimolar $[H^+]$ and $[A^-]$ (where HA is CH_3COOH, and A^- is CH_3COO^-), we can write

$$3.0 = 4.76 + \log ([\text{A}^-]/[\text{HA}])$$

$$-1.76 = \log ([\text{A}^-]/[\text{HA}]) = -\log ([\text{HA}]/[\text{A}^-])$$

$$[\text{HA}]/[\text{A}^-] = 10^{1.76} = 58$$

Thus, $[HA] = 58[A^-]$. At pH 3.0, $[H^+] = [A^-] = 10^{-3}$, so

$$[\text{HA}] = 58 \times 10^{-3}\ \text{M} = 0.058\ \text{mol/L}$$

Dividing density (g/mL) by molecular weight (g/mol) for acetic acid gives

$$\frac{1.049\ \text{g/mL}}{60\ \text{g/mol}} = 0.017\ \text{mol/mL}$$

Dividing this answer into 0.058 mol/L gives the volume of acetic acid needed to prepare 1.0 L of a 0.058 M solution:

$$\frac{0.058\ \text{mol/L}}{0.017\ \text{mol/mL}} = 3.3\ \text{mL/L}$$

10. Identifying the Conjugate Base Which is the conjugate base in each of the pairs below?

(a) $RCOOH$, $RCOO^-$
(b) RNH_2, RNH_3^+
(c) $H_2PO_4^-$, H_3PO_4
(d) H_2CO_3, HCO_3^-

Answer In each pair, the acid is the species that gives up a proton; the conjugate base is the deprotonated species. By inspection, the conjugate base is the species with fewer hydrogen atoms. **(a)** $RCOO^-$ **(b)** RNH_2 **(c)** $H_2PO_4^-$ **(d)** HCO_3^-

11. Calculation of the pH of a Mixture of a Weak Acid and Its Conjugate Base Calculate the pH of a dilute solution that contains a molar ratio of potassium acetate to acetic acid (pK_a = 4.76) of **(a)** 2:1; **(b)** 1:3; **(c)** 5:1; **(d)** 1:1; **(e)** 1:10.

Answer Using the Henderson-Hasselbalch equation,

$$\text{pH} = \text{p}K_a + \log \frac{[\text{A}^-]}{[\text{HA}]}$$

pH = 4.76 + log ([acetate]/[acetic acid]), where [acetate]/[acetic acid] is the ratio given for each part of the question.

(a) log (2/1) = 0.30; pH = 4.76 + 0.30 = 5.06
(b) log (1/3) = −0.48; pH = 4.76 + (−0.48) = 4.28

(c) log (5/1) = 0.70; pH = 4.76 + 0.70 = 5.46
(d) log (1/1) = 0; pH = 4.76
(e) log (1/10) = −1.00; pH = 4.76 + (−1.00) = 3.76

12. Effect of pH on Solubility The strongly polar hydrogen-bonding properties of water make it an excellent solvent for ionic (charged) species. By contrast, nonionized, nonpolar organic molecules, such as benzene, are relatively insoluble in water. In principle, the aqueous solubility of an organic acid or base can be increased by converting the molecules to charged species. For example, the solubility of benzoic acid in water is low. The addition of sodium bicarbonate to a mixture of water and benzoic acid raises the pH and deprotonates the benzoic acid to form benzoate ion, which is quite soluble in water.

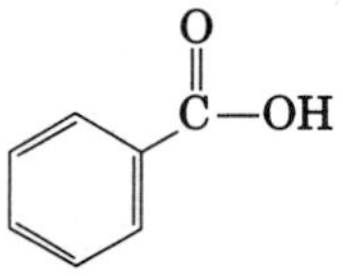

Benzoic acid
$pK_a \approx 5$

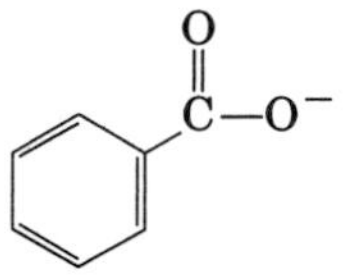

Benzoate ion

Are the following compounds more soluble in an aqueous solution of 0.1 M NaOH or 0.1 M HCl? (The dissociable proton in **(c)** is that of the —OH group.)

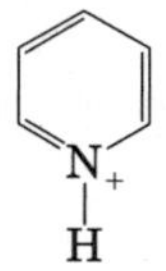

Pyridine ion
$pK_a \approx 5$
(a)

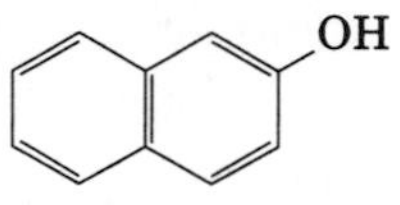

β-Naphthol
$pK_a \approx 10$
(b)

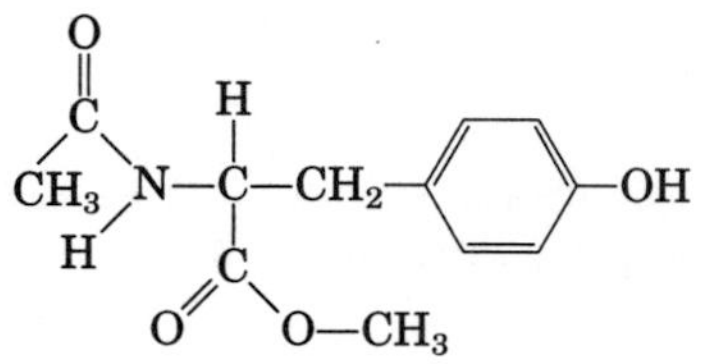

N-Acetyltyrosine methyl ester
$pK_a \approx 10$
(c)

Answer

(a) Pyridine is ionic in its protonated form and therefore more soluble at the lower pH, in 0.1 M HCl.

(b) *β*-Naphthol is ionic when *de*protonated and thus more soluble at the higher pH, in 0.1 M NaOH.

(c) *N*-Acetyltyrosine methyl ester is ionic when *de*protonated and thus more soluble in 0.1 M NaOH.

13. Treatment of Poison Ivy Rash The components of poison ivy and poison oak that produce the characteristic itchy rash are catechols substituted with long-chain alkyl groups.

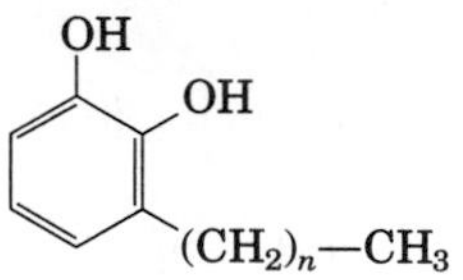

OH OH $(CH_2)_n$—CH_3
$pK_a \approx 8$

If you were exposed to poison ivy, which of the treatments below would you apply to the affected area? Justify your choice.

(a) Wash the area with cold water.

(b) Wash the area with dilute vinegar or lemon juice.

(c) Wash the area with soap and water.

(d) Wash the area with soap, water, and baking soda (sodium bicarbonate).

Answer The best choice is **(d).** Soap helps to emulsify and dissolve the hydrophobic alkyl group of an alkylcatechol. Given that the pK_a of an alkylcatechol is about 8, in a mildly alkaline solution of bicarbonate ($NaHCO_3$) its —OH group ionizes, making the compound much more water-soluble. A neutral or acidic solution, as in **(a)** or **(b),** would not be effective.

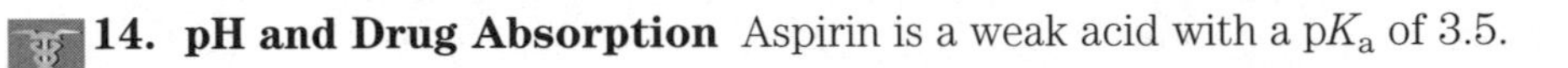

14. pH and Drug Absorption Aspirin is a weak acid with a pK_a of 3.5.

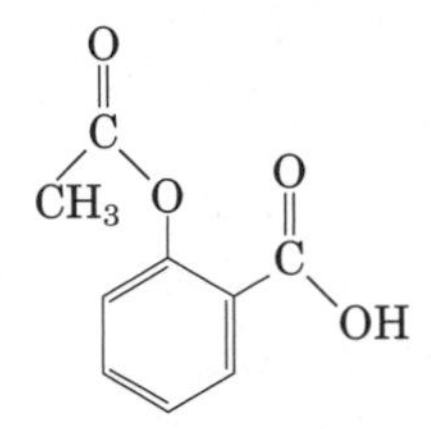

It is absorbed into the blood through the cells lining the stomach and the small intestine. Absorption requires passage through the plasma membrane, the rate of which is determined by the polarity of the molecule: charged and highly polar molecules pass slowly, whereas neutral hydrophobic ones pass rapidly. The pH of the stomach contents is about 1.5, and the pH of the contents of the small intestine is about 6. Is more aspirin absorbed into the bloodstream from the stomach or from the small intestine? Clearly justify your choice.

Answer With a pK_a of 3.5, aspirin is in its protonated (neutral) form at pH below 2.5. At higher pH, it becomes increasingly deprotonated (anionic). Thus, aspirin is better absorbed in the more acidic environment of the stomach.

15. Calculation of pH from Molar Concentrations What is the pH of a solution containing 0.12 mol/L of NH_4Cl and 0.03 mol/L of NaOH (pK_a of NH_4^+/NH_3 is 9.25)?

Answer For the equilibrium

$$NH_4^+ \rightleftharpoons NH_3 + H^+$$

$$pH = pK_a + \log([NH_3]/[NH_4^+])$$

we know that $[NH_4^+] + [NH_3] = 0.12$ mol/L, and that NaOH completely dissociates to give $[OH^-] = 0.03$ mol/L. Thus, $[NH_3] = 0.03$ mol/L and $[NH_4^+] = 0.09$ mol/L, and

$$pH = 9.25 + \log(0.03/0.09) = 9.25 - 0.48 = 8.77, \text{ which rounds to 9.}$$

16. Calculation of pH after Titration of Weak Acid A compound has a pK_a of 7.4. To 100 mL of a 1.0 M solution of this compound at pH 8.0 is added 30 mL of 1.0 M hydrochloric acid. What is the pH of the resulting solution?

Answer Begin by calculating the ratio of conjugate base to acid in the starting solution, using the Henderson-Hasselbalch equation:

$$pH = pK_a + \log([A^-]/[HA])$$

$$8.0 = 7.4 + \log([A^-]/[HA])$$

$$\log([A^-]/[HA]) = 0.6$$

$$[A^-]/[HA] = 10^{0.6} = 4$$

The solution contains 100 meq of the compound (conjugate base plus acid), so 80 meq are in the conjugate base form and 20 meq are in the acid form, for a [base]/[acid] ratio of 4.

Because HCl is a strong acid and dissociates completely, adding 30 mL of 1.0 M HCl adds 30 meq of H^+ to the solution. These 30 meq titrate 30 meq of the conjugate base, so the [base]/[acid] ratio is 1. Solving the Henderson-Hasselbalch equation for pH:

$$\begin{aligned} pH &= pK_a + \log([A^-]/[HA]) \\ &= 7.4 + \log 1 = 7.4 \end{aligned}$$

17. Properties of a Buffer The amino acid glycine is often used as the main ingredient of a buffer in biochemical experiments. The amino group of glycine, which has a pK_a of 9.6, can exist either in the protonated form ($—NH_3^+$) or as the free base ($—NH_2$), because of the reversible equilibrium

$$R—NH_3^+ \rightleftharpoons R—NH_2 + H^+$$

(a) In what pH range can glycine be used as an effective buffer due to its amino group?

(b) In a 0.1 M solution of glycine at pH 9.0, what fraction of glycine has its amino group in the $—NH_3^+$ form?

(c) How much 5 M KOH must be added to 1.0 L of 0.1 M glycine at pH 9.0 to bring its pH to exactly 10.0?

(d) When 99% of the glycine is in its $—NH_3^+$ form, what is the numerical relation between the pH of the solution and the pK_a of the amino group?

Answer

(a) In general, a buffer functions best in the zone from about one pH unit below to one pH unit above its pK_a. Thus, glycine is a good buffer (through ionization of its amino group) between pH 8.6 and pH 10.6.

(b) Using the Henderson-Hasselbalch equation

$$pH = pK_a + \log \frac{[A^-]}{[HA]}$$

we can write

$$9.0 = 9.6 + \log \frac{[A^-]}{[HA]}$$

$$\frac{[A^-]}{[HA]} = 10^{-0.6} = 0.25$$

which corresponds to a ratio of 1/4. This indicates that the amino group of glycine is about 1/5 deprotonated and 4/5 protonated at pH 9.0.

(c) From **(b)** we know that the amino group is about 1/5, or 20%, deprotonated at pH 9.0. Thus, in moving from pH 9.0 to pH 9.6 (at which, by definition, the amino group is 50% deprotonated), 30%, or 0.3, of the glycine is titrated. We can now calculate from the Henderson-Hasselbalch equation the percentage protonation at pH 10.0:

$$10.0 = 9.6 + \log \frac{[A^-]}{[HA]}$$

$$\frac{[A^-]}{[HA]} = 10^{0.4} = 2.5 = 5/2$$

This ratio indicates that glycine is 5/7, or 71%, deprotonated at pH 10.0, an additional 21%, or 0.21, deprotonation above that (50%, or 0.5) at the pK_a. Thus, the total fractional deprotonation in moving from pH 9.0 to 10.0 is 0.30 + 0.21 = 0.51, which corresponds to

$$0.51 \times 0.1 \text{ mol} = 0.05 \text{ mol of KOH}$$

Thus, the volume of 5 M KOH solution required is (0.5 mol)/(5 mol/L) = 0.01 L, or 10 mL.

(d) From the Henderson-Hasselbalch equation,

$$\begin{aligned} \text{pH} &= \text{p}K_a + \log ([\text{—NH}_2]/[\text{—NH}_3^+]) \\ &= \text{p}K_a + \log (0.01/0.99) \\ &= \text{p}K_a + (-2) = \text{p}K_a - 2 \end{aligned}$$

In general, any group with an ionizable hydrogen is almost completely protonated at a pH at least two pH units below its pK_a value.

18. Preparation of a Phosphate Buffer What molar ratio of HPO_4^{2-} to $H_2PO_4^-$ in solution would produce a pH of 7.0? Phosphoric acid (H_3PO_4), a triprotic acid, has 3 pK_a values: 2.14, 6.86, and 12.4. Hint: Only one of the pK_a values is relevant here.

Answer Only the pK_a close to the pH is relevant here, because the concentrations of the other species (H_3PO_4 and PO_4^{3-}) are insignificant compared with the concentrations of HPO_4^{2-} and $H_2PO_4^-$. Begin with the Henderson-Hasselbalch equation:

$$\text{pH} = \text{p}K_a + \log ([\text{conjugate base}]/[\text{acid}])$$

$$\log ([HPO_4^{2-}]/[H_2PO_4^-]) = \text{pH} - \text{p}K_a = 7.0 - 6.86 = 0.14$$

$$[HPO_4^{2-}]/[H_2PO_4^-] = 10^{0.14} = 1.38 = 1.4 \text{ (two significant figures)}$$

19. Preparation of Standard Buffer for Calibration of a pH Meter The glass electrode used in commercial pH meters gives an electrical response proportional to the concentration of hydrogen ion. To convert these responses to a pH reading, the electrode must be calibrated against standard solutions of known H^+ concentration. Determine the weight in grams of sodium dihydrogen phosphate ($NaH_2PO_4 \cdot H_2O$; FW 138) and disodium hydrogen phosphate (Na_2HPO_4; FW 142) needed to prepare 1 L of a standard buffer at pH 7.00 with a total phosphate concentration of 0.100 M (see Fig. 2–15). See Problem 18 for the pK_a values of phosphoric acid.

Answer In solution, the two salts ionize as indicated below.

Sodium dihydrogen phosphate
(sodium phosphate, monobasic)
$NaH_2PO_4 \cdot H_2O$

Disodium hydrogen phosphate
(sodium phosphate, dibasic)
Na_2HPO_4

The buffering capacity of the solution is determined by the concentration ratio of proton acceptor (A^-) to proton donor (HA), or $[HPO_4^{2-}]/[H_2PO_4^-]$. From Figure 2–15, the pK_a for the dissociation of the ionizable hydrogen of dihydrogen phosphate

$$H_2PO_4^- \rightleftharpoons HPO_4^{2-} + H^+$$

is 6.86. Using the Henderson-Hasselbalch equation,

$$\text{pH} = \text{p}K_a + \log \frac{[A^-]}{[HA]}$$

we calculate:

$$7.00 - 6.86 = \log \frac{[A^-]}{[HA]}$$

$$\frac{[A^-]}{[HA]} = 10^{0.14} = 1.38$$

This ratio is approximately 7/5; that is, 7 parts Na_2HPO_4 to 5 parts $NaH_2PO_4 \cdot H_2O$. Because $[HPO_4^{2-}] + [H_2PO_4^-] = 0.100$ M, $[H_2PO_4^-] = 0.100$ M $- [HPO_4^{2-}]$, and we can now calculate the amount of each salt required for a 0.100 M solution:

$$\frac{[HPO_4^{2-}]}{0.100 \text{ M} - [HPO_4^{2-}]} = 1.38$$

Solving for $[HPO_4^{2-}]$,

$$[HPO_4^{2-}] = \frac{0.138}{2.38} \text{ M} = 0.058 \text{ M} = 0.058 \text{ mol/L}$$

and $[H_2PO_4^-] = 0.100$ M $- 0.058$ M $= 0.042$ M $= 0.042$ mol/L.

The amount needed for 1 L of solution = FW (mol/L).

$$\text{For } NaH_2PO_4 \cdot H_2O\text{: } (138 \text{ g/mol})(0.042 \text{ mol/L}) = 5.8 \text{ g/L}$$

$$\text{For } Na_2HPO_4\text{: } (142 \text{ g/mol})(0.058 \text{ mol/L}) = 8.2 \text{ g/L}$$

20. Calculation of Molar Ratios of Conjugate Base to Weak Acid from pH For a weak acid with a pK_a of 6.0, calculate the ratio of conjugate base to acid at a pH of 5.0.

Answer Using the Henderson-Hasselbalch equation,

$$pH = pK_a + \log([A^-]/[HA])$$

$$5.0 = 6.0 + \log([A^-]/[HA])$$

$$\log([A^-]/[HA]) = -1.0$$

$$[A^-]/[HA] = 10^{-1.0} = 0.10$$

21. Preparation of Buffer of Known pH and Strength Given 0.10 M solutions of acetic acid (pK_a = 4.76) and sodium acetate, describe how you would go about preparing 1.0 L of 0.10 M acetate buffer of pH 4.00.

Answer Use the Henderson-Hasselbalch equation to calculate the ratio $[Ac^-]/[HAc]$ in the final buffer.

$$pH = pK_a + \log([Ac^-]/[HAc])$$

$$\log([Ac^-]/[HAc]) = pH - pK_a = 4.00 - 4.76 = -0.76$$

$$[Ac^-]/[HAc] = 10^{-0.76}$$

The fraction of the solution that is $Ac^- = [Ac^-]/[HAc + Ac^-] = 10^{-0.76}/(1 + 10^{-0.76}) = 0.148$, which must be rounded to 0.15 (two significant figures). Therefore, to make 1.0 L of acetate buffer, use 150 mL of sodium acetate and 850 mL of acetic acid.

22. Choice of Weak Acid for a Buffer Which of these compounds would be the best buffer at pH 5.0: formic acid (pK_a = 3.8), acetic acid (pK_a = 4.76), or ethylamine (pK_a = 9.0)? Briefly justify your answer.

Answer Acetic acid; its pK_a is closest to the desired pH.

23. Working with Buffers A buffer contains 0.010 mol of lactic acid ($pK_a = 3.86$) and 0.050 mol of sodium lactate per liter. **(a)** Calculate the pH of the buffer. **(b)** Calculate the change in pH when 5 mL of 0.5 M HCl is added to 1 L of the buffer. **(c)** What pH change would you expect if you added the same quantity of HCl to 1 L of pure water?

Answer Using the Henderson-Hasselbalch equation,

$$pH = pK_a + \log \frac{[A^-]}{[HA]}$$

(a) $pH = pK_a + \log$ ([lactate]/[lactic acid]) = 3.86 + log (0.050 M/0.010 M) = 3.86 + 0.70 = 4.56. Thus, the pH is 4.6.

(b) Strong acids ionize completely, so 0.005 L × 0.5 mol/L = 0.002 mol of H^+ is added. The added acid will convert some of the salt form to the acid form. Thus, the final pH is

$$pH = 3.86 + \log [(0.050 - 0.0025)/(0.010 - 0.0025)]$$

$$= 3.86 + 0.58 = 4.44$$

The change in pH = 4.56 − 4.44 = 0.12, which rounds to 0.1 pH unit.

(c) HCl completely dissociates. So, when 5 mL of 0.5 M HCl is added to 1 L of water,

$$[H^+] = (0.002 \text{ mol})/(1 \text{ L}) = 0.002 \text{ mol/L} = 0.002 \text{ M}$$

$$pH = -\log 0.002 = 2.7$$

The pH of pure water is 7.0, so the change in pH = 7.0 – 2.7 = 4.3, which rounds to 4 pH units.

24. Use of Molar Concentrations to Calculate pH What is the pH of a solution that contains 0.20 M sodium acetate and 0.60 M acetic acid ($pK_a = 4.76$)?

Answer $pH = pK_a + \log$ ([base]/[acid])

$= pK_a + \log$ ([acetate]/[acetic acid])

= 4.76 + log (0.20/0.60)

= 4.76 + (−0.48) = 4.3 (two significant figures, based on precision of concentrations)

25. Preparation of an Acetate Buffer Calculate the concentrations of acetic acid ($pK_a = 4.76$) and sodium acetate necessary to prepare a 0.2 M buffer solution at pH 5.0.

Answer First, calculate the required ratio of conjugate base to acid.

$$pH = pK_a + \log ([\text{acetate}]/[\text{acetic acid}])$$

$$\log ([\text{acetate}]/[\text{acetic acid}]) = pH - pK_a = 5.0 - 4.76 = 0.24$$

$$[\text{acetate}]/[\text{acetic acid}] = 10^{0.24} = 1.7$$

$$[\text{acetate}]/[\text{acetate + acetic acid}] = 1.7/2.7 = 0.63 \text{ (two significant figures)}$$

Thus, 63% of the 0.2 M buffer is acetate and 27% is acetic acid. So at pH 5.0 the buffer has 0.13 M acetate and 0.07 M acetic acid.

26. pH of Insect Defensive Secretion You have been observing an insect that defends itself from enemies by secreting a caustic liquid. Analysis of the liquid shows it to have a total concentration of formate plus formic acid ($K_a = 1.8 \times 10^{-4}$) of 1.45 M; the concentration of formate ion is 0.015 M. What is the pH of the secretion?

Answer Solve the Henderson-Hasselbalch equation for pH.

$$pH = pK_a + \log\,([\text{conjugate base}]/[\text{acid}])$$

Given the K_a of formic acid ($K_a = 1.8 \times 10^{-4}$), you can calculate pK_a as $-\log K_a = 3.7$. If the concentration of formate + formic acid = 1.45 M and the concentration of formate is 0.015 M, then the concentration of formic acid is 1.45 M − 0.015 M + 1.435 M.

$$\log\,([\text{formate}]/[\text{formic acid}]) = \log\,(0.015/1.435) = -2.0$$

$$pH = 3.7 - 2.0 = 1.7 \text{ (two significant figures)}$$

27. Calculation of pK_a An unknown compound, X, is thought to have a carboxyl group with a pK_a of 2.0 and another ionizable group with a pK_a between 5 and 8. When 75 mL of 0.1 M NaOH is added to 100 mL of a 0.1 M solution of X at pH 2.0, the pH increases to 6.72. Calculate the pK_a of the second ionizable group of X.

Answer At the first pH (pH = 2), 50% of the carboxyl group is dissociated (pK_a = pH). Then

$$\text{Amount of NaOH added} = 0.075 \text{ L} \times 0.1 \text{ mol/L} = 0.0075 \text{ mol}$$

$$\text{Amount of X present} = 0.1 \text{ L} \times 0.1 \text{ mol/L} = 0.01 \text{ mol}$$

At the new pH of 6.72, the carboxyl group is completely dissociated (because pH is much greater than the pK_a). The amount of NaOH required to titrate this remaining 50% of the carboxyl group is 0.5 × 0.01 mol = 0.005 mol.

Thus, 0.0075 mol − 0.005 mol = 0.0025 mol of NaOH is available to titrate the other group, and, using the Henderson-Hasselbalch equation,

$$pH = pK_a + \log \frac{[A^-]}{[HA]}$$

we can find the pK_a of the second ionizable group of X:

$$6.72 = pK_a + \log\,[0.0025/(0.01 - 0.0025)]$$

$$pK_a = 6.72 - (-0.48) = 7.20, \text{ which rounds to 7.}$$

28. Ionic Forms of Alanine Alanine is a diprotic acid that can undergo two dissociation reactions (see Table 3–1 for pK_a values). **(a)** Given the structure of the partially protonated form (or zwitterion; see Fig. 3–9) below, draw the chemical structures of the other two forms of alanine that predominate in aqueous solution: the fully protonated form and the fully deprotonated form.

$$\begin{array}{c} COO^- \\ | \\ H_3\overset{+}{N}—C—H \\ | \\ CH_3 \end{array}$$

Alanine

Of the three possible forms of alanine, which would be present at the highest concentration in solutions of the following pH: **(b)** 1.0; **(c)** 6.2; **(d)** 8.02; **(e)** 11.9. Explain your answers in terms of pH relative to the two pK_a values.

Answer

(a)

$$\begin{array}{c} COOH \\ | \\ H_3\overset{+}{N}—C—H \\ | \\ CH_3 \end{array}$$

Fully protonated

$$\begin{array}{c} COO^- \\ | \\ H_2N—C—H \\ | \\ CH_3 \end{array}$$

Fully deprotonated

(b) At pH 1.0, 1.3 pH units below the pK_a of the carboxyl group, more than 90% of the carboxyl groups are protonated, and protonated amino groups predominate by a factor of more than 107.

(c) At pH 6.2 the zwitterion predominates. This is 4 pH units above the pK_a of the carboxyl group, so the vast majority of carboxyl groups are deprotonated. It is 3.5 pH units below the pK_a of the amino group, so the vast majority of amino groups are protonated.

(d) At pH 8.02 the zwitterion still predominates. The carboxyl groups are deprotonated and, with the pH still 1.6 units below the pK_a of the amino group, the vast majority of amino groups are protonated.

(e) At pH 11.9, 2.2 pH units above the pK_a of the amino group, the vast majority of amino groups are deprotonated; and the carboxyl groups, at 9.6 pH units above their pK_a, remain deprotonated.

29. Control of Blood pH by Respiration Rate

(a) The partial pressure of CO_2 in the lungs can be varied rapidly by the rate and depth of breathing. For example, a common remedy to alleviate hiccups is to increase the concentration of CO_2 in the lungs. This can be achieved by holding one's breath, by very slow and shallow breathing (hypoventilation), or by breathing in and out of a paper bag. Under such conditions, pCO_2 in the air space of the lungs rises above normal. Qualitatively explain the effect of these procedures on the blood pH.

(b) A common practice of competitive short-distance runners is to breathe rapidly and deeply (hyperventilate) for about half a minute to remove CO_2 from their lungs just before the race begins. Blood pH may rise to 7.60. Explain why the blood pH increases.

(c) During a short-distance run, the muscles produce a large amount of lactic acid ($CH_3CH(OH)COOH$, $K_a = 1.38 \times 10^{-4}$) from their glucose stores. In view of this fact, why might hyperventilation before a dash be useful?

Answer

(a) Blood pH is controlled by the carbon dioxide–bicarbonate buffer system, as shown by the net equation

$$CO_2 + H_2O \rightleftharpoons H^+ + HCO_3^-$$

During *hypoventilation,* the concentration of CO_2 in the lungs and arterial blood increases, driving the equilibrium to the right and raising the [H^+]; that is, the pH is lowered.

(b) During *hyperventilation,* the concentration of CO_2 in the lungs and arterial blood falls. This drives the equilibrium to the left, which requires the consumption of hydrogen ions, reducing [H^+] and increasing pH.

(c) Lactate is a moderately strong acid (pK_a = 3.86) that completely dissociates under physiological conditions:

$$CH_3CH(OH)COOH \rightleftharpoons CH_3CH(OH)COO^- + H^+$$

This lowers the pH of the blood and muscle tissue. Hyperventilation is useful because it removes hydrogen ions, raising the pH of the blood and tissues in anticipation of the acid buildup.

30. Calculation of Blood pH from CO_2 and Bicarbonate Levels Calculate the pH of a blood plasma sample with a total CO_2 concentration of 26.9 mM and bicarbonate concentration of 25.6 mM. Recall from page 63 that the relevant pK_a of carbonic acid is 6.1.

Answer Use the Henderson-Hasselbalch equation:

$$pH = pK_a + \log([\text{bicarbonate}]/[\text{carbonic acid}])$$

If total $[CO_2]$ = 26.9 M and [bicarbonate] = 25.6 M, then the concentration of carbonic acid is 26.9 M − 25.6 M = 1.3 M.

$$pH = 6.1 + \log(25.6/1.3) = 7.4 \text{ (two significant figures)}$$

31. Effect of Holding One's Breath on Blood pH The pH of the extracellular fluid is buffered by the bicarbonate/carbonic acid system. Holding your breath can increase the concentration of $CO_2(g)$ in the blood. What effect might this have on the pH of the extracellular fluid? Explain by showing the relevant equilibrium equation(s) for this buffer system.

Answer Dissolving more CO_2 in the blood increases $[H^+]$ in blood and extracellular fluids, lowering pH: $CO_2(d) + H_2O \rightleftharpoons H_2CO_3 \rightleftharpoons H^+ + HCO_3^-$

Data Analysis Problem

32. "Switchable" Surfactants Hydrophobic molecules do not dissolve well in water. Given that water is a very commonly used solvent, this makes certain processes very difficult: washing oily food residue off dishes, cleaning up spilled oil, keeping the oil and water phases of salad dressings well mixed, and carrying out chemical reactions that involve both hydrophobic and hydrophilic components.

Surfactants are a class of amphipathic compounds that includes soaps, detergents, and emulsifiers. With the use of surfactants, hydrophobic compounds can be suspended in aqueous solution by forming micelles (see Fig. 2–7). A micelle has a hydrophobic core consisting of the hydrophobic compound and the hydrophobic "tails" of the surfactant; the hydrophilic "heads" of the surfactant cover the surface of the micelle. A suspension of micelles is called an emulsion. The more hydrophilic the head group of the surfactant, the more powerful it is—that is, the greater its capacity to emulsify hydrophobic material.

When you use soap to remove grease from dirty dishes, the soap forms an emulsion with the grease that is easily removed by water through interaction with the hydrophilic head of the soap molecules. Likewise, a detergent can be used to emulsify spilled oil for removal by water. And emulsifiers in commercial salad dressings keep the oil suspended evenly throughout the water-based mixture.

There are some situations in which it would be very useful to have a "switchable" surfactant: a molecule that could be reversibly converted between a surfactant and a nonsurfactant.

(a) Imagine such a "switchable" surfactant existed. How would you use it to clean up and then recover the oil from an oil spill?

Liu et al. describe a prototypical switchable surfactant in their 2006 article "Switchable Surfactants." The switching is based on the following reaction:

$$R{-}N{=}C(CH_3){-}N(CH_3)_2 + CO_2 + H_2O \rightleftharpoons R{-}NH{-}C(CH_3){=}N^+(CH_3)_2 + HCO_3^-$$

Amidine form — Amidinium form

(b) Given that the pK_a of a typical amidinium ion is 12.4, in which direction (left or right) would you expect the equilibrium of the above reaction to lie? (See Fig. 2–16 for relevant pK_a values.) Justify your answer. Hint: Remember the reaction $H_2O + CO_2 \rightleftharpoons H_2CO_3$.

Liu and colleagues produced a switchable surfactant for which R = $C_{16}H_{33}$. They do not name the molecule in their article; for brevity, we'll call it s-surf.

(c) The amidinium form of s-surf is a powerful surfactant; the amidine form is not. Explain this observation.

Liu and colleagues found that they could switch between the two forms of s-surf by changing the gas that they bubbled through a solution of the surfactant. They demonstrated this switch by measuring the electrical conductivity of the s-surf solution; aqueous solutions of ionic compounds have higher conductivity than solutions of nonionic compounds. They started with a solution of the amidine form of s-surf in water. Their results are shown below; dotted lines indicate the switch from one gas to another.

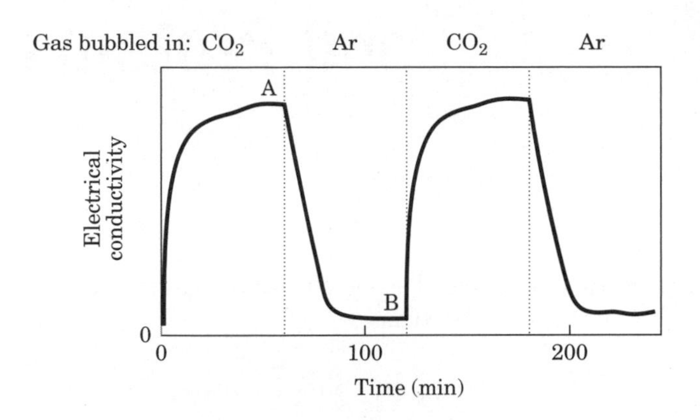

(d) In which form is the majority of s-surf at point A? At point B?

(e) Why does the electrical conductivity rise from time 0 to point A?

(f) Why does the electrical conductivity fall from point A to point B?

(g) Explain how you would use s-surf to clean up and recover the oil from an oil spill.

Answer

(a) Use the substance in its surfactant form to emulsify the spilled oil, collect the emulsified oil, then switch to the nonsurfactant form. The oil and water will separate and the oil can be collected for further use.

(b) The equilibrium lies strongly to the right. The stronger acid (lower pK_a), H_2CO_3, donates a proton to the conjugate base of the weaker acid (higher pK_a), amidine.

(c) The strength of a surfactant depends on the hydrophilicity of its head groups: the more hydrophilic, the more powerful the surfactant. The amidinium form of s-surf is much more hydrophilic than the amidine form, so it is a more powerful surfactant.

(d) *Point A:* amidinium; the CO_2 has had plenty of time to react with the amidine to produce the amidinium form. *Point B:* amidine; Ar has removed CO_2 from the solution, leaving the amidine form.

(e) The conductivity rises as uncharged amidine reacts with CO_2 to produce the charged amidinium form.

(f) The conductivity falls as Ar removes CO_2, shifting the equilibrium to the uncharged amidine form.

(g) Treat s-surf with CO_2 to produce the surfactant amidinium form and use this to emulsify the spill. Treat the emulsion with Ar to remove the CO_2 and produce the nonsurfactant amidine from. The oil will separate from the water and can be recovered.

Reference

Liu, Y., Jessop, P.G., Cunningham, M., Eckert, C.A., & Liotta, C.L. (2006) *Science* **313,** 958–960.

chapter

3 Amino Acids, Peptides, and Proteins

1. **Absolute Configuration of Citrulline** The citrulline isolated from watermelons has the structure shown below. Is it a D- or L-amino acid? Explain.

$$\begin{array}{c} CH_2(CH_2)_2NH-\underset{\underset{O}{\|}}{C}-NH_2 \\ H-C-\overset{+}{N}H_3 \\ COO^- \end{array}$$

Answer Rotating the structural representation 180° in the plane of the page puts the most highly oxidized group—the carboxyl (—COO$^-$) group—at the top, in the same position as the —CHO group of glyceraldehyde in Figure 3–4. In this orientation, the α-amino group is on the left, and thus the absolute configuration of the citrulline is L.

2. **Relationship between the Titration Curve and the Acid-Base Properties of Glycine** A 100 mL solution of 0.1 M glycine at pH 1.72 was titrated with 2 M NaOH solution. The pH was monitored and the results were plotted as shown in the following graph. The key points in the titration are designated I to V. For each of the statements **(a)** to **(o)**, *identify* the appropriate key point in the titration and *justify* your choice.

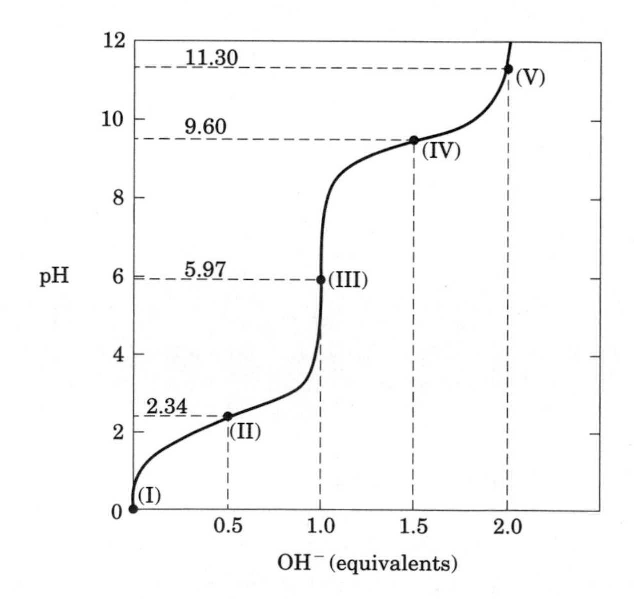

Note: before considering statements **(a)** through **(o)**, refer to Figure 3–10. The three species involved in the titration of glycine can be considered in terms of a useful physical analogy. Each ionic species can be viewed as a different floor of a building, each with a different net charge:

$^{+}H_3N—CH_2—COOH$	+1
$^{+}H_3N—CH_2—COO^-$	0 (zwitterion)
$H_2N—CH_2—COO^-$	−1

The floors are connected by steep stairways, and each stairway has a landing halfway between the floors. A titration curve traces the path one would follow between the different floors as the pH changes in response to added OH^-. Recall that the pK_a of an acid (on a halfway landing) represents the pH at which half of the acid is deprotonated. The isoelectric point (pI) is the pH at which the average net charge is zero. Now you are ready to consider statements **(a)** through **(o)**.

(a) Glycine is present predominantly as the species $^{+}H_3N—CH_2—COOH$.

(b) The *average* net charge of glycine is $+\frac{1}{2}$.

(c) Half of the amino groups are ionized.

(d) The pH is equal to the pK_a of the carboxyl group.

(e) The pH is equal to the pK_a of the protonated amino group.

(f) Glycine has its maximum buffering capacity.

(g) The *average* net charge of glycine is zero.

(h) The carboxyl group has been completely titrated (first equivalence point).

(i) Glycine is completely titrated (second equivalence point).

(j) The predominant species is $^{+}H_3N—CH_2—COO^-$.

(k) The *average* net charge of glycine is −1.

(l) Glycine is present predominantly as a 50:50 mixture of $^{+}H_3N—CH_2—COOH$ and $^{+}H_3N—CH_2—COO^-$.

(m) This is the isoelectric point.

(n) This is the end of the titration.

(o) These are the *worst* pH regions for buffering power.

Answer

(a) I; maximum protonation occurs at the lowest pH (the highest [H^+]).

(b) II; at the first pK_a, or pK_1 (2.34), half of the protons are removed from the α-carboxyl group (i.e., it is half deprotonated), changing its charge from 0 to $-\frac{1}{2}$. The average net charge of glycine is $(-\frac{1}{2}) + 1 = \frac{1}{2}$.

(c) IV; the α-amino group is half-deprotonated at its pK_a, or pK_2 (9.60).

(d) II; from the Henderson-Hasselbalch equation, pH = pK_a + log ([A^-]/[HA]). If [A^-]/[HA] = 1, or [A^-] = [HA], then pH = pK_a. (Recall that log 1 = 0.)

(e) IV; see answers **(c)** and **(d)**.

(f) II and IV; in the pK_a regions, acid donates protons to or base abstracts protons from glycine, with minimal pH changes.

(g) III; this occurs at the isoelectric point; pI = (pK_1 + pK_2)/2 = (2.34 + 9.60)/2 = 5.97.

(h) III; the pH at which 1.0 equivalent of OH^- has been added, pH 5.97 (3.6 pH units away from either pK_a).

(i) V; pH 11.3 (1.7 pH units above pK_2).

(j) III; at pI (5.97) the carboxyl group is fully negatively charged (deprotonated) and the amino group is fully positively charged (protonated).

(k) V; both groups are fully deprotonated, with a neutral amino group and a negatively charged carboxyl group (net charge = −1).

(l) II; the carboxyl group is half ionized at pH = pK_1.

(m) III; see answers **(g)** and **(j)**.

(n) V; glycine is fully titrated after 2.0 equivalents of OH^- have been added.

(o) I, III, and V; each is several pH units removed from either pK_a, where the best pH buffering action occurs.

3. How Much Alanine Is Present as the Completely Uncharged Species? At a pH equal to the isoelectric point of alanine, the *net* charge on alanine is zero. Two structures can be drawn that have a net charge of zero, but the predominant form of alanine at its pI is zwitterionic.

$$H_3\overset{+}{N}-C(CH_3)(H)-C(=O)O^- \qquad H_2N-C(CH_3)(H)-C(=O)OH$$

Zwitterionic   Uncharged

(a) Why is alanine predominantly zwitterionic rather than completely uncharged at its pI?

(b) What fraction of alanine is in the completely uncharged form at its pI? Justify your assumptions.

Answer

(a) The pI of alanine is well above the pK_a of the α-carboxyl group and well below the pK_a of the α-amino group. Hence, at pH = pI, both groups are present predominantly in their charged (ionized) forms.

(b) From Table 3–1, the pI of alanine is 6.01, midway between the two pK_a values 2.34 and 9.69. From the Henderson-Hasselbalch equation, pH − pK_a = log ([A^-]/[HA]). For the carboxyl group:

$$\log \frac{[A^-]}{[HA]} = 6.01 - 2.34 = 3.67$$

$$\frac{[HA]}{[A^-]} = 10^{-3.67} = \frac{1}{4.68 \times 10^3}$$

That is, one molecule in 4,680 is still in the form —COOH. Similarly, at pH = pI, one molecule in 4,680 is in the form $—NH_2$. Thus, the fraction of molecules with both groups uncharged (—COOH *and* $—NH_2$) is 1 in 4,680 × 4,680, or 1 in 2.19×10^7.

4. Ionization State of Histidine Each ionizable group of an amino acid can exist in one of two states, charged or neutral. The electric charge on the functional group is determined by the relationship between its pK_a and the pH of the solution. This relationship is described by the Henderson-Hasselbalch equation.

(a) Histidine has three ionizable functional groups. Write the equilibrium equations for its three ionizations and assign the proper pK_a for each ionization. Draw the structure of histidine in each ionization state. What is the net charge on the histidine molecule in each ionization state?

(b) Draw the structures of the predominant ionization state of histidine at pH 1, 4, 8, and 12. Note that the ionization state can be approximated by treating each ionizable group independently.

(c) What is the net charge of histidine at pH 1, 4, 8, and 12? For each pH, will histidine migrate toward the anode (+) or cathode (−) when placed in an electric field?

Answer

(a)

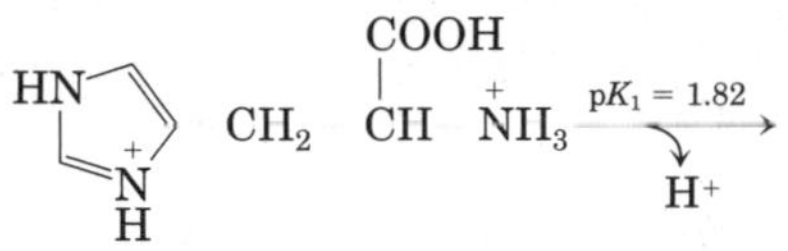

HN (imidazolium, $\overset{+}{N}$H) CII_2 — CII(COOH) — $\overset{+}{N}II_3$ $\xrightarrow{pK_1 = 1.82}$ H^+

1

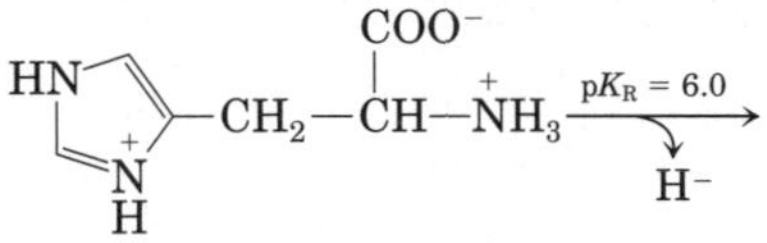

HN (imidazolium, $\overset{+}{N}$H) —CH_2—CH(COO^-)—$\overset{+}{N}H_3$ $\xrightarrow{pK_R = 6.0}$ H^-

2

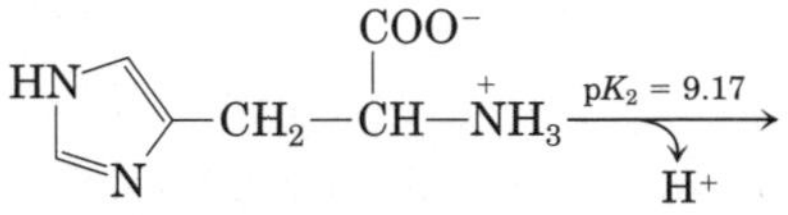

HN (imidazole, N) —CH_2—CH(COO^-)—$\overset{+}{N}H_3$ $\xrightarrow{pK_2 = 9.17}$ H^+

3

HN (imidazole, N) —CH_2—CH(COO^-)—NH_2

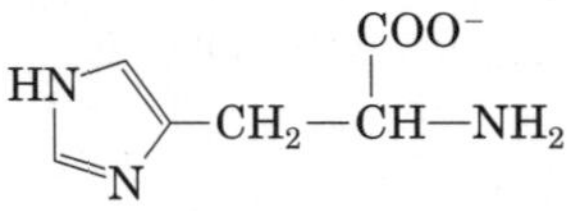

4

We start with the most highly protonated species of histidine (structure **1,** found at the most acidic pH). The pK_a values are from Table 3–1. As base is added, the group with the lowest pK_a loses its proton first, followed by the group with the next lowest pK_a, then the group with the highest pK_a. (In the following table, R = —CH_2–imidazole.)

Structure	Net charge
1 ^+H_3N—CH(RH^+)—COOH	+2
2 ^+H_3N—CH(RH^+)—COO^-	+1
3 ^+H_3N—CH(R)—COO^-	0
4 H_2N—CH(R)—COO^-	−1

(b) and **(c)** See the structures in **(a).**

pH	Structure	Net charge	Migrates toward:
1	**1**	+2	Cathode
4	**2**	+1	Cathode
8	**3**	0	Does not migrate
12	**4**	−1	Anode

5. Separation of Amino Acids by Ion-Exchange Chromatography Mixtures of amino acids can be analyzed by first separating the mixture into its components through ion-exchange chromatography. Amino acids placed on a cation-exchange resin (see Fig. 3–17a) containing sulfonate ($—SO_3^-$) groups flow down the column at different rates because of two factors that influence their movement: (1) ionic attraction between the sulfonate residues on the column and positively charged functional groups on the amino acids, and (2) hydrophobic interactions between amino acid side chains and the strongly hydrophobic backbone of the polystyrene resin. For each pair of amino acids listed, determine which will be eluted first from an ion-exchange column by a pH 7.0 buffer.

(a) Asp and Lys
(b) Arg and Met
(c) Glu and Val
(d) Gly and Leu
(e) Ser and Ala

Answer See Table 3–1 for pK_a values for the amino acid side chains. At pH < pI, an amino acid has a net positive charge; at pH > pI, it has a net negative charge. For any pair of amino acids, the more negatively charged one passes through the sulfonated resin faster. For two neutral amino acids, the less polar one passes through more slowly because of its stronger hydrophobic interactions with the polystyrene.

	pI values	Net charge (pH 7)	Elution order	Basis for separation
(a) Asp, Lys	2.77, 9.74	−1, +1	Asp, Lys	Charge
(b) Arg, Met	10.76, 5.74	+1, 0	Met, Arg	Charge
(c) Glu, Val	3.22, 5.97	−1, 0	Glu, Val	Charge
(d) Gly, Leu	5.97, 5.98	0, 0	Gly, Leu	Polarity
(e) Ser, Ala	5.68, 6.01	0, 0	Ser, Ala	Polarity

6. Naming the Stereoisomers of Isoleucine The structure of the amino acid isoleucine is

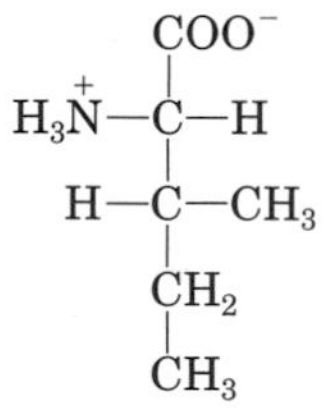

$$\begin{array}{c} COO^- \\ | \\ H_3\overset{+}{N}—C—H \\ | \\ H—C—CH_3 \\ | \\ CH_2 \\ | \\ CH_3 \end{array}$$

(a) How many chiral centers does it have?
(b) How many optical isomers?
(c) Draw perspective formulas for all the optical isomers of isoleucine.

Answer

(a) Two; at C-2 and C-3 (the α and β carbons).
(b) Four; the two chiral centers permit four possible diastereoisomers: (*S*,*S*), (*S*,*R*), (*R*,*R*), and (*R*,*S*).

(c)

```
    COO⁻              COO⁻             COO⁻              COO⁻
    |                 |                |                 |
 +                 +                                  
H3N—C—H          H3N—C—H          H—C—NH3 +          H—C—NH3 +
    |                 |                |                 |
 H—C—CH3         H3C—C—H          H—C—CH3          H3C—C—H
    |                 |                |                 |
    CH2               CH2              CH2               CH2
    |                 |                |                 |
    CH3               CH3              CH3               CH3
```

7. Comparing the pK_a Values of Alanine and Polyalanine The titration curve of alanine shows the ionization of two functional groups with pK_a values of 2.34 and 9.69, corresponding to the ionization of the carboxyl and the protonated amino groups, respectively. The titration of di-, tri-, and larger oligopeptides of alanine also shows the ionization of only two functional groups, although the experimental pK_a values are different. The trend in pK_a values is summarized in the table.

Amino acid or peptide	pK_1	pK_2
Ala	2.34	9.69
Ala–Ala	3.12	8.30
Ala–Ala–Ala	3.39	8.03
Ala–(Ala)$_n$–Ala, $n \geq 4$	3.42	7.94

(a) Draw the structure of Ala–Ala–Ala. Identify the functional groups associated with pK_1 and pK_2.

(b) Why does the value of pK_1 *increase* with each additional Ala residue in the Ala oligopeptide?

(c) Why does the value of pK_2 *decrease* with each additional Ala residue in the Ala oligopeptide?

Answer

(a) The structure at pH 7 is:

```
                  O            O            O
                  ‖            ‖            ‖
 +
H3N—CH—C—N—CH—C—N—CH—C—O⁻
     |       |  |        |  |
     CH3     H  CH3      H  CH3
pK2 = 8.03                          pK1 = 3.39
```

Note that only the amino- and carboxyl-terminal groups ionize.

(b) As the length of poly(Ala) increases, the two terminal groups move farther apart, separated by an intervening sequence with an "insulating" nonpolar structure. The carboxyl group becomes a weaker acid, as reflected in its higher pK_a, because the electrostatic repulsion between the carboxyl proton and the positive charge on the NH_3^+ group diminishes as the groups become more distant.

(c) The negative charge on the terminal carboxyl group has a stabilizing effect on the positively charged (protonated) terminal amino group. With increasing numbers of intervening Ala residues, this stabilizing effect is diminished and the NH_3^+ group loses its proton more easily. The lower pK_2 indicates that the terminal amino group has become a weaker base (stronger acid). The intramolecular effects of the amide (peptide bond) linkages keep pK_a values lower than they would be for an alkyl-substituted amine.

8. The Size of Proteins What is the approximate molecular weight of a protein with 682 amino acid residues in a single polypeptide chain?

Answer Assuming that the average M_r per residue is 110 (corrected for loss of water in formation of the peptide bond), a protein containing 682 residues has an M_r of approximately $682 \times 110 = 75{,}000$.

9. The Number of Tryptophan Residues in Bovine Serum Albumin A quantitative amino acid analysis reveals that bovine serum albumin (BSA) contains 0.58% tryptophan (M_r 204) by weight.

(a) Calculate the *minimum* molecular weight of BSA (i.e., assuming there is only one tryptophan residue per protein molecule).

(b) Gel filtration of BSA gives a molecular weight estimate of 70,000. How many tryptophan residues are present in a molecule of serum albumin?

Answer

(a) The M_r of a Trp residue must be adjusted to account for the removal of water during peptide bond formation: $M_r = 204 - 18 = 186$. The molecular weight of BSA can be calculated using the following proportionality, where n is the number of Trp residues in the protein:

$$\frac{\text{wt Trp}}{\text{wt BSA}} = \frac{n(M_r\ \text{Trp})}{M_r\ \text{BSA}}$$

$$\frac{0.58\ \text{g}}{100\ \text{g}} = \frac{n(186)}{M_r\ \text{BSA}}$$

A *minimum* molecular weight can be found by assuming only one Trp residue per BSA molecule ($n = 1$).

$$\frac{(100\ \text{g})(186)(1)}{0.58\text{g}} = 32{,}000$$

(b) Given that the M_r of BSA is approximately 70,000, BSA has ~ 70,000/32,000 = 2.2, or 2 Trp residues per molecule. (The remainder from this division suggests that the estimate of M_r 70,000 for BSA is somewhat high.)

10. Subunit Composition of a Protein A protein has a molecular mass of 400 kDa when measured by gel filtration. When subjected to gel electrophoresis in the presence of sodium dodecyl sulfate (SDS), the protein gives three bands with molecular masses of 180, 160, and 60 kDa. When electrophoresis is carried out in the presence of SDS and dithiothreitol, three bands are again formed, this time with molecular masses of 160, 90, and 60 kDa. Determine the subunit composition of the protein.

Answer The protein has four subunits, with molecular masses of 160, 90, 90, and 60 kDa. The two 90 kDa subunits (possibly identical) are linked by one or more disulfide bonds.

11. Net Electric Charge of Peptides A peptide has the sequence

Glu–His–Trp–Ser–Gly–Leu–Arg–Pro–Gly

(a) What is the net charge of the molecule at pH 3, 8, and 11? (Use pK_a values for side chains and terminal amino and carboxyl groups as given in Table 3–1.)

(b) Estimate the pI for this peptide.

Answer

(a) When pH > pK_a, ionizing groups lose their protons. The pK_a values of importance here are those of the amino-terminal (9.67) and carboxyl-terminal (2.34) groups and those of the R groups of the Glu (4.25), His (6.00), and Arg (12.48) residues. **Note:** here we are using the available pK_a values: those for the *free* amino acids as given in Table 3–1. As demonstrated in Problem 7, however, the pK_a values of the α-amino and α-carboxyl groups shift when the amino acid is at the amino or carboxyl terminus, respectively, of a peptide, and this would affect the net charge and the pI of the peptide.

pH	^+H_3N	Glu	His	Arg	COO^-	Net charge
3	+1	0	+1	+1	−1	+2
8	+1	−1	0	+1	−1	0
11	0	−1	0	+1	−1	−1

(b) Two different methods can be used to estimate pI. Find the two ionizable groups with pK_a values that "straddle" the point at which net peptide charge = 0 (here, two groups that ionize near pH 8): the amino-terminal α-amino group of Glu and the His imidazole group. Thus, we can estimate

$$\text{pI} = \frac{9.67 + 6.00}{2} = 7.8$$

Alternatively, plot the calculated net charges as a function of pH, and determine graphically the pH at which the net charge is zero on the vertical axis. More data points are needed to use this method accurately.

Note: although at any instant an individual amino acid molecule will have an integral charge, it is possible for a population of amino acid molecules in solution to have a fractional charge. For example, at pH 1.0 glycine exists entirely as the form $^+H_3N—CH_2—COOH$ with a net positive charge of 1.0. However, at pH 2.34, where there is an equal mixture of $^+H_3N—CH_2—COOH$ and $^+H_3N—CH_2—COO^-$, the average or net charge on the population of glycine molecules is 0.5 (see the discussion on pp. 80–81). You can use the Henderson-Hasselbalch equation to calculate the exact ratio of charged and uncharged species at equilibrium at various pH values.

12. Isoelectric Point of Pepsin Pepsin is the name given to a mix of several digestive enzymes secreted (as larger precursor proteins) by glands that line the stomach. These glands also secrete hydrochloric acid, which dissolves the particulate matter in food, allowing pepsin to enzymatically cleave individual protein molecules. The resulting mixture of food, HCl, and digestive enzymes is known as chyme and has a pH near 1.5. What pI would you predict for the pepsin proteins? What functional groups must be present to confer this pI on pepsin? Which amino acids in the proteins would contribute such groups?

Answer Pepsin proteins have a relatively low pI (near the pH of gastric juice) in order to remain soluble and thus functional in the stomach. (Pepsin—the mixture of enzymes—has a pI of ~1.) As pH increases, pepsins acquire a net charge and undergo ionic interactions with oppositely charged molecules (such as dissolved salts), causing the pepsin proteins to precipitate. Pepsin is active only in the stomach. In the relatively high pH of the intestine, pepsin proteins precipitate and become inactive.

A low pI requires large numbers of negatively charged (low pK_a) groups. These are contributed by the carboxylate groups of Asp and Glu residues.

13. The Isoelectric Point of Histones Histones are proteins found in eukaryotic cell nuclei, tightly bound to DNA, which has many phosphate groups. The pI of histones is very high, about 10.8. What amino acid residues must be present in relatively large numbers in histones? In what way do these residues contribute to the strong binding of histones to DNA?

Answer Large numbers of positively charged (high pK_a) groups in a protein give it a high pI. In histones, the positively charged R groups of Lys, Arg, and His residues interact strongly with the negatively charged phosphate groups of DNA through ionic interactions.

14. Solubility of Polypeptides One method for separating polypeptides makes use of their different solubilities. The solubility of large polypeptides in water depends upon the relative polarity of their R groups, particularly on the number of ionized groups: the more ionized groups there are, the more soluble the polypeptide. Which of each pair of the polypeptides that follow is more soluble at the indicated pH?

(a) $(Gly)_{20}$ or $(Glu)_{20}$ at pH 7.0
(b) $(Lys–Ala)_3$ or $(Phe–Met)_3$ at pH 7.0
(c) $(Ala–Ser–Gly)_5$ or $(Asn–Ser–His)_5$ at pH 6.0
(d) $(Ala–Asp–Gly)_5$ or $(Asn–Ser–His)_5$ at pH 3.0

Answer

(a) $(Glu)_{20}$; it is highly negatively charged (polar) at pH 7. $(Gly)_{20}$ is uncharged except for the amino- and carboxyl-terminal groups.
(b) $(Lys–Ala)_3$; this is highly positively charged (polar) at pH 7. $(Phe–Met)_3$ is much less polar and hence less soluble.
(c) $(Asn–Ser–His)_5$; both polymers have polar Ser side chains, but $(Asn–Ser–His)_5$ also has the polar Asn side chains and partially protonated His side chains.
(d) $(Asn–Ser–His)_5$; at pH 3, the carboxylate groups of Asp residues are partially protonated and neutral, whereas the imidazole groups of His residues are fully protonated and positively charged.

15. Purification of an Enzyme A biochemist discovers and purifies a new enzyme, generating the purification table below.

Procedure	Total protein (mg)	Activity (units)
1. Crude extract	20,000	4,000,000
2. Precipitation (salt)	5,000	3,000,000
3. Precipitation (pH)	4,000	1,000,000
4. Ion-exchange chromatography	200	800,000
5. Affinity chromatography	50	750,000
6. Size-exclusion chromatography	45	675,000

(a) From the information given in the table, calculate the specific activity of the enzyme after each purification procedure.
(b) Which of the purification procedures used for this enzyme is most effective (i.e., gives the greatest relative increase in purity)?
(c) Which of the purification procedures is least effective?
(d) Is there any indication based on the results shown in the table that the enzyme after step 6 is now pure? What else could be done to estimate the purity of the enzyme preparation?

Answer

(a) From the percentage recovery of activity (units), we calculate percentage yield and specific activity (units/mg).

Procedure	Protein (mg)	Activity (units)	% Yield	Specific activity (units/mg)	Purification factor (overall)
1	20,000	4,000,000	(100)	200	(1.0)
2	5,000	3,000,000	75	600	× 3.0
3	4,000	1,000,000	25	250	× 1.25
4	200	800,000	20	4,000	× 20
5	50	750,000	19	15,000	× 75
6	45	675,000	17	15,000	× 75

(b) Step 4, ion-exchange chromatography; this gives the greatest increase in specific activity (an index of purity and degree of increase in purification).

(c) Step 3, pH precipitation; two-thirds of the total activity from the previous step was lost here.

(d) Yes. The specific activity did not increase further after step 5. SDS polyacrylamide gel electrophoresis is an excellent, standard way of checking homogeneity and purity.

16. Dialysis A purified protein is in a Hepes (*N*-(2-hydroxyethyl)piperazine-*N'*-(2-ethanesulfonic acid)) buffer at pH 7 with 500 mM NaCl. A sample (1 mL) of the protein solution is placed in a tube made of dialysis membrane and dialyzed against 1 L of the same Hepes buffer with 0 mM NaCl. Small molecules and ions (such as Na^+, Cl^-, and Hepes) can diffuse across the dialysis membrane, but the protein cannot.

(a) Once the dialysis has come to equilibrium, what is the concentration of NaCl in the protein sample? Assume no volume changes occur in the sample during the dialysis.

(b) If the original 1 mL sample were dialyzed twice, successively, against 100 mL of the same Hepes buffer with 0 mM NaCl, what would be the final NaCl concentration in the sample?

Answer

(a) [NaCl] = 0.5 mM

(b) [NaCl] = 0.05 mM.

17. Peptide Purification At pH 7.0, in what order would the following three peptides be eluted from a column filled with a cation-exchange polymer? Their amino acid compositions are:

Protein A: Ala 10%, Glu 5%, Ser 5%, Leu 10%, Arg 10%, His 5%, Ile 10%, Phe 5%, Tyr 5%, Lys 10%, Gly 10%, Pro 5%, and Trp 10%.

Protein B: Ala 5%, Val 5%, Gly 10%, Asp 5%, Leu 5%, Arg 5%, Ile 5%, Phe 5%, Tyr 5%, Lys 5%, Trp 5%, Ser 5%, Thr 5%, Glu 5%, Asn 5%, Pro 10%, Met 5%, and Cys 5%.

Protein C: Ala 10%, Glu 10%, Gly 5%, Leu 5%, Asp 10%, Arg 5%, Met 5%, Cys 5%, Tyr 5%, Phe 5%, His 5%, Val 5%, Pro 5%, Thr 5%, Ser 5%, Asn 5%, and Gln 5%.

Answer Protein C has a net negative charge because there are more Glu and Asp residues than Lys, Arg, and His residues. Protein A has a net positive charge. Protein B has no net charge at neutral pH. A cation-exchange column has a negatively charged polymer, so protein C interacts most weakly with the column and is eluted first, followed by B, then A.

18. Sequence Determination of the Brain Peptide Leucine Enkephalin A group of peptides that influence nerve transmission in certain parts of the brain has been isolated from normal brain tissue. These peptides are known as opioids because they bind to specific receptors that also bind opiate drugs, such as morphine and naloxone. Opioids thus mimic some of the properties of opiates. Some researchers consider these peptides to be the brain's own painkillers. Using the information below, determine the amino acid sequence of the opioid leucine enkephalin. Explain how your structure is consistent with each piece of information.

(a) Complete hydrolysis by 6 M HCl at 110 °C followed by amino acid analysis indicated the presence of Gly, Leu, Phe, and Tyr in a 2:1:1:1 molar ratio.

(b) Treatment of the peptide with 1-fluoro-2,4-dinitrobenzene followed by complete hydrolysis and chromatography indicated the presence of the 2,4-dinitrophenyl derivative of tyrosine. No free tyrosine could be found.

(c) Complete digestion of the peptide with chymotrypsin followed by chromatography yielded free tyrosine and leucine, plus a tripeptide containing Phe and Gly in a 1:2 ratio.

Answer

(a) The empirical composition is (2 Gly, Leu, Phe, Tyr)$_n$.

(b) Tyr is the amino-terminal residue, and there are no other Tyr residues, so $n = 1$ and the sequence is Tyr–(2 Gly, Leu, Phe).

(c) As shown in Table 3–7, chymotrypsin cleaves on the carboxyl side of aromatic residues (Phe, Trp, and Tyr). The peptide has only two aromatic residues, Tyr at the amino terminus and a Phe. Because there are three cleavage products, the Phe residue cannot be at the carboxyl terminus. Rather, release of free leucine means that Leu must be at the carboxyl terminus and must be on the carboxyl side of Phe in the peptide. Thus the sequence must be

Tyr–(2 Gly)–Phe–Leu = Tyr–Gly–Gly–Phe–Leu

19. Structure of a Peptide Antibiotic from *Bacillus brevis* Extracts from the bacterium *Bacillus brevis* contain a peptide with antibiotic properties. This peptide forms complexes with metal ions and seems to disrupt ion transport across the cell membranes of other bacterial species, killing them. The structure of the peptide has been determined from the following observations.

(a) Complete acid hydrolysis of the peptide followed by amino acid analysis yielded equimolar amounts of Leu, Orn, Phe, Pro, and Val. Orn is ornithine, an amino acid not present in proteins but present in some peptides. It has the structure

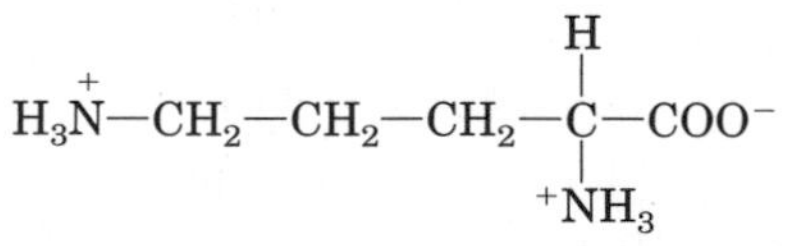

$$H_3\overset{+}{N}-CH_2-CH_2-CH_2-\underset{^+NH_3}{\overset{H}{\underset{|}{\overset{|}{C}}}}-COO^-$$

(b) The molecular weight of the peptide was estimated as about 1,200.

(c) The peptide failed to undergo hydrolysis when treated with the enzyme carboxypeptidase. This enzyme catalyzes the hydrolysis of the carboxyl-terminal residue of a polypeptide unless the residue is Pro or, for some reason, does not contain a free carboxyl group.

(d) Treatment of the intact peptide with 1-fluoro-2,4-dinitrobenzene, followed by complete hydrolysis and chromatography, yielded only free amino acids and the following derivative:

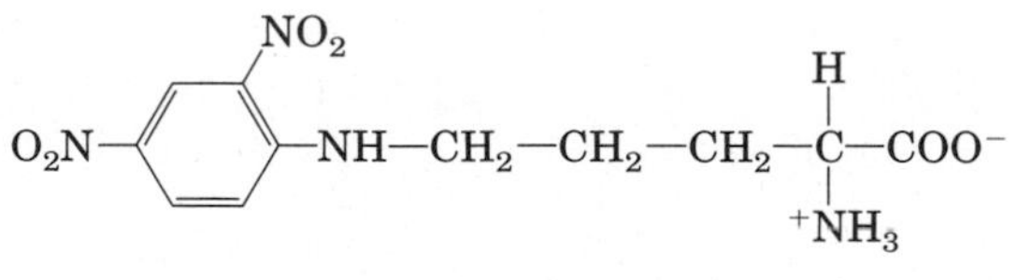

$$O_2N-C_6H_3(NO_2)-NH-CH_2-CH_2-CH_2-\underset{^+NH_3}{\overset{H}{\underset{|}{\overset{|}{C}}}}-COO^-$$

(Hint: note that the 2,4-dinitrophenyl derivative involves the amino group of a side chain rather than the α-amino group.)

(e) Partial hydrolysis of the peptide followed by chromatographic separation and sequence analysis yielded the following di- and tripeptides (the amino-terminal amino acid is always at the left):

Leu–Phe Phe–Pro Orn–Leu Val–Orn

Val–Orn–Leu Phe–Pro–Val Pro–Val–Orn

Given the above information, deduce the amino acid sequence of the peptide antibiotic. Show your reasoning. When you have arrived at a structure, demonstrate that it is consistent with *each* experimental observation.

Answer The information obtained from each experiment is as follows.

(a) The simplest empirical formula for the peptide is (Leu, Orn, Phe, Pro, Val)$_n$.

(b) Assuming an average residue M_r of 110, the minimum molecular weight for the peptide is 550. Because 1,200/550 ≈ 2, the empirical formula is (Leu, Orn, Phe, Pro, Val)$_2$.

(c) Failure of carboxypeptidase to cleave the peptide could result from Pro at the carboxyl terminus *or* the absence of a carboxyl-terminal residue—as in a cyclic peptide.

(d) Failure of FDNB to derivatize an α-amino group indicates either the absence of a free amino-terminal group or that Pro (an imino acid) is at the amino-terminal position. (The derivative formed is 2,4 dinitrophenyl-ε-ornithine.)

(e) The presence of Pro at an internal position in the peptide Phe–Pro–Val indicates that it is *not* at the amino or carboxyl terminus. The information from these experiments suggests that the peptide is cyclic. The alignment of overlapping sequences is

Leu–Phe

Phe–Pro

Phe–Pro–Val

Pro–Val–Orn

Val–Orn

Orn–(Leu)

Thus, the peptide is a cyclic dimer of Leu–Phe–Pro–Val–Orn:

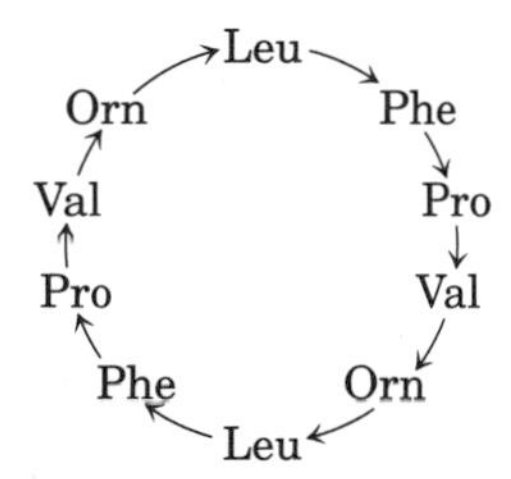

where the arrows indicate the –CO → NH–, or C → N, direction of the peptide bonds. This structure is consistent with all the data.

20. Efficiency in Peptide Sequencing A peptide with the primary structure Lys–Arg–Pro–Leu–Ile–Asp–Gly–Ala is sequenced by the Edman procedure. If each Edman cycle is 96% efficient, what percentage of the amino acids liberated in the fourth cycle will be leucine? Do the calculation a second time, but assume a 99% efficiency for each cycle.

Answer 88%, 97%. The formula for calculating the percentage of correct amino acid liberated after sequencing cycle n, given an efficiency x, is x^n/x, or x^{n-1}. If the efficiency is 0.96, the fraction of correct amino acid liberated in the fourth cycle is $(0.96)^3 = 0.88$. If the efficiency is 0.99, the fraction is $(0.99)^3 = 0.97$.

The $n - 1$ term can be explained by considering what happens in each cycle. For example, in the first cycle, a Lys residue is liberated from 96% of the ends, so that 96% of the termini now have Arg and the remaining 4% still have Lys. However, 100% of the residues actually removed in this cycle are Lys. In the second cycle, Arg is removed from 96% of the ends that contain it, or from $0.96 \times 0.96 = 0.92 = 92\%$ of the ends. Lys is removed from $0.96 \times 0.04 = 0.04$ of the ends. However, the fraction of liberated residues that are Arg is greater than 92% because only 96% of the ends had residues removed. Hence, the fraction of residues liberated as Arg in the second cycle is $0.92/0.96 = 0.96 = 96\%$—and so forth.

21. Sequence Comparisons Proteins called molecular chaperones (described in Chapter 4) assist in the process of protein folding. One class of chaperone found in organisms from bacteria to mammals is heat shock protein 90 (Hsp90). All Hsp90 chaperones contain a 10 amino acid "signature sequence," which allows for ready identification of these proteins in sequence databases. Two representations of this signature sequence are shown below.

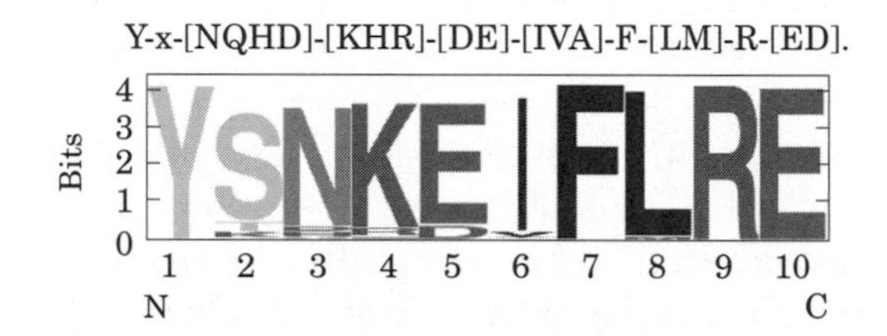

(a) In this sequence, which amino acid residues are invariant (conserved across all species)?

(b) At which position(s) are amino acids limited to those with positively charged side chains? For each position, which amino acid is more commonly found?

(c) At which positions are substitutions restricted to amino acids with negatively charged side chains? For each position, which amino acid predominates?

(d) There is one position that can be any amino acid, although one amino acid appears much more often than any other. What position is this, and which amino acid appears most often?

Answer

(a) Y (1), F (7), and R (9)

(b) Positions 4 and 9; K (Lys) is more common at 4, R (Arg) is invariant at 9

(c) Positions 5 and 10; E (Glu) is more common at both positions

(d) Position 2; S (Ser)

22. Biochemistry Protocols: Your First Protein Purification As the newest and least experienced student in a biochemistry research lab, your first few weeks are spent washing glassware and labeling test tubes. You then graduate to making buffers and stock solutions for use in various laboratory procedures. Finally, you are given responsibility for purifying a protein. It is citrate synthase (an enzyme of the citric acid cycle, to be discussed in Chapter 16), which is located in the mitochondrial matrix. Following a protocol for the purification, you proceed through the steps below. As you work, a more experienced student questions you about the rationale for each procedure. Supply the answers. (Hint: see Chapter 2 for information about osmolarity; see p. 7 for information on separation of organelles from cells.)

(a) You pick up 20 kg of beef hearts from a nearby slaughterhouse (muscle cells are rich in mitochondria, which supply energy for muscle contraction). You transport the hearts on ice, and perform each step of the purification on ice or in a walk-in cold room. You homogenize the beef heart tissue in a high-speed blender in a medium containing 0.2 M sucrose, buffered to a pH of 7.2. *Why do you use beef heart tissue, and in such large quantity? What is the purpose of keeping the tissue cold and suspending it in 0.2 M sucrose, at pH 7.2? What happens to the tissue when it is homogenized?*

(b) You subject the resulting heart homogenate, which is dense and opaque, to a series of differential centrifugation steps. *What does this accomplish?*

(c) You proceed with the purification using the supernatant fraction that contains mostly intact mitochondria. Next, you osmotically lyse the mitochondria. The lysate, which is less dense than the homogenate, but still opaque, consists primarily of mitochondrial membranes and internal mitochondrial contents. To this lysate you add ammonium sulfate, a highly soluble salt, to a specific concentration. You centrifuge the solution, decant the supernatant, and discard the pellet. To the supernatant, which is clearer than the lysate, you add *more* ammonium sulfate. Once again, you centrifuge the sample, but this time you save the pellet because it contains the citrate synthase. *What is the rationale for the two-step addition of the salt?*

(d) You solubilize the ammonium sulfate pellet containing the mitochondrial proteins and dialyze it overnight against large volumes of buffered (pH 7.2) solution. *Why isn't ammonium sulfate included in the dialysis buffer? Why do you use the buffer solution instead of water?*

(e) You run the dialyzed solution over a size-exclusion chromatographic column. Following the protocol, you collect the *first* protein fraction that exits the column and discard the fractions that elute from the column later. You detect the protein by measuring UV absorbance (at 280 nm) in the fractions. *What does the instruction to collect the first fraction tell you about the protein? Why is UV absorbance at 280 nm a good way to monitor for the presence of protein in the eluted fractions?*

(f) You place the fraction collected in (e) on a cation-exchange chromatographic column. After discarding the initial solution that exits the column (the flowthrough), you add a washing solution of higher pH to the column and collect the protein fraction that immediately elutes. *Explain what you are doing.*

(g) You run a small sample of your fraction, now very reduced in volume and quite clear (though tinged pink), on an isoelectric focusing gel. When stained, the gel shows three sharp bands. According to the protocol, the citrate synthase is the protein with a pI of 5.6, but you decide to do one more assay of the protein's purity. You cut out the pI 5.6 band and subject it to SDS polyacrylamide gel electrophoresis. The protein resolves as a single band. *Why were you unconvinced of the purity of the "single" protein band on your isoelectric focusing gel? What did the results of the SDS gel tell you? Why is it important to do the SDS gel electrophoresis* after *the isoelectric focusing?*

Answer

(a) *Why do you use beef heart tissue, and why do you need so much of it?* The protein you are to isolate and purify (citrate synthase [CS]) is found in mitochondria, which are abundant in cells with high metabolic activity such as heart muscle cells. Beef hearts are relatively cheap and easy to get at the local slaughterhouse. You begin with a large amount of tissue because cells contain thousands of different proteins, and no single protein is present in high concentration—that is, the specific activity of CS is low. To purify a significant quantity, you must start with a large excess of tissue.

Why do you need to keep the tissues cold? The cold temperatures inhibit the action of lysosomal enzymes that would destroy the sample.

Why is the tissue suspended in 0.2 M sucrose, at pH 7.2? Sucrose is used in the homogenization buffer to create a medium that is isotonic with the organelles. This prevents diffusion of water into the organelles, causing them to swell, burst, and spill their contents. A pH of 7.2 helps to decrease the activity of lysosomal enzymes and maintain the native structure of proteins in the sample.

What happens to the tissue when it is homogenized? Homogenization breaks open the heart muscle cells, releasing the organelles and cytosol.

(b) *What does differential centrifugation accomplish?* Organelles differ in size and therefore sediment at different rates during centrifugation. Larger organelles and pieces of cell debris sediment first, and progressively smaller cellular components can be isolated in a series of centrifugation steps at increasing speed. The contents of each fraction can be determined microscopically or by enzyme assay.

(c) *What is the rationale for the two-step addition of ammonium sulfate?* Proteins have characteristic solubilities at different salt concentrations, depending on the functional groups in the protein. In a concentration of ammonium sulfate just below the precipitation point of CS, some unwanted proteins can be precipitated (salted out). The ammonium sulfate concentration is then increased so that CS is salted out. It can then be recovered by centrifugation.

(d) *Why is a buffer solution without ammonium sulfate used for the dialysis step?* Osmolarity (as well as pH and temperature) affects the conformation and stability of proteins. To solubilize and renature the protein, the ammonium sulfate must be removed. In dialysis against a buffered solution containing no ammonium sulfate, the ammonium sulfate in the sample moves into the buffer until its concentration is equal in both solutions. By dialyzing against large volumes of buffer that are changed frequently, the concentration of ammonium sulfate in the sample can be reduced to almost zero. This procedure usually takes a long time (typically overnight). The dialysate must be buffered to keep the pH (and ionic strength) of the sample in a range that promotes the native conformation of the protein.

(e) *What does the instruction to collect the first fraction tell you about the protein?* The CS molecule is larger than the pore size of the chromatographic gel. Size-exclusion columns retard the flow of smaller molecules, which enter the pores of the column matrix material. Larger molecules flow around the matrix, taking a direct route through the column.

Why is UV absorbance at 280 nm a good way to monitor for the presence of protein in the eluted fractions? The aromatic side chains of Tyr and Trp residues strongly absorb at 280 nm.

(f) *Explain the procedure on the cation-exchange chromatography column.* CS has a positive charge (at the pH of the separation) and binds to the negatively charged beads of the cation-exchange column, while negatively charged and neutral proteins pass through. CS is displaced from the column by raising the pH of the mobile phase and thus altering the charge on the CS molecules.

(g) *Why were you unconvinced of the purity of the "single" protein band on your isoelectric focusing gel?* Several different proteins, all with the same pI, could be focused in the "single" band. SDS polyacrylamide gel electrophoresis separates on the basis of mass and therefore would separate any polypeptides in the pI 5.6 band.

Why is it important to do the SDS gel electrophoresis after the isoelectric focusing? SDS is a highly negatively charged detergent that binds tightly and uniformly along the length of a polypeptide. Removing SDS from a protein is difficult, and a protein with only traces of SDS no longer has its native acid-base properties, including its native pI.

Data Analysis Problem

23. Determining the Amino Acid Sequence of Insulin Figure 3–24 shows the amino acid sequence of the hormone insulin. This structure was determined by Frederick Sanger and his coworkers. Most of this work is described in a series of articles published in the *Biochemical Journal* from 1945 to 1955.

When Sanger and colleagues began their work in 1945, it was known that insulin was a small protein consisting of two or four polypeptide chains linked by disulfide bonds. Sanger and his coworkers had developed a few simple methods for studying protein sequences.

Treatment with FDNB. FDNB (1-fluoro-2,4-dinitrobenzene) reacted with free amino (but not amido or guanidino) groups in proteins to produce dinitrophenyl (DNP) derivatives of amino acids:

$$R{-}NH_2 + F{-}C_6H_3(NO_2){-}NO_2 \longrightarrow R{-}N(H){-}C_6H_3(NO_2){-}NO_2 + HF$$

Amine FDNB DNP-amine

Acid Hydrolysis. Boiling a protein with 10% HCl for several hours hydrolyzed all of its peptide and amide bonds. Short treatments produced short polypeptides; the longer the treatment, the more complete the breakdown of the protein into its amino acids.

Oxidation of Cysteines. Treatment of a protein with performic acid cleaved all the disulfide bonds and converted all Cys residues to cysteic acid residues (Fig. 3–26).

Paper Chromatography. This more primitive version of thin-layer chromatography (see Fig. 10–24) separated compounds based on their chemical properties, allowing identification of single amino acids and, in some cases, dipeptides. Thin-layer chromatography also separates larger peptides.

As reported in his first paper (1945), Sanger reacted insulin with FDNB and hydrolyzed the resulting protein. He found many free amino acids, but only three DNP–amino acids: α-DNP-glycine (DNP group attached to the α-amino group); α-DNP-phenylalanine; and ε-DNP-lysine (DNP attached to the ε-amino group). Sanger interpreted these results as showing that insulin had two protein chains: one with Gly at its amino terminus and one with Phe at its amino terminus. One of the two chains also contained a Lys residue, not at the amino terminus. He named the chain beginning with a Gly residue "A" and the chain beginning with Phe "B."

(a) Explain how Sanger's results support his conclusions.

(b) Are the results consistent with the known structure of insulin (Fig. 3–24)?

In a later paper (1949), Sanger described how he used these techniques to determine the first few amino acids (amino-terminal end) of each insulin chain. To analyze the B chain, for example, he carried out the following steps:

1. Oxidized insulin to separate the A and B chains.
2. Prepared a sample of pure B chain with paper chromatography.
3. Reacted the B chain with FDNB.
4. Gently acid-hydrolyzed the protein so that some small peptides would be produced.
5. Separated the DNP-peptides from the peptides that did not contain DNP groups.
6. Isolated four of the DNP-peptides, which were named B1 through B4.
7. Strongly hydrolyzed each DNP-peptide to give free amino acids.
8. Identified the amino acids in each peptide with paper chromatography.

The results were as follows:

B1: α-DNP-phenylalanine only
B2: α-DNP-phenylalanine; valine
B3: aspartic acid; α-DNP-phenylalanine; valine
B4: aspartic acid; glutamic acid; α-DNP-phenylalanine; valine

(c) Based on these data, what are the first four (amino-terminal) amino acids of the B chain? Explain your reasoning.

(d) Does this result match the known sequence of insulin (Fig. 3–24)? Explain any discrepancies.

Sanger and colleagues used these and related methods to determine the entire sequence of the A and B chains. Their sequence for the A chain was as follows (amino terminus on left):

1 5 10 15 20
Gly–Ile–Val–Glx–Glx–Cys–Cys–Ala–Ser–Val– Cys–Ser–Leu–Tyr–Glx–Leu–Glx–Asx–Tyr–Cys–Asx

Because acid hydrolysis had converted all Asn to Asp and all Gln to Glu, these residues had to be designated Asx and Glx, respectively (exact identity in the peptide unknown). Sanger solved this problem by using protease enzymes that cleave peptide bonds, but not the amide bonds in Asn and Gln residues, to prepare short peptides. He then determined the number of amide groups present in each peptide by measuring the NH_4^+ released when the peptide was acid-hydrolyzed. Some of the results for the A chain are shown below. The peptides may not have been completely pure, so the numbers were approximate—but good enough for Sanger's purposes.

Peptide name	Peptide sequence	Number of amide groups in peptide
Ac1	Cys–Asx	0.7
Ap15	Tyr–Glx–Leu	0.98
Ap14	Tyr–Glx–Leu–Glx	1.06
Ap3	Asx–Tyr–Cys–Asx	2.10
Ap1	Glx–Asx–Tyr–Cys–Asx	1.94
Ap5pa1	Gly–Ile–Val–Glx	0.15
Ap5	Gly–Ile–Val–Glx–Glx–Cys–Cys–Ala–Ser–Val–Cys–Ser–Leu	1.16

(e) Based on these data, determine the amino acid sequence of the A chain. Explain how you reached your answer. Compare it with Figure 3–24.

Answer

(a) Any linear polypeptide chain has only two kinds of free amino groups: a single α-amino group at the amino terminus, and an ε-amino group on each Lys residue present. These amino groups react with FDNB to form a DNP–amino acid derivative. Insulin gave two different α-amino-DNP derivatives, suggesting that it has two amino termini and thus two polypeptide chains—one with an amino-terminal Gly and the other with an amino-terminal Phe. Because the DNP-lysine product is ε-DNP-lysine, the Lys is not at an amino terminus.

(b) Yes. The A chain has amino-terminal Gly; the B chain has amino-terminal Phe; and (non-terminal) residue 29 in the B chain is Lys.

(c) Phe–Val–Asp–Glu–. Peptide B1 shows that the amino-terminal residue is Phe. Peptide B2 also includes Val, but since no DNP-Val is formed, Val is not at the amino terminus; it must be on the carboxyl side of Phe. Thus the sequence of B2 is DNP-Phe–Val. Similarly, the sequence of B3 must be DNP-Phe–Val–Asp, and the sequence of the A chain must begin Phe–Val–Asp–Glu–.

(d) No. The known amino-terminal sequence of the A chain is Phe–Val–Asn–Gln–. The Asn and Gln appear in Sanger's analysis as Asp and Glu because the vigorous hydrolysis in step ⑦ hydrolyzed the amide bonds in Asn and Gln (as well as the peptide bonds), forming Asp and Glu. Sanger et al. could not distinguish Asp from Asn or Glu from Gln at this stage in their analysis.

(e) The sequence exactly matches that in Figure 3–24. Each peptide in the table gives specific information about which Asx residues are Asn or Asp and which Glx residues are Glu or Gln.

Ac1: residues 20–21. This is the only Cys–Asx sequence in the A chain; there is ~1 amido group in this peptide, so it must be Cys–Asn:

N–Gly–Ile–Val–Glx–Glx–Cys–Cys–Ala–Ser–Val–Cys–Ser–Leu–Tyr–Glx–Leu–Glx–Asx–Tyr–Cys–**Asn**–*C*
1 5 10 15 20

Ap15: residues 14–15–16. This is the only Tyr–Glx–Leu in the A chain; there is ~1 amido group, so the peptide must be Tyr–Gln–Leu:

N–Gly–Ile–Val–Glx–Glx–Cys–Cys–Ala–Ser–Val–Cys–Ser–Leu–Tyr–**Gln**–Leu–Glx–Asx–Tyr–Cys–Asn–*C*
1 5 10 15 20

Ap14: residues 14–15–16–17. It has ~1 amido group, and we already know that residue 15 is Gln, so residue 17 must be Glu:

N–Gly–Ile–Val–Glx–Glx–Cys–Cys–Ala–Ser–Val–Cys–Ser–Leu–Tyr–Gln–Leu–**Glu**–Asx–Tyr–Cys–Asn–*C*
1 5 10 15 20

Ap3: residues 18–19–20–21. It has ~2 amido groups, and we know that residue 21 is Asn, so residue 18 must be Asn:

N–Gly–Ile–Val–Glx–Glx–Cys–Cys–Ala–Ser–Val–Cys–Ser–Leu–Tyr–Gln–Leu–Glu–**Asn**–Tyr–Cys–Asn–*C*
1 5 10 15 20

Ap1: residues 17–18–19–20–21, which is consistent with residues 18 and 21 being Asn.

Ap5pa1: residues 1–2–3–4. It has ~0 amido group, so residue 4 must be Glu:

N–Gly–Ile–Val–**Glu**–Glx–Cys–Cys–Ala–Ser–Val–Cys–Ser–Leu–Tyr–Gln–Leu–Glu–Asn–Tyr–Cys–Asn–*C*
1 5 10 15 20

Ap5: residues 1 through 13. It has ~1 amido group, and we know that residue 14 is Glu, so residue 5 must be Gln:

N–Gly–Ile–Val–Glu–**Gln**–Cys–Cys–Ala–Ser–Val–Cys–Ser–Leu–Tyr–Gln–Leu–Glu–Asn–Tyr–Cys–Asn–*C*
1 5 10 15 20

References

Sanger, F. (1945) The free amino groups of insulin. *Biochem. J.* **39,** 507–515.

Sanger, F. (1949) The terminal peptides of insulin. *Biochem. J.* **45,** 563–574.

chapter

4 The Three-Dimensional Structure of Proteins

1. **Properties of the Peptide Bond** In x-ray studies of crystalline peptides, Linus Pauling and Robert Corey found that the C—N bond in the peptide link is intermediate in length (1.32 Å) between a typical C—N single bond (1.49 Å) and a C═N double bond (1.27 Å). They also found that the peptide bond is planar (all four atoms attached to the C—N group are located in the same plane) and that the two α-carbon atoms attached to the C—N are always trans to each other (on opposite sides of the peptide bond).
   (a) What does the length of the C—N bond in the peptide linkage indicate about its strength and its bond order (i.e., whether it is single, double, or triple)?
   (b) What do the observations of Pauling and Corey tell us about the ease of rotation about the C—N peptide bond?

   Answer
   (a) The higher the bond order (double or triple vs. single), the shorter and stronger are the bonds. Thus, bond length is an indication of bond order. For example, the C═N bond is shorter (1.27 Å) and has a higher order ($n = 2.0$) than a typical C—N bond (length = 1.49 Å, $n = 1.0$). The length of the C—N bond of the peptide link (1.32 Å) indicates that it is intermediate in strength and bond order between a single and double bond.
   (b) Rotation about a double bond is generally impossible at physiological temperatures, and the steric relationship of the groups attached to the two atoms involved in the double bond is spatially "fixed." Since the peptide bond has considerable double-bond character, there is essentially no rotation, and the —C═O and —N—H groups are fixed in the trans configuration.

2. **Structural and Functional Relationships in Fibrous Proteins** William Astbury discovered that the x-ray diffraction pattern of wool shows a repeating structural unit spaced about 5.2 Å along the length of the wool fiber. When he steamed and stretched the wool, the x-ray pattern showed a new repeating structural unit at a spacing of 7.0 Å. Steaming and stretching the wool and then letting it shrink gave an x-ray pattern consistent with the original spacing of about 5.2 Å. Although these observations provided important clues to the molecular structure of wool, Astbury was unable to interpret them at the time.
   (a) Given our current understanding of the structure of wool, interpret Astbury's observations.
   (b) When wool sweaters or socks are washed in hot water or heated in a dryer, they shrink. Silk, on the other hand, does not shrink under the same conditions. Explain.

Answer

(a) The principal structural units in the wool fiber polypeptide, α-keratin, are successive turns of the α helix, which are spaced at 5.4 Å intervals; two α-keratin strands twisted into a coiled coil produce the 5.2 Å spacing. The intrinsic stability of the helix (and thus the fiber) results from *intra* chain hydrogen bonds (see Fig. 4–4a). Steaming and stretching the fiber yields an extended polypeptide chain with the β conformation, in which the distance between adjacent R groups is about 7.0 Å. Upon resteaming, the polypeptide chains again assume the less-extended α-helix conformation.

(b) Freshly sheared wool is primarily in its α-keratin (α-helical coiled coil) form (see Fig. 4–10). Because raw wool is crimped or curly, it is combed and stretched to straighten it before being spun into fibers for clothing. This processing converts the wool from its native α-helical conformation to a more extended β form. Moist heat triggers a conformational change back to the native α-helical structure, which shrinks both the fiber and the clothing. Under conditions of mechanical tension and moist heat, wool can be stretched back to a fully extended form. In silk, by contrast, the polypeptide chains have a very stable β-pleated sheet structure, fully extended along the axis of the fiber (see Fig. 4–6), and have small, closely packed amino acid side chains (see Fig. 4–13). These characteristics make silk resistant to stretching and shrinking.

3. Rate of Synthesis of Hair α-Keratin Hair grows at a rate of 15 to 20 cm/yr. All this growth is concentrated at the base of the hair fiber, where α-keratin filaments are synthesized inside living epidermal cells and assembled into ropelike structures (see Fig. 4–10). The fundamental structural element of α-keratin is the α helix, which has 3.6 amino acid residues per turn and a rise of 5.4 Å per turn (see Fig. 4–4a). Assuming that the biosynthesis of α-helical keratin chains is the rate-limiting factor in the growth of hair, calculate the rate at which peptide bonds of α-keratin chains must be synthesized (peptide bonds per second) to account for the observed yearly growth of hair.

Answer Because there are 3.6 amino acids (AAs) per turn and the rise is 5.4 Å/turn, the length per AA of the α helix is

$$\frac{5.4 \text{ Å/turn}}{3.6 \text{ AA/turn}} = 1.5 \text{ Å/AA} = 1.5 \times 10^{-10} \text{ m/AA}$$

A growth rate of 20 cm/yr is equivalent to

$$\frac{20 \text{ cm/year}}{(365 \text{ days/yr})(24 \text{ h/day})(60 \text{ min/h})(60 \text{ s/min})} = 6.3 \times 10^{-7} \text{ cm/s} = 6.3 \times 10^{-9} \text{ m/s}$$

Thus, the rate at which amino acids are added is

$$\frac{6.3 \times 10^{-9} \text{ m/s}}{1.5 \times 10^{-10} \text{ m/AA}} = 42 \text{ AA/s} = 42 \text{ peptide bonds per second}$$

4. Effect of pH on the Conformation of α-Helical Secondary Structures The unfolding of the α helix of a polypeptide to a randomly coiled conformation is accompanied by a large decrease in a property called specific rotation, a measure of a solution's capacity to rotate plane-polarized light. Polyglutamate, a polypeptide made up of only L-Glu residues, has the α-helical conformation at pH 3.

When the pH is raised to 7, there is a large decrease in the specific rotation of the solution. Similarly, polylysine (L-Lys residues) is an α helix at pH 10, but when the pH is lowered to 7 the specific rotation also decreases, as shown by the following graph.

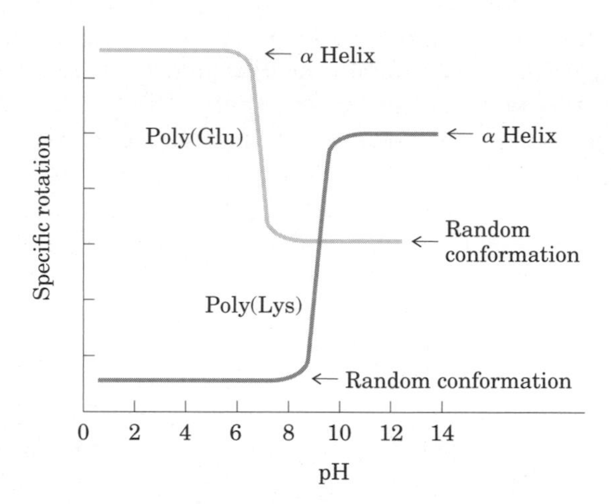

What is the explanation for the effect of the pH changes on the conformations of poly(Glu) and poly(Lys)? Why does the transition occur over such a narrow range of pH?

Answer At pH values above 6, deprotonation of the carboxylate side chains of poly(Glu) leads to repulsion between adjacent negatively charged groups, which destabilizes the α helix and results in unfolding. Similarly, at pH 7 protonation of the amino-group side chains of poly(Lys) causes repulsion between positively charged groups, which leads to unfolding.

5. **Disulfide Bonds Determine the Properties of Many Proteins** Some natural proteins are rich in disulfide bonds, and their mechanical properties (tensile strength, viscosity, hardness, etc.) are correlated with the degree of disulfide bonding.

(a) Glutenin, a wheat protein rich in disulfide bonds, is responsible for the cohesive and elastic character of dough made from wheat flour. Similarly, the hard, tough nature of tortoise shell is due to the extensive disulfide bonding in its α-keratin. What is the molecular basis for the correlation between disulfide-bond content and mechanical properties of the protein?

(b) Most globular proteins are denatured and lose their activity when briefly heated to 65 °C. However, globular proteins that contain multiple disulfide bonds often must be heated longer at higher temperatures to denature them. One such protein is bovine pancreatic trypsin inhibitor (BPTI), which has 58 amino acid residues in a single chain and contains three disulfide bonds. On cooling a solution of denatured BPTI, the activity of the protein is restored. What is the molecular basis for this property?

Answer

(a) Disulfide bonds are covalent bonds, which are much stronger than the noncovalent interactions (hydrogen bonds, hydrophobic interactions, van der Waals interactions) that stabilize the three-dimensional structure of most proteins. Disulfide bonds serve to cross-link protein chains, increasing stiffness, hardness, and mechanical strength.

(b) As the temperature is raised, the increased thermal motion of the polypeptide chains and the vibrational motions of hydrogen bonds ultimately lead to thermal denaturation (unfolding) of a protein. Cystine residues (disulfide bonds) can, depending on their location in the protein structure, prevent or restrict the movement of folded protein domains, block access of solvent water to the interior of the protein, and prevent the complete unfolding of the protein. Refolding to the native structure from a random conformation is seldom spontaneous, owing to the very large number of conformations possible. Disulfide bonds limit the number of conformations by allowing only a few minimally unfolded structures, and hence the protein returns to its native conformation more easily upon cooling.

6. Amino Acid Sequence and Protein Structure Our growing understanding of how proteins fold allows researchers to make predictions about protein structure based on primary amino acid sequence data. Consider the following amino acid sequence.

Ile–	Ala–	His–	Thr–	Tyr–	Gly–	Pro–	Phe–	Glu–	Ala–	Ala–	Met–	Cys–	Lys–	Trp–	Glu–	Ala–	Gln–	Pro–	Asp–	Gly–	Met–	Glu–	Cys–	Ala–	Phe–	His–	Arg
1	2	3	4	5	6	7	8	9	10	11	12	13	14	15	16	17	18	19	20	21	22	23	24	25	26	27	28

(a) Where might bends or β turns occur?
(b) Where might intrachain disulfide cross-linkages be formed?
(c) Assuming that this sequence is part of a larger globular protein, indicate the probable location (the external surface or interior of the protein) of the following amino acid residues: Asp, Ile, Thr, Ala, Gln, Lys. Explain your reasoning. (Hint: See the hydropathy index in Table 3–1.)

Answer

(a) Bends or turns are most likely to occur at residues 7 and 19 because Pro residues are often (but not always) found at bends in globular folded proteins. A bend may also occur at the Thr residue (residue 4) and, assuming that this is a portion of a larger polypeptide, at the Ile residue (residue 1).
(b) Intrachain disulfide cross-linkages can form only between residues 13 and 24 (Cys residues).
(c) Amino acids with ionic (charged) or strongly polar neutral groups (e.g., Asp, Gln, and Lys in this protein) are located on the external surface, where they interact optimally with solvent water. Residues with nonpolar side chains (such as Ala and Ile) are situated in the interior, where they escape the polar environment. Thr is of intermediate polarity and could be found either in the interior or on the exterior surface (see Table 3–1).

7. Bacteriorhodopsin in Purple Membrane Proteins Under the proper environmental conditions, the salt-loving bacterium *Halobacterium halobium* synthesizes a membrane protein (M_r 26,000) known as bacteriorhodopsin, which is purple because it contains retinal (see Fig. 10–21). Molecules of this protein aggregate into "purple patches" in the cell membrane. Bacteriorhodopsin acts as a light-activated proton pump that provides energy for cell functions. X-ray analysis of this protein reveals that it consists of seven parallel α-helical segments, each of which traverses the bacterial cell membrane (thickness 45 Å). Calculate the minimum number of amino acids necessary for one segment of α helix to traverse the membrane completely. Estimate the fraction of the bacteriorhodopsin protein that is involved in membrane-spanning helices. (Use an average amino acid residue weight of 110.)

Answer Using the parameters from Problem 3 (3.6 AA/turn, 5.4 Å/turn), we can calculate that there are 0.67 AA/Å along the axis of a helix. Thus, a helix of length 45 Å (sufficient to span the membrane) requires a minimum of (45 Å)(0.67 AA/Å) = 30 amino acid residues.

The membrane protein has an M_r of 26,000 and average AA M_r of 110. Thus the protein contains 26,000/110 = 240 AA. Of these, (30 AA/helix)(7 helices) = 210 AA are involved in membrane-spanning helices, which is 210/240 = 0.87, or 87%, of the protein.

8. Protein Structure Terminology Is myoglobin a motif, a domain, or a complete three-dimensional structure?

Answer Myoglobin is all three. The folded structure, the "globin fold," is a motif found in all globins. The polypeptide folds into a single domain, which for this protein represents the entire three-dimensional structure.

9. Pathogenic Action of Bacteria That Cause Gas Gangrene The highly pathogenic anaerobic bacterium *Clostridium perfringens* is responsible for gas gangrene, a condition in which animal tissue structure is destroyed. This bacterium secretes an enzyme that efficiently catalyzes the hydrolysis of the peptide bond indicated by an asterisk:

$$—X\overset{*}{—}Gly—Pro—Y— \xrightarrow{H_2O} —X—COO^- + H_3\overset{+}{N}—Gly—Pro—Y—$$

where X and Y are any of the 20 common amino acids. How does the secretion of this enzyme contribute to the invasiveness of this bacterium in human tissues? Why does this enzyme not affect the bacterium itself?

Answer Collagen is distinctive in its amino acid composition, having a very high proportion of Gly (35%) and Pro residues. The enzyme secreted by the bacterium is a collagenase, which breaks down collagen at the X–Gly bonds and damages the connective-tissue barrier (skin, hide, etc.) of the host; this allows the bacterium to invade the host tissues. Bacteria do not contain collagen and thus are unaffected by collagenase.

10. Number of Polypeptide Chains in a Multisubunit Protein A sample (660 mg) of an oligomeric protein of M_r 132,000 was treated with an excess of 1-fluoro-2,4-dinitrobenzene (Sanger's reagent) under slightly alkaline conditions until the chemical reaction was complete. The peptide bonds of the protein were then completely hydrolyzed by heating it with concentrated HCl. The hydrolysate was found to contain 5.5 mg of the following compound:

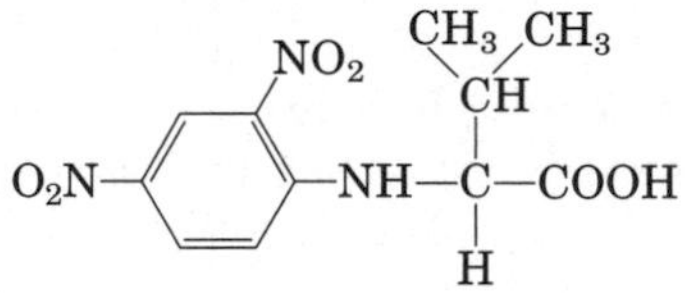

NO_2 ; CH_3 CH_3 ; CH

O_2N—(benzene ring)—NH—C—COOH

H

2,4-Dinitrophenyl derivatives of the α-amino groups of other amino acids could not be found.

(a) Explain how this information can be used to determine the number of polypeptide chains in an oligomeric protein.

(b) Calculate the number of polypeptide chains in this protein.

(c) What other protein analysis technique could you employ to determine whether the polypeptide chains in this protein are similar or different?

Answer

(a) Because only a single 2,4-dinitrophenyl (DNP) amino acid derivative is found, there is only one kind of amino acid at the amino terminus (i.e., all the polypeptide chains have the same amino-terminal residue). Comparing the number of moles of this derivative to the number of moles of protein gives the number of polypeptide chains.

(b) The amount of protein = (0.66 g)/(132,000 g/mol) = 5×10^{-6} mol.
Because M_r for DNP-Val ($C_{11}H_{13}O_6N_3$) = 283, the amount of DNP-Val = (0.0055 g)/(283 g/mol) = 1.9×10^{-5} mol.

The ratio of moles of DNP-Val to moles of protein gives the number of amino-terminal residues and thus the number of chains per oligomer:

$$\frac{1.9 \times 10^{-5} \text{ mol DNP-Val}}{5 \times 10^{-6} \text{ mol protein}} = 4 \text{ polypeptide chains}$$

An alternative approach to the problem is through the proportionality (n = number of polypeptide chains):

$$\frac{n(283 \text{ g/mol})}{132{,}000 \text{ g/mol}} = \frac{5.5 \text{ mg}}{660 \text{ mg}}$$

$$n = \frac{(5.5 \text{ mg})(132{,}000 \text{ g/mol})}{(660 \text{ mg})(283 \text{ g/mol})} = 3.9 \approx 4$$

(c) Polyacrylamide gel electrophoresis in the presence of a detergent (such as sodium dodecylsulfate [SDS]) and an agent that prevents the formation of disulfide bonds (such as β-mercaptoethanol) would provide information on subunit structure of a protein. In the example here, an oligomeric protein of M_r 132,000 that had four *identical* subunits would produce a single band on the electrophoretic gel, with apparent M_r ~33,000 (132,000/33,000 = 4). If the protein were made up of different polypeptide subunits, they would likely appear as multiple discrete bands on the gel.

11. Predicting Secondary Structure Which of the following peptides is more likely to take up an α-helical structure, and why?

(a) LKAENDEAARAMSEA

(b) CRAGGFPWDQPGTSN

Answer By cursory inspection, peptide **(a)** has five Ala residues (most likely to take up an α-helical conformation), and peptide **(b)** has five Pro and Gly residues (least often found in an α helix). This suggests that **(a)** is more likely than **(b)** to form an α helix. Referring to Table 4–1, **(a)** has 15 residues with a total $\Delta\Delta G^\circ$ of 13 kJ/mol, and **(b)** has 15 residues with a total $\Delta\Delta G^\circ$ of 41 kJ/mol. Given that a lower $\Delta\Delta G^\circ$ indicates a greater tendency to take up an α-helical structure, this confirms that peptide **(a)** is much more likely to form an α helix.

12. Amyloid Fibers in Disease Several small aromatic molecules, such as phenol red (used as a non-toxic drug model), have been shown to inhibit the formation of amyloid in laboratory model systems. A goal of the research on these small aromatic compounds is to find a drug that would efficiently inhibit the formation of amyloid in the brain in people with incipient Alzheimer's disease.

(a) Suggest why molecules with aromatic substituents would disrupt the formation of amyloid.

(b) Some researchers have suggested that a drug used to treat Alzheimer's disease may also be effective in treating type 2 (adult onset) diabetes mellitus. Why might a single drug be effective in treating these two different conditions?

Answer

(a) Aromatic residues seem to play an important role in stabilizing amyloid fibrils. Thus, molecules with aromatic substituents may inhibit amyloid formation by interfering with the stacking or association of the aromatic side chains.

(b) Amyloid is formed in the pancreas in association with type 2 diabetes, as it is in the brain in Alzheimer's disease. Although the amyloid fibrils in the two diseases involve different proteins, the fundamental structure of the amyloid is similar and similarly stabilized in both, and thus they are potential targets for similar drugs designed to disrupt this structure.

Biochemistry on the Internet

13. Protein Modeling on the Internet A group of patients with Crohn's disease (an inflammatory bowel disease) underwent biopsies of their intestinal mucosa in an attempt to identify the causative agent. Researchers identified a protein that was present at higher levels in patients with Crohn's disease than in patients with an unrelated inflammatory bowel disease or in unaffected controls. The protein was isolated, and the following *partial* amino acid sequence was obtained (reads left to right):

EAELCPDRCI HSFQNLGIQC VKKRDLEQAI SQRIQTNNNP FQVPIEEQRG
DYDLNAVRLC FQVTVRDPSG RPLRLPPVLP HPIFDNRAPN TAELKICRVN
RNSGSCLGGD EIFLLCDKVQ KEDIEVYFTG PGWEARGSFS QADVHRQVAI
VFRTPPYADP SLQAPVRVSM QLRRPSDREL SEPMEFQYLP DTDDRHRIEE
KRKRTYETFK SIMKKSPFSG PTDPRPPPRR IAVPSRSSAS VPKPAPQPYP

(a) You can identify this protein using a protein database on the Internet. Some good places to start include Protein Information Resource (PIR; http://pir.georgetown.edu), Structural Classification of Proteins (SCOP; http://scop.mrc-lmb.cam.ac.uk/scop), and Prosite (http://expasy.org/prosite).
At your selected database site, follow links to the sequence comparison engine. Enter about 30 residues from the protein sequence in the appropriate search field and submit it for analysis. What does this analysis tell you about the identity of the protein?

(b) Try using different portions of the amino acid sequence. Do you always get the same result?

(c) A variety of websites provide information about the three-dimensional structure of proteins. Find information about the protein's secondary, tertiary, and quaternary structures using database sites, such as the Protein Data Bank (PDB; www.rcsb.org) or SCOP.

(d) In the course of your Web searches, what did you learn about the cellular function of the protein?

Answer

(a) At the PIR—International Protein Sequence Database (http://pir.georgetown.edu), click on "Search/Analysis" and choose "BLAST search." Paste the first 30 amino acid residues of the sequence into the search box and submit the sequence for comparison. The table that returns includes many proteins that have 100% sequence identity with these 30 residues. Among the human proteins are RelA and the transcription factor NFκB. Proteins from other species match as well. Click on the "Help" button for explanations of the various options and table items.

(b) As more proteins are sequenced, the number of hits returned from a 30-residue sequence increases. Sequence matching based on the first 30 residues brings up several proteins that contain this sequence (identity = 100%). Using sequence segments from different parts of the protein will return some different results, but the proteins with high sequence identities will likely be similar. Even when the *entire* sequence is entered in the search field, similar proteins from cattle, mouse, and rat match with very high scores, and several hundred hits are returned. When the entire sequence is used, the human protein with the best match is the p65 subunit of nuclear transcription factor kappa B (NFκB). A synonym for this protein is RelA transforming protein.

(c) At the PDB (rcsb.org) search on "NF-kappa-B p65." You will get more than a dozen hits. Adding "human" to the search limits the results further. Go back to the more general search on "NF-kappa-B p65" and scan through the returned items. NFκB has two subunits. There are multiple variants of the subunits, with the best-characterized being 50, 52, and 65 kDa (p50, p52, and p65, respectively). These pair with each other to form a variety of homodimers and heterodimers.

(d) The various proteins that predominate in this search are eukaryotic transcription factors, which stimulate transcription of genes involved in development and some immune responses. The proteins have two distinct domains, including an amino-terminal Rel homology domain 300 amino acid residues long and a carboxyl-terminal domain involved in gene activation. A search of the various links in the databases will reveal much additional information about the proteins' structure and function.

Data Analysis Problem

14. Mirror-Image Proteins As noted in Chapter 3, "The amino acid residues in protein molecules are exclusively L stereoisomers." It is not clear whether this selectivity is necessary for proper protein function or is an accident of evolution. To explore this question, Milton and colleagues (1992) published a study of an enzyme made entirely of D stereoisomers. The enzyme they chose was HIV protease, a proteolytic enzyme made by HIV that converts inactive viral pre-proteins to their active forms.

Previously, Wlodawer and coworkers (1989) had reported the complete chemical synthesis of HIV protease from L-amino acids (the L-enzyme), using the process shown in Figure 3–29. Normal HIV protease contains two Cys residues at positions 67 and 95. Because chemical synthesis of proteins containing Cys is technically difficult, Wlodawer and colleagues substituted the synthetic amino acid L-α-amino-n-butyric acid (Aba) for the two Cys residues in the protein. In the authors' words, this was done so as to "reduce synthetic difficulties associated with Cys deprotection and ease product handling."

(a) The structure of Aba is shown below. Why was this a suitable substitution for a Cys residue? Under what circumstances would it not be suitable?

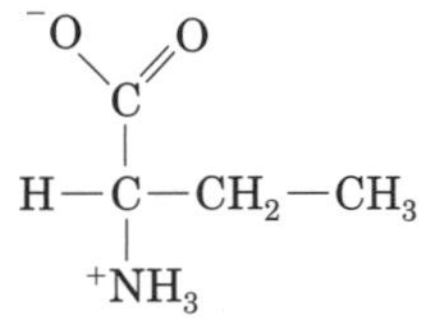

$$\begin{array}{c} {}^-O \quad\;\; O \\ \diagdown \;\; /\!\!/ \\ C \\ | \\ H-C-CH_2-CH_3 \\ | \\ {}^+NH_3 \end{array}$$

L-α-Amino-n-butyric acid

Wlodawer and coworkers denatured the newly synthesized protein by dissolving it in 6 M guanidine HCl, and then allowed it to fold slowly by dialyzing away the guanidine against a neutral buffer (10% glycerol, 25 mM $NaPO_4$, pH 7).

(b) There are many reasons to predict that a protein synthesized, denatured, and folded in this manner would not be active. Give three such reasons.

(c) Interestingly, the resulting L-protease was active. What does this finding tell you about the role of disulfide bonds in the native HIV protease molecule?

In their new study, Milton and coworkers synthesized HIV protease from D-amino acids, using the same protocol as the earlier study (Wlodawer et al.). Formally, there are three possibilities for the folding of the D-protease: it would give (1) the same shape as the L-protease, (2) the mirror image of the L-protease, or (3) something else, possibly inactive.

(d) For each possibility, decide whether or not it is a likely outcome and defend your position.

In fact, the D-protease was active: it cleaved a particular synthetic substrate and was inhibited by specific inhibitors. To examine the structure of the D- and L-enzymes, Milton and coworkers tested both forms for activity with D and L forms of a chiral peptide substrate and for inhibition by D and L forms of a chiral peptide-analog inhibitor. Both forms were also tested for inhibition by the achiral inhibitor Evans blue. The findings are given in the table.

HIV protease	Substrate hydrolysis		Inhibition		
			Peptide inhibitor		Evans blue (achiral)
	D-substrate	L-protease	D-inhibitor	L-inhibitor	
L-protease	−	+	−	+	+
D-protease	+	−	+	−	+

(e) Which of the three models proposed above is supported by these data? Explain your reasoning.

(f) Why does Evans blue inhibit both forms of the protease?

(g) Would you expect chymotrypsin to digest the D-protease? Explain your reasoning.

(h) Would you expect total synthesis from D-amino acids followed by renaturation to yield active enzyme for any enzyme? Explain your reasoning.

Answer

(a) Aba is a suitable replacement because Aba and Cys have approximately the same sized side chain and are similarly hydrophobic. However, Aba cannot form disulfide bonds so it will not be a suitable replacement if these are required.

(b) There are many important differences between the synthesized protein and HIV protease produced by a human cell, any of which could result in an inactive synthetic enzyme: (1) Although Aba and Cys have similar size and hydrophobicity, Aba may not be similar enough for the protein to fold properly. (2) HIV protease may require disulfide bonds for proper functioning. (3) Many proteins synthesized by ribosomes fold as they are produced; the protein in this study folded only after the chain was complete. (4) Proteins synthesized by ribosomes may interact with the ribosomes as they fold; this is not possible for the protein in the study. (5) Cytosol is a more complex solution than the buffer used in the study; some proteins may require specific, unknown proteins for proper folding. (6) Proteins synthesized in cells often require chaperones for proper folding; these are not present in the study buffer. (7) In cells, HIV protease is synthesized as part of a larger chain that is then proteolytically processed; the protein in the study was synthesized as a single molecule.

(c) Because the enzyme *is* functional with Aba substituted for Cys, disulfide bonds do not play an important role in the structure of HIV protease.

(d) *Model 1*: it would fold like the L-protease. *Argument for*: the covalent structure is the same (except for chirality), so it should fold like the L-protease. *Argument against*: chirality is not a trivial detail; three-dimensional shape is a key feature of biological molecules. The synthetic enzyme will not fold like the L-protease. *Model 2*: it would fold to the mirror image of the L-protease. *For*: because the individual components are mirror images of those in the biological protein, it will fold in the mirror-image shape. *Against*: the interactions involved in protein folding are very complex, so the synthetic protein will most likely fold in another form. *Model* 3: it would fold to something else. *For*: the interactions involved in protein folding are very complex, so the synthetic protein will most likely fold in another form. *Against*: because the individual components are mirror images of those in the biological protein, it will fold in the mirror-image shape.

(e) Model 1. The enzyme is active, but with the enantiomeric form of the biological substrate, and it is inhibited by the enantiomeric form of the biological inhibitor. This is consistent with the D-protease being the mirror image of the L-protease.

(f) Evans blue is achiral; it binds to both forms of the enzyme.

(g) No. Because proteases contain only L-amino acids and recognize only L-peptides, chymotrypsin would not digest the D-protease.

(h) Not necessarily. Depending on the individual enzyme, any of the problems listed in **(b)** could result in an inactive enzyme.

References

Milton, R. C., Milton, S. C., & Kent, S. B. (1992) Total chemical synthesis of a D-enzyme: the enantiomers of HIV-1 protease show demonstration of reciprocal chiral substrate specificity. *Science* **256,** 1445–1448.

Wlodawer, A., Miller, M., Jaskólski, M., Sathyanarayana, B. K., Baldwin, E., Weber, I. T., Selk, L. M., Clawson, L., Schneider, J., & Kent, S. B. (1989) Conserved folding in retroviral proteases: crystal structure of a synthetic HIV-1 protease. *Science* **245,** 616–621.

chapter

5 Protein Function

1. **Relationship between Affinity and Dissociation Constant** Protein A has a binding site for ligand X with a K_d of 10^{-6} M. Protein B has a binding site for ligand X with a K_d of 10^{-9} M. Which protein has a higher affinity for ligand X? Explain your reasoning. Convert the K_d to K_a for both proteins.

    Answer Protein B has a higher affinity for ligand X. The lower K_d indicates that protein B will be half-saturated with bound ligand X at a much lower concentration of X than will protein A. Because $K_a = 1/K_d$, protein A has $K_a = 10^6$ M^{-1}; protein B has $K_a = 10^9$ M^{-1}.

2. **Negative Cooperativity** Which of the following situations would produce a Hill plot with $n_H < 1.0$? Explain your reasoning in each case.
    (a) The protein has multiple subunits, each with a single ligand-binding site. Binding of ligand to one site decreases the binding affinity of other sites for the ligand.
    (b) The protein is a single polypeptide with two ligand-binding sites, each having a different affinity for the ligand.
    (c) The protein is a single polypeptide with a single ligand-binding site. As purified, the protein preparation is heterogeneous, containing some protein molecules that are partially denatured and thus have a lower binding affinity for the ligand.

    Answer All three situations would produce $n_H < 1.0$. An n_H (Hill coefficient) of <1.0 generally suggests situation **(a)**—the classic case of negative cooperativity. However, closer examination of the properties of a protein exhibiting apparent negative cooperativity in ligand binding often reveals situation **(b)** or **(c).** When two or more types of ligand-binding sites with different affinities for the ligand are present on the same or different proteins in the same solution, apparent negative cooperativity is observed. In **(b),** the higher-affinity ligand-binding sites bind the ligand first. As the ligand concentration is increased, binding to the lower-affinity sites produces an $n_H < 1.0$, even though binding to the two ligand-binding sites is completely independent. Even more common is situation **(c),** in which the protein preparation is heterogeneous. Unsuspected proteolytic digestion by contaminating proteases and partial denaturation of the protein under certain solvent conditions are common artifacts of protein purification. There are few well-documented cases of *true* negative cooperativity.

3. **Affinity for Oxygen of Hemoglobin** What is the effect of the following changes on the O_2 affinity of hemoglobin? **(a)** A drop in the pH of blood plasma from 7.4 to 7.2. **(b)** A decrease in the partial pressure of CO_2 in the lungs from 6 kPa (holding one's breath) to 2 kPa (normal). **(c)** An increase in the BPG level from 5 mM (normal altitudes) to 8 mM (high altitudes). **(d)** An increase in CO from 1.0 parts per million (ppm) in a normal indoor atmosphere to 30 ppm in a home that has a malfunctioning or leaking furnace.

Answer The affinity of hemoglobin for O_2 is regulated by the binding of the ligands H^+, CO_2, and BPG. The binding of each ligand shifts the O_2-saturation curve to the right—that is, the O_2 affinity of hemoglobin is reduced in the presence of ligand. **(a)** decreases the affinity; **(b)** increases the affinity; **(c)** decreases the affinity; **(d)** decreases the affinity.

4. **Reversible Ligand Binding** The protein calcineurin binds to the protein calmodulin with an association rate of $8.9 \times 10^3\ \text{M}^{-1}\text{s}^{-1}$ and an overall dissociation constant, K_d, of 10 nM. Calculate the dissociation rate, k_d, including appropriate units.

Answer K_d, the dissociation constant, is the ratio of k_d, the rate constant for the dissociation reaction, to k_a, the rate constant for the association reaction.

$$K_d = k_d/k_a$$

Rearrange to solve for k_d and substitute the known values.

$$k_d = K_d \times k_a = (10 \times 10^{-9}\ \text{M})(8.9 \times 10^3\ \text{M}^{-1}\text{s}^{-1}) = 8.9 \times 10^{-5}\ \text{s}^{-1}$$

5. **Cooperativity in Hemoglobin** Under appropriate conditions, hemoglobin dissociates into its four subunits. The isolated α subunit binds oxygen, but the O_2-saturation curve is hyperbolic rather than sigmoid. In addition, the binding of oxygen to the isolated α subunit is not affected by the presence of H^+, CO_2, or BPG. What do these observations indicate about the source of the cooperativity in hemoglobin?

Answer These observations indicate that the cooperative behavior—the sigmoid O_2-binding curve and the positive cooperativity in ligand binding—of hemoglobin arises from interaction between subunits.

6. **Comparison of Fetal and Maternal Hemoglobins** Studies of oxygen transport in pregnant mammals show that the O_2-saturation curves of fetal and maternal blood are markedly different when measured under the same conditions. Fetal erythrocytes contain a structural variant of hemoglobin, HbF, consisting of two α and two γ subunits ($\alpha_2\gamma_2$), whereas maternal erythrocytes contain HbA ($\alpha_2\beta_2$).

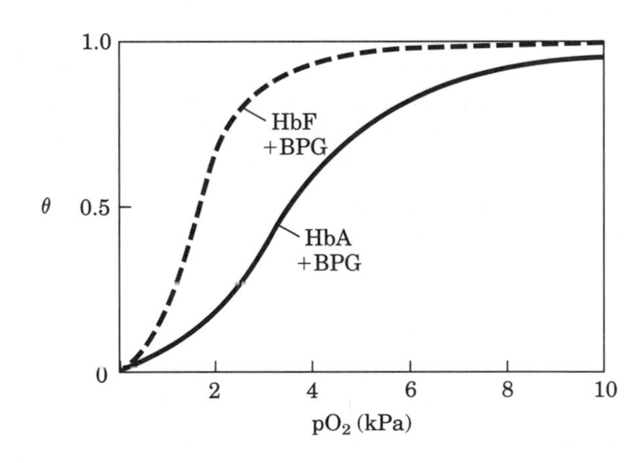

(a) Which hemoglobin has a higher affinity for oxygen under physiological conditions, HbA or HbF? Explain.

(b) What is the physiological significance of the different O_2 affinities?

(c) When all the BPG is carefully removed from samples of HbA and HbF, the measured O_2-saturation curves (and consequently the O_2 affinities) are displaced to the left. However, HbA now has a greater affinity for oxygen than does HbF. When BPG is reintroduced, the O_2-saturation curves return to normal, as shown in the graph. What is the effect of BPG on the O_2 affinity of hemoglobin? How can the above information be used to explain the different O_2 affinities of fetal and maternal hemoglobin?

Answer

(a) The observation that hemoglobin A (HbA; maternal) is about 60% saturated at $pO_2 = 4$ kPa (the pO_2 in tissues), whereas hemoglobin F (HbF; fetal) is more than 90% saturated under the same physiological conditions, indicates that HbF has a higher O_2 affinity than HbA. In other words, at identical O_2 concentrations, HbF binds more oxygen than does HbA. Thus, HbF must bind oxygen more tightly (with higher affinity) than HbA under physiological conditions.

(b) The higher O_2 affinity of HbF ensures that oxygen will flow from maternal blood to fetal blood in the placenta. For maximal O_2 transport, the oxygen pressure at which fetal blood approaches full saturation must be in the region where the O_2 affinity of HbA is low. This is indeed the case.

(c)

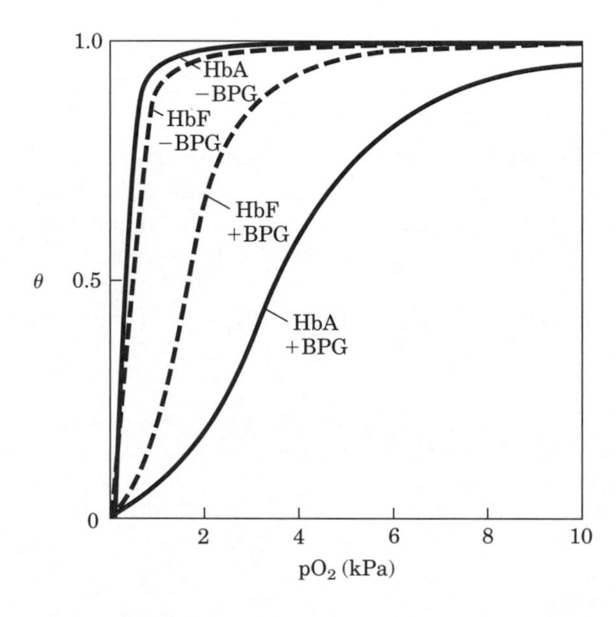

Binding of BPG to hemoglobin reduces the affinity of hemoglobin for O_2, as shown in the graph. The O_2-saturation curve for HbA shifts far to the right when BPG binds (solid curves)—that is, the O_2 affinity is dramatically lowered. The O_2-saturation curve for HbF also shifts to the right when BPG binds (dashed curves), but not as far. Because the O_2-saturation curve of HbA undergoes a larger shift on BPG binding than does that of HbF, we can conclude that HbA binds BPG more tightly than does HbF. Differential binding of BPG to the two hemoglobins may determine the difference in their O_2 affinities.

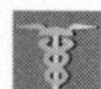

7. **Hemoglobin Variants** There are almost 500 naturally occurring variants of hemoglobin. Most are the result of a single amino acid substitution in a globin polypeptide chain. Some variants produce clinical illness, though not all variants have deleterious effects. A brief sample follows:

HbS (sickle-cell Hb): substitutes a Val for a Glu on the surface

Hb Cowtown: eliminates an ion pair involved in T-state stabilization

Hb Memphis: substitutes one uncharged polar residue for another of similar size on the surface

Hb Bibba: substitutes a Pro for a Leu involved in an α helix

Hb Milwaukee: substitutes a Glu for a Val

Hb Providence: substitutes an Asn for a Lys that normally projects into the central cavity of the tetramer

Hb Philly: substitutes a Phe for a Tyr, disrupting hydrogen bonding at the $\alpha_1\beta_1$ interface

Explain your choices for each of the following:

(a) The Hb variant *least* likely to cause pathological symptoms.

(b) The variant(s) most likely to show pI values different from that of HbA on an isoelectric focusing gel.

(c) The variant(s) most likely to show a decrease in BPG binding and an increase in the overall affinity of the hemoglobin for oxygen.

Answer

(a) Hb Memphis; it has a conservative substitution that is unlikely to have a significant effect on function.

(b) HbS, Hb Milwaukee, and Hb Providence; all have substitutions that alter the net charge on the protein, which will change the pI. The loss of an ion pair in Hb Cowtown may indicate loss of a charged residue, which would also change the pI, but there is not enough information to be sure.

(c) Hb Providence; it has an Asn residue in place of a Lys that normally projects into the central cavity of hemoglobin. Loss of the positively charged Lys that normally interacts with the negative charges on BPG results in Hb Providence having lower affinity for BPG and thus higher affinity for O_2.

8. Oxygen Binding and Hemoglobin Structure A team of biochemists uses genetic engineering to modify the interface region between hemoglobin subunits. The resulting hemoglobin variants exist in solution primarily as $\alpha\beta$ dimers (few, if any, $\alpha_2\beta_2$ tetramers form). Are these variants likely to bind oxygen more weakly or more tightly? Explain your answer.

Answer More tightly. An inability to form tetramers would limit the cooperativity of these variants, and the binding curve would become more hyperbolic. Also, the BPG-binding site would be disrupted. Oxygen binding would probably be tighter, because the default state in the absence of bound BPG is the tight-binding R state.

9. Reversible (but Tight) Binding to an Antibody An antibody binds to an antigen with a K_d of 5×10^{-8} M. At what concentration of antigen will θ be **(a)** 0.2, **(b)** 0.5, **(c)** 0.6, **(d)** 0.8?

Answer **(a)** 1×10^{-8} M, **(b)** 5×10^{-8} M, **(c)** 8×10^{-8} M, **(d)** 2×10^{-7} M. These are calculated from a rearrangement of Equation 5–8 to give $[L] = \theta K_d/(1 - \theta)$, and for this antigen-antibody binding, $[L] = \theta(5 \times 10^{-8}\ \text{M})/(1 - \theta)$. For example, for **(a)** $[L] = 0.2(5 \times 10^{-8}\ \text{M})/(0.8) = 1 \times 10^{-8}$ M.

10. Using Antibodies to Probe Structure-Function Relationships in Proteins A monoclonal antibody binds to G-actin but not to F-actin. What does this tell you about the epitope recognized by the antibody?

Answer The epitope is likely to be a structure that is buried when G-actin polymerizes to form F-actin.

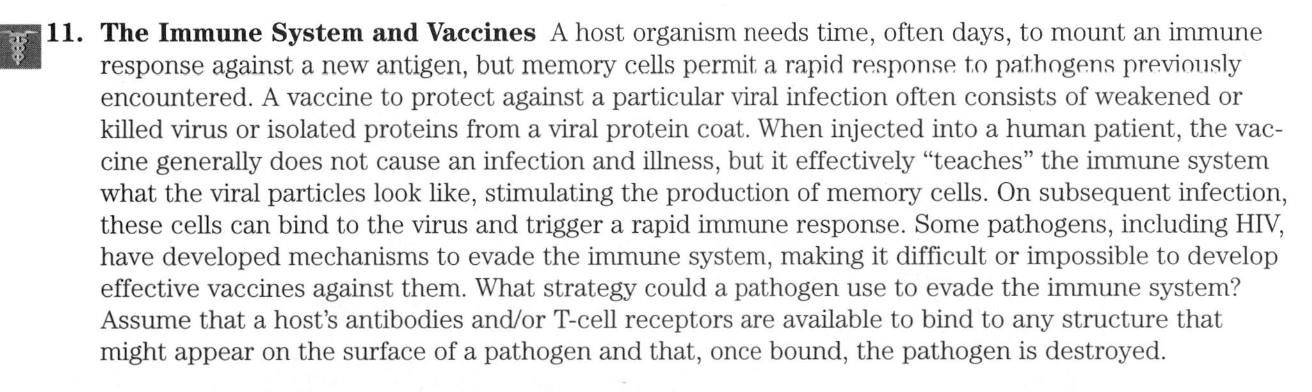

11. The Immune System and Vaccines A host organism needs time, often days, to mount an immune response against a new antigen, but memory cells permit a rapid response to pathogens previously encountered. A vaccine to protect against a particular viral infection often consists of weakened or killed virus or isolated proteins from a viral protein coat. When injected into a human patient, the vaccine generally does not cause an infection and illness, but it effectively "teaches" the immune system what the viral particles look like, stimulating the production of memory cells. On subsequent infection, these cells can bind to the virus and trigger a rapid immune response. Some pathogens, including HIV, have developed mechanisms to evade the immune system, making it difficult or impossible to develop effective vaccines against them. What strategy could a pathogen use to evade the immune system? Assume that a host's antibodies and/or T-cell receptors are available to bind to any structure that might appear on the surface of a pathogen and that, once bound, the pathogen is destroyed.

Answer Many pathogens, including HIV, have evolved mechanisms by which they can repeatedly alter the surface proteins to which immune system components initially bind. Thus the host organism regularly faces new antigens and requires time to mount an immune response to each one. As the immune system responds to one variant, new variants are created. Some molecular mechanisms that are used to vary viral surface proteins are described in Part III of the text. HIV uses an additional strategy to evade the immune system: it actively infects and destroys immune system cells.

12. **How We Become a "Stiff"** When a vertebrate dies, its muscles stiffen as they are deprived of ATP, a state called rigor mortis. Explain the molecular basis of the rigor state.

    Answer Binding of ATP to myosin triggers dissociation of myosin from the actin thin filament. In the absence of ATP, actin and myosin bind tightly to each other.

13. **Sarcomeres from Another Point of View** The symmetry of thick and thin filaments in a sarcomere is such that six thin filaments ordinarily surround each thick filament in a hexagonal array. Draw a cross section (transverse cut) of a myofibril at the following points: **(a)** at the M line; **(b)** through the I band; **(c)** through the dense region of the A band; **(d)** through the less dense region of the A band, adjacent to the M line (see Fig. 5–29b, c).

    Answer

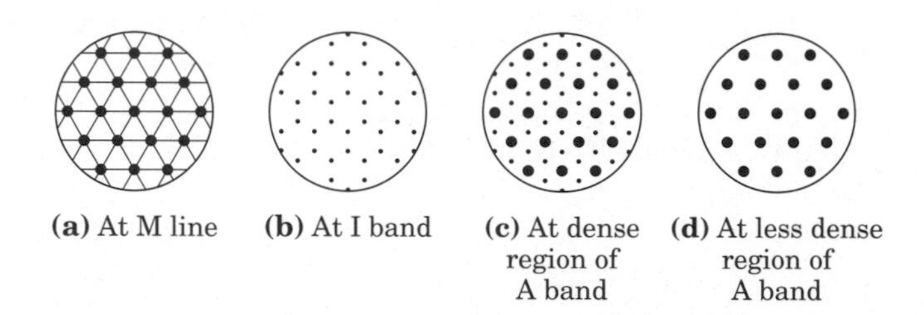

The less dense region of the A band, also known as the H zone (not shown in Fig. 5–29b), is the region in which the myosin thick filaments do not overlap the actin thin filaments. When the sarcomere contracts (see Fig. 5–29c), the H zone and the I band decrease in width.

Biochemistry on the Internet

14. **Lysozyme and Antibodies** To fully appreciate how proteins function in a cell, it is helpful to have a three-dimensional view of how proteins interact with other cellular components. Fortunately, this is possible using Web-based protein databases and three-dimensional molecular viewing utilities. Some molecular viewers require that you download a program or plug-in; some can be problematic when used with certain operating systems or browsers; some require the use of command-line code; some have a more user-friendly interface. We suggest you go to www.umass.edu/microbio/rasmol and look at the information about RasMol, Protein Explorer, and Jmol FirstGlance. Choose the viewer most compatible with your operating system, browser, and level of expertise. Then download and install any software or plug-ins you may need.

    In this exercise you will examine the interactions between the enzyme lysozyme (Chapter 4) and the Fab portion of the anti-lysozyme antibody. Use the PDB identifier 1FDL to explore the structure of the IgG1 Fab fragment–lysozyme complex (antibody-antigen complex). To answer the following questions, use the information on the Structure Summary page at the Protein Data Bank (www.rcsb.org), and view the structure using RasMol, Protein Explorer, or FirstGlance in Jmol.

    (a) Which chains in the three-dimensional model correspond to the antibody fragment and which correspond to the antigen, lysozyme?

    (b) What type of secondary structure predominates in this Fab fragment?

    (c) How many amino acid residues are in the heavy and light chains of the Fab fragment? In lysozyme? Estimate the percentage of the lysozyme that interacts with the antigen-binding site of the antibody fragment.

    (d) Identify the specific amino acid residues in lysozyme and in the variable regions of the Fab heavy and light chains that are situated at the antigen-antibody interface. Are the residues contiguous in the primary sequence of the polypeptide chains?

Answer

(a) Chain L is the light chain and chain H is the heavy chain of the Fab fragment of this antibody molecule. Chain Y is lysozyme.

(b) At the PDB, the SCOP and CATH data show that the proteins have predominantly β secondary structure forming immunoglobulin-like β-sandwich folds. Use the Jmol viewing utility at the PDB to view the complex. You should be able to identify the β structures in the variable and constant regions of both the light and heavy chains.

(c) The heavy chain of the Fab fragment has 218 amino acid residues, the light chain fragment has 214, and lysozyme has 129. Viewing the structure in the spacefill mode shows that less than 15% of the total lysozyme molecule is in contact with the combined V_L and V_H domains of the antibody fragment.

(d) To answer this question you may wish to use FirstGlance in Jmol (http://firstglance.jmol.org). Enter the PDB ID 1FDL. When the molecule appears, check the "Spin" box to stop the molecule from spinning. Next, click "Contacts." With "Chains" selected as the target, click on the lysozyme portion of the complex (Chain Y). The atoms will have asterisks when they are selected. Click "Show Atoms Contacting Target." Only the atoms (in the immunoglobulin chains) that are in contact with lysozyme will remain in space-filling mode. A quick click on each atom will bring up identifying information. Repeat the process with each of the immunoglobulin chains selected to find the lysozyme residues at the interface. In the H chain these residues include Gly^{31}, Tyr^{32}, Asp^{100}, and Tyr^{101}; in the L chain, Tyr^{32}, Tyr^{49}, Tyr^{50}, and Trp^{92}. In lysozyme, residues Asn^{19}, Gly^{22}, Tyr^{23}, Ser^{24}, Lys^{116}, Gly^{117}, Thr^{118}, Asp^{119}, Gln^{121}, and Arg^{125} appear to be situated at the antigen-antibody interface. Not all these residues are adjacent in the primary structure. In any antibody, the residues in the V_L and V_H domains that come into contact with the antigen are located primarily in the loops connecting the β strands of the β-sandwich supersecondary structure. Folding of the polypeptide chain into higher levels of structure brings the nonconsecutive residues together to form the antigen-binding site.

15. Exploring Reversible Interactions of Proteins and Ligands with Living Graphs Use the living graphs for Equations 5–8, 5–11, 5–14, and 5–16 to work through the following exercises.

(a) Reversible binding of a ligand to a simple protein, without cooperativity. For Equation 5–8, set up a plot of θ versus [L] (vertical and horizontal axes, respectively). Examine the plots generated when K_d is set at 5, 10, 20, and 100 μM. Higher affinity of the protein for the ligand means more binding at lower ligand concentrations. Suppose that four different proteins exhibit these four different K_d values for ligand L. Which protein would have the highest affinity for L?

Examine the plot generated when $K_d = 10$ μM. How much does θ increase when [L] increases from 0.2 to 0.4 μM? How much does θ increase when [L] increases from 40 to 80 μM?

You can do the same exercise for Equation 5–11. Convert [L] to pO_2 and K_d to P_{50}. Examine the curves generated when P_{50} is set at 0.5, 1, 2, and 10 kPa. For the curve generated when $P_{50} = 1$ kPa, how much does θ change when the pO_2 increases from 0.02 to 0.04 kPa? From 4 to 8 kPa?

(b) Cooperative binding of a ligand to a multisubunit protein. Using Equation 5–14, generate a binding curve for a protein and ligand with $K_d = 10$ μM and $n = 3$. Note the altered definition of K_d in Equation 5–16. On the same plot, add a curve for a protein with $K_d = 20$ μM and $n = 3$. Now see how both curves change when you change to $n = 4$. Generate Hill plots (Eqn 5–16) for each of these cases. For $K_d = 10$ μM and $n = 3$, what is θ when [L] = 20 μM?

(c) Explore these equations further by varying all the parameters used above.

Answer

(a) The plots should be a series of hyperbolic curves, with $\theta = 1.0$ as the limit. Each curve passes through $\theta = 0.5$ at the point on the x axis where [L] = K_d. The protein with $K_d = 5$ μM has the highest affinity for ligand L. When $K_d = 10$ μM, doubling [L] from 0.2 to 0.4 μM (values well below K_d) nearly doubles θ (the actual increase factor is 1.96).

This is a property of the hyperbolic curve; at low ligand concentrations, θ is an almost linear function of [L]. By contrast, doubling [L] from 40 to 80 μM (well above K_d, where the binding curve is approaching its asymptotic limit) increases θ by a factor of only 1.1. The increase factors are identical for the curves generated from Equation 5–11.

(b) The curves generated from Equation 5–14 should be sigmoidal. Increasing the Hill coefficient (n) increases the slope of the curves at the inflection point. Using Equation 5–14, with [L] = 20 μM, K_d = 10 μM, and n = 3, you will find that θ = 0.998.

(c) A variety of answers will be obtained depending on the values entered for the different parameters.

Data Analysis Problem

16. Protein Function During the 1980s, the structures of actin and myosin were known only at the resolution shown in Figure 5–28a, b. Although researchers knew that the S1 portion of myosin binds to actin and hydrolyzes ATP, there was a substantial debate about where in the myosin molecule the contractile force was generated. At the time, two competing models were proposed for the mechanism of force generation in myosin.

In the "hinge" model, S1 bound to actin, but the pulling force was generated by contraction of the "hinge region" in the myosin tail. The hinge region is in the heavy meromyosin portion of the myosin molecule, near where trypsin cleaves off light meromyosin (see Fig. 5–27b). This is roughly the point labeled "Two supercoiled α helices" in Figure 5–27a. In the "S1" model, the pulling force was generated in the S1 "head" itself and the tail was just for structural support.

Many experiments had been performed but provided no conclusive evidence. In 1987, James Spudich and his colleagues at Stanford University published a study that, although not conclusive, went a long way toward resolving this controversy.

Recombinant DNA techniques were not sufficiently developed to address this issue in vivo, so Spudich and colleagues used an interesting in vitro motility assay. The alga *Nitella* has extremely long cells, often several centimeters in length and about 1 mm in diameter. These cells have actin fibers that run along their long axes, and the cells can be cut open along their length to expose the actin fibers. Spudich and his group had observed that plastic beads coated with myosin would "walk" along these fibers in the presence of ATP, just as myosin would do in contracting muscle.

For these experiments, they used a more well-defined method for attaching the myosin to the beads. The "beads" were clumps of killed bacterial (*Staphylococcus aureus*) cells. These cells have a protein on their surface that binds to the Fc region of antibody molecules (Fig. 5–21a). The antibodies, in turn, bind to several (unknown) places along the tail of the myosin molecule. When bead-antibody-myosin complexes were prepared with intact myosin molecules, they would move along *Nitella* actin fibers in the presence of ATP.

(a) Sketch a diagram showing what a bead-antibody-myosin complex might look like at the molecular level.

(b) Why was ATP required for the beads to move along the actin fibers?

(c) Spudich and coworkers used antibodies that bound to the myosin tail. Why would this experiment have failed if they had used an antibody that bound to the part of S1 that normally binds to actin? Why would this experiment have failed if they had used an antibody that bound to actin?

To help focus in on the part of myosin responsible for force production, Spudich and his colleagues used trypsin to produce two partial myosin molecules (see Fig. 5–27): (1) heavy meromyosin (HMM), made by briefly digesting myosin with trypsin; HMM consists of S1 and the part of the tail that includes the hinge; and (2) short heavy meromyosin (SHMM), made from a more extensive digestion of HMM with trypsin; SHMM consists of S1 and a shorter part of the tail that does not include the hinge. Brief digestion of myosin with trypsin produces HMM and light meromyosin (Fig. 5–27), by cleavage of a single specific peptide bond in the myosin molecule.

(d) Why might trypsin attack this peptide bond first rather than other peptide bonds in myosin?

Spudich and colleagues prepared bead-antibody-myosin complexes with varying amounts of myosin, HMM, and SHMM, and measured their speeds along *Nitella* actin fibers in the presence of ATP. The graph below sketches their results.

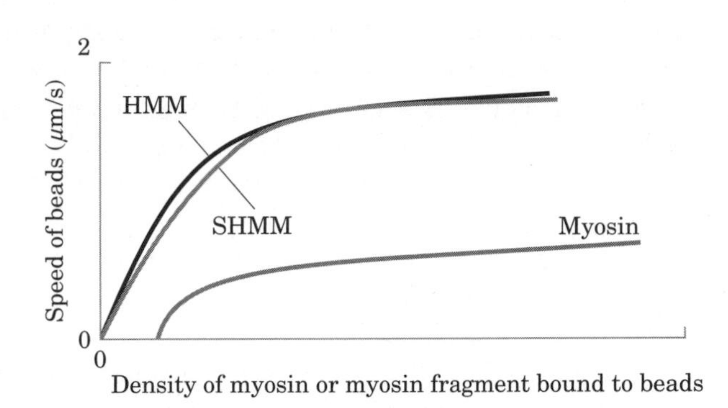

(e) Which model ("S1" or "hinge") is consistent with these results? Explain your reasoning.

(f) Provide a plausible explanation for why the speed of the beads increased with increasing myosin density.

(g) Provide a plausible explanation for why the speed of the beads reached a plateau at high myosin density.

The more extensive trypsin digestion required to produce SHMM had a side effect: another specific cleavage of the myosin polypeptide backbone in addition to the cleavage in the tail. This second cleavage was in the S1 head.

(h) Based on this information, why is it surprising that SHMM was still capable of moving beads along actin fibers?

(i) As it turns out, the tertiary structure of the S1 head remains intact in SHMM. Provide a plausible explanation of how the protein remains intact and functional even though the polypeptide backbone has been cleaved and is no longer continuous.

Answer

(a)

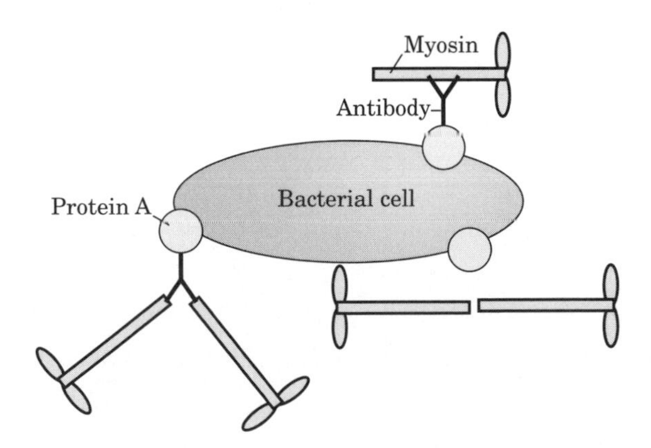

The drawing is not to scale; any given cell would have many more myosin molecules on its surface.

(b) ATP is needed to provide the chemical energy to drive the motion (see Chapter 13).

(c) An antibody that bound to the myosin tail, the actin-binding site, would block actin binding and prevent movement. An antibody that bound to actin would also prevent actin-myosin interaction and thus movement.

(d) There are two possible explanations: (1) Trypsin cleaves only at Lys and Arg residues (see Table 3–7) so would not cleave at many sites in the protein. (2) Not all Arg or Lys residues are equally accessible to trypsin; the most-exposed sites would be cleaved first.

(e) The S1 model. The hinge model predicts that bead-antibody-HMM complexes (with the hinge) would move, but bead-antibody-SHMM complexes (no hinge) would not. The S1 model predicts that because both complexes include S1, both would move. The finding that the beads move with SHMM (no hinge) is consistent only with the S1 model.

(f) With fewer myosin molecules bound, the beads could temporarily fall off the actin as a myosin let go of it. The beads would then move more slowly, as time is required for a second myosin to bind. At higher myosin density, as one myosin lets go another quickly binds, leading to faster motion.

(g) Above a certain density, what limits the rate of movement is the intrinsic speed with which myosin molecules move the beads. The myosin molecules are moving at a maximum rate and adding more will not increase speed.

(h) Because the force is produced in the S1 head, damaging the S1 head would probably inactivate the resulting molecule, and SHMM would be incapable of producing movement.

(i) The S1 head must be held together by noncovalent interactions that are strong enough to retain the active shape of the molecule.

Reference

Hynes, T.R., Block, S.M., White, B.T., & Spudich, J.A. (1987) Movement of myosin fragments in vitro: domains involved in force production. *Cell* **48,** 953–963.

chapter

6

Enzymes

1. **Keeping the Sweet Taste of Corn** The sweet taste of freshly picked corn (maize) is due to the high level of sugar in the kernels. Store-bought corn (several days after picking) is not as sweet, because about 50% of the free sugar is converted to starch within one day of picking. To preserve the sweetness of fresh corn, the husked ears can be immersed in boiling water for a few minutes ("blanched") then cooled in cold water. Corn processed in this way and stored in a freezer maintains its sweetness. What is the biochemical basis for this procedure?

   Answer After an ear of corn has been removed from the plant, the enzyme-catalyzed conversion of sugar to starch continues. Inactivation of these enzymes slows down the conversion to an imperceptible rate. One of the simplest techniques for inactivating enzymes is heat denaturation. Freezing the corn lowers any remaining enzyme activity to an insignificant level.

2. **Intracellular Concentration of Enzymes** To approximate the actual concentration of enzymes in a bacterial cell, assume that the cell contains equal concentrations of 1,000 different enzymes in solution in the cytosol and that each protein has a molecular weight of 100,000. Assume also that the bacterial cell is a cylinder (diameter 1.0 μm, height 2.0 μm), that the cytosol (specific gravity 1.20) is 20% soluble protein by weight, and that the soluble protein consists entirely of enzymes. Calculate the *average* molar concentration of each enzyme in this hypothetical cell.

   Answer There are three different ways to approach this problem.

   (i) The concentration of total protein in the cytosol is

$$\frac{(1.2\text{ g/mL})(0.20)}{100{,}000\text{ g/mol}} = 0.24 \times 10^{-5}\text{ mol/mL} = 2.4 \times 10^{-3}\text{ M}$$

   Thus, for 1 enzyme in 1,000, the enzyme concentration is

$$\frac{2.4 \times 10^{-3}\text{ M}}{1000} = 2.4 \times 10^{-6}\text{ M}$$

   (ii) The average molar concentration $= \dfrac{\text{moles of each enzyme in cell}}{\text{volume of cell in liters}}$

$$\begin{aligned}\text{Volume of bacterial cytosol} &= \pi r^2 h = (3.14)(0.50\ \mu\text{m})^2(2.0\ \mu\text{m}) = 1.6\ \mu\text{m}^3\\ &= 1.6 \times 10^{-12}\text{ cm}^3 = 1.6 \times 10^{-12}\text{ mL}\\ &= 1.6 \times 10^{-15}\text{ L}\end{aligned}$$

   Amount (in moles) of each enzyme in cell is

$$\frac{(0.20)(1.2\text{ g/cm}^3)(1.6\ \mu\text{m}^3)(10^{-12}\text{ cm}^3/\mu\text{m}^3)}{(100{,}000\text{ g/mol})(1000)} = 3.8 \times 10^{-21}\text{ mol}$$

$$\begin{aligned}\text{Average molar concentration} &= \frac{3.8 \times 10^{-21}\text{ mol}}{1.6 \times 10^{-15}\text{ L}}\\ &= 2.4 \times 10^{-6}\text{ mol/L} = 2.4 \times 10^{-6}\text{ M}\end{aligned}$$

(iii) Volume of bacterial cytosol $= \pi r^2 h$

$= (3.14)(0.50\ \mu\text{m})^2(2.0\ \mu\text{m}) = 1.6\ \mu\text{m}^3 = 1.6 \times 10^{-12}$ mL

Weight of cytosol = (specific gravity)(volume)

$= (1.2\ \text{g/mL})(1.6 \times 10^{-12}\ \text{mL}) = 1.9 \times 10^{-12}$ g

Average weight of each protein (1 in 1,000, 20% wt/wt protein)

$= (1.9 \times 10^{-12}\ \text{g})(0.20)/(1{,}000) = 3.8 \times 10^{-16}$ g

Average molar concentration of each protein

$= (\text{average weight})/(M_r)(\text{volume})$

$= (3.8 \times 10^{-16}\ \text{g})/(10^5\ \text{g/mol})(1.6 \times 10^{-12}\ \text{mL})(1\ \text{L}/1000\ \text{mL})$

$= 2.4 \times 10^{-6}\ \text{mol/L} = 2.4 \times 10^{-6}$ M

3. **Rate Enhancement by Urease** The enzyme urease enhances the rate of urea hydrolysis at pH 8.0 and 20 °C by a factor of 10^{14}. If a given quantity of urease can completely hydrolyze a given quantity of urea in 5.0 min at 20 °C and pH 8.0, how long would it take for this amount of urea to be hydrolyzed under the same conditions in the absence of urease? Assume that both reactions take place in sterile systems so that bacteria cannot attack the urea.

Answer

Time to hydrolyze urea

$$= \frac{(5.0\ \text{min})(10^{14})}{(60\ \text{min/hr})(24\ \text{hr/day})(365\ \text{days/yr})}$$

$= 9.5 \times 10^8$ yr

= 950 million years!

4. **Protection of an Enzyme against Denaturation by Heat** When enzyme solutions are heated, there is a progressive loss of catalytic activity over time due to denaturation of the enzyme. A solution of the enzyme hexokinase incubated at 45 °C lost 50% of its activity in 12 min, but when incubated at 45 °C in the presence of a very large concentration of one of its substrates, it lost only 3% of its activity in 12 min. Suggest why thermal denaturation of hexokinase was retarded in the presence of one of its substrates.

Answer One possibility is that the ES complex is more stable than the free enzyme. This implies that the ground state for the ES complex is at a lower energy level than that for the free enzyme, thus *increasing the height of the energy barrier* to be crossed in passing from the native to the denatured or unfolded state.

An alternative view is that an enzyme denatures in two stages: reversible conversion of active native enzyme (N) to an inactive unfolded state (U), followed by irreversible conversion to inactivated enzyme (I):

$$\text{N} \rightleftharpoons \text{U} \longrightarrow \text{I}$$

If substrate, S, binds only to N, saturation with S to form NS would leave less free N available for conversion to U or I, as the N $\rightleftharpoons$ U equilibrium is perturbed toward N. If N but not NS is converted to U or I, then substrate binding will cause stabilization.

5. **Requirements of Active Sites in Enzymes** Carboxypeptidase, which sequentially removes carboxyl-terminal amino acid residues from its peptide substrates, is a single polypeptide of 307 amino acids. The two essential catalytic groups in the active site are furnished by Arg^{145} and Glu^{270}.

(a) If the carboxypeptidase chain were a perfect α helix, how far apart (in Å) would Arg^{145} and Glu^{270} be? (Hint: see Fig. 4–4a.)

(b) Explain how the two amino acid residues can catalyze a reaction occurring in the space of a few angstroms.

Answer

(a) Arg^{145} is separated from Glu^{270} by 270 − 145 = 125 amino acid (AA) residues. From Figure 4–4a we see that the α helix has 3.6 AA/turn and increases in length along the major axis by 5.4 Å/turn. Thus, the distance between the two residues is

$$\frac{(125 \text{ AA})(5.4 \text{ Å/turn})}{3.6 \text{ AA/turn}} = 190 \text{ Å}$$

(b) Three-dimensional folding of the enzyme brings the two amino acid residues into close proximity.

6. Quantitative Assay for Lactate Dehydrogenase The muscle enzyme lactate dehydrogenase catalyzes the reaction

$$\underset{\text{Pyruvate}}{CH_3-\overset{\overset{\displaystyle O}{\|}}{C}-COO^-} + NADH + H^+ \longrightarrow \underset{\text{Lactate}}{CH_3-\overset{\overset{\displaystyle OH}{|}}{\underset{\underset{\displaystyle H}{|}}{C}}-COO^-} + NAD^+$$

NADH and NAD^+ are the reduced and oxidized forms, respectively, of the coenzyme NAD. Solutions of NADH, but *not* NAD^+, absorb light at 340 nm. This property is used to determine the concentration of NADH in solution by measuring spectrophotometrically the amount of light absorbed at 340 nm by the solution. Explain how these properties of NADH can be used to design a quantitative assay for lactate dehydrogenase.

Answer The reaction rate can be measured by following the decrease in absorption at 340 nm (as NADH is converted to NAD^+) as the reaction proceeds. The researcher needs to obtain three pieces of information to develop a good quantitative assay for lactate dehydrogenase:

(i) Determine K_m values (see Box 6–1).

(ii) Measure the initial rate at several known concentrations of enzyme with saturating concentrations of NADH and pyruvate.

(iii) Plot the initial rates as a function of [E]; the plot should be linear, with a slope that provides a measure of lactate dehydrogenase concentration.

7. Effect of Enzymes on Reactions Which of the following effects would be brought about by any enzyme catalyzing the simple reaction

$$S \underset{k_2}{\overset{k_1}{\rightleftharpoons}} P \quad \text{where} \quad K'_{eq} = \frac{[P]}{[S]}?$$

(a) Decreased K'_{eq}; (b) Increased k_1; (c) Increased K'_{eq}; (d) Increased $\Delta G^\ddagger$; (e) Decreased $\Delta G^\ddagger$; (f) More negative $\Delta G'^\circ$; (g) Increased k_2.

Answer (b), (e), (g). Enzymes do not change a reaction's equilibrium constant and thus catalyze the reaction in both directions, making (b) and (g) correct. Enzymes increase the rate of a reaction by lowering the activation energy, hence (e) is correct.

8. Relation between Reaction Velocity and Substrate Concentration: Michaelis-Menten Equation

(a) At what substrate concentration would an enzyme with a k_{cat} of 30.0 s^{-1} and a K_m of 0.0050 M operate at one-quarter of its maximum rate?

(b) Determine the fraction of V_{max} that would be obtained at the following substrate concentrations [S]: $\frac{1}{2}K_m$, $2K_m$, and $10K_m$.

(c) An enzyme that catalyzes the reaction X $\rightleftharpoons$ Y is isolated from two bacterial species. The enzymes have the same V_{max}, but different K_m values for the substrate X. Enzyme A has a K_m of 2.0 μM, while enzyme B has a K_m of 0.5 μM. The plot below shows the kinetics of reactions carried out with the same concentration of each enzyme and with [X] = 1 μM. Which curve corresponds to which enzyme?

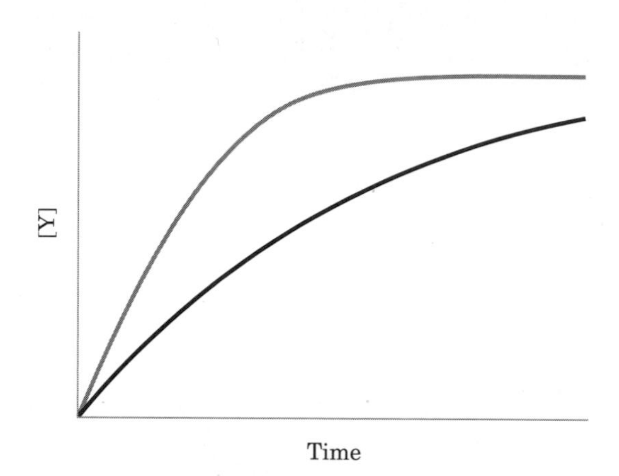

Answer

(a) Here we want to find the value of [S] when $V_0 = 0.25\ V_{max}$. The Michaelis-Menten equation is

$$V_0 = V_{max}[S]/(K_m + [S])$$

so $V_0 = V_{max}$ when $[S]/(K_m + [S]) = 0.25$; or

$$[S] = 0.33K_m = 0.33(0.0050\ \text{M}) = 1.7 \times 10^{-3}\ \text{M}$$

(b) The Michaelis-Menten equation can be rearranged to

$$V_0/V_{max} = [S]/(K_m + [S])$$

Substituting $[S] = \frac{1}{2} K_m$ into the equation gives

$$V_0/V_{max} = 0.5\ K_m/1.5K_m = 0.33$$

Similarly, substituting $[S] = 2K_m$ gives

$$V_0/V_{max} = 0.67$$

And substituting $[S] = 10K_m$ gives

$$V_0/V_{max} = 0.91$$

(c) The upper curve corresponds to enzyme B ([X] is greater than the K_m for this enzyme), and the lower curve corresponds to enzyme A. When the initial concentration of substrate is greater than K_m, the rate of the reaction is less sensitive to the depletion of substrate at early stages of the reaction and the rate remains approximately linear for a longer time.

9. Applying the Michaelis-Menten Equation I A research group discovers a new version of happyase, which they call happyase*, that catalyzes the chemical reaction

$$\text{HAPPY} \rightleftharpoons \text{SAD}$$

The researchers begin to characterize the enzyme.

(a) In the first experiment, with [E_t] at 4 nM, they find that the V_{max} is 1.6 μM s^{-1}. Based on this experiment, what is the k_{cat} for happyase*? (Include appropriate units.)

(b) In another experiment, with $[E_t]$ at 1 nM and [HAPPY] at 30 μM, the researchers find that $V_0 =$ 300 nM s^{-1}. What is the measured K_m of happyase* for its substrate HAPPY? (Include appropriate units.)

(c) Further research shows that the purified happyase* used in the first two experiments was actually contaminated with a reversible inhibitor called ANGER. When ANGER is carefully removed from the happyase* preparation, and the two experiments repeated, the measured V_{max} in **(a)** is increased to 4.8 μM s^{-1}, and the measured K_m in **(b)** is now 15 μM. For the inhibitor ANGER, calculate the values of α and α'.

(d) Based on the information given above, what type of inhibitor is ANGER?

Answer

(a) Use the equation $k_{cat} = V_{max}/[E_t]$. $k_{cat} = 1600 \text{ nM s}^{-1}/4 \text{ nM} = 400 \text{ s}^{-1}$.

(b) Use the equation $V_{max} = k_{cat}[E_t]$. When $[E_t] = 1$ nM, $V_{max} = 400 \text{ nM s}^{-1}$.

$$V_0/V_{max} = 300 \text{ nM s}^{-1}/400 \text{ nM s}^{-1} = \frac{3}{4}$$

Rearrange the Michaelis-Menten equation, substitute for V_0/V_{max}, and solve for K_m.

$$V_0/V_{max} = [S]/(K_m + [S])$$
$$\frac{3}{4} = [S]/(K_m + [S])$$
$$K_m = [S]/3$$

In this experiment, the concentration of the substrate, HAPPY, was 30 μM, so $K_m = 10$ μM.

(c) As shown in Table 6–9, V_{max} varies as a function of V_{max}/α'. Because V_{max} increased by a factor of 3, $\alpha' = 3$. Similarly, K_m varies as a function of $\alpha K_m/\alpha'$. Given that K_m increased by a factor of 1.5 when ANGER was removed (that is, the inhibitor decreased the observed K_m by $\frac{2}{3}$) and $\alpha' = 3$, then $\alpha = 2$.

(d) Because both α and α' are affected, ANGER is a mixed inhibitor.

10. Applying the Michaelis-Menten Equation II Another enzyme is found that catalyzes the reaction

$$A \rightleftharpoons B$$

Researchers find that the K_m for the substrate A is 4 μM, and the k_{cat} is 20 min^{-1}.

(a) In an experiment, [A] = 6 mM, and the initial velocity, V_0 was 480 nM min^{-1}. What was the $[E_t]$ used in the experiment?

(b) In another experiment, $[E_t] = 0.5$ μM, and the measured $V_0 = 5$ μM min^{-1}. What was the [A] used in the experiment?

(c) The compound Z is found to be a very strong competitive inhibitor of the enzyme, with an α of 10. In an experiment with the same $[E_t]$ as in part **(a),** but a different [A], an amount of Z is added that reduces the rate V_0 to 240 nM min^{-1}. What is the [A] in this experiment?

(d) Based on the kinetic parameters given above, has this enzyme evolved to achieve catalytic perfection? Explain your answer briefly, using the kinetic parameter(s) that define catalytic perfection.

Answer

(a) Because [S] is much greater than (more than 1000-fold) K_m, assume that the measured rate of the reaction reflects V_{max}. Use the equation $V_{max} = k_{cat}[E_t]$, and solve for $[E_t]$. $[E_t] = V_{max}/k_{cat} = 480 \text{ nM min}^{-1}/20 \text{ min}^{-1} = 24$ nM.

(b) At this $[E_t]$, the calculated $V_{max} = k_{cat}[E_t] = 20 \text{ min}^{-1} \times 0.5\ \mu\text{M} = 10\ \mu\text{M min}^{-1}$. Recall that K_m equals the substrate concentration at which $V_0 = \frac{1}{2}V_{max}$. The measured V_0 is exactly half V_{max}, so $[A] = K_m = 4$ μM.

(c) Given the same [E_t] as in **(a),** V_{max} = 480 nM min^{-1}. The V_0 is again exactly half V_{max} (V_0 = 240 nM min^{-1}), so [A] = the apparent or measured K_m. In the presence of an inhibitor with α = 10, the measured K_m = 40 μM = [S].

(d) No. $k_{cat}/K_m = 0.33/(4 \times 10^{-6}\ M^{-1}\ s^{-1}) = 8.25 \times 10^4\ M^{-1}\ s^{-1}$, well below the diffusion-controlled limit.

11. Estimation of V_{max} and K_m by Inspection Although graphical methods are available for accurate determination of the V_{max} and K_m of an enzyme-catalyzed reaction (see Box 6–1), sometimes these quantities can be quickly estimated by inspecting values of V_0 at increasing [S]. Estimate the V_{max} and K_m of the enzyme-catalyzed reaction for which the following data were obtained.

[S] (M)	V_0 (μM/min)	[S] (M)	V_0 (μM/min)
2.5×10^{-6}	28	4×10^{-5}	112
4.0×10^{-6}	40	1×10^{-4}	128
1×10^{-5}	70	2×10^{-3}	139
2×10^{-5}	95	1×10^{-2}	140

Answer Notice how little the velocity changes as the substrate concentration increases by fivefold from 2 to 10 mM. Thus, we can estimate a V_{max} of 140 μM/min. K_m is defined as the substrate concentration that produces a velocity of $\frac{1}{2}V_{max}$, or 70 μM/min. Inspection of the table indicates that this V_0 occurs at [S] = 1×10^{-5} M, thus $K_m \approx 1 \times 10^{-5}$ M.

12. Properties of an Enzyme of Prostaglandin Synthesis Prostaglandins are a class of eicosanoids, fatty acid derivatives with a variety of extremely potent actions on vertebrate tissues. They are responsible for producing fever and inflammation and its associated pain. Prostaglandins are derived from the 20-carbon fatty acid arachidonic acid in a reaction catalyzed by the enzyme prostaglandin endoperoxide synthase. This enzyme, a cyclooxygenase, uses oxygen to convert arachidonic acid to PGG_2, the immediate precursor of many different prostaglandins (prostaglandin synthesis is described in Chapter 21).

(a) The kinetic data given below are for the reaction catalyzed by prostaglandin endoperoxide synthase. Focusing here on the first two columns, determine the V_{max} and K_m of the enzyme.

[Arachidonic acid] (mM)	Rate of formation of PGG_2 (mM/min)	Rate of formation of PGG_2 with 10 mg/mL ibuprofen (mM/min)
0.5	23.5	16.67
1.0	32.2	25.25
1.5	36.9	30.49
2.5	41.8	37.04
3.5	44.0	38.91

(b) Ibuprofen is an inhibitor of prostaglandin endoperoxide synthase. By inhibiting the synthesis of prostaglandins, ibuprofen reduces inflammation and pain. Using the data in the first and third columns of the table, determine the type of inhibition that ibuprofen exerts on prostaglandin endoperoxide synthase.

Answer

(a) Calculate the reciprocal values for the data, as in parentheses below, and prepare a double-reciprocal plot to determine the kinetic parameters.

[S] (mM) (1/[S] (mM^{-1}))	V_0 (mM/min) (1/V_0 (min/mM))	V_0 with 10 mg/mL ibuprofen (mM/min) (1/V_0 (min/mM))
0.5 (2.0)	23.5 (0.043)	16.67 (0.0600)
1.0 (1.0)	32.2 (0.031)	25.25 (0.0396)
1.5 (0.67)	36.9 (0.027)	30.49 (0.0328)
2.5 (0.40)	41.8 (0.024)	37.04 (0.0270)
3.5 (0.28)	44.0 (0.023)	38.91 (0.0257)

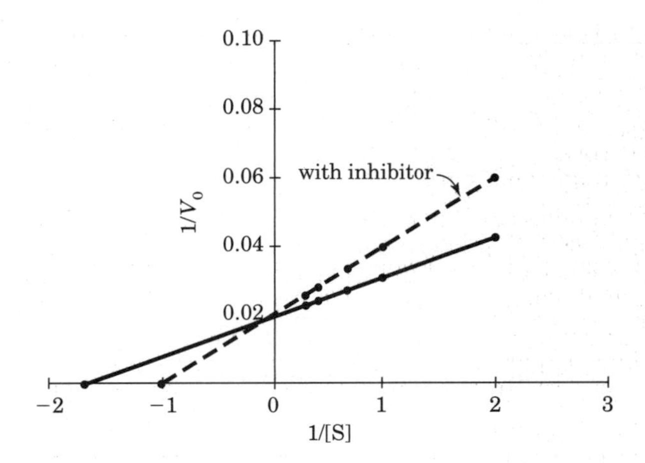

The intercept on the vertical axis = $-1/V_{max}$ and the intercept on the horizontal axis = $-1/K_m$. From these values, we can calculate V_{max} and K_m.

$-1/V_{max} = -0.0194$, and $V_{max} = 51.5$ mM/min

$-1/K_m = -1.7$, and $K_m = 0.59$ mM

(b) Ibuprofen acts as a competitive inhibitor. The double-reciprocal plot (with inhibitor) shows that, in the presence of ibuprofen, the V_{max} of the reaction is unchanged (the intercept on the $1/V_0$ axis is the same) and K_m is increased ($-1/K_m$ is closer to the origin).

13. Graphical Analysis of V_{max} and K_m The following experimental data were collected during a study of the catalytic activity of an intestinal peptidase with the substrate glycylglycine:

$$\text{Glycylglycine} + H_2O \longrightarrow 2 \text{ glycine}$$

[S] (mM)	Product formed (μmol/min)	[S] (mM)	Product formed (μmol/min)
1.5	0.21	4.0	0.33
2.0	0.24	8.0	0.40
3.0	0.28	16.0	0.45

Use graphical analysis (see Box 6–1) to determine the K_m and V_{max} for this enzyme preparation and substrate.

Answer As described in Box 6–1, the standard method is to use V_0 versus [S] data to calculate $1/V_0$ and 1/[S].

V_0 (μmol/min)	$1/V_0$ (min/μmol)	[S] (mM)	1/[S] (mM^{-1})
0.21	4.8	1.5	0.67
0.24	4.2	2.0	0.50
0.28	3.6	3.0	0.33
0.33	3.0	4.0	0.25
0.40	2.5	8.0	0.13
0.45	2.2	16.0	0.06

Graphing these values gives a Lineweaver-Burk plot. From the best straight line through the data, the intercept on the horizontal axis = $-1/K_m$ and the intercept on the vertical axis = $1/V_{max}$. From these values, we can calculate K_m and V_{max}:

$-1/K_m = -0.45$, and $K_m = 2.2$ mM

$-1/V_{max} = -2.0$, and $V_{max} = 0.50$ μmol/min

14. The Eadie-Hofstee Equation One transformation of the Michaelis-Menten equation is the Lineweaver-Burk, or double-reciprocal, equation. Multiplying both sides of the Lineweaver-Burk equation by V_{max} and rearranging gives the Eadie-Hofstee equation:

$$V_0 = (-K_m)\frac{V_0}{[S]} + V_{max}$$

A plot of V_0 vs. V_0/[S] for an enzyme-catalyzed reaction is shown below. The curve labeled "Slope = $-K_m$" was obtained in the absence of inhibitor. Which of the other curves (A, B, or C) shows the enzyme activity when a competitive inhibitor is added to the reaction mixture? Hint: See Equation 6–30.

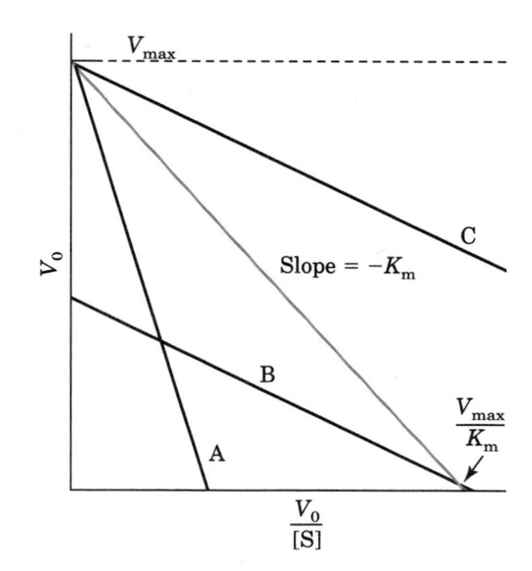

Answer Curve A shows competitive inhibition. V_{max} for A is the same as for the normal curve, as seen by the identical intercepts on the V_0 axis. And, for every value of [S] (until maximal velocity is reached at saturating substrate levels), V_0 is lower for curve A than for the normal curve, indicating competitive inhibition. Note that as [S] increases, V_0/[S] decreases, so that V_{max}—that is, the V_0 at the highest (saturating) [S]—is found at the intersection of the curve at the y axis. Curve C, while also having an identical V_{max}, shows higher V_0 values for every [S] (and for every V_0/[S]) than the normal curve, which is not indicative of inhibition. The lower V_{max} for curve B rules out competitive inhibition.

15. The Turnover Number of Carbonic Anhydrase Carbonic anhydrase of erythrocytes (M_r 30,000) has one of the highest turnover numbers we know of. It catalyzes the reversible hydration of CO_2:

$$H_2O + CO_2 \rightleftharpoons H_2CO_3$$

This is an important process in the transport of CO_2 from the tissues to the lungs. If 10.0 μg of pure carbonic anhydrase catalyzes the hydration of 0.30 g of CO_2 in 1 min at 37 °C at V_{max}, what is the turnover number (k_{cat}) of carbonic anhydrase (in units of min^{-1})?

Answer The turnover number of an enzyme is the number of substrate molecules transformed per unit time by a single enzyme molecule (or a single catalytic site) when the enzyme is saturated with substrate:

$$k_{cat} = V_{max}/E_t$$

where E_t = total moles of active sites.

We can convert the values given in the problem into a turnover number (min^{-1}) by converting the weights of enzyme and substrate to molar amounts:

$$V_{max} \text{ (moles of CO}_2\text{/min)} = \frac{0.30 \text{ g/min}}{44 \text{ g/mol}} = 6.8 \times 10^{-3} \text{ mol/min}$$

$$\text{Amount of enzyme (moles)} = \frac{(10.0\ \mu\text{g})(1 \text{ g}/10^6\ \mu\text{g})}{30{,}000 \text{ g/mol}} = 3.3 \times 10^{-10} \text{ mol}$$

The turnover number is obtained by dividing moles of CO_2/min by moles of enzyme:

$$k_{cat} = \frac{6.8 \times 10^{-3} \text{ mol/min}}{3.3 \times 10^{-10} \text{ mol}} = 2.0 \times 10^7 \text{ min}^{-1}$$

16. Deriving a Rate Equation for Competitive Inhibition The rate equation for an enzyme subject to competitive inhibition is

$$V_0 = \frac{V_{max}[S]}{\alpha K_m + [S]}$$

Beginning with a new definition of total enzyme as

$$[E_t] = [E] + [ES] + [EI]$$

and the definitions of α and K_I provided in the text, derive the rate equation above. Use the derivation of the Michaelis-Menten equation as a guide.

Answer The basic assumptions used to derive the Michaelis-Menten equation still hold. The reaction is at steady state, and the overall rate is determined by

$$V_0 = k_2[ES] \qquad \text{(a)}$$

With the competitive inhibitor, I, now to be added, the goal again is to describe V_0 in terms of the measurable quantities $[E_t]$, [S], and [I]. In the presence of inhibitor,

$$[E_t] = [ES] + [E] + [EI] \qquad (b)$$

We first solve for [EI]. As we have seen,

$$K_I = \frac{[E][I]}{[EI]}; \text{ so } [EI] = \frac{[E][I]}{K_I}$$

Substituting for [EI] in (b) gives

$$[E_t] = [ES] + [E] + \frac{[E][I]}{K_I} \qquad (c)$$

and simplifying gives

$$[E_t] = [ES] + [E]\left(1 + \frac{[I]}{K_I}\right) = [ES] + [E]\alpha \qquad (d)$$

where α describes the effect of the competitive inhibitor. [E] in the absence of inhibitor can be obtained from a rearrangement of Equation 6–19 (remembering that $[E_t] = [ES] + [E]$), to give

$$[E] = \frac{[ES]K_m}{[S]} \qquad (e)$$

Substituting (e) into (d) gives

$$[E_t] = [ES] + \left(\frac{[ES]K_m}{[S]}\right)\alpha \qquad (f)$$

and rearranging and solving for [ES] gives

$$[ES] = \frac{[E_t][S]}{\alpha K_m + [S]} \qquad (g)$$

Next, substituting (g) into (a), and defining $k_2[E_t] = V_{max}$, we get the final equation for reaction velocity in the presence of a competitive inhibitor:

$$V_0 = \frac{V_{max}[S]}{\alpha K_m + [S]}$$

17. Irreversible Inhibition of an Enzyme Many enzymes are inhibited irreversibly by heavy metal ions such as Hg^{2+}, Cu^{2+}, or Ag^+, which can react with essential sulfhydryl groups to form mercaptides:

$$\text{Enz—SH} + Ag^+ \longrightarrow \text{Enz—S—Ag} + H^+$$

The affinity of Ag^+ for sulfhydryl groups is so great that Ag^+ can be used to titrate —SH groups quantitatively. To 10.0 mL of a solution containing 1.0 mg/mL of a pure enzyme, an investigator added just enough $AgNO_3$ to completely inactivate the enzyme. A total of 0.342 μmol of $AgNO_3$ was required. Calculate the minimum molecular weight of the enzyme. Why does the value obtained in this way give only the *minimum* molecular weight?

Answer An equivalency exists between millimoles of $AgNO_3$ required for inactivation and millimoles of —SH group and thus, assuming one —SH group per enzyme molecule, millimoles of enzyme:

$$0.342 \times 10^{-3} \text{ mmol} = \frac{(1.0 \text{ mg/mL})(10.0 \text{ mL})}{(\text{minimum } M_r)(\text{mg/mmol})}$$

Thus, the minimum $M_r = \dfrac{(1.0 \text{ mg/mL})(10.0 \text{ mL})}{0.342 \times 10^{-3} \text{ mmol}} = 2.9 \times 10^4 = 29{,}000$

This is the *minimum* molecular weight because it assumes only one titratable —SH group per enzyme molecule.

18. Clinical Application of Differential Enzyme Inhibition Human blood serum contains a class of enzymes known as acid phosphatases, which hydrolyze biological phosphate esters under slightly acidic conditions (pH 5.0):

$$R{-}O{-}PO_3^{2-} + H_2O \longrightarrow R{-}OH + HO{-}PO_3^{2-}$$

Acid phosphatases are produced by erythrocytes, the liver, kidney, spleen, and prostate gland. The enzyme of the prostate gland is clinically important because its increased activity in the blood can be an indication of prostate cancer. The phosphatase from the prostate gland is strongly inhibited by tartrate ion, but acid phosphatases from other tissues are not. How can this information be used to develop a specific procedure for measuring the activity of the acid phosphatase of the prostate gland in human blood serum?

Answer First, measure the *total* acid phosphatase activity in a blood sample in units of μmol of phosphate ester hydrolyzed per mL of serum. Next, remeasure this activity in the presence of tartrate ion at a concentration sufficient to completely inhibit the enzyme from the prostate gland. The difference between the two activities represents the activity of acid phosphatase from the prostate gland.

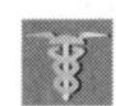

19. Inhibition of Carbonic Anhydrase by Acetazolamide Carbonic anhydrase is strongly inhibited by the drug acetazolamide, which is used as a diuretic (i.e., to increase the production of urine) and to lower excessively high pressure in the eye (due to accumulation of intraocular fluid) in glaucoma. Carbonic anhydrase plays an important role in these and other secretory processes because it participates in regulating the pH and bicarbonate content of several body fluids. The experimental curve of initial reaction velocity (as percentage of V_{max}) versus [S] for the carbonic anhydrase reaction is illustrated below (upper curve). When the experiment is repeated in the presence of acetazolamide, the lower curve is obtained. From an inspection of the curves and your knowledge of the kinetic properties of competitive and mixed enzyme inhibitors, determine the nature of the inhibition by acetazolamide. Explain your reasoning.

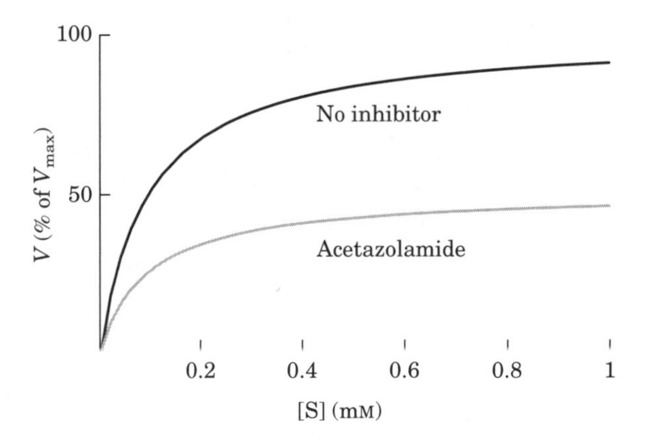

Answer The graph gives us several pieces of information. First, the inhibitor prevents the enzyme from achieving the same V_{max} as in the absence of inhibitor. Second, the overall shape of the two curves is very similar: at any [S] the ratio of the two velocities (±inhibitor) is the same. Third, the velocity does not change very much above [S] = 1 mM, so at much higher [S] the observed velocity is essentially V_{max} for each curve. Fourth, if we estimate the [S] at which $\frac{1}{2}V_{max}$ is achieved, this value is nearly identical for both curves. Noncompetitive inhibition, a special form of mixed inhibition that is rarely observed, alters the V_{max} of enzymes but leaves K_m unchanged. Thus, acetazolamide acts as a noncompetitive (mixed) inhibitor of carbonic anhydrase.

20. The Effects of Reversible Inhibitors Derive the expression for the effect of a reversible inhibitor on observed K_m (apparent $K_m = \alpha K_m/\alpha'$). Start with Equation 6–30 and the statement that apparent K_m is equivalent to the [S] at which $V_0 = V_{max}/2\alpha'$.

Answer Equation 6–30 is

$$V_0 = \frac{V_{max}[S]}{\alpha K_m + \alpha'[S]}$$

Or $V_0 = V_{max} \times [S]/(\alpha K_m + \alpha'[S])$. Thus, the [S] at which $V_0 = V_{max}/2\alpha'$ is obtained when all the terms on the right side of the equation except V_{max} equal $\frac{1}{2}\alpha'$:

$$[S]/(\alpha K_m + \alpha'[S]) = \tfrac{1}{2}\alpha'$$

We can now solve this equation for [S]:

$2\alpha'[S] = \alpha K_m + \alpha'[S]$

$2\alpha'[S] - \alpha'[S] = \alpha K_m$

$\alpha'[S] = \alpha K_m$

$[S] = \alpha K_m/\alpha'$

Thus, observed $K_m = \alpha K_m/\alpha'$.

21. pH Optimum of Lysozyme The active site of lysozyme contains two amino acid residues essential for catalysis: Glu^{35} and Asp^{52}. The pK_a values of the carboxyl side chains of these residues are 5.9 and 4.5, respectively. What is the ionization state (protonated or deprotonated) of each residue at pH 5.2, the pH optimum of lysozyme? How can the ionization states of these residues explain the pH-activity profile of lysozyme shown below?

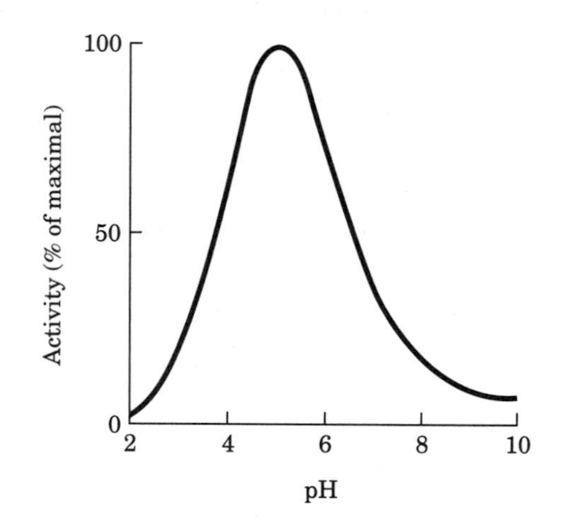

Answer At a pH midway between the two pK_a values (pH 5.2), the side-chain carboxyl group of Asp^{52}, with the lower pK_a (4.5), is mainly deprotonated (—COO^-), whereas Glu^{35}, with the higher pK_a (5.9; the stronger base), is protonated (—COOH). At pH values below 5.2, Asp^{52} becomes protonated and the activity decreases. Similarly, at pH values above 5.2, Glu^{35} becomes deprotonated and the activity also decreases. The pH-activity profile suggests that maximum catalytic activity occurs at a pH midway between the pK_a values of the two acidic groups, when Glu^{35} is protonated and Asp^{52} is deprotonated.

22. Working with Kinetics Go to the Living Graphs for Chapter 6.

(a) Using the Living Graph for Equation 6–9, create a V versus [S] plot. Use $V_{max} = 100\ \mu\text{M s}^{-1}$, and $K_m = 10\ \mu\text{M}$. How much does V_0 increase when [S] is doubled, from 0.2 to 0.4 μM? What is V_0 when [S] = 10 μM? How much does the V_0 increase when [S] increases from 100 to 200 μM? Observe how the graph changes when the values for V_{max} or K_m are halved or doubled.

(b) Using the Living Graph for Equation 6–30 and the kinetic parameters in **(a),** create a plot in which both α and α' are 1.0. Now observe how the plot changes when $\alpha = 2.0$; when $\alpha' = 3.0$; and when $\alpha = 2.0$ and $\alpha' = 3.0$.

(c) Using the Living Graphs for Equation 6–30 and the Lineweaver-Burk equation in Box 6–1, create Lineweaver-Burk (double-reciprocal) plots for all the cases in **(a)** and **(b).** When $\alpha = 2.0$, does the x intercept move to the right or to the left? If $\alpha = 2.0$ and $\alpha' = 3.0$, does the x intercept move to the right or to the left?

Answer

(a) When [S] increases from 0.2 to 0.4 μM, V_0 increases by a factor of 1.96. When [S] = 10 μM, $V_0 = 50\ \mu\text{M s}^{-1}$. When [S] increases from 100 to 200 μM, V_0 increases by a factor of 1.048.

(b) When $\alpha = 2.0$, the curve is shifted to the right as the K_m is increased by a factor of 2. When $\alpha' = 3.0$, the asymptote of the curve (the V_{max}) declines by a factor of 3. When $\alpha = 2.0$ and $\alpha' = 3.0$, the curve briefly rises above the curve where both α and $\alpha' = 1.0$, due to a decline in K_m. However, the asymptote is lower because V_{max} declines by a factor of 3.

(c) When $\alpha = 2.0$, the x intercept moves to the right. When $\alpha = 2.0$ and $\alpha' = 3.0$, the x intercept moves to the left.

Data Analysis Problem

23. Exploring and Engineering Lactate Dehydrogenase Examining the structure of an enzyme results in hypotheses about the relationship between different amino acids in the protein's structure and the protein's function. One way to test these hypotheses is to use recombinant DNA technology to generate mutant versions of the enzyme and then examine the structure and function of these altered forms. The technology used to do this is described in Chapter 9.

One example of this kind of analysis is the work of Clarke and colleagues on the enzyme lactate dehydrogenase, published in 1989. Lactate dehydrogenase (LDH) catalyzes the reduction of pyruvate with NADH to form lactate (see Section 14.3). A schematic of the enzyme's active site is shown below; the pyruvate is in the center:

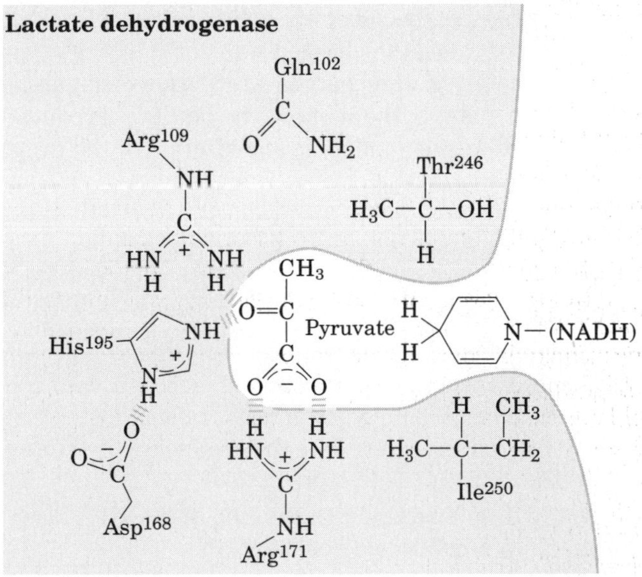

Lactate dehydrogenase

The reaction mechanism is similar to many NADH reductions (Fig. 13–24); it is approximately the reverse of steps 2 and 3 of Figure 14–7. The transition state involves a strongly polarized carbonyl group of the pyruvate molecule as shown below:

$$
\begin{array}{c}
CH_3 \\
| \\
{}^{-}O-C^{+} \\
| \\
C \\
O \overset{-}{\diagup\diagdown} O
\end{array}
$$

(a) A mutant form of LDH in which Arg^{109} is replaced with Gln shows only 5% of the pyruvate binding and 0.07% of the activity of wild-type enzyme. Provide a plausible explanation for the effects of this mutation.

(b) A mutant form of LDH in which Arg^{171} is replaced with Lys shows only 0.05% of the wild-type level of substrate binding. Why is this dramatic effect surprising?

(c) In the crystal structure of LDH, the guanidinium group of Arg^{171} and the carboxyl group of pyruvate are aligned as shown in a co-planar "forked" configuration. Based on this, provide a plausible explanation for the dramatic effect of substituting Arg^{171} with Lys.

(d) A mutant form of LDH in which Ile^{250} is replaced with Gln shows reduced binding of NADH. Provide a plausible explanation for this result.

Clarke and colleagues also set out to engineer a mutant version of LDH that would bind and reduce oxaloacetate rather than pyruvate. They made a single substitution, replacing Gln^{102} with Arg; the resulting enzyme would reduce oxaloacetate to malate and would no longer reduce pyruvate to lactate. They had therefore converted LDH to malate dehydrogenase.

(e) Sketch the active site of this mutant LDH with oxaloacetate bound.

(f) Provide a plausible explanation for why this mutant enzyme now "prefers" oxaloacetate instead of pyruvate.

(g) The authors were surprised that substituting a larger amino acid in the active site allowed a larger substrate to bind. Provide a plausible explanation for this result.

Answer

(a) In the wild-type enzyme, the substrate is held in place by a hydrogen bond and an ion-dipole interaction between the charged side chain of Arg^{109} and the polar carbonyl of pyruvate. During catalysis, the charged Arg^{109} side chain also stabilizes the polarized carbonyl transition state. In the mutant, the binding is reduced to just a hydrogen bond, substrate binding is weaker, and ionic stabilization of the transition state is lost, reducing catalytic activity.

(b) Because Lys and Arg are roughly the same size and have a similar positive charge, they probably have very similar properties. Furthermore, because pyruvate binds to Arg^{171} by (presumably) an ionic interaction, an Arg to Lys mutation would probably have little effect on substrate binding.

(c) The "forked" arrangement aligns two positively charged groups of Arg residues with the negatively charged oxygens of pyruvate and facilitates two combined hydrogen-bond and ion-dipole interactions. When Lys is present, only one such combined hydrogen-bond and ion-dipole interaction is possible, thus reducing the strength of the interaction. The positioning of the substrate is less precise.

(d) Ile^{250} interacts hydrophobically with the ring of NADH. This type of interaction is not possible with the hydrophilic side chain of Gln.

(e) The structure is shown below.

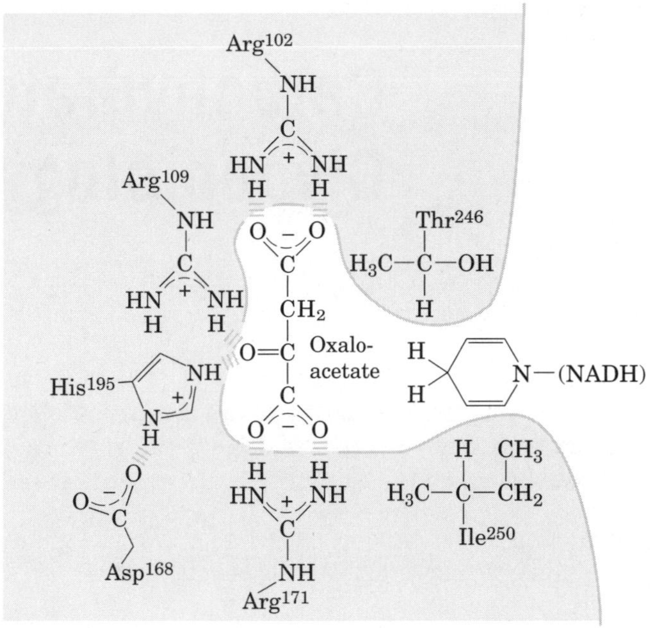

(f) The mutant enzyme rejects pyruvate because pyruvate's hydrophobic methyl group will not interact with the highly hydrophilic guanidinium group of Arg^{102}. The mutant binds oxaloacetate because of the strong ionic interaction between the Arg^{102} side chain and the carboxyl of oxaloacetate.

(g) The protein must be flexible enough to accommodate the added bulk of the side chain and the larger substrate.

References

Clarke, A.R., Atkinson, T., & Holbrook, J.J. (1989) From analysis to synthesis: new ligand binding sites on the lactate dehydrogenase framework, Part I. *Trends Biochem. Sci.* **14,** 101–105.

Clarke, A.R., Atkinson, T., & Holbrook, J.J. (1989) From analysis to synthesis: new ligand binding sites on the lactate dehydrogenase framework, Part II. *Trends Biochem. Sci.* **14,** 145–148.

chapter

7 Carbohydrates and Glycobiology

1. **Sugar Alcohols** In the monosaccharide derivatives known as sugar alcohols, the carbonyl oxygen is reduced to a hydroxyl group. For example, D-glyceraldehyde can be reduced to glycerol. However, this sugar alcohol is no longer designated D or L. Why?

   Answer With reduction of the carbonyl oxygen to a hydroxyl group, the stereochemistry at C-1 and C-3 is the same; the glycerol molecule is not chiral.

2. **Recognizing Epimers** Using Figure 7–3, identify the epimers of **(a)** D-allose, **(b)** D-gulose, and **(c)** D-ribose at C-2, C-3, and C-4.

   Answer Epimers differ by the configuration about only *one* carbon.
   (a) D-altrose (C-2), D-glucose (C-3), D-gulose (C-4)
   (b) D-idose (C-2), D-galactose (C-3), D-allose (C-4)
   (c) D-arabinose (C-2), D-xylose (C-3)

3. **Melting Points of Monosaccharide Osazone Derivatives** Many carbohydrates react with phenylhydrazine ($C_6H_5NHNH_2$) to form bright yellow crystalline derivatives known as osazones:

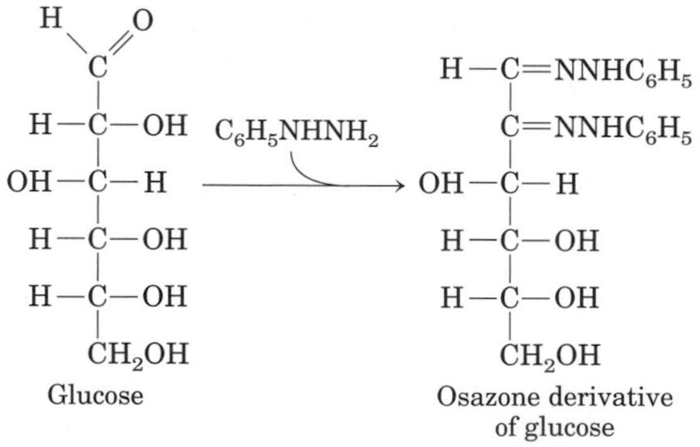

   Glucose: H–C=O; H—C—OH; OH—C—H; H—C—OH; H—C—OH; CH_2OH

   $\xrightarrow{C_6H_5NHNH_2}$

   Osazone derivative of glucose: H—C=$NNHC_6H_5$; C=$NNHC_6H_5$; OH—C—H; H—C—OH; H—C—OH; CH_2OH

   The melting temperatures of these derivatives are easily determined and are characteristic for each osazone. This information was used to help identify monosaccharides before the development of HPLC or gas-liquid chromatography. Listed below are the melting points (MPs) of some aldose-osazone derivatives:

Monosaccharide	MP of anhydrous monosaccharide (°C)	MP of osazone derivative (°C)
Glucose	146	205
Mannose	132	205
Galactose	165–168	201
Talose	128–130	201

As the table shows, certain pairs of derivatives have the same melting points, although the underivatized monosaccharides do not. Why do glucose and mannose, and similarly galactose and talose, form osazone derivatives with the same melting points?

Answer The configuration at C-2 of an aldose is lost in its osazone derivative, so aldoses differing only at the C-2 configuration (C-2 epimers) give the same derivative, with the same melting point. Glucose and mannose are C-2 epimers and thus form the same osazone; the same is true for galactose and talose (see Fig. 7–3).

4. Interconversion of D-Glucose Forms A solution of one enantiomer of a given monosaccharide rotates plane-polarized light to the left (counterclockwise) and is called the levorotatory isomer, designated (−); the other enantiomer rotates plane-polarized light to the same extent but to the right (clockwise) and is called the dextrorotatory isomer, designated (+). An equimolar mixture of the (+) and (−) forms does not rotate plane-polarized light.

The optical activity of a stereoisomer is expressed quantitatively by its *optical rotation*, the number of degrees by which plane-polarized light is rotated on passage through a given path length of a solution of the compound at a given concentration. The *specific rotation* $[\alpha]_\lambda^t$ of an optically active compound is defined thus:

$$[\alpha]_\lambda^t = \frac{\text{observed optical rotation }(^\circ)}{\text{optical path length (dm)} \times \text{concentration (g/mL)}}$$

The temperature (t) and the wavelength of the light (λ) employed (usually, as here, the D line of sodium, 589 nm) must be specified.

A freshly prepared solution of α-D-glucose shows a specific rotation of +112°. Over time, the rotation of the solution gradually decreases and reaches an equilibrium value corresponding to $[\alpha]_D^{25^\circ C}$ = +52.5°. In contrast, a freshly prepared solution of β-D-glucose has a specific rotation of +19°. The rotation of this solution increases over time to the same equilibrium value as that shown by the α anomer.

(a) Draw the Haworth perspective formulas of the α and β forms of D-glucose. What feature distinguishes the two forms?

(b) Why does the specific rotation of a freshly prepared solution of the α form gradually decrease with time? Why do solutions of the α and β forms reach the same specific rotation at equilibrium?

(c) Calculate the percentage of each of the two forms of D-glucose present at equilibrium.

Answer

(a)

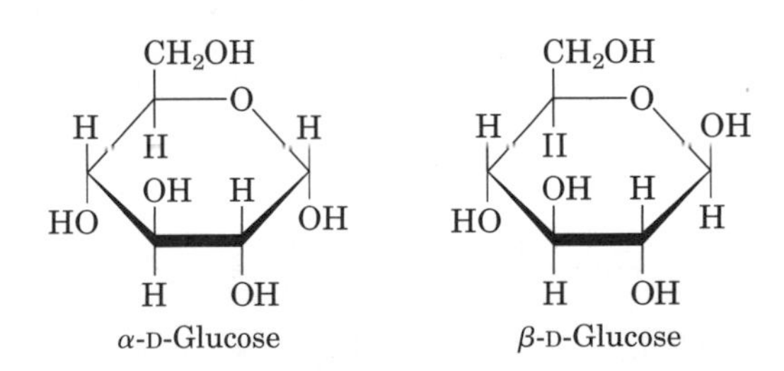

The α and β forms of D-glucose differ only at the hemiacetal carbon (C-1; the anomeric carbon).

(b) A fresh solution of the α form of glucose undergoes mutarotation to an equilibrium mixture containing both the α and β forms. The same applies to a fresh solution of the β form.

(c) The change in specific rotation of a solution in changing from 100% α form ($[\alpha]_D^{25^\circ C}$ 112°) to 100% β form ($[\alpha]_D^{25^\circ C}$ 19°) is 93°. For an equilibrium mixture having $[\alpha]_D^{25^\circ C}$ 52.5°, the fraction of D-glucose in the α form is

$$\frac{52.5^\circ - 19^\circ}{112^\circ - 19^\circ} = \frac{33.5^\circ}{93^\circ} = 0.36 = 36\%$$

Thus, ignoring the small portions of furanose forms (~0.5% each), the mixture contains about 36% α-D-glucose and 64% β-D-glucose.

5. Configuration and Conformation Which bond(s) in α-D-glucose must be broken to change its configuration to β-D-glucose? Which bond(s) to convert D-glucose to D-mannose? Which bond(s) to convert one "chair" form of D-glucose to the other?

Answer To convert α-D-glucose to β-D-glucose, the bond between C-1 and the hydroxyl on C-5 must be broken and reformed in the opposite configuration (as in Fig. 7–6). To convert D-glucose to D-mannose, either the —H or the —OH on C-2 must be broken and reformed in the opposite configuration. Conversion between chair conformations does not require bond breakage; this is the critical distinction between configuration and conformation.

6. Deoxysugars Is D-2-deoxygalactose the same chemical as D-2-deoxyglucose? Explain.

Answer No; glucose and galactose differ in their configuration at C-4.

7. Sugar Structures Describe the common structural features and the differences for each pair: **(a)** cellulose and glycogen; **(b)** D-glucose and D-fructose; **(c)** maltose and sucrose.

Answer (a) Both are polymers of D-glucose, but they differ in the glycosidic linkage: (β1→4) for cellulose, (α1→4) for glycogen. **(b)** Both are hexoses, but glucose is an aldohexose, fructose a ketohexose. **(c)** Both are disaccharides, but maltose has two (α1→4)-linked D-glucose units; sucrose has (α1↔2β)-linked D-glucose and D-fructose.

8. Reducing Sugars Draw the structural formula for α-D-glucosyl-(1→6)-D-mannosamine and circle the part of this structure that makes the compound a reducing sugar.

Answer

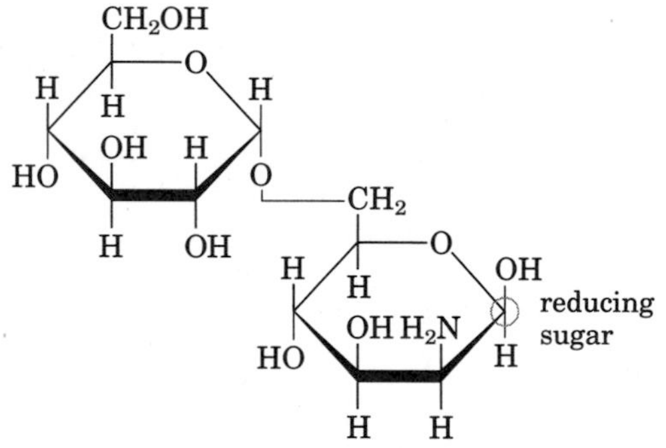

9. Hemiacetal and Glycosidic Linkages Explain the difference between a hemiacetal and a glycoside.

Answer A hemiacetal is formed when an aldose or ketose condenses with an alcohol; a glycoside is formed when a hemiacetal condenses with an alcohol (see Fig. 7–5, p. 238).

10. A Taste of Honey The fructose in honey is mainly in the β-D-pyranose form. This is one of the sweetest carbohydrates known, about twice as sweet as glucose; the β-D-furanose form of fructose is much less sweet. The sweetness of honey gradually decreases at a high temperature. Also, high-fructose corn syrup (a commercial product in which much of the glucose in corn syrup is converted to fructose) is used for sweetening *cold* but not *hot* drinks. What chemical property of fructose could account for both these observations?

Answer Straight-chain fructose can cyclize to yield either the pyranose or the furanose structure. Increasing the temperature shifts the equilibrium in the direction of the furanose form, reducing the sweetness of the solution. The higher the temperature, the less sweet is the fructose solution.

11. Reducing Disaccharide A disaccharide, which you know to be either maltose or sucrose, is treated with Fehling's solution, and a red color is formed. Which sugar is it, and how do you know?

Answer Maltose; sucrose has no reducing (oxidizable) group, as the anomeric carbons of both monosaccharides are involved in the glycosidic bond.

12. Glucose Oxidase in Determination of Blood Glucose The enzyme glucose oxidase isolated from the mold *Penicillium notatum* catalyzes the oxidation of β-D-glucose to D-glucono-δ-lactone. This enzyme is highly specific for the β anomer of glucose and does not affect the α anomer. In spite of this specificity, the reaction catalyzed by glucose oxidase is commonly used in a clinical assay for total blood glucose—that is, for solutions consisting of a mixture of β- and α-D-glucose. What are the circumstances required to make this possible? Aside from allowing the detection of smaller quantities of glucose, what advantage does glucose oxidase offer over Fehling's reagent for the determination of blood glucose?

Answer The rate of mutarotation (interconversion of the α and β anomers) is sufficiently high that, as the enzyme consumes β-D-glucose, more α-D-glucose is converted to the β form, and, eventually, all the glucose is oxidized. Glucose oxidase is specific for glucose and does not detect other reducing sugars (such as galactose). Fehling's reagent reacts with any reducing sugar.

13. Invertase "Inverts" Sucrose The hydrolysis of sucrose (specific rotation +66.5°) yields an equimolar mixture of D-glucose (specific rotation +52.5°) and D-fructose (specific rotation −92°). (See Problem 4 for details of specific rotation.)

(a) Suggest a convenient way to determine the rate of hydrolysis of sucrose by an enzyme preparation extracted from the lining of the small intestine.

(b) Explain why, in the food industry, an equimolar mixture of D-glucose and D-fructose formed by hydrolysis of sucrose is called invert sugar.

(c) The enzyme invertase (now commonly called sucrase) is allowed to act on a 10% (0.1 g/mL) solution of sucrose until hydrolysis is complete. What will be the observed optical rotation of the solution in a 10 cm cell? (Ignore a possible small contribution from the enzyme.)

Answer

(a) An equimolar mixture of D-glucose and D-fructose, such as that formed from sucrose hydrolysis, has optical rotation = 52.5° + (−92.0°) = −39.5°. Enzyme (sucrase) activity can be assayed by observing the change in optical rotation of a solution of 100% sucrose (specific rotation = +66.5°) as it is converted to a 1:1 mixture of D-glucose and D-fructose.

(b) The optical rotation of the hydrolysis mixture is negative (inverted) relative to that of the unhydrolyzed sucrose solution.

(c) The addition of 1 mol of water (M_r 18) in the hydrolysis of 1 mol of sucrose (M_r 342) gives the products an increase in weight of (18/342)(100%) = 5.26% with respect to the starting sugar. Accordingly, a 10% sucrose solution yields a [10 + (0.053 × 10)]% = 10.5% solution of invert sugar. Of this 10.5%, 5.25% (0.0525 g/mL) is D-glucose and 5.25% is D-fructose. By rearranging the equation in Problem 4,

$$[\alpha]_D^{25°C} = \frac{\text{observed optical rotation (°)}}{\text{optical path length (dm)} \times \text{concentration (g/mL)}}$$

we can determine the optical rotation of each sugar in the mixture in a 10 cm cell:
Optical rotation of glucose = (52.5°)(1 dm)(0.0525 g/mL) = 2.76°
Optical rotation of fructose = (92°)(1 dm)(0.0525 g/mL) = −4.8°
The observed optical rotation of the solution is 2.76° + (−4.8°) = −2.0°

14. Manufacture of Liquid-Filled Chocolates The manufacture of chocolates containing a liquid center is an interesting application of enzyme engineering. The flavored liquid center consists largely of an aqueous solution of sugars rich in fructose to provide sweetness. The technical dilemma is the following: the chocolate coating must be prepared by pouring hot melted chocolate over a solid (or almost solid) core, yet the final product must have a liquid, fructose-rich center. Suggest a way to solve this problem. (Hint: Sucrose is much less soluble than a mixture of glucose and fructose.)

Answer Prepare the core as a semisolid slurry of sucrose and water. Add a small amount of sucrase (invertase), and quickly coat the semisolid mixture with chocolate. After the chocolate coat has cooled and hardened, the sucrase hydrolyzes enough of the sucrose to form a more liquid center: a mixture of fructose, glucose, and sucrose.

15. Anomers of Sucrose? Lactose exists in two anomeric forms, but no anomeric forms of sucrose have been reported. Why?

Answer Lactose (Gal(β1→4)Glc) has a free anomeric carbon (on the glucose residue). In sucrose (Glc(α1↔2β)Fru), the anomeric carbons of both monosaccharide units are involved in the glycosidic bond, and the disaccharide has no free anomeric carbon to undergo mutarotation.

16. Gentiobiose Gentiobiose (D-Glc(β1→6)D-Glc) is a disaccharide found in some plant glycosides. Draw the structure of gentobiose based on its abbreviated name. Is it a reducing sugar? Does it undergo mutarotation?

Answer

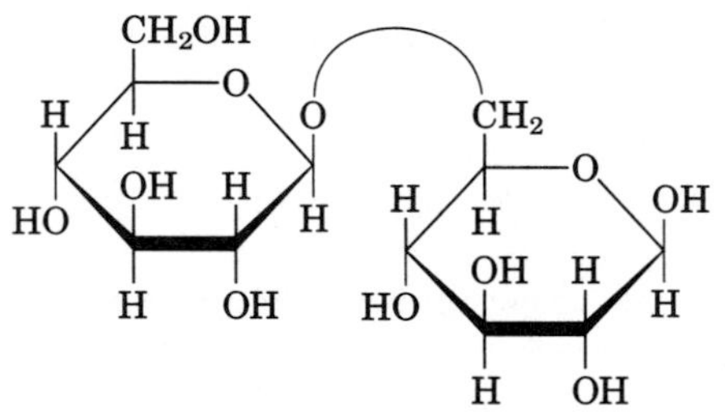

It is a reducing sugar; it undergoes mutarotation.

17. Identifying Reducing Sugars Is *N*-acetyl-β-D-glucosamine (Fig. 7–9) a reducing sugar? What about D-gluconate? Is the disaccharide GlcN(α1↔1α)Glc a reducing sugar?

Answer *N*-Acetyl-β-D-glucosamine is a reducing sugar; its C-1 can be oxidized (see Fig. 7–10, p. 241). D-Gluconate is not a reducing sugar; its C-1 is already at the oxidation state of a carboxylic acid. GlcN(α1↔1α)Glc is not a reducing sugar; the anomeric carbons of both monosaccharides are involved in the glycosidic bond.

18. Cellulose Digestion Cellulose could provide a widely available and cheap form of glucose, but humans cannot digest it. Why not? If you were offered a procedure that allowed you to acquire this ability, would you accept? Why or why not?

Answer Humans cannot break down cellulose to its monosaccharides because they lack cellulases, a family of enzymes, produced chiefly by fungi, bacteria, and protozoans, that catalyze the hydrolysis of cellulose to glucose. In ruminant animals (such as cows and sheep), the rumen (one of four stomach compartments) acts as an anaerobic fermenter in which bacteria and protozoa degrade cellulose, making its glucose available as a nutrient to the animal. If cellulase were present in the human digestive tract, we could use foods rich in cellulose as nutrients. This would greatly increase the forms of biomass that could be used for human nutrition. This change might require some changes in the teeth that would allow cellulosic materials to be ground into small pieces to serve as cellulase substrates.

19. Physical Properties of Cellulose and Glycogen The almost pure cellulose obtained from the seed threads of *Gossypium* (cotton) is tough, fibrous, and completely insoluble in water. In contrast, glycogen obtained from muscle or liver disperses readily in hot water to make a turbid solution. Despite their markedly different physical properties, both substances are (1→4)-linked D-glucose polymers of comparable molecular weight. What structural features of these two polysaccharides underlie their different physical properties? Explain the biological advantages of their respective properties.

Answer Native cellulose consists of glucose units linked by (β1→4) glycosidic bonds. The β linkages force the polymer chain into an extended conformation. Parallel series of these extended chains can form *intermolecular* hydrogen bonds, thus aggregating into long, tough, insoluble fibers. Glycogen consists of glucose units linked by (α1→4) glycosidic bonds. The α linkages cause bends in the chain, and glycogen forms helical structures with *intramolecular* hydrogen bonding; it cannot form long fibers. In addition, glycogen is highly branched and, because many of its hydroxyl groups are exposed to water, is highly hydrated and therefore very water-soluble. It can be extracted as a dispersion in hot water.

The physical properties of the two polymers are well suited to their biological roles. Cellulose serves as a structural material in plants, consistent with the side-by-side aggregation of long molecules into tough, insoluble fibers. Glycogen is a storage fuel in animals. The highly hydrated glycogen granules, with their abundance of free, nonreducing ends, can be rapidly hydrolyzed by glycogen phosphorylase to release glucose 1-phosphate, available for oxidation and energy production.

20. Dimensions of a Polysaccharide Compare the dimensions of a molecule of cellulose and a molecule of amylose, each of M_r 200,000.

Answer Cellulose is several times longer; it assumes an extended conformation, whereas amylose has a helical structure.

21. Growth Rate of Bamboo The stems of bamboo, a tropical grass, can grow at the phenomenal rate of 0.3 m/day under optimal conditions. Given that the stems are composed almost entirely of cellulose fibers oriented in the direction of growth, calculate the number of sugar residues per second that must be added enzymatically to growing cellulose chains to account for the growth rate. Each D-glucose unit contributes ~0.5 nm to the length of a cellulose molecule.

Answer First, calculate the growth per second:

$$\frac{0.3 \text{ m/day}}{(24 \text{ h/day})(60 \text{ min/h})(60 \text{ s/min})} = 3 \times 10^{-6} \text{ m/s}$$

Given that each glucose residue increases the length of the cellulose chain by 0.5 nm (5×10^{-10} m), the number of residues added per second is

$$\frac{3 \times 10^{-6} \text{ m/s}}{5 \times 10^{-10} \text{ m/residue}} = 6{,}000 \text{ residues/s}$$

22. Glycogen as Energy Storage: How Long Can a Game Bird Fly? Since ancient times it has been observed that certain game birds, such as grouse, quail, and pheasants, are easily fatigued. The Greek historian Xenophon wrote, "The bustards . . . can be caught if one is quick in starting them up, for they will fly only a short distance, like partridges, and soon tire; and their flesh is delicious." The flight muscles of game birds rely almost entirely on the use of glucose 1-phosphate for energy, in the form of ATP (Chapter 14). The glucose 1-phosphate is formed by the breakdown of stored muscle glycogen, catalyzed by the enzyme glycogen phosphorylase. The rate of ATP production is limited by the rate at which glycogen can be broken down. During a "panic flight," the game bird's rate of glycogen breakdown is quite high, approximately 120 μmol/min of glucose 1-phosphate produced per gram of fresh tissue. Given that the flight muscles usually contain about 0.35% glycogen by weight, calculate how long a game bird can fly. (Assume the average molecular weight of a glucose residue in glycogen is 162 g/mol.)

Answer Given the average molecular weight of a glucose residue = 162, the amount of usable glucose (as glycogen) in 1 g of tissue is

$$\frac{3.5 \times 10^{-3}\ \text{g}}{162\ \text{g/mol}} = 2.2 \times 10^{-5}\ \text{mol}$$

In 1 min, 120 μmol of glucose 1-phosphate is produced, so 120 μmol of glucose is hydrolyzed. Thus, depletion of the glycogen would occur in

$$\frac{(2.2 \times 10^{-5}\ \text{mol})(60\ \text{s/min})}{120 \times 10^{-6}\ \text{mol/min}} = 11\ \text{s}$$

23. Relative Stability of Two Conformers Explain why the two structures shown in Figure 7–19 are so different in energy (stability). Hint: See Figure 1–21.

Answer The ball-and-stick model of the disaccharide in Figure 7–19 shows no steric interactions, but a space-filling model, showing atoms with their real relative sizes, would show several strong steric hindrances in the −170°,−170° conformer that are not present in the 30°,−40° conformer.

24. Volume of Chondroitin Sulfate in Solution One critical function of chondroitin sulfate is to act as a lubricant in skeletal joints by creating a gel-like medium that is resilient to friction and shock. This function seems to be related to a distinctive property of chondroitin sulfate: the volume occupied by the molecule is much greater in solution than in the dehydrated solid. Why is the volume so much larger in solution?

Answer In solution, the negative charges on chondroitin sulfate repel each other and force the molecule into an extended conformation. The polar molecule also attracts many water molecules (water of hydration), further increasing the molecular volume. In the dehydrated solid, each negative charge is counterbalanced by a counterion, such as Na^+, and the molecule collapses into its condensed form.

25. Heparin Interactions Heparin, a highly negatively charged glycosaminoglycan, is used clinically as an anticoagulant. It acts by binding several plasma proteins, including antithrombin III, an inhibitor of blood clotting. The 1:1 binding of heparin to antithrombin III seems to cause a conformational change in the protein that greatly increases its ability to inhibit clotting. What amino acid residues of antithrombin III are likely to interact with heparin?

Answer Positively charged amino acid residues would be the best candidates to bind to the highly negatively charged groups on heparin. In fact, Lys residues of antithrombin III interact with heparin.

26. Permutations of a Trisaccharide Think about how one might estimate the number of possible trisaccharides composed of *N*-acetylglucosamine 4-sulfate (GlcNAc4S) and glucuronic acid (GlcA), and draw 10 of them.

Answer If GlcNAc4S is represented as A, and GlcA as B, the trimer could have any of these sequences: AAA, AAB, ABB, ABA, BBB, BBA, BAA, or BAB (8 possible sequences). The connections between each pair of monosaccharides could be 1→6, 1→4, 1→3, or 1→1 (4 possibilities for each of two bonds, or 4 × 4 = 16 possible linkages in all), and each linkage could involve either the α or the β anomer of each sugar (2 possibilities for each of two bonds, so 2 × 2 = 4 stereochemical possibilities). Therefore there are 8 × 16 × 4 = 512 possible permutations!

27. Effect of Sialic Acid on SDS Polyacrylamide Gel Electrophoresis Suppose you have four forms of a protein, all with identical amino acid sequence but containing zero, one, two, or three oligosaccharide chains, each ending in a single sialic acid residue. Draw the gel pattern you would expect when a mixture of these four glycoproteins is subjected to SDS polyacrylamide gel electrophoresis (see Fig. 3–18) and stained for protein. Identify any bands in your drawing.

Answer

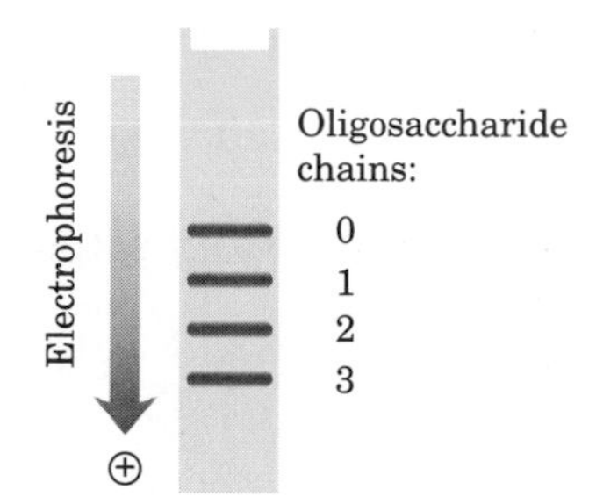

The significant feature of sialic acids is the negative charge of their carboxyl group. The four glycoproteins would have the same charge except for the additional 1, 2, or 3 negative charges of the sialic acid residues. In SDS gel electrophoresis, the proteins are coated uniformly with a layer of sodium dodecyl sulfate (which is negatively charged, p. 89) and thus move toward the positive electrode. The glycoproteins with 1, 2, or 3 extra negative charges will move progressively faster than the form without sialic acid.

28. Information Content of Oligosaccharides The carbohydrate portion of some glycoproteins may serve as a cellular recognition site. In order to perform this function, the oligosaccharide moiety of glycoproteins must have the potential to exist in a large variety of forms. Which can produce a greater variety of structures: oligopeptides composed of five different amino acid residues or oligosaccharides composed of five different monosaccharide residues? Explain.

Answer Oligosaccharides; their monosaccharide residues can be combined in more ways than the amino acid residues of oligopeptides. Each of the several hydroxyl groups of each monosaccharide can participate in glycosidic bonds, and the configuration of each glycosidic bond can be either α or β. Furthermore, the polymer can be linear or branched. Oligopeptides are unbranched polymers, with all amino acid residues linked through identical peptide bonds.

29. Determination of the Extent of Branching in Amylopectin The amount of branching (number of (α1→6) glycosidic bonds) in amylopectin can be determined by the following procedure. A sample of amylopectin is exhaustively methylated—treated with a methylating agent (methyl iodide) that replaces the hydrogen of every sugar hydroxyl with a methyl group, converting —OH to —OCH_3. All the glycosidic bonds in the treated sample are then hydrolyzed in aqueous acid, and the amount of 2,3-di-*O*-methylglucose so formed is determined.

CH_2OH, O, H, H, H, HO, OCH_3, H, OH, H, OCH_3

2,3-Di-*O*-methylglucose

(a) Explain the basis of this procedure for determining the number of (α1→6) branch points in amylopectin. What happens to the unbranched glucose residues in amylopectin during the methylation and hydrolysis procedure?

(b) A 258 mg sample of amylopectin treated as described above yielded 12.4 mg of 2,3-di-*O*-methylglucose. Determine what percentage of the glucose residues in amylopectin contain an (α1→6) branch. (Assume the average molecular weight of a glucose residue in amylopectin is 162 g/mol.)

Answer

(a) In glucose residues at branch points, the hydroxyl of C-6 is protected from methylation because it is involved in a glycosidic linkage. During complete methylation and subsequent hydrolysis, the branch-point residues yield 2,3-di-*O*-methylglucose and the unbranched residues yield 2,3,6-tri-*O*-methylglucose.

(b) Given the average molecular weight of a glucose residue = 162, then 258 mg of amylopectin contains

$$\frac{258 \times 10^{-3}\ \text{g}}{162\ \text{g/mol}} = 1.59 \times 10^{-3}\ \text{mol of glucose}$$

The 12.4 mg yield of 2,3-di-*O*-methylglucose (M_r 208) is equivalent to

$$\frac{12.4 \times 10^{-3}\ \text{g}}{208\ \text{g/mol}} = 5.96 \times 10^{-5}\ \text{mol of glucose}$$

Thus, the percentage of glucose residues in amylopectin that yield 2,3-di-*O*-methylglucose is

$$\frac{(5.96 \times 10^{-5}\ \text{mol})(100\%)}{1.59 \times 10^{-3}\ \text{mol}} = 3.75\%$$

30. Structural Analysis of a Polysaccharide A polysaccharide of unknown structure was isolated, subjected to exhaustive methylation, and hydrolyzed. Analysis of the products revealed three methylated sugars: 2,3,4-tri-*O*-methyl-D-glucose, 2,4-di-*O*-methyl-D-glucose, and 2,3,4,6-tetra-*O*-methyl-D-glucose, in the ratio 20:1:1. What is the structure of the polysaccharide?

Answer The polysaccharide is a branched glucose polymer. Because the predominant product is 2,3,4-tri-*O*-methyl-D-glucose, the predominant glycosidic linkage must be (1→6). The formation of 2,4-di-*O*-methyl-D-glucose indicates that branch points occur through C-3. The ratio of these two methylated sugars indicates that a branch occurs at an average frequency of once every 20 residues. The 2,3,4,6-tetra-*O*-methyl-D-glucose is derived from nonreducing chain ends, which compose about $\frac{1}{20}$, or 5%, of the residues, consistent with a high degree of branching. Thus, the polysaccharide has chains of (1→6)-linked D-glucose residues with (1→3)-linked branches, about one branch every 20 residues.

Data Analysis Problem

31. Determining the Structure of ABO Blood Group Antigens The human ABO blood group system was first discovered in 1901, and in 1924 this trait was shown to be inherited at a single gene locus with three alleles. In 1960, W. T. J. Morgan published a paper summarizing what was known at that time about the structure of the ABO antigen molecules. When the paper was published, the complete structures of the A, B, and O antigens were not yet known; this paper is an example of what scientific knowledge looks like "in the making."

In any attempt to determine the structure of an unknown biological compound, researchers must deal with two fundamental problems: (1) If you don't know what *it* is, how do you know if *it* is pure? (2) If you don't know what *it* is, how do you know that your extraction and purification conditions have not changed *its* structure? Morgan addressed problem 1 through several methods. One method is described in his paper as observing "constant analytical values after fractional solubility tests" (p. 312). In this case, "analytical values" are measurements of chemical composition, melting point, and so forth.

(a) Based on your understanding of chemical techniques, what could Morgan mean by "fractional solubility tests"?

(b) Why would the analytical values obtained from fractional solubility tests of a *pure* substance be constant, and those of an *impure* substance not be constant?

Morgan addressed problem 2 by using an assay to measure the immunological activity of the substance present in different samples.

(c) Why was it important for Morgan's studies, and especially for addressing problem 2, that this activity assay be quantitative (measuring a level of activity) rather than simply qualitative (measuring only the presence or absence of a substance)?

The structure of the blood group antigens is shown in Figure 10–15. In his paper (p. 314), Morgan listed several properties of the three antigens, A, B, and O, that were known at that time:

1. Type B antigen has a higher content of galactose than A or O.
2. Type A antigen contains more total amino sugars than B or O.
3. The glucosamine/galactosamine ratio for the A antigen is roughly 1.2; for B, it is roughly 2.5.

(d) Which of these findings is (are) consistent with the known structures of the blood group antigens?

(e) How do you explain the discrepancies between Morgan's data and the known structures?

In later work, Morgan and his colleagues used a clever technique to obtain structural information about the blood group antigens. Enzymes had been found that would specifically degrade the antigens. However, these were available only as crude enzyme preparations, perhaps containing more than one enzyme of unknown specificity. Degradation of the blood type antigens by these crude enzymes could be inhibited by the addition of particular sugar molecules to the reaction. Only sugars found in the blood type antigens would cause this inhibition. One enzyme preparation, isolated from the protozoan *Trichomonas foetus*, would degrade all three antigens and was inhibited by the addition of particular sugars. The results of these studies are summarized in the table below, showing the percentage of substrate remaining unchanged when the *T. foetus* enzyme acted on the blood group antigens in the presence of sugars.

	Unchanged substrate (%)		
Sugar added	**A antigen**	**B antigen**	**O antigen**
Control—no sugar	3	1	1
L-Fucose	3	1	100
D-Fucose	3	1	1
L-Galactose	3	1	3
D-Galactose	6	100	1
N-Acetylglucosamine	3	1	1
N-Acetylgalactosamine	100	6	1

For the O antigen, a comparison of the control and L-fucose results shows that L-fucose inhibits the degradation of the antigen. This is an example of product inhibition, in which an excess of reaction product shifts the equilibrium of the reaction, preventing further breakdown of substrate.

(f) Although the O antigen contains galactose, *N*-acetylglucosamine, and *N*-acetylgalactosamine, none of these sugars inhibited the degradation of this antigen. Based on these data, is the enzyme preparation from *T. foetus* an endo- or exoglycosidase? (Endoglycosidases cut bonds between interior residues; exoglycosidases remove one residue at a time from the end of a polymer.) Explain your reasoning.

(g) Fucose is also present in the A and B antigens. Based on the structure of these antigens, why does fucose fail to prevent their degradation by the *T. foetus* enzyme? What structure would be produced?

(h) Which of the results in **(f)** and **(g)** are consistent with the structures shown in Figure 10–15? Explain your reasoning.

Answer

(a) The tests involve trying to dissolve only part of the sample in a variety of solvents, then analyzing both dissolved and undissolved materials to see whether their compositions differ.

(b) For a pure substance, all molecules are the same and any dissolved fraction will have the same composition as any undissolved fraction. An impure substance is a mixture of more than one compound. When treated with a particular solvent, more of one component may dissolve, leaving more of the other component(s) behind. As a result, the dissolved and undissolved fractions have different compositions.

(c) A quantitative assay allows researchers to be sure that none of the activity has been lost through degradation. When determining the structure of a molecule, it is important that the sample under analysis consist only of intact (undegraded) molecules. If the sample is contaminated with degraded material, this will give confusing and perhaps uninterpretable structural results. A qualitative assay would detect the presence of activity, even if it had become significantly degraded.

(d) Results 1 and 2. Result 1 is consistent with the known structure, because type B antigen has three molecules of galactose; types A and O each have only two. Result 2 is also consistent, because type A has two amino sugars (*N*-acetylgalactosamine and *N*-acetylglucosamine); types B and O have only one (*N*-acetylglucosamine). Result 3 is *not* consistent with the known structure: for type A, the glucosamine:galactosamine ratio is 1:1; for type B, it is 1:0.

(e) The samples were probably impure and/or partly degraded. The first two results were correct possibly because the method was only roughly quantitative and thus not as sensitive to inaccuracies in measurement. The third result is more quantitative and thus more likely to differ from predicted values, because of impure or degraded samples.

(f) An exoglycosidase. If it were an endoglycosidase, one of the products of its action on O antigen would include galactose, *N*-acetylglucosamine, or *N*-acetylgalactosamine, and at least one of those sugars would be able to inhibit the degradation. Given that the enzyme is not inhibited by any of these sugars, it must be an exoglycosidase, removing only the terminal sugar from the chain. The terminal sugar of O antigen is fucose, so fucose is the only sugar that could inhibit the degradation of O antigen.

(g) The exoglycosidase removes *N*-acetylgalactosamine from A antigen and galactose from B antigen. Because fucose is not a product of either reaction, it will not prevent removal of these sugars, and the resulting substances will no longer be active as A or B antigen. However, the products should be active as O antigen, because degradation stops at fucose.

(h) All the results are consistent with Figure 10–15. (1) D-Fucose and L-galactose, which would protect against degradation, are not present in any of the antigens. (2) The terminal sugar of A antigen is *N*-acetylgalactosamine, and this sugar alone protects this antigen from degradation. (3) The terminal sugar of B antigen is galactose, which is the only sugar capable of protecting this antigen.

Reference

Morgan, W.T. (1960) The Croonian Lecture: a contribution to human biochemical genetics; the chemical basis of blood-group specificity. *Proc. R. Soc. Lond. B Biol. Sci.* **151,** 308–347.

chapter

8

Nucleotides and Nucleic Acids

1. **Nucleotide Structure** Which positions in the purine ring of a purine nucleotide in DNA have the potential to form hydrogen bonds but are not involved in Watson-Crick base pairing?

   Answer All purine ring nitrogens (N-1, N-3, N-7, and N-9) have the potential to form hydrogen bonds (see Figs. 8–1, 8–11, and 2–3). However, N-1 is involved in Watson-Crick hydrogen bonding with a pyrimidine, and N-9 is involved in the *N*-glycosyl linkage with deoxyribose and has very limited hydrogen-bonding capacity. Thus, N-3 and N-7 are available to form further hydrogen bonds.

2. **Base Sequence of Complementary DNA Strands** One strand of a double-helical DNA has the sequence (5′)GCGCAATATTTCTCAAAATATTGCGC(3′). Write the base sequence of the complementary strand. What special type of sequence is contained in this DNA segment? Does the double-stranded DNA have the potential to form any alternative structures?

   Answer The complementary strand is

   (5′)GCGCAATATTTTGAGAAATATTGCGC(3′)

   (Note that the sequence of a single strand is always written in the 5′→3′ direction.) This sequence has a palindrome, an inverted repeat with twofold symmetry:

   (5′)<u>GCGCAATATTT</u>CTCA<u>AAATATTGCGC</u>(3′)
   (3′)<u>CGCGTTATAAA</u>GAGT<u>TTTATAACGCG</u>(5′)

   Because this sequence is self-complementary, the individual strands have the potential to form hairpin structures. The two strands together may also form a cruciform.

3. **DNA of the Human Body** Calculate the weight in grams of a double-helical DNA molecule stretching from the Earth to the moon (~320,000 km). The DNA double helix weighs about 1×10^{-18} g per 1,000 nucleotide pairs; each base pair extends 3.4 Å. For an interesting comparison, your body contains about 0.5 g of DNA!

   Answer
   The length of the DNA is

   $$(3.2 \times 10^5 \text{ km})(10^{12} \text{ nm/km})(10 \text{ Å/nm}) = 3.2 \times 10^{18} \text{ Å}$$

   The number of base pairs (bp) is

   $$\frac{3.2 \times 10^{18} \text{Å}}{3.4 \text{ Å/bp}} = 9.4 \times 10^{17} \text{ bp}$$

   Thus, the weight of the DNA molecule is

   $$(9.4 \times 10^{17} \text{ bp})(1 \times 10^{-18} \text{ g}/10^3 \text{ bp}) = 9.4 \times 10^{-4} \text{ g} = 0.00094 \text{ g}$$

4. **DNA Bending** Assume that a poly(A) tract five base pairs long produces a 20° bend in a DNA strand. Calculate the total (net) bend produced in a DNA if the center base pairs (the third of five) of two successive $(dA)_5$ tracts are located **(a)** 10 base pairs apart; **(b)** 15 base pairs apart. Assume 10 base pairs per turn in the DNA double helix.

   Answer When bending elements are repeated in phase with the helix turn (i.e., every 10 base pairs) as in **(a),** the total bend is additive; when bending elements are repeated out of phase by one half-turn as in **(b),** they cancel each other out. Thus, the net bend is **(a)** 40°; **(b)** 0°.

5. **Distinction between DNA Structure and RNA Structure** Hairpins may form at palindromic sequences in single strands of either RNA or DNA. How is the helical structure of a long and fully base-paired (except at the end) hairpin in RNA different from that of a similar hairpin in DNA?

   Answer The RNA helix assumes the A conformation; the DNA helix generally assumes the B conformation. (The presence of the 2′-OH group on ribose makes it sterically impossible for double-helical RNA to assume the B-form helix.)

6. **Nucleotide Chemistry** The cells of many eukaryotic organisms have highly specialized systems that specifically repair G–T mismatches in DNA. The mismatch is repaired to form a G≡C (not A=T) base pair. This G–T mismatch repair mechanism occurs in addition to a more general system that repairs virtually all mismatches. Can you suggest why cells might require a specialized system to repair G–T mismatches?

   Answer Many C residues of CpG sequences in eukaryotic DNA are methylated at the 5′ position to 5-methylcytosine. (About 5% of all C residues are methylated.) Spontaneous deamination of 5-methylcytosine yields thymine, T, and a G–T mismatch resulting from spontaneous deamination of 5-methylcytosine in a G≡C base pair is one of the most common mismatches in eukaryotic cells. The specialized repair mechanism to convert G–T back to G≡C is directed at this common class of mismatch.

7. **Spontaneous DNA Damage** Hydrolysis of the *N*-glycosyl bond between deoxyribose and a purine in DNA creates an AP site. An AP site generates a thermodynamic destabilization greater than that created by any DNA mismatched base pair. This effect is not completely understood. Examine the structure of an AP site (see Fig. 8–33b) and describe some chemical consequences of base loss.

   Answer Without the base, the ribose ring can be opened to generate the noncyclic aldehyde form. This, and the loss of base-stacking interactions, could contribute significant flexibility to the DNA backbone.

8. **Nucleic Acid Structure** Explain why the absorption of UV light by double-stranded DNA increases (the hyperchromic effect) when the DNA is denatured.

   Answer The double-helical structure is stabilized by hydrogen bonding between complementary bases on opposite strands and by base stacking between adjacent bases on the same strand. Base stacking in nucleic acids causes a decrease in the absorption of UV light (relative to the non-stacked structure). On denaturation of DNA, the base stacking is lost and UV absorption increases.

9. **Determination of Protein Concentration in a Solution Containing Proteins and Nucleic Acids** The concentration of protein or nucleic acid in a solution containing both can be estimated by using their different light absorption properties: proteins absorb most strongly at 280 nm and nucleic acids at 260 nm. Estimates of their respective concentrations in a mixture can be made by measuring the absorbance (A) of the solution at 280 nm and 260 nm and using the table that follows, which gives $R_{280/260}$, the ratio of absorbances at 280 and 260 nm; the percentage of total mass that is nucleic acid; and a factor, F, that corrects the A_{280} reading and gives a more accurate protein estimate. The protein concentration (in mg/ml) $= F \times A_{280}$ (assuming the cuvette is 1 cm wide). Calculate the protein concentration in a solution of $A_{280} = 0.69$ and $A_{260} = 0.94$.

$R_{280/260}$	Proportion of nucleic acid (%)	F
1.75	0.00	1.116
1.63	0.25	1.081
1.52	0.50	1.054
1.40	0.75	1.023
1.36	1.00	0.994
1.30	1.25	0.970
1.25	1.50	0.944
1.16	2.00	0.899
1.09	2.50	0.852
1.03	3.00	0.814
0.979	3.50	0.776
0.939	4.00	0.743
0.874	5.00	0.682
0.846	5.50	0.656
0.822	6.00	0.632
0.804	6.50	0.607
0.784	7.00	0.585
0.767	7.50	0.565
0.753	8.00	0.545
0.730	9.00	0.508
0.705	10.00	0.478
0.671	12.00	0.422
0.644	14.00	0.377
0.615	17.00	0.322
0.595	20.00	0.278

Answer For this protein solution, $R_{280/260} = 0.69/0.94 = 0.73$, so (from the table) $F = 0.508$. The concentration of protein is $F \times A_{280} = (0.508 \times 0.69)$ mg/mL $= 0.35$ mg/mL. **Note:** the table applies to mixtures of proteins, such as might be found in a crude cellular extract, and reflects the absorption properties of average proteins. For a purified protein, the values of F would have to be altered to reflect the unique molar extinction coefficient of that protein.

10. Solubility of the Components of DNA Draw the following structures and rate their relative solubilities in water (most soluble to least soluble): deoxyribose, guanine, phosphate. How are these solubilities consistent with the three-dimensional structure of double-stranded DNA?

Answer

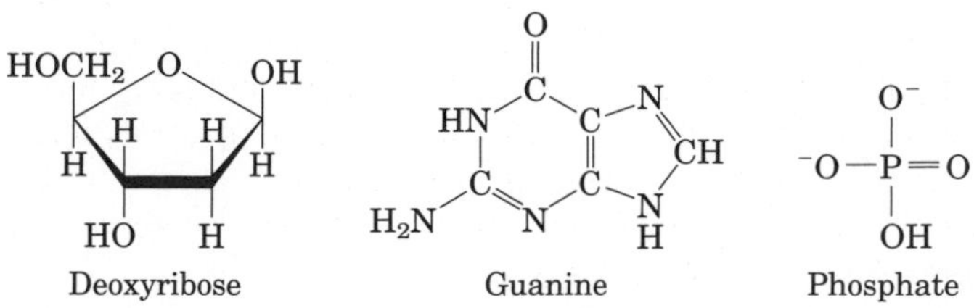

Deoxyribose Guanine Phosphate

Solubilities: phosphate > deoxyribose > guanine. The negatively charged phosphate is the most water-soluble; the deoxyribose, with several hydroxyl groups, is quite water-soluble; and guanine, a hydrophobic base, is relatively insoluble in water. The polar phosphate groups and sugars are on the outside of the DNA double helix, exposed to water. The hydrophobic bases are located in the interior of the double helix, away from water.

11. Sanger Sequencing Logic In the Sanger (dideoxy) method for DNA sequencing, a small amount of a dideoxynucleotide triphosphate—say, ddCTP—is added to the sequencing reaction along with a larger amount of the corresponding dCTP. What result would be observed if the dCTP were omitted?

Answer If dCTP is omitted from the reaction mixture, when the first G residue is encountered in the template, ddCTP is added and polymerization halts. Only one band will appear in the sequencing gel.

12. DNA Sequencing The following DNA fragment was sequenced by the Sanger method. The asterisk indicates a fluorescent label.

*5′ ——— 3′-OH
3′ ——— ATTACGCAAGGACATTAGAC---5′

A sample of the DNA was reacted with DNA polymerase and each of the nucleotide mixtures (in an appropriate buffer) listed below. Dideoxynucleotides (ddNTPs) were added in relatively small amounts.

1. dATP, dTTP, dCTP, dGTP, ddTTP
2. dATP, dTTP, dCTP, dGTP, ddGTP
3. dATP, dCTP, dGTP, ddTTP
4. dATP, dTTP, dCTP, dGTP

The resulting DNA was separated by electrophoresis on an agarose gel, and the fluorescent bands on the gel were located. The band pattern resulting from nucleotide mixture 1 is shown below. Assuming that all mixtures were run on the same gel, what did the remaining lanes of the gel look like?

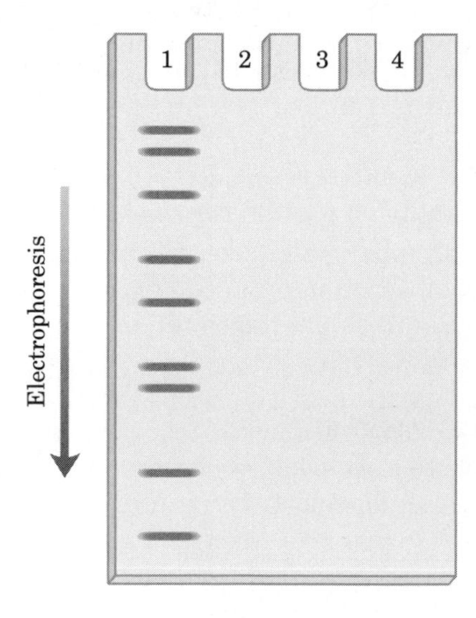

Answer

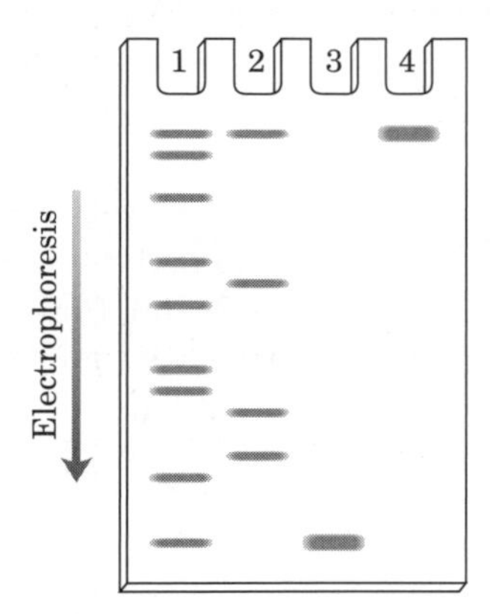

Lane 1: The reaction mixture that generated these bands included all the deoxynucleotides, plus dideoxythymidine. The fragments are of various lengths, all terminating where a ddTTP was substituted for a dTTP. For a small portion of the strands synthesized in the experiment, ddTTP would not be inserted and the strand would thus extend to the final G. Thus, the nine products are (from top to bottom of the gel):

5′-primer-TAATGCGTTCCTGTAATCTG
5′-primer-TAATGCGTTCCTGTAATCT
5′-primer-TAATGCGTTCCTGTAAT
5′-primer-TAATGCGTTCCTGT
5′-primer-TAATGCGTTCCT
5′-primer-TAATGCGTT
5′-primer-TAATGCGT
5′-primer-TAAT
5′-primer-T

Lane 2: Similarly, this lane will have four bands (top to bottom), for the following fragments, each terminating where ddGTP was inserted in place of dGTP:

5′-primer-TAATGCGTTCCTGTAATCTG
5′-primer-TAATGCGTTCCTG
5′-primer-TAATGCG
5′-primer-TAATG

Lane 3: Because mixture 3 lacked dTTP, every fragment was terminated immediately after the primer as ddTTP was inserted, to form 5′-primer-T. The result will be a single thick band near the bottom of the gel.

Lane 4: When all the deoxynucleotides were provided, but no dideoxynucleotide, a single labeled product formed: 5′-primer-TAATGCGTTCCTGTAATCTG. This will appear as a single thick band at the top of the gel.

13. Snake Venom Phosphodiesterase An exonuclease is an enzyme that sequentially cleaves nucleotides from the end of a polynucleotide strand. Snake venom phosphodiesterase, which hydrolyzes nucleotides from the 3′ end of any oligonucleotide with a free 3′-hydroxyl group, cleaves between the 3′ hydroxyl of the ribose or deoxyribose and the phosphoryl group of the next nucleotide. It acts on single-stranded DNA or RNA and has no base specificity. This enzyme was used in sequence determination experiments before the development of modern nucleic acid sequencing techniques. What are the products of partial digestion by snake venom phosphodiesterase of an oligonucleotide with the following sequence?

(5′)GCGCCAUUGC(3′)–OH

Answer When snake venom phosphodiesterase cleaves a nucleotide from a nucleic acid strand, it leaves the phosphoryl group attached to the 5′ position of the released nucleotide and a free 3′-OH group on the remaining strand. Partial digestion of the oligonucleotide gives a mixture of fragments of all lengths, as well as some of the original, undigested strand, so the products are (P represents the phosphate group):

(5′)P–GCGCCAUUGC(3′)–OH
(5′)P–GCGCCAUUG(3′)–OH
(5′)P–GCGCCAUU(3′)–OH
(5′)P–GCGCCAU(3′)–OH
(5′)P–GCGCCA(3′)–OH
(5′)P–GCGCC(3′)–OH
(5′)P–GCGC(3′)–OH
(5′)P–GCG(3′)–OH
(5′)P–GC(3′)–OH

and the released nucleoside 5′-phosphates, GMP, UMP, AMP, and CMP.

14. Preserving DNA in Bacterial Endospores Bacterial endospores form when the environment is no longer conducive to active cell metabolism. The soil bacterium *Bacillus subtilis*, for example, begins the process of sporulation when one or more nutrients are depleted. The end product is a small, metabolically dormant structure that can survive almost indefinitely with no detectable metabolism. Spores have mechanisms to prevent accumulation of potentially lethal mutations in their DNA over periods of dormancy that can exceed 1,000 years. *B. subtilis* spores are much more resistant than are the organism's growing cells to heat, UV radiation, and oxidizing agents, all of which promote mutations.

(a) One factor that prevents potential DNA damage in spores is their greatly decreased water content. How would this affect some types of mutations?

(b) Endospores have a category of proteins called small acid-soluble proteins (SASPs) that bind to their DNA, preventing formation of cyclobutane-type dimers. What causes cyclobutane dimers, and why do bacterial endospores need mechanisms to prevent their formation?

Answer

(a) Water is a participant in most biological reactions, including those that cause mutations. The low water content in endospores reduces the activity of mutation-causing enzymes and slows the rate of nonenzymatic depurination reactions, which are hydrolysis reactions.

(b) UV light induces the condensation of adjacent pyrimidine bases to form cyclobutane pyrimidine dimers. The spores of *B. subtilis,* a soil organism, are at constant risk of being lofted to the top of the soil or into the air, where they are subject to UV exposure, possibly for prolonged periods. Protection from UV-induced mutation is critical to spore DNA integrity.

15. Oligonucleotide Synthesis In the scheme of Figure 8–35, each new base to be added to the growing oligonucleotide is modified so that its 3′ hydroxyl is activated and the 5′ hydroxyl has a dimethoxytrityl (DMT) group attached. What is the function of the DMT group on the incoming base?

Answer DMT is a blocking group that prevents reaction of the incoming base with itself.

Biochemistry on the Internet

16. The Structure of DNA Elucidation of the three-dimensional structure of DNA helped researchers understand how this molecule conveys information that can be faithfully replicated from one generation to the next. To see the secondary structure of double-stranded DNA, go to the Protein Data Bank website (www.rcsb.org). Use the PDB identifiers listed below to retrieve the structure summaries for the two forms of DNA. Open the structures using Jmol (linked under the Display Options), and use the controls in the Jmol menu (accessed with a control-click or by clicking on the Jmol logo in the lower right corner of the image screen) to complete the following exercises. Refer to the Jmol help links as needed.

(a) Obtain the file for 141D, a highly conserved, repeated DNA sequence from the end of the HIV-1 (the virus that causes AIDS) genome. Display the molecule as a ball-and-stick structure (in the control menu, choose Select > All, then Render > Scheme > Ball and Stick). Identify the sugar–phosphate backbone for each strand of the DNA duplex. Locate and identify individual bases. Identify the 5′ end of each strand. Locate the major and minor grooves. Is this a right- or left-handed helix?

(b) Obtain the file for 145D, a DNA with the Z conformation. Display the molecule as a ball-and-stick structure. Identify the sugar–phosphate backbone for each strand of the DNA duplex. Is this a right- or left-handed helix?

(c) To fully appreciate the secondary structure of DNA, view the molecules in stereo. On the control menu, Select > All, then Render > Stereographic > Cross-eyed or Wall-eyed. You will see two images of the DNA molecule. Sit with your nose approximately 10 inches from the monitor and focus on the tip of your nose (cross-eyed) or the opposite edges of the screen (wall-eyed). In the background you should see three images of the DNA helix. Shift your focus to the middle image, which should appear three-dimensional. (Note that only one of the two authors can make this work.)

Answer

(a) The DNA fragment modeled in file 141D, from the human immunodeficiency virus, is the B form, the standard Watson-Crick structure (although this particular structure is a bent B-form DNA). This fragment has an adenine at the 5′ end and a guanine at the 3′ end; click on the bases at each end of the helix to identify which is the 5′ end. When the helix is oriented with the 5′ adenine at the upper left-hand side of the model, the *minor groove* is in the center of the model. Rotating the model so that the 5′ adenine is at the upper right-hand side positions the *major groove* in the center. The spiral of this helix runs upward in a counterclockwise direction, so this is a right-handed helix.

(b) The model of DNA in the Z conformation includes a shell of water molecules around the helix. The water molecules are visible when the complex is viewed in ball-and-stick mode. Turn off the display of the water molecules using the console controls Select > Nucleic > DNA. Then Select > Display Selected Only. The backbone of DNA in the Z conformation is very different from that in the B conformation. The helix spiral runs upward in a clockwise direction, so this is a left-handed helix.

(c) Viewing the structures in stereo takes a bit of practice, but perseverance will be rewarded! Here are some tips for successful three-dimensional viewing:

(1) Turn off or lower the room lighting.
(2) Sit directly in front of the screen.
(3) Use a ruler to make sure you are 10 to 11 inches from the screen.
(4) Position your head so that when you focus on the tip of your nose, the screen images are on either side of the tip (i.e., look down your nose at the structures).
(5) Move your head slightly closer to or farther away from the screen to bring the middle image into focus. Don't look directly at the middle image as you try to bring it into focus.

(6) If you find it uncomfortable to focus on the tip of your nose, try using the tip of a finger (positioned just beyond the tip of your nose) instead.

(7) Relax as you attempt to view the three-dimensional image.

Note that many people, including one of the text authors, have some trouble making this work!

Data Analysis Problem

17. Chargaff's Studies of DNA Structure The chapter section "DNA Is a Double Helix that Stores Genetic Information" includes a summary of the main findings of Erwin Chargaff and his coworkers, listed as four conclusions ("Chargaff's rules"; p. 278). In this problem, you will examine the data Chargaff collected in support of these conclusions.

In one paper, Chargaff (1950) described his analytical methods and some early results. Briefly, he treated DNA samples with acid to remove the bases, separated the bases by paper chromatography, and measured the amount of each base with UV spectroscopy. His results are shown in the three tables below. The *molar ratio* is the ratio of the number of moles of each base in the sample to the number of moles of phosphate in the sample—this gives the fraction of the total number of bases represented by each particular base. The *recovery* is the sum of all four bases (the sum of the molar ratios); full recovery of all bases in the DNA would give a recovery of 1.0.

	Molar ratios in ox DNA					
	Thymus			Spleen		Liver
Base	Prep. 1	Prep. 2	Prep. 3	Prep. 1	Prep. 2	Prep. 1
Adenine	0.26	0.28	0.30	0.25	0.26	0.26
Guanine	0.21	0.24	0.22	0.20	0.21	0.20
Cytosine	0.16	0.18	0.17	0.15	0.17	
Thymine	0.25	0.24	0.25	0.24	0.24	
Recovery	*0.88*	*0.94*	*0.94*	*0.84*	*0.88*	

	Molar ratios in human DNA				
	Sperm		Thymus	Liver	
Base	Prep. 1	Prep. 2	Prep. 1	Normal	Carcinoma
Adenine	0.29	0.27	0.28	0.27	0.27
Guanine	0.18	0.17	0.19	0.19	0.18
Cytosine	0.18	0.18	0.16		0.15
Thymine	0.31	0.30	0.28		0.27
Recovery	*0.96*	*0.92*	*0.91*		*0.87*

	Molar ratios in DNA of microorganisms		
	Yeast		Avian tubercle bacilli
Base	Prep. 1	Prep. 2	Prep. 1
Adenine	0.24	0.30	0.12
Guanine	0.14	0.18	0.28
Cytosine	0.13	0.15	0.26
Thymine	0.25	0.29	0.11
Recovery	*0.76*	*0.92*	*0.77*

(a) Based on these data, Chargaff concluded that "no differences in composition have so far been found in DNA from different tissues of the same species." This corresponds to conclusion 2 in this chapter. However, a skeptic looking at the data above might say, "They certainly look different to me!" If you were Chargaff, how would you use the data to convince the skeptic to change her mind?

(b) The base composition of DNA from normal and cancerous liver cells (hepatocarcinoma) was not distinguishably different. Would you expect Chargaff's technique to be capable of detecting a difference between the DNA of normal and cancerous cells? Explain your reasoning.

As you might expect, Chargaff's data were not completely convincing. He went on to improve his techniques, as described in a later paper (Chargaff, 1951), in which he reported molar ratios of bases in DNA from a variety of organisms:

Source	A:G	T:C	A:T	G:C	Purine:pyrimidine
Ox	1.29	1.43	1.04	1.00	1.1
Human	1.56	1.75	1.00	1.00	1.0
Hen	1.45	1.29	1.06	0.91	0.99
Salmon	1.43	1.43	1.02	1.02	1.02
Wheat	1.22	1.18	1.00	0.97	0.99
Yeast	1.67	1.92	1.03	1.20	1.0
Haemophilus influenzae type c	1.74	1.54	1.07	0.91	1.0
E. coli K-12	1.05	0.95	1.09	0.99	1.0
Avian tubercle bacillus	0.4	0.4	1.09	1.08	1.1
Serratia marcescens	0.7	0.7	0.95	0.86	0.9
Bacillus schatz	0.7	0.6	1.12	0.89	1.0

(c) According to Chargaff, as stated in conclusion 1 in this chapter, "The base composition of DNA generally varies from one species to another." Provide an argument, based on the data presented so far, that supports this conclusion.

(d) According to conclusion 4, "In *all* cellular DNAs, regardless of the species . . . A + G = T + C." Provide an argument, based on the data presented so far, that supports this conclusion.

Part of Chargaff's intent was to disprove the "tetranucleotide hypothesis"; this was the idea that DNA was a monotonous tetranucleotide polymer $(AGCT)_n$ and therefore not capable of containing sequence information. Although the data presented above show that DNA cannot be simply a tetranucleotide—if so, all samples would have molar ratios of 0.25 for each base—it was still possible that the DNA from different organisms was a slightly more complex, but still monotonous, repeating sequence.

To address this issue, Chargaff took DNA from wheat germ and treated it with the enzyme deoxyribonuclease for different time intervals. At each time interval, some of the DNA was converted to small fragments; the remaining, larger fragments he called the "core." In the table below, the "19% core" corresponds to the larger fragments left behind when 81% of the DNA was degraded; the "8% core" corresponds to the larger fragments left after 92% degradation.

Base	Intact DNA	19% Core	8% Core
Adenine	0.27	0.33	0.35
Guanine	0.22	0.20	0.20
Cytosine	0.22	0.16	0.14
Thymine	0.27	0.26	0.23
Recovery	*0.98*	*0.95*	*0.92*

(e) How would you use these data to argue that wheat germ DNA is not a monotonous repeating sequence?

Answer

(a) It would not be easy! The data for different samples from the same organism show significant variation, and the recovery is never 100%. The numbers for C and T show much more consistency than those for A and G, so for C and T it is much easier to make the case that samples from the same organism have the same composition. But even with the less consistent values for A and G, (1) the range of values for different tissues does overlap substantially; (2) the difference between different preparations of the same tissue is about the same as the difference between samples from different tissues; and (3) in samples for which recovery is high, the numbers are more consistent.

(b) This technique would not be sensitive enough to detect a difference between normal and cancerous cells. Cancer is caused by mutations, but these changes in DNA—a few base pairs out of several billion—would be too small to detect with these techniques.

(c) The ratios of A:G and T:C vary widely among different species. For example, in the bacterium *Serratia marcescens*, both ratios are 0.4, meaning that the DNA contains mostly G and C. In *Haemophilus influenzae*, by contrast, the ratios are 1.74 and 1.54, meaning that the DNA is mostly A and T.

(d) Conclusion 4 has three requirements:

A = T: The table shows an A:T ratio very close to 1 in all cases. Certainly, the variation in this ratio is substantially less than the variation in the A:G and T:C ratios.

G = C: Again, the G:C ratio is very close to 1, and the other ratios vary widely.

(A + G) = (T + C): This is the purine:pyrimidine ratio, which also is very close to 1.

(e) The different "core" fractions represent different regions of the wheat germ DNA. If the DNA were a monotonous repeating sequence, the base composition of all regions would be the same. Because different core regions have different sequences, the DNA sequence must be more complex.

References

Chargaff, E. (1950) Chemical specificity of nucleic acids and mechanism of their enzymic degradation. *Experientia* **6,** 201–209.

Chargaff, E. (1951) Structure and function of nucleic acids as cell constituents. *Fed. Proc.* **10,** 654–659.

chapter

DNA-Based Information Technologies

9

1. **Cloning** When joining two or more DNA fragments, a researcher can adjust the sequence at the junction in a variety of subtle ways, as seen in the following exercises.

   (a) Draw the structure of each end of a linear DNA fragment produced by an *Eco*RI restriction digest (include those sequences remaining from the *Eco*RI recognition sequence).

   (b) Draw the structure resulting from the reaction of this end sequence with DNA polymerase I and the four deoxynucleoside triphosphates (see Fig. 8–33).

   (c) Draw the sequence produced at the junction that arises if two ends with the structure derived in **(b)** are ligated (see Fig. 25–17).

   (d) Draw the structure produced if the structure derived in **(a)** is treated with a nuclease that degrades only single-stranded DNA.

   (e) Draw the sequence of the junction produced if an end with structure **(b)** is ligated to an end with structure **(d).**

   (f) Draw the structure of the end of a linear DNA fragment that was produced by a *Pvu*II restriction digest (include those sequences remaining from the *Pvu*II recognition sequence).

   (g) Draw the sequence of the junction produced if an end with structure **(b)** is ligated to an end with structure **(f).**

   (h) Suppose you can synthesize a short duplex DNA fragment with any sequence you desire. With this synthetic fragment and the procedures described in **(a)** through **(g),** design a protocol that would remove an *Eco*RI restriction site from a DNA molecule and incorporate a new *Bam*HI restriction site at approximately the same location. (See Fig. 9–2.)

   (i) Design four different short synthetic double-stranded DNA fragments that would permit ligation of structure **(a)** with a DNA fragment produced by a *Pst*I restriction digest. In one of these fragments, design the sequence so that the final junction contains the recognition sequences for both *Eco*RI and *Pst*I. In the second and third fragments, design the sequence so that the junction contains only the *Eco*RI and only the *Pst*I recognition sequence, respectively. Design the sequence of the fourth fragment so that neither the *Eco*RI nor the *Pst*I sequence appears in the junction.

Answer Type II restriction enzymes cleave double-stranded DNA within recognition sequences to create either blunt-ended or sticky-ended fragments. Blunt-ended DNA fragments can be joined by the action of T4 DNA ligase. Sticky-ended DNA fragments can be joined by either *E. coli* or T4 DNA ligases, provided that the sticky ends are complementary. Sticky-ended fragments without complementary ends can be joined only after the ends are made blunt, either by exonucleases or by *E. coli* DNA polymerase I.

(a) The recognition sequence for *Eco*RI is (5′)GAATTC(3′), with the cleavage site between G and A (see Table 9–2). Thus, digestion of a DNA molecule with one *Eco*RI site

(5′)---GAATTC---(3′)
(3′)---CTTAAG---(5′)

would yield two fragments:

(5′)---G(3′) and (5′)AATTC---(3′)

(3′)---CTTAA(5′) (3′)G---(5′)

(b) DNA polymerase I catalyzes the synthesis of DNA in the 5′→3′ direction in the presence of the four deoxyribonucleoside triphosphates. Therefore, both fragments generated in **(a)** will be made blunt ended:

(5′)---GAATT(3′) and (5′)AATTC---(3′)
(3′)---CTTAA(5′) (3′)TTAAG---(5′)

(c) The two fragments generated in **(b)** can be ligated by T4 DNA ligase to form

(5′)---GAATTAATTC---(3′)
(3′)---CTTAATTAAG---(5′)

(d) The fragments in **(a)** have sticky ends, with a protruding single-stranded region. Treatment of these DNA fragments with a single-strand-specific nuclease will yield DNA fragments with blunt ends:

(5′)---G(3′) and (5′)C---(3′)
(3′)---C(5′) (3′)G---(5′)

(e) The left-hand DNA fragment in **(b)** can be joined to the right-hand fragment in **(d)** to yield

(5′)---GAATTC---(3′)
(3′)---CTTAAG---(5′)

The same recombinant DNA molecule is produced by joining the right-hand fragment in **(b)** to the left-hand fragment in **(d).**

(f) The recognition sequence for *Pvu*II is (5′)CAGCTG(3′), with the cleavage site between G and C (see Table 9–2). Thus, a DNA molecule with a *Pvu*II site will yield two fragments when digested with *Pvu*II:

(5′)---CAG(3′) and (5′)CTG---(3′)
(3′)---GTC(5′) (3′)GAC---(5′)

(g) The left-hand DNA fragment in **(b)** can be joined to the right-hand fragment in **(f)** to yield

(5′)---GAATTCTG---(3′)
(3′)---CTTAAGAC---(5′)

The same recombinant DNA is produced by joining the right-hand fragment in **(b)** to the left-hand fragment in **(f):**

(5′)---CAGAATTC---(3′)
(3′)---GTCTTAAG---(5′)

(h) There are two ways to convert an *Eco*RI restriction site to a *Bam*HI restriction site.

Method 1: Digest DNA with *Eco*RI, and then create blunt ends by using either DNA polymerase I to fill in the single-stranded region as in **(b)** or a single-strand-specific nuclease to remove the single-stranded region as in **(d).** Ligate a synthetic linker that contains the *Bam*HI recognition sequence (5′)GGATCC(3′) (see Table 9–2),

(5′)GCGGATCCCG(3′)
(3′)CGCCTAGGGC(5′)

between the two blunt-ended DNA fragments to yield, if the *Eco*RI-digested DNA is treated as in **(b),**

(5′)---GAATTGCGGATCCCGAATTC---(3′)
(3′)---CTTAACGCCTAGGGCTTAAG---(5′)

or, if the *Eco*RI-digested DNA is treated as in **(d),**

(5′)---GGCGGATCCCG(3′)
(3′)---CCGCCTAGGGC(5′)

Notice that the *Eco*RI site is not regenerated after ligation of the linker.

Method 2: This method uses a "conversion adaptor" to introduce a *Bam*HI site into the DNA molecule. A synthetic oligonucleotide with the sequence (5′)AATTGGATCC(3′) is partially self-complementary, and it spontaneously forms the structure

(5′)AATTGGATCC(3′)
(3′)   CCTAGGTTAA(5′)

The sticky ends of this adaptor are complementary to the sticky ends generated by *Eco*RI digestion, so the adaptor can be ligated between the two *Eco*RI fragments to form

(5′)---GAATTGGATCCAATT---(3′)
(3′)---CTTAACCTAGGTTAA---(5′)

Because ligation between DNA molecules with compatible sticky ends is more efficient than ligation between DNA molecules with blunt ends, Method 2 is preferred over Method 1.

(i) Joining of the DNA fragments in **(a)** to a fragment generated by *Pst*I digestion requires a conversion adaptor. This adaptor should contain a single-stranded region complementary to the sticky end of an *Eco*RI-generated DNA fragment, and a single-stranded region complementary to the sticky end generated by *Pst*I digestion. The four adaptor sequences that fulfill this requirement are shown below, in order of discussion in the problem (N = any nucleotide):

(5′)AATTCNNNNCTGCA(3′)
(3′)GNNNNG(5′)
(5′)AATTCNNNNGTGCA(3′)
(3′)GNNNNC(5′)
(5′)AATTGNNNNCTGCA(3′)
(3′)CNNNNG(5′)
(5′)AATTGNNNNGTGCA(3′)
(3′)CNNNNC(5′)

For the first adaptor: Ligation of the adaptor to the *Eco*RI-digested DNA molecule would yield

(5′)---GAATTCNNNNCTGCA(3′)
(3′)---CTTAAGNNNNG(5′)

This product can now be ligated to a DNA fragment produced by a *Pst*I digest, which has the terminal sequence

(5′)G---(3′)
(3′)ACGTC---(5′)

to yield

(5′)---GAATTCNNNNCTGCAG---(3′)
(3′)---CTTAAGNNNNGACGTC---(5′)

Notice that the *Eco*RI and *Pst*I sites are retained.

In a similar fashion, each of the other three adaptors can be ligated to the *Eco*RI-digested DNA molecule, and the ligated molecule joined to a DNA fragment produced by a *Pst*I digest. The final products are as follows.

For the second adaptor:

(5′)---GAATTCNNNNGTGCAG---(3′)
(3′)---CTTAAGNNNNCACGTC---(5′)

The *Eco*RI site is retained, but not the *Pst*I site.

For the third adaptor:

(5′)---GAATTGNNNNCTGCAG---(3′)
(3′)---CTTAACNNNNGACGTC---(5′)

The *Pst*I site is retained, but not the *Eco*RI site.

For the fourth adaptor:

(5′)---GAATTGNNNNGTGCAG---(3′)
(3′)---CTTAACNNNNCACGTC---(5′)

Neither the *Eco*RI nor the *Pst*I site is retained.

2. **Selecting for Recombinant Plasmids** When cloning a foreign DNA fragment into a plasmid, it is often useful to insert the fragment at a site that interrupts a selectable marker (such as the tetracycline-resistance gene of pBR322). The loss of function of the interrupted gene can be used to identify clones containing recombinant plasmids with foreign DNA. With a bacteriophage λ vector it is not necessary to do this, yet one can easily distinguish vectors that incorporate large foreign DNA fragments from those that do not. How are these recombinant vectors identified?

Answer Bacteriophage λ DNA can be packaged into infectious phage particles only if it is between 40,000 and 53,000 bp long. The two essential pieces of the bacteriophage λ vector have about 30,000 bp in all, so the vector is not packaged into phage particles unless the additional, foreign DNA is of sufficient length: 10,000 to 23,000 bp.

3. **DNA Cloning** The plasmid cloning vector pBR322 (see Fig. 9–3) is cleaved with the restriction endonuclease *Pst*I. An isolated DNA fragment from a eukaryotic genome (also produced by *Pst*I cleavage) is added to the prepared vector and ligated. The mixture of ligated DNAs is then used to transform bacteria, and plasmid-containing bacteria are selected by growth in the presence of tetracycline.

(a) In addition to the desired recombinant plasmid, what other types of plasmids might be found among the transformed bacteria that are tetracycline resistant? How can the types be distinguished?

(b) The cloned DNA fragment is 1,000 bp long and has an *Eco*RI site 250 bp from one end. Three different recombinant plasmids are cleaved with *Eco*RI and analyzed by gel electrophoresis, giving the patterns shown. What does each pattern say about the cloned DNA? Note that in pBR322, the *Pst*I and *Eco*RI restriction sites are about 750 bp apart. The entire plasmid with no cloned insert is 4,361 bp. Size markers in lane 4 have the number of nucleotides noted.

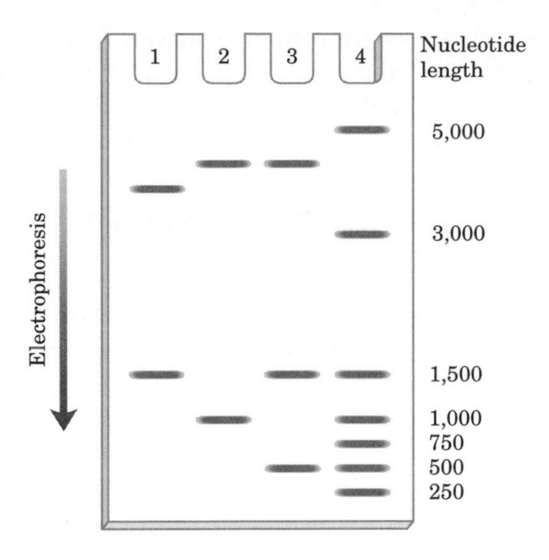

Answer

(a) Ligation of the linear pBR322 to regenerate circular pBR322 is a unimolecular process and thus occurs more efficiently than the ligation of a foreign DNA fragment to the linear pBR322, which is a bimolecular process (assuming equimolar amounts of linear pBR322 and foreign DNA in the reaction mixture). The tetracycline-resistant bacteria would include recombinant plasmids and plasmids in which the original pBR322 was regenerated without insertion of a foreign DNA fragment. (These would also retain resistance to ampicillin.) In addition, two or more molecules of pBR322 might be ligated together with or without insertion of foreign DNA.

(b) The clones giving rise to the patterns in lanes 1 and 2 each have one DNA fragment inserted, but in different orientations (see the diagrams, which are not drawn to scale; keep in mind that the products on the gel are from *Eco*RI cleavage). The clone producing the pattern in lane 3 has two DNA fragments, ligated such that the *Eco*RI-site proximal ends are joined.

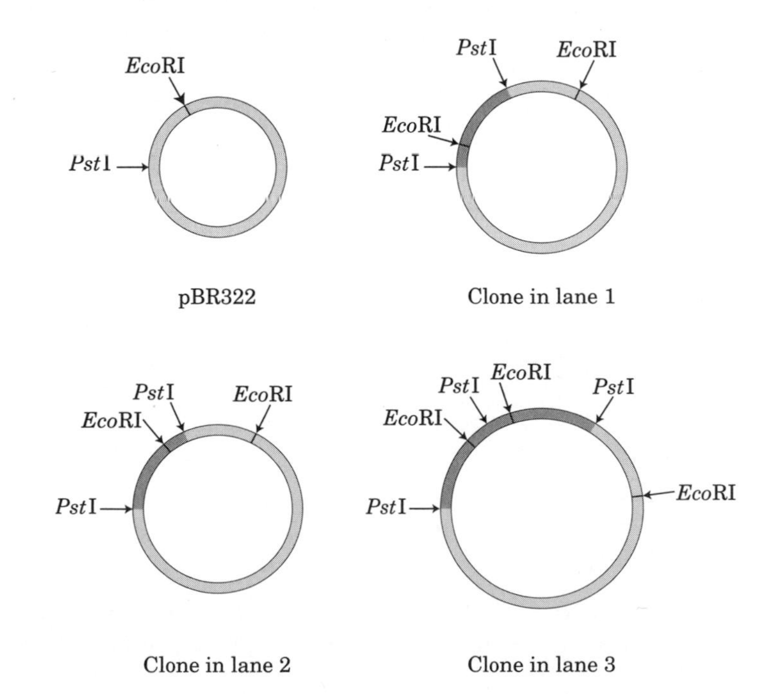

4. Identifying the Gene for a Protein with a Known Amino Acid Sequence Using Figure 27–7 to translate the genetic code, design a DNA probe that would allow you to identify the gene for a protein with the following amino-terminal amino acid sequence. The probe should be 18 to 20 nucleotides long, a size that provides adequate specificity if there is sufficient homology between the probe and the gene.

H_3N^+–Ala–Pro–Met–Thr–Trp–Tyr–Cys–Met–Asp–Trp–Ile–
Ala–Gly–Gly–Pro–Trp–Phe–Arg–Lys–Asn–Thr–Lys–

Answer Most amino acids are encoded by two or more codons (see Fig. 27–7). To minimize the ambiguity in codon assignment for a given peptide sequence, we must select a region of the peptide that contains amino acids specified by the smallest number of codons. Focus on the amino acids with the fewest codons: Met and Trp (see Fig. 27–7 and Table 27–3). The best possibility for a probe is a span of DNA from the codon for the first Trp residue to the first two nucleotides of the codon for Ile. The sequence of the probe would be

(5′)UGG UA(U/C) UG(U/C) AUG GA(U/C) UGG AU

The synthesis would be designed to incorporate either U or C where indicated, producing a mixture of eight 20-nucleotide probes.

5. Designing a Diagnostic Test for a Genetic Disease Huntington's disease (HD) is an inherited neurodegenerative disorder, characterized by the gradual, irreversible impairment of psychological, motor, and cognitive functions. Symptoms typically appear in middle age, but onset can occur at almost any age. The course of the disease can last 15 to 20 years. The molecular basis of the disease is becoming better understood. The genetic mutation underlying HD has been traced to a gene encoding a protein (M_r 350,000) of unknown function. In individuals who will not develop HD, a region of the gene that encodes the amino terminus of the protein has a sequence of CAG codons (for glutamine) that is repeated 6 to 39 times in succession. In individuals with adult-onset HD, this codon is typically repeated 40 to 55 times. In individuals with childhood-onset HD, this codon is repeated more than 70 times. The length of this simple trinucleotide repeat indicates whether an individual will develop HD, and at approximately what age the first symptoms will occur.

A small portion of the amino-terminal coding sequence of the 3,143-codon HD gene is given below. The nucleotide sequence of the DNA is shown, with the amino acid sequence corresponding to the gene below it, and the CAG repeat shaded. Using Figure 27–7 to translate the genetic code, outline a PCR-based test for HD that could be carried out using a blood sample. Assume the PCR primer must be 25 nucleotides long. By convention, unless otherwise specified a DNA sequence encoding a protein is displayed with the coding strand (the sequence identical to the mRNA transcribed from the gene) on top such that it is read 5′ to 3′, left to right.

```
307 ATGGCGACCCTGGAAAAGCTGATGAAGGCCTTCGAGTCCCTCAAGTCCTTC
1   M  A  T  L  E  K  L  M  K  A  F  E  S  L  K  S  F

358 CAGCAGTTCCAGCAGCAGCAGCAGCAGCAGCAGCAGCAGCAGCAGCAGCAG
18  Q  Q  F  Q  Q  Q  Q  Q  Q  Q  Q  Q  Q  Q  Q  Q  Q

409 CAGCAGCAGCAGCAGCAGCAGCAACAGCCGCCACCGCCGCCGCCGCCGCCG
35  Q  Q  Q  Q  Q  Q  Q  Q  Q  P  P  P  P  P  P  P  P

460 CCGCCTCCTCAGCTTCCTCAGCCGCCGCCG
52  P  P  P  Q  L  P  Q  P  P  P
```

Source: The Huntington's Disease Collaborative Research Group. (1993) A novel gene containing a trinucleotide repeat that is expanded and unstable on Huntington's disease chromosomes. *Cell* **72,** 971–983.

Answer Your test would require DNA primers, a heat-stable DNA polymerase, deoxynucleoside triphosphates, and a PCR machine (thermal cycler). The primers would be designed to amplify a DNA segment encompassing the CAG repeat. The DNA strand shown is the coding strand, oriented $5' \rightarrow 3'$ left to right. The primer targeted to DNA to the left of the repeat would be identical to any 25-nucleotide sequence shown in the region to the left of the CAG repeat. Such a primer will direct synthesis of DNA across the repeat from left to right. The primer on the right side must be *complementary* and *antiparallel* to a 25-nucleotide sequence to the right of the CAG repeat. Such a primer will direct $5' \rightarrow 3'$ synthesis of DNA across the repeat from right to left. Choosing unique sequences relatively close to the CAG repeat will make the amplified region smaller and the test more sensitive to small changes in size. Using the primers, DNA including the CAG repeat would be amplified by PCR, and its size would be determined by comparison to size markers after electrophoresis. The length of the DNA would reflect the length of the CAG repeat, providing a simple test for the disease. Such a test could be carried out on a blood sample and completed in less than a day.

6. Using PCR to Detect Circular DNA Molecules In a species of ciliated protist, a segment of genomic DNA is sometimes deleted. The deletion is a genetically programmed reaction associated with cellular mating. A researcher proposes that the DNA is deleted in a type of recombination called site-specific recombination, with the DNA on either end of the segment joined together and the deleted DNA ending up as a circular DNA reaction product.

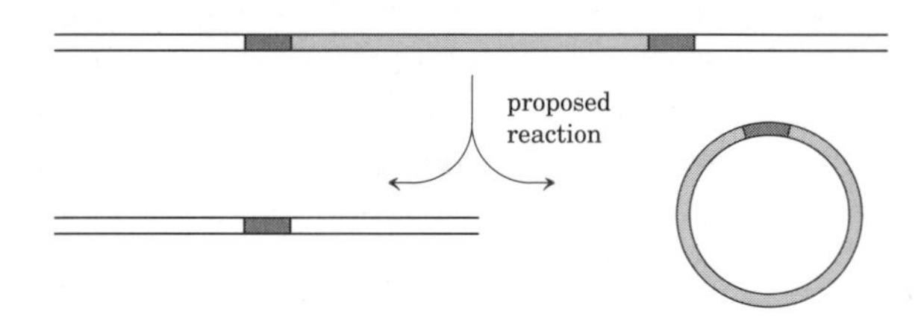

Suggest how the researcher might use the polymerase chain reaction (PCR) to detect the presence of the circular form of the deleted DNA in an extract of the protist.

Answer Design PCR primers complementary to DNA in the deleted segment, but which would direct DNA synthesis away from each other. No PCR product will be generated unless the ends of the deleted segment are joined to create a circle.

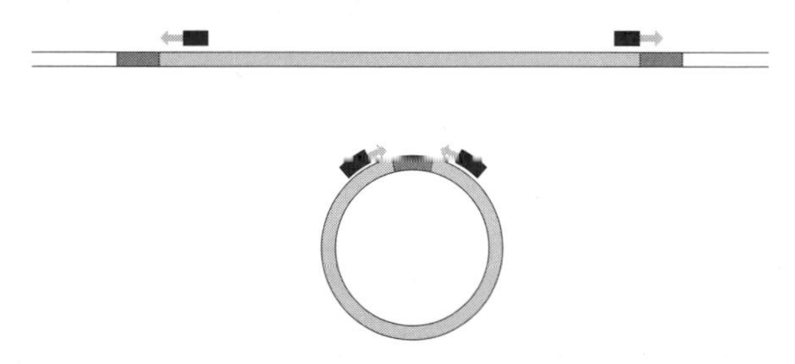

7. Glowing Plants When grown in ordinary garden soil and watered normally, a plant engineered to express green fluorescent protein (see Fig. 9–15a) will glow in the dark, whereas a plant engineered to express firefly luciferase (see Fig. 9–29) will not. Explain these observations.

Answer The plant expressing firefly luciferase must take up luciferin, the substrate of luciferase, before it can "glow" (albeit weakly). The plant expressing green fluorescent protein glows without requiring any other compound.

8. RFLP Analysis for Paternity Testing DNA fingerprinting and RFLP analysis are often used to test for paternity. A child inherits chromosomes from the mother and the father, so DNA from a child displays restriction fragments derived from each parent. In the gel shown here, which child, if any, can be excluded as being the biological offspring of the putative father? Explain your reasoning. Lane M is the sample from the mother, F from the putative father, and C1, C2, and C3 from the children.

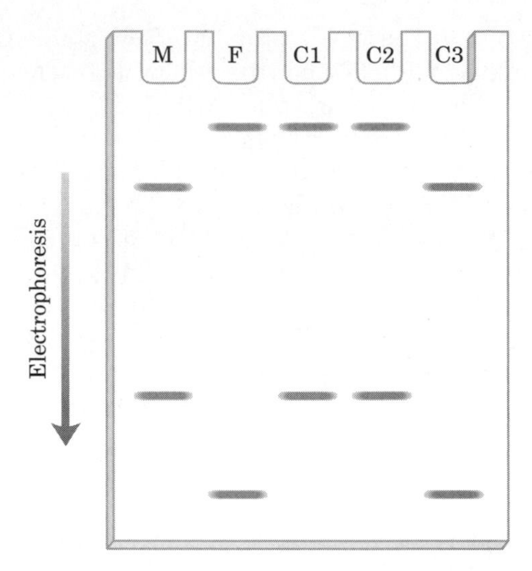

Answer None of the children can be excluded. Each child has one band that could be derived from the father.

9. **Mapping a Chromosome Segment** A group of overlapping clones, designated A through F, is isolated from one region of a chromosome. Each of the clones is separately cleaved by a restriction enzyme and the pieces resolved by agarose gel electrophoresis, with the results shown in the figure below. There are nine different restriction fragments in this chromosomal region, with a subset appearing in each clone. Using this information, deduce the order of the restriction fragments in the chromosome.

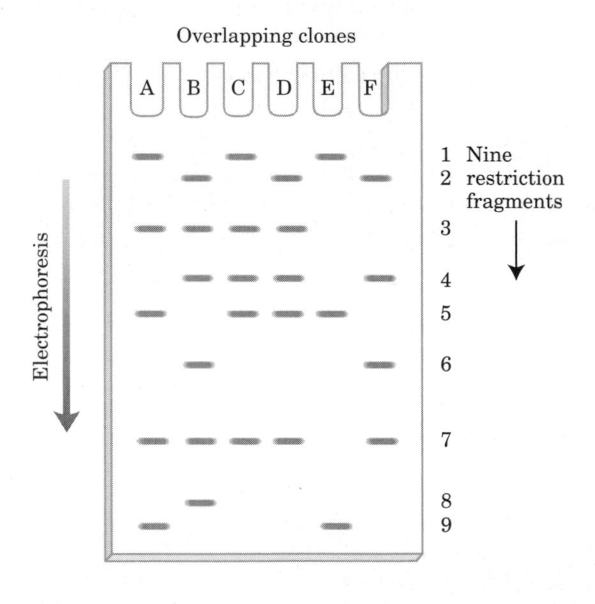

Answer Solving a restriction fragment map is a logic puzzle. The agarose gel shows which fragments are part of each clone, but deducing their order on the chromosome takes some work.

Clone A: fragments 1, 3, 5, 7, and 9
Clone B: fragments 2, 3, 4, 6, 7, and 8
Clone C: fragments 1, 3, 4, 5, and 7
Clone D: fragments 2, 3, 4, 5, and 7
Clone E: fragments 1, 5, and 9
Clone F: fragments 2, 4, 6, and 7

Begin with clone E, which has the fewest fragments. Fragments 1, 5, and 9 must be adjacent, but may be in any order. However, clone C includes fragments 1 and 5, but not 9, so we can conclude that fragments 1 and 5 are adjacent; 9 could be adjacent to either 1 or 5, but is not between them (i.e., the order is 9 1 5 or 1 5 9). Clones C and D are identical except that C also includes fragment 1, and D also includes fragment 2; from this we can deduce that fragments 1 and 2 must be at opposite ends of the overlapping region, which includes fragments 3, 4, 5, and 7 (in an as yet undetermined order). Because we have already concluded that fragments 1 and 5 are adjacent, we can propose the following sequence: 1-5-(3,4,7)-2.

To find the order of fragments 3, 4, and 7, look for fragments that contain a subset of these with a flanking fragment. Clone A includes fragments 5, 3, and 7, but not 4 (thus, 5-(3,7)-4); clone F includes fragments 4, 7, and 2 (thus, (4,7)-2). Combining these possibilities allows us to deduce the order as 1-5-3-7-4-2. We concluded earlier that 9 is adjacent to either 1 or 5. If it were adjacent to 5, it would be a part of clones C and D; because it is not in C or D, it must be adjacent to 1.

We can now propose the following sequence: 9-1-5-3-7-4-2. Because clone F includes these last three fragments and fragment 6, we can append 6 after 2: 9-1-5-3-7-4-2-6. Finally, clone B is the only one that includes fragment 8, so it must occur at either end of our deduced sequence. Given the other fragments in clone B, fragment 8 must be adjacent to fragment 6. Thus, we have the order 9-1-5-3-7-4-2-6-8.

The fragments were numbered based on their migration distance in the gel, which correlates inversely with the size of the fragment. Fragment 1 is the longest; fragment 9 the shortest. The relative sizes and the positions of the fragments on the chromosome are shown below. Molecular weight markers in the gel would allow a better estimation of the sizes of the various fragments.

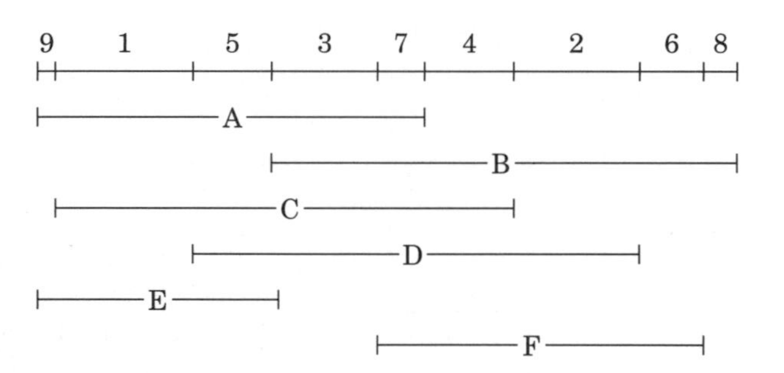

10. Cloning in Plants The strategy outlined in Figure 9–28 employs *Agrobacterium* cells that contain two separate plasmids. Suggest why the sequences on the two plasmids are not combined on one plasmid.

Answer Simply for convenience; the 200,000 bp Ti plasmid, even when the T DNA is removed, is too large to isolate in quantity and manipulate in vitro. It is also too large to reintroduce into a cell by standard transformation techniques. Single-plasmid systems in which the T DNA of a Ti plasmid has been replaced by foreign DNA (by low-efficiency recombination in vivo) have been used successfully, but this approach is very laborious. The *vir* genes can facilitate transfer of any DNA between the T DNA repeats, even if they are on a separate plasmid. The second plasmid in the two-plasmid system, because it requires only the T DNA repeats and a few sequences necessary for plasmid selection and propagation, is relatively small, easily isolated, and easily manipulated (foreign DNA is easily added and/or altered). It can be propagated in *E. coli* or *Agrobacterium* and is readily reintroduced into either bacterium.

11. DNA Fingerprinting and RFLP Analysis DNA is extracted from the blood cells of two humans, individuals 1 and 2. In separate experiments, the DNA from each individual is cleaved by restriction endonucleases A, B, and C, and the fragments separated by electrophoresis. A hypothetical map of a 10,000 bp segment of a human chromosome is shown (1 kbp = 1,000 bp). Individual 2 has point mutations that eliminate restriction recognition sites B* and C*. You probe the gel with a radioactive oligonucleotide complementary to the indicated sequence and expose a piece of x-ray film to the gel. Indicate where you would expect to see bands on the film. The lanes of the gel are marked in the accompanying diagram.

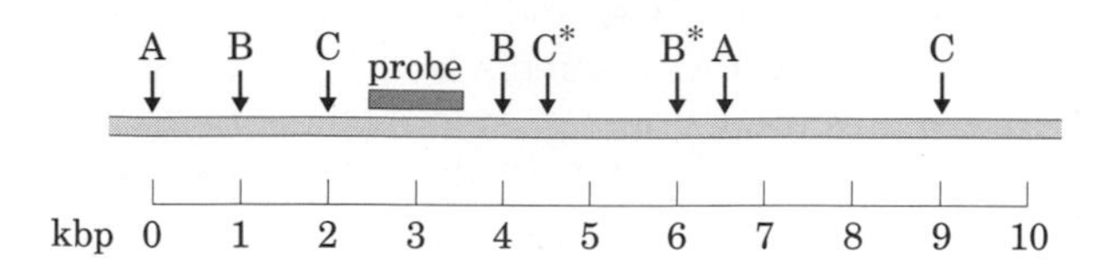

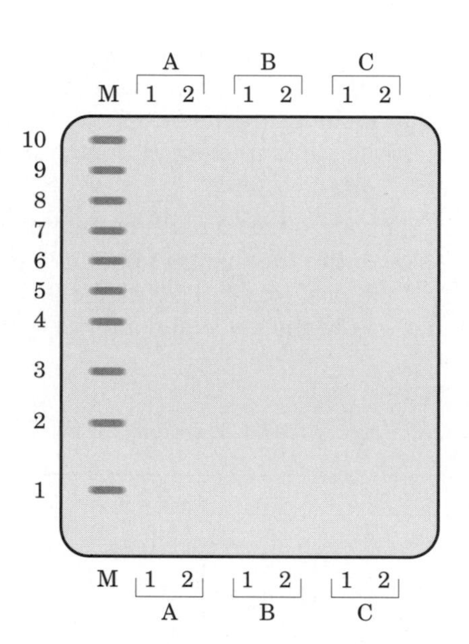

Answer Cleaving DNA with restriction enzyme A produces identical fragments in both individuals: 6.5 kbp and 3.5 kbp fragments. The probe hybridizes to both 6.5 kbp fragments, resulting in two identical bands in column A. Restriction enzyme B produces different cleavage products: DNA from individual 1 is cleaved into 3, 2, and 4 kbp fragments; that from individual 2 (who has an altered B recognition sequence) into 3 and 6 kbp fragments. However, the probe binds to the 3 kbp fragments from both individuals and therefore produces the same pattern of bands on the gel (in column B). Restriction enzyme C cleaves DNA from individual 1 into 2.5 and 4.5 kbp fragments, and the probe labels the 2.5 kbp piece. DNA from individual 2, however, is cleaved to produce a single 7 kbp fragment, which hybridizes with the probe. Thus, only in column C does a difference in DNA sequence between individuals 1 and 2 become apparent. This exercise points out the importance of the choice of restriction enzymes, as well as the choice of probes, when performing DNA fingerprinting and RFLP analysis.

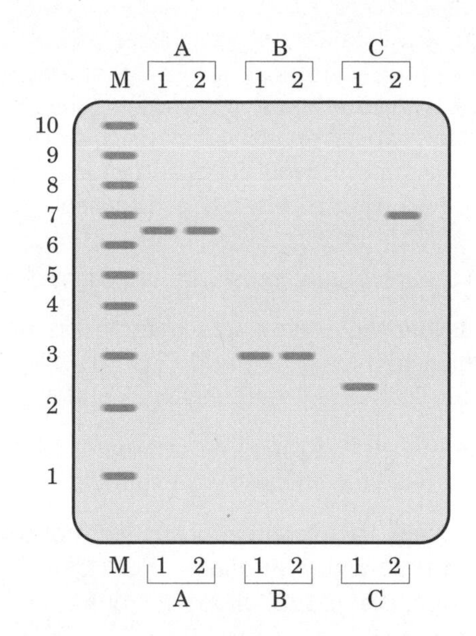

12. Use of Photolithography to Make a DNA Microarray Figure 9–21 shows the first steps in the process of making a DNA microarray, or DNA chip, using photolithography. Describe the remaining steps needed to obtain the desired sequences (a different four-nucleotide sequence on each of the four spots) shown in the first panel of the figure. After each step, give the resulting nucleotide sequence attached at each spot.

Answer Cover spot 4, add solution containing activated T, irradiate, and wash. The resulting sequences are now

1. A-T	2. G-T	3. A-T	4. G-C

Cover spots 2 and 4, add solution containing activated G, irradiate, and wash.

1. A-T-G	2. G-T	3. A-T-G	4. G-C

Cover spot 3, add solution containing activated C, irradiate, and wash.

1. A-T-G-C	2. G-T-C	3. A-T-G	4. G-C-C

Cover spots 1, 3, and 4, add solution containing activated C, irradiate, and wash.

1. A-T-G-C	2. G-T-C-C	3. A-T-G	4. G-C-C

Cover spots 1 and 2, add solution containing activated G, irradiate, and wash.

1. A-T-G-C	2. G-T-C-C	3. A-T-G-C	4. G-C-C-C

13. Cloning in Mammals The retroviral vectors described in Figure 9–32 make possible the efficient integration of foreign DNA into a mammalian genome. Explain how these vectors, which lack genes for replication and viral packaging (*gag, pol, env*), are assembled into infectious viral particles. Suggest why it is important that these vectors lack the replication and packaging genes.

Answer The retroviral vectors must be introduced into a cell infected with a helper virus that can provide the necessary replication and packaging functions but cannot itself be packaged. The vectors packaged into infectious viral particles are then used to introduce the recombinant DNA into a mammalian cell. Once this DNA is integrated into the target cell's chromosome, the absence of replication and packaging functions makes the integration very stable by preventing deletion or replication of the integrated DNA.

Data Analysis Problem

14. *Hinc*II: The First Restriction Endonuclease Discovery of the first restriction endonuclease to be of practical use was reported in two papers published in 1970. In the first paper, Smith and Wilcox described the isolation of an enzyme that cleaved double-stranded DNA. They initially demonstrated the enzyme's nuclease activity by measuring the decrease in viscosity of DNA samples treated with the enzyme.

(a) Why does treatment with a nuclease decrease the viscosity of a solution of DNA?

The authors determined whether the enzyme was an endo- or an exonuclease by treating ^{32}P-labeled DNA with the enzyme, then adding trichloroacetic acid (TCA). Under the conditions used in their experiment, single nucleotides would be TCA-soluble and oligonucleotides would precipitate.

(b) No TCA-soluble ^{32}P-labeled material formed on treatment of ^{32}P-labeled DNA with the nuclease. Based on this finding, is the enzyme an endo- or exonuclease? Explain your reasoning.

When a polynucleotide is cleaved, the phosphate usually is not removed but remains attached to the 5′ or 3′ end of the resulting DNA fragment. Smith and Wilcox determined the location of the phosphate on the fragment formed by the nuclease in the following steps:

1. Treat unlabeled DNA with the nuclease.
2. Treat a sample (A) of the product with γ-^{32}P-labeled ATP and polynucleotide kinase (which can attach the γ-phosphate of ATP to a 5′ OH but not to a 5′ phosphate or to a 3′ OH or 3′ phosphate). Measure the amount of ^{32}P incorporated into the DNA.
3. Treat another sample (B) of the product of step 1 with alkaline phosphatase (which removes phosphate groups from free 5′ and 3′ ends), followed by polynucleotide kinase and γ-^{32}P-labeled ATP. Measure the amount of ^{32}P incorporated into the DNA.

(c) Smith and Wilcox found that sample A had 136 counts/min of ^{32}P; sample B had 3,740 counts/min. Did the nuclease cleavage leave the phosphate on the 5′ or the 3′ end of the DNA fragments? Explain your reasoning.

(d) Treatment of bacteriophage T7 DNA with the nuclease gave approximately 40 specific fragments of various lengths. How is this result consistent with the enzyme's recognizing a specific sequence in the DNA as opposed to making random double-strand breaks?

At this point, there were two possibilities for the site-specific cleavage: the cleavage occurred either (1) at the site of recognition or (2) near the site of recognition but not within the sequence recognized. To address this issue, Kelly and Smith determined the sequence of the 5′ ends of the DNA fragments generated by the nuclease, in the following steps:

1. Treat phage T7 DNA with the enzyme.
2. Treat the resulting fragments with alkaline phosphatase to remove the 5′ phosphates.
3. Treat the dephosphorylated fragments with polynucleotide kinase and γ-^{32}P-labeled ATP to label the 5′ ends.
4. Treat the labeled molecules with DNases to break them into a mixture of mono-, di-, and trinucleotides.
5. Determine the sequence of the labeled mono-, di-, and trinucleotides by comparing them with oligonucleotides of known sequence on thin-layer chromatography.

The labeled products were identified as follows: mononucleotides: A and G; dinucleotides: 5′-pApA-3′ and 5′-pGpA-3′; trinucleotides: 5′-pApApC-3′ and 5′-pGpApC-3′.

(e) Which model of cleavage is consistent with these results? Explain your reasoning.

Kelly and Smith went on to determine the sequence of the 3′ ends of the fragments. They found a mixture of 5′-pTpC-3′ and 5′-pTpT-3′. They did not determine the sequence of any trinucleotides at the 3′ end.

(f) Based on these data, what is the recognition sequence for the nuclease and where in the sequence is the DNA backbone cleaved? Use Table 9–2 as a model for your answer.

Answer

(a) DNA solutions are highly viscous because the very long molecules are tangled in solution. Shorter molecules tend to tangle less and form a less viscous solution, so decreased viscosity corresponds to shortening of the polymers—as caused by nuclease activity.

(b) An endonuclease. An exonuclease removes single nucleotides from the 5′ or 3′ end and would produce TCA-soluble ^{32}P-labeled nucleotides. An endonuclease cuts DNA into oligonucleotide fragments and produces little or no TCA-soluble ^{32}P-labeled material.

(c) The 5′ end. If the phosphate were left on the 3′ end, the kinase would incorporate significant ^{32}P as it added phosphate to the 5′ end; treatment with the phosphatase would have no effect on this. In this case, samples A and B would incorporate significant amounts of ^{32}P. When the phosphate is left on the 5′ end, the kinase does not incorporate any ^{32}P: it cannot add a phosphate if one is already present. Treatment with the phosphatase removes 5′ phosphate, and the kinase then incorporates significant amounts of ^{32}P. Sample A will have little or no ^{32}P, and B will show substantial ^{32}P incorporation—as was observed.

(d) Random breaks would produce a distribution of fragments of random size. The production of specific fragments indicates that the enzyme is site-specific.

(e) Cleavage at the site of recognition. This produces a specific sequence at the 5′ end of the fragments. If cleavage occurred near but not within the recognition site, the sequence at the 5′ end of the fragments would be random.

(f) The results are consistent with two recognition sequences, as shown below, cleaved where shown by the arrows:

```
             ↓
(5′)---GTT AAC---(3′)
(3′)---CAA TTG---(5′)
             ↑
```

which gives the (5′)pApApC and (3′)TpTp fragments; and

```
             ↓
(5′)---GTC GAC---(3′)
(3′)---CAG CTG---(5′)
             ↑
```

which gives the (5′)pGpApC and (3′)CpTp fragments.

References

Kelly, T.J. & Smith, H.O. (1970) A restriction enzyme from *Haemophilus influenzae*: II. Base sequence of the recognition site. *J. Mol. Biol.* **51,** 393–409.

Smith, H.O. & Wilcox, K.W. (1970) A restriction enzyme from *Haemophilus influenzae*: I. Purification and general properties. *J. Mol. Biol.* **51,** 379–391.

chapter 10 Lipids

1. **Operational Definition of Lipids** How is the definition of "lipid" different from the types of definitions used for other biomolecules that we have considered, such as amino acids, nucleic acids, and proteins?

   Answer The term "lipid" does not specify a particular chemical structure. Whereas one can write a general formula for an amino acid, nucleic acid, or protein, lipids are much more chemically diverse. Compounds are categorized as lipids based on their greater solubility in organic solvents than in water.

2. **Melting Points of Lipids** The melting points of a series of 18-carbon fatty acids are: stearic acid, 69.6 °C; oleic acid, 13.4 °C; linoleic acid, −5 °C; and linolenic acid, −11 °C.
   (a) What structural aspect of these 18-carbon fatty acids can be correlated with the melting point?
   (b) Draw all the possible triacylglycerols that can be constructed from glycerol, palmitic acid, and oleic acid. Rank them in order of increasing melting point.
   (c) Branched-chain fatty acids are found in some bacterial membrane lipids. Would their presence increase or decrease the fluidity of the membranes (that is, give them a lower or higher melting point)? Why?

   Answer
   (a) The number of cis double bonds (stearic acid, 18:0; oleic, 18:1; linoleic, 18:2; linolenic, 18:3). Each cis double bond causes a bend in the hydrocarbon chain, and bent chains are less well packed than straight chains in a crystal lattice. The lower the extent of packing, the lower the melting temperature.
   (b) Six different triacylglycerols are possible: one with glycerol and only palmitic acid (PPP); one with glycerol and only oleic acid (OOO); and four with glycerol and a mixture of oleic and palmitic acids. Four mixed triacylglycerols are possible, because the three carbons of glycerol are not equivalent: thus OOP and OPO are positional isomers, as are POP and OPP. The greater the content of saturated fatty acid (P), the higher the melting point. Thus, the order of melting points is OOO < OOP = OPO < POP = OPP < PPP. See Table 10–1 and Figure 10–3 for information on how to draw the triacylglycerols.
   (c) Branched-chain fatty acids will increase the fluidity of membranes (i.e., lower their melting point) because they decrease the extent of packing possible within the membrane. The effect of branches is similar to that of bends caused by double bonds.

3. **Preparation of Béarnaise Sauce** During the preparation of béarnaise sauce, egg yolks are incorporated into melted butter to stabilize the sauce and avoid separation. The stabilizing agent in the egg yolks is lecithin (phosphatidylcholine). Suggest why this works.

   Answer Lecithin (see Fig. 10–14 for structure), an amphipathic molecule, is an emulsifying agent, solubilizing the fat (triacylglycerols) in butter. Lecithin is such a good emulsifying agent that it can be used to create a stable emulsion in a mixture that contains up to 75% oil. Mayonnaise, too, is an emulsion created with egg yolks, with an oil:vinegar mixture in a 3:1 ratio.

4. Isoprene Units in Isoprenoids Geraniol, farnesol, and squalene are called isoprenoids, because they are synthesized from five-carbon isoprene units. In each compound, circle the five-carbon units representing isoprene units (see Fig. 10–22).

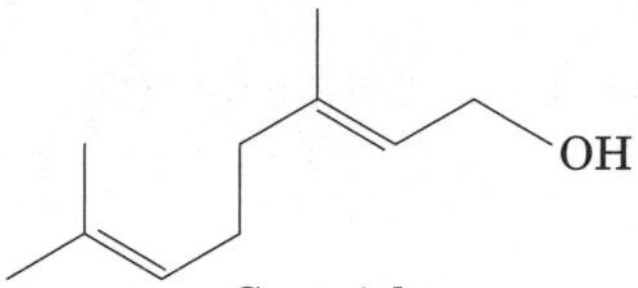

Geraniol

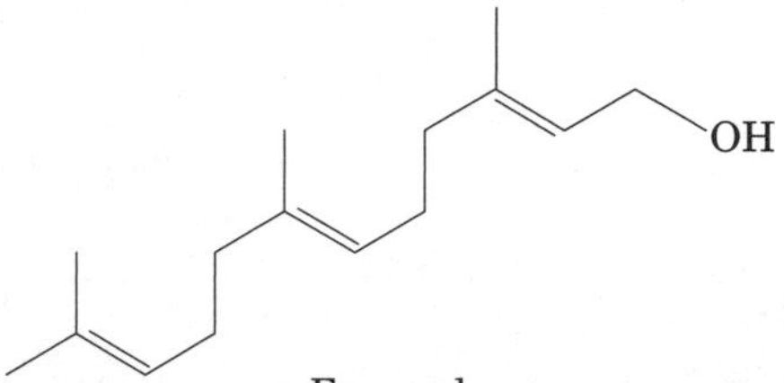

Farnesol

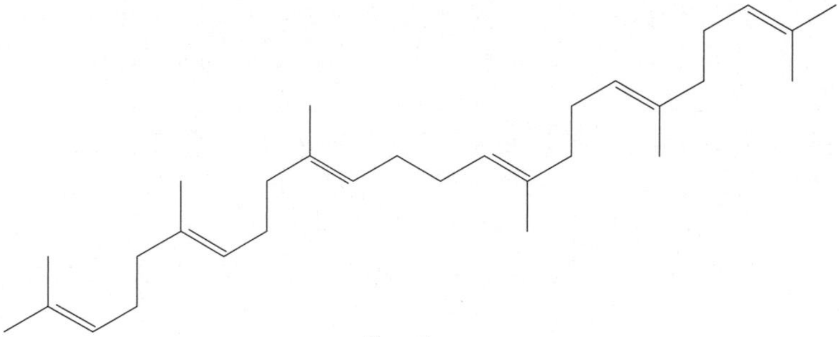

Squalene

Answer

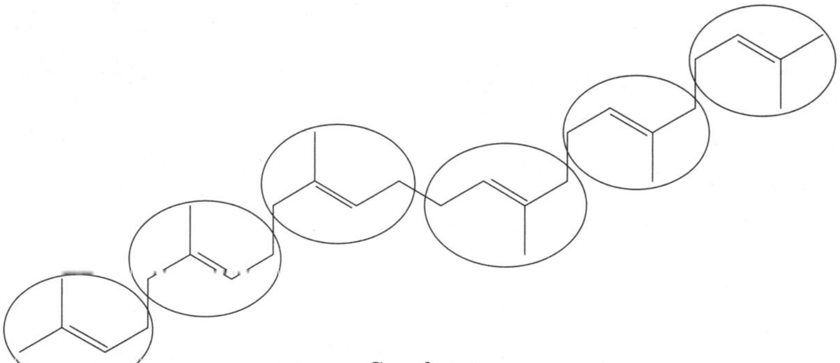

Squalene

5. Naming Lipid Stereoisomers The two compounds below are stereoisomers of carvone with quite different properties; the one on the left smells like spearmint, and that on the right, like caraway. Name the compounds using the RS system.

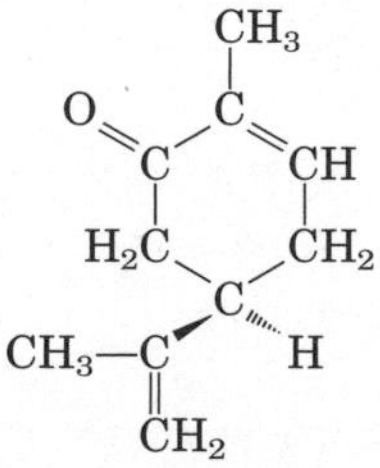

Spearmint

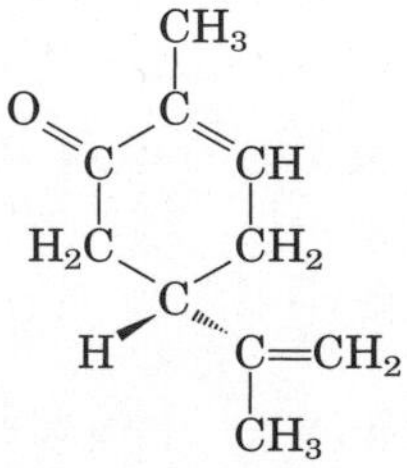

Caraway

Answer Spearmint is (*R*)-carvone; caraway is (*S*)-carvone.

6. **RS Designations for Alanine and Lactate** Draw (using wedge-bond notation) and label the (*R*) and (*S*) isomers of 2-aminopropanoic acid (alanine) and 2-hydroxypropanoic acid (lactic acid).

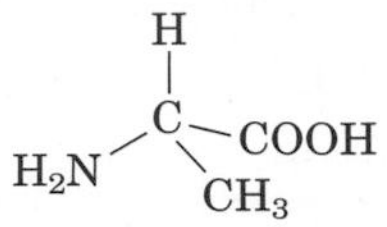

2-Aminopropanoic acid (alanine)

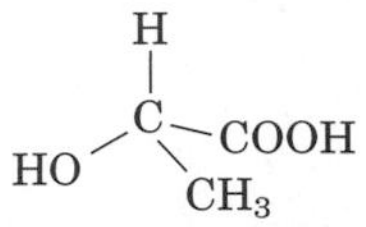

2-Hydroxypropanoic acid (lactic acid)

Answer

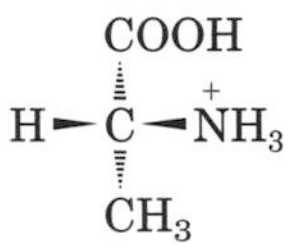

(*R*)-2-Aminopropanoic acid

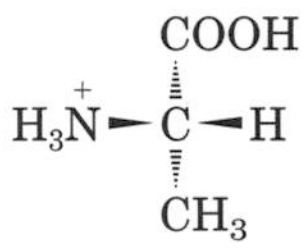

(*S*)-2-Aminopropanoic acid

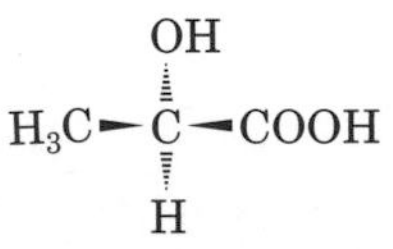

(*R*)-2-Hydroxypropanoic acid

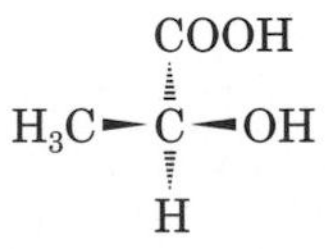

(*S*)-2-Hydroxypropanoic acid

7. **Hydrophobic and Hydrophilic Components of Membrane Lipids** A common structural feature of membrane lipids is their amphipathic nature. For example, in phosphatidylcholine, the two fatty acid chains are hydrophobic and the phosphocholine head group is hydrophilic. For each of the following membrane lipids, name the components that serve as the hydrophobic and hydrophilic units: **(a)** phosphatidylethanolamine; **(b)** sphingomyelin; **(c)** galactosylcerebroside; **(d)** ganglioside; **(e)** cholesterol.

Answer

	Hydrophobic unit(s)	Hydrophilic unit(s)
(a)	2 Fatty acids	Phosphoethanolamine
(b)	1 Fatty acid and the hydrocarbon chain of sphingosine	Phosphocholine
(c)	1 Fatty acid and the hydrocarbon chain of sphingosine	D-Galactose
(d)	1 Fatty acid and the hydrocarbon chain of sphingosine	Several sugar molecules
(e)	Steroid nucleus and acyl side chain	Alcohol group

8. **Structure of Omega-6 Fatty Acid** Draw the structure of the omega-6 fatty acid 16:1.

Answer

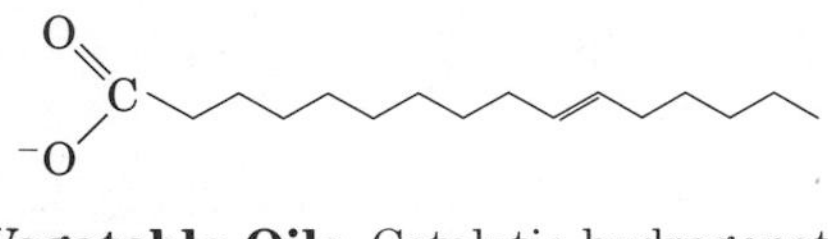

9. **Catalytic Hydrogenation of Vegetable Oils** Catalytic hydrogenation, used in the food industry, converts double bonds in the fatty acids of the oil triacylglycerols to $—CH_2—CH_2—$. How does this affect the physical properties of the oils?

Answer It reduces double bonds, which increases the melting point of lipids containing the fatty acids.

10. **Alkali Lability of Triacylglycerols** A common procedure for cleaning the grease trap in a sink is to add a product that contains sodium hydroxide. Explain why this works.

Answer Triacylglycerols, a component of grease (consisting largely of animal fats), are hydrolyzed by NaOH to form glycerol and the sodium salts of free fatty acids, a process known as saponification. The fatty acids form micelles, which are more water-soluble than triacylglycerols.

11. Deducing Lipid Structure from Composition Compositional analysis of a certain lipid shows that it has exactly one mole of fatty acid per mole of inorganic phosphate. Could this be a glycerophospholipid? A ganglioside? A sphingomyelin?

Answer It could only be a sphingolipid (sphingomyelin). Sphingomyelin has one fatty acid molecule and a phosphocholine molecule attached to the sphingosine backbone, for a ratio of fatty acid to inorganic phosphate of 1:1. Glycerophospholipids have two fatty acyl chains and a head group attached to a phosphoglycerol molecule. Unless the head group included additional phosphate groups, the ratio of fatty acid to inorganic phosphate would be 2:1. (Phosphatidylinositol 4,5-bisphosphate would have a ratio of 2:3; cardiolipin, a ratio of 4:2.) Gangliosides do not contain inorganic phosphate.

12. Deducing Lipid Structure from Molar Ratio of Components Complete hydrolysis of a glycerophospholipid yields glycerol, two fatty acids (16:1(Δ^9) and 16:0), phosphoric acid, and serine in the molar ratio 1:1:1:1:1. Name this lipid and draw its structure.

Answer

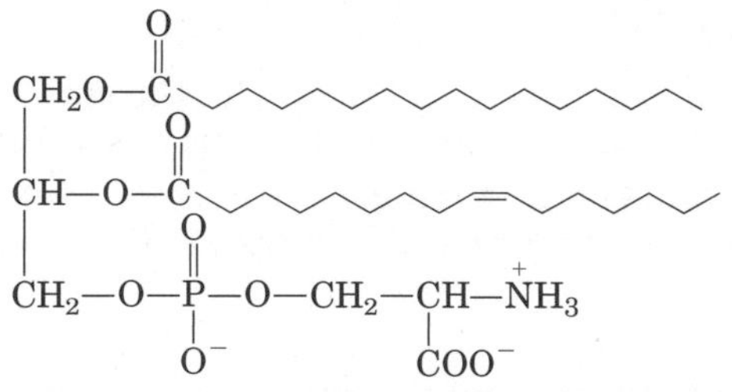

$CH_2O-C(=O)-$ (16:0 chain)
$CH-O-C(=O)-$ (16:1 chain)
$CH_2-O-P(=O)(O^-)-O-CH_2-CH(COO^-)-\overset{+}{N}H_3$

Phosphatidylserine

13. Impermeability of Waxes What property of the waxy cuticles that cover plant leaves makes the cuticles impermeable to water?

Answer Long, saturated acyl chains, nearly solid at air temperature, form a hydrophobic layer in which a polar compound such as H_2O cannot dissolve or diffuse.

14. The Action of Phospholipases The venom of the Eastern diamondback rattler and the Indian cobra contains phospholipase A_2, which catalyzes the hydrolysis of fatty acids at the C-2 position of glycerophospholipids. The phospholipid breakdown product of this reaction is lysolecithin (lecithin is phosphatidylcholine). At high concentrations, this and other lysophospholipids act as detergents, dissolving the membranes of erythrocytes and lysing the cells. Extensive hemolysis may be life-threatening.

(a) All detergents are amphipathic. What are the hydrophilic and hydrophobic portions of lysolecithin?

(b) The pain and inflammation caused by a snake bite can be treated with certain steroids. What is the basis of this treatment?

(c) Though the high levels of phospholipase A_2 can be deadly, this enzyme is necessary for a variety of normal metabolic processes. What are these processes?

Answer

(a) The free —OH group on C-2 and the phosphocholine head group on C-3 are the hydrophilic portions; the fatty acid on C-1 of the lysolecithin is the hydrophobic portion.

(b) Certain steroids such as prednisone inhibit the action of phospholipase A_2, the enzyme that releases the fatty acid arachidonate from the C-2 position of some membrane glycerophospholipids. Arachidonate is converted to a variety of eicosanoids, some of which cause inflammation and pain.

(c) Phospholipase A_2 is necessary to release arachidonate from certain membrane glycerophopholipids. Arachidonate is a precursor of other eicosanoids that have vital protective functions in the body. The enzyme is also important in digestion, breaking down dietary glycerophospholipids.

15. Lipids in Blood Group Determination We note in Figure 10–15 that the structure of glycosphingolipids determines the blood groups A, B, and O in humans. It is also true that glycoproteins determine blood groups. How can both statements be true?

Answer The part of the membrane lipid that determines blood type is the oligosaccharide in the head group of the membrane sphingolipids (see Fig. 10–15, p. 355). This same oligosaccharide is attached to certain membrane glycoproteins, which also serve as points of recognition by the antibodies that distinguish blood groups.

16. Intracellular Messengers from Phosphatidylinositols When the hormone vasopressin stimulates cleavage of phosphatidylinositol 4,5-bisphosphate by hormone-sensitive phospholipase C, two products are formed. What are they? Compare their properties and their solubilities in water, and predict whether either would diffuse readily through the cytosol.

Answer Phosphatidylinositol 4,5-bisphosphate is a membrane lipid. The two products of cleavage are a diacylglycerol and inositol 1,4,5-trisphosphate (IP_3). Diacylglycerol is not water-soluble and remains in the membrane, acting as a second messenger. The IP_3 is highly polar and very soluble in water; it readily diffuses in the cytosol, acting as a soluble second messenger.

17. Storage of Fat-Soluble Vitamins In contrast to water-soluble vitamins, which must be part of our daily diet, fat-soluble vitamins can be stored in the body in amounts sufficient for many months. Suggest an explanation for this difference.

Answer Unlike water-soluble compounds, lipid-soluble compounds are not readily mobilized—that is, they do not readily pass into aqueous solution. The body's lipids provide a reservoir for storage of lipid-soluble vitamins. Water-soluble vitamins cannot be stored and are rapidly removed from the blood by the kidneys.

18. Hydrolysis of Lipids Name the products of mild hydrolysis with dilute NaOH of **(a)** 1-stearoyl-2,3-dipalmitoylglycerol; **(b)** 1-palmitoyl-2-oleoylphosphatidylcholine.

Answer Mild hydrolysis cleaves the ester linkages between glycerol and fatty acids, forming **(a)** glycerol and the sodium salts of palmitic and stearic acids; **(b)** D-glycerol 3-phosphocholine and the sodium salts of palmitic and oleic acids.

19. Effect of Polarity on Solubility Rank the following in order of increasing solubility in water: a triacylglycerol, a diacylglycerol, and a monoacylglycerol, all containing only palmitic acid.

Answer Solubilities: monoacylglycerol > diacylglycerol > triacylglycerol. Increasing the number of palmitic acid moieties increases the proportion of the molecule that is hydrophobic.

20. Chromatographic Separation of Lipids A mixture of lipids is applied to a silica gel column, and the column is then washed with increasingly polar solvents. The mixture consists of phosphatidylserine, phosphatidylethanolamine, phosphatidylcholine, cholesteryl palmitate (a sterol ester), sphingomyelin, palmitate, *n*-tetradecanol, triacylglycerol, and cholesterol. In what order will the lipids elute from the column? Explain your reasoning.

Answer Because silica gel is polar, the most hydrophobic lipids elute first, the most hydrophilic last. The neutral lipids elute first: cholesteryl palmitate and triacylglycerol. Cholesterol and *n*-tetradecanol, neutral but somewhat more polar, elute next. The neutral phospholipids phosphatidylcholine and phosphatidylethanolamine follow. Sphingomyelin, neutral but slightly more polar, elutes after the neutral phospholipids. The negatively charged phosphatidylserine and palmitate elute last—phosphatidylserine first because it is larger and has a lower charge-to-mass ratio.

21. Identification of Unknown Lipids Johann Thudichum, who practiced medicine in London about 100 years ago, also dabbled in lipid chemistry in his spare time. He isolated a variety of lipids from neural tissue, and characterized and named many of them. His carefully sealed and labeled vials of isolated lipids were rediscovered many years later.

(a) How would you confirm, using techniques not available to Thudichum, that the vials labeled "sphingomyelin" and "cerebroside" actually contain these compounds?

(b) How would you distinguish sphingomyelin from phosphatidylcholine by chemical, physical, or enzymatic tests?

Answer

(a) First, create an acid hydrolysate of each compound. Sphingomyelin yields sphingosine, fatty acids, phosphocholine, choline, and phosphate. Cerebroside yields sphingosine, fatty acids, and sugars, but no phosphate. Subject each hydrolysate to chromatography (gas-liquid or silica gel thin-layer chromatography) and compare the result with known standards.

(b) On strong alkaline hydrolysis, sphingomyelin yields sphingosine, fatty acids, and phosphocholine; phosphatidylcholine yields glycerol, fatty acids, and phosphocholine. The distinguishing features are the presence of *sphingosine* in sphingomyelin and *glycerol* in phosphatidylcholine, which can be detected on thin-layer chromatograms of each hydrolysate compared against known standards. The hydrolysates could also be distinguished by their reaction with the Sanger reagent (1-fluoro-2,4-dinitrobenzene, FDNB); only the sphingosine in the sphingomyelin hydrolysate has a primary amine that would react with FDNB to form a colored product. Alternatively, enzymatic treatment of the two samples with phospholipase A_1 or A_2 would release free fatty acids from phosphatidylcholine, but not from sphingomyelin.

22. Ninhydrin to Detect Lipids on TLC Plates Ninhydrin reacts specifically with primary amines to form a purplish-blue product. A thin-layer chromatogram of rat liver phospholipids is sprayed with ninhydrin, and the color is allowed to develop. Which phospholipids can be detected in this way?

Answer Phosphatidylethanolamine and phosphatidylserine; they are the only phospholipids that have primary amine groups that can react with ninhydrin.

Data Analysis Problem

23. Determining the Structure of the Abnormal Lipid in Tay-Sachs Disease Box 10–2, Figure 1, shows the pathway of breakdown of gangliosides in healthy (normal) individuals and individuals with certain genetic diseases. Some of the data on which the figure is based were presented in a paper by Lars Svennerholm (1962). Note that the sugar Neu5Ac, *N*-acetylneuraminic acid, represented in the Box 10–2 figure as a purple ◆, is a sialic acid.

Svennerholm reported that "about 90% of the monosialiogangliosides isolated from normal human brain" consisted of a compound with ceramide, hexose, *N*-acetylgalactosamine, and *N*-acetylneuraminic acid in the molar ratio 1:3:1:1.

(a) Which of the gangliosides (GM1 through GM3 and globoside) in Box 10–2, Figure 1, fits this description? Explain your reasoning.

(b) Svennerholm reported that 90% of the gangliosides from a patient with Tay-Sachs had a molar ratio (of the same four components given above) of 1:2:1:1. Is this consistent with the Box 10–2 figure? Explain your reasoning.

To determine the structure in more detail, Svennerholm treated the gangliosides with neuraminidase to remove the *N*-acetylneuraminic acid. This resulted in an asialoganglioside that was much easier to analyze. He hydrolyzed it with acid, collected the ceramide-containing products, and determined the molar ratio of the sugars in each product. He did this for both the normal and the Tay-Sachs gangliosides. His results are shown below.

Ganglioside	Ceramide	Glucose	Galactose	Galactosamine
Normal				
Fragment 1	1	1	0	0
Fragment 2	1	1	1	0
Fragment 3	1	1	1	1
Fragment 4	1	1	2	1
Tay-Sachs				
Fragment 1	1	1	0	0
Fragment 2	1	1	1	0
Fragment 3	1	1	1	1

(c) Based on these data, what can you conclude about the structure of the normal ganglioside? Is this consistent with the structure in Box 10–2? Explain your reasoning.

(d) What can you conclude about the structure of the Tay-Sachs ganglioside? Is this consistent with the structure in Box 10–2? Explain your reasoning.

Svennerholm also reported the work of other researchers who "permethylated" the normal asialoganglioside. Permethylation is the same as exhaustive methylation: a methyl group is added to every free hydroxyl group on a sugar. They found the following permethylated sugars: 2,3,6-trimethylglycopyranose; 2,3,4,6-tetramethylgalactopyranose; 2,4,6-trimethylgalactopyranose; and 4,6-dimethyl-2-deoxy-2-aminogalactopyranose.

(e) To which sugar of GM1 does each of the permethylated sugars correspond? Explain your reasoning.

(f) Based on all the data presented so far, what pieces of information about normal ganglioside structure are missing?

Answer

(a) GM1 and globoside. Both glucose and galactose are hexoses, so "hexose" in the molar ratio refers to glucose + galactose. The ratios for the four gangliosides are: GM1, 1:3:1:1; GM2, 1:2:1:1; GM3, 1:2:0:1; globoside, 1:3:1:0.

(b) Yes. The ratio matches GM2, the ganglioside expected to build up in Tay-Sachs disease (see Box 10–2, Fig. 1).

(c) This analysis is similar to that used by Sanger to determine the amino acid sequence of insulin. The analysis of each fragment reveals only its *composition,* not its *sequence,* but because each fragment is formed by sequential removal of one sugar, we can draw conclusions about sequence. The structure of the normal asialoganglioside is ceramide–glucose–galactose–galactosamine–galactose, consistent with Box 10–2 (excluding Neu5Ac, removed before hydrolysis).

(d) The Tay-Sachs asialoganglioside is ceramide–glucose–galactose–galactosamine, consistent with Box 10–2.

(e) The structure of the normal asialoganglioside, GM1, is: *ceramide–glucose* (2 —OH involved in glycosidic links; 1 —OH involved in ring structure; 3 —OH (2,3,6) free for methylation)– *galactose* (2 —OH in links; 1 —OH in ring; 3 —OH (2,4,6) free for methylation)–*galactosamine* (2 —OH in links; 1 —OH in ring; 1 —NH_2 instead of an —OH; 2 —OH (4,6) free for methylation)–*galactose* (1 —OH in link; 1 —OH in ring; 4 —OH (2,3,4,6) free for methylation).

(f) Two key pieces of information are missing: What are the linkages between the sugars? Where is Neu5Ac attached?

Reference

Svennerholm, L. (1962) The chemical structure of normal human brain and Tay-Sachs gangliosides. *Biochem. Biophys. Res. Comm.* **9,** 436–441.

chapter

11 Biological Membranes and Transport

1. **Determining the Cross-Sectional Area of a Lipid Molecule** When phospholipids are layered gently onto the surface of water, they orient at the air-water interface with their head groups in the water and their hydrophobic tails in the air. An experimental apparatus **(a)** has been devised that reduces the surface area available to a layer of lipids. By measuring the force necessary to push the lipids together, it is possible to determine when the molecules are packed tightly in a continuous monolayer; as that area is approached, the force needed to further reduce the surface area increases sharply **(b).** How would you use this apparatus to determine the average area occupied by a single lipid molecule in the monolayer?

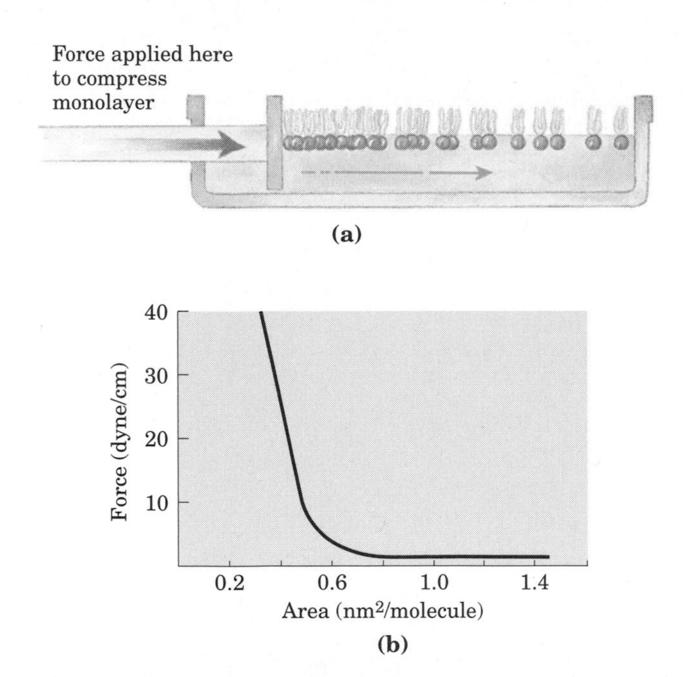

Answer Determine the surface area of the water at which the pressure increases sharply. Divide this surface area by the number of lipid molecules on the surface, which is calculated by multiplying the number of moles (calculated from the concentration and the molecular weight) by Avogadro's number.

2. **Evidence for a Lipid Bilayer** In 1925, E. Gorter and F. Grendel used an apparatus like that described in Problem 1 to determine the surface area of a lipid monolayer formed by lipids extracted from erythrocytes of several animal species. They used a microscope to measure the dimensions of individual cells, from which they calculated the average surface area of one erythrocyte. They obtained the data shown in the following table. Were these investigators justified in concluding that "chromocytes [erythrocytes] are covered by a layer of fatty substances that is two molecules thick" (i.e., a lipid bilayer)?

Animal	Volume of packed cells (mL)	Number of cells (per mm^3)	Total surface area of lipid monolayer from cells (m^2)	Total surface area of one cell (μm^2)
Dog	40	8,000,000	62	98
Sheep	10	9,900,000	6.0	29.8
Human	1	4,740,000	0.92	99.4

Source: Data from Gorter, E. & Grendel, F. (1925) On bimolecular layers of lipoids on the chromocytes of the blood. *J. Exp. Med.* **41,** 439–443.

Answer The conclusions are justified for dog erythrocytes but not for sheep or human erythrocytes. The table provides the total surface area of the lipid monolayer. To determine the monolayer surface area per cell, first calculate the total number of cells. For example, for dog erythrocytes, the number of cells is 8×10^6 per $mm^3 = 8 \times 10^9$ per cm^3 (or per mL). In 40 mL, there is a total of $(40 \text{ mL})(8 \times 10^9 \text{ cells/mL}) = 3 \times 10^{11}$ cells. From the table, this number of cells yielded a monolayer surface area of $62 \text{ m}^2 = 6.2 \times 10^5 \text{ cm}^2$. Dividing the surface area by the number of cells gives

$$\frac{6.2 \times 10^5 \text{ cm}^2}{3 \times 10^{11} \text{ cells}} = 2 \times 10^{-6} \text{ cm}^2\text{/cell}$$

Comparing this number to the total surface area of one erythrocyte ($98 \ \mu m^2 = 0.98 \times 10^{-6} \text{ cm}^2$), we find a 2-to-1 relationship. This result justifies the investigators' conclusion of a lipid bilayer in dog erythrocytes. Similar calculations for the sheep and human erythrocytes reveal a 1-to-1 relationship, suggesting a lipid monolayer. However, there were significant experimental errors in these early experiments; recent, more accurate measurements support a bilayer in all cases.

3. Number of Detergent Molecules per Micelle When a small amount of the detergent sodium dodecyl sulfate (SDS; $Na^+CH_3(CH_2)_{11}OSO_3^-$) is dissolved in water, the detergent ions enter the solution as monomeric species. As more detergent is added, a concentration is reached (the critical micelle concentration) at which the monomers associate to form micelles. The critical micelle concentration of SDS is 8.2 mM. The micelles have an average particle weight (the sum of the molecular weights of the constituent monomers) of 18,000. Calculate the number of detergent molecules in the average micelle.

Answer The molecular weight of sodium dodecyl sulfate is 288. Given an average micelle particle weight of 18,000, there are 18,000/288 = 63 SDS molecules per micelle.

4. Properties of Lipids and Lipid Bilayers Lipid bilayers formed between two aqueous phases have this important property: they form two-dimensional sheets, the edges of which close upon each other and undergo self-sealing to form liposomes.

(a) What properties of lipids are responsible for this property of bilayers? Explain.

(b) What are the consequences of this property for the structure of biological membranes?

Answer

(a) Lipids that form bilayers are amphipathic molecules: they contain hydrophilic and hydrophobic regions. In order to minimize the hydrophobic area exposed to the water surface, these lipids form two-dimensional sheets, with the hydrophilic regions exposed to water and the hydrophobic regions buried in the interior of the sheet. Furthermore, to avoid exposing the hydrophobic edges of the sheet to water, lipid bilayers close upon themselves. Similarly, if the sheet is perforated, the hole will seal because the membrane is semifluid.

(b) These sheets form the closed membrane surfaces that envelop cells and compartments within cells (organelles).

5. Length of a Fatty Acid Molecule The carbon–carbon bond distance for single-bonded carbons such as those in a saturated fatty acyl chain is about 1.5 Å. Estimate the length of a single molecule of palmitate in its fully extended form. If two molecules of palmitate were placed end to end, how would their total length compare with the thickness of the lipid bilayer in a biological membrane?

Answer Given that the C–C bond length is 0.15 nm and that the bond angle of tetrahedral carbon is 109°, the distance between the first and third carbons in an acyl chain (calculated by trigonometry) is about 0.24 nm. For palmitate (16:0), the length of the extended chain is about 8×0.24 nm = 2 nm. Two palmitate chains end to end (as in a bilayer) would extend 4 nm. This is about the thickness of a lipid bilayer.

6. Temperature Dependence of Lateral Diffusion The experiment described in Figure 11–17 was performed at 37 °C. If the experiment were carried out at 10 °C, what effect would you expect on the rate of diffusion? Why?

Answer When the temperature drops, the fluidity of a membrane decreases. This is caused by a decrease in the rate of diffusion of lipids. Consequently, all processes depending on diffusion, such as the lateral diffusion experiment shown in Figure 11–17, would slow down.

7. Synthesis of Gastric Juice: Energetics Gastric juice (pH 1.5) is produced by pumping HCl from blood plasma (pH 7.4) into the stomach. Calculate the amount of free energy required to concentrate the H^+ in 1 L of gastric juice at 37 °C. Under cellular conditions, how many moles of ATP must be hydrolyzed to provide this amount of free energy? The free-energy change for ATP hydrolysis under cellular conditions is about −58 kJ/mol (as explained in Chapter 13). Ignore the effects of the transmembrane electrical potential.

Answer Given that pH = $-\log [H^+]$, then $[H^+] = 10^{-pH}$.

At pH 1.5, $[H^+] = 10^{-1.5} = 3.2 \times 10^{-2}$ M.

At pH 7.4, $[H^+] = 10^{-7.4} = 4.0 \times 10^{-8}$ M.

Because $\Delta G_t = RT \ln (C_2/C_1)$, and at 37 °C, $RT = 2.58$ kJ/mol,

$$\Delta G_t = (2.58 \text{ kJ/mol}) \ln \left(\frac{3.2 \times 10^{-2}}{4.0 \times 10^{-8}} \right) = 35 \text{ kJ/mol.}$$

The amount of ATP required to provide 35 kJ is

$$\frac{35 \text{ kJ}}{58 \text{ kJ/mol}} = 0.60 \text{ mol}$$

8. Energetics of the Na^+K^+ ATPase For a typical vertebrate cell with a membrane potential of −0.070 V (inside negative), what is the free-energy change for transporting 1 mol of Na^+ out of the cell and into the blood at 37 °C? Assume the concentration of Na^+ inside the cell is 12 mM, and that in blood plasma is 145 mM.

Answer

$$\Delta G_t = RT \ln (C_2/C_1) + Z\mathcal{J}\Delta\psi$$

$$= (2.58 \text{ kJ/mol}) \ln \left(\frac{145}{12} \right) + (1)(96{,}480 \text{ J/V} \cdot \text{mol})(0.070 \text{ V})$$

$$= 6.4 \text{ kJ/mol} + 6.8 \text{ kJ/mol} = 13 \text{ kJ/mol}$$

Note that 6.8 kJ/mol is the membrane potential portion.

9. Action of Ouabain on Kidney Tissue Ouabain specifically inhibits the Na^+K^+ ATPase activity of animal tissues but is not known to inhibit any other enzyme. When ouabain is added to thin slices of living kidney tissue, it inhibits oxygen consumption by 66%. Why? What does this observation tell us about the use of respiratory energy by kidney tissue?

Answer Oxidative phosphorylation to supply the cell with ATP accounts for the vast majority of oxygen consumption. A decrease in oxygen consumption by 66% on addition of ouabain indicates that consumption of ATP by the Na^+K^+ ATPase in kidney cells accounts for about two-thirds of the tissue's ATP requirements, and thus of its use of respiratory energy.

10. **Energetics of Symport** Suppose that you determined experimentally that a cellular transport system for glucose, driven by symport of Na^+, could accumulate glucose to concentrations 25 times greater than in the external medium, while the external [Na^+] was only 10 times greater than the intracellular [Na^+]. Would this violate the laws of thermodynamics? If not, how could you explain this observation?

Answer No; the symport may be able to transport more than one equivalent of glucose per Na^+.

11. **Location of a Membrane Protein** The following observations are made on an unknown membrane protein, X. It can be extracted from disrupted erythrocyte membranes into a concentrated salt solution, and it can be cleaved into fragments by proteolytic enzymes. Treatment of erythrocytes with proteolytic enzymes followed by disruption and extraction of membrane components yields intact X. However, treatment of erythrocyte "ghosts" (which consist of just plasma membranes, produced by disrupting the cells and washing out the hemoglobin) with proteolytic enzymes followed by disruption and extraction yields extensively fragmented X. What do these observations indicate about the location of X in the plasma membrane? Do the properties of X resemble those of an integral or peripheral membrane protein?

Answer Because protein X can be removed by salt treatment, it must be a peripheral membrane protein. Inability to digest the protein with proteases unless the membrane has been disrupted indicates that protein X is located internally, bound to the inner surface of the erythrocyte plasma membrane.

12. **Membrane Self-sealing** Cellular membranes are self-sealing—if they are punctured or disrupted mechanically, they quickly and automatically reseal. What properties of membranes are responsible for this important feature?

Answer Hydrophobic interactions are the driving force for membrane formation. Because these forces are noncovalent and reversible, membranes can easily anneal after disruption.

13. **Lipid Melting Temperatures** Membrane lipids in tissue samples obtained from different parts of the leg of a reindeer have different fatty acid compositions. Membrane lipids from tissue near the hooves contain a larger proportion of unsaturated fatty acids than those from tissue in the upper leg. What is the significance of this observation?

Answer The temperature of body tissues at the extremities, such as near the hooves, is generally lower than that of tissues closer to the center of the body. To maintain fluidity, as required by the fluid-mosaic model, membranes at lower temperatures must contain a higher percentage of polyunsaturated fatty acids: a higher content of unsaturated fatty acids lowers the melting point of lipid mixtures.

14. **Flip-Flop Diffusion** The inner leaflet (monolayer) of the human erythrocyte membrane consists predominantly of phosphatidylethanolamine and phosphatidylserine. The outer leaflet consists predominantly of phosphatidylcholine and sphingomyelin. Although the phospholipid components of the membrane can diffuse in the fluid bilayer, this sidedness is preserved at all times. How?

Answer The energy required to flip a charged polar head group through a hydrophobic lipid bilayer is prohibitively high.

15. Membrane Permeability At pH 7, tryptophan crosses a lipid bilayer at about one-thousandth the rate of indole, a closely related compound:

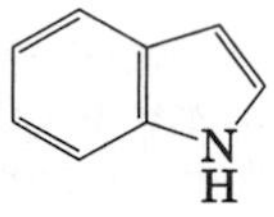

Suggest an explanation for this observation.

Answer At pH 7, tryptophan exists as a zwitterion (having a positive and negative charge), whereas indole is uncharged. The movement of the less polar indole through the hydrophobic core of the bilayer is more energetically favorable.

16. Water Flow through an Aquaporin A human erythrocyte has about 2×10^5 AQP-1 monomers. If water molecules flow through the plasma membrane at a rate of 5×10^8 per AQP-1 tetramer per second, and the volume of an erythrocyte is 5×10^{-11} mL, how rapidly could an erythrocyte halve its volume as it encountered the high osmolarity (1 M) in the interstitial fluid of the renal medulla? Assume that the erythrocyte consists entirely of water.

Answer First, calculate the number of water molecules that must leave the erythrocyte to halve its volume. The volume of the cell is 5×10^{-11} mL. For $[H_2O] = 55$ M, the number of water molecules in the cell is

$$(5 \times 10^{-11} \text{mL/cell})(6.02 \times 10^{20} \text{ molecules/mmol})(55 \text{ mmol } H_2O\text{/mL}) = 1.7 \times 10^{12}$$

Half of these molecules (8.5×10^{11}) must leave to halve the cell volume.

Next, calculate how fast the cell can lose water molecules. The cell has 2×10^5 aquaporin monomers, or 5×10^4 tetramers. Each tetramer allows passage of 5×10^8 H_2O molecules per second, so the flux of water molecules through the plasma membrane is

$$(5 \times 10^8 \text{ } H_2O \text{ molecules/s/aquaporin tetramer})(5 \times 10^4 \text{ aquaporin tetramers/cell})$$
$$= 2.5 \times 10^{13} \text{ } H_2O \text{ molecules/s}$$

Removal of half the volume of water would take

$$(8.5 \times 10^{11} \text{ } H_2O \text{ molecules})/(2.5 \times 10^{13} \text{ } H_2O \text{ molecules/s}) = 3 \times 10^{-2} \text{ s}$$

17. Labeling the Lactose Transporter A bacterial lactose transporter, which is highly specific for lactose, contains a Cys residue that is essential to its transport activity. Covalent reaction of *N*-ethylmaleimide (NEM) with this Cys residue irreversibly inactivates the transporter. A high concentration of lactose in the medium prevents inactivation by NEM, presumably by sterically protecting the Cys residue, which is in or near the lactose-binding site. You know nothing else about the transporter protein. Suggest an experiment that might allow you to determine the M_r of the Cys-containing transporter polypeptide.

Answer Treat a suspension of the bacteria as follows: Add lactose at a concentration well above the K_t, so that virtually every molecule of galactoside transporter binds lactose. Next, add nonradiolabeled NEM and allow it to react with all available —SH groups on the cell surface. Remove excess lactose by centrifuging and resuspending the cells, then add radiolabeled NEM. The only Cys residues now available to react with NEM are those in the transporter protein. Dissolve the membrane proteins in sodium dodecylsulfate (SDS), and separate them on the basis of size by SDS gel electrophoresis. The M_r of the labeled band should represent that of the galactoside transporter.

18. Predicting Membrane Protein Topology from Sequence You have cloned the gene for a human erythrocyte protein, which you suspect is a membrane protein. From the nucleotide sequence of the gene, you know the amino acid sequence. From this sequence alone, how would you evaluate the possibility that the protein is an integral protein? Suppose the protein proves to be an integral protein, either type I or type II. Suggest biochemical or chemical experiments that might allow you to determine which type it is.

Answer Construct and analyze a hydropathy plot for the protein. You can assume that any hydrophobic regions of more than 20 consecutive residues are transmembrane segments of an integral protein. To determine whether the external domain is carboxyl- or amino-terminal, treat intact erythrocytes with a membrane-impermeant reagent known to react with primary amines and determine whether the protein reacts. If it does, the amino terminus is on the external surface of the erythrocyte membrane and this is a type I protein (see Fig. 11–8). If it does not, a type II protein is indicated.

19. Intestinal Uptake of Leucine You are studying the uptake of L-leucine by epithelial cells of the mouse intestine. Measurements of the rate of uptake of L-leucine and several of its analogs, with and without Na^+ in the assay buffer, yield the results given in the table. What can you conclude about the properties and mechanism of the leucine transporter? Would you expect L-leucine uptake to be inhibited by ouabain?

	Uptake in presence of Na^+		Uptake in absence of Na^+	
Substrate	V_{max}	K_t (mM)	V_{max}	K_t (mM)
L-Leucine	420	0.24	23	0.2
D-Leucine	310	4.7	5	4.7
L-Valine	225	0.31	19	0.31

Answer The similar K_t values for L-leucine and L-valine indicate that the transporter binding site can accommodate the side chains of both amino acids equally well; it is probably a hydrophobic pocket of suitable size for either R group. The 20-fold higher K_t for D- than for L-leucine indicates that the binding site recognizes differences of configuration about the α carbon. Based on the lower V_{max} in the absence of Na^+ for all three substrates, we know that Na^+ entry is essential for amino acid uptake; the transporter acts by symport of leucine (or valine) and Na^+.

20. Effect of an Ionophore on Active Transport Consider the leucine transporter described in Problem 19. Would V_{max} and/or K_t change if you added a Na^+ ionophore to the assay solution containing Na^+? Explain.

Answer By dissipating the transmembrane Na^+ gradient, the ionophore would prevent symport of L-leucine and reduce the rate of uptake, measured as V_{max}. The value of K_t, a measure of the transporter's affinity for the substrate (L-leucine), should not change. Valinomycin (the likely ionophore here) does not resemble L-leucine in structure and almost certainly would not bind the transporter to affect K_t.

21. Surface Density of a Membrane Protein *E. coli* can be induced to make about 10,000 copies of the lactose transporter (M_r 31,000) per cell. Assume that *E. coli* is a cylinder 1 μm in diameter and 2 μm long. What fraction of the plasma membrane surface is occupied by the lactose transporter molecules? Explain how you arrived at this conclusion.

Answer The surface area of a cylinder is $2\pi r^2 + \pi dh$, where r = radius, d = diameter, and h = height. For a cylinder 2 μm high and 1 μm in diameter, the surface area is $2\pi(0.5\ \mu\text{m})^2 + \pi(1\ \mu\text{m})(2\ \mu\text{m}) = 2.5\pi\ \mu\text{m}^2 = 8\ \mu\text{m}^2$. This is the *E. coli* surface area.

To estimate the cross-sectional area of a globular protein of M_r 31,000, we can use the dimensions for hemoglobin (M_r = 64,500; diameter = 5.5 nm; see text p. 159), thus a protein of M_r 31,000 has a diameter of about 3 nm, assuming the proteins have the same density. The cross-sectional area of a sphere of diameter 3 nm (0.003 μm)—or of a single transporter molecule—is

$$\pi r^2 = 3.14(1.5 \times 10^{-3}\ \mu\text{m})^2 = 7 \times 10^{-6}\ \mu\text{m}^2$$

and the total cross-sectional area of 10,000 transporter molecules is $7 \times 10^{-2}\ \mu m^2$. Thus, the fraction of an *E. coli* cell surface covered by transporter molecules is

$$(7 \times 10^{-2}\ \mu m^2)/(8\ \mu m^2) = 0.009, \text{ or about } 1\%.$$

This answer is clearly an approximation, given the method of estimating the diameter of the transporter molecule, but it is certainly of the right order of magnitude.

22. Use of the Helical Wheel Diagram A helical wheel is a two-dimensional representation of a helix, a view along its central axis (see Fig. 11–29b; see also Fig. 4–4d). Use the helical wheel diagram below to determine the distribution of amino acid residues in a helical segment with the sequence –Val–Asp–Arg–Val–Phe–Ser–Asn–Val–Cys–Thr–His–Leu–Lys–Thr–Leu–Gln–Asp–Lys–

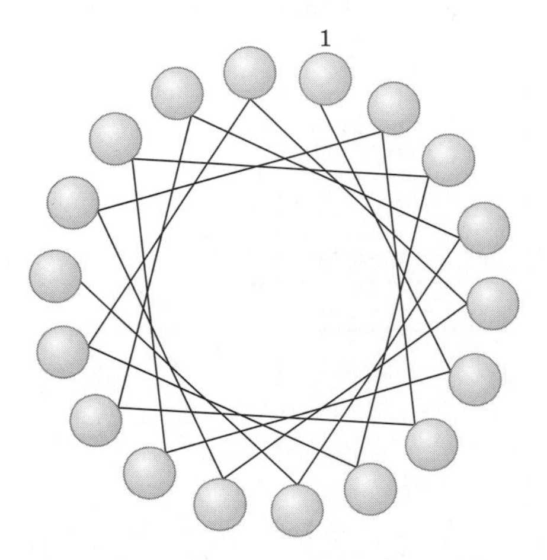

What can you say about the surface properties of this helix? How would you expect the helix to be oriented in the tertiary structure of an integral membrane protein?

Answer A helical wheel is a two-dimensional representation of a helix obtained by projecting the helix down its central axis. An α helix contains 3.6 residues per turn, so each amino acid in the helix lies 100° around the axis from the previous residue: (360°/turn)/(3.6 residues/turn) = 100° per residue. For the 18 amino acid helix considered here, the 18 vertices are separated by 20° increments. If there were a 19th residue, it would lie under the first residue on the projection, but five turns down the helix: 5 turns × 0.54 nm/turn (pitch for an α helix) = 2.70 nm "behind" residue 1. To complete the diagram, follow the lines from residue 1 to residue 2, and so on, numbering the residues. Then, using the sequence given, label each residue with its one-letter abbreviation and a characterization of its R group properties—**P** for polar, and **N** for nonpolar.

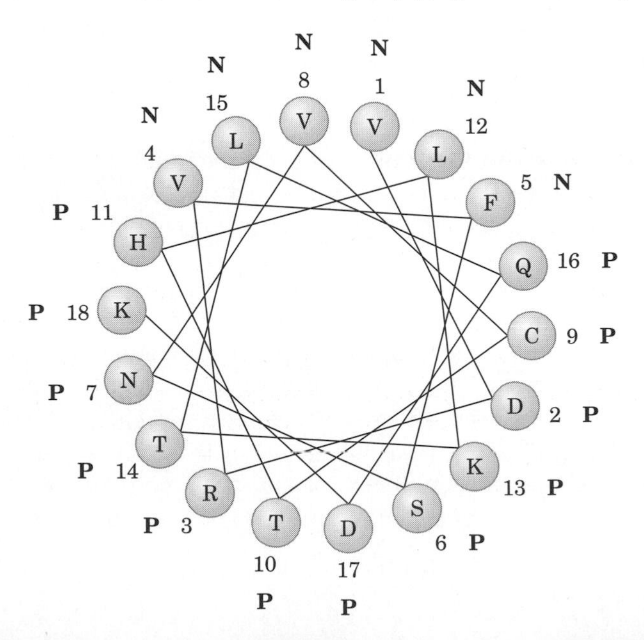

The "top" side of the helix contains only hydrophobic side chains, while the other surfaces are polar or charged; this is an amphipathic helix. As an integral membrane protein, it is likely to dip its hydrophobic surface into the lipid bilayer but expose the other surfaces to the aqueous phase. An alternative arrangement might be to cluster, say, 10 helices, one from each of 10 subunits, around a central hydrophilic core, while exposing only the hydrophobic surface to the lipid bilayer.

23. Molecular Species in the *E. coli* Membrane The plasma membrane of *E. coli* is about 75% protein and 25% phospholipid by weight. How many molecules of membrane lipid are present for each molecule of membrane protein? Assume an average protein M_r of 50,000 and an average phospholipid M_r of 750. What more would you need to know to estimate the fraction of the membrane surface that is covered by lipids?

Answer Consider a sample that contains 1 g of membrane, of which 0.75 g is protein (M_r = 50,000) and 0.25 g is phospholipid (M_r = 750).

$$(0.75 \text{ g protein})(1 \text{ mol}/5 \times 10^4 \text{ g}) = 1.5 \times 10^{-5} \text{ mol protein in 1 g of membrane}$$

$$(0.25 \text{ g phospholipid})(1 \text{ mol}/750 \text{ g}) = 3.3 \times 10^{-4} \text{ mol phospholipid in 1 g of membrane}$$

$$3.3 \times 10^{-4} \text{ mol phospholipid}/1.5 \times 10^{-5} \text{ mol protein} = 22 \text{ mol phospholipid/mol protein}$$

To estimate the percentage of the surface covered by phospholipid, you would need to know (or estimate) the average cross-sectional area of a phospholipid in a bilayer (which you might learn from an experiment such as that diagrammed in Problem 1, above) and the average cross-sectional area of a 50 kDa protein.

Biochemistry on the Internet

24. Membrane Protein Topology The receptor for the hormone epinephrine in animal cells is an integral membrane protein (M_r 64,000) that is believed to have seven membrane-spanning regions.

(a) Show that a protein of this size is capable of spanning the membrane seven times.

(b) Given the amino acid sequence of this protein, how would you predict which regions of the protein form the membrane-spanning helices?

(c) Go to the Protein Data Bank (www.rcsb.org). Use the PDB identifier 1DEP to retrieve the data page for a portion of the β-adrenergic receptor (one type of epinephrine receptor) from a turkey. Using Jmol to explore the structure, predict where this portion of the receptor is located: within the membrane or at the membrane surface. Explain.

(d) Retrieve the data for a portion of another receptor, the acetylcholine receptor of neurons and myocytes, using the PDB identifier 1A11. As in **(c)**, predict where this portion of the receptor is located and explain your answer.

If you have not used the PDB, see Box 4–4 (p. 129) for more information.

Answer

(a) Assume that the transmembrane portion of the peptide is an α helix. The rise per amino acid (AA) residue of an α helix is 1.5Å/AA = 1.5 nm/AA. Assume the lipid bilayer is 4.0 nm thick; thus (4.0 nm)/(0.15 nm/AA) = 27 AA are needed to span the bilayer, and seven spans require 7 × 27 = 189 residues. For an average AA residue M_r of 110, a protein of M_r 64,000 has approximately 64,000/110 = 580 AA residues.

(b) Hydropathy plots are used to identify potential transmembrane regions. The most hydrophobic (hydropathic) stretches are those most likely to pass through the apolar lipid bilayer.

(c) This portion of the epinephrine receptor is an intracellular loop that connects adjacent membrane-spanning regions of the protein. You can predict that this α helix is *not* located in the membrane on the basis of the properties of its amino acids. At the PDB, click on the icon to display the PDB file. The helix has only 15 amino acid residues—RSPDFRKAFKRLLCF—too few to span the bilayer. About half of these residues are charged. Look at the structure in Jmol, using the console controls to display the side chains as ball-and-stick structures. Rotate the molecule so that you look directly down the axis of the helix, with the side chains spiraling around the outside. Note that most of the nonpolar residues fall along one side of this amphipathic helix.

(d) This portion of the acetylcholine receptor is one of the membrane-spanning regions of the protein. You can predict that this α helix is membrane-spanning on the basis of the properties of its amino acid residues. In the PDB file you can see that the amino acid sequence is GSEKMSTAISVLLAQAVFLLLTSQR. In Jmol, display the side chains in space-filling mode. Notice that the charged residues occur primarily at the ends. The long run of hydrophobic or uncharged residues between Lys at position 4 and Arg at position 25 is enough to span a lipid bilayer.

Data Analysis Problem

25. The Fluid Mosaic Model of Biological Membrane Structure Figure 11–3 shows the currently accepted fluid mosaic model of biological membrane structure. This model was presented in detail in a review article by S. J. Singer in 1971. In the article, Singer presented the three models of membrane structure that had been proposed by that time:

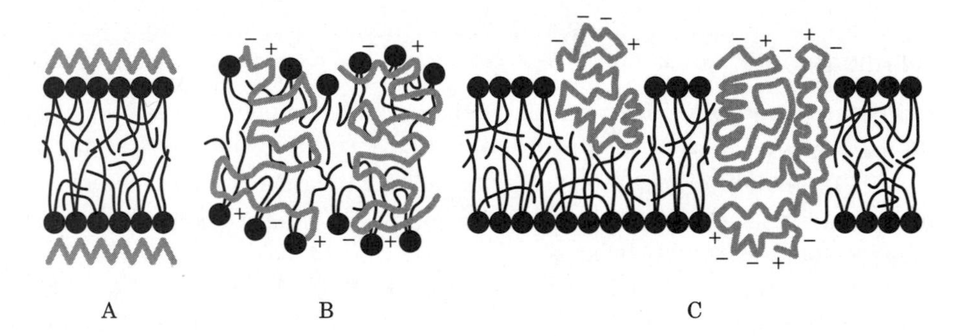

A. The Davson-Danielli-Robertson Model. This was the most widely accepted model in 1971, when Singer's review was published. In this model, the phospholipids are arranged as a bilayer. Proteins are found on both surfaces of the bilayer, attached to it by ionic interactions between the charged head groups of the phospholipids and charged groups in the proteins. Crucially, there is no protein in the interior of the bilayer.

B. The Benson Lipoprotein Subunit Model. Here, the proteins are globular and the membrane is a protein-lipid mixture. The hydrophobic tails of the lipids are embedded in the hydrophobic parts of the proteins. The lipid head groups are exposed to the solvent. There is no lipid bilayer.

C. The Lipid-Globular Protein Mosaic Model. This is the model shown in Figure 11–3. The lipids form a bilayer and proteins are embedded in it, some extending through the bilayer and others not. Proteins are anchored in the bilayer by hydrophobic interactions between the hydrophobic tails of the lipids and hydrophobic portions of the protein.

For the data given below, consider how each piece of information aligns with each of the three models of membrane structure. Which model(s) are supported, which are not supported, and what reservations do you have about the data or their interpretation? Explain your reasoning.

(a) When cells were fixed, stained with osmium tetroxide, and examined in the electron microscope, they gave images like that in Figure 11–1: the membranes showed a "railroad track" appearance, with two dark-staining lines separated by a light space.

(b) The thickness of membranes in cells fixed and stained in the same way was found to be 5 to 9 nm. The thickness of a "naked" phospholipid bilayer, without proteins, was 4 to 4.5 nm. The thickness of a single monolayer of proteins was about 1 nm.

(c) In Singer's words: "The average amino acid composition of membrane proteins is not distinguishable from that of soluble proteins. In particular, a substantial fraction of the residues is hydrophobic" (p. 165).

(d) As described in Problems 1 and 2 of this chapter, researchers had extracted membranes from cells, extracted the lipids, and compared the area of the lipid monolayer with the area of the original cell membrane. The interpretation of the results was complicated by the issue illustrated in the graph of Problem 1: the area of the monolayer depended on how hard it was pushed. With very light pressures, the ratio of monolayer area to cell membrane area was about 2.0. At higher pressures—thought to be more like those found in cells—the ratio was substantially lower.

(e) Circular dichroism spectroscopy uses changes in polarization of UV light to make inferences about protein secondary structure (see Fig. 4–9). On average, this technique showed that membrane proteins have a large amount of α helix and little or no β sheet. This finding was consistent with most membrane proteins having a globular structure.

(f) Phospholipase C is an enzyme that removes the polar head group (including the phosphate) from phospholipids. In several studies, treatment of intact membranes with phospholipase C removed about 70% of the head groups without disrupting the "railroad track" structure of the membrane.

(g) Singer described a study in which "a glycoprotein of molecular weight about 31,000 in human red blood cell membranes is cleaved by tryptic treatment of the membranes into soluble glycopeptides of about 10,000 molecular weight, while the remaining portions are quite hydrophobic" (p. 199). Trypsin treatment did not cause gross changes in the membranes, which remained intact.

Singer's review also included many more studies in this area. In the end, though, the data available in 1971 did not conclusively prove Model C was correct. As more data have accumulated, this model of membrane structure has been accepted by the scientific community.

Answer

(a) *Model A*: supported. The two dark lines are either the protein layers or the phospholipid heads, and the clear space is either the bilayer or the hydrophobic core, respectively. *Model B*: not supported. This model requires a more-or-less uniformly stained band surrounding the cell. *Model C*: supported, with one reservation. The two dark lines are the phospholipid heads; the clear zone is the tails. This assumes that the membrane proteins are not visible, because they do not stain with osmium or do not happen to be in the sections viewed.

(b) *Model A*: supported. A "naked" bilayer (4.5 nm) + two layers of protein (2 nm) sums to 6.5 nm, which is within the observed range of thickness. *Model B*: neither. This model makes no predictions about membrane thickness. *Model C*: unclear. The result is hard to reconcile with this model, which predicts a membrane as thick as, or slightly thicker than (due to the projecting ends of embedded proteins), a "naked" bilayer. The model is supported only if the smallest values for membrane thickness are correct or if a substantial amount of protein projects from the bilayer.

(c) *Model A*: unclear. The result is hard to reconcile with this model. If the proteins are bound to the membrane by ionic interactions, the model predicts that the proteins contain a high proportion of charged amino acids, in contrast to what was observed. Also, because the protein layer must be very thin (see **(b)**), there would not be much room for a hydrophobic protein core, so hydrophobic residues would be exposed to the solvent. *Model B*: supported. The proteins have a mixture of hydrophobic residues (interacting with lipids) and charged residues (interacting with water). *Model C*: supported. The proteins have a mixture of hydrophobic residues (anchoring in the membrane) and charged residues (interacting with water).

(d) *Model A*: unclear. The result is hard to reconcile with this model, which predicts a ratio of exactly 2.0; this would be hard to achieve under physiologically relevant pressures. *Model B*: neither. This model makes no predictions about amount of lipid in the membrane. *Model C*: supported. Some membrane surface area is taken up with proteins, so the ratio would be less than 2.0, as was observed under more physiologically relevant conditions.

(e) *Model A*: unclear. The model predicts proteins in extended conformations rather than globular, so supported only if one assumes that proteins layered on the surfaces include helical segments. *Model B*: supported. The model predicts mostly globular proteins (containing some helical segments). *Model C*: supported. The model predicts mostly globular proteins.

(f) *Model A*: unclear. The phosphorylamine head groups are protected by the protein layer, but only if the proteins completely cover the surface will the phospholipids be completely protected from phospholipase. *Model B*: supported. Most head groups are accessible to phospholipase. *Model C*: supported. All head groups are accessible to phospholipase.

(g) *Model A*: not supported. Proteins are entirely accessible to trypsin digestion and virtually all will undergo multiple cleavage, with no protected hydrophobic segments. *Model B*: not supported. Virtually all proteins are in the bilayer and inaccessible to trypsin. *Model C*: supported. Segments of protein that penetrate or span the bilayer are protected from trypsin; those exposed at the surfaces will be cleaved. The trypsin-resistant portions have a high proportion of hydrophobic residues.

Reference

Singer, S.J. (1971) The molecular organization of biological membranes. In *Structure and Function of Biological Membranes* (Rothfield, L.I., ed.), pp. 145–222, Academic Press, Inc., New York.

chapter

12

Biosignaling

1. **Hormone Experiments in Cell-Free Systems** In the 1950s, Earl W. Sutherland, Jr., and his colleagues carried out pioneering experiments to elucidate the mechanism of action of epinephrine and glucagon. Given what you have learned in this chapter about hormone action, interpret each of the experiments described below. Identify substance X and indicate the significance of the results.
   (a) Addition of epinephrine to a homogenate of normal liver resulted in an increase in the activity of glycogen phosphorylase. However, if the homogenate was first centrifuged at a high speed and epinephrine or glucagon was added to the clear supernatant fraction that contains phosphorylase, no increase in the phosphorylase activity occurred.
   (b) When the particulate fraction from the centrifugation in **(a)** was treated with epinephrine, substance X was produced. The substance was isolated and purified. Unlike epinephrine, substance X activated glycogen phosphorylase when added to the clear supernatant fraction of the centrifuged homogenate.
   (c) Substance X was heat-stable; that is, heat treatment did not affect its capacity to activate phosphorylase. (Hint: Would this be the case if substance X were a protein?) Substance X was nearly identical to a compound obtained when pure ATP was treated with barium hydroxide. (Fig. 8–6 will be helpful.)

   Answer Substance X is cyclic AMP. Epinephrine stimulates glycogen phosphorylase by activating the enzyme adenylyl cyclase, which catalyzes formation of cAMP, the second messenger.
   (a) Adenylyl cyclase is a membrane-bound protein; centrifugation sediments it into the particulate fraction.
   (b) Cyclic AMP directly stimulates glycogen phosphorylase.
   (c) Cyclic AMP is heat-stable; it can be prepared by treating ATP with barium hydroxide.

2. **Effect of Dibutyryl cAMP versus cAMP on Intact Cells** The physiological effects of epinephrine should in principle be mimicked by addition of cAMP to the target cells. In practice, addition of cAMP to intact target cells elicits only a minimal physiological response. Why? When the structurally related derivative dibutyryl cAMP (shown below) is added to intact cells, the expected physiological response is readily apparent. Explain the basis for the difference in cellular response to these two substances. Dibutyryl cAMP is widely used in studies of cAMP function.

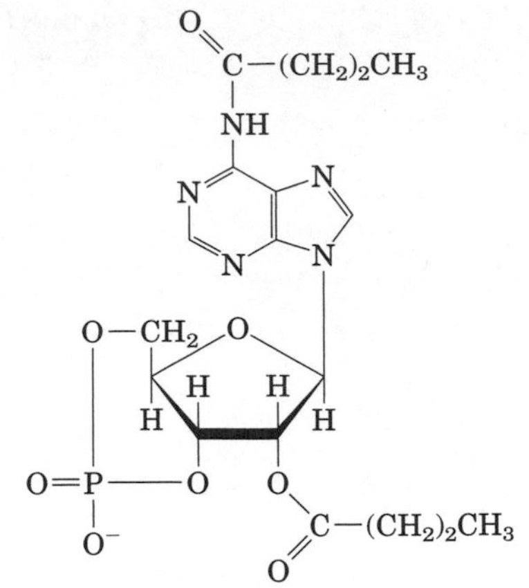

Dibutyryl cAMP
(N^6,$O^{2'}$-Dibutyryl adenosine 3′,5′-cyclic monophosphate)

Answer Unlike cAMP, dibutyryl cAMP passes readily through the plasma membrane.

3. Effect of Cholera Toxin on Adenylyl Cyclase The gram-negative bacterium *Vibrio cholerae* produces a protein, cholera toxin (M_r 90,000), that is responsible for the characteristic symptoms of cholera: extensive loss of body water and Na^+ through continuous, debilitating diarrhea. If body fluids and Na^+ are not replaced, severe dehydration results; untreated, the disease is often fatal. When the cholera toxin gains access to the human intestinal tract it binds tightly to specific sites in the plasma membrane of the epithelial cells lining the small intestine, causing adenylyl cyclase to undergo prolonged activation (hours or days).

(a) What is the effect of cholera toxin on [cAMP] in the intestinal cells?

(b) Based on the information above, suggest how cAMP normally functions in intestinal epithelial cells.

(c) Suggest a possible treatment for cholera.

Answer (a) It increases [cAMP]. **(b)** The observations suggest that cAMP regulates Na^+ permeability. **(c)** Replace lost body fluids and electrolytes.

4. Mutations in PKA Explain how mutations in the R or C subunit of cAMP-dependent protein kinase (PKA) might lead to **(a)** a constantly active PKA or **(b)** a constantly inactive PKA.

Answer

(a) If a mutation in the R subunit makes it unable to bind to the C subunit, the C subunit is never inhibited; it is constantly active.

(b) If a mutation prevents the binding of cAMP to the R subunit but still allows normal R-C interaction, the inhibition of C by R cannot be relieved by elevated [cAMP], so the enzyme is constantly inactive.

5. Therapeutic Effects of Albuterol The respiratory symptoms of asthma result from constriction of the bronchi and bronchioles of the lungs caused by contraction of the smooth muscle of their walls. This constriction can be reversed by raising the [cAMP] in the smooth muscle. Explain the therapeutic effects of albuterol, a β-adrenergic agonist taken (by inhalation) for asthma. Would you expect this drug to have any side effects? How might one design a better drug that did not have these effects?

Answer By mimicking the actions of epinephrine on smooth muscle, albuterol raises [cAMP], leading to relaxation and enlargement (dilation) of the bronchi and bronchioles. Because β-adrenergic receptors also control many other processes, drugs that act as β-adrenergic agonists generally have other, undesirable effects. To minimize such side effects, the goal is to find an agonist that is specific for the subtype of β-adrenergic receptors found in bronchial smooth muscle.

6. **Termination of Hormonal Signals** Signals carried by hormones must eventually be terminated. Describe several different mechanisms for signal termination.

Answer A hormone can be degraded by extracellular enzymes (such as acetylcholinesterase). The GTP bound to a G protein can be hydrolyzed to GDP. A second messenger can be degraded (cAMP, cGMP), further metabolized (IP_3), or resequestered (Ca^{2+}, in the endoplasmic reticulum). A receptor can be desensitized (acetylcholine receptor/channel), phosphorylated/inactivated, bound to an arrestin, or removed from the surface (β-adrenergic receptor, rhodopsin).

7. **Using FRET to Explore Protein-Protein Interactions in Vivo** Figure 12–8 shows the interaction between β-arrestin and the β-adrenergic receptor. How would you use FRET (see Box 12–3) to demonstrate this interaction in living cells? Which proteins would you fuse? Which wavelengths would you use to illuminate the cells, and which would you monitor? What would you expect to observe if the interaction occurred? If it did not occur? How might you explain the failure of this approach to demonstrate this interaction?

Answer Fuse CFP to β-arrestin and YFP to the cytoplasmic domain of the β-adrenergic receptor, or vice versa. In either case, illuminate at 433 nm and observe at both 476 and 527 nm. If the interaction occurs, emitted light intensity will decrease at 476 nm and increase at 527 nm on addition of epinephrine to cells expressing the fusion proteins. If the interaction does not occur, the wavelength of the emitted light will remain at 476 nm. There are several reasons why this might fail; for example, the fusion proteins (1) are inactive or otherwise unable to interact, (2) are not translocated to their normal subcellular location, or (3) are not stable to proteolytic breakdown.

8. **EGTA Injection** EGTA (ethylene glycol-bis(β-aminoethyl ether)-*N,N,N′,N′*-tetraacetic acid) is a chelating agent with high affinity and specificity for Ca^{2+}. By microinjecting a cell with an appropriate Ca^{2+}-EGTA solution, an experimenter can prevent cytosolic [Ca^{2+}] from rising above 10^{-7} M. How would EGTA microinjection affect a cell's response to vasopressin (see Table 12–4)? To glucagon?

Answer Vasopressin acts through a PLC-coupled GPCR. The IP_3 released by PLC normally elevates cytosolic [Ca^{2+}] to 10^{-6} M, activating (with diacylglycerol) protein kinase C. Preventing this elevation of [Ca^{2+}] by using EGTA to "buffer" the internal [Ca^{2+}] would block vasopressin action, but should not directly affect the response to glucagon, which uses cAMP, *not* Ca^{2+}, as its intracellular second messenger.

9. **Amplification of Hormonal Signals** Describe all the sources of amplification in the insulin receptor system.

Answer The amplification results from catalysts activating catalysts—including protein kinases that act in enzyme cascades. Two molecules of insulin activate an *insulin receptor* dimer for a finite period, during which the receptor phosphorylates many molecules of *IRS-1*. Through a series of interactions with other proteins (including Grb2, Sos, Ras), IRS-1 activates *Raf,* which phosphorylates and activates many molecules of *MEK,* each of which phosphorylates and activates many molecules of *ERK.* Each activated ERK phosphorylates and activates several molecules of a *transcription factor,* and each of these stimulates the transcription of multiple copies of *mRNA* for specific genes. Each mRNA can direct the synthesis of many copies of the protein it encodes. (See Fig. 12–15.)

10. Mutations in *ras* How does a mutation in *ras* that leads to formation of a Ras protein with no GTPase activity affect a cell's response to insulin?

Answer When active, Ras activates the protein kinase Raf, which initiates the MAPK cascade that leads to the phosphorylation of nuclear proteins. A Ras protein without GTPase activity would, once activated by the binding of GTP, remain active, continuing to produce an insulin response.

11. Differences among G Proteins Compare the G proteins G_s, which acts in transducing the signal from β-adrenergic receptors, and Ras. What properties do they share? How do they differ? What is the functional difference between G_s and G_i?

Answer *Shared properties of Ras and G_s*: both can bind either GDP or GTP; both are activated by GTP; both can, when active, activate a downstream enzyme; both have intrinsic GTPase activity that shuts them off after a short period of activation. *Differences between Ras and G_s*: Ras is a small, monomeric protein; G_s is heterotrimeric. *Functional differences between G_s and G_i*: G_s activates adenylyl cyclase, G_i inhibits it.

12. Mechanisms for Regulating Protein Kinases Identify eight general types of protein kinases found in eukaryotic cells, and explain what factor is *directly* responsible for activating each type.

Answer Kinase (factor(s)): PKA (cAMP); PKG (cGMP); PKC (Ca^{2+}, DAG); Ca^{2+}/CaM kinase (Ca^{2+}, CaM); cyclin-dependent kinase (cyclin); protein Tyr kinase (ligand for the receptor, such as insulin); MAPK (Raf); Raf (Ras); glycogen phosphorylase kinase (PKA).

13. Nonhydrolyzable GTP Analogs Many enzymes can hydrolyze GTP between the β and γ phosphates. The GTP analog β,γ-imidoguanosine 5′-triphosphate Gpp(NH)p, shown below, cannot be hydrolyzed between the β and γ phosphates. Predict the effect of microinjection of Gpp(NH)p into a myocyte on the cell's response to β-adrenergic stimulation.

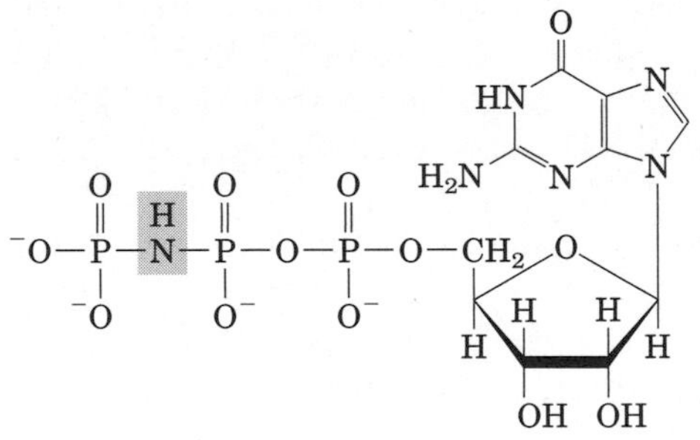

Gpp(NH)p
(β,γ-imidoguanosine 5′-triphosphate)

Answer Nonhydrolyzable analogs of GTP have the effect of keeping the stimulatory G protein (G_s) in its activated form once it has encountered the receptor-hormone complex; it cannot shut itself off by converting the bound GTP (analog) to GDP. Injection of the analog would therefore be expected to prolong the effect of epinephrine on the injected cell.

14. Use of Toxin Binding to Purify a Channel Protein α-Bungarotoxin is a powerful neurotoxin found in the venom of a poisonous snake (*Bungarus multicinctus*). It binds with high specificity to the nicotinic acetylcholine receptor (AChR) protein and prevents the ion channel from opening. This interaction was used to purify AChR from the electric organ of torpedo fish.

(a) Outline a strategy for using α-bungarotoxin covalently bound to chromatography beads to purify the AChR protein. (Hint: See Fig. 3–17c.)

(b) Outline a strategy for the use of $[^{125}I]\alpha$-bungarotoxin to purify the AChR protein.

Answer

(a) Use the (α-bungarotoxin–bound beads for affinity purification (see Fig. 3–17c, p. 87) of AChR. Extract proteins from the electric organs and pass the mixture through the chromatography column; the AChR binds selectively to the beads. Elute the AChR with a solution of NaCl or a solvent of lower pH, which weakens its interaction with α-bungarotoxin.

(b) Use binding of [^{125}I]α-bungarotoxin as a *quantitative assay* for AChR during purification by various techniques. At each step, assay AChR by measuring [^{125}I]α-bungarotoxin binding to the proteins in the sample. Optimize purification for the highest specific activity of AChR (counts/min of bound [^{125}I]α-bungarotoxin per mg of protein) in the final material.

15. Resting Membrane Potential A variety of unusual invertebrates, including giant clams, mussels, and polychaete worms, live on the fringes of deep-sea hydrothermal vents, where the temperature is 60 °C.

(a) The adductor muscle of a giant clam has a resting membrane potential of −95 mV. Given the intracellular and extracellular ionic compositions shown below, would you have predicted this membrane potential? Why or why not?

	Concentration (mM)	
Ion	Intracellular	Extracellular
Na^+	50	440
K^+	400	20
Cl^-	21	560
Ca^{2+}	0.4	10

(b) Assume that the adductor muscle membrane is permeable to only one of the ions listed above. Which ion could determine the V_m?

Answer

(a) In most myocytes at rest, the plasma membrane is permeable primarily to K^+ ions. V_m is a function of the distribution of K^+ ions across the membrane. If V_m in the clam adductor muscle is determined primarily by K^+, then V_m at rest would be predicted by the Nernst equation

$$E_{ion} = (RT/Z\mathcal{J})\ \ln\ (C_{out}/C_{in})$$

using the values for [K^+] given in the table.

$$E_{K^+} = [(8.315\ \text{J/mol}\cdot\text{K})(333\ \text{K})/(1)(96{,}480\ \text{J/V}\cdot\text{mol})]\ \ln\ (20/400)$$

$$= 0.0287\ \text{V} \times (-3.0) = -0.09\ \text{V, or } -90\ \text{mV}$$

Because the experimentally observed V_m is −95 mV, the plasma membrane in the adductor muscle must be permeable to some other ion or combination of ions.

(b) Use the Nernst equation to calculate E for each ion. The ion with an E value closest to the membrane potential is the permeant ion that influences V_m.

$$E_{K^+} = -90\ \text{mV (see above)}$$

$$E_{Na^+} = 0.0287\ \text{V} \times \ln\ (440/50) = 0.06\ \text{V, or } 60\ \text{mV}$$

$$E_{Cl^-} = (0.0287\ \text{V}/-1) \times \ln\ (560/21) = -0.094\ \text{V, or } -94\ \text{mV (note that } Z \text{ for } Cl^- \text{ is } -1)$$

$$E_{Ca^{2+}} = (0.0287\ \text{V}/2) \times \ln\ (10/0.4) = 0.05\ \text{V, or } 50\ \text{mV } (Z \text{ for } Ca^{2+} \text{ is } 2)$$

Thus, because $E_{Cl^-} = -94$ mV is very close to the resting V_m of -95 mV, it is likely that the membrane is permeable only to Cl^- ions at rest. You could verify this experimentally by changing the extracellular $[Cl^-]$, then measuring the effect on resting membrane potential. If this potential does depend only on Cl^- ions, the Nernst equation should predict how the membrane potential will change.

16. Membrane Potentials in Frog Eggs Fertilization of a frog oocyte by a sperm cell triggers ionic changes similar to those observed in neurons (during movement of the action potential) and initiates the events that result in cell division and development of the embryo. Oocytes can be stimulated to divide without fertilization by suspending them in 80 mM KCl (normal pond water contains 9 mM KCl).

(a) Calculate how much the change in extracellular [KCl] changes the resting membrane potential of the oocyte. (Hint: Assume the oocyte contains 120 mM K^+ and is permeable *only* to K^+.) Assume a temperature of 20 °C.

(b) When the experiment is repeated in Ca^{2+}-free water, elevated [KCl] has no effect. What does this suggest about the mechanism of the KCl effect?

Answer

(a) $$V_m = \frac{RT}{Z\mathcal{J}} \ln\left(\frac{[K^+]_{out}}{[K^+]_{in}}\right)$$

$$= [(8.315 \text{ J/mol} \cdot \text{K})(293 \text{ K})/(1)(96{,}480 \text{ J/V} \cdot \text{mol})] \ln([K^+]_{out}/[K^+]_{in})$$

$$= (0.025 \text{ V}) \ln([K^+]_{out}/[K^+]_{in})$$

V_m in pond water $= 0.025$ V ln $(9/120) = -0.06$ V, or -60 mV

V_m in 80 mM KCl $= 0.025$ V ln $(80/120) = -0.01$ V, or -10 mV

The membrane of the oocyte has been *depolarized*—the resting membrane potential has become less negative—by exposure to elevated extracellular $[K^+]$.

(b) This observation suggests that the effect of increased [KCl] depends on an influx of Ca^{2+} from the extracellular medium, which is required to stimulate cell division. High [KCl] treatment must depolarize the oocyte sufficiently to open voltage-dependent Ca^{2+} channels in the plasma membrane.

17. Excitation Triggered by Hyperpolarization In most neurons, membrane *depolarization* leads to the opening of voltage-dependent ion channels, generation of an action potential, and ultimately an influx of Ca^{2+}, which causes release of neurotransmitter at the axon terminus. Devise a cellular strategy by which *hyperpolarization* in rod cells could produce excitation of the visual pathway and passage of visual signals to the brain. (Hint: The neuronal signaling pathway in higher organisms consists of a *series* of neurons that relay information to the brain (see Fig. 12–35). The signal released by one neuron can be either excitatory or inhibitory to the following, postsynaptic neuron.)

Answer Hyperpolarization of rod cells in the retina occurs when the membrane potential, V_m, becomes more negative. This results in the closing of voltage-dependent Ca^{2+} channels in the presynaptic region of the rod cell. The resulting decrease in intracellular $[Ca^{2+}]$ causes a corresponding decrease in the release of neurotransmitter by exocytosis. The neurotransmitter released by rod cells is actually an inhibitory neurotransmitter, which leads to suppression of activity in the next neuron of the visual circuit. When this inhibition is removed in response to a light stimulus, the circuit becomes active and visual centers in the brain are excited.

18. Genetic "Channelopathies" There are many genetic diseases that result from defects in ion channels. For each of the following, explain how the molecular defect might lead to the symptoms described.

(a) A loss of function mutation in the gene encoding the α subunit of the cGMP-gated cation channel of retinal cone cells leads to a complete inability to distinguish colors.

(b) Loss-of-function alleles of the gene encoding the α subunit of the ATP-gated K^+ channel shown in Figure 23–29 lead to a condition known as congenital hyperinsulinism—persistently high levels of insulin in the blood.

(c) Mutations affecting the β subunit of the ATP-gated K^+ channel that prevent ATP binding lead to neonatal diabetes—persistently low levels of insulin in the blood in newborn babies.

Answer

(a) Loss of function of the cGMP-gated channel prevents influx of Na^+ and Ca^{2+} into cone cells in response to light; consequently, the cells fail to signal the brain that light had been received. Because rod cells are unaffected, the individual can see but does not have color vision.

(b) A loss-of-function mutation in the ATP-gated cation channel prevents efflux of K^+ through these channels, leading to continuous depolarization of the β-cell membrane and constitutive release of insulin into the blood.

(c) ATP is responsible for closing this channel, so in an individual with the mutant protein, the channels will remain open, preventing depolarization of the β-cell membrane and thereby preventing release of insulin, resulting in diabetes.

19. Visual Desensitization Oguchi's disease is an inherited form of night blindness. Affected individuals are slow to recover vision after a flash of bright light against a dark background, such as the headlights of a car on the freeway. Suggest what the molecular defect(s) might be in Oguchi's disease. Explain in molecular terms how this defect would account for night blindness.

Answer Some individuals with Oguchi's disease have a defective rhodopsin kinase that slows the recycling of rhodopsin after its conversion to the all-trans form on illumination. This defect leaves retinal rod and cone cells insensitive for some time after a bright flash. Other individuals have genetic defects in arrestin that prevent it from interacting with phosphorylated rhodopsin to trigger the process that leads to replacement of all-*trans*-retinal with 11-*cis*-retinal.

20. Effect of a Permeant cGMP Analog on Rod Cells An analog of cGMP, 8-Br-cGMP, will permeate cellular membranes, is only slowly degraded by a rod cell's PDE activity, and is as effective as cGMP in opening the gated channel in the cell's outer segment. If you suspended rod cells in a buffer containing a relatively high [8-Br-cGMP], then illuminated the cells while measuring their membrane potential, what would you observe?

Answer Rod cells would no longer show any change in membrane potential in response to light. This experiment has been done. Illumination did activate PDE, but the enzyme could not significantly reduce the 8-Br-cGMP level, which remained well above that needed to keep the gated ion channels open. Thus, light had no impact on membrane potential.

21. Hot and Cool Taste Sensations The sensations of heat and cold are transduced by a group of temperature-gated cation channels. For example, TRPV1, TRPV3, and TRPM8 are usually closed, but open under the following conditions: TRPV1 at $\geq$43 °C; TRPV3 at $\geq$33 °C; and TRPM8 at $<$25 °C. These channels are expressed in sensory neurons known to be responsible for temperature sensation.

(a) Propose a reasonable model to explain how exposing a sensory neuron containing TRPV1 to high temperature leads to a sensation of heat.

(b) Capsaicin, one of the active ingredients in "hot" peppers, is an agonist of TRPV1. Capsaicin shows 50% activation of the TRPV1 response at a concentration (i.e., it has an EC_{50}) of 32 nM. Explain why even a very few drops of hot pepper sauce can taste very "hot" without actually burning you.

(c) Menthol, one of the active ingredients in mint, is an agonist of TRPM8 (EC_{50} = 30 μM) and TRPV3 (EC_{50} = 20 μM). What sensation would you expect from contact with low levels of menthol? With high levels?

Answer

(a) On exposure to heat, TRPV1 channels open, causing an influx of Na^+ and Ca^{2+} into the sensory neuron. This depolarizes the neuron, triggering an action potential. When the action potential reaches the axon terminus, neurotransmitter is released, signaling the nervous system that heat has been sensed.

(b) Capsaicin mimics the effects of heat by binding to and opening the TRPV1 channel at low temperature, leading to the false sensation of heat. The extremely low EC_{50} indicates that even very small amounts of capsaicin will have dramatic sensory effects.

(c) At low levels, menthol should open the TRPM8 channel, leading to a sensation of cool; at high levels, both TRPM8 and TRPV3 will open, leading to a mixed sensation of cool and heat, such as you may have experienced with very strong peppermints.

22. Oncogenes, Tumor-Suppressor Genes, and Tumors For each of the following situations, provide a plausible explanation for how it could lead to unrestricted cell division.

(a) Colon cancer cells often contain mutations in the gene encoding the prostaglandin E_2 receptor. PGE_2 is a growth factor required for the division of cells in the gastrointestinal tract.

(b) Kaposi sarcoma, a common tumor in people with untreated AIDS, is caused by a virus carrying a gene for a protein similar to the chemokine receptors CXCR1 and CXCR2. Chemokines are cell-specific growth factors.

(c) Adenovirus, a tumor virus, carries a gene for the protein E1A, which binds to the retinoblastoma protein, pRb. (Hint: See Fig. 12–48.)

(d) An important feature of many oncogenes and tumor suppressor genes is their cell-type specificity. For example, mutations in the PGE_2 receptor are not typically found in lung tumors. Explain this observation. (Note that PGE_2 acts through a GPCR in the plasma membrane.)

Answer

(a) These mutations might lead to permanent activation of the PGE_2 receptor. The mutant cells would behave as though stimulatory levels of PGE_2 were always present, leading to unregulated cell division and tumor formation.

(b) The viral gene might encode a constitutively active form of the receptor, such that the cells send a constant signal for cell division. This unrestrained division would lead to tumor formation.

(c) E1A protein might bind to pRb and prevent E2F from binding, so E2F is constantly active as a transcription factor. It constantly activates genes that trigger cell division, so cells divide uncontrollably.

(d) Lung cells do not normally respond to PGE_2 because they do not express the PGE_2 receptor; mutations resulting in a constitutively active PGE_2 receptor do not affect lung cells.

23. Mutations in Tumor Suppressor Genes and Oncogenes Explain why mutations in tumor suppressor genes are recessive (both copies of the gene must be defective for the regulation of cell division to be defective), whereas mutations in oncogenes are dominant.

Answer A *tumor suppressor gene* in its normal cellular form encodes a protein that restrains cell division. Mutant forms of the protein fail to suppress cell division, but if either of the two alleles of the gene present in the individual encodes a normal protein, normal function will continue. Only if both alleles are defective will the suppression of cell division fail, leading to unregulated division. An *oncogene* in its normal form encodes a regulatory protein that signals the cell to divide, but only when other, external or internal factors (such as growth factors) signal cell division. If a defective oncogene product is formed by either of the two alleles, unregulated cell growth and division will occur: the mutant protein sends the signal for cell division, whether or not growth factors are present.

**24. Retinoblastoma in Children** Explain why some children with retinoblastoma develop multiple tumors of the retina in both eyes, whereas others have a single tumor in only one eye.

Answer Children who develop multiple tumors in both eyes were born with a defective copy of the *Rb* gene, occurring in every cell of the retina. Early in their lives, as retinal cells divided, one or several cells independently underwent a second mutation that damaged the remaining good copy of the *Rb* gene. Each cell with two defective *Rb* alleles develops into a tumor. In the later onset, single-tumor form of the disease, children were born with two good copies of the *Rb* gene. A tumor develops when mutation in a single retinal cell damages one allele of the *Rb* gene, then a second mutation damages the second allele in the same cell. Two mutations in the same gene in the same cell are extremely rare, and when this does happen, it occurs in only one cell and develops into a single tumor.

25. Specificity of a Signal for a Single Cell Type Discuss the validity of the following proposition. A signaling molecule (hormone, growth factor, or neurotransmitter) elicits identical responses in different types of target cells if they contain identical receptors.

Answer The proposition is invalid. Two cells expressing the same surface receptor for a given hormone may have different complements of target proteins for phosphorylation by protein kinases, resulting in different physiological and biochemical responses in different cells.

Data Analysis Problem

26. Exploring Taste Sensation in Mice Figure 12–41 shows the signal-transduction pathway for sweet taste in mammals. Pleasing tastes are an evolutionary adaptation to encourage animals to consume nutritious foods. Zhao and coauthors (2003) examined the two major pleasurable taste sensations: sweet and umami. Umami is a "distinct savory taste" triggered by amino acids, especially aspartate and glutamate, and probably encourages animals to consume protein-rich foods. Monosodium glutamate (MSG) is a flavor enhancer that exploits this sensitivity.

At the time the article was published, specific taste receptor proteins (labeled SR in Fig. 12–41) for sweet and umami had been tentatively characterized. Three such proteins were known—T1R1, T1R2, and T1R3—which function as heterodimeric receptor complexes: T1R1-T1R3 was tentatively identified as the umami receptor, and T1R2-T1R3 as the sweet receptor. It was not clear how taste sensation was encoded and sent to the brain, and two possible models had been suggested. In the cell-based model, individual taste-sensing cells express only one kind of receptor; that is, there are "sweet cells," "bitter cells," "umami cells," and so on, and each type of cell sends its information to the brain via a different nerve. The brain "knows" which taste is detected by the identity of the nerve fiber that transmits the message. In the receptor-based model, individual taste-sensing cells have several kinds of receptors and send different messages along the same nerve fiber to the brain, the message depending on which receptor is activated. Also unclear at the time was whether there was any interaction between the different taste sensations, or whether parts of one taste-sensing system were required for other taste sensations.

(a) Previous work had shown that different taste receptor proteins are expressed in nonoverlapping sets of taste receptor cells. Which model does this support? Explain your reasoning.

Zhao and colleagues constructed a set of "knockout mice"—mice homozygous for loss-of-function alleles for one of the three receptor proteins, T1R1, T1R2, or T1R3—and double-knockout mice with nonfunctioning T1R2 and T1R3. The researchers measured the taste perception of these mice by measuring their "lick rate" of solutions containing different taste molecules. Mice will lick the spout of a feeding bottle with a pleasant-tasting solution more often than one with an unpleasant-tasting solution. The researchers measured relative lick rates: how often the mice licked a sample solution compared with water. A relative lick rate of 1 indicated no preference; <1, an aversion; and >1, a preference.

(b) All four types of knockout strains had the same responses to salt and bitter tastes as did wild-type mice. Which of the above issues did this experiment address? What do you conclude from these results?

The researchers then studied umami taste reception by measuring the relative lick rates of the different mouse strains with different quantities of MSG in the feeding solution. Note that the solutions also contained inosine monophosphate (IMP), a strong potentiator of umami taste reception (and a common ingredient in ramen soups, along with MSG), and ameloride, which suppresses the pleasant salty taste imparted by the sodium of MSG. The results are shown in the graph.

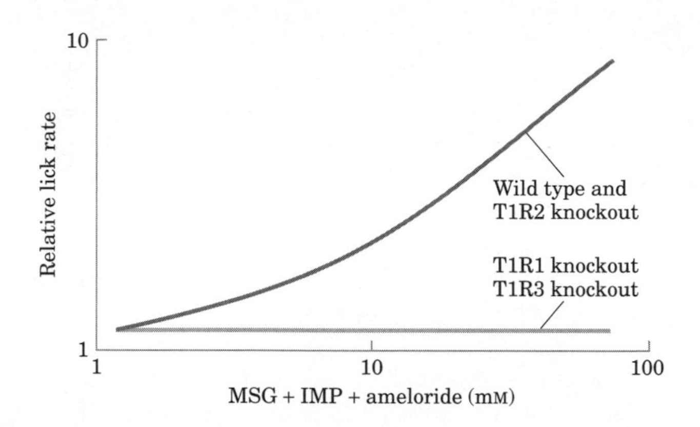

(c) Are these data consistent with the umami taste receptor consisting of a heterodimer of T1R1 and T1R3? Why or why not?

(d) Which model(s) of taste encoding does this result support? Explain your reasoning.

Zhao and coworkers then performed a series of similar experiments using sucrose as a sweet taste. These results are shown below.

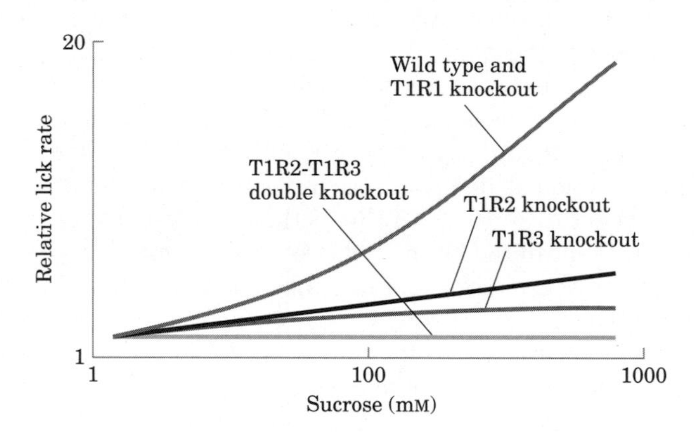

(e) Are these data consistent with the sweet taste receptor consisting of a heterodimer of T1R2 and T1R3? Why or why not?

(f) There were some unexpected responses at very high sucrose concentrations. How do these complicate the idea of a heterodimeric system as presented above?

In addition to sugars, humans also taste other compounds (e.g., the peptides monellin and aspartame) as sweet; mice do not taste these as sweet. Zhao and coworkers inserted into T1R2 knockout mice a copy of the human T1R2 gene under the control of the mouse T1R2 promoter. These modified mice now tasted monellin and saccharin as sweet. The researchers then went further, adding to T1R1 knockout mice the RASSL protein—a G protein–linked receptor for the synthetic opiate spiradoline; the RASSL gene was under the control of a promoter that could be induced by feeding the mice tetracycline. These mice did not prefer spiradoline in the absence of tetracycline; in the presence of tetracycline, they showed a strong preference for nanomolar concentrations of spiradoline.

(g) How do these results strengthen Zhao and coauthors' conclusions about the mechanism of taste sensation?

Answer

(a) The cell-based model, which predicts different receptors present on different cells.

(b) This experiment addresses the issue of the independence of different taste sensations. Even though the receptors for sweet and/or umami are missing, the animals' other taste sensations are normal; thus, pleasant and unpleasant taste sensations are independent.

(c) Yes. Loss of either T1R1 or T1R3 subunits abolishes umami taste sensation.

(d) Both models. With either model, removing one receptor would abolish that taste sensation.

(e) Yes. Loss of either the T1R2 or T1R3 subunits almost completely abolishes the sweet taste sensation; complete elimination of sweet taste requires deletion of both subunits.

(f) At very high sucrose concentrations, T1R2 and, to a lesser extent, T1R3 receptors, as homodimers, can detect sweet taste.

(g) The results are consistent with either model of taste encoding, but do strengthen the researchers' conclusions. Ligand binding can be completely separated from taste sensation. If the ligand for the receptor in "sweet-tasting cells" binds a molecule, mice prefer that molecule as a sweet compound.

Reference

Zhao, G.Q., Zhang, Y., Hoon, M.A., Chandrashekar, J., Erlenbach, I., Ryba, N.J.P., & Zuker, C. (2003) The receptors for mammalian sweet and umami taste. *Cell* **115,** 255–266.

chapter

13 Bioenergetics and Biochemical Reaction Types

1. **Entropy Changes during Egg Development** Consider a system consisting of an egg in an incubator. The white and yolk of the egg contain proteins, carbohydrates, and lipids. If fertilized, the egg is transformed from a single cell to a complex organism. Discuss this irreversible process in terms of the entropy changes in the system, surroundings, and universe. Be sure that you first clearly define the system and surroundings.

   Answer Consider the developing chick as the system. The nutrients, egg shell, and outside world are the surroundings. Transformation of the single cell into a chick drastically reduces the entropy of the system (increases the order). Initially, the parts of the egg outside the embryo (within the surroundings) contain complex fuel molecules (a low-entropy condition). During incubation, some of these complex molecules are converted to large numbers of CO_2 and H_2O molecules (high entropy). This increase in entropy of the surroundings is larger than the decrease in entropy of the chick (the system). Thus, the entropy of the universe (the system + surroundings) increases.

2. **Calculation of $\Delta G'^{\circ}$ from an Equilibrium Constant** Calculate the standard free-energy change for each of the following metabolically important enzyme-catalyzed reactions, using the equilibrium constants given for the reactions at 25 °C and pH 7.0.

   (a) $\text{Glutamate} + \text{oxaloacetate} \xrightleftharpoons[]{\text{aspartate aminotransferase}} \text{aspartate} + \alpha\text{-ketoglutarate} \qquad K'_{eq} = 6.8$

   (b) $\text{Dihydroxyacetone phosphate} \xrightleftharpoons[]{\text{triose phosphate isomerase}} \text{glyceraldehyde 3-phosphate} \qquad K'_{eq} = 0.0475$

   (c) $\text{Fructose 6-phosphate} + \text{ATP} \xrightleftharpoons[]{\text{phosphofructokinase}} \text{fructose 1,6-bisphosphate} + \text{ADP} \qquad K'_{eq} = 254$

   Answer

   $$\Delta G = \Delta G'^{\circ} + RT \ln [\text{products}]/[\text{reactants}]$$

   and [products]/[reactants] is the mass-action ratio, Q. At equilibrium, $\Delta G = 0$ and $Q = K'_{eq}$, so

   $$\Delta G'^{\circ} = -RT \ln K'_{eq}$$

   where $R = 8.315$ J/mol · K and $T = 25$ °C $= 298$ K. Using the value $RT = 2.48$ kJ/mol, we can calculate the $\Delta G'^{\circ}$ values from the K'_{eq} for each reaction.

   (a) $\Delta G'^{\circ} = -(2.48 \text{ kJ/mol}) \ln 6.8 = -4.8 \text{ kJ/mol}$
   (b) $\Delta G'^{\circ} = -(2.48 \text{ kJ/mol}) \ln 0.0475 = 7.56 \text{ kJ/mol}$
   (c) $\Delta G'^{\circ} = -(2.48 \text{ kJ/mol}) \ln 254 = -13.7 \text{ kJ/mol}$

3. Calculation of the Equilibrium Constant from $\Delta G'^\circ$ Calculate the equilibrium constant K'_{eq} for each of the following reactions at pH 7.0 and 25 °C, using the $\Delta G'^\circ$ values in Table 13–4.

(a) Glucose 6-phosphate + $H_2O \xrightleftharpoons{\text{glucose 6-phosphatase}}$ glucose + P_i

(b) Lactose + $H_2O \xrightleftharpoons{\beta\text{-galactosidase}}$ glucose + galactose

(c) Malate $\xrightleftharpoons{\text{fumarase}}$ fumarate + H_2O

Answer

As noted in Problem 2, $\Delta G = \Delta G'^\circ + RT \ln Q$, and at equilibrium, $Q = K'_{eq}$, $\Delta G = 0$, and

$$\Delta G'^\circ = -RT \ln K'_{eq}$$

So, at equilibrium, $\ln K'_{eq} = -\Delta G'^\circ/RT$, or $K'_{eq} = e^{-(\Delta G'^\circ/RT)}$; at 25 °C, $RT = 2.48$ kJ/mol.

From these relationships, we can calculate K'_{eq} for each reaction using the values of $\Delta G'^\circ$ in Table 13–4.

(a) For glucose 6-phosphatase:

$\Delta G'^\circ = -13.8$ kJ/mol

$\ln K'_{eq} = -(-13.8 \text{ kJ/mol})/(2.48 \text{ kJ/mol}) = 5.57$

$K'_{eq} = e^{5.57} = 262$

(b) For β-galactosidase:

$\Delta G'^\circ = -15.9$ kJ/mol

$\ln K'_{eq} = -(-15.9 \text{ kJ/mol})/(2.48 \text{ kJ/mol}) = 6.41$

$K'_{eq} = e^{6.41} = 608$

(c) For fumarase:

$\Delta G'^\circ = 3.1$ kJ/mol

$\ln K'_{eq} = -(3.1 \text{ kJ/mol})/(2.48 \text{ kJ/mol}) = -1.2$

$K'_{eq} = e^{-1.2} = 0.30$

4. Experimental Determination of K'_{eq} and $\Delta G'^\circ$ If a 0.1 M solution of glucose 1-phosphate at 25 °C is incubated with a catalytic amount of phosphoglucomutase, the glucose 1-phosphate is transformed to glucose 6-phosphate. At equilibrium, the concentrations of the reaction components are

$$\underset{4.5 \times 10^{-3}\ \text{M}}{\text{Glucose 1-phosphate}} \rightleftharpoons \underset{9.6 \times 10^{-2}\ \text{M}}{\text{glucose 6-phosphate}}$$

Calculate K'_{eq} and $\Delta G'^\circ$ for this reaction.

Answer

$K'_{eq} = [\text{G6P}]/[\text{G1P}] = (9.6 \times 10^{-2}\ \text{M})/(4.5 \times 10^{-3}\ \text{M})$

$= 21$

$\Delta G'^\circ = -RT \ln K'_{eq}$

$= -(2.48 \text{ kJ/mol})(\ln 21) = -7.6$ kJ/mol

5. Experimental Determination of $\Delta G'^{\circ}$ for ATP Hydrolysis A direct measurement of the standard free-energy change associated with the hydrolysis of ATP is technically demanding because the minute amount of ATP remaining at equilibrium is difficult to measure accurately. The value of $\Delta G'^{\circ}$ can be calculated indirectly, however, from the equilibrium constants of two other enzymatic reactions having less favorable equilibrium constants:

$$\text{Glucose 6-phosphate} + H_2O \longrightarrow \text{glucose} + P_i \qquad K'_{eq} = 270$$

$$\text{ATP} + \text{glucose} \longrightarrow \text{ADP} + \text{glucose 6-phosphate} \qquad K'_{eq} = 890$$

Using this information for equilibrium constants determined at 25 °C, calculate the standard free energy of hydrolysis of ATP.

Answer The reactions, if coupled together, constitute a "futile cycle" that results in the net hydrolysis of ATP:

(1) $\text{G6P} + H_2O \longrightarrow \text{glucose} + P_i$
(2) $\text{ATP} + \text{glucose} \longrightarrow \text{ADP} + \text{G6P}$

Sum: $\text{ATP} + H_2O \longrightarrow \text{ADP} + P_i$

Calculating from $\Delta G'^{\circ} = -RT \ln K'_{eq}$:

$$\Delta G_1'^{\circ} = (-2.48 \text{ kJ/mol})(\ln 270) = -14 \text{ kJ/mol}$$
$$\Delta G_2'^{\circ} = (-2.48 \text{ kJ/mol})(\ln 890) = -17 \text{ kJ/mol}$$
$$\Delta G_{sum}'^{\circ} = \Delta G_1'^{\circ} + \Delta G_2'^{\circ} = -31 \text{ kJ/mol}$$

6. Difference between $\Delta G'^{\circ}$ and ΔG Consider the following interconversion, which occurs in glycolysis (Chapter 14):

$$\text{Fructose 6-phosphate} \rightleftharpoons \text{glucose 6-phosphate} \qquad K'_{eq} = 1.97$$

(a) What is $\Delta G'^{\circ}$ for the reaction (K'_{eq} measured at 25 °C)?

(b) If the concentration of fructose 6-phosphate is adjusted to 1.5 M and that of glucose 6-phosphate is adjusted to 0.50 M, what is ΔG?

(c) Why are $\Delta G'^{\circ}$ and ΔG different?

Answer

(a) At equilibrium, $\Delta G'^{\circ} = -RT \ln K'_{eq}$
$= -(2.48 \text{ kJ/mol}) \ln 1.97$
$= -1.68 \text{ kJ/mol}$

(b) $\Delta G = \Delta G'^{\circ} + RT \ln Q$
$Q = [\text{G6P}]/[\text{F6P}] = 0.5 \text{ M}/1.5 \text{ M} = 0.33$
$\Delta G = -1.68 \text{ kJ/mol} + (2.48 \text{ kJ/mol}) \ln 0.33$
$= -4.4 \text{ kJ/mol}$

(c) $\Delta G'^{\circ}$ for any reaction is a fixed parameter because it is defined for standard conditions of temperature (25 °C = 298 K) and concentration (both F6P and G6P = 1 M). In contrast, ΔG is a variable and can be calculated for any set of product and reactant concentrations. ΔG is defined as $\Delta G'^{\circ}$ (standard conditions) plus whatever difference occurs in ΔG on moving to nonstandard conditions.

7. Free Energy of Hydrolysis of CTP Compare the structure of the nucleoside triphosphate CTP with the structure of ATP.

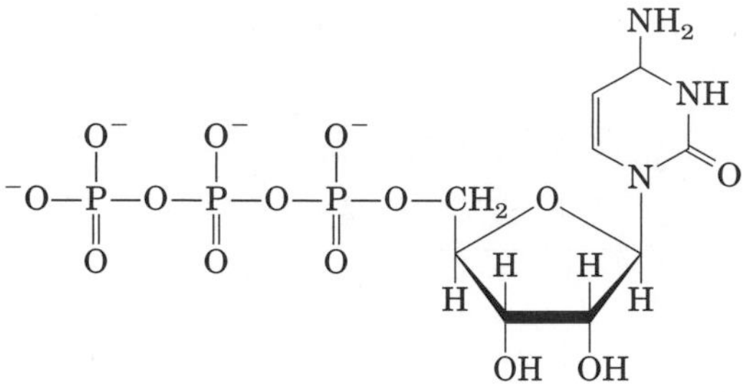

Cytidine triphosphate (CTP)

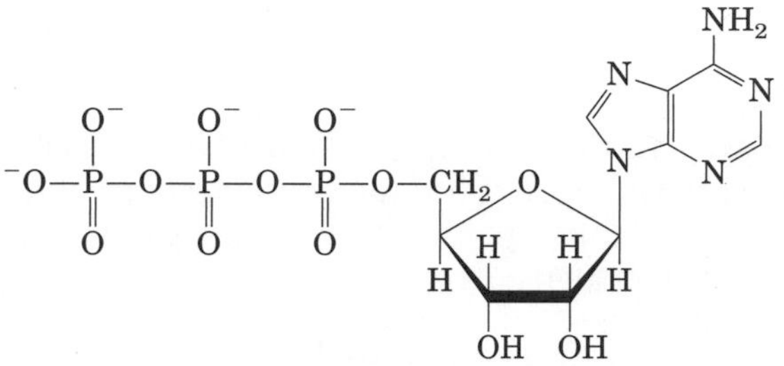

Adenosine triphosphate (ATP)

Now predict the K'_{eq} and $\Delta G'^{\circ}$ for the following reaction:

$$\text{ATP} + \text{CDP} \longrightarrow \text{ADP} + \text{CTP}$$

Answer $\Delta G'^{\circ}$ near 0; K'_{eq} near 1. The high $\Delta G'^{\circ}$ of ATP is related to structural features not of the base or the sugar, but primarily of the anhydride linkages between phosphate groups. In this structural feature, CTP is equivalent to ATP, and thus it most likely has about the same $\Delta G'^{\circ}$ as ATP. If this is the case, the reaction ATP + CDP $\longrightarrow$ ADP + CTP has a $\Delta G'^{\circ}$ very close to zero, and a K'_{eq} close to 1 (see Table 13–3).

8. Dependence of ΔG on pH The free energy released by the hydrolysis of ATP under standard conditions at pH 7.0 is -30.5 kJ/mol. If ATP is hydrolyzed under standard conditions except at pH 5.0, is more or less free energy released? Explain. Use the Living Graph to explore this relationship.

Answer Less; the overall equation for ATP hydrolysis can be approximated as

$$\text{ATP}^{4-} + \text{H}_2\text{O} \rightleftharpoons \text{ADP}^{3-} + \text{HPO}_4^{2-} + \text{H}^+$$

(This is only an approximation, because the ionized species shown here are the major, but not the only, forms present.) Under standard conditions (i.e., [ATP] = [ADP] = [P_i] = 1 M), the concentration of water is 55 M and does not change during the reaction. Because H^+ ions are produced in the reaction, the lower the pH at which the reaction proceeds—that is, the higher the [H^+]—the more the equilibrium shifts toward reactants. As a result, at lower pH the reaction does not proceed as far toward products, and less free energy is released.

9. The $\Delta G'^{\circ}$ for Coupled Reactions Glucose 1-phosphate is converted into fructose 6-phosphate in two successive reactions:

$$\text{Glucose 1-phosphate} \longrightarrow \text{glucose 6-phosphate}$$

$$\text{Glucose 6-phosphate} \longrightarrow \text{fructose 6-phosphate}$$

Using the $\Delta G'^{\circ}$ values in Table 13–4, calculate the equilibrium constant, K'_{eq}, for the sum of the two reactions:

$$\text{Glucose 1-phosphate} \longrightarrow \text{fructose 6-phosphate}$$

Answer

$$\begin{array}{llll} (1) & \text{G1P} \longrightarrow \text{G6P} & \Delta G_1'^{\circ} = & -7.3 \text{ kJ/mol} \\ (2) & \text{G6P} \longrightarrow \text{F6P} & \Delta G_2'^{\circ} = & 1.7 \text{ kJ/mol} \\ \hline \text{Sum:} & \text{G1P} \longrightarrow \text{F6P} & \Delta G_{sum}'^{\circ} = & -5.6 \text{ kJ/mol} \end{array}$$

$$\begin{aligned} \ln K'_{eq} &= -\Delta G'^{\circ}/RT \\ &= -(-5.6 \text{ kJ/mol})/(2.48 \text{ kJ/mol}) \\ &= 2.3 \\ K'_{eq} &= 10 \end{aligned}$$

10. Effect of [ATP]/[ADP] Ratio on Free Energy of Hydrolysis of ATP Using Equation 13–4, plot ΔG against $\ln Q$ (mass-action ratio) at 25 °C for the concentrations of ATP, ADP, and P_i in the table below. $\Delta G'^{\circ}$ for the reaction is –30.5 kJ/mol. Use the resulting plot to explain why metabolism is regulated to keep the ratio [ATP]/[ADP] high.

	Concentration (mM)				
ATP	5	3	1	0.2	5
ADP	0.2	2.2	4.2	5.0	25
P_i	10	12.1	14.1	14.9	10

Answer The reaction is ATP $\longrightarrow$ ADP + P_i. From Equation 13–4, with Q (the mass action ratio) = [ADP][P_i]/[ATP], expressed as molar concentrations, the free-energy change for this reaction is:

$$\Delta G = \Delta G'^{\circ} + RT \ln ([\text{ADP}][\text{P}_i]/[\text{ATP}])$$

Calculate $\ln Q$ for each of the five cases:

$$\ln Q_1 = \ln [(2 \times 10^{-4})(1.0 \times 10^{-2})/(5 \times 10^{-3})] = -7.8$$

$$\ln Q_2 = \ln [(2.2 \times 10^{-3})(1.21 \times 10^{-2})/(3 \times 10^{-3})] = -4.7$$

$$\ln Q_3 = \ln [(4.2 \times 10^{-3})(1.41 \times 10^{-2})/(1 \times 10^{-3})] = -2.8$$

$$\ln Q_4 = \ln [(5.0 \times 10^{-3})(1.49 \times 10^{-2})/(2 \times 10^{-4})] = -1.0$$

$$\ln Q_5 = \ln [(2.5 \times 10^{-2})(1.0 \times 10^{-2})/(5 \times 10^{-3})] = -3.0$$

Substitute each of these values for $\ln Q$, −30.5 kJ/mol for $\Delta G'^{\circ}$, and 2.48 kJ/mol for RT in Equation 13–4:

$$\Delta G_1 = -30.5 \text{ kJ/mol} + (2.48 \text{ kJ/mol})(-7.8) = -50 \text{ kJ/mol}$$

$$\Delta G_2 = -30.5 \text{ kJ/mol} + (2.48 \text{ kJ/mol})(-4.7) = -42 \text{ kJ/mol}$$

$$\Delta G_3 = -30.5 \text{ kJ/mol} + (2.48 \text{ kJ/mol})(-2.8) = -38 \text{ kJ/mol}$$

$$\Delta G_4 = -30.5 \text{ kJ/mol} + (2.48 \text{ kJ/mol})(-1.0) = -33 \text{ kJ/mol}$$

$$\Delta G_5 = -30.5 \text{ kJ/mol} + (2.48 \text{ kJ/mol})(-3.0) = -38 \text{ kJ/mol}$$

Now plot ΔG versus $\ln Q$ for each case:

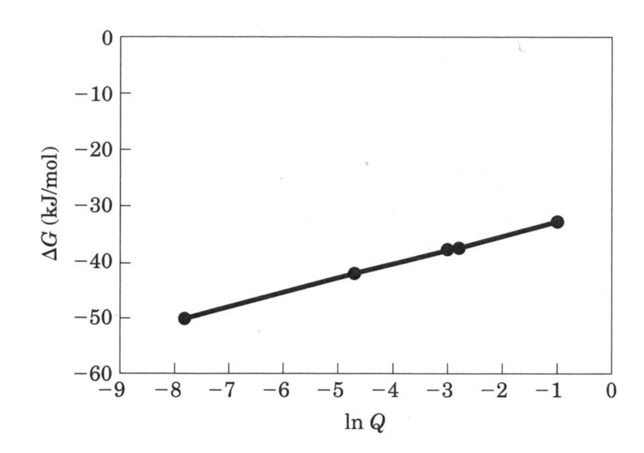

The ΔG for ATP hydrolysis is smaller when [ATP]/[ADP] is low (<<1) than when [ATP]/[ADP] is high. The energy available to a cell from a given amount of ATP is smaller when [ATP]/[ADP] falls and greater when this ratio rises.

11. Strategy for Overcoming an Unfavorable Reaction: ATP-Dependent Chemical Coupling The phosphorylation of glucose to glucose 6-phosphate is the initial step in the catabolism of glucose. The direct phosphorylation of glucose by P_i is described by the equation

$$\text{Glucose} + P_i \longrightarrow \text{glucose 6-phosphate} + H_2O \qquad \Delta G'^\circ = 13.8 \text{ kJ/mol}$$

(a) Calculate the equilibrium constant for the above reaction at 37 °C. In the rat hepatocyte the physiological concentrations of glucose and P_i are maintained at approximately 4.8 mM. What is the equilibrium concentration of glucose 6-phosphate obtained by the direct phosphorylation of glucose by P_i? Does this reaction represent a reasonable metabolic step for the catabolism of glucose? Explain.

(b) In principle, at least, one way to increase the concentration of glucose 6-phosphate is to drive the equilibrium reaction to the right by increasing the intracellular concentrations of glucose and P_i. Assuming a fixed concentration of P_i at 4.8 mM, how high would the intracellular concentration of glucose have to be to give an equilibrium concentration of glucose 6-phosphate of 250 μM (the normal physiological concentration)? Would this route be physiologically reasonable, given that the maximum solubility of glucose is less than 1 M?

(c) The phosphorylation of glucose in the cell is coupled to the hydrolysis of ATP; that is, part of the free energy of ATP hydrolysis is used to phosphorylate glucose:

(1) $\text{Glucose} + P_i \longrightarrow \text{glucose 6-phosphate} + H_2O \qquad \Delta G'^\circ = 13.8 \text{ kJ/mol}$

(2) $\text{ATP} + H_2O \longrightarrow \text{ADP} + P_i \qquad \Delta G'^\circ = -30.5 \text{ kJ/mol}$

Sum: $\text{Glucose} + \text{ATP} \longrightarrow \text{glucose 6-phosphate} + \text{ADP}$

Calculate K'_{eq} at 37 °C for the overall reaction. For the ATP-dependent phosphorylation of glucose, what concentration of glucose is needed to achieve a 250 μM intracellular concentration of glucose 6-phosphate when the concentrations of ATP and ADP are 3.38 mM and 1.32 mM, respectively? Does this coupling process provide a feasible route, at least in principle, for the phosphorylation of glucose in the cell? Explain.

(d) Although coupling ATP hydrolysis to glucose phosphorylation makes thermodynamic sense, we have not yet specified how this coupling is to take place. Given that coupling requires a common intermediate, one conceivable route is to use ATP hydrolysis to raise the intracellular concentration of P_i and thus drive the unfavorable phosphorylation of glucose by P_i. Is this a reasonable route? (Think about the solubility products of metabolic intermediates.)

(e) The ATP-coupled phosphorylation of glucose is catalyzed in hepatocytes by the enzyme glucokinase. This enzyme binds ATP and glucose to form a glucose-ATP-enzyme complex, and the phosphoryl group is transferred directly from ATP to glucose. Explain the advantages of this route.

Answer

(a) $\Delta G'^\circ = -RT \ln K'_{eq}$

$\ln K'_{eq} = -\Delta G'^\circ/RT$

$= -(13.8 \text{ kJ/mol})/(2.48 \text{ kJ/mol})$

$K'_{eq} = e^{-5.56}$

$= 3.85 \times 10^{-3} \text{ M}^{-1}$

(Note: this value has units M^{-1} because the expression for K'_{eq} from the chemical equilibrium includes H_2O; see below.)

$$K'_{eq} = \frac{[\text{G6P}]}{[\text{Glc}][\text{P}_i]}$$

$[\text{G6P}] = K'_{eq}\,[\text{Glc}][\text{P}_i]$

$= (3.85 \times 10^{-3} \text{ M}^{-1})(4.8 \times 10^{-3} \text{ M})(4.8 \times 10^{-3} \text{ M})$

$= 8.9 \times 10^{-8} \text{ M}$

This would not be a reasonable route for glucose catabolism because the cellular [G6P] is likely to be much higher than 8.9×10^{-8} M, and the reaction would be unfavorable.

(b) Because $K'_{eq} = \dfrac{[\text{G6P}]}{[\text{Glc}][\text{P}_i]}$

then $[\text{Glc}] = \dfrac{[\text{G6P}]}{K'_{eq}[\text{P}_i]}$

$$= \frac{250 \times 10^{-6} \text{ M}}{(3.85 \times 10^{-3} \text{ M}^{-1})(4.8 \times 10^{-3} \text{ M})} = 14 \text{ M}$$

This would not be a reasonable route because the maximum solubility of glucose is less than 1 M.

(c) (1) $\text{Glc} + \text{P}_i \longrightarrow \text{G6P} + H_2O$ $\qquad \Delta G_1'^\circ = 13.8 \text{ kJ/mol}$

(2) $\text{ATP} + H_2O \longrightarrow \text{ADP} + \text{P}_i$ $\qquad \Delta G_2'^\circ = -30.5 \text{ kJ/mol}$

Sum: $\text{Glc} + \text{ATP} \longrightarrow \text{G6P} + \text{ADP}$ $\qquad \Delta G'^\circ_{sum} = -16.7 \text{ kJ/mol}$

$\ln K'_{eq} = -\Delta G'^\circ/RT$

$= -(-16.7 \text{ kJ/mol})/(2.48 \text{ kJ/mol})$

$= 6.73$

$K'_{eq} = 837$

Because $K'_{eq} = \dfrac{[\text{G6P}][\text{ADP}]}{[\text{Glc}][\text{ATP}]}$

then $[\text{Glc}] = \dfrac{[\text{G6P}][\text{ADP}]}{K'_{eq}\,[\text{ATP}]}$

$$= \frac{(250 \times 10^{-6} \text{ M})(1.32 \times 10^{-3} \text{ M})}{(837)(3.38 \times 10^{-3} \text{ M})}$$

$= 1.2 \times 10^{-7} \text{ M}$

This route is feasible because the glucose concentration is reasonable.

(d) No; this is not reasonable. When glucose is at its physiological level, the required P_i concentration would be so high that phosphate salts of divalent cations would precipitate out.

(e) Direct transfer of the phosphoryl group from ATP to glucose takes advantage of the high phosphoryl group transfer potential of ATP and does not demand that the concentration of intermediates be very high, unlike the mechanism proposed in **(d).** In addition, the usual benefits of enzymatic catalysis apply, including binding interactions between the enzyme and its substrates; induced fit leading to the exclusion of water from the active site, so that only glucose is phosphorylated; and stabilization of the transition state.

12. Calculations of $\Delta G'^\circ$ for ATP-Coupled Reactions From data in Table 13–6 calculate the $\Delta G'^\circ$ value for the reactions.

(a) Phosphocreatine + ADP $\longrightarrow$ creatine + ATP

(b) ATP + fructose $\longrightarrow$ ADP + fructose 6-phosphate

Answer

(a) The $\Delta G'^\circ$ value for the overall reaction is calculated from the sum of the $\Delta G'^\circ$ values for the coupled reactions.

	Reaction	
(1)	Phosphocreatine + $H_2O \longrightarrow$ creatine + P_i	$\Delta G_1'^\circ = -43.0$ kJ/mol
(2)	ADP + $P_i \longrightarrow$ ATP + H_2O	$\Delta G_2'^\circ = 30.5$ kJ/mol
Sum:	Phosphocreatine + ADP $\longrightarrow$ creatine + ATP	$\Delta G_{sum}'^\circ = -12.5$ kJ/mol

(b)

	Reaction	
(1)	ATP + $H_2O \longrightarrow$ ADP + P_i	$\Delta G_1'^\circ = -30.5$ kJ/mol
(2)	Fructose + $P_i \longrightarrow$ F6P + H_2O	$\Delta G_2'^\circ = 15.9$ kJ/mol
Sum:	ATP + fructose $\longrightarrow$ ADP + F6P	$\Delta G_{sum}'^\circ = -14.6$ kJ/mol

13. Coupling ATP Cleavage to an Unfavorable Reaction To explore the consequences of coupling ATP hydrolysis under physiological conditions to a thermodynamically unfavorable biochemical reaction, consider the hypothetical transformation X $\rightarrow$ Y, for which $\Delta G'^\circ = 20$ kJ/mol.

(a) What is the ratio [Y]/[X] at equilibrium?

(b) Suppose X and Y participate in a sequence of reactions during which ATP is hydrolyzed to ADP and P_i. The overall reaction is

$$X + ATP + H_2O \longrightarrow Y + ADP + P_i$$

Calculate [Y]/[X] for this reaction at equilibrium. Assume that the temperature is 25 °C and the equilibrium concentrations of ATP, ADP, and P_i are all 1 M.

(c) We know that [ATP], [ADP], and [P_i] are *not* 1 M under physiological conditions. Calculate [Y]/[X] for the ATP-coupled reaction when the values of [ATP], [ADP], and [P_i] are those found in rat myocytes (Table 13–5).

Answer

(a) The ratio $[Y]_{eq}/[X]_{eq}$ is equal to the equilibrium constant, K'_{eq}.

$$\begin{aligned} \ln K'_{eq} &= -\Delta G'^\circ/RT \\ &= -(20 \text{ kJ/mol})/(2.48 \text{ kJ/mol}) \\ &= -8 \\ K'_{eq} &= e^{-8} = 3 \times 10^{-4} = [Y]_{eq}/[X]_{eq} \end{aligned}$$

This is a very small value of K'_{eq}; consequently, $\Delta G'^\circ$ is large and positive, making the reaction energetically unfavorable as written.

(b) First, we need to calculate $\Delta G'^\circ$ for the overall reaction.

	Reaction	
(1)	X $\longrightarrow$ Y	$\Delta G_1'^\circ = 20$ kJ/mol
(2)	ATP + $H_2O \longrightarrow$ ADP + P_i	$\Delta G_2'^\circ = -30.5$ kJ/mol
Sum:	X + ATP + $H_2O \longrightarrow$ ADP + P_i + Y	$\Delta G_{sum}'^\circ = -10.5$ kJ/mol

$$K'_{eq} = \frac{[Y]_{eq}[P_i]_{eq}[ADP]_{eq}}{[X]_{eq}[ATP]_{eq}}\text{; note: water is omitted.}$$

Because [ADP], [ATP], and [P_i] are 1 M, this simplifies to K'_{eq} = [Y]/[X] in units of M.

$$\ln K'_{eq} = -\Delta G'^{\circ}/RT$$
$$= -(-10.5 \text{ kJ/mol})/(2.48 \text{ kJ/mol}) = 4.23$$
$$K'_{eq} = e^{4.23} = 68.7 = [\text{Y}]/[\text{X}]$$

$\Delta G'^{\circ}$ is fairly large and negative; the coupled reaction is favorable as written.

(c) Here we are dealing with the nonstandard conditions of the cell. Under physiological conditions, a favorable reaction (under standard conditions) becomes even more favorable.

$$K'_{eq} = \frac{[\text{Y}]_{eq}[\text{P}_i]_{eq}[\text{ADP}]_{eq}}{[\text{X}]_{eq}[\text{ATP}]_{eq}}$$

If we hold the values of [P_i], [ADP], and [ATP] at the values known to exist in the cell, we can calculate the values of [X] and [Y] that meet the equilibrium expression above, giving the equilibrium constant we calculated in **(b).**

$$[\text{Y}]/[\text{X}] = \frac{K'_{eq}[\text{ATP}]}{[\text{P}_i][\text{ADP}]}$$
$$= \frac{(68.7 \text{ M})(8.05 \times 10^{-3} \text{ M})}{(8.05 \times 10^{-3} \text{ M})(0.93 \times 10^{-3} \text{ M})}$$
$$= 7.4 \times 10^4$$

So by coupling the conversion X → Y to ATP hydrolysis, and by holding [ATP], [ADP], and [P_i] far from their equilibrium levels, the cell can greatly increase the ratio [product]/[reactant]; the reaction goes essentially to completion.

14. Calculations of ΔG at Physiological Concentrations Calculate the actual, physiological ΔG for the reaction

$$\text{Phosphocreatine} + \text{ADP} \longrightarrow \text{creatine} + \text{ATP}$$

at 37 °C, as it occurs in the cytosol of neurons, with phosphocreatine at 4.7 mM, creatine at 1.0 mM, ADP at 0.73 mM, and ATP at 2.6 mM.

Answer

Using $\Delta G'^{\circ}$ values from Table 13–6:

(1) Phosphocreatine + H_2O ⟶ creatine + P_i $\Delta G_1'^{\circ} = -43.0$ kJ/mol

(2) ADP + P_i ⟶ ATP + H_2O $\Delta G_2'^{\circ} = 30.5$ kJ/mol

Sum: Phosphocreatine = ADP ⟶ creatine + ATP $\Delta G_{sum}'^{\circ} = -12.5$ kJ/mol

$$\text{Mass-action ratio, } Q = \frac{[\text{products}]}{[\text{reactants}]} = \frac{[\text{creatine}][\text{ATP}]}{[\text{phosphocreatine}][\text{ADP}]}$$
$$= \frac{(1 \times 10^{-3} \text{ M})(2.6 \times 10^{-3} \text{ M})}{(4.7 \times 10^{-3} \text{ M})(7.3 \times 10^{-4} \text{ M})}$$
$$= 0.75$$
$$\Delta G = \Delta G'^{\circ} + RT \ln Q$$
$$= -12.5 \text{ kJ/mol} + (8.315 \text{ J/mol} \cdot \text{K})(310 \text{ K}) \ln 0.75$$
$$= -13 \text{ kJ/mol}$$

15. Free Energy Required for ATP Synthesis under Physiological Conditions In the cytosol of rat hepatocytes, the temperature is 37 °C and the mass-action ratio, Q, is

$$\frac{[\text{ATP}]}{[\text{ADP}][\text{P}_\text{i}]} = 5.33 \times 10^2\ \text{M}^{-1}$$

Calculate the free energy required to synthesize ATP in a rat hepatocyte.

Answer The reaction for the synthesis of ATP is

$$\text{ADP} + \text{P}_\text{i} \longrightarrow \text{ATP} + \text{H}_2\text{O} \qquad \Delta G'^\circ = 30.5\ \text{kJ/mol}$$

The mass-action ratio is

$$\frac{[\text{products}]}{[\text{reactants}]} = \frac{[\text{ATP}]}{[\text{P}_\text{i}][\text{ADP}]} = 5.33 \times 10^2\ \text{M}^{-1}$$

Because $\Delta G = \Delta G'^\circ + RT \ln [\text{products}]/[\text{reactants}]$,

$$\begin{aligned}\Delta G &= 30.5\ \text{kJ/mol} + (8.315\ \text{J/mol} \cdot \text{K})(310\ \text{K}) \ln 5.33 \times 10^2\ \text{M}^{-1} \\ &= 46.7\ \text{kJ/mol}\end{aligned}$$

16. Chemical Logic In the glycolytic pathway, a six-carbon sugar (fructose 1,6-bisphosphate) is cleaved to form two three-carbon sugars, which undergo further metabolism (see Fig. 14–5). In this pathway, an isomerization of glucose 6-phosphate to fructose 6-phosphate (shown below) occurs two steps before the cleavage reaction (the intervening step is phosphorylation of fructose 6-phosphate to fructose 1,6-bisphosphate (p. 532)).

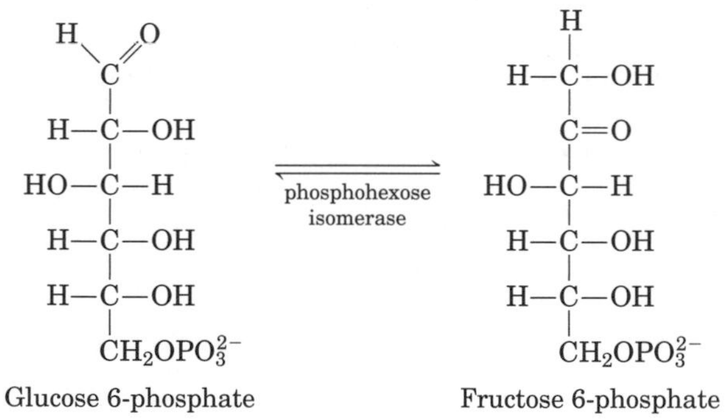

Glucose 6-phosphate: H–C(=O) / H—C—OH / HO—C—H / H—C—OH / H—C—OH / $CH_2OPO_3^{2-}$ ⇌ (phosphohexose isomerase) Fructose 6-phosphate: H—C(H)—OH / C=O / HO—C—H / H—C—OH / H—C—OH / $CH_2OPO_3^{2-}$

Glucose 6-phosphate   Fructose 6-phosphate

What does the isomerization step accomplish from a chemical perspective? (Hint: Consider what might happen if the C—C bond cleavage were to proceed without the preceding isomerization.)

Answer C—C bond cleavage is facilitated by the presence of a carbonyl group one carbon removed from the bond being cleaved. Isomerization moves the carbonyl group from C-1 to C-2, setting up a carbon–carbon bond cleavage between C-3 and C-4. Without isomerization, bond cleavage would occur between C-2 and C-3, generating one two-carbon and one four-carbon compound.

17. Enzymatic Reaction Mechanisms I Lactate dehydrogenase is one of the many enzymes that require NADH as coenzyme. It catalyzes the conversion of pyruvate to lactate:

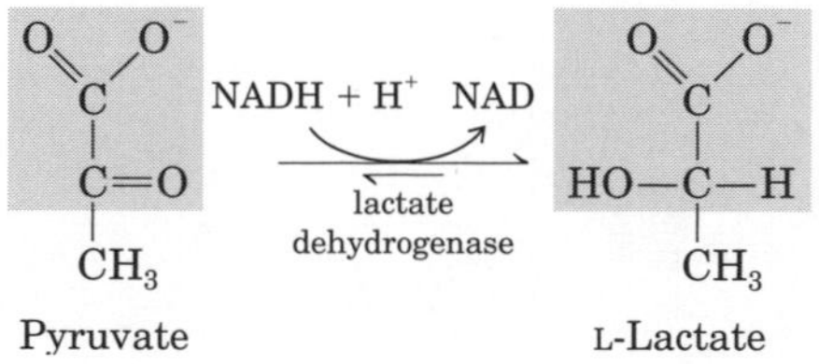

Pyruvate: COO^- / C=O / CH_3 → (NADH + H^+ → NAD; lactate dehydrogenase) L-Lactate: COO^- / HO—C—H / CH_3

Pyruvate   L-Lactate

Draw the mechanism of this reaction (show electron-pushing arrows). (Hint: This is a common reaction throughout metabolism; the mechanism is similar to that catalyzed by other dehydrogenases that use NADH, such as alcohol dehydrogenase.)

Answer The mechanism is the same as that of the alcohol dehydrogenase reaction (Fig. 14–13, p. 547).

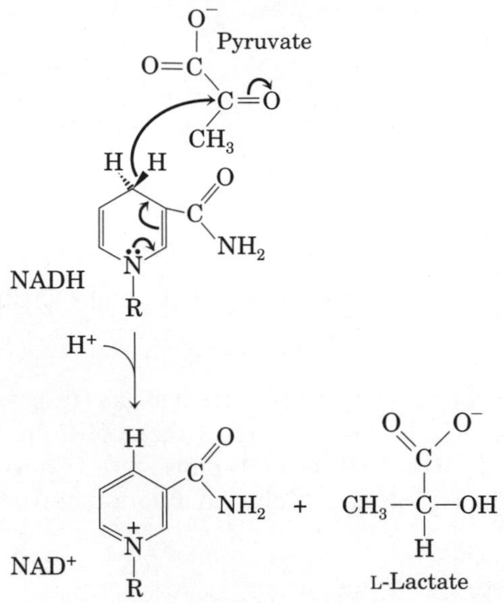

Pyruvate + NADH + H^+ → NAD^+ + CH_3—C(H)(OH)—COO^-

NADH

H^+

NAD^+

L-Lactate

18. Enzymatic Reaction Mechanisms II Biochemical reactions often look more complex than they really are. In the pentose phosphate pathway (Chapter 14), sedoheptulose 7-phosphate and glyceraldehyde 3-phosphate react to form erythrose 4-phosphate and fructose 6-phosphate in a reaction catalyzed by transaldolase.

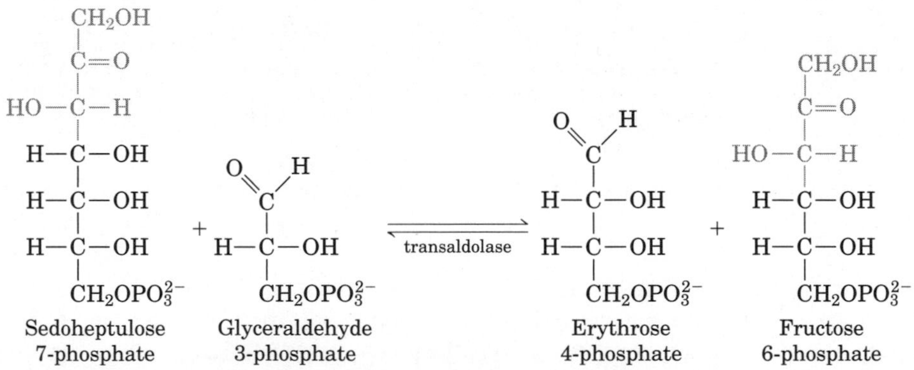

Sedoheptulose 7-phosphate + Glyceraldehyde 3-phosphate $\rightleftharpoons$ (transaldolase) Erythrose 4-phosphate + Fructose 6-phosphate

Draw a mechanism for this reaction (show electron-pushing arrows). (Hint: Take another look at aldol condensations, then consider the name of this enzyme.)

Answer The first step is the reverse of an aldol condensation (see the aldolase mechanism, Fig. 14–5, p. 534); the second step is an aldol condensation (see Fig. 13–4, p. 497).

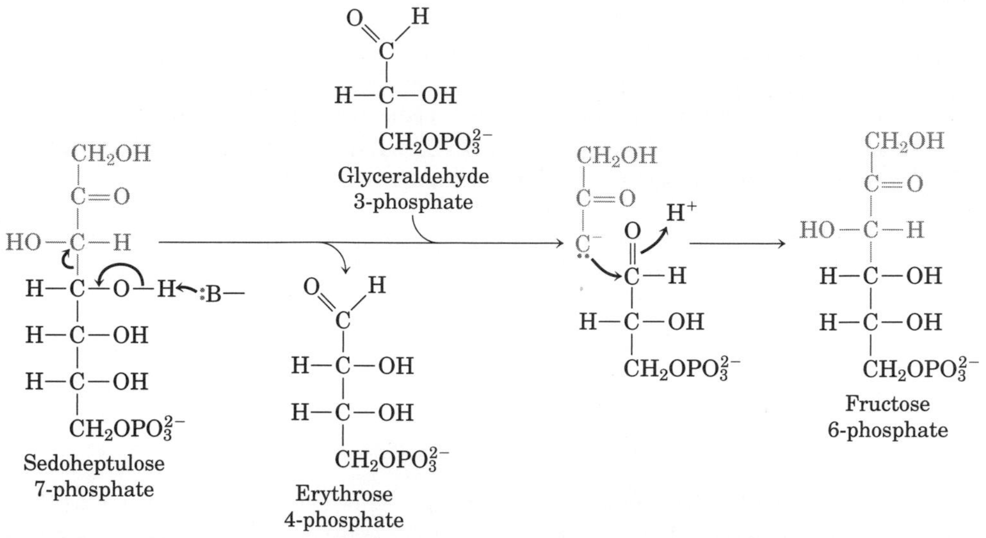

Sedoheptulose 7-phosphate → Erythrose 4-phosphate + Glyceraldehyde 3-phosphate → Fructose 6-phosphate

19. Daily ATP Utilization by Human Adults

(a) A total of 30.5 kJ/mol of free energy is needed to synthesize ATP from ADP and P_i when the reactants and products are at 1 M concentrations and the temperature is 25 °C (standard state). Because the actual physiological concentrations of ATP, ADP, and P_i are not 1 M, and the temperature is 37 °C, the free energy required to synthesize ATP under physiological conditions is different from $\Delta G'^{\circ}$. Calculate the free energy required to synthesize ATP in the human hepatocyte when the physiological concentrations of ATP, ADP, and P_i are 3.5, 1.50, and 5.0 mM, respectively.

(b) A 68 kg (150 lb) adult requires a caloric intake of 2,000 kcal (8,360 kJ) of food per day (24 hours). The food is metabolized and the free energy is used to synthesize ATP, which then provides energy for the body's daily chemical and mechanical work. Assuming that the efficiency of converting food energy into ATP is 50%, calculate the weight of ATP used by a human adult in 24 hours. What percentage of the body weight does this represent?

(c) Although adults synthesize large amounts of ATP daily, their body weight, structure, and composition do not change significantly during this period. Explain this apparent contradiction.

Answer

(a) $\text{ADP} + \text{P}_i \longrightarrow \text{ATP} + \text{H}_2\text{O} \qquad \Delta G'^{\circ} = 30.5 \text{ kJ/mol}$

$$\text{Mass action ratio, } Q = \frac{[\text{ATP}]}{[\text{P}_i][\text{ADP}]} = \frac{[3.5 \times 10^{-3}\ \text{M}]}{[1.5 \times 10^{-3}\ \text{M}]\,[5.0 \times 10^{-3}\ \text{M}]} = 4.7 \times 10^{2}\ \text{M}^{-1}$$

$$\begin{aligned}\Delta G &= \Delta G'^{\circ} + RT \ln Q \\ &= 30.5 \text{ kJ/mol} + (2.58 \text{ kJ/mol}) \ln (4.7 \times 10^{2}\ \text{M}^{-1}) \\ &= 46 \text{ kJ/mol}\end{aligned}$$

(b) The energy going into ATP synthesis in 24 hr is 8,360 kJ × 50% = 4,180 kJ. Using the value of ΔG from **(a),** the amount of ATP synthesized is

$$(4{,}180 \text{ kJ})/(46 \text{ kJ/mol}) = 91 \text{ mol}$$

The molecular weight of ATP is 503 (calculated by summing atomic weights). Thus, the weight of ATP synthesized is

$$(91 \text{ mol ATP})(503 \text{ g/mol}) = 46 \text{ kg}$$

As a percentage of body weight:

$$100\%(46 \text{ kg ATP})/(68 \text{ kg body weight}) = 68\%$$

(c) The concentration of ATP in a healthy body is maintained in a steady state; this is an example of homeostasis, a condition in which the body synthesizes and breaks down ATP as needed.

20. Rates of Turnover of γ and β Phosphates of ATP If a small amount of ATP labeled with radioactive phosphorus in the terminal position, [γ-^{32}P]ATP, is added to a yeast extract, about half of the ^{32}P activity is found in P_i within a few minutes, but the concentration of ATP remains unchanged. Explain. If the same experiment is carried out using ATP labeled with ^{32}P in the central position, [β-^{32}P]ATP, the ^{32}P does not appear in P_i within such a short time. Why?

Answer We can represent ATP as A-P-P-P (the P farthest from A is the γ P) and a radiolabeled phosphate group as *P. One possible reaction for γ-labeled ATP would be phosphorylation of glucose:

$$\text{A-P-P-*P} + \text{Glc} \longrightarrow \text{A-P-P} + \text{G6*P} \rightarrow \rightarrow \rightarrow \text{*P}_i$$

or, more generally:

$$\text{A-P-P-*P} + \text{H}_2\text{O} \longrightarrow \text{A-P-P} + \text{*P}_i$$

The ATP system is in a dynamic steady state; [ATP] remains constant because the rate of ATP consumption, as depicted above, equals its rate of synthesis. ATP consumption involves release of the terminal (γ) phosphoryl group; synthesis of ATP from ADP involves replacement of this phosphoryl group. Hence, the terminal phosphate undergoes rapid turnover.

The reaction

$$\text{A-P-*P-P} + \text{H}_2\text{O} \longrightarrow \text{A-P} + \text{*P}_i + \text{P}_i$$

occurs more slowly: the central (β) phosphate undergoes only relatively slow turnover.

21. Cleavage of ATP to AMP and PP_i during Metabolism Synthesis of the activated form of acetate (acetyl-CoA) is carried out in an ATP-dependent process:

$$\text{Acetate} + \text{CoA} + \text{ATP} \longrightarrow \text{acetyl-CoA} + \text{AMP} + \text{PP}_i$$

(a) The $\Delta G'^{\circ}$ for the hydrolysis of acetyl-CoA to acetate and CoA is -32.2 kJ/mol and that for hydrolysis of ATP to AMP and PP_i is -30.5 kJ/mol. Calculate $\Delta G'^{\circ}$ for the ATP-dependent synthesis of acetyl-CoA.

(b) Almost all cells contain the enzyme inorganic pyrophosphatase, which catalyzes the hydrolysis of PP_i to P_i. What effect does the presence of this enzyme have on the synthesis of acetyl-CoA? Explain.

Answer

(a) The $\Delta G'^{\circ}$ can be determined for the coupled reactions:

	Reaction	
(1)	$\text{Acetate} + \text{CoA} \longrightarrow \text{acetyl-CoA} + \text{H}_2\text{O}$	$\Delta G_1'^{\circ} = 32.2$ kJ/mol
(2)	$\text{ATP} + \text{H}_2\text{O} \longrightarrow \text{AMP} + \text{PP}_i$	$\Delta G_2'^{\circ} = -30.5$ kJ/mol
Sum:	$\text{Acetate} + \text{CoA} + \text{ATP} \longrightarrow \text{acetyl-CoA} + \text{AMP} + \text{PP}_i$	$\Delta G_{sum}'^{\circ} = 1.7$ kJ/mol

(b) Hydrolysis of PP_i would drive the reaction forward, favoring the synthesis of acetyl-CoA.

22. Energy for H^+ Pumping The parietal cells of the stomach lining contain membrane "pumps" that transport hydrogen ions from the cytosol (pH 7.0) into the stomach, contributing to the acidity of gastric juice (pH 1.0). Calculate the free energy required to transport 1 mol of hydrogen ions through these pumps. (Hint: see Chapter 11.) Assume a temperature of 37 °C.

Answer The free energy required to transport 1 mol of H^+ from the interior of the cell, where $[H^+]$ is 10^{-7} M, across the membrane to where $[H^+]$ is 10^{-1} M is

$$\begin{aligned}\Delta G_t &= RT \ln (C_2/C_1)\\ &= RT \ln (10^{-1}/10^{-7})\\ &= (8.315 \text{ J/mol} \cdot \text{K})(310 \text{ K}) \ln 10^6\\ &= 36 \text{ kJ/mol}\end{aligned}$$

23. Standard Reduction Potentials The standard reduction potential, E'°, of any redox pair is defined for the half-cell reaction:

$$\text{Oxidizing agent} + n \text{ electrons} \longrightarrow \text{reducing agent}$$

The E'° values for the NAD^+/NADH and pyruvate/lactate conjugate redox pairs are −0.32 V and −0.19 V, respectively.

(a) Which redox pair has the greater tendency to lose electrons? Explain.

(b) Which pair is the stronger oxidizing agent? Explain.

(c) Beginning with 1 M concentrations of each reactant and product at pH 7 and 25 °C, in which direction will the following reaction proceed?

$$\text{Pyruvate} + \text{NADH} + H^+ \rightleftharpoons \text{lactate} + NAD^+$$

(d) What is the standard free-energy change ($\Delta G'^\circ$) for the conversion of pyruvate to lactate?

(e) What is the equilibrium constant (K'_{eq}) for this reaction?

Answer

(a) The NAD^+/NADH pair is more likely to lose electrons. The equations in Table 13–7 are written in the direction of reduction (gain of electrons). E'° is positive if the oxidized member of a conjugate pair has a tendency to accept electrons. E'° is negative if the oxidized member of a conjugate pair does *not* have a tendency to accept electrons. Both NAD^+/NADH and pyruvate/lactate have negative E'° values. The E'° of NAD^+/NADH (−0.0320 V) is more negative than that for pyruvate/lactate (−0.185 V), so this pair has the greater tendency to accept electrons and is thus the stronger oxidizing system.

(b) The pyruvate/lactate pair is the more likely to accept electrons and thus is the stronger oxidizing agent. For the same reason that NADH tends to donate electrons to pyruvate, pyruvate tends to accept electrons from NADH. Pyruvate is reduced to lactate; NADH is oxidized to NAD^+. Pyruvate is the oxidizing agent; NADH is the reducing agent.

(c) From the answers to **(a)** and **(b),** it is evident that the reaction will tend to go in the direction of lactate formation.

(d) The first step is to calculate $\Delta E'^\circ$ for the reaction, using the E'° values in Table 13–7. Recall that, by convention, $\Delta E'^\circ = (E'^\circ \text{ of electron acceptor}) - (E'^\circ \text{ of electron donor})$. For

$$\text{NADH} + \text{pyruvate} \longrightarrow NAD^+ + \text{lactate}$$

$$\begin{aligned}\Delta E'^\circ &= (E'^\circ \text{ for pyruvate/lactate}) - (E'^\circ \text{ for } NAD^+/\text{NADH})\\ &= -0.185 \text{ V} - (-0.320 \text{ V}) = 0.135 \text{ V}\\ \Delta G'^\circ &= -n\mathcal{F}\,\Delta E'^\circ\\ &= -2(96.5 \text{ kJ/V} \cdot \text{mol})(0.135 \text{ V})\\ &= -26.1 \text{ kJ/mol}\end{aligned}$$

(e) $\ln K'_{eq} = -\Delta G'^{\circ}/RT$

$= -\ (-26.1 \text{ kJ/mol})/(2.48 \text{ kJ/mol})$

$= -\ 10.5$

$K'_{eq} = e^{10.5} = 3.63 \times 10^4$

24. Energy Span of the Respiratory Chain Electron transfer in the mitochondrial respiratory chain may be represented by the net reaction equation

$$\text{NADH} + \text{H}^+ + \tfrac{1}{2}\text{O}_2 \rightleftharpoons \text{H}_2\text{O} + \text{NAD}^+$$

(a) Calculate $\Delta E'^{\circ}$ for the net reaction of mitochondrial electron transfer. Use E'° values from Table 13–7.

(b) Calculate $\Delta G'^{\circ}$ for this reaction.

(c) How many ATP molecules can *theoretically* be generated by this reaction if the free energy of ATP synthesis under cellular conditions is 52 kJ/mol?

Answer

(a) Using E'° values from Table 13–7:

For

$\text{NADH} + \text{H}^+ + \tfrac{1}{2}\,\text{O}_2 \longrightarrow \text{H}_2\text{O} + \text{NAD}^+$

$\Delta E'^{\circ} = (E'^{\circ} \text{ for } \tfrac{1}{2}\,\text{O}_2/\text{H}_2\text{O}) - (E'^{\circ} \text{ for NAD}^+/\text{NADH})$

$= 0.816 \text{ V} - (-0.320 \text{ V}) = 1.14 \text{ V}$

(b) $\Delta G'^{\circ} = -n\mathcal{J}\,\Delta E'^{\circ}$

$= -2(96.5 \text{ kJ/V} \cdot \text{mol})(1.14 \text{ V})$

$= -220 \text{ kJ/mol}$

(c) For ATP synthesis, the reaction is

$$\text{ADP} + \text{P}_i \longrightarrow \text{ATP}$$

The free energy required for this reaction in the cell is 52 kJ/mol. Thus, the number of ATP molecules that could, in theory, be generated is

$$\frac{220 \text{ kJ/mol}}{52 \text{ kJ/mol}} = 4.2 \approx 4$$

25. Dependence of Electromotive Force on Concentrations Calculate the electromotive force (in volts) registered by an electrode immersed in a solution containing the following mixtures of NAD^+ and NADH at pH 7.0 and 25 °C, with reference to a half-cell of E'° 0.00 V.

(a) 1.0 mM NAD^+ and 10 mM NADH

(b) 1.0 mM NAD^+ and 1.0 mM NADH

(c) 10 mM NAD^+ and 1.0 mM NADH

Answer The relevant equation for calculating E for this system is

$$E = E'^{\circ} + \frac{RT}{n\mathcal{J}} \ln \frac{[\text{NAD}^+]}{[\text{NADH}]}$$

At 25 °C, the $RT/n\mathcal{J}$ term simplifies to 0.026 V/n.

(a) From Table 13–7, E'° for the NAD^+/NADH redox pair is −0.320 V. Because two electrons are transferred, $n = 2$. Thus,

$E = (-0.320 \text{ V}) + (0.026 \text{ V/2}) \ln (1 \times 10^{-3})/(10 \times 10^{-3})$

$= -0.320 \text{ V} + (-\ 0.03 \text{ V}) = -0.35 \text{ V}$

(b) The conditions specified here are "standard conditions," so we expect that $E = E'^\circ$. As proof, we know that ln 1 = 0, so under standard conditions the term $[(0.026 \text{ V}/n) \ln 1] = 0$, and $E = E'^\circ = -0.320$ V.

(c) Here the concentration of NAD^+ (the electron acceptor) is 10 times that of NADH (the electron donor). This affects the value of E:

$$E = (-0.320 \text{ V}) + (0.026/2 \text{ V}) \ln (10 \times 10^{-3})/(1 \times 10^{-3})$$
$$= -0.320 \text{ V} + 0.03 \text{ V} = -0.29 \text{ V}$$

26. Electron Affinity of Compounds List the following in order of increasing tendency to accept electrons: (a), α-ketoglutarate + CO_2 (yielding isocitrate); (b), oxaloacetate; (c), O_2; (d), $NADP^+$.

Answer To solve this problem, first write the half-reactions as in Table 13–7, and then find the value for E'° for each. Pay attention to the sign!

Half-reaction	E'° (V)
(a) α-Ketoglutarate + CO_2 + $2H^+$ + $2e^-$ $\longrightarrow$ isocitrate	−0.38
(b) Oxaloacetate + $2H^+$ + $2e^-$ $\longrightarrow$ malate	−0.166
(c) $\frac{1}{2}O_2$ + $2H^+$ + $2e^-$ $\longrightarrow$ H_2O	+0.816
(d) $NADP^+$ + H^+ + $2e^-$ $\longrightarrow$ NADPH	−0.324

The more positive the E'°, the more likely the substance will accept electrons; thus, we can list the substances in order of increasing tendency to accept electrons: (a), (d), (b), (c).

27. Direction of Oxidation-Reduction Reactions Which of the following reactions would you expect to proceed in the direction shown, under standard conditions, assuming that the appropriate enzymes are present to catalyze them?

(a) Malate + NAD^+ $\longrightarrow$ oxaloacetate + NADH + H^+
(b) Acetoacetate + NADH + H^+ $\longrightarrow$ β-hydroxybutyrate + NAD^+
(c) Pyruvate + NADH + H^+ $\longrightarrow$ lactate + NAD^+
(d) Pyruvate + β-hydroxybutyrate $\longrightarrow$ lactate + acetoacetate
(e) Malate + pyruvate $\longrightarrow$ oxaloacetate + lactate
(f) Acetaldehyde + succinate $\longrightarrow$ ethanol + fumarate

Answer It is important to note that standard conditions do not exist in the cell. The value of $\Delta E'^\circ$, as calculated in this problem, gives an indication of whether a reaction would or would not occur in a cell without additional energy being added (usually from ATP); but $\Delta E'^\circ$ does not tell the entire story. The actual cellular concentrations of the electron donors and electron acceptors contribute significantly to the value of E'° (e.g., see Problem 25). Under nonstandard conditions, the potential can either add to an already favorable $\Delta E'^\circ$ or be such a large positive number as to "overwhelm" an unfavorable $\Delta E'^\circ$, making ΔE favorable.

To solve this problem, calculate the $\Delta E'^\circ$ for each reaction. $\Delta E'^\circ$ = (E'° of electron acceptor in the reaction) − (E'° of electron donor in the reaction). Use E'° values in Table 13–7.

(a) Not favorable.

$$\Delta E'^\circ = (E'^\circ \text{ for oxaloacetate/malate}) - (E'^\circ \text{ for NAD}^+/\text{NADH})$$
$$= -0.320 \text{ V} - (-0.166 \text{ V})$$
$$= -0.154 \text{ V}$$

(b) Not favorable.

$$\begin{aligned}\Delta E'^\circ &= (E'^\circ \text{ for acetoacetate}/\beta\text{-hydroxybutyrate}) - (E'^\circ \text{ for NAD}^+/\text{NADH})\\ &= (-0.346 \text{ V}) - (-0.320 \text{ V})\\ &= -0.026 \text{ V}\end{aligned}$$

(c) Favorable.

$$\begin{aligned}\Delta E'^\circ &= (E'^\circ \text{ for pyruvate/lactate}) - (E'^\circ \text{ for NAD}^+/\text{NADH})\\ &= -0.185 \text{ V} - (-0.320 \text{ V})\\ &= 0.135 \text{ V}\end{aligned}$$

(d) Favorable.

$$\begin{aligned}\Delta E'^\circ &= (E'^\circ \text{ for pyruvate/lactate}) - (E'^\circ \text{ for acetoacetate}/\beta\text{-hydroxybutyrate})\\ &= -0.185 \text{ V} - (-0.346 \text{ V})\\ &= 0.161 \text{ V}\end{aligned}$$

(e) Not favorable.

$$\begin{aligned}\Delta E'^\circ &= (E'^\circ \text{ for pyruvate/lactate}) - (E'^\circ \text{ for oxaloacetate/malate})\\ &= -0.185 \text{ V} - (-0.166 \text{ V})\\ &= -0.019 \text{ V}\end{aligned}$$

(f) Not favorable.

$$\begin{aligned}\Delta E'^\circ &= (E'^\circ \text{ for acetaldehyde/ethanol}) - (E'^\circ \text{for fumarate/succinate})\\ &= -0.197 \text{ V} - (+0.031 \text{ V})\\ &= -0.228 \text{ V}\end{aligned}$$

Data Analysis Problem

28. Thermodynamics Can Be Tricky Thermodynamics is a challenging area of study and one with many opportunities for confusion. An interesting example is found in an article by Robinson, Hampson, Munro, and Vaney, published in *Science* in 1993. Robinson and colleagues studied the movement of small molecules between neighboring cells of the nervous system through cell-to-cell channels (gap junctions). They found that the dyes Lucifer yellow (a small, negatively charged molecule) and biocytin (a small zwitterionic molecule) moved in only one direction between two particular types of glia (nonneuronal cells of the nervous system). Dye injected into astrocytes would rapidly pass into adjacent astrocytes, oligodendrocytes, or Müller cells, but dye injected into oligodendrocytes or Müller cells passed slowly if at all into astrocytes. All of these cell types are connected by gap junctions.

Although it was not a central point of their article, the authors presented a molecular model for how this unidirectional transport might occur, as shown in their Figure 3:

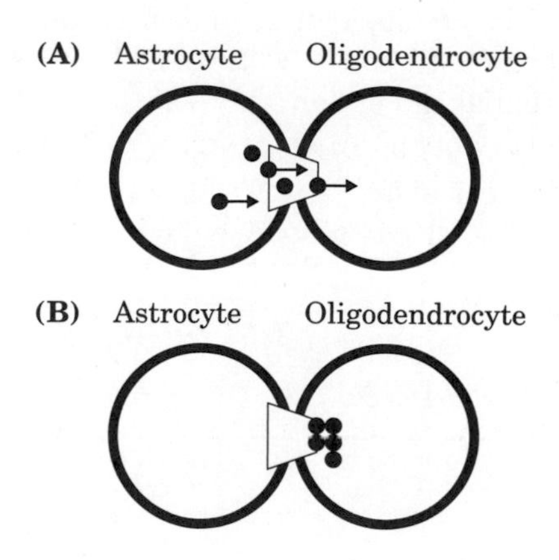

The figure legend reads: "Model of the unidirectional diffusion of dye between coupled oligodendrocytes and astrocytes, based on differences in connection pore diameter. Like a fish in a fish trap, dye molecules (black circles) can pass from an astrocyte to an oligodendrocyte (A) but not back in the other direction (B)."

Although this article clearly passed review at a well-respected journal, several letters to the editor (1994) followed, showing that Robinson and coauthors' model violated the second law of thermodynamics.

(a) Explain how the model violates the second law. Hint: Consider what would happen to the entropy of the system if one started with equal concentrations of dye in the astrocyte and oligodendrocyte connected by the "fish trap" type of gap junctions.

(b) Explain why this model cannot work for small molecules, although it may allow one to catch fish.

(c) Explain why a fish trap *does* work for fish.

(d) Provide two plausible mechanisms for the unidirectional transport of dye molecules between the cells that do not violate the second law of thermodynamics.

Answer

(a) The lowest-energy, highest-entropy state occurs when the dye concentration is the same in both cells. If a "fish trap" gap junction allowed unidirectional transport, more of the dye would end up in the oligodendrocyte and less in the astrocyte. This would be a higher-energy, lower-entropy state than the starting state, violating the second law of thermodynamics. Robinson et al.'s model requires an impossible spontaneous decrease in entropy. In terms of energy, the model entails a spontaneous change from a lower-energy to a higher-energy state without an energy input—again, thermodynamically impossible.

(b) Molecules, unlike fish, do not exhibit *directed behavior*; they move randomly by Brownian motion. Diffusion results in *net* movement of molecules from a region of higher concentration to a region of lower concentration simply because it is more likely that a molecule on the high-concentration side will enter the connecting channel. Look at this as a pathway with a rate-limiting step: the narrow end of the channel. The narrower end limits the rate at which molecules pass through because random motion of the molecules is less likely to move them through the smaller cross section. The wide end of the channel does *not* act like a funnel for molecules, although it may for fish, because molecules are not "crowded" by the sides of the narrowing funnel as fish would be. The narrow end limits the rate of movement equally in both directions. When the concentrations on both sides are equal, the rates of movement in both directions are equal and there will be no change in concentration.

(c) Fish exhibit *nonrandom behavior*, adjusting their actions in response to the environment. Fish that enter the large opening of the channel tend to move forward because fish have behavior that tends to make them prefer forward movement, and they experience "crowding" as they move through the narrowing channel. It is easy for fish to enter the large opening, but they don't move out of the trap as readily because they are less likely to enter the small opening.

(d) There are many possible explanations, some of which were proposed by the letter-writers who criticized the article. Here are two. (1) *The dye could bind to a molecule in the oligodendrocyte*. Binding effectively removes the dye from the bulk solvent, so it doesn't "count" as a solute for thermodynamic considerations yet remains visible in the fluorescence microscope. (2) *The dye could be sequestered in a subcellular organelle of the oligodendrocyte*, either actively pumped at the expense of ATP or drawn in by its attraction to other molecules in that organelle.

References

Letters to the editor. (1994) *Science* **265,** 1017–1019.

Robinson, S.R., Hampson, E.C.G.M., Munro, M.N., & Vaney, D.I. (1993) Unidirectional coupling of gap junctions between neuroglia. *Science* **262,** 1072–1074.

chapter

14 Glycolysis, Gluconeogenesis, and the Pentose Phosphate Pathway

1. **Equation for the Preparatory Phase of Glycolysis** Write balanced biochemical equations for all the reactions in the catabolism of glucose to two molecules of glyceraldehyde 3-phosphate (the preparatory phase of glycolysis), including the standard free-energy change for each reaction. Then write the overall or net equation for the preparatory phase of glycolysis, with the net standard free-energy change.

   Answer The initial phase of glycolysis requires ATP; it is endergonic. There are five reactions in this phase:

   1. Glucose + ATP $\longrightarrow$ glucose 6-phosphate + ADP $\quad \Delta G'^\circ = -16.7$ kJ/mol
   2. Glucose 6-phosphate $\longrightarrow$ fructose 6-phosphate $\quad \Delta G'^\circ = 1.7$ kJ/mol
   3. Fructose 6-phosphate + ATP $\longrightarrow$ fructose 1,6-bisphosphate $\quad \Delta G'^\circ = 14.2$ kJ/mol
   4. Fructose 1,6-bisphosphate $\longrightarrow$ dihydroxyacetone phosphate + glyceraldehyde 3-phosphate $\quad \Delta G'^\circ = 23.8$ kJ/mol
   5. Dihydroxyacetone phosphate $\longrightarrow$ glyceraldehyde 3-phosphate $\quad \Delta G'^\circ = 7.5$ kJ/mol

   The net equation for this phase is

   $$\text{Glucose} + 2\text{ATP} \longrightarrow 2 \text{ glyceraldehyde 3-phosphate} + 2\text{ADP} + 2\text{H}^+$$

   The overall standard free-energy change can be calculated by summing the individual reactions: $\Delta G'^\circ = 2.1$ kJ/mol (endergonic).

2. **The Payoff Phase of Glycolysis in Skeletal Muscle** In working skeletal muscle under anaerobic conditions, glyceraldehyde 3-phosphate is converted to pyruvate (the payoff phase of glycolysis), and the pyruvate is reduced to lactate. Write balanced biochemical equations for all the reactions in this process, with the standard free-energy change for each reaction. Then write the overall or net equation for the payoff phase of glycolysis (with lactate as the end product), including the net standard free-energy change.

   Answer The payoff phase of glycolysis produces ATP, and thus is exergonic. This phase consists of five reactions, designated 6 to 10 in the text:

   6. Glyceraldehyde 3-phosphate + P_i + NAD^+ $\longrightarrow$ 1,3-bisphosphoglycerate + NADH + H^+ $\quad \Delta G'^\circ = 6.3$ kJ/mol
   7. 1,3-Bisphosphoglycerate + ADP $\longrightarrow$ 3-phosphoglycerate + ATP $\quad \Delta G'^\circ = -185$ kJ/mol
   8. 3-Phosphoglycerate $\longrightarrow$ 2-phosphoglycerate $\quad \Delta G'^\circ = 4.4$ kJ/mol
   9. 2-Phosphoglycerate $\longrightarrow$ phosphoenolpyruvate $\quad \Delta G'^\circ = 7.5$ kJ/mol
   10. Phosphoenolpyruvate + ADP $\longrightarrow$ pyruvate + ATP $\quad \Delta G'^\circ = -31.4$ kJ/mol

   The pyruvate is then converted to lactate:

   $$\text{Pyruvate} + \text{NADH} + \text{H}^+ \longrightarrow \text{lactate} + \text{NAD}^+ \qquad \Delta G'^\circ = -25.1 \text{ kJ/mol}$$

The net equation is

$$\text{Glyceraldehyde 3-phosphate} + 2\text{ADP} + \text{P}_\text{i} \longrightarrow \text{lactate} + \text{NAD}^+ \qquad \Delta G'^\circ = -57 \text{ kJ/mol}$$

Because the payoff phase uses two glyceraldehyde 3-phosphate molecules from each glucose entering glycolysis, the net equation is

$$2 \text{ Glyceraldehyde 3-phosphate} + 4\text{ADP} + 2\text{P}_\text{i} \longrightarrow 2 \text{ lactate} + 2\text{NAD}^+$$

and the energetic payoff for the net reaction is $\Delta G'^\circ = -114$ kJ/mol.

3. GLUT Transporters Compare the localization of GLUT4 with that of GLUT2 and GLUT3, and explain why these localizations are important in the response of muscle, adipose tissue, brain, and liver to insulin.

Answer GLUT2 (and GLUT1) is found in liver and is always present in the plasma membrane of hepatocytes. GLUT3 is always present in the plasma membrane of certain brain cells. GLUT4 is normally sequestered in vesicles in cells of muscle and adipose tissue and enters the plasma membrane only in response to insulin. Thus, liver and brain can take up glucose from blood regardless of insulin level, but muscle and adipose tissue take up glucose only when insulin levels are elevated in response to high blood glucose.

4. Ethanol Production in Yeast When grown anaerobically on glucose, yeast (*S. cerevisiae*) converts pyruvate to acetaldehyde, then reduces acetaldehyde to ethanol using electrons from NADH. Write the equation for the second reaction, and calculate its equilibrium constant at 25 °C, given the standard reduction potentials in Table 13–7.

Answer

$$\underset{\text{Acetaldehyde}}{\text{CH}_3\text{CHO}} + \text{NADH} + \text{H}^+ \rightleftharpoons \underset{\text{Ethanol}}{\text{CH}_3\text{CH}_2\text{OH}} + \text{NAD}^+$$

Solve for K'_{eq} using the E'° values in Table 13–7 and Equations 13–3 and 13–7.

$$\Delta G'^\circ = -RT \ln K'_{eq}$$

$$\Delta G'^\circ = -n\mathcal{F}\Delta E'^\circ$$

$$RT \ln K'_{eq} = n\mathcal{F}\Delta E'^\circ$$

$$\ln K'_{eq} = \frac{n\mathcal{F}\Delta E'^\circ}{RT}$$

In this reaction, $n = 2$, and $\Delta E'^\circ = 0.123$ V (calculated from values in Table 13–7 as shown in Worked Example 13–3). Substitute the standard values for the faraday and R, and 298 K for the temperature:

$$\ln K'_{eq} = \frac{2(96{,}480 \text{ J/V} \cdot \text{mol})(0.123 \text{ V})}{(8.315 \text{ J/mol} \cdot \text{K})(298 \text{ K})} = 9.58$$

$$K'_{eq} = e^{9.58} = 1.45 \times 10^4$$

5. Energetics of the Aldolase Reaction Aldolase catalyzes the glycolytic reaction

$$\text{Fructose 1,6-bisphosphate} \longrightarrow \text{glyceraldehyde 3-phosphate} + \text{dihydroxyacetone phosphate}$$

The standard free-energy change for this reaction in the direction written is +23.8 kJ/mol. The concentrations of the three intermediates in the hepatocyte of a mammal are: fructose 1,6-bisphosphate, 1.4×10^{-5} M; glyceraldehyde 3-phosphate, 3×10^{-6} M; and dihydroxyacetone phosphate, 1.6×10^{-5} M. At body temperature (37 °C), what is the actual free-energy change for the reaction?

Answer For this reaction, $\Delta G' = \Delta G'^{\circ} + RT \ln$ [glyceraldehyde 3-phosphate][dihydroxyacetone phosphate]/[fructose 1,6-bisphosphate]:

$$\Delta G' = 23.8 \text{ kJ/mol} + (8.315 \times 10^{-3} \text{ kJ/mol} \cdot \text{K})(310 \text{ K}) \ln [(3 \times 10^{-6})(1.6 \times 10^{-5})/(1.4 \times 10^{-5})]$$
$$= 23.8 \text{ kJ/mol} + (2.578 \text{ kJ/mol}) \ln (3.43 \times 10^{-6})$$
$$= 23.8 \text{ kJ/mol} + (2.578 \text{ kJ/mol})(-12.58)$$
$$= 23.8 \text{ kJ/mol} + (-32.4 \text{ kJ/mol}) = -8.6 \text{ kJ/mol}$$

6. **Pathway of Atoms in Fermentation** A "pulse-chase" experiment using ^{14}C-labeled carbon sources is carried out on a yeast extract maintained under strictly anaerobic conditions to produce ethanol. The experiment consists of incubating a small amount of ^{14}C-labeled substrate (the pulse) with the yeast extract just long enough for each intermediate in the fermentation pathway to become labeled. The label is then "chased" through the pathway by the addition of excess unlabeled glucose. The chase effectively prevents any further entry of labeled glucose into the pathway.
   (a) If [1-^{14}C]glucose (glucose labeled at C-1 with ^{14}C) is used as a substrate, what is the location of ^{14}C in the product ethanol? Explain.
   (b) Where would ^{14}C have to be located in the starting glucose to ensure that all the ^{14}C activity is liberated as $^{14}CO_2$ during fermentation to ethanol? Explain.

Answer Anaerobiosis requires the regeneration of NAD^+ from NADH in order to allow glycolysis to continue.

(a) Figure 14–6 illustrates the fate of the carbon atoms of glucose. C-1 (or C-6) becomes C-3 of glyceraldehyde 3-phosphate and subsequently pyruvate. When pyruvate is decarboxylated and reduced to ethanol, C-3 of pyruvate becomes the C-2 of ethanol ($^{14}CH_3$—CH_2—OH).

(b) If all the labeled carbon from glucose is converted to $^{14}CO_2$ during ethanol fermentation, the original label must have been on C-3 and/or C-4 of glucose, because these are converted to the carboxyl group of pyruvate.

7. **Heat from Fermentations** Large-scale industrial fermenters generally require constant, vigorous cooling. Why?

Answer Fermentation releases energy, some conserved in the form of ATP but much of it dissipated as heat. Unless the fermenter contents are cooled to counterbalance this heat production, the temperature would become high enough to kill the microorganisms.

8. **Fermentation to Produce Soy Sauce** Soy sauce is prepared by fermenting a salted mixture of soybeans and wheat with several microorganisms, including yeast, over a period of 8 to 12 months. The resulting sauce (after solids are removed) is rich in lactate and ethanol. How are these two compounds produced? To prevent the soy sauce from having a strong vinegar taste (vinegar is dilute acetic acid), oxygen must be kept out of the fermentation tank. Why?

Answer Soybeans and wheat contain starch, a polymer of glucose, which is broken down to glucose by the microorganisms. The glucose is then degraded to pyruvate via glycolysis. Because the process is carried out in the absence of oxygen (i.e., it is a fermentation), pyruvate is reduced to lactic acid and ethanol. If oxygen were present, pyruvate would be oxidized to acetyl-CoA and then to CO_2 and H_2O. Some of the acetyl-CoA, however, would also be hydrolyzed to acetic acid (vinegar) in the presence of oxygen.

9. Equivalence of Triose Phosphates ^{14}C-Labeled glyceraldehyde 3-phosphate was added to a yeast extract. After a short time, fructose 1,6-bisphosphate labeled with ^{14}C at C-3 and C-4 was isolated. What was the location of the ^{14}C label in the starting glyceraldehyde 3-phosphate? Where did the second ^{14}C label in fructose 1,6-bisphosphate come from? Explain.

Answer Problem 1 outlines the steps in glycolysis involving fructose 1,6-bisphosphate, glyceraldehyde 3-phosphate, and dihydroxyacetone phosphate. Keep in mind that the aldolase reaction is readily reversible and the triose phosphate isomerase reaction catalyzes extremely rapid interconversion of its substrates. Thus, the label at C-1 of glyceraldehyde 3-phosphate would equilibrate with C-1 of dihydroxyacetone phosphate ($\Delta G'^\circ = 7.5$ kJ/mol). Because the aldolase reaction has $\Delta G'^\circ = -23.8$ kJ/mol in the direction of hexose formation, fructose 1,6-bisphosphate would be readily formed, and labeled in C-3 and C-4 (see Fig. 14–6).

10. Glycolysis Shortcut Suppose you discovered a mutant yeast whose glycolytic pathway was shorter because of the presence of a new enzyme catalyzing the reaction

$$\text{Glyceraldehyde 3-phosphate} + H_2 \xrightarrow{NAD^+ \;\rightarrow\; NADH + H^+} \text{3-phosphoglycerate}$$

Would shortening the glycolytic pathway in this way benefit the cell? Explain.

Answer Under anaerobic conditions, the phosphoglycerate kinase and pyruvate kinase reactions are essential. The shortcut in the mutant yeast would bypass the formation of an acyl phosphate by glyceraldehyde 3-phosphate dehydrogenase and therefore would not allow the formation of 1,3-bisphosphoglycerate. Without the formation of a substrate for 3-phosphoglycerate kinase, no ATP would be formed. Under anaerobic conditions, the net reaction for glycolysis normally produces 2 ATP per glucose. In the mutant yeast, net production of ATP would be zero and growth could not occur. Under aerobic conditions, however, because the majority of ATP formation occurs via oxidative phosphorylation, the mutation would have no observable effect.

11. Role of Lactate Dehydrogenase During strenuous activity, the demand for ATP in muscle tissue is vastly increased. In rabbit leg muscle or turkey flight muscle, the ATP is produced almost exclusively by lactic acid fermentation. ATP is formed in the payoff phase of glycolysis by two reactions, promoted by phosphoglycerate kinase and pyruvate kinase. Suppose skeletal muscle were devoid of lactate dehydrogenase. Could it carry out strenuous physical activity; that is, could it generate ATP at a high rate by glycolysis? Explain.

Answer The key point here is that NAD^+ must be regenerated from NADH in order for glycolysis to continue. Some tissues, such as skeletal muscle, obtain almost all their ATP through the glycolytic pathway and are capable of short-term exercise only (see Box 14–2). In order to generate ATP at a high rate, the NADH formed during glycolysis must be oxidized. In the absence of significant amounts of O_2 in the tissues, lactate dehydrogenase converts pyruvate and NADH to lactate and NAD^+. In the absence of this enzyme, NAD^+ could not be regenerated and glycolytic production of ATP would stop—and as a consequence, muscle activity could not be maintained.

12. Efficiency of ATP Production in Muscle The transformation of glucose to lactate in myocytes releases only about 7% of the free energy released when glucose is completely oxidized to CO_2 and H_2O. Does this mean that anaerobic glycolysis in muscle is a wasteful use of glucose? Explain.

Answer The transformation of glucose to lactate occurs when myocytes are low in oxygen, and it provides a means of generating ATP under oxygen-deficient conditions. Because lactate can be transformed to pyruvate, glucose is not wasted: the pyruvate can be oxidized by aerobic reactions when oxygen becomes plentiful. This metabolic flexibility gives the organism a greater capacity to adapt to its environment.

13. Free-Energy Change for Triose Phosphate Oxidation The oxidation of glyceraldehyde 3-phosphate to 1,3-bisphosphoglycerate, catalyzed by glyceraldehyde 3-phosphate dehydrogenase, proceeds with an unfavorable equilibrium constant ($K'_{eq} = 0.08$; $\Delta G'^{\circ} = 6.3$ kJ/mol), yet the flow through this point in the glycolytic pathway proceeds smoothly. How does the cell overcome the unfavorable equilibrium?

Answer In organisms, where directional flow in a pathway is required, exergonic reactions are coupled to endergonic reactions to overcome unfavorable free-energy changes. The endergonic glyceraldehyde 3-phosphate dehydrogenase reaction is followed by the phosphoglycerate kinase reaction, which rapidly removes the product of the former reaction. Consequently, the dehydrogenase reaction does not reach equilibrium and its unfavorable free-energy change is thus circumvented. The net $\Delta G'^{\circ}$ of the two reactions, when coupled, is -18.5 kJ/mol $+ 6.3$ kJ/mol $= -12.2$ kJ/mol.

14. Arsenate Poisoning Arsenate is structurally and chemically similar to inorganic phosphate (P_i), and many enzymes that require phosphate will also use arsenate. Organic compounds of arsenate are less stable than analogous phosphate compounds, however. For example, acyl *arsenates* decompose rapidly by hydrolysis:

$$\mathrm{R{-}\overset{\overset{\displaystyle O}{\|}}{C}{-}O{-}\underset{\underset{\displaystyle O^-}{|}}{\overset{\overset{\displaystyle O}{\|}}{As}}{-}O^- + H_2O \longrightarrow R{-}\overset{\overset{\displaystyle O}{\|}}{C}{-}O^- + HO{-}\underset{\underset{\displaystyle O^-}{|}}{\overset{\overset{\displaystyle O}{\|}}{As}}{-}O^- + H^+}$$

On the other hand, acyl *phosphates,* such as 1,3-bisphosphoglycerate, are more stable and undergo further enzyme-catalyzed transformation in cells.

(a) Predict the effect on the net reaction catalyzed by glyceraldehyde 3-phosphate dehydrogenase if phosphate were replaced by arsenate.

(b) What would be the consequence to an organism if arsenate were substituted for phosphate? Arsenate is very toxic to most organisms. Explain why.

Answer

(a) In the presence of arsenate, the product of the glyceraldehyde 3-phosphate dehydrogenase reaction is 1-arseno-3-phosphoglycerate, which nonenzymatically decomposes to 3-phosphoglycerate and arsenate; the substrate for the phosphoglycerate kinase is therefore bypassed.

(b) No ATP can be formed in the presence of arsenate because 1,3-bisphosphoglycerate is not formed. Under anaerobic conditions, this would result in no net glycolytic synthesis of ATP. Arsenate poisoning can be used as a test for the presence of an acyl phosphate intermediate in a reaction pathway.

15. Requirement for Phosphate in Ethanol Fermentation In 1906 Harden and Young, in a series of classic studies on the fermentation of glucose to ethanol and CO_2 by extracts of brewer's yeast, made the following observations. (1) Inorganic phosphate was essential to fermentation; when the supply of phosphate was exhausted, fermentation ceased before all the glucose was used. (2) During fermentation under these conditions, ethanol, CO_2, and a hexose bisphosphate accumulated. (3) When arsenate was substituted for phosphate, no hexose bisphosphate accumulated, but the fermentation proceeded until all the glucose was converted to ethanol and CO_2.

(a) Why did fermentation cease when the supply of phosphate was exhausted?

(b) Why did ethanol and CO_2 accumulate? Was the conversion of pyruvate to ethanol and CO_2 essential? Why? Identify the hexose bisphosphate that accumulated. Why did it accumulate?

(c) Why did the substitution of arsenate for phosphate prevent the accumulation of the hexose bisphosphate yet allow fermentation to ethanol and CO_2 to go to completion? (See Problem 14.)

Answer Ethanol fermentation in yeast has the following overall equation

$$\text{Glucose} + 2\text{ADP} + 2\text{P}_\text{i} \longrightarrow 2\text{ ethanol} + 2\text{CO}_2 + 2\text{ATP} + 2\text{H}_2\text{O}$$

It is clear that phosphate is required for the continued operation of glycolysis and ethanol formation. In extracts to which glucose is added, fermentation proceeds until ADP and P_i (present in the extracts) are exhausted.

(a) Phosphate is required in the glyceraldehyde 3-phosphate dehydrogenase reaction, and glycolysis will stop at this step when P_i is exhausted. Because glucose remains, it will be phosphorylated by ATP, but P_i will not be released.

(b) Fermentation in yeast cells produces ethanol and CO_2 rather than lactate (see Box 14–3). Without these reactions (in the absence of oxygen), NADH would accumulate and no new NAD^+ would be available for further glycolysis (see Problem 11). The hexose bisphosphate that accumulates is fructose 1,6-bisphosphate; in terms of energetics, this intermediate lies at a "low point" or valley in the pathway, between the energy-input reactions that precede it and the energy-payoff reactions that follow.

(c) Arsenate replaces P_i in the glyceraldehyde 3-phosphate dehydrogenase reaction to yield an acyl arsenate, which spontaneously hydrolyzes. This prevents formation of fructose 1,6-bisphosphate and ATP but allows formation of 3-phosphoglycerate, which continues through the pathway.

16. Role of the Vitamin Niacin Adults engaged in strenuous physical activity require an intake of about 160 g of carbohydrate daily but only about 20 mg of niacin for optimal nutrition. Given the role of niacin in glycolysis, how do you explain the observation?

Answer Dietary niacin is used to synthesize NAD^+. Oxidations carried out by NAD^+ are part of cyclic oxidation-reduction processes, with NAD^+/NADH as an electron carrier. Because of this cycling, one molecule of NAD^+ can oxidize many thousands of molecules of glucose, and thus the dietary requirement for the precursor vitamin (niacin) is relatively small.

17. Synthesis of Glycerol Phosphate The glycerol 3-phosphate required for the synthesis of glycerophospholipids can be synthesized from a glycolytic intermediate. Propose a reaction sequence for this conversion.

Answer Glycerol 3-phosphate and dihydroxyacetone 3-phosphate differ only at C-2. A dehydrogenase with the cofactor NADH acting on dihydroxyacetone 3-phosphate would form glycerol 3-phospate.

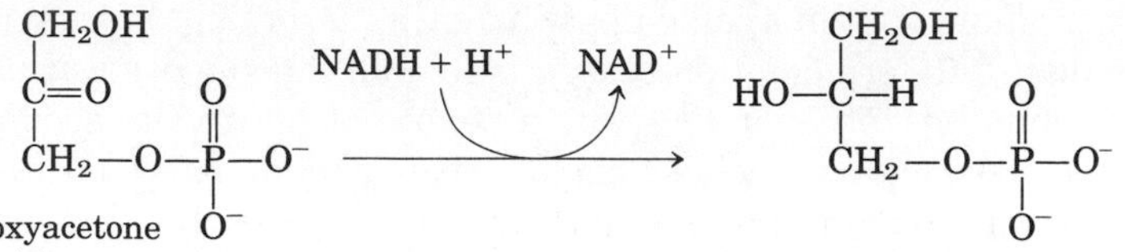

$$\underset{\text{Dihydroxyacetone phosphate}}{\begin{array}{l}CH_2OH\\ |\\ C{=}O\quad\ \ O\\ |\qquad\quad\ \ \|\\ CH_2{-}O{-}P{-}O^-\\ \qquad\quad\ \ |\\ \qquad\quad\ \ O^-\end{array}} \xrightarrow{NADH + H^+ \quad NAD^+} \underset{\text{Glycerol 3-phosphate}}{\begin{array}{l}\quad\ \ CH_2OH\\ \qquad |\\ HO{-}C{-}H\quad\ \ O\\ \qquad |\qquad\quad\ \ \|\\ \quad\ \ CH_2{-}O{-}P{-}O^-\\ \qquad\qquad\quad\ \ |\\ \qquad\qquad\quad\ \ O^-\end{array}}$$

In fact, the enzyme glycerol 3-phosphate dehydrogenase catalyzes this reaction (see Fig. 21–17).

18. Severity of Clinical Symptoms Due to Enzyme Deficiency The clinical symptoms of two forms of galactosemia—deficiency of galactokinase or of UDP-glucose:galactose 1-phosphate uridylyltransferase—show radically different severity. Although both types produce gastric discomfort after milk ingestion, deficiency of the transferase also leads to liver, kidney, spleen, and brain dysfunction and eventual death. What products accumulate in the blood and tissues with each type of enzyme deficiency? Estimate the relative toxicities of these products from the above information.

Answer In galactokinase deficiency, galactose accumulates; in UDP-glucose:galactose 1-phosphate uridylyltransferase deficiency, galactose 1-phosphate accumulates (see Fig. 14–12). The latter metabolite is clearly more toxic.

19. Muscle-Wasting in Starvation One consequence of starvation is a reduction in muscle mass. What happens to the muscle proteins?

Answer Muscle proteins are selectively degraded by proteases in myocytes, and the resulting amino acids move, in the bloodstream, from muscle to liver. In the liver, glucogenic amino acids are the starting materials for gluconeogenesis, to provide glucose for export to the brain (which cannot use fatty acids as fuel).

20. Pathway of Atoms in Gluconeogenesis A liver extract capable of carrying out all the normal metabolic reactions of the liver is briefly incubated in separate experiments with the following ^{14}C-labeled precursors:

(a) [^{14}C]Bicarbonate, $HO{-}^{14}C(=O){-}O^-$

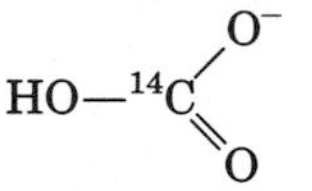

(b) [1-^{14}C]Pyruvate, $CH_3{-}C(=O){-}^{14}COO^-$

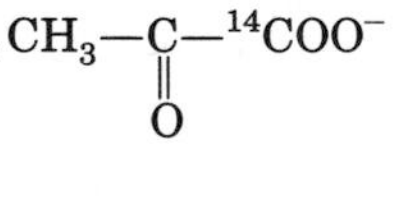

Trace the pathway of each precursor through gluconeogenesis. Indicate the location of ^{14}C in all intermediates and in the product, glucose.

Answer

(a) In the pyruvate carboxylase reaction, $^{14}CO_2$ is added to pyruvate to form [4-^{14}C]oxaloacetate, but the phosphoenolpyruvate carboxykinase reaction removes the *same* CO_2 in the next step. Thus, ^{14}C is not (initially) incorporated into glucose.

(b)

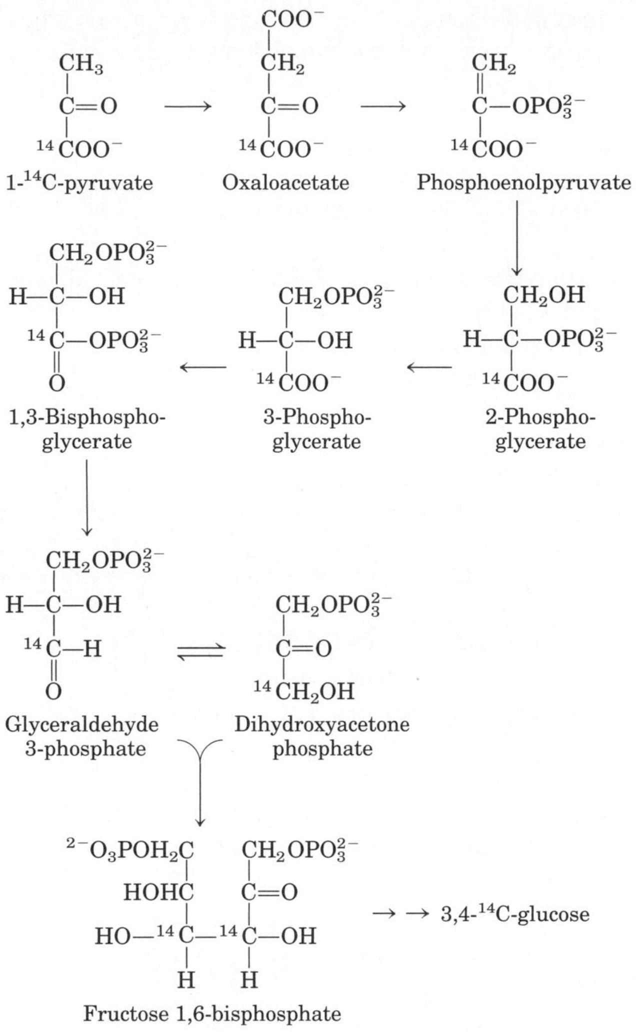

CH_3—C=O—$^{14}COO^-$
1-^{14}C-pyruvate
⟶
COO^-—CH_2—C=O—$^{14}COO^-$
Oxaloacetate
⟶
CH_2=C—OPO_3^{2-}—$^{14}COO^-$
Phosphoenolpyruvate
↓
CH_2OH—H—C—OPO_3^{2-}—$^{14}COO^-$
2-Phospho-glycerate
⟵
$CH_2OPO_3^{2-}$—H—C—OH—$^{14}COO^-$
3-Phospho-glycerate
⟵
$CH_2OPO_3^{2-}$—H—C—OH—^{14}C—OPO_3^{2-} (=O)
1,3-Bisphospho-glycerate
↓
$CH_2OPO_3^{2-}$—H—C—OH—^{14}C—H (=O)
Glyceraldehyde 3-phosphate
⇌
$CH_2OPO_3^{2-}$—C=O—$^{14}CH_2OH$
Dihydroxyacetone phosphate
↓
$^{2-}O_3POH_2C$—HOHC—HO—^{14}C(H)—^{14}C(H)—OH—C=O—$CH_2OPO_3^{2-}$
Fructose 1,6-bisphosphate
→ → 3,4-^{14}C-glucose

21. Energy Cost of a Cycle of Glycolysis and Gluconeogenesis What is the cost (in ATP equivalents) of transforming glucose to pyruvate via glycolysis and back again to glucose via gluconeogenesis?

Answer The overall reaction of glycolysis is

$$\text{Glucose} + 2\text{ADP} + 2\text{P}_i + 2\text{NAD}^+ \longrightarrow 2\text{ pyruvate} + 2\text{ATP} + 2\text{NADH} + 2\text{H}^+ + 2\text{H}_2\text{O}$$

The overall reaction of gluconeogenesis is

$$2\text{ Pyruvate} + 4\text{ATP} + 2\text{GTP} + 2\text{NADH} + 2\text{H}^+ + 4\text{H}_2\text{O} \longrightarrow \text{glucose} + 2\text{NAD}^+ + 4\text{ADP} + 2\text{GDP} + 6\text{P}_i$$

The cost of transforming glucose to pyruvate and back to glucose is given by the difference between these two equations:

$$2\text{ATP} + 2\text{GTP} + 2\text{H}_2\text{O} \longrightarrow 2\text{ADP} + 2\text{GDP} + 4\text{P}_i$$

The energy cost is four ATP equivalents per glucose molecule.

22. Relationship between Gluconeogenesis and Glycolysis Why is it important that gluconeogenesis is not the exact reversal of glycolysis?

Answer If gluconeogenesis were simply the reactions of glycolysis in reverse, the process would be energetically unfeasible (highly endergonic), because of the three reactions with large, negative standard free-energy changes in the catabolic (glycolytic) direction. Furthermore, if the same enzymes were used for all reactions in the two pathways, it would be impossible to regulate the two processes separately; anything that stimulated (or inhibited) the forward reaction for a given enzyme would stimulate (or inhibit) the reverse reaction to the same extent.

23. Energetics of the Pyruvate Kinase Reaction Explain in bioenergetic terms how the conversion of pyruvate to phosphoenolpyruvate in gluconeogenesis overcomes the large, negative standard free-energy change of the pyruvate kinase reaction in glycolysis.

Answer In converting pyruvate to PEP, the cell invests two ATP equivalents: ATP in the pyruvate carboxylase reaction, then GTP (equivalent to ATP) in the PEP carboxykinase reaction. By coupling the expenditure of two ATP equivalents to the conversion of pyruvate to PEP, the gluconeogenic process is made exergonic.

24. Glucogenic Substrates A common procedure for determining the effectiveness of compounds as precursors of glucose in mammals is to starve the animal until the liver glycogen stores are depleted and then administer the compound in question. A substrate that leads to a *net* increase in liver glycogen is termed glucogenic because it must first be converted to glucose 6-phosphate. Show by means of known enzymatic reactions which of the following substances are glucogenic:

(a) Succinate, $^{-}OOC{-}CH_2{-}CH_2{-}COO^{-}$

(b) Glycerol, $\underset{OH}{CH_2}{-}\underset{OH}{\overset{H}{C}}{-}\underset{OH}{CH_2}$ (OH on each carbon; H on the central C)

(c) Acetyl-CoA, $CH_3{-}\overset{O}{\overset{\|}{C}}{-}S\text{-}CoA$

(d) Pyruvate, $CH_3{-}\overset{O}{\overset{\|}{C}}{-}COO^{-}$

(e) Butyrate, $CH_3{-}CH_2{-}CH_2{-}COO^{-}$

Answer

(a) Glucogenic. In the citric acid cycle, succinate is converted to fumarate by succinate dehydrogenase, then to malate by fumarase, then to oxaloacetate by malate dehydrogenase. OAA can then leave the mitochondrion via the malate-aspartate shuttle, and in the cytosol is converted to PEP, which is glucogenic.

(b) Glucogenic. Glycerol kinase converts glycerol to glycerol 1-phosphate, which is then converted by a dehydrogenase (using NAD^+) to dihydroxyacetone phosphate, which is glucogenic.

(c) Not glucogenic. Higher animals do not have the enzymes to convert acetyl-CoA to pyruvate.

(d) Glucogenic. Pyruvate carboxylase converts pyruvate to oxaloacetate, which is used for gluconeogenesis as in **(a).**

(e) Not glucogenic. Butyrate is converted to butyryl-CoA by an acyl-CoA synthetase, and a single turn of the β-oxidation pathway converts butyryl-CoA to two molecules of acetyl-CoA, which is not glucogenic.

25. Ethanol Affects Blood Glucose Levels The consumption of alcohol (ethanol), especially after periods of strenuous activity or after not eating for several hours, results in a deficiency of glucose in the blood, a condition known as hypoglycemia. The first step in the metabolism of ethanol by the liver is oxidation to acetaldehyde, catalyzed by liver alcohol dehydrogenase:

$$CH_3CH_2OH + NAD^+ \longrightarrow CH_3CHO + NADH + H^+$$

Explain how this reaction inhibits the transformation of lactate to pyruvate. Why does this lead to hypoglycemia?

Answer The first step in the synthesis of glucose from lactate in the liver is oxidation of the lactate to pyruvate; like the oxidation of ethanol to acetaldehyde, this requires NAD^+. Consumption of alcohol forces a competition for NAD^+ between ethanol metabolism and gluconeogenesis, reducing the conversion of lactate to glucose and resulting in hypoglycemia. The problem is compounded by strenuous exercise and lack of food because at these times the level of blood glucose is already low.

26. Blood Lactate Levels during Vigorous Exercise The concentrations of lactate in blood plasma before, during, and after a 400 m sprint are shown in the graph.

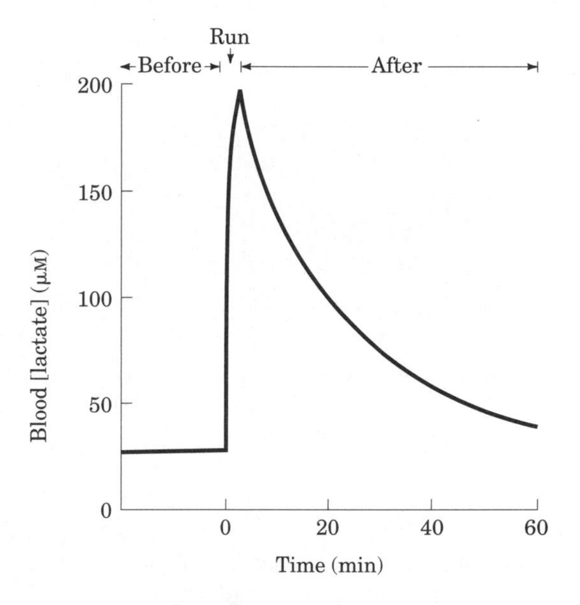

(a) What causes the rapid rise in lactate concentration?

(b) What causes the decline in lactate concentration after completion of the sprint? Why does the decline occur more slowly than the increase?

(c) Why is the concentration of lactate not zero during the resting state?

Answer

(a) Rapid depletion of ATP during strenuous muscular exertion causes the rate of glycolysis to increase dramatically, producing higher cytosolic concentrations of pyruvate and NADH; lactate dehydrogenase converts these to lactate and NAD^+ (lactic acid fermentation).

(b) When energy demands are reduced, the oxidative capacity of the mitochondria is again adequate, and lactate is transformed to pyruvate by lactate dehydrogenase, and the pyruvate is converted to glucose. The rate of the dehydrogenase reaction is slower in this direction because of the limited availability of NAD^+ and because the equilibrium of the reaction is strongly in favor of lactate (conversion of lactate to pyruvate is energy-requiring).

(c) The equilibrium of the lactate dehydrogenase reaction

$$\text{Pyruvate} + \text{NADH} + \text{H}^+ \longrightarrow \text{lactate} + \text{NAD}^+$$

is *strongly* in favor of lactate. Thus, even at very low concentrations of NADH and pyruvate, there is a significant concentration of lactate.

27. Relationship between Fructose 1,6-Bisphosphatase and Blood Lactate Levels A congenital defect in the liver enzyme fructose 1,6-bisphosphatase results in abnormally high levels of lactate in the blood plasma. Explain.

Answer In the liver, lactate is converted to pyruvate and then to glucose by gluconeogenesis (see Figs 14–15, 14–16). This pathway includes the glycolytic bypass step catalyzed by fructose 1,6-bisphosphatase (FBPase-1). A defect in this enzyme would prevent the entry of lactate into the gluconeogenic pathway in hepatocytes, causing lactate to accumulate in the blood.

28. Effect of Phloridzin on Carbohydrate Metabolism Phloridzin, a toxic glycoside from the bark of the pear tree, blocks the normal reabsorption of glucose from the kidney tubule, thus causing blood glucose to be almost completely excreted in the urine. In an experiment, rats fed phloridzin and sodium succinate excreted about 0.5 mol of glucose (made by gluconeogenesis) for every 1 mol of sodium succinate ingested. How is the succinate transformed to glucose? Explain the stoichiometry.

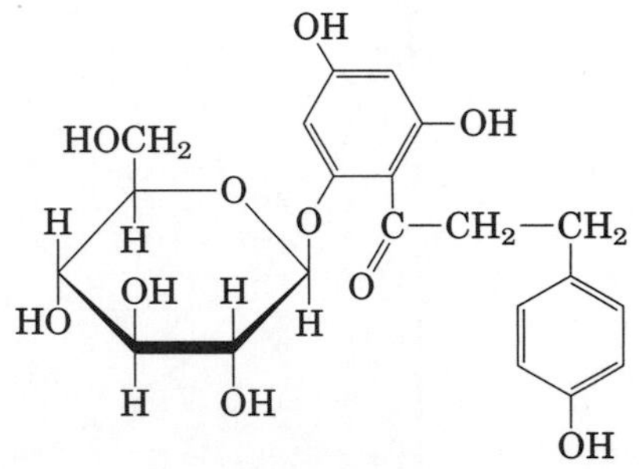

Phloridzin

Answer Excretion of glucose promoted by phloridzin causes a drop in blood glucose, which stimulates gluconeogenesis. The ingested succinate enters the mitochondrion via the dicarboxylate transport system and is transformed to oxaloacetate by enzymes of the citric acid cycle. The oxaloacetate passes into the cytosol and is transformed to phosphoenolpyruvate by PEP carboxykinase. Two moles of PEP are then required to produce a mole of glucose by the route outlined in Figure 14–16, consistent with the observed stoichiometry. Note that the rate of glucose production must be much higher than the rate of utilization by tissues because almost 100% of the glucose is excreted.

29. Excess O_2 Uptake during Gluconeogenesis Lactate absorbed by the liver is converted to glucose, with the input of 6 mol of ATP for every mole of glucose produced. The extent of this process in a rat liver preparation can be monitored by administering [^{14}C]lactate and measuring the amount of [^{14}C]glucose produced. Because the stoichiometry of O_2 consumption and ATP production is known (about 5 ATP per O_2), we can predict the extra O_2 consumption above the normal rate when a given amount of lactate is administered. However, when the extra O_2 used in the synthesis of glucose from lactate is actually measured, it is always higher than predicted by known stoichiometric relationships. Suggest a possible explanation for this observation.

Answer If the catabolic and biosynthetic pathways operate simultaneously, a certain amount of ATP is consumed in "futile cycles" (or "substrate cycles") in which no useful work is done. Examples of such cycles are that between glucose and glucose 6-phosphate and that between fructose 6-phosphate and fructose 1,6-bisphosphate. The net hydrolysis of ATP to ADP and P_i increases the consumption of oxygen, the terminal electron acceptor in oxidative phosphorylation.

30. Role of the Pentose Phosphate Pathway If the oxidation of glucose 6-phosphate via the pentose phosphate pathway were being used primarily to generate NADPH for biosynthesis, the other product, ribose 5-phosphate, would accumulate. What problems might this cause?

> **Answer** At the very least, accumulation of ribose 5-phosphate would tend to force this reaction in the reverse direction by mass action (see Eqn 13–4, p. 493). It might also affect other metabolic reactions that involve ribose 5-phosphate as a substrate or product—such as the pathways of nucleotide synthesis.

Data Analysis Problem

31. Engineering a Fermentation System Fermentation of plant matter to produce ethanol for fuel is one potential method for reducing the use of fossil fuels and thus the CO_2 emissions that lead to global warming. Many microorganisms can break down cellulose then ferment the glucose to ethanol. However, many potential cellulose sources, including agricultural residues and switchgrass, also contain substantial amounts of arabinose, which is not as easily fermented.

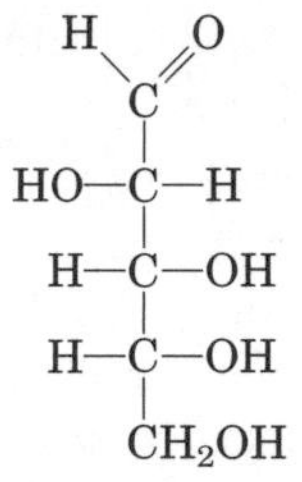

```
  H   O
   \ //
    C
    |
HO—C—H
    |
 H—C—OH
    |
 H—C—OH
    |
   CH2OH
```

D-Arabinose

Escherichia coli is capable of fermenting arabinose to ethanol, but it is not naturally tolerant of high ethanol levels, thus limiting its utility for commercial ethanol production. Another bacterium, *Zymomonas mobilis*, is naturally tolerant of high levels of ethanol but cannot ferment arabinose. Deanda, Zhang, Eddy, and Picataggio (1996) described their efforts to combine the most useful features of these two organisms by introducing the *E. coli* genes for the arabinose-metabolizing enzymes into *Z. mobilis*.

(a) Why is this a simpler strategy than the reverse: engineering *E. coli* to be more ethanol-tolerant?

Deanda and colleagues inserted five *E. coli* genes into the *Z. mobilis* genome: *araA*, coding for L-arabinose isomerase, which interconverts L-arabinose and L-ribulose; *araB*, L-ribulokinase, which uses ATP to phosphorylate L-ribulose at C-5; *araD*, L-ribulose 5-phosphate epimerase, which interconverts L-ribulose 5-phosphate and L-xylulose 5-phosphate; *talB*, transaldolase; and *tktA*, transketolase.

(b) For each of the three *ara* enzymes, briefly describe the chemical transformation it catalyzes and, where possible, name an enzyme discussed in this chapter that carries out an analogous reaction.

The five *E. coli* genes inserted in *Z. mobilis* allowed the entry of arabinose into the nonoxidative phase of the pentose phosphate pathway (Fig. 14–22), where it was converted to glucose 6-phosphate and fermented to ethanol.

(c) The three *ara* enzymes eventually converted arabinose into which sugar?

(d) The product from part **(c)** feeds into the pathway shown in Figure 14–22. Combining the five *E. coli* enzymes listed above with the enzymes of this pathway, describe the overall pathway for the fermentation of 6 molecules of arabinose to ethanol.

(e) What is the stoichiometry of the fermentation of 6 molecules of arabinose to ethanol and CO_2? How many ATP molecules would you expect this reaction to generate?

(f) *Z. mobilis* uses a slightly different pathway for ethanol fermentation from the one described in this chapter. As a result, the expected ATP yield is only 1 ATP per molecule of arabinose. Although this is less beneficial for the bacterium, it is better for ethanol production. Why?

Another sugar commonly found in plant matter is xylose.

```
  H   O
   \ //
    C
    |
 H—C—OH
    |
HO—C—H
    |
 H—C—OH
    |
   CH2OH
```

D-Xylose

(g) What additional enzymes would you need to introduce into the modified *Z. mobilis* strain described above to enable it to use xylose as well as arabinose to produce ethanol? You don't need to name the enzymes (they may not even exist in the real world!); just give the reactions they would need to catalyze.

Answer

(a) Ethanol tolerance is likely to involve many more genes, and thus the engineering would be a much more involved project.

(b) L-Arabinose isomerase (the *araA* enzyme) converts an aldose to a ketose by moving the carbonyl of a nonphosphorylated sugar from C-1 to C-2. No analogous enzyme is discussed in this chapter; all the enzymes described here act on phosphorylated sugars. An enzyme that carries out a similar transformation with phosphorylated sugars is phosphohexose isomerase. L-Ribulokinase (*araB*) phosphorylates a sugar at C-5 by transferring the γ phosphate from ATP. Many such reactions are described in this chapter, including the hexokinase reaction. L-Ribulose 5-phosphate epimerase (*araD*) switches the —H and —OH groups on a chiral carbon of a sugar. No analogous reaction is described in the chapter, but it is described in Chapter 20 (see Fig. 20–1, p. 774).

(c) The three *ara* enzymes would convert arabinose to xylulose 5-phosphate by the following pathway:

$$\text{Arabinose} \xrightarrow{\text{L-arabinose isomerase}} \text{L-ribulose} \xrightarrow{\text{L-ribulokinase}} \text{L-ribulose 5-phosphate} \xrightarrow{\text{epimerase}} \text{xylulose 5-phosphate.}$$

(d) The arabinose is converted to xylulose 5-phosphate as in **(c),** which enters the pathway in Figure 14–22; the glucose 6-phosphate product is then fermented to ethanol and CO_2.

(e) 6 molecules of arabinose + 6 molecules of ATP are converted to 6 molecules of xylulose 5-phosphate, which feed into the pathway in Figure 14–22 to yield 5 molecules of glucose 6-phosphate, each of which is fermented to yield 3 ATP (they enter as glucose 6-phosphate, not glucose)—15 ATP in all. Overall, you would expect a yield of 15 ATP – 6 ATP = 9 ATP from the 6 arabinose molecules. The other products are 10 molecules of ethanol and 10 molecules of CO_2.

(f) Given the lower ATP yield, for an amount of growth (i.e., of available ATP) equivalent to growth without the added genes the engineered *Z. mobilis* must ferment more arabinose, and thus it produces more ethanol.

(g) One way to allow the use of xylose would be to add the genes for two enzymes: an analog of the *araD* enzyme that converts xylose to ribose by switching the —H and —OH on C-3, and an analog of the *araB* enzyme that phosphorylates ribose at C-5. The resulting ribose 5-phosphate would feed into the existing pathway.

Reference

Deanda, K., Zhang, M., Eddy, C., & Picataggio, S. (1996) Development of an arabinose-fermenting *Zymomonas mobilis* strain by metabolic pathway engineering. *Appl. Environ. Microbiol.* **62,** 4465–4470.

chapter

15

Principles of Metabolic Regulation

1. **Measurement of Intracellular Metabolite Concentrations** Measuring the concentrations of metabolic intermediates in a living cell presents great experimental difficulties—usually a cell must be destroyed before metabolite concentrations can be measured. Yet enzymes catalyze metabolic interconversions very rapidly, so a common problem associated with these types of measurements is that the findings reflect not the physiological concentrations of metabolites but the equilibrium concentrations. A reliable experimental technique requires all enzyme-catalyzed reactions to be instantaneously stopped in the intact tissue so that the metabolic intermediates do not undergo change. This objective is accomplished by rapidly compressing the tissue between large aluminum plates cooled with liquid nitrogen (-190 °C), a process called **freeze-clamping.** After freezing, which stops enzyme action instantly, the tissue is powdered and the enzymes are inactivated by precipitation with perchloric acid. The precipitate is removed by centrifugation, and the clear supernatant extract is analyzed for metabolites. To calculate intracellular concentrations, the intracellular volume is determined from the total water content of the tissue and a measurement of the extracellular volume.

   The intracellular concentrations of the substrates and products of the phosphofructokinase-1 reaction in isolated rat heart tissue are given in the table below.

Metabolite	Concentration μM*
Fructose 6-phosphate	87.0
Fructose 1,6-bisphosphate	22.0
ATP	11,400
ADP	1,320

Source: From Williamson, J.R. (1965) Glycolytic control mechanisms I: inhibition of glycolysis by acetate and pyruvate in the isolated, perfused rat heart. *J. Biol. Chem.* **240,** 2308–2321.

*Calculated as μmol/mL of intracellular water.

(a) Calculate Q, [fructose 1,6-bisphosphate][ADP]/[fructose 6-phosphate][ATP], for the PFK-1 reaction under physiological conditions.

(b) Given a $\Delta G'^{\circ}$ for the PFK-1 reaction of -14.2 kJ/mol, calculate the equilibrium constant for this reaction.

(c) Compare the values of Q and K'_{eq}. Is the physiological reaction near or far from equilibrium? Explain. What does this experiment suggest about the role of PFK-1 as a regulatory enzyme?

Answer

(a) The mass-action ratio, $Q = \dfrac{(22.0\ \mu\text{M})(1{,}320\ \mu\text{M})}{(87.0\ \mu\text{M})(11{,}400\ \mu\text{M})} = 0.0293$

(b) $\Delta G'^\circ = -RT \ln K'_{eq}$ (RT at 25 °C = 2.48 kJ/mol)

$$\ln K'_{eq} = -\Delta G'^\circ/RT$$
$$= -(-14.2 \text{ kJ/mol})/(2.48 \text{ kJ/mol})$$
$$= 5.73$$
$$K'_{eq} = e^{-5.73} = 308$$

(c) Because K'_{eq} is much greater than Q, it is clear that the PFK-1 reaction does not approach equilibrium in vivo, and thus the product of the reaction, fructose 1,6-bisphosphate, does not approximate an equilibrium concentration. Metabolic pathways are "open systems," operating under near–steady state conditions, with substrates flowing in and products flowing out at all times. Thus, all the fructose 1,6-bisphosphate formed is rapidly used or turned over. The PFK-1–catalyzed reaction, which is the rate-limiting step in glycolysis, is an excellent candidate for the critical regulatory point of the pathway.

2. Are All Metabolic Reactions at Equilibrium?

(a) Phosphoenolpyruvate (PEP) is one of the two phosphoryl group donors in the synthesis of ATP during glycolysis. In human erythrocytes, the steady-state concentration of ATP is 2.24 mM, that of ADP is 0.25 mM, and that of pyruvate is 0.051 mM. Calculate the concentration of PEP at 25 °C, assuming that the pyruvate kinase reaction (see Fig. 13–13) is at equilibrium in the cell.

(b) The physiological concentration of PEP in human erythrocytes is 0.023 mM. Compare this with the value obtained in **(a).** Explain the significance of this difference.

Answer

(a) First, we must calculate the overall $\Delta G'^\circ$ for the pyruvate kinase reaction by breaking this process into two reactions and summing the $\Delta G'^\circ$ values (from Table 13–6):

$$(1)\ \text{PEP} + H_2O \longrightarrow \text{pyruvate} + P_i \qquad \Delta G_1'^\circ = -61.9 \text{ kJ/mol}$$
$$(2)\ \text{ADP} + P_i \longrightarrow \text{ATP} + H_2O \qquad \Delta G_2'^\circ = 30.5 \text{ kJ/mol}$$

$$\text{Sum: PEP} + \text{ADP} \longrightarrow \text{pyruvate} + \text{ATP} \qquad \Delta G'^\circ_{sum} = -31.4 \text{ kJ/mol}$$

Assuming the reaction is at equilibrium, we can calculate K'_{eq}:

$$\Delta G'^\circ = -RT \ln K'_{eq}$$
$$\ln K'_{eq} = -\Delta G'^\circ/RT$$
$$= -(-31.4 \text{ kJ/mol})/(2.48 \text{ kJ/mol})$$
$$= 12.7$$
$$K'_{eq} = 3.28 \times 10^5$$

Because $K'_{eq} = \dfrac{[\text{pyruvate}][\text{ATP}]}{[\text{ADP}][\text{PEP}]}$, where all reactants and products are at their equilibrium concentrations,

$$[\text{PEP}] = \frac{[\text{pyruvate}][\text{ATP}]}{[\text{ADP}]\,K'_{eq}}$$
$$= \frac{(5.1 \times 10^{-5} \text{ M})(2.24 \times 10^{-3} \text{ M})}{(2.5 \times 10^{-4} \text{ M})(3.28 \times 10^5)}$$
$$= 1.4 \times 10^{-9} \text{M}$$

(b) The physiological [PEP] of 0.023 mM is

$$\frac{0.023 \times 10^{-3} \text{ M}}{1.4 \times 10^{-9} \text{ M}} = 16{,}000 \text{ times the equilibrium concentration}$$

This reaction, like many others in the cell, is *not* at equilibrium.

3. Effect of O_2 Supply on Glycolytic Rates The regulated steps of glycolysis in intact cells can be identified by studying the catabolism of glucose in whole tissues or organs. For example, the glucose consumption by heart muscle can be measured by artificially circulating blood through an isolated intact heart and measuring the concentration of glucose before and after the blood passes through the heart. If the circulating blood is deoxygenated, heart muscle consumes glucose at a steady rate. When oxygen is added to the blood, the rate of glucose consumption drops dramatically, then is maintained at the new, lower rate. Explain.

Answer In the absence of O_2, the ATP needs of the cell are met by anaerobic glucose metabolism (fermentation) to form lactate. This produces a maximum of 2 ATP per glucose. Because the aerobic metabolism of glucose produces far more ATP per glucose (by oxidative phosphorylation), far less glucose is needed to produce the same amount of ATP. The Pasteur effect was the first demonstration of the primacy of energy production—that is, of ATP levels—in controlling the rate of glycolysis.

4. Regulation of PFK-1 The effect of ATP on the allosteric enzyme PFK-1 is shown below. For a given concentration of fructose 6-phosphate, the PFK-1 activity increases with increasing concentrations of ATP, but a point is reached beyond which increasing the concentration of ATP inhibits the enzyme.

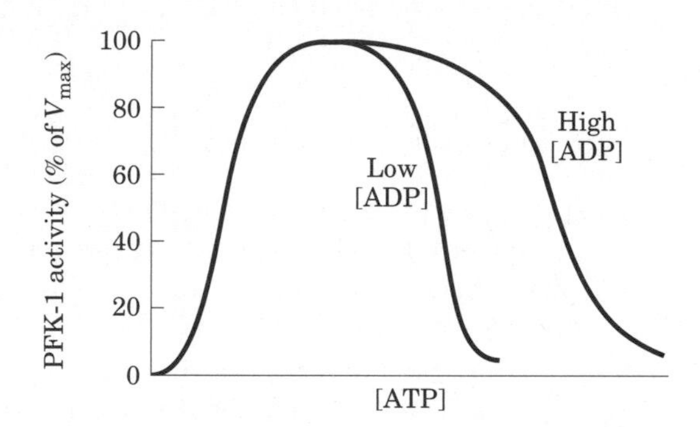

(a) Explain how ATP can be both a substrate and an inhibitor of PFK-1. How is the enzyme regulated by ATP?

(b) In what ways is glycolysis regulated by ATP levels?

(c) The inhibition of PFK-1 by ATP is diminished when the ADP concentration is high, as shown in the illustration. How can this observation be explained?

Answer

(a) In addition to their binding sites for substrate(s), allosteric enzymes have binding sites for regulatory metabolites. Binding of effectors to these regulatory sites modifies the enzyme's activity by altering its V_{max} or K_m value. ATP is a substrate and an allosteric inhibitor of PFK-1. Binding of ATP to the catalytic site increases activity, whereas binding to the allosteric site inhibits activity.

(b) Because ATP is a negative regulator of PFK-1, elevation of [ATP] when energy is abundant inhibits the enzyme and thus the flux of metabolites through the glycolytic pathway.

(c) The graph indicates that increased [ADP] suppresses the inhibition of PFK-1 by ATP. Because the total adenine nucleotide pool is more or less constant in all cells, utilization of ATP leads to an increase in [ADP]. The data show that the activity of the enzyme may be regulated in vivo by the ratio [ATP]/[ADP].

5. Cellular Glucose Concentration The concentration of glucose in human blood plasma is maintained at about 5 mM. The concentration of free glucose inside a myocyte is much lower. Why is the concentration so low in the cell? What happens to glucose after entry into the cell? Glucose is administered intravenously as a food source in certain clinical situations. Given that the transformation of glucose to glucose 6-phosphate consumes ATP, why not administer intravenous glucose 6-phosphate instead?

Answer Glucose enters cells and is immediately exposed to hexokinase, which converts glucose to glucose 6-phosphate using the energy of ATP. This reaction is highly exergonic ($\Delta G'^{\circ} = -16.7$ kJ/mol), and formation of glucose 6-phosphate is strongly favored. Glucose 6-phosphate is negatively charged and cannot diffuse across the membrane. The glucose transporter is specific for glucose; glucose 6-phosphate cannot leave the cell on this transporter and must be stored (by conversion to glycogen) or metabolized via glycolysis. Intravenous administration of glucose 6-phosphate is not useful because the phosphorylated glucose cannot enter cells on the glucose transporter or by diffusion.

6. Enzyme Activity and Physiological Function The V_{max} of the enzyme glycogen phosphorylase from skeletal muscle is much greater than the V_{max} of the same enzyme from liver tissue.

(a) What is the physiological function of glycogen phosphorylase in skeletal muscle? In liver tissue?

(b) Why does the V_{max} of the muscle enzyme need to be larger than that of the liver enzyme?

Answer

(a) The role of glycogen and glycogen metabolism differs in muscle and liver. *In muscle,* glycogen is broken down to supply energy (ATP), via glycolysis and lactic acid fermentation. Glycogen phosphorylase catalyzes the conversion of stored glycogen to glucose 1-phosphate, which is converted to glucose 6-phosphate and thus enters glycolysis. During strenuous activity, muscle becomes anaerobic and large quantities of glucose 6-phosphate undergo lactic acid fermentation to form the necessary ATP.

In the liver, glycogen is used to maintain the level of glucose in the blood (primarily between meals). In this case, the glucose 6-phosphate is converted to glucose and exported into the bloodstream.

(b) Strenuous muscular activity requires large amounts of ATP, which must be formed rapidly and efficiently, so glycogen phosphorylase must have a high ratio of V_{max}/K_m in muscle. This is not a critical requirement in liver tissue.

7. Glycogen Phosphorylase Equilibrium Glycogen phosphorylase catalyzes the removal of glucose from glycogen. The $\Delta G'^{\circ}$ for this reaction is 3.1 kJ/mol. **(a)** Calculate the ratio of [P_i] to [glucose 1-phosphate] when the reaction is at equilibrium. (Hint: The removal of glucose units from glycogen does not change the glycogen concentration.) **(b)** The measured ratio [P_i]/[glucose 1-phosphate] in myocytes under physiological conditions is more than 100:1. What does this indicate about the direction of metabolite flow through the glycogen phosphorylase reaction in muscle? **(c)** Why are the equilibrium and physiological ratios different? What is the possible significance of this difference?

Answer

(a) First, we need to calculate the equilibrium constant (assuming a temperature of 25 °C):

$$\begin{aligned} \Delta G'^{\circ} &= -RT \ln K'_{eq} \\ \ln K'_{eq} &= -\Delta G'^{\circ}/RT \\ &= -(3.1 \text{ kJ/mol})/(2.48 \text{ kJ/mol}) \\ &= -1.2 \\ K'_{eq} &= e^{-1.2} = 0.30 \end{aligned}$$

For the glycogen phosphorylase reaction:

$$\text{Glycogen}_n + P_i \longrightarrow \text{glycogen}_{n-1} + \text{glucose 1-phosphate}$$

$$K'_{eq} = \frac{[\text{glycogen}_{n-1}][\text{glucose 1-phosphate}]}{[\text{glycogen}_n][\text{P}_i]}$$

where all reactants are at their equilibrium concentrations. Because the concentration of glycogen remains constant, these terms cancel and the expression becomes

$$K'_{eq} = \frac{[\text{glucose 1-phosphate}]}{[\text{P}_i]}$$

This may be arranged to

$$\frac{[\text{P}_i]}{[\text{glucose 1-phosphate}]} = \frac{1}{K'_{eq}} = \frac{1}{0.3} = 3.3$$

(b) and **(c)** The high $[P_i]$/[glucose 1-phosphate] ratio in myocytes (>100:1) means that [glucose 1-phosphate] is far below the equilibrium value. The rate at which phosphoglucomutase removes glucose 1-phosphate (by conversion to glucose 6-phosphate, which enters glycolysis) is greater than the rate at which glycogen phosphorylase can produce it. This indicates that the direction of metabolite flow is from glycogen to glucose-1-phosphate, and that the glycogen phosphorylase reaction is the rate-limiting step in glycogen breakdown.

8. **Regulation of Glycogen Phosphorylase** In muscle tissue, the rate of conversion of glycogen to glucose 6-phosphate is determined by the ratio of phosphorylase *a* (active) to phosphorylase *b* (less active). Determine what happens to the rate of glycogen breakdown if a muscle preparation containing glycogen phosphorylase is treated with **(a)** phosphorylase kinase and ATP; **(b)** PP1; **(c)** epinephrine.

Answer

(a) Treatment with the kinase and ATP converts glycogen phosphorylase to the more active, phosphorylated form, phosphorylase *a;* glycogen breakdown accelerates.

(b) Treatment with the phosphatase PP1 converts the active phosphorylase *a* to the less active phosphorylase *b;* glycogen breakdown slows.

(c) Addition of epinephrine to muscle tissue causes the synthesis of cyclic AMP, which activates phosphorylase kinase. The kinase converts phosphorylase *b* (less active) to phosphorylase *a* (more active); glycogen breakdown accelerates.

9. **Glycogen Breakdown in Rabbit Muscle** The intracellular use of glucose and glycogen is tightly regulated at four points. To compare the regulation of glycolysis when oxygen is plentiful and when it is depleted, consider the utilization of glucose and glycogen by rabbit leg muscle in two physiological settings: a resting rabbit, with low ATP demands, and a rabbit that sights its mortal enemy, the coyote, and dashes into its burrow. For each setting, determine the relative levels (high, intermediate, or low) of AMP, ATP, citrate, and acetyl-CoA and describe how these levels affect the flow of metabolites through glycolysis by regulating specific enzymes. In periods of stress, rabbit leg muscle produces much of its ATP by anaerobic glycolysis (lactate fermentation) and very little by oxidation of acetyl-CoA derived from fat breakdown.

Answer A primary role of glycolysis is the production of ATP, and the pathway is regulated to ensure efficient ATP formation. The utilization of glycogen and glucose to supply energy is regulated at the following steps: glycogen phosphorylase, phosphofructokinase-1, pyruvate kinase, and entry of acetyl-CoA into the citric acid cycle. In muscle, the primary regulatory metabolites are ATP, AMP, citrate, and acetyl-CoA. ATP is an inhibitor of glycogen phosphorylase and PFK-1; AMP stimulates both. Citrate inhibits PFK-1, and acetyl-CoA inhibits pyruvate kinase. Lack of O_2 leads to elevated levels of NADH, inhibiting pyruvate dehydrogenase and promoting fermentation of pyruvate to lactate.

Under resting conditions, [ATP] is high and [AMP] low because the total adenine nucleotide pool is constant. [Citrate] and [acetyl-CoA] are intermediate because O_2 is not limiting and the citric acid cycle is functioning. Under conditions of active exertion (running), O_2 becomes limiting and ATP synthesis decreases. Consequently, [ATP] is relatively low and [AMP] relatively high, compared with aerobic conditions. [Citrate] and [acetyl-CoA] are low. These changes release the inhibition of glycolysis and stimulate lactic acid production.

10. Glycogen Breakdown in Migrating Birds Unlike the rabbit with its short dash, migratory birds require energy for extended periods of time. For example, ducks generally fly several thousand miles during their annual migration. The flight muscles of migratory birds have a high oxidative capacity and obtain the necessary ATP through the oxidation of acetyl-CoA (obtained from fats) via the citric acid cycle. Compare the regulation of muscle glycolysis during short-term intense activity, as in the fleeing rabbit, and during extended activity, as in the migrating duck. Why must the regulation in these two settings be different?

Answer Migratory birds have a very efficient respiratory system to ensure that O_2 is available to flight muscles under stress (see Box 14–2). Birds also rely on the aerobic oxidation of fat, which produces the greatest amount of energy per gram of fuel. The migratory bird must regulate glycolysis so that glycogen is used only for short bursts of energy, not for the long-term stress of prolonged flight. The sprinting rabbit relies on breakdown of stored (liver) glycogen and anaerobic glycolysis for short-term production of ATP for muscle activity.

The regulation of these two means of ATP production is very different. Under aerobic conditions (see answer to Problem 9), glycolysis is inhibited by the relatively high [ATP], as acetyl-CoA units derived from fat feed into the citric acid cycle and ATP is produced by oxidative phosphorylation. Under anaerobic conditions, glycolysis is stimulated and metabolism of fats does not occur at an appreciable rate because [citrate] and [acetyl-CoA] are low and O_2 (the final acceptor of electrons in oxidative phosphorylation) is absent.

11. Enzyme Defects in Carbohydrate Metabolism Summaries of four clinical case studies follow. For each case, determine which enzyme is defective and designate the appropriate treatment, from the lists provided at the end of the problem. Justify your choices. Answer the questions contained in each case study. (You may need to refer to information in Chapter 14.)

Case A The patient develops vomiting and diarrhea shortly after milk ingestion. A lactose tolerance test is administered. (The patient ingests a standard amount of lactose, and the glucose and galactose concentrations of blood plasma are measured at intervals. In individuals with normal carbohydrate metabolism, the levels increase to a maximum in about 1 hour, then decline.) The patient's blood glucose and galactose concentrations do not increase during the test. Why do blood glucose and galactose increase and then decrease during the test in healthy individuals? Why do they fail to rise in the patient?

Case B The patient develops vomiting and diarrhea after ingestion of milk. His blood is found to have a low concentration of glucose but a much higher than normal concentration of reducing sugars. The urine tests positive for galactose. Why is the concentration of reducing sugar in the blood high? Why does galactose appear in the urine?

Case C The patient complains of painful muscle cramps when performing strenuous physical exercise but has no other symptoms. A muscle biopsy indicates a muscle glycogen concentration much higher than normal. Why does glycogen accumulate?

Case D The patient is lethargic, her liver is enlarged, and a biopsy of the liver shows large amounts of excess glycogen. She also has a lower than normal blood glucose level. What is the reason for the low blood glucose in this patient?

Defective Enzyme

(a) Muscle PFK-1

(b) Phosphomannose isomerase

(c) Galactose 1-phosphate uridylyltransferase

(d) Liver glycogen phosphorylase

(e) Triose kinase
(f) Lactase in intestinal mucosa
(g) Maltase in intestinal mucosa
(h) Muscle debranching enzyme

Treatment
1. Jogging 5 km each day
2. Fat-free diet
3. Low-lactose diet
4. Avoiding strenuous exercise
5. Large doses of niacin (the precursor of NAD^+)
6. Frequent feedings (smaller portions) of a normal diet

Answer

***Case A:* (f)** Lactase in the intestinal mucosa hydrolyzes milk lactose to glucose and galactose, so the levels of these sugars increase transiently after milk ingestion. This increase would not occur in a person lacking lactase; the patient would show symptoms of lactose toxicity. The patient should exclude lactose (milk) from the diet (treatment **3**).

***Case B:* (c)** Galactose 1-phosphate uridylyltransferase is an enzyme involved in conversion of galactose to glucose, which then can enter glycolysis. Absence of this enzyme leads to accumulation of galactose in the blood and excretion in the urine. A patient with this deficiency should be on a low-lactose diet (treatment **3**).

***Case C:* (h)** Without the muscle form of glycogen phosphorylase, glycogen cannot be mobilized to supply ATP during extended exercise, and the result is weakness and cramping during the exercise. Because glycogen continues to be synthesized but is not mobilized, the muscle glycogen level is higher than normal. The patient should avoid strenuous exercise (treatment **4**).

***Case D:* (d)** Liver glycogen functions as a source of blood glucose. Accumulation of glycogen and low blood glucose suggest that the liver glycogen phosphorylase is defective or of low activity. The patient should eat light meals regularly and frequently (treatment **6**).

12. Effects of Insufficient Insulin in a Person with Diabetes A man with insulin-dependent diabetes is brought to the emergency room in a near-comatose state. While vacationing in an isolated place, he lost his insulin medication and has not taken any insulin for two days.

(a) For each tissue listed below, is each pathway faster, slower, or unchanged in this patient, compared with the normal level when he is getting appropriate amounts of insulin?

(b) For each pathway, describe at least one control mechanism responsible for the change you predict.

Tissue and Pathways
1. Adipose: fatty acid synthesis
2. Muscle: glycolysis; fatty acid synthesis; glycogen synthesis
3. Liver: glycolysis; gluconeogenesis; glycogen synthesis; fatty acid synthesis; pentose phosphate pathway

Answer

(a) (1) Adipose: fatty acid synthesis slower. (2) Muscle: glycolysis, fatty acid synthesis, and glycogen synthesis slower. (3) Liver: glycolysis faster; gluconeogenesis, glycogen synthesis, and fatty acid synthesis slower; pentose phosphate pathway unchanged.

(b) (1) Adipose and (3) liver: fatty acid synthesis slower because lack of insulin results in inactive acetyl-CoA carboxylase, the first enzyme of fatty acid synthesis. Glycogen synthesis inhibited by cAMP-dependent phosphorylation (thus activation) of glycogen synthase. (2) Muscle: glycolysis slower because GLUT4 is inactive, so glucose uptake is inhibited. (3) Liver: glycolysis slower because the bifunctional PFK-2/FBPase-2 is converted to the form with active FBPase-2, decreasing [fructose 2,6-bisphosphate], which allosterically stimulates phosphofructokinase and inhibits FBPase-1; this also accounts for the stimulation of gluconeogenesis.

13. Blood Metabolites in Insulin Insufficiency For the patient described in Problem 12, predict the levels of the following metabolites in his blood *before* treatment in the emergency room, relative to levels maintained during adequate insulin treatment: **(a)** glucose; **(b)** ketone bodies; **(c)** free fatty acids.

Answer

(a) Elevated; without insulin, muscle and other tissues cannot take up glucose, because the GLUT4 transporters do not move to the plasma membrane and ingested glucose remains in the blood.

(b) Elevated; accumulation of acetyl-CoA from fatty acid oxidation shifts the thiolase equilibrium toward acetoacetyl-CoA formation, the first step in ketone body formation.

(c) Elevated; to provide fuel for muscle and other tissues, fatty acids are mobilized from adipose and move through the blood to muscle and other tissues.

14. Metabolic Effects of Mutant Enzymes Predict and explain the effect on glycogen metabolism of each of the following defects caused by mutation: **(a)** loss of the cAMP-binding site on the regulatory subunit of protein kinase A (PKA); **(b)** loss of the protein phosphatase inhibitor (inhibitor 1 in Fig. 15–40); **(c)** overexpression of phosphorylase *b* kinase in liver; **(d)** defective glucagon receptors in liver.

Answer

(a) PKA cannot be activated in response to glucagon or epinephrine, and glycogen phosphorylase is not activated.

(b) PP1 remains active, allowing it to dephosphorylate glycogen synthase (activating it) and glycogen phosphorylase (inhibiting it). Glycogen synthesis is stimulated, and glycogen breakdown inhibited.

(c) Phosphorylase remains phosphorylated (active), increasing the breakdown of glycogen and thus depleting the store of liver glycogen.

(d) Gluconeogenesis cannot be stimulated when blood glucose is low, leading to dangerously low blood glucose during periods of fasting.

15. Hormonal Control of Metabolic Fuel Between your evening meal and breakfast, your blood glucose drops and your liver becomes a net producer rather than consumer of glucose. Describe the hormonal basis for this switch, and explain how the hormonal change triggers glucose production by the liver.

Answer The drop in blood glucose triggers release of glucagon by the pancreas. In the liver, glucagon activates glycogen phosphorylase by stimulating its cAMP-dependent phosphorylation and stimulates gluconeogenesis by lowering [fructose 2,6-bisphosphate], thus stimulating FBPase-1, a key enzyme in gluconeogenesis.

16. Altered Metabolism in Genetically Manipulated Mice Researchers can manipulate the genes of a mouse so that a single gene in a single tissue either produces an inactive protein (a "knockout" mouse) or produces a protein that is always (constitutively) active. What effects on metabolism would you predict for mice with the following genetic changes: **(a)** knockout of glycogen debranching enzyme in the liver; **(b)** knockout of hexokinase IV in liver; **(c)** knockout of FBPase-2 in liver; **(d)** constitutively active FBPase-2 in liver; **(e)** constitutively active AMPK in muscle; **(f)** constitutively active ChREBP in liver?

Answer

(a) Reduced capacity to mobilize glycogen; lowered blood glucose between meals

(b) Reduced capacity to lower blood glucose after a carbohydrate meal; elevated blood glucose (see Fig. 15–12)

(c) Reduced fructose 2,6-bisphosphate (F26BP) in liver, stimulating glycolysis and inhibiting gluconeogenesis (see Fig. 15–15)

(d) Reduced F26BP, stimulating gluconeogenesis and inhibiting glycolysis

(e) Increased uptake of fatty acids and glucose; increased oxidation of both (see Fig. 15–6)

(f) Increased conversion of pyruvate to acetyl-CoA; increased fatty acid synthesis

Data Analysis Problem

17. Optimal Glycogen Structure Muscle cells need rapid access to large amounts of glucose during heavy exercise. This glucose is stored in liver and skeletal muscle in polymeric form as particles of glycogen. The typical glycogen particle contains about 55,000 glucose residues (see Fig. 15–33b). Meléndez-Hevia, Waddell, and Shelton (1993) explored some theoretical aspects of the structure of glycogen, as described in this problem.

(a) The cellular concentration of glycogen in liver is about 0.01 μM. What cellular concentration of free glucose would be required to store an equivalent amount of glucose? Why would this concentration of free glucose present a problem for the cell?

Glucose is released from glycogen by glycogen phosphorylase, an enzyme that can remove glucose molecules, one at a time, from one end of a glycogen chain. Glycogen chains are branched (see Figs 15–26 and 15–33b), and the degree of branching—the number of branches per chain—has a powerful influence on the rate at which glycogen phosphorylase can release glucose.

(b) Why would a degree of branching that was too low (i.e., below an optimum level) reduce the rate of glucose release? (Hint: Consider the extreme case of no branches in a chain of 55,000 glucose residues.)

(c) Why would a degree of branching that was too high also reduce the rate of glucose release? (Hint: Think of the physical constraints.)

Meléndez-Hevia and colleagues did a series of calculations and found that two branches per chain (see Fig. 15–33b) was optimal for the constraints described above. This is what is found in glycogen stored in muscle and liver.

To determine the optimum number of glucose residues per chain, Meléndez-Hevia and coauthors considered two key parameters that define the structure of a glycogen particle: t = the number of tiers of glucose chains in a particle (the molecule in Fig. 15–33b has five tiers); g_c = the number of glucose residues in each chain. They set out to find the values of t and g_c that would maximize three quantities: (1) the amount of glucose stored in the particle (G_T) per unit volume, (2) the number of unbranched glucose chains (C_A) per unit volume (i.e., number of chains in the outermost tier, readily accessible to glycogen phosphorylase); and (3) the amount of glucose available to phosphorylase in these unbranched chains (G_{PT}).

(d) Show that $C_A = 2^{t-1}$. This is the number of chains available to glycogen phosphorylase before the action of the debranching enzyme.

(e) Show that C_T, the total number of chains in the particle, is given by $C_T = 2^t - 1$. Thus $G_T = g_c(C_T) = g_c(2^t - 1)$, the total number of glucose residues in the particle.

(f) Glycogen phosphorylase cannot remove glucose from glycogen chains that are shorter than five glucose residues. Show that $G_{PT} = (g_c - 4)(2^{t-1})$. This is the amount of glucose readily available to glycogen phosphorylase.

(g) Based on the size of a glucose residue and the location of branches, the thickness of one tier of glycogen is $0.12g_c$ nm + 0.35 nm. Show that the volume of a particle, V_s, is given by the equation $V_s = \frac{4}{3}\pi t^3(0.12g_c + 0.35)^3$ nm^3.

Meléndez-Hevia and coauthors then determined the optimum values of t and g_c—those that gave the maximum value of a quality function, f, that maximizes G_T, C_A, and G_{PT}, while minimizing V_S: $f = \frac{G_T C_A G_{PT}}{V_s}$. They found that the optimum value of g_c is independent of t.

(h) Choose a value of t between 5 and 15 and find the optimum value of g_c. How does this compare with the g_c found in liver glycogen (see Fig. 15–33b)? (Hint: You may find it useful to use a spreadsheet program.)

Answer

(a) Given that each particle contains about 55,000 glucose residues, the equivalent free glucose concentration would be 55,000 × 0.01 μM = 550 mM, or 0.55 M. This would present a serious osmotic challenge for the cell! (Body fluids have a substantially lower osmolarity.)

(b) The lower the number of branches, the lower the number of free ends available for glycogen phosphorylase activity, and the slower the rate of glucose release. With no branches, there would be just one site for phosphorylase to act.

(c) The outer tier of the particle would be too crowded with glucose residues for the enzyme to gain access to cleave bonds and release glucose.

(d) The number of chains doubles in each succeeding tier: tier 1 has one chain (2^0), tier 2 has two (2^1), tier 3 has four (2^2), and so on. Thus, for t tiers, the number of chains in the outermost tier, C_A, is 2^{t-1}.

(e) The total number of chains is $2^0 + 2^1 + 2^2 + \ldots 2^{t-1} = 2^t - 1$. Each chain contains g_c glucose molecules, so the total number of glucose molecules, C_T, is $g_c(2^t - 1)$.

(f) Glycogen phosphorylase can release all but four of the glucose residues in a chain of length g_c. Therefore, from each chain in the outer tier it can release ($g_c - 4$) glucose molecules. Given that there are 2^{t-1} chains in the outer tier, the number of glucose molecules the enzyme can release, G_{PT}, is $(g_c - 4)(2^{t-1})$.

(g) The volume of a sphere is $\frac{4}{3}\pi r^3$. In this case, r is the thickness of one tier times the number of tiers, or $(0.12g_c + 0.35)t$ nm. Thus $V_s = \frac{4}{3}\pi t^3(0.12g_c + 0.35)^3$ nm^3.

(h) You can show algebraically that the value of g_c that maximizes f is independent of t. Choosing $t = 3$:

g_c	C_A	G_T	G_{PT}	V_s	f
5	4	35	4	11	5.8
6	4	42	8	19	9.7
7	4	49	12	24	12
8	4	56	16	28	14
9	4	63	20	32	15
10	4	70	24	34	16
11	4	77	28	36	16
12	4	84	32	38	17
13	4	91	36	40	17
14	4	98	40	41	17
15	4	100	44	42	16
16	4	110	48	43	16

The optimum value of g_c (i.e., at maximum f) is 13. In nature, g_c varies from 12 to 14, which corresponds to f values very close to the optimum. If you choose another value for t, the numbers will differ but the optimal g_c will still be 13.

Reference

Meléndez-Hevia, E., Waddell, T.G., & Shelton, E.D. (1993) Optimization of molecular design in the evolution of metabolism: the glycogen molecule. *Biochem.* J. **295,** 477–483.

chapter

16

The Citric Acid Cycle

1. **Balance Sheet for the Citric Acid Cycle** The citric acid cycle has eight enzymes: citrate synthase, aconitase, isocitrate dehydrogenase, α-ketoglutarate dehydrogenase, succinyl-CoA synthetase, succinate dehydrogenase, fumarase, and malate dehydrogenase.

   (a) Write a balanced equation for the reaction catalyzed by each enzyme.

   (b) Name the cofactor(s) required by each enzyme reaction.

   (c) For each enzyme determine which of the following describes the type of reaction(s) catalyzed: condensation (carbon–carbon bond formation); dehydration (loss of water); hydration (addition of water); decarboxylation (loss of CO_2); oxidation-reduction; substrate-level phosphorylation; isomerization.

   (d) Write a balanced net equation for the catabolism of acetyl-CoA to CO_2.

Answer

Citrate synthase

(a) Acetyl-CoA + oxaloacetate + $H_2O \longrightarrow$ citrate + CoA

(b) CoA

(c) Condensation

Aconitase

(a) Citrate $\longrightarrow$ isocitrate

(b) No cofactors

(c) Isomerization

Isocitrate dehydrogenase

(a) Isocitrate + $NAD^+ \longrightarrow$ α-ketoglutarate + CO_2 + NADH

(b) NAD^+

(c) Oxidative decarboxylation

α-Ketoglutarate dehydrogenase

(a) α-Ketoglutarate + NAD^+ + CoA $\longrightarrow$ succinyl-CoA + CO_2 + NADH

(b) NAD^+, CoA, thiamine pyrophosphate

(c) Oxidative decarboxylation

Succinyl-CoA synthetase

(a) Succinyl-CoA + P_i + GDP $\longrightarrow$ succinate + CoA + GTP

(b) CoA

(c) Substrate-level phosphorylation and acyl transfer

Succinate dehydrogenase

(a) Succinate + FAD $\longrightarrow$ fumarate + $FADH_2$

(b) FAD

(c) Oxidation

Fumarase

(a) Fumarate + $H_2O \longrightarrow$ malate

(b) No cofactors

(c) Hydration

Malate dehydrogenase

(a) Malate + $NAD^+ \longrightarrow$ oxaloacetate + NADH + H^+

(b) NAD^+

(c) Oxidation

(d) The net equation for the catabolism of acetyl-CoA is

$$\text{Acetyl-CoA} + 3NAD^+ + FAD + GDP + P_i + 2H_2O \longrightarrow 2CO_2 + CoA + 3NADH + FADH_2 + GTP + 2H^+$$

2. Net Equation for Glycolysis and the Citric Acid Cycle Write the net biochemical equation for the metabolism of a molecule of glucose by glycolysis and the citric acid cycle, including all cofactors.

Answer

Glycolysis:

$$\text{Glucose} + 2ADP + 2P_i + 2NAD^+ \longrightarrow 2ATP + 2NADH + 2\text{ pyruvate}$$

Pyruvate dehydrogenase:

$$2\text{ Pyruvate} + 2NAD^+ + 2CoASH \longrightarrow 2\text{ acetyl-CoA} + 2CO_2 + 2NADH$$

Citric acid cycle:

$$2\text{ Acetyl-CoA} + 2FAD + 6NAD^+ + 2ADP + 2P_i \longrightarrow 2CoASH + 2FADH_2 + 6NADH + 2ATP + 4CO_2$$

Overall:

$$\text{Glucose} + 4ADP + 4P_i + 10NAD^+ + 2FAD \longrightarrow 4ATP + 10NADH + 2FADH_2 + 6CO_2$$

3. Recognizing Oxidation and Reduction Reactions One biochemical strategy of many living organisms is the stepwise oxidation of organic compounds to CO_2 and H_2O and the conservation of a major part of the energy thus produced in the form of ATP. It is important to be able to recognize oxidation-reduction processes in metabolism. Reduction of an organic molecule results from the hydrogenation of a double bond (Eqn 1, below) or of a single bond with accompanying cleavage (Eqn 2). Conversely, oxidation results from dehydrogenation. In biochemical redox reactions, the coenzymes NAD and FAD dehydrogenate/hydrogenate organic molecules in the presence of the proper enzymes.

$$\underset{\text{Acetaldehyde}}{CH_3{-}C({=}O){-}H} + H{-}H \underset{\text{oxidation}}{\overset{\text{reduction}}{\rightleftharpoons}} \left[CH_3{-}C(H)(\overset{\|}{O}\cdots H)\cdots H\right] \underset{\text{oxidation}}{\overset{\text{reduction}}{\rightleftharpoons}} \underset{\text{Ethanol}}{CH_3{-}C(H)(H){-}O{-}H} \qquad (1)$$

$$\underset{\text{Acetate}}{CH_3{-}C({=}O){-}O^-} + H^+ + H{-}H \underset{\text{oxidation}}{\overset{\text{reduction}}{\rightleftharpoons}} \left[CH_3{-}C({=}O)\cdots O^- \;\; H\cdots H \;\; H^+\right] \underset{\text{oxidation}}{\overset{\text{reduction}}{\rightleftharpoons}} \underset{\text{Acetaldehyde}}{CH_3{-}C({=}O){-}H} + H{-}O{-}H \qquad (2)$$

For each of the metabolic transformations in **(a)** through **(h)**, determine whether oxidation or reduction has occurred. Balance each transformation by inserting H—H and, where necessary, H_2O.

(a) $CH_3—OH \longrightarrow H—C(=O)—H$
Methanol → Formaldehyde

(b) $H—C(=O)—H \longrightarrow H—C(=O)—O^- + H^+$
Formaldehyde → Formate

(c) $O{=}C{=}O \longrightarrow H—C(=O)—O^- + H^+$
Carbon dioxide → Formate

(d) $CH_2(OH)—CH(OH)—C(=O)—O^- + H^+ \longrightarrow CH_2(OH)—CH(OH)—C(=O)—H$
Glycerate → Glyceraldehyde

(e) $CH_2(OH)—CH(OH)—CH_2(OH) \longrightarrow CH_2(OH)—C(=O)—CH_2(OH)$
Glycerol → Dihydroxyacetone

(f) $C_6H_5—CH_3 \longrightarrow C_6H_5—C(=O)—O^- + H^+$
Toluene → Benzoate

(g) $^-O—C(=O)—CH_2—CH_2—C(=O)—O^- \longrightarrow {}^-O—C(=O)—CH{=}CH—C(=O)—O^-$
Succinate → Fumarate

(h) $CH_3—C(=O)—C(=O)—O^- \longrightarrow CH_3—C(=O)—O^- + CO_2$
Pyruvate → Acetate

Answer Keep in mind that oxidation is the loss of electrons and accompanying H^+, whereas reduction is the gain of electrons (or H—H).

(a) Oxidation: Methanol ⟶ formaldehyde + H—H
(b) Oxidation: Formaldehyde ⟶ formate + H—H
(c) Reduction: CO_2 + H—H ⟶ formate + H^+
(d) Reduction: Glycerate + H—H + H^+ ⟶ glyceraldehyde + H_2O
(e) Oxidation: Glycerol ⟶ dihydroxyacetone + H—H
(f) Oxidation: Toluene + $2H_2O$ ⟶ benzoate + H^+ + 3H—H
(g) Oxidation: Succinate ⟶ fumarate + H—H
(h) Oxidation: Pyruvate + H_2O ⟶ acetate + CO_2 + H—H

4. **Relationship between Energy Release and the Oxidation State of Carbon** A eukaryotic cell can use glucose ($C_6H_{12}O_6$) and hexanoic acid ($C_6H_{14}O_2$) as fuels for cellular respiration. On the basis of their structural formulas, which substance releases more energy per gram on complete combustion to CO_2 and H_2O?

Answer From the structural formulas, we see that the carbon-bound H/C ratio of hexanoic acid (11/6) is higher than that of glucose (7/6). Hexanoic acid is more reduced and yields more energy upon complete combustion to CO_2 and H_2O.

5. **Nicotinamide Coenzymes as Reversible Redox Carriers** The nicotinamide coenzymes (see Fig. 13–24) can undergo reversible oxidation-reduction reactions with specific substrates in the presence of the appropriate dehydrogenase. In these reactions, NADH + H^+ serves as the hydrogen

source, as described in Problem 3. Whenever the coenzyme is oxidized, a substrate must be simultaneously reduced:

$$\underset{\text{Oxidized}}{\text{Substrate}} + \underset{\text{Reduced}}{\text{NADH}} + H^+ \rightleftharpoons \underset{\text{Reduced}}{\text{product}} + \underset{\text{Oxidized}}{NAD^+}$$

For each of the reactions in **(a)** through **(f)**, determine whether the substrate has been oxidized or reduced or is unchanged in oxidation state (see Problem 3). If a redox change has occurred, balance the reaction with the necessary amount of NAD^+, NADH, H^+, and H_2O. The objective is to recognize when a redox coenzyme is necessary in a metabolic reaction.

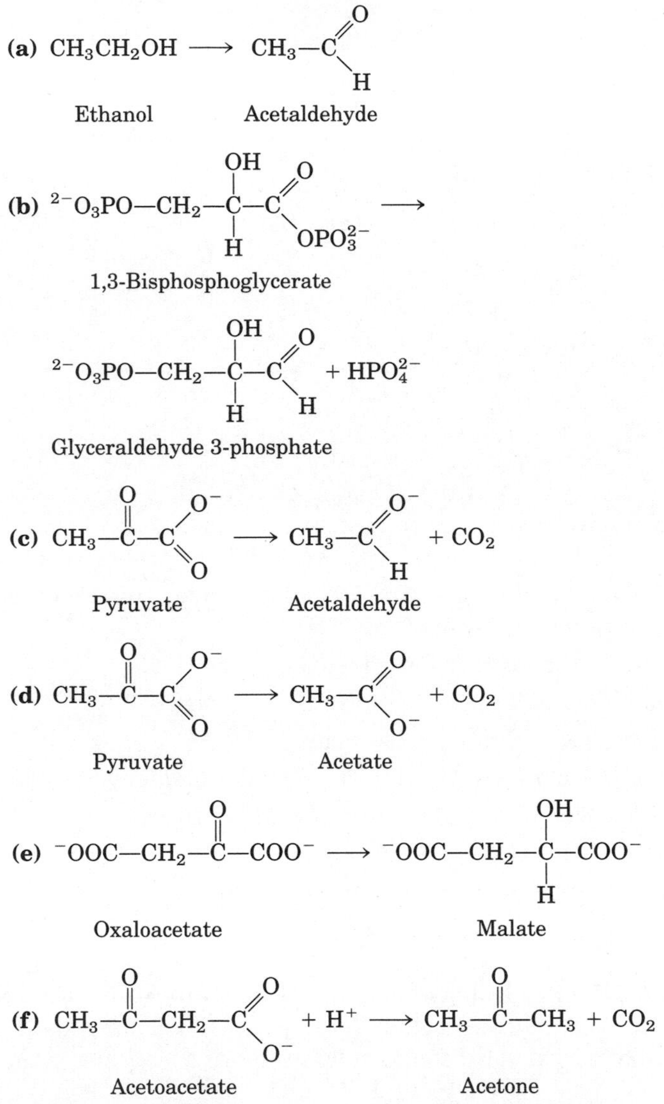

(a) $CH_3CH_2OH \longrightarrow CH_3{-}C({=}O){-}H$

Ethanol → Acetaldehyde

(b) $^{2-}O_3PO{-}CH_2{-}CH(OH){-}C({=}O){-}OPO_3^{2-} \longrightarrow$

1,3-Bisphosphoglycerate

$^{2-}O_3PO{-}CH_2{-}CH(OH){-}C({=}O){-}H + HPO_4^{2-}$

Glyceraldehyde 3-phosphate

(c) $CH_3{-}C({=}O){-}C({=}O){-}O^- \longrightarrow CH_3{-}C({=}O){-}H + CO_2$

Pyruvate → Acetaldehyde

(d) $CH_3{-}C({=}O){-}C({=}O){-}O^- \longrightarrow CH_3{-}C({=}O){-}O^- + CO_2$

Pyruvate → Acetate

(e) $^{-}OOC{-}CH_2{-}C({=}O){-}COO^- \longrightarrow {}^{-}OOC{-}CH_2{-}CH(OH){-}COO^-$

Oxaloacetate → Malate

(f) $CH_3{-}C({=}O){-}CH_2{-}C({=}O){-}O^- + H^+ \longrightarrow CH_3{-}C({=}O){-}CH_3 + CO_2$

Acetoacetate → Acetone

Answer

(a) Oxidized: Ethanol + $NAD^+ \longrightarrow$ acetaldehyde + NADH + H^+

(b) Reduced: 1,3-Bisphosphoglycerate + NADH + $H^+ \longrightarrow$ glyceraldehyde 3-phosphate + NAD^+ + HPO_4^{2-}

(c) Unchanged: Pyruvate + $H^+ \longrightarrow$ acetaldehyde + CO_2

(d) Oxidized: Pyruvate + $NAD^+ \longrightarrow$ acetate + CO_2 + NADH + H^+

(e) Reduced: Oxaloacetate + NADH + $H^+ \longrightarrow$ malate + NAD^+

(f) Unchanged: Acetoacetate + $H^+ \longrightarrow$ acetone + CO_2

6. Pyruvate Dehydrogenase Cofactors and Mechanism Describe the role of each cofactor involved in the reaction catalyzed by the pyruvate dehydrogenase complex.

Answer *TPP:* thiazolium ring adds to α carbon of pyruvate, then stabilizes the resulting carbanion by acting as an electron sink. *Lipoic acid:* oxidizes pyruvate to level of acetate (acetyl-CoA), and activates acetate as a thioester. *CoA-SH:* activates acetate as thioester. *FAD*: oxidizes lipoic acid. *NAD*$^+$: oxidizes FAD. (See Fig. 16–6.)

7. Thiamine Deficiency Individuals with a thiamine-deficient diet have relatively high levels of pyruvate in their blood. Explain this in biochemical terms.

Answer Thiamine is essential for the formation of thiamine pyrophosphate (TPP), one of the cofactors in the pyruvate dehydrogenase reaction. Without TPP, the pyruvate generated by glycolysis accumulates in cells and enters the blood.

8. Isocitrate Dehydrogenase Reaction What type of chemical reaction is involved in the conversion of isocitrate to α-ketoglutarate? Name and describe the role of any cofactors. What other reaction(s) of the citric acid cycle are of this same type?

Answer Oxidative decarboxylation involving $NADP^+$ or NAD^+ as the electron acceptor; the α-ketoglutarate dehydrogenase reaction is also an oxidative decarboxylation, but its mechanism is different and involves different cofactors: TPP, lipoate, FAD, NAD^+, and CoA-SH.

9. Stimulation of Oxygen Consumption by Oxaloacetate and Malate In the early 1930s, Albert Szent-Györgyi reported the interesting observation that the addition of small amounts of oxaloacetate or malate to suspensions of minced pigeon breast muscle stimulated the oxygen consumption of the preparation. Surprisingly, the amount of oxygen consumed was about seven times more than the amount necessary for complete oxidation (to CO_2 and H_2O) of the added oxaloacetate or malate. Why did the addition of oxaloacetate or malate stimulate oxygen consumption? Why was the amount of oxygen consumed so much greater than the amount necessary to completely oxidize the added oxaloacetate or malate?

Answer Oxygen consumption is a measure of the activity of the first two stages of cellular respiration: glycolysis and the citric acid cycle. Initial nutrients being oxidized are carbohydrates and lipids. Because several intermediates of the citric acid cycle can be siphoned off into biosynthetic pathways, the cycle may slow down for lack of oxaloacetate in the citrate synthase reaction, and acetyl-CoA will accumulate. Addition of oxaloacetate or malate (converted to oxaloacetate by malate dehydrogenase) will stimulate the cycle and allow it to use the accumulated acetyl-CoA. This stimulates respiration. Oxaloacetate is regenerated in the cycle, so addition of oxaloacetate (or malate) stimulates the oxidation of a much larger amount of acetyl-CoA

10. Formation of Oxaloacetate in a Mitochondrion In the last reaction of the citric acid cycle, malate is dehydrogenated to regenerate the oxaloacetate necessary for the entry of acetyl-CoA into the cycle:

$$\text{L-Malate} + NAD^+ \longrightarrow \text{oxaloacetate} + NADH + H^+ \qquad \Delta G'^\circ = 30.0 \text{ kJ/mol}$$

(a) Calculate the equilibrium constant for this reaction at 25 °C.

(b) Because $\Delta G'^\circ$ assumes a standard pH of 7, the equilibrium constant calculated in **(a)** corresponds to

$$K'_{eq} = \frac{[\text{oxaloacetate}][\text{NADH}]}{[\text{L-malate}][\text{NAD}^+]}$$

The measured concentration of L-malate in rat liver mitochondria is about 0.20 mM when $[NAD^+]/[NADH]$ is 10. Calculate the concentration of oxaloacetate at pH 7 in these mitochondria.

(c) To appreciate the magnitude of the mitochondrial oxaloacetate concentration, calculate the number of oxaloacetate molecules in a single rat liver mitochondrion. Assume the mitochondrion is a sphere of diameter 2.0 μm.

Answer

(a)
$$\Delta G'^{\circ} = -RT \ln K'_{eq}$$
$$\ln K'_{eq} = -\Delta G'^{\circ}/RT$$
$$= -(30.0 \text{ kJ/mol})/(2.48 \text{ kJ/mol})$$
$$= -12.1$$
$$K'_{eq} = e^{-12.1} = 5.6 \times 10^{-6}$$

(b) Given that

$$K'_{eq} = ([\text{OAA}]_{eq}[\text{NADH}]_{eq})/([\text{malate}]_{eq}[\text{NAD}^+]_{eq})$$

if we hold the values of [malate], [NADH], and [NAD^+] at the values that exist in the cell, we can calculate what [oxaloacetate] must be at equilibrium to give the equilibrium constant calculated in **(a):**

$$[\text{oxaloacetate}] = K'_{eq}\,[\text{malate}][\text{NAD}^+]/[\text{NADH}]$$
$$= (5.6 \times 10^{-6})(0.20 \text{ mM})(10)$$
$$= 1.1 \times 10^{-5} \text{ mM} = 1.1 \times 10^{-8} \text{ M}$$

This predicts that [oxaloacetate] at equilibrium would be very low, and the measured concentration is indeed low: less than 10^{-7} M.

(c) The volume of a sphere is $\frac{4}{3}\pi r^3$, thus the volume of a mitochondrion ($r = 1.0 \times 10^{-3}$ mm) is

$$\tfrac{4}{3}(3.14)(1.0 \times 10^{-3} \text{ mm})^3 = 4.2 \times 10^{-9} \text{ mm}^3 = 4.2 \times 10^{-15} \text{ L}$$

Given the concentration of oxaloacetate and Avogadro's number, we can calculate the number of molecules in a mitochondrion:

$$(1.1 \times 10^{-8} \text{ mol/L})(6.02 \times 10^{23} \text{ molecules/mol})(4.2 \times 10^{-15} \text{ L}) = 28 \text{ molecules}$$

11. Cofactors for the Citric Acid Cycle Suppose you have prepared a mitochondrial extract that contains all of the soluble enzymes of the matrix but has lost (by dialysis) all the low molecular weight cofactors. What must you add to the extract so that the preparation will oxidize acetyl-CoA to CO_2?

Answer ADP (or GDP), P_i, CoA-SH, TPP, NAD^+; *not* lipoic acid, which is covalently attached to the isolated enzymes that use it (see Fig. 16–7).

12. Riboflavin Deficiency How would a riboflavin deficiency affect the functioning of the citric acid cycle? Explain your answer.

Answer The flavin nucleotides, FMN and FAD, would not be synthesized. Because FAD is required by the citric acid cycle enzyme succinate dehydrogenase, flavin deficiency would strongly inhibit the cycle.

13. Oxaloacetate Pool What factors might decrease the pool of oxaloacetate available for the activity of the citric acid cycle? How can the pool of oxaloacetate be replenished?

Answer Oxaloacetate might be withdrawn for aspartate synthesis or for gluconeogenesis. Oxaloacetate is replenished by the anaplerotic reactions catalyzed by PEP carboxykinase, PEP carboxylase, malic enzyme, or pyruvate carboxylase (see Fig. 16–15, p. 632).

14. Energy Yield from the Citric Acid Cycle The reaction catalyzed by succinyl-CoA synthetase produces the high-energy compound GTP. How is the free energy contained in GTP incorporated into the cellular ATP pool?

Answer The terminal phosphoryl group in GTP can be transferred to ADP in a reaction catalyzed by nucleoside diphosphate kinase, with an equilibrium constant of 1.0:

$$\text{GTP} + \text{ADP} \longrightarrow \text{GDP} + \text{ATP}$$

15. Respiration Studies in Isolated Mitochondria Cellular respiration can be studied in isolated mitochondria by measuring oxygen consumption under different conditions. If 0.01 M sodium malonate is added to actively respiring mitochondria that are using pyruvate as fuel source, respiration soon stops and a metabolic intermediate accumulates.

(a) What is the structure of this intermediate?

(b) Explain why it accumulates.

(c) Explain why oxygen consumption stops.

(d) Aside from removal of the malonate, how can this inhibition of respiration be overcome? Explain.

Answer Malonate is a structural analog of succinate and a competitive inhibitor of succinate dehydrogenase.

(a) Succinate: $^{-}OOC—CH_2—CH_2—COO^{-}$

(b) When succinate dehydrogenase is inhibited, succinate accumulates.

(c) Inhibition of any reaction in a pathway causes the substrate of that reaction to accumulate. Because this substrate is also the product of the preceding reaction, its accumulation changes the effective ΔG of that reaction, and so on for all the preceding steps in the pathway. The net rate of the pathway (or cycle) slows and eventually becomes almost negligible. In the case of the citric acid cycle, ceasing to produce the primary product, NADH, has the effect of stopping electron transfer and consumption of oxygen, the final acceptor of electrons derived from NADH.

(d) Because malonate is a competitive inhibitor, the addition of large amounts of succinate will overcome the inhibition.

16. Labeling Studies in Isolated Mitochondria The metabolic pathways of organic compounds have often been delineated by using a radioactively labeled substrate and following the fate of the label.

(a) How can you determine whether glucose added to a suspension of isolated mitochondria is metabolized to CO_2 and H_2O?

(b) Suppose you add a brief pulse of [3-^{14}C] pyruvate (labeled in the methyl position) to the mitochondria. After one turn of the citric acid cycle, what is the location of the ^{14}C in the oxaloacetate? Explain by tracing the ^{14}C label through the pathway. How many turns of the cycle are required to release all the [3-^{14}C] pyruvate as CO_2?

Answer

(a) If you added uniformly labeled glucose (^{14}C in all carbon atoms), release of labeled CO_2 would indicate that the glucose is metabolized to CO_2 and H_2O.

(b) One turn of the cycle produces oxaloacetate with label equally distributed between C-2 and C-3. The route of the label is from C-3 in pyruvate, to C-2 in acetyl-CoA, to a methylene ($—CH_2—$) carbon, C-2 or C-4 (see Fig. 16–7), in intermediates to succinate, which is symmetric; from succinate, the label is in C-2 or C-3. The second turn of the cycle releases half the label, and every subsequent turn releases half of what remains, so an infinite number of turns are required to release *all* the labeled carbon.

17. Pathway of CO_2 in Gluconeogenesis In the first bypass step of gluconeogenesis, the conversion of pyruvate to phosphoenolpyruvate (PEP), pyruvate is carboxylated by pyruvate carboxylase to oxaloacetate, which is subsequently decarboxylated to PEP by PEP carboxykinase (Chapter 14). Because the addition of CO_2 is directly followed by the loss of CO_2, you might expect that in tracer experiments, the ^{14}C of $^{14}CO_2$ would not be incorporated into PEP, glucose, or any intermediates in gluconeogenesis.

However, investigators find that when a rat liver preparation synthesizes glucose in the presence of $^{14}CO_2$, ^{14}C slowly appears in PEP and eventually at C-3 and C-4 of glucose. How does the ^{14}C label get into the PEP and glucose? (Hint: During gluconeogenesis in the presence of $^{14}CO_2$, several of the four-carbon citric acid cycle intermediates also become labeled.)

Answer Because pyruvate carboxylase is a mitochondrial enzyme, the [^{14}C]oxaloacetate (OAA) formed by this reaction mixes with the OAA pool of the citric acid cycle. A mixture of [1-^{14}C] and [4-^{14}C] OAA eventually forms by randomization of the C-1 and C-4 positions in the reversible conversions OAA $\rightarrow$ malate $\rightarrow$ succinate. [1-^{14}C] OAA leads to formation of [3,4-^{14}C]glucose.

18. [1-^{14}C]Glucose Catabolism An actively respiring bacterial culture is briefly incubated with [1-^{14}C] glucose, and the glycolytic and citric acid cycle intermediates are isolated. Where is the ^{14}C in each of the intermediates listed below? Consider only the initial incorporation of ^{14}C, in the first pass of labeled glucose through the pathways.

(a) Fructose 1,6-bisphosphate
(b) Glyceraldehyde 3-phosphate
(c) Phosphoenolpyruvate
(d) Acetyl-CoA
(e) Citrate
(f) α-Ketoglutarate
(g) Oxaloacetate

Answer Figures 14–2, 14–6, and 16–7 and Box 16–3 outline the fate of all the carbon atoms of glucose. In one pass through the pathways, the label appears at:

(a) C-1
(b) C-3
(c) C-3
(d) C-2 (methyl group)
(e) C-2 (see Box 16–3)
(f) C-4
(g) Equally distributed in C-2 and C-3

19. Role of the Vitamin Thiamine People with beriberi, a disease caused by thiamine deficiency, have elevated levels of blood pyruvate and α-ketoglutarate, especially after consuming a meal rich in glucose. How are these effects related to a deficiency of thiamine?

Answer Thiamine is required for the synthesis of thiamin pyrophosphate (TPP), a prosthetic group in the pyruvate dehydrogenase and α-ketoglutarate dehydrogenase complexes. A thiamin deficiency reduces the activity of these enzyme complexes and causes the observed accumulation of precursors.

20. Synthesis of Oxaloacetate by the Citric Acid Cycle Oxaloacetate is formed in the last step of the citric acid cycle by the NAD^+-dependent oxidation of L-malate. Can a net synthesis of oxaloacetate from acetyl-CoA occur using only the enzymes and cofactors of the citric acid cycle, without depleting the intermediates of the cycle? Explain. How is oxaloacetate that is lost from the cycle (to biosynthetic reactions) replenished?

Answer In the citric acid cycle, the entering acetyl-CoA combines with oxaloacetate to form citrate. One turn of the cycle regenerates oxaloacetate and produces two CO_2 molecules. There is *no* net synthesis of oxaloacetate in the cycle. If any cycle intermediates are channeled into biosynthetic reactions, replenishment of oxaloacetate is essential. Four enzymes can produce oxaloacetate (or malate) from pyruvate or phosphoenolpyruvate. Pyruvate carboxylase (liver, kidney) and PEP carboxykinase (heart, skeletal muscle) are the most important in animals, and PEP carboxylase is most important in plants, yeast and bacteria. Malic enzyme produces malate from pyruvate in many organisms (see Table 16–2).

21. Oxaloacetate Depletion Mammalian liver can carry out gluconeogenesis using oxaloacetate as the starting material (Chapter 14). Would the operation of the citric acid cycle be affected by extensive use of oxaloacetate for gluconeogenesis? Explain your answer.

Answer Oxaloacetate depletion would tend to inhibit the citric acid cycle. Oxaloacetate is present at relatively low concentrations in mitochondria, and removing it for gluconeogenesis would tend to shift the equilibrium for the citrate synthase reaction toward oxaloacetate. However, anaplerotic reactions (see Fig. 16–15) counter this effect by replacing oxaloacetate.

22. Mode of Action of the Rodenticide Fluoroacetate Fluoroacetate, prepared commercially for rodent control, is also produced by a South African plant. After entering a cell, fluoroacetate is converted to fluoroacetyl-CoA in a reaction catalyzed by the enzyme acetate thiokinase:

$$F{-}CH_2COO^- + \text{CoA-SH} + \text{ATP} \longrightarrow F{-}CH_2\underset{\underset{O}{\|}}{C}{-}\text{S-CoA} + \text{AMP} + PP_i$$

The toxic effect of fluoroacetate was studied in an experiment using intact isolated rat heart. After the heart was perfused with 0.22 mM fluoroacetate, the measured rate of glucose uptake and glycolysis decreased, and glucose 6-phosphate and fructose 6-phosphate accumulated. Examination of the citric acid cycle intermediates revealed that their concentrations were below normal, except for citrate, with a concentration 10 times higher than normal.

(a) Where did the block in the citric acid cycle occur? What caused citrate to accumulate and the other cycle intermediates to be depleted?

(b) Fluoroacetyl-CoA is enzymatically transformed in the citric acid cycle. What is the structure of the end product of fluoroacetate metabolism? Why does it block the citric acid cycle? How might the inhibition be overcome?

(c) In the heart perfusion experiments, why did glucose uptake and glycolysis decrease? Why did hexose monophosphates accumulate?

(d) Why is fluoroacetate poisoning fatal?

Answer

(a) The block occurs at the aconitase reaction, which normally converts citrate to isocitrate.

(b) Fluoroacetate, an analog of acetate, can be activated to fluoroacetyl-CoA, which condenses with oxaloacetate to form fluorocitrate—the end product of fluoroacetate metabolism. Fluorocitrate is a structural analog of citrate and a strong competitive inhibitor of aconitase. The inhibition can be overcome by addition of large amounts of citrate.

(c) Citrate and fluorocitrate are allosteric inhibitors of phosphofructokinase-1, and as their concentration increases, glycolysis and glucose uptake slow down. Inhibition of PFK-1 causes the accumulation of glucose 6-phosphate and fructose 6-phosphate.

(d) The net effect of fluoroacetate poisoning is to shut down ATP synthesis, aerobic (oxidative) and anaerobic (fermentative).

23. Synthesis of L-Malate in Wine Making The tartness of some wines is due to high concentrations of L-malate. Write a sequence of reactions showing how yeast cells synthesize L-malate from glucose under anaerobic conditions in the presence of dissolved CO_2 (HCO_3^-). Note that the overall reaction for this fermentation cannot involve the consumption of nicotinamide coenzymes or citric acid cycle intermediates.

Answer The glycolytic reactions

$$\text{Glucose} + 2\text{P}_i + 2\text{ADP} + 2\text{NAD}^+ \longrightarrow 2\text{ pyruvate} + 2\text{ATP} + 2\text{NADH} + 2\text{H}^+ + 2\text{H}_2\text{O}$$

are followed by the pyruvate carboxylase reaction

$$2\text{ Pyruvate} + 2\text{CO}_2 + 2\text{ATP} + 2\text{H}_2\text{O} \longrightarrow 2\text{ oxaloacetate} + 2\text{ADP} + 2\text{P}_i + 4\text{H}^+$$

In the citric acid cycle, the malate dehydrogenase reaction

$$2\text{ Oxaloacetate} + 2\text{NADH} + 2\text{H}^+ \longrightarrow 2\text{ L-malate} + 2\text{NAD}^+$$

recycles nicotinamide coenzymes under anaerobic conditions. The overall reaction is

$$\text{Glucose} + 2\text{CO}_2 \longrightarrow 2\text{ L-malate} + 4\text{H}^+$$

which produces four H^+ per glucose, increasing the acidity and thus the tartness of the wine.

24. Net Synthesis of α-Ketoglutarate α-Ketoglutarate plays a central role in the biosynthesis of several amino acids. Write a sequence of enzymatic reactions that could result in the net synthesis of α-ketoglutarate from pyruvate. Your proposed sequence must not involve the net consumption of other citric acid cycle intermediates. Write an equation for the overall reaction and identify the source of each reactant.

Answer Anaplerotic reactions replenish intermediates in the citric acid cycle. Net synthesis of α-ketoglutarate from pyruvate occurs by the sequential actions of (1) pyruvate carboxylase (which makes extra molecules of oxaloacetate), (2) pyruvate dehydrogenase, and the citric acid cycle enzymes (3) citrate synthase, (4) aconitase, and (5) isocitrate dehydrogenase:

(1) $\text{Pyruvate} + \text{ATP} + \text{CO}_2 + \text{H}_2\text{O} \longrightarrow \text{oxaloacetate} + \text{ADP} + \text{P}_i + \text{H}^+$

(2) $\text{Pyruvate} + \text{CoA} + \text{NAD}^+ \longrightarrow \text{acetyl-CoA} + \text{CO}_2 + \text{NADH} + \text{H}^+$

(3) $\text{Oxaloacetate} + \text{acetyl-CoA} \longrightarrow \text{citrate} + \text{CoA}$

(4) $\text{Citrate} \longrightarrow \text{isocitrate}$

(5) $\text{Isocitrate} + \text{NAD}^+ \longrightarrow \alpha\text{-ketoglutarate} + \text{CO}_2 + \text{NADH} + \text{H}^+$

Net reaction: $2\text{ Pyruvate} + \text{ATP} + 2\text{NAD}^+ + \text{H}_2\text{O} \longrightarrow$
$\alpha\text{-ketoglutarate} + \text{CO}_2 + \text{ADP} + \text{P}_i + 2\text{NADH} + 3\text{H}^+$

25. Amphibolic Pathways Explain, giving examples, what is meant by the statement that the citric acid cycle is amphibolic.

Answer Amphibolic pathways can serve either in energy-yielding catabolic or in energy-requiring biosynthetic processes, depending on the cellular circumstances. For example, the citric acid cycle generates NADH and $FADH_2$ when functioning catabolically. But it can also provide precursors for the synthesis of such products as glutamate and aspartate (from α-ketoglutarate and oxaloacetate, respectively), which in turn serve as precursors for other products, such as glutamine, proline, and asparagine (see Fig. 16–15).

26. Regulation of the Pyruvate Dehydrogenase Complex In animal tissues, the rate of conversion of pyruvate to acetyl-CoA is regulated by the ratio of active, phosphorylated to inactive, unphosphorylated PDH complex. Determine what happens to the rate of this reaction when a preparation of rabbit muscle mitochondria containing the PDH complex is treated with **(a)** pyruvate dehydrogenase kinase, ATP, and NADH; **(b)** pyruvate dehydrogenase phosphatase and Ca^{2+}; **(c)** malonate.

Answer Pyruvate dehydrogenase is regulated by covalent modification and by allosteric inhibitors. The mitochondrial preparation responds as follows: **(a)** Active pyruvate dehydrogenase (dephosphorylated) is converted to inactive pyruvate dehydrogenase (phosphorylated) and the rate of conversion of pyruvate to acetyl-CoA decreases. **(b)** The phosphoryl group on pyruvate dehydrogenase phosphate is removed enzymatically to yield active pyruvate dehydrogenase, which increases the rate of conversion of pyruvate to acetyl-CoA. **(c)** Malonate inhibits succinate dehydrogenase, and citrate accumulates. The accumulated citrate inhibits citrate synthase, and acetyl-CoA accumulates. High levels of acetyl-CoA inhibit pyruvate dehydrogenase, and the rate of conversion of pyruvate to acetyl-CoA is reduced.

27. Commercial Synthesis of Citric Acid Citric acid is used as a flavoring agent in soft drinks, fruit juices, and many other foods. Worldwide, the market for citric acid is valued at hundreds of millions of dollars per year. Commercial production uses the mold *Aspergillus niger,* which metabolizes sucrose under carefully controlled conditions.

(a) The yield of citric acid is strongly dependent on the concentration of $FeCl_3$ in the culture medium, as indicated in the graph. Why does the yield decrease when the concentration of Fe^{3+} is above or below the optimal value of 0.5 mg/L?

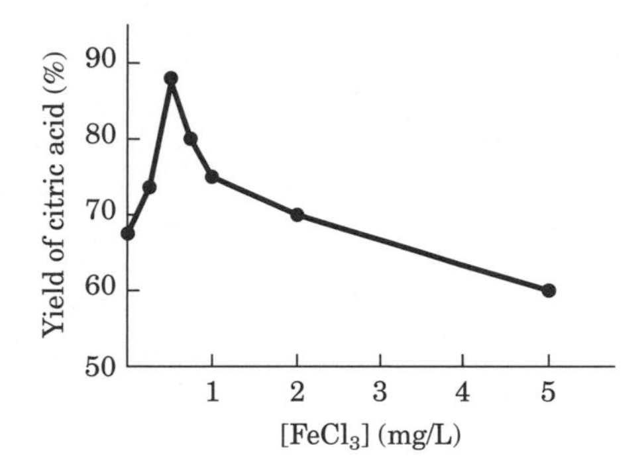

(b) Write the sequence of reactions by which *A. niger* synthesizes citric acid from sucrose. Write an equation for the overall reaction.

(c) Does the commercial process require the culture medium to be aerated—that is, is this a fermentation or an aerobic process? Explain.

Answer

(a) Citrate is produced through the action of citrate synthase on oxaloacetate and acetyl-CoA. Although the citric acid cycle does not normally result in an accumulation of intermediates, citrate synthase can be used for net synthesis of citrate when (1) there is a continuous influx of new oxaloacetate and acetyl-CoA, and (2) the transformation of citrate to isocitrate is blocked or at least restricted. *A. niger* grown in a medium rich in sucrose but low in Fe^{3+} meets both requirements. Citrate is transformed to isocitrate by aconitase, an Fe^{3+}-containing enzyme. In an Fe^{3+}-restricted medium, synthesis of aconitase is restricted and thus the breakdown of citrate is partially blocked; citrate accumulates and can be isolated in commercial quantities. Note that *some* aconitase activity is necessary—the mold will not thrive at [Fe^{3+}] below 0.5 mg/L. At higher [Fe^{3+}], however, aconitase is synthesized in increasing amounts; this will lead to a decrease in the yield of citrate as it cycles through the citric acid cycle.

(b) Sucrose + $H_2O \longrightarrow$ glucose + fructose

Glucose + $2P_i$ + 2ADP + $2NAD^+ \longrightarrow$ 2 pyruvate + 2ATP + 2NADH + $2H^+$ + $2H_2O$

Fructose + $2P_i$ + 2ADP + $2NAD^+ \longrightarrow$ 2 pyruvate + 2ATP + 2NADH + $2H^+$ + $2H_2O$

2 Pyruvate + $2NAD^+$ + 2CoA $\longrightarrow$ 2 acetyl-CoA + 2NADH + $2H^+$ + $2CO_2$

2 Pyruvate + $2CO_2$ + 2ATP + $2H_2O \longrightarrow$ 2 oxaloacetate + 2ADP + $2P_i$ + $4H^+$

2 Acetyl-CoA + 2 oxaloacetate + $2H_2O \longrightarrow$ 2 citrate + 2CoA

The overall reaction is

Sucrose + H_2O + $2P_i$ + 2ADP + $6NAD^+$ $\longrightarrow$ 2 citrate + 2ATP + 6NADH + $10H^+$

(c) Note that the overall reaction consumes NAD^+. Because the cellular pool of this oxidized coenzyme is limited, it must be recycled by oxidation of NADH via the electron-transfer chain, with consumption of oxygen. Consequently, the overall conversion of sucrose to citrate is an aerobic process and requires molecular oxygen.

28. Regulation of Citrate Synthase In the presence of saturating amounts of oxaloacetate, the activity of citrate synthase from pig heart tissue shows a sigmoid dependence on the concentration of acetyl-CoA, as shown in the graph. When succinyl-CoA is added, the curve shifts to the right and the sigmoid dependence is more pronounced.

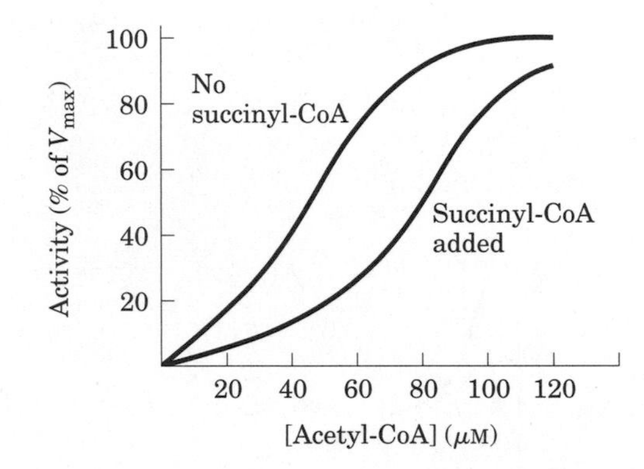

On the basis of these observations, suggest how succinyl-CoA regulates the activity of citrate synthase. (Hint: see Fig. 6–34) Why is succinyl-CoA an appropriate signal for regulation of the citric acid cycle? How does the regulation of citrate synthase control the rate of cellular respiration in pig heart tissue?

Answer Succinyl-CoA is an intermediate of the citric acid cycle—the first four-carbon intermediate, formed in the α-ketoglutarate dehydrogenase reaction. Its accumulation signals reduced flux through the cycle, and thus the need for reduced entry of acetyl-CoA into the cycle.

As seen in the graph, succinyl-CoA shifts the half-saturation point, $[S]_{0.5}$ (or $K_{0.5}$), for acetyl-CoA to the right but does not alter V_{max}. This indicates that succinyl-CoA acts as a negative modulator, either directly as a competitive inhibitor with acetyl-CoA or by binding to a site separate from the active site.

Citrate synthase catalyzes the step at which acetyl-CoA enters the cycle, and thus regulation of this enzyme controls the activity of the cycle, the rate of production of reduced coenzymes, and thus the rate of cellular respiration.

29. Regulation of Pyruvate Carboxylase The carboxylation of pyruvate by pyruvate carboxylase occurs at a very low rate unless acetyl-CoA, a positive allosteric modulator, is present. If you have just eaten a meal rich in fatty acids (triacylglycerols) but low in carbohydrates (glucose), how does this regulatory property shut down the oxidation of glucose to CO_2 and H_2O but increase the oxidation of acetyl-CoA derived from fatty acids?

Answer Fatty acid catabolism increases the level of acetyl-CoA, which stimulates pyruvate carboxylase. The resulting increase in oxaloacetate concentration stimulates acetyl-CoA consumption through the citric acid cycle, causing the citrate and ATP concentrations to rise. These metabolites inhibit glycolysis at PFK-1 and inhibit pyruvate dehydrogenase, effectively slowing the utilization of sugars and pyruvate.

30. Relationship between Respiration and the Citric Acid Cycle Although oxygen does not participate directly in the citric acid cycle, the cycle operates only when O_2 is present. Why?

Answer Oxygen is the terminal electron acceptor in oxidative phosphorylation, and thus is needed to recycle NAD^+ from NADH. NADH is produced in greatest quantities by the oxidative reactions of the citric acid cycle. In the absence of O_2, the supply of NAD^+ is depleted, and the accumulated NADH allosterically inhibits pyruvate dehydrogenase and α-ketoglutarate dehydrogenase (see Fig. 16–18).

31. Effect of [NADH]/[NAD^+] on the Citric Acid Cycle How would you expect the operation of the citric acid cycle to respond to a rapid increase in the [NADH]/[NAD^+] ratio in the mitochondrial matrix? Why?

Answer Increased [NADH]/[NAD^+] inhibits the citric acid cycle by mass action at each of the three steps that involve reduction of NAD^+; high [NADH] shifts the equilibrium toward NAD^+. Another way to look at this effect is to consider how an increased ratio of product (NADH) to reactant (NAD^+) affects the free-energy change for any of the three NAD^+-dependent steps of the citric acid cycle. Look, for example, at Equation 13–4 (p. 493).

32. Thermodynamics of Citrate Synthase Reaction in Cells Citrate is formed by the condensation of acetyl-CoA with oxaloacetate, catalyzed by citrate synthase:

$$\text{Oxaloacetate} + \text{acetyl-CoA} + H_2O \longrightarrow \text{citrate} + \text{CoA} + H^+$$

In rat heart mitochondria at pH 7.0 and 25° C, the concentrations of reactants and products are: oxaloacetate, 1 μM; acetyl-CoA, 1 μM; citrate, 220 μM; and CoA, 65 μM. The standard free-energy change for the citrate synthase reaction is −32.2 kJ/mol. What is the direction of metabolite flow through the citrate synthase reaction in rat heart cells? Explain.

Answer The free-energy change of the citrate synthase reaction in the cell is

$$\Delta G = \Delta G'^\circ + RT \ln \frac{[\text{citrate}][\text{CoA}]}{[\text{OAA}][\text{acetyl-CoA}]}$$

$$= -32.2 \text{ kJ/mol} + (2.48 \text{ kJ/mol}) \ln \frac{(220 \times 10^{-6})(65 \times 10^{-6})}{(1 \times 10^{-6})(1 \times 10^{-6})}$$

$$= -8 \text{ kJ/mol}$$

Thus, the citrate synthase reaction is exergonic and proceeds in the direction of citrate formation.

33. Reactions of the Pyruvate Dehydrogenase Complex Two of the steps in the oxidative decarboxylation of pyruvate (steps ④ and ⑤ in Fig. 16–6) do not involve any of the three carbons of pyruvate yet are essential to the operation of the PDH complex. Explain.

Answer The pyruvate dehydrogenase complex can be thought of as performing five enzymatic reactions. The first three (see Fig. 16–6) catalyze the oxidation of pyruvate to acetyl-CoA and reduction of the enzyme. The last two reactions are essential to reoxidize the reduced enzyme, reducing NAD^+ to NADH + H^+. The moiety on the enzyme that is oxidized/reduced is the lipoamide cofactor.

34. Citric Acid Cycle Mutants There are many cases of human disease in which one or another enzyme activity is lacking due to genetic mutation. However, cases in which individuals lack one of the enzymes of the citric acid cycle are extremely rare. Why?

Answer The citric acid cycle is so central to metabolism that a serious defect in any cycle enzyme would probably be lethal to the embryo.

35. Partitioning between the Citric Acid and Glyoxylate Cycles In an organism (such as *E. coli*) that has both the citric acid cycle and the glyoxylate cycle, what determines which of these pathways isocitrate will enter?

Answer Isocitrate can be metabolized by the citric acid cycle or by the glyoxylate cycle. The first enzyme in each pathway is allosterically regulated, so that the accumulation of citric acid cycle intermediates stimulate that cycle while inhibiting the glyoxylate cycle. AMP and ADP, which signal an inadequate reserve of ATP, inhibit the glyoxylate cycle, shifting the use of isocitrate to the energy-producing citric acid cycle. This reciprocal regulation of the two enzymes at the branch point determine which pathway isocitrate will enter.

Data Analysis Problem

36. How the Citric Acid Cycle Was Determined The detailed biochemistry of the citric acid cycle was determined by several researchers over a period of decades. In a 1937 article, Krebs and Johnson summarized their work and the work of others in the first published description of this pathway.

The methods used by these researchers were very different from those of modern biochemistry. Radioactive tracers were not commonly available until the 1940s, so Krebs and other researchers had to use nontracer techniques to work out the pathway. Using freshly prepared samples of pigeon breast muscle, they determined oxygen consumption by suspending minced muscle in buffer in a sealed flask and measuring the volume (in μL) of oxygen consumed under different conditions. They measured levels of substrates (intermediates) by treating samples with acid to remove contaminating proteins, then assaying the quantities of various small organic molecules. The two key observations that led Krebs and colleagues to propose a citric acid *cycle* as opposed to a *linear pathway* (like that of glycolysis) were made in the following experiments.

Experiment I. They incubated 460 mg of minced muscle in 3 mL of buffer at 40 °C for 150 minutes. Addition of *citrate* increased O_2 consumption by 893 μL compared with samples without added citrate. They calculated, based on the O_2 consumed during respiration of other carbon-containing compounds, that the expected O_2 consumption for complete respiration of this quantity of citrate was only 302 μL.

Experiment II. They measured O_2 consumption by 460 mg of minced muscle in 3 mL of buffer when incubated with *citrate* and/or with *1-phosphoglycerol* (glycerol 1-phosphate; this was known to be readily oxidized by cellular respiration) at 40 °C for 140 minutes. The results are shown in the table.

Sample	Substrate(s) added	μL O_2 absorbed
1	No extra	342
2	0.3 mL 0.2 M 1-phosphoglycerol	757
3	0.15 mL 0.02 M citrate	431
4	0.3 mL 0.2 M 1-phosphoglycerol and 0.15 mL 0.02 M citrate	1,385

(a) Why is O_2 consumption a good measure of cellular respiration?

(b) Why does sample 1 (unsupplemented muscle tissue) consume some oxygen?

(c) Based on the results for samples 2 and 3, can you conclude that 1-phosphoglycerol and citrate serve as substrates for cellular respiration in this system? Explain your reasoning.

(d) Krebs and colleagues used the results from these experiments to argue that citrate was "catalytic"—that it helped the muscle tissue samples metabolize 1-phosphoglycerol more completely. How would you use their data to make this argument?

(e) Krebs and colleagues further argued that citrate was not simply consumed by these reactions, but had to be *regenerated*. Therefore, the reactions had to be a *cycle* rather than a linear pathway. How would you make this argument?

Other researchers had found that *arsenate* (AsO_4^{3-}) inhibits α-ketoglutarate dehydrogenase and that *malonate* inhibits succinate dehydrogenase.

(f) Krebs and coworkers found that muscle tissue samples treated with arsenate and citrate would consume citrate only in the presence of oxygen; and under these conditions, oxygen was consumed. Based on the pathway in Figure 16–7, what was the citrate converted to in this experiment, and why did the samples consume oxygen?

In their article, Krebs and Johnson further reported the following. (1) In the presence of arsenate, 5.48 mmol of citrate was converted to 5.07 mmol of α-ketoglutarate. (2) In the presence of malonate, citrate was quantitatively converted to large amounts of succinate and small amounts of α-ketoglutarate. (3) Addition of oxaloacetate in the absence of oxygen led to production of a large amount of citrate; the amount was increased if glucose was also added.

Other workers had found the following pathway in similar muscle tissue preparations:

$$\text{Succinate} \longrightarrow \text{fumarate} \longrightarrow \text{malate} \longrightarrow \text{oxaloacetate} \longrightarrow \text{pyruvate}$$

(g) Based only on the data presented in this problem, what is the order of the intermediates in the citric acid cycle? How does this compare with Figure 16–7? Explain your reasoning.

(h) Why was it important to show the *quantitative* conversion of citrate to α-ketoglutarate?

The Krebs and Johnson article also contains other data that filled in most of the missing components of the cycle. The only component left unresolved was the molecule that reacted with oxaloacetate to form citrate.

Answer

(a) The only reaction in muscle tissue that consumes significant amounts of oxygen is cellular respiration, so O_2 consumption is a good proxy for respiration.

(b) Freshly prepared muscle tissue contains some residual glucose; O_2 consumption is due to oxidation of this glucose.

(c) Yes. Because the amount of O_2 consumed increased when citrate or 1-phosphoglycerol was added, both can serve as substrate for cellular respiration in this system.

(d) *Experiment I*: citrate is causing much more O_2 consumption than would be expected from its complete oxidation. Each molecule of citrate seems to be acting as though it were more than one molecule. The only possible explanation is that each molecule of citrate functions more than once in the reaction—which is how a catalyst operates. *Experiment II*: the key is to calculate the excess O_2 consumed by each sample compared with the control (sample 1).

Sample	Substrate(s) added	μL O_2 absorbed	Excess μL O_2 consumed
1	No extra	342	0
2	0.3 mL 0.2 M 1-phosphoglycerol	757	415
3	0.15 mL 0.02 M citrate	431	89
4	0.3 mL 0.2 M 1-phosphoglycerol + 0.15 mL 0.02 M citrate	1,385	1,043

If both citrate and 1-phosphoglycerol were simply substrates for the reaction, you would expect the excess O_2 consumption by sample 4 to be the sum of the individual excess consumptions by samples 2 and 3 (415 μL + 89 μL = 504 μL). However, the excess consumption when both substrates are present is roughly twice this amount (1,043 μL). Thus citrate increases the ability of the tissue to metabolize 1-phosphoglycerol. This behavior is typical of a catalyst. Both experiments (I and II) are required to make this case convincing. Based on experiment I only, citrate is somehow accelerating the reaction, but it is not clear whether it acts by helping substrate metabolism or by some other mechanism. Based on experiment II only, it is not clear which molecule is the catalyst, citrate or 1-phosphoglycerol. Together, the experiments show that citrate is acting as a "catalyst" for the oxidation of 1-phosphoglycerol.

(e) Given that the pathway can consume citrate (see sample 3), if citrate is to act as a catalyst it must be regenerated. If the set of reactions first consumes then regenerates citrate, it must be a circular rather than a linear pathway.

(f) When the pathway is blocked at α-ketoglutarate dehydrogenase, citrate is converted to α-ketoglutarate but the pathway goes no further. Oxygen is consumed by reoxidation of the NADH produced by isocitrate dehydrogenase.

(g)

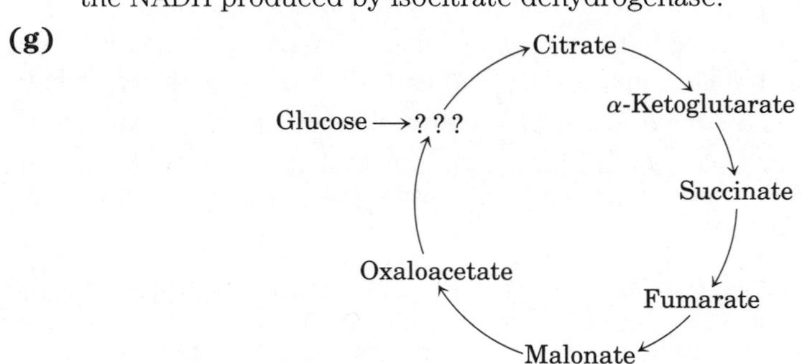

This differs from Figure 16–7 in that it does not include *cis*-aconitate and isocitrate (between citrate and α-ketoglutarate), or succinyl-CoA, or acetyl-CoA.

(h) Establishing a quantitative conversion was essential to rule out a branched or other, more complex pathway.

Reference

Krebs, H.A. & Johnson, W.A. (1937) The role of citric acid in intermediate metabolism in animal tissues. *Enzymologia* **4,** 148–156. [Reprinted (1980) in *FEBS Lett.* **117** (Suppl.), K2–K10.]

chapter

17

Fatty Acid Catabolism

1. **Energy in Triacylglycerols** On a per-carbon basis, where does the largest amount of biologically available energy in triacylglycerols reside: in the fatty acid portions or the glycerol portion? Indicate how knowledge of the chemical structure of triacylglycerols provides the answer.

   Answer The fatty acids of triacylglycerols are hydrocarbons, with a single carboxyl group. Glycerol, on the other hand, has an —OH group on each carbon and is thus much more highly oxidized than a fatty acid. On oxidation, fatty acids therefore produce far more energy per carbon than does glycerol. Triacylglycerols have an energy of oxidation more than twice that of the same weight of carbohydrates or proteins.

2. **Fuel Reserves in Adipose Tissue** Triacylglycerols, with their hydrocarbon-like fatty acids, have the highest energy content of the major nutrients.
   (a) If 15% of the body mass of a 70.0 kg adult consists of triacylglycerols, what is the total available fuel reserve, in kilojoules and kilocalories, in the form of triacylglycerols? Recall that 1.00 kcal = 4.18 kJ.
   (b) If the basal energy requirement is approximately 8,400 kJ/day (2,000 kcal/day), how long could this person survive if the oxidation of fatty acids stored as triacylglycerols were the only source of energy?
   (c) What would be the weight loss in pounds per day under such starvation conditions (1 lb = 0.454 kg)?

   Answer
   (a) Given (in the text) that the energy value of stored triacylglycerol is 38 kJ/g, the available fuel reserve is

$$(0.15)(70.0 \times 10^3 \text{ g})(38 \text{ kJ/g}) = 4.0 \times 10^5 \text{ kJ}$$
$$= 9.6 \times 10^4 \text{ kcal}$$

   (b) At a rate of 8.4×10^3 kJ/day, the fuel supply would last

$$(4.0 \times 10^5 \text{ kJ})/(8.4 \times 10^3 \text{ kJ/day}) = 48 \text{ days}$$

   (c) If all the triacylglycerol is used over a 48-day period, this represents a rate of weight loss of

$$\frac{(0.15)(70.0 \text{ kg})}{48 \text{ days}} = 0.22 \text{ kg/day}$$

$$\text{or } (0.22 \text{ kg/day})/(0.454 \text{ kg/lb}) = 0.48 \text{ lb/day}$$

3. **Common Reaction Steps in the Fatty Acid Oxidation Cycle and Citric Acid Cycle** Cells often use the same enzyme reaction pattern for analogous metabolic conversions. For example, the steps in the oxidation of pyruvate to acetyl-CoA and of α-ketoglutarate to succinyl-CoA, although catalyzed by

different enzymes, are very similar. The first stage of fatty acid oxidation follows a reaction sequence closely resembling a sequence in the citric acid cycle. Use equations to show the analogous reaction sequences in the two pathways.

Answer The first three reactions in the β oxidation of fatty acyl–CoA molecules are analogous to three reactions of the citric acid cycle.

The fatty acyl–CoA dehydrogenase reaction is analogous to the succinate dehydrogenase reaction; both are FAD-requiring oxidations:

$$\text{Succinate} + \text{FAD} \longrightarrow \text{fumarate} + \text{FADH}_2$$

$$\text{Fatty acyl–CoA} + \text{FAD} \longrightarrow \textit{trans}\text{-}\Delta^2\text{-enoyl-CoA} + \text{FADH}_2$$

The enoyl-CoA hydratase reaction is analogous to the fumarase reaction; both add water to an olefinic bond:

$$\text{Fumarate} + \text{H}_2\text{O} \longrightarrow \text{malate}$$

$$\textit{trans}\text{-}\Delta^2\text{-Enoyl-CoA} + \text{H}_2\text{O} \longrightarrow \text{L-}\beta\text{-hydroxyacyl-CoA}$$

The β-hydroxyacyl-CoA dehydrogenase reaction is analogous to the malate dehydrogenase reaction; both are NAD-requiring and act on β-hydroxyacyl compounds:

$$\text{Malate} + \text{NAD}^+ \longrightarrow \text{oxaloacetate} + \text{NADH}$$

$$\text{L-}\beta\text{-Hydroxyacyl-CoA} + \text{NAD}^+ \longrightarrow \beta\text{-ketoacyl-CoA} + \text{NADH}$$

4. **β Oxidation: How Many Cycles?** How many cycles of β oxidation are required for the complete oxidation of activated oleic acid, 18:1(Δ^9)?

Answer 7 cycles; the last releases 2 acetyl-CoA.

5. **Chemistry of the Acyl-CoA Synthetase Reaction** Fatty acids are converted to their coenzyme A esters in a reversible reaction catalyzed by acyl-CoA synthetase:

$$\text{R—COO}^- + \text{ATP} + \text{CoA} \rightleftharpoons \text{R—}\overset{\overset{\displaystyle \text{O}}{\|}}{\text{C}}\text{—CoA} + \text{AMP} + \text{PP}_i$$

(a) The enzyme-bound intermediate in this reaction has been identified as the mixed anhydride of the fatty acid and adenosine monophosphate (AMP), acyl-AMP:

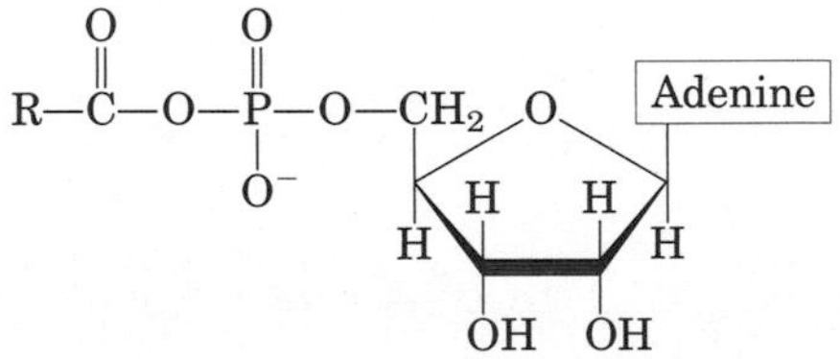

$$\text{R—}\overset{\overset{\displaystyle \text{O}}{\|}}{\text{C}}\text{—O—}\underset{\underset{\displaystyle \text{O}^-}{|}}{\overset{\overset{\displaystyle \text{O}}{\|}}{\text{P}}}\text{—O—CH}_2\text{—(ribose: H, H, H, H, OH, OH)—Adenine}$$

Write two equations corresponding to the two steps of the reaction catalyzed by acyl-CoA synthetase.

(b) The acyl-CoA synthetase reaction is readily reversible, with an equilibrium constant near 1. How can this reaction be made to favor formation of fatty acyl–CoA?

Answer Activation of carboxyl groups by ATP could in theory be accomplished by three types of reactions: the formation of acyl-phosphate + ADP; of acyl-ADP + P_i; or of acyl-AMP + PP_i. All these reactions are readily reversible. To create an activation reaction with a highly negative $\Delta G'^\circ$ (effectively irreversible), the third type of reaction can be coupled to a pyrophosphatase reaction, as in the synthesis of fatty acyl–CoA molecules.

(a) $R—COO^- + ATP \longrightarrow$ acyl-AMP + PP_i

Acyl-AMP + CoA $\longrightarrow$ acyl-CoA + AMP

(b) Hydrolysis of PP_i by an inorganic pyrophosphatase pulls the reaction in the direction of fatty acyl–CoA formation.

6. Intermediates in Oleic Acid Oxidation What is the structure of the partially oxidized fatty acyl group that is formed when oleic acid, 18:1(Δ^9), has undergone three cycles of β oxidation? What are the next two steps in the continued oxidation of this intermediate?

Answer After three rounds of β oxidation, the fatty acyl–CoA has been shortened by six carbons with the removal of three acetyl-CoAs. The resulting 12-carbon intermediate is *cis*-Δ^3-dodecanoyl-CoA, with the double bond between the third and fourth carbons from the carboxyl end of the chain. Before another round of β oxidation can occur, that double bond must be moved from Δ^3 to Δ^2, which is catalyzed by Δ^3,Δ^2-enoyl isomerase (see Fig. 17–9). Water is then added to the double bond to form the β-hydroxydodecanoyl-CoA derivative, which can undergo further β oxidation.

7. β Oxidation of an Odd-Chain Fatty Acid What are the direct products of β oxidation of a fully saturated, straight-chain fatty acid of 11 carbons?

Answer 4 acetyl-CoA and 1 propionyl-CoA

8. Oxidation of Tritiated Palmitate Palmitate uniformly labeled with tritium (^3H) to a specific activity of 2.48×10^8 counts per minute (cpm) per micromole of palmitate is added to a mitochondrial preparation that oxidizes it to acetyl-CoA. The acetyl-CoA is isolated and hydrolyzed to acetate. The specific activity of the isolated acetate is 1.00×10^7 cpm/μmol. Is this result consistent with the β-oxidation pathway? Explain. What is the final fate of the removed tritium?

Answer The β-oxidation pathway includes two dehydrogenase enzymes that remove hydrogen (H–H) from the fatty acyl–CoA chain, first at a $—CH_2—CH_2—$ and then at a $—CH_2—CH(OH)—$. The net result of the two reactions is removal of one of the two hydrogens at the point of formation of the enoyl-CoA intermediate. The two other hydrogens in the methyl group of acetyl-CoA come from water.

Palmitate contains 16 carbons, with $(14 \times 2) + 3 = 31$ hydrogens, so each two-carbon unit contains about 4/31 or about 1/8 of the total ^3H present. Thus, the counts per minute expected per acetyl-CoA, with two of the four acetyl hydrogens labeled (the other two arising from unlabeled water), is $(2/4)(2.48 \times 10^8 \text{ cpm}/\mu\text{mol})(1/8) = 1.6 \times 10^7 \text{ cpm}/\mu\text{mol}$, somewhat higher than observed. Exchange between β-ketoacyl-CoA and solvent water could cause loss of ^3H.

The final fate of the tritium removed from palmitate is its appearance in water, as reduced carriers ($FADH_2$, NADH) are reoxidized by the mitochondria.

9. Compartmentation in β Oxidation Free palmitate is activated to its coenzyme A derivative (palmitoyl-CoA) in the cytosol before it can be oxidized in the mitochondrion. If palmitate and [^{14}C]coenzyme A are added to a liver homogenate, palmitoyl-CoA isolated from the cytosolic fraction is radioactive, but that isolated from the mitochondrial fraction is not. Explain.

Answer The transport of fatty acid molecules into mitochondria requires a shuttle system involving a fatty acyl–carnitine intermediate. Fatty acids are first converted to fatty acyl–CoA molecules in the cytosol (by the action of acyl–CoA synthetases) then, at the outer mitochondrial membrane, the fatty acyl group is transferred to carnitine (by the action of carnitine acyltransferase I). After transport of fatty acyl–carnitine through the inner membrane, the fatty acyl group is transferred to mitochondrial CoA. The cytosolic and mitochondrial pools of CoA are thus kept separate, and no labeled CoA from the cytosolic pool enters the mitochondrion.

10. Comparative Biochemistry: Energy-Generating Pathways in Birds One indication of the relative importance of various ATP-producing pathways is the V_{max} of certain enzymes of these pathways. The values of V_{max} of several enzymes from the pectoral muscles (chest muscles used for flying) of pigeon and pheasant are listed below.

Enzyme	V_{max} (μmol substrate/min/g tissue)	
	Pigeon	**Pheasant**
Hexokinase	3.0	2.3
Glycogen phosphorylase	18.0	120.0
Phosphofructokinase-1	24.0	143.0
Citrate synthase	100.0	15.0
Triacylglycerol lipase	0.07	0.01

(a) Discuss the relative importance of glycogen metabolism and fat metabolism in generating ATP in the pectoral muscles of these birds.
(b) Compare oxygen consumption in the two birds.
(c) Judging from the data in the table, which bird is the long-distance flyer? Justify your answer.
(d) Why were these particular enzymes selected for comparison? Would the activities of triose phosphate isomerase and malate dehydrogenase be equally good bases for comparison? Explain.

Answer
(a) In the pigeon, aerobic oxidation of fatty acids—β oxidation and oxidative phosphorylation—predominates; in the pheasant, anaerobic glycolysis of glycogen predominates. Note the high citrate synthase activity in the pigeon, and the high glycogen phosphorylase and PFK-1 activities in the pheasant.
(b) Using aerobic oxidation, pigeon muscle consumes more oxygen during flight.
(c) The energy available per gram is higher for fat than for glycogen. In addition, anaerobic breakdown of glycogen is limited by tolerance to lactate buildup. Thus the pigeon, using predominantly the oxidative catabolism of fats, is the long-distance flyer.
(d) The enzymes listed in the table (unlike triose phosphate isomerase and malate dehydrogenase) are the regulatory enzymes of their respective pathways and thus limit ATP production rates.

11. Mutant Carnitine Acyltransferase What changes in metabolic pattern would result from a mutation in the muscle carnitine acyltransferase I in which the mutant protein has lost its affinity for malonyl-CoA but not its catalytic activity?

Answer Malonyl-CoA would no longer inhibit fatty acid entry into the mitochondrion and β oxidation, so there might be a futile cycle of simultaneous fatty acid synthesis in the cytosol and fatty acid breakdown in mitochondria. (See Fig. 17–12.)

12. Effect of Carnitine Deficiency An individual developed a condition characterized by progressive muscular weakness and aching muscle cramps. The symptoms were aggravated by fasting, exercise, and a high-fat diet. The homogenate of a skeletal muscle specimen from the patient oxidized added oleate more slowly than did control homogenates, consisting of muscle specimens from healthy individuals. When carnitine was added to the patient's muscle homogenate, the rate of oleate oxidation equaled that in the control homogenates. The patient was diagnosed as having a carnitine deficiency.
(a) Why did added carnitine increase the rate of oleate oxidation in the patient's muscle homogenate?

(b) Why were the patient's symptoms aggravated by fasting, exercise, and a high-fat diet?

(c) Suggest two possible reasons for the deficiency of muscle carnitine in this individual.

Answer

(a) The carnitine-mediated transport of fatty acids into mitochondria is the rate-limiting step in β oxidation (see Fig. 17–6). Carnitine deficiency decreases the rate of transport of fatty acids into mitochondria and thus the rate of β oxidation, so addition of carnitine would increase the rate of oxidation.

(b) Fasting, exercise, and a high-fat diet all cause an increased need for β oxidation of fatty acids and thus an increased demand for carnitine shuttle activity. The symptoms of carnitine deficiency would therefore become more severe under these conditions.

(c) The deficiency of carnitine may result from a dietary deficiency of its precursor, lysine, or from a defect in one of the enzymes that synthesize carnitine from this precursor.

13. Fatty Acids as a Source of Water Contrary to legend, camels do not store water in their humps, which actually consist of large fat deposits. How can these fat deposits serve as a source of water? Calculate the amount of water (in liters) that a camel can produce from 1.0 kg of fat. Assume for simplicity that the fat consists entirely of tripalmitoylglycerol.

Answer Oxidation of fatty acids produces water in significant amounts. From Equation 17–6

$$\text{Palmitoyl-CoA} + 23\text{O}_2 + 108\text{P}_i + 108\text{ADP} \longrightarrow \text{CoA} + 16\text{CO}_2 + 108\text{ATP} + 23\text{H}_2\text{O}$$

we know that the oxidation of 1 mol of palmitoyl-CoA produces 23 mol of water.

Tripalmitoin (glycerol plus three palmitates in ester linkage) has a molecular weight of 885, so 1 kg of tripalmitoin contains (1.0 kg)(1,000 g/kg)/(885 g/mol) = 1.1 mol. Complete oxidation of the three palmitoyl groups will produce

$$(1.1 \text{ mol tripalmitoin})(3 \text{ mol palmitate/mol tripalmitoin})(23 \text{ mol H}_2\text{O/mol palmitate}) = 76 \text{ mol H}_2\text{O}$$

Thus, the volume of water produced (ignoring the contribution of glycerol oxidation) is

$$(76 \text{ mol})(18 \text{ g/mol})(1 \text{ kg/1,000 g})(1 \text{ L/kg}) = 1.4 \text{ L}$$

Note: in reality, this may be an overestimate. The fatty acyl groups of the triacylglycerol in the camel's fat may be less highly reduced than palmitate.

14. Petroleum as a Microbial Food Source Some microorganisms of the genera *Nocardia* and *Pseudomonas* can grow in an environment where hydrocarbons are the only food source. These bacteria oxidize straight-chain aliphatic hydrocarbons, such as octane, to their corresponding carboxylic acids:

$$\text{CH}_3(\text{CH}_2)_6\text{CH}_3 + \text{NAD}^+ + \text{O}_2 \rightleftharpoons \text{CH}_3(\text{CH}_2)_6\text{COOH} + \text{NADH} + \text{H}^+$$

How could these bacteria be used to clean up oil spills? What would be some of the limiting factors to the efficiency of this process?

Answer By oxidizing hydrocarbons to their corresponding fatty acids, these microbes can obtain all their energy from β oxidation and oxidative phosphorylation, converting the hydrocarbons to CO_2 and H_2O. Theoretically, oil spills could be broken down by treatment with these microbes.

Because of the extreme hydrophobicity of hydrocarbons, close contact between substrate and bacterial enzymes might be difficult to achieve; under field conditions (e.g., an oil spill), detergents are often added to improve this contact. In addition, other nutrients, such as nitrogen or phosphorus, may be limiting for the bacterial populations, and these elements are often added to foster the growth of the hydrocarbon-oxidizers.

15. Metabolism of a Straight-Chain Phenylated Fatty Acid A crystalline metabolite was isolated from the urine of a rabbit that had been fed a straight-chain fatty acid containing a terminal phenyl group:

$$C_6H_5{-}CH_2{-}(CH_2)_n{-}COO^-$$

A 302 mg sample of the metabolite in aqueous solution was completely neutralized by 22.2 mL of 0.100 M NaOH.

(a) What is the probable molecular weight and structure of the metabolite?

(b) Did the straight-chain fatty acid contain an even or an odd number of methylene ($—CH_2—$) groups (i.e., is n even or odd)? Explain.

Answer

(a) 22.2 mL of 0.1 M NaOH is equivalent to $(22.2 \times 10^{-3}\text{ L})\ (0.100\text{ mol/L}) = 22.2 = 10^{-4}$ mol of unknown metabolite (assuming that it contains only one carboxyl group) in the 302 mg sample. Thus, the M_r of the metabolite is

$$\frac{302 \times 10^{-3}\text{ g}}{22.2 \times 10^{-4}\text{ mol}} = 136$$

This is the M_r of phenylacetic acid.

(b) Because β oxidation removes two-carbon units, and the end product is a two-carbon unit, the original fatty acyl chain must have had an even number of methylene groups (with the phenyl group counted as equivalent to a terminal methyl group). An odd-numbered fatty acid would have produced phenylpropionate.

16. Fatty Acid Oxidation in Uncontrolled Diabetes When the acetyl-CoA produced during β oxidation in the liver exceeds the capacity of the citric acid cycle, the excess acetyl-CoA forms ketone bodies—acetone, acetoacetate, and D-β-hydroxybutyrate. This occurs in severe, uncontrolled diabetes: because the tissues cannot use glucose, they oxidize large amounts of fatty acids instead. Although acetyl-CoA is not toxic, the mitochondrion must divert the acetyl-CoA to ketone bodies. What problem would arise if acetyl-CoA were not converted to ketone bodies? How does the diversion to ketone bodies solve the problem?

Answer Individuals with uncontrolled diabetes oxidize large quantities of fat because they cannot use glucose efficiently. This leads to a decrease in activity of the citric acid cycle (see Problem 17) and an increase in the pool of acetyl-CoA. If acetyl-CoA were not converted to ketone bodies, the CoA pool would become depleted. Because the mitochondrial CoA pool is small, liver mitochondria recycle CoA by condensing two acetyl-CoA molecules to form acetoacetyl-CoA + CoA (see Fig. 17–18). The acetoacetyl-CoA is converted to other ketones, and the CoA is recycled for use in the β-oxidation pathway and energy production.

17. Consequences of a High-Fat Diet with No Carbohydrates Suppose you had to subsist on a diet of whale blubber and seal blubber, with little or no carbohydrate.

(a) What would be the effect of carbohydrate deprivation on the utilization of fats for energy?

(b) If your diet were totally devoid of carbohydrate, would it be better to consume odd- or even-numbered fatty acids? Explain.

Answer

(a) Pyruvate, formed from glucose via glycolysis, is the main source of the oxaloacetate needed to replenish citric acid cycle intermediates (see Table 16–2). In the absence of carbohydrate in the diet, the oxaloacetate level drops and the citric acid cycle slows. This increases the rate of β oxidation of fatty acids and leads to ketosis.

(b) The last cycle of β oxidation produces two acetyl-CoA molecules from an even-numbered fatty acid, or propionyl-CoA + acetyl-CoA from an odd-numbered fatty acid. Propionyl-CoA can be converted to succinyl-CoA (see Fig. 17–11), which when converted to oxaloacetate stimulates the citric acid cycle and relieves the conditions leading to ketosis. Thus, it would be better to consume odd-numbered fatty acids.

18. Even- and Odd-Chain Fatty Acids in the Diet In a laboratory experiment, two groups of rats are fed two different fatty acids as their sole source of carbon for a month. The first group gets heptanoic acid (7:0), and the second gets octanoic acid (8:0). After the experiment, a striking difference is seen between the two groups. Those in the first group are healthy and have gained weight, whereas those in the second group are weak and have lost weight as a result of losing muscle mass. What is the biochemical basis for this difference?

Answer The β oxidation of heptanoic acid (which has an odd number of carbons) produces the three-carbon intermediate propionyl-CoA, which can be converted by propionyl-CoA carboxylase to methylmalonyl-CoA, then to succinyl-CoA. This four-carbon product of fatty acid oxidation can then be converted to oxaloacetate in the citric acid cycle, and the oxaloacetate can be used for gluconeogenesis—thus providing the animal with carbohydrate as well as energy from fatty acid oxidation. Animals fed octanoic acid (with an even number of carbons) degrade it completely to acetyl-CoA by three rounds of β oxidation. This provides energy via the citric acid cycle but does not provide starting material for gluconeogenesis. These animals are therefore deficient in glucose, the primary fuel for the brain and an intermediate in many biosynthetic pathways.

19. Metabolic Consequences of Ingesting ω-Fluorooleate The shrub *Dichapetalum toxicarium*, native to Sierra Leone, produces ω-fluorooleate, which is highly toxic to warm-blooded animals.

$$F-CH_2-(CH_2)_7-\overset{H}{\overset{|}{C}}=\overset{H}{\overset{|}{C}}-(CH_2)_7-COO^-$$

ω-Fluorooleate

This substance has been used as an arrow poison, and powdered fruit from the plant is sometimes used as a rat poison (hence the plant's common name, ratsbane). Why is this substance so toxic? (Hint: review Chapter 16, Problem 22.)

Answer Oxidation of ω-fluorooleate in the β-oxidation pathway forms fluoroacetyl-CoA in the last pass through the sequence. Entry of fluoroacetyl-CoA into the citric acid cycle produces fluorocitrate, a powerful inhibitor of the enzyme aconitase. As a result of this inhibition, the citric acid cycle shuts down and the flow of reducing equivalents to oxidative phosphorylation is fatally impaired.

20. Mutant Acetyl-CoA Carboxylase What would be the consequences for fat metabolism of a mutation in acetyl-CoA carboxylase that replaced the Ser residue normally phosphorylated by AMPK to an Ala residue? What might happen if the same Ser were replaced by Asp? (Hint: See Fig. 17–12.)

Answer The Ser-to-Ala change would produce an enzyme that could not be inhibited by phosphorylation by AMPK. The first step in fatty acid synthesis would be constantly turned on, and the malonyl-CoA produced by acetyl-CoA carboxylase would inhibit entry of fatty acids into mitochondria, shutting down β oxidation. The Ser-to-Asp mutation would put a negatively charged Asp residue in the position occupied by Ⓟ-Ser in the inhibited wild-type enzyme. This might mimic the effect of a phosphorylated Ser residue, shutting down acetyl-CoA carboxylase, inhibiting fatty acid synthesis, and stimulating β oxidation.

21. Effect of PDE Inhibitor on Adipocytes How would an adipocyte's response to epinephrine be affected by the addition of an inhibitor of cAMP phosphodiesterase (PDE)? (Hint: See Fig. 12–4.)

Answer Response to glucagon or epinephrine would be prolonged because cAMP, once formed, would persist, stimulating protein kinase A for a longer period and leading to longer-lasting mobilization of fatty acids in adipocytes.

22. Role of FAD as Electron Acceptor Acyl-CoA dehydrogenase uses enzyme-bound FAD as a prosthetic group to dehydrogenate the α and β carbons of fatty acyl–CoA. What is the advantage of using FAD as an electron acceptor rather than NAD^+? Explain in terms of the standard reduction potentials for the Enz-FAD/$FADH_2$ ($E'° = -0.219$ V) and NAD^+/NADH ($E'° = -0.320$ V) half-reactions.

Answer Enz-FAD, having a more positive standard reduction potential, is a better electron acceptor than NAD^+, and the reaction is driven in the direction of fatty acyl–CoA oxidation (a negative free-energy change). This more favorable free-energy change is obtained at the expense of 1 ATP; only 1.5 ATP molecules are formed per $FADH_2$ oxidized in the respiratory chain, compared with 2.5 ATP per NADH.

23. β Oxidation of Arachidic Acid How many turns of the fatty acid oxidation cycle are required for complete oxidation of arachidic acid (see Table 10–1) to acetyl-CoA?

Answer Arachidic acid is a 20-carbon saturated fatty acid. Nine cycles of the β-oxidation pathway are required for its oxidation, producing 10 molecules of acetyl-CoA, the last two in the ninth turn.

24. Fate of Labeled Propionate If [3-^{14}C]propionate (^{14}C in the methyl group) is added to a liver homogenate, ^{14}C-labeled oxaloacetate is rapidly produced. Draw a flow chart for the pathway by which propionate is transformed to oxaloacetate, and indicate the location of the ^{14}C in oxaloacetate.

Answer Propionate is first converted to the CoA derivative. Figure 17–11 shows the three-step pathway that converts propionyl-CoA to succinyl-CoA, which can be summarized as follows. Use these descriptions to prepare your own flow diagram.

1. Propionyl-CoA carboxylase uses CO_2 and ATP to form D-methylmalonyl-CoA by carboxylation at C-2 of the propionyl group.
2. Methylmalonyl-CoA epimerase shifts the CoA thioester from C-1 (of the original propionyl group) to the newly added carboxylate, making the product L-methylmalonyl-CoA.
3. Methylmalonyl-CoA mutase moves the carboxy-CoA group from C-2 to C-3 within the original propionyl unit, forming succinyl-CoA.
4. Once succinyl-CoA is formed, the citric acid cycle can convert it to oxaloacetate.
   Given the stereochemistry of these reactions, the [^{14}C]-label is equilibrated at C-2 and C-3 of the oxaloacetate.

25. Phytanic Acid Metabolism When phytanic acid uniformly labeled with ^{14}C is fed to a mouse, radioactivity can be detected in malate, a citric acid cycle intermediate, within minutes. Draw a metabolic pathway that could account for this. Which of the carbon atoms in malate would contain ^{14}C label?

Answer Phytanic acid is degraded to pristanic acid by the pathway shown in Figure 17–17. Pristanic acid undergoes β oxidation, with each round yielding propionyl-CoA (not acetyl-CoA, as for a straight-chain fatty acid). Degradation of uniformly labeled phytanic acid produces propionyl-CoA labeled in all three carbons of propionate. Propionyl-CoA is converted to succinyl-CoA by the series of reactions shown in Figure 17–11. The C-2 and C-3 of the succinyl moiety are labeled, and either C-1 or C-4 as well. When this succinate is converted to malate in the citric acid cycle, the malate is labeled at C-2 and C-3, and labeled half as much at C-1 and C-4.

26. Sources of H_2O Produced in β Oxidation The complete oxidation of palmitoyl-CoA to carbon dioxide and water is represented by the overall equation

$$\text{Palmitoyl-CoA} + 23O_2 + 108P_i + 108\text{ADP} \longrightarrow \text{CoA} + 16CO_2 + 108\text{ATP} + 23H_2O$$

Water is also produced in the reaction

$$\text{ADP} + P_i \longrightarrow \text{ATP} + H_2O$$

but is not included as a product in the overall equation. Why?

Answer ATP hydrolysis in the cell's energy-requiring reactions uses water, in the reaction

$$\text{ATP} + H_2O \longrightarrow \text{ADP} + P_i$$

In a cell at steady state, for every mole of ATP hydrolyzed, a mole of ATP is formed by condensation of ADP + P_i. There is no *net* change in [ATP] and thus no *net* production of H_2O.

27. Biological Importance of Cobalt In cattle, deer, sheep, and other ruminant animals, large amounts of propionate are produced in the rumen through the bacterial fermentation of ingested plant matter. Propionate is the principal source of glucose for these animals, via the route propionate → oxaloacetate → glucose. In some areas of the world, notably Australia, ruminant animals sometimes show symptoms of anemia with concomitant loss of appetite and retarded growth, resulting from an inability to transform propionate to oxaloacetate. This condition is due to a cobalt deficiency caused by very low cobalt levels in the soil and thus in plant matter. Explain.

Answer One of the enzymes necessary for the conversion of propionate to oxaloacetate is methylmalonyl-CoA mutase (see Fig. 17–11). This enzyme requires as an essential cofactor the cobalt-containing coenzyme B_{12}, which is synthesized from vitamin B_{12}. A cobalt deficiency in animals would result in coenzyme B_{12} deficiency.

28. Fat Loss during Hibernation Bears expend about 25×10^6 J/day during periods of hibernation, which may last as long as seven months. The energy required to sustain life is obtained from fatty acid oxidation. How much weight loss (in kilograms) has occurred after seven months? How might ketosis be minimized during hibernation? (Assume the oxidation of fat yields 38 kJ/g.)

Answer If the catabolism of fat yields 38 kJ/g, or 3.8×10^4 kJ/kg, and the bear expends 25×10^6 J/day, or 2.5×10^4 kJ/day, then the bear will lose

$$(2.5 \times 10^4 \text{ kJ/day})/(3.8 \times 10^4 \text{ kJ/kg}) = 0.66 \text{ kg/day}$$

and in 7 months, or 210 days, will lose

$$0.66 \text{ kg/day} \times 210 \text{ days} = 140 \text{ kg}$$

To minimize ketosis, a slow but steady degradation of nonessential proteins would provide three-, four-, and five-carbon products essential to the formation of glucose by gluconeogenesis. This would avoid the inhibition of the citric acid cycle that occurs when oxaloacetate is withdrawn from the cycle to be used for gluconeogenesis. The citric acid cycle could continue to degrade acetyl-CoA, rather than shunting it into ketone body formation.

Data Analysis Problem

29. β Oxidation of Trans Fats Unsaturated fats with trans double bonds are commonly referred to as "trans fats." There has been much discussion about the effects of dietary trans fats on health. In their investigations of the effects of trans fatty acid metabolism on health, Yu and colleagues (2004) showed that a model trans fatty acid was processed differently from its cis isomer. They used three related 18-carbon fatty acids to explore the difference in β oxidation between cis and trans isomers of the same-size fatty acid.

O
OH

Stearic acid
(octadecenoic acid)

O
OH

Oleic acid
(*cis*-Δ^9-octadecenoic acid)

O
OH

Elaidic acid
(*trans*-Δ^9-octadecenoic acid)

The researchers incubated the coenzyme A derivative of each acid with rat liver mitochondria for 5 minutes, then separated the remaining CoA derivatives in each mixture by HPLC (high-performance liquid chromatography). The results are shown below, with separate panels for the three experiments.

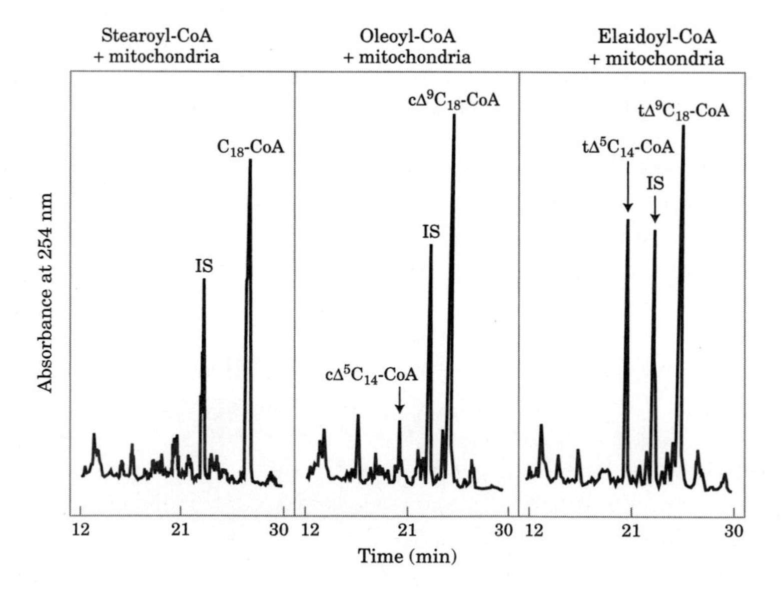

In the figure, IS indicates an internal standard (pentadecanoyl-CoA) added to the mixture, after the reaction, as a molecular marker. The researchers abbreviated the CoA derivatives as follows: stearoyl-CoA, C_{18}-CoA; *cis*-Δ^5-tetradecenoyl-CoA, cΔ^5C_{14}-CoA; oleoyl-CoA, cΔ^9C_{18}-CoA; *trans*-Δ^5-tetradecenoyl-CoA, tΔ^5C_{14}-CoA; and elaidoyl-CoA, tΔ^9C_{18}- CoA.

cis-Δ^5-Tetradecenoyl-CoA

trans-Δ^5-Tetradecenoyl-CoA

(a) Why did Yu and colleagues need to use CoA derivatives rather than the free fatty acids in these experiments?

(b) Why were no lower molecular weight CoA derivatives found in the reaction with stearoyl-CoA?

(c) How many rounds of β oxidation would be required to convert the oleoyl-CoA and the elaidoyl-CoA to *cis*-Δ^5-tetradecenoyl-CoA and *trans*-Δ^5-tetradecenoyl-CoA, respectively?

There are two forms of the enzyme acyl-CoA dehydrogenase (see Fig. 17–8a): long-chain acyl-CoA dehydrogenase (LCAD) and very-long-chain acyl-CoA dehydrogenase (VLCAD). Yu and coworkers measured the kinetic parameters of both enzymes. They used the CoA derivatives of three fatty acids: tetradecanoyl-CoA (C_{14}-CoA), *cis*-Δ^5-tetradecenoyl-CoA (cΔ^5C_{14}-CoA), and *trans*-Δ^5-tetradecenoyl-CoA (tΔ^5C_{14}-CoA). The results are shown below. (See Chapter 6 for definitions of the kinetic parameters.)

	LCAD			VLCAD		
	C_{14}-CoA	cΔ^5C_{14}-CoA	tΔ^5C_{14}-CoA	C_{14}-CoA	cΔ^5C_{14}-CoA	tΔ^5C_{14}-CoA
V_{max}	3.3	3.0	2.9	1.4	0.32	0.88
K_m	0.41	0.40	1.6	0.57	0.44	0.97
k_{cat}	9.9	8.9	8.5	2.0	0.42	1.12
k_{cat}/K_m	24	22	5	4	1	1

(d) For LCAD, the K_m differs dramatically for the cis and trans substrates. Provide a plausible explanation for this observation in terms of the structures of the substrate molecules. (Hint: You may want to refer to Fig. 10–2.)

(e) The kinetic parameters of the two enzymes are relevant to the differential processing of these fatty acids *only* if the LCAD or VLCAD reaction (or both) is the rate-limiting step in the pathway. What evidence is there to support this assumption?

(f) How do these different kinetic parameters explain the different levels of the CoA derivatives found after incubation of rat liver mitochondria with stearoyl-CoA, oleoyl-CoA, and elaidoyl-CoA (shown in the three-panel figure)?

Yu and coworkers measured the substrate specificity of rat liver mitochondrial thioesterase, which hydrolyzes acyl-CoA to CoA and free fatty acid (see Chapter 21). This enzyme was approximately twice as active with C_{14}-CoA thioesters as with C_{18}-CoA thioesters.

(g) Other research has suggested that free fatty acids can pass through membranes. In their experiments, Yu and colleagues found *trans*-Δ^5-tetradecenoic acid outside mitochondria (i.e., in the medium) that had been incubated with elaidoyl-CoA. Describe the pathway that led to this extramitochondrial *trans*-Δ^5-tetradecenoic acid. Be sure to indicate where in the cell the various transformations take place, as well as the enzymes that catalyze the transformations.

(h) It is often said in the popular press that "trans fats are not broken down by your cells and instead accumulate in your body." In what sense is this statement correct and in what sense is it an oversimplification?

Answer

(a) Fatty acids are converted to their CoA derivatives by enzymes in the cytoplasm; the acyl-CoAs are then imported into mitochondria for oxidation. Given that the researchers were using isolated mitochondria, they had to use CoA derivatives.

(b) Stearoyl-CoA was rapidly converted to 9 acetyl-CoA by the β-oxidation pathway. All intermediates reacted rapidly and none were detectable at significant levels.

(c) Two rounds. Each round removes two carbon atoms, thus two rounds convert an 18-carbon to a 14-carbon fatty acid and 2 acetyl-CoA.

(d) The K_m is higher for the trans isomer than for the cis, so a higher concentration of trans isomer is required for the same rate of breakdown. Roughly speaking, the trans isomer binds less well than the cis, probably because differences in shape, even though not at the target site for the enzyme, affect substrate binding to the enzyme.

(e) The substrate for LCAD/VLCAD builds up differently, depending on the particular substrate; this is expected for the rate-limiting step in a pathway.

(f) The kinetic parameters show that the trans isomer is a poorer substrate than the cis for LCAD, but there is little difference for VLCAD. Because it is a poorer substrate, the trans isomer accumulates to higher levels than the cis.

(g) One possible pathway is shown below (indicating "inside" and "outside" mitochondria).

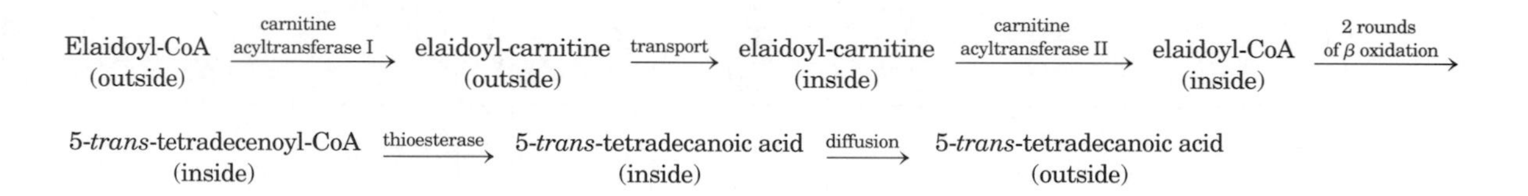

(h) It is correct insofar as trans fats are broken down less efficiently than cis fats, and thus trans fats may "leak" out of mitochondria. It is incorrect to say that trans fats are not broken down by cells; they are broken down, but at a slower rate than cis fats.

Reference

Yu, W., Liang, X., Ensenauer, R., Vockley, J., Sweetman, L., & Schultz, H. (2004) Leaky β-oxidation of a *trans*-fatty acid. *J. Biol. Chem.* **279,** 52,160–52,167.

Amino Acid Oxidation and the Production of Urea

chapter **18**

1. **Products of Amino Acid Transamination** Name and draw the structure of the α-keto acid resulting when each of the following amino acids undergoes transamination with α-ketoglutarate: **(a)** aspartate, **(b)** glutamate, **(c)** alanine, **(d)** phenylalanine.

Answer

(a) $^{-}OOC—CH_2—C(=O)—COO^{-}$ Oxaloacetate

(b) $^{-}OOC—CH_2—CH_2—C(=O)—COO^{-}$ α-Ketoglutarate

(c) $CH_3—C(=O)—COO^{-}$ Pyruvate

(d) 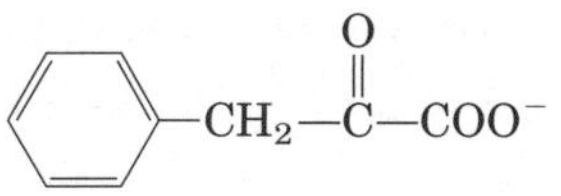 $C_6H_5—CH_2—C(=O)—COO^{-}$ Phenylpyruvate

2. **Measurement of Alanine Aminotransferase Activity** The activity (reaction rate) of alanine aminotransferase is usually measured by including an excess of pure lactate dehydrogenase and NADH in the reaction system. The rate of alanine disappearance is equal to the rate of NADH disappearance measured spectrophotometrically. Explain how this assay works.

Answer The measurement of the activity of alanine aminotransferase by measurement of the reaction of its product with lactate dehydrogenase is an example of a "coupled" assay. The product of the transamination (pyruvate) is rapidly consumed in the subsequent "indicator reaction," catalyzed by an excess of lactate dehydrogenase. The dehydrogenase uses the cofactor NADH, the disappearance of which is conveniently measured by observing the rate of decrease in NADH absorption at 340 nm. Thus, the rate of disappearance of NADH is a measure of the rate of the aminotransferase reaction, *if NADH and lactate dehydrogenase are added in excess.*

3. **Alanine and Glutamine in the Blood** Normal human blood plasma contains all the amino acids required for the synthesis of body proteins, but not in equal concentrations. Alanine and glutamine are present in much higher concentrations than any other amino acids. Suggest why.

Answer Muscle tissue can convert amino acids to their keto acids plus ammonia, then oxidize the keto acids to produce ATP for muscle contraction. However, urea cannot be formed in muscle. Alanine and glutamine transport amino groups in the bloodstream to the liver (see Fig. 18–2) from muscle and other nonhepatic tissues. In muscle, amino groups from all amino acids are transferred to pyruvate or glutamate to form alanine or glutamine, and these latter amino acids are transported to the liver.

4. **Distribution of Amino Nitrogen** If your diet is rich in alanine but deficient in aspartate, will you show signs of aspartate deficiency? Explain.

Answer No; aspartate is readily formed by the transfer of the amino group of alanine to oxaloacetate. Cellular levels of aminotransferases are sufficient to provide all of the amino acids in this fashion, if the α-keto acids are available.

5. **Lactate versus Alanine as Metabolic Fuel: The Cost of Nitrogen Removal** The three carbons in lactate and alanine have identical oxidation states, and animals can use either carbon source as a metabolic fuel. Compare the net ATP yield (moles of ATP per mole of substrate) for the complete oxidation (to CO_2 and H_2O) of lactate versus alanine when the cost of nitrogen excretion as urea is included.

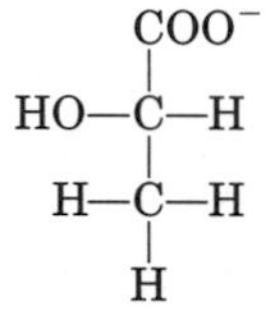

Lactate	Alanine
COO^-	COO^-
HO—C—H	$H_3\overset{+}{N}$—C—H
H—C—H	H—C—H
H	H

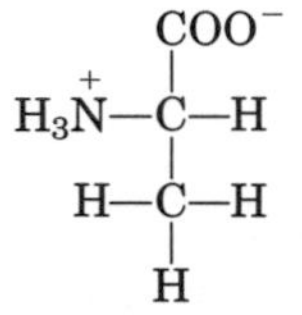

Answer Lactate and alanine are converted to pyruvate by their respective dehydrogenases, lactate dehydrogenase and alanine dehydrogenase, producing pyruvate and NADH + H^+ and, in the case of alanine, NH_4^+. Complete oxidation of 1 mol of pyruvate to CO_2 and H_2O produces 12.5 mol of ATP via the citric acid cycle and oxidative phosphorylation (see Table 16–1). In addition, the NADH from each dehydrogenase reaction produces 2.5 mol of ATP per mole of NADH reoxidized. Thus oxidation produces 15 mol of ATP per mole of lactate. Urea formation uses the equivalent of 4 mol of ATP per mole of urea formed (Fig. 18–10), or 2 mol of ATP per mol of NH_4^+. Subtracting this value from the energy yield of alanine results in 13 mol of ATP per mole of alanine oxidized.

6. **Ammonia Toxicity Resulting from an Arginine-Deficient Diet** In a study conducted some years ago, cats were fasted overnight then given a single meal complete in all amino acids except arginine. Within 2 hours, blood ammonia levels increased from a normal level of 18 μg/L to 140 μg/L, and the cats showed the clinical symptoms of ammonia toxicity. A control group fed a complete amino acid diet or an amino acid diet in which arginine was replaced by ornithine showed no unusual clinical symptoms.

(a) What was the role of fasting in the experiment?

(b) What caused the ammonia levels to rise in the experimental group? Why did the absence of arginine lead to ammonia toxicity? Is arginine an essential amino acid in cats? Why or why not?

(c) Why can ornithine be substituted for arginine?

Answer

(a) Fasting resulted in lowering of blood glucose levels. Subsequent feeding of an arginine-free diet led to a rapid catabolism of all the ingested amino acids, especially the glucogenic ones. This catabolism was exacerbated by the lack of an essential amino acid, which prevented protein synthesis.

(b) Oxidative deamination of amino acids caused the elevation of ammonia levels. In addition, the lack of arginine (an intermediate in the urea cycle) slowed the conversion of ammonia to urea. Arginine (or ornithine) synthesis in the cat is not sufficient to meet the needs imposed by the stress of this experiment, suggesting that arginine is an essential amino acid.

(c) Ornithine (or citrulline) can be substituted for arginine because it also is an intermediate in the urea cycle.

7. Oxidation of Glutamate Write a series of balanced equations, and an overall equation for the net reaction, describing the oxidation of 2 mol of glutamate to 2 mol of α-ketoglutarate and 1 mol of urea.

Answer

H_2O + glutamate + NAD^+ $\longrightarrow$ α-ketoglutarate + NH_4^+ + NADH + H^+
NH_4^+ + 2ATP + H_2O + CO_2 $\longrightarrow$ carbamoyl phosphate + 2ADP + P_i + $3H^+$
Carbamoyl phosphate + ornithine $\longrightarrow$ citrulline + P_i + H^+
Citrulline + aspartate + ATP $\longrightarrow$ argininosuccinate + AMP + PP_i + H^+
Argininosuccinate $\longrightarrow$ arginine + fumarate
Fumarate + H_2O $\longrightarrow$ malate
Malate + NAD^+ $\longrightarrow$ oxaloacetate + NADH + H^+
Oxaloacetate + glutamate $\longrightarrow$ aspartate + α-ketoglutarate
Arginine + H_2O $\longrightarrow$ urea + ornithine

The sum of these reactions is

$$\text{2 Glutamate} + CO_2 + 4H_2O + 2NAD^+ + 3ATP \longrightarrow \text{2 } \alpha\text{-ketoglutarate} + 2NADH + 7H^+ + \text{urea} + 2ADP + AMP + PP_i + 2P_i \quad (1)$$

Three additional reactions need to be considered:

$$AMP + ATP \longrightarrow 2ADP \quad (2)$$
$$O_2 + 8H^+ + 2NADH + 6ADP + 6P_i \longrightarrow 2NAD^+ + 6ATP + 8H_2O \quad (3)$$
$$H_2O + PP_i \longrightarrow 2P_i + H^+ \quad (4)$$

Summing the last four equations:

$$\text{2 Glutamate} + CO_2 + O_2 + 2ADP + 2P_i \longrightarrow \text{2 } \alpha\text{-ketoglutarate} + \text{urea} + 3H_2O + 2ATP$$

8. Transamination and the Urea Cycle Aspartate aminotransferase has the highest activity of all the mammalian liver aminotransferases. Why?

Answer The second amino group introduced into urea is transferred from aspartate. This amino acid is generated in large quantities by transamination between oxaloacetate and glutamate (and many other amino acids), catalyzed by aspartate aminotransferase. Approximately one half of all the amino groups excreted as urea must pass through the aspartate aminotransferase reaction, and liver contains higher levels of this aminotransferase than of any other.

9. The Case against the Liquid Protein Diet A weight-reducing diet heavily promoted some years ago required the daily intake of "liquid protein" (soup of hydrolyzed gelatin), water, and an assortment of vitamins. All other food and drink were to be avoided. People on this diet typically lost 10 to 14 lb in the first week.

(a) Opponents argued that the weight loss was almost entirely due to water loss and would be regained very soon after a normal diet was resumed. What is the biochemical basis for this argument?

(b) A number of people on this diet died. What are some of the dangers inherent in the diet and how can they lead to death?

Answer

(a) A person on a diet consisting only of protein must use amino acids as the principal source of metabolic fuel. Because the catabolism of amino acids requires the removal of nitrogen as urea, the process consumes large quantities of water to dilute and excrete the urea in the urine. Furthermore, electrolytes in the "liquid protein" must be diluted with water and excreted. If this abnormally large daily water loss through the kidney is not balanced by a sufficient water intake, a net loss of body water results.

(b) When considering the nutritional benefits of protein, keep in mind the total amount of amino acids needed for protein synthesis and the distribution of amino acids in the dietary protein. Gelatin contains a nutritionally unbalanced distribution of amino acids. As large amounts of gelatin are ingested and the excess amino acids are catabolized, the capacity of the urea cycle may be exceeded, leading to ammonia toxicity. This is further complicated by the dehydration that may result from excretion of large quantities of urea. A combination of these two factors could produce coma and death.

10. Ketogenic Amino Acids Which amino acids are exclusively ketogenic?

Answer Lysine and leucine are exclusively ketogenic. These amino acids are degraded entirely to acetyl-CoA and acetoacetyl-CoA, and no parts of their carbon skeletons can be used for glucose synthesis. Leucine is especially common in proteins. Its degradation makes a substantial contribution to ketosis under starvation conditions.

11. A Genetic Defect in Amino Acid Metabolism: A Case History A two-year-old child was taken to the hospital. His mother said that he vomited frequently, especially after feedings. The child's weight and physical development were below normal. His hair, although dark, contained patches of white. A urine sample treated with ferric chloride ($FeCl_3$) gave a green color characteristic of the presence of phenylpyruvate. Quantitative analysis of urine samples gave the results shown in the table.

	Concentration (mM)	
Substance	**Patient's urine**	**Normal urine**
Phenylalanine	7.0	0.01
Phenylpyruvate	4.8	0
Phenyllactate	10.3	0

(a) Suggest which enzyme might be deficient in this child. Propose a treatment.

(b) Why does phenylalanine appear in the urine in large amounts?

(c) What is the source of phenylpyruvate and phenyllactate? Why does this pathway (normally not functional) come into play when the concentration of phenylalanine rises?

(d) Why does the boy's hair contain patches of white?

Answer

(a) Because phenylalanine (and its related phenylketones) accumulate in this patient, it is likely that the first enzyme in phenylalanine catabolism, phenylalanine hydroxylase (also called phenylalanine-4-monooxygenase), is defective or missing (see Fig. 18–23). The most appropriate treatment for patients with this disease, known as phenylketonuria (PKU), is to establish a low-phenylalanine diet that provides just enough of the amino acid to meet the needs for protein synthesis.

(b) Phenylalanine appears in the urine because high levels of this amino acid accumulate in the bloodstream and the body attempts to dispose of it.

(c) Phenylalanine is converted to phenylpyruvate by transamination, a reaction that has an equilibrium constant of about 1.0. Phenyllactate is formed from phenylpyruvate by reduction (see Fig. 18–25). This pathway is of importance only when phenylalanine hydroxylase is defective.

(d) The normal catabolic pathway of phenylalanine is through tyrosine, a precursor of melanin, the dark pigment normally present in hair. Decreased tyrosine levels in patients with phenylketonuria result in varying degrees of pigment loss.

12. Role of Cobalamin in Amino Acid Catabolism Pernicious anemia is caused by impaired absorption of vitamin B_{12}. What is the effect of this impairment on the catabolism of amino acids? Are all amino acids equally affected? (Hint: see Box 17–2.)

Answer The catabolism of the carbon skeletons of valine, isoleucine, and methionine is impaired because of the absence of a functional methylmalonyl-CoA mutase. This enzyme requires coenzyme B_{12} as a cofactor, and a deficiency of this vitamin leads to elevated methylmalonic acid levels (methylmalonic acidemia). The symptoms and effects of this deficiency are severe (see Table 18–2 and Box 18–2).

13. Vegetarian Diets Vegetarian diets can provide high levels of antioxidants and a lipid profile that can help prevent coronary disease. However, there can be some associated problems. Blood samples were taken from a large group of volunteer subjects who were vegans (strict vegetarians: no animal products), lactovegetarians (vegetarians who eat dairy products), or omnivores (individuals with a normal, varied diet including meat). In each case, the volunteers had followed the diet for several years. The blood levels of both homocysteine and methylmalonate were elevated in the vegan group, somewhat lower in the lactovegetarian group, and much lower in the omnivore group. Explain.

Answer The vegan diet lacks vitamin B_{12}, leading to the increase in homocysteine and methylmalonate (reflecting the deficiencies in methionine synthase and methylmalonic acid mutase, respectively) in individuals on the diet for several years. Dairy products provide some vitamin B_{12} in the lactovegetarian diet.

14. Pernicious Anemia Vitamin B_{12} deficiency can arise from a few rare genetic diseases that lead to low B_{12} levels despite a normal diet that includes B_{12}-rich meat and dairy sources. These conditions cannot be treated with dietary B_{12} supplements. Explain.

Answer The genetic forms of pernicious anemia generally arise as a result of defects in the pathway that mediates absorption of dietary vitamin B_{12} (see Box 17–2, p. 658). Because dietary supplements are not absorbed in the intestine, these conditions are treated by injecting supplementary B_{12} directly into the bloodstream.

15. Pyridoxal Phosphate Reaction Mechanisms Threonine can be broken down by the enzyme threonine dehydratase, which catalyzes the conversion of threonine to α-ketobutyrate and ammonia. The enzyme uses PLP as a cofactor. Suggest a mechanism for this reaction, based on the mechanisms in Figure 18–6. Note that this reaction includes an elimination at the β carbon of threonine.

$$\underset{\text{Threonine}}{CH_3{-}\overset{\overset{\displaystyle OH}{|}}{CH}{-}\overset{\overset{\displaystyle \overset{+}{N}H_3}{|}}{CH}{-}COO^-} \xrightarrow[\text{threonine dehydratase}]{\text{PLP}} \underset{\alpha\text{-Ketobutyrate}}{CH_3{-}CH_2{-}\overset{\overset{\displaystyle O}{\|}}{C}{-}COO^-} + NH_3 + H_2O$$

Answer The mechanism is identical to that for serine dehydratase (see Fig. 18–20a, p. 693) except that the extra methyl group of threonine is retained, yielding α-ketobutyrate instead of pyruvate.

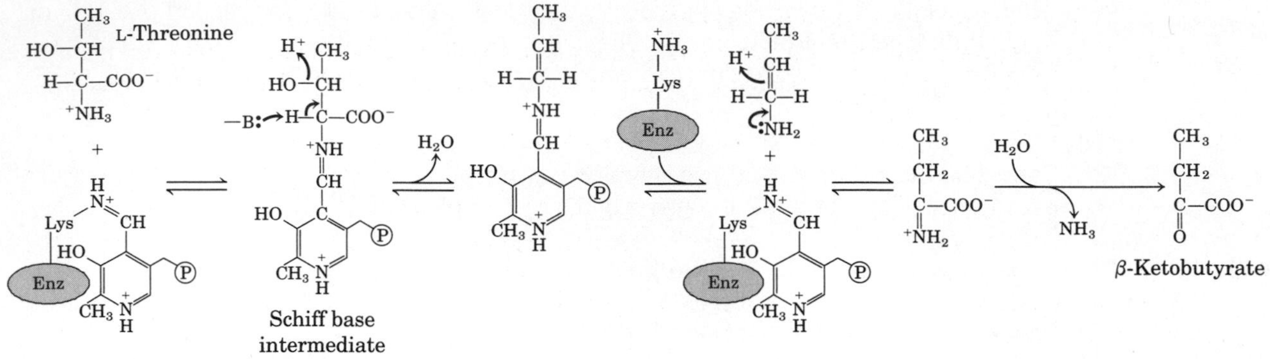

L-Threonine

Schiff base intermediate

β-Ketobutyrate

16. Pathway of Carbon and Nitrogen in Glutamate Metabolism When [2-^{14}C, ^{15}N] glutamate undergoes oxidative degradation in the liver of a rat, in which atoms of the following metabolites will each isotope be found: **(a)** urea, **(b)** succinate, **(c)** arginine, **(d)** citrulline, **(e)** ornithine, **(f)** aspartate?

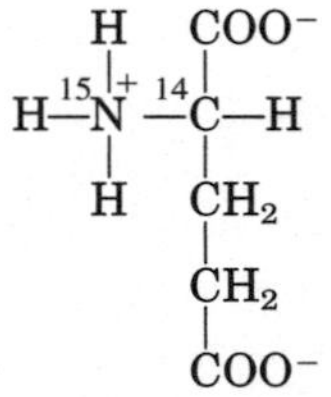

$$H_3{}^{15}\overset{+}{N}-{}^{14}CH(COO^-)-CH_2-CH_2-COO^-$$

Labeled glutamate

Answer

(a) The amino groups of urea contain ^{15}N, a result of glutamate dehydrogenase producing $^{15}NH_4^+$ or of a transaminase producing ^{15}N-labeled aspartate.

$$^{15}NH_2-CO-{}^{15}NH_2$$

(b) After loss of the amino group, the [2-^{14}C] α-ketoglutarate is metabolized in the citric acid cycle. Succinate thus formed is labeled in the carboxyl groups.

$$^-OO^{14}C-CH_2-CH_2-{}^{14}COO^-$$

(c) The arginine formed in the urea cycle contains ^{15}N in both guanidino nitrogens.

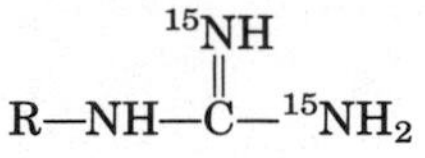

$$R-NH-C(={}^{15}NH)-{}^{15}NH_2$$

(d) Citrulline formed in the urea cycle contains ^{15}N in the carboxamide group.

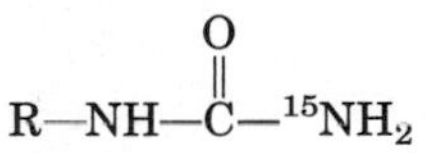

$$R-NH-C(=O)-{}^{15}NH_2$$

(e) No labeled N is found in ornithine.

(f) Aspartate contains ^{15}N in its amino group as a result of transamination from glutamate. It also contains ^{14}C in its carboxyl groups as a result of succinate conversion to oxaloacetate (as in **(b)**).

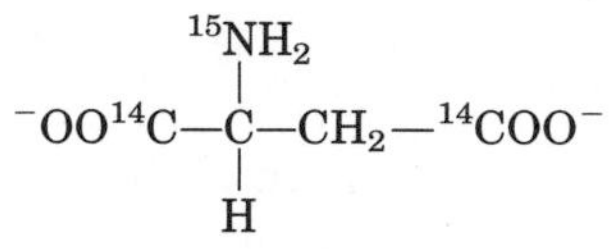

$$^{-}OO^{14}C-\underset{\displaystyle H}{\overset{\displaystyle ^{15}NH_2}{C}}-CH_2-^{14}COO^{-}$$

Note: in **(c)**, **(d)**, and **(e)**, these urea cycle intermediates will contain low levels of ^{14}C as a result of a very weak synthesis of ornithine from glutamate.

17. Chemical Strategy of Isoleucine Catabolism Isoleucine is degraded in six steps to propionyl-CoA and acetyl-CoA.

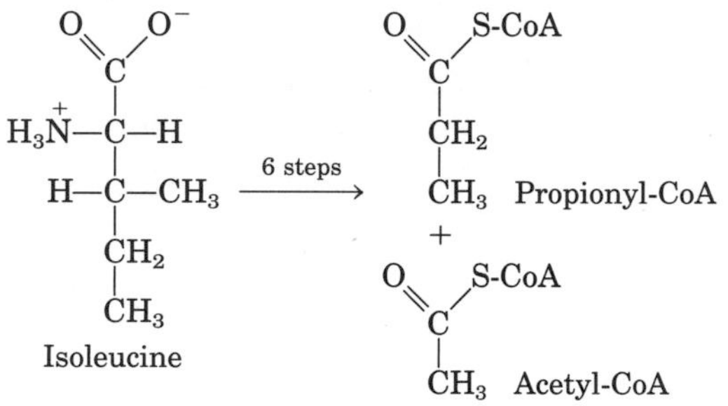

Isoleucine: $^{-}OOC-CH(\overset{+}{N}H_3)-CH(CH_3)-CH_2-CH_3$ $\xrightarrow{\text{6 steps}}$ Propionyl-CoA: $CoA\text{-}S-C(=O)-CH_2-CH_3$ + Acetyl-CoA: $CoA\text{-}S-C(=O)-CH_3$

(a) The chemical process of isoleucine degradation includes strategies analogous to those used in the citric acid cycle and the β oxidation of fatty acids. The intermediates of isoleucine degradation (I to V) shown below are not in the proper order. Use your knowledge and understanding of the citric acid cycle and β-oxidation pathway to arrange the intermediates in the proper metabolic sequence for isoleucine degradation.

I: $CoA\text{-}S-C(=O)-C(CH_3)=CH-CH_3$

II: $^{-}O-C(=O)-C(=O)-CH(CH_3)-CH_2-CH_3$

III: $CoA\text{-}S-C(=O)-CH(CH_3)-C(=O)-CH_3$

IV: $CoA\text{-}S-C(=O)-CH(CH_3)-CH_2-CH_3$

V: $CoA\text{-}S-C(=O)-CH(CH_3)-CH(OH)-CH_3$

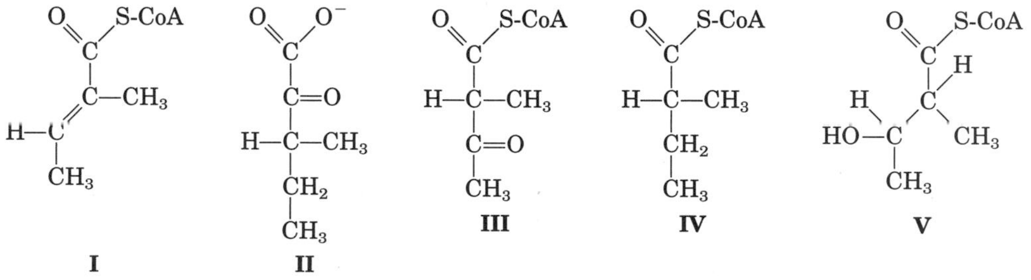

(b) For each step you propose, describe the chemical process, provide an analogous example from the citric acid cycle or β-oxidation pathway (where possible), and indicate any necessary cofactors.

Answer

(a) Isoleucine $\xrightarrow{1}$ II $\xrightarrow{2}$ IV $\xrightarrow{3}$ I $\xrightarrow{4}$ V $\xrightarrow{5}$ III $\xrightarrow{6}$ acetyl-CoA + propionyl-CoA

(b) Step 1 is a transamination that has no analogous reaction; it requires PLP. Step 2 is an oxidative decarboxylation similar to the pyruvate dehydrogenase reaction; it requires NAD^+, TPP, lipoate, FAD. Step 3 is an oxidation similar to the succinate dehydrogenase reaction; it requires FAD. Step 4 is a hydration analogous to the fumarase reaction; no cofactor is required. Step 5 is an oxidation analogous to the malate dehyrogenase reaction of the citric acid cycle; it requires NAD^+. Step 6 is a thiolysis analogous to the final cleavage step of β oxidation catalyzed by thiolase; it requires CoA.

18. Role of Pyridoxal Phosphate in Glycine Metabolism The enzyme serine hydroxymethyltransferase requires pyridoxal phosphate as cofactor. Propose a mechanism for the reaction catalyzed by this enzyme, in the direction of serine degradation (glycine production). (Hint: see Figs 18–19 and 18–20b.)

Answer See the mechanism below. The formaldehyde produced in the second step reacts rapidly with tetrahydrofolate at the enzyme active site to produce N^5,N^{10}-methylene tetrahydrofolate (see Fig. 18–17).

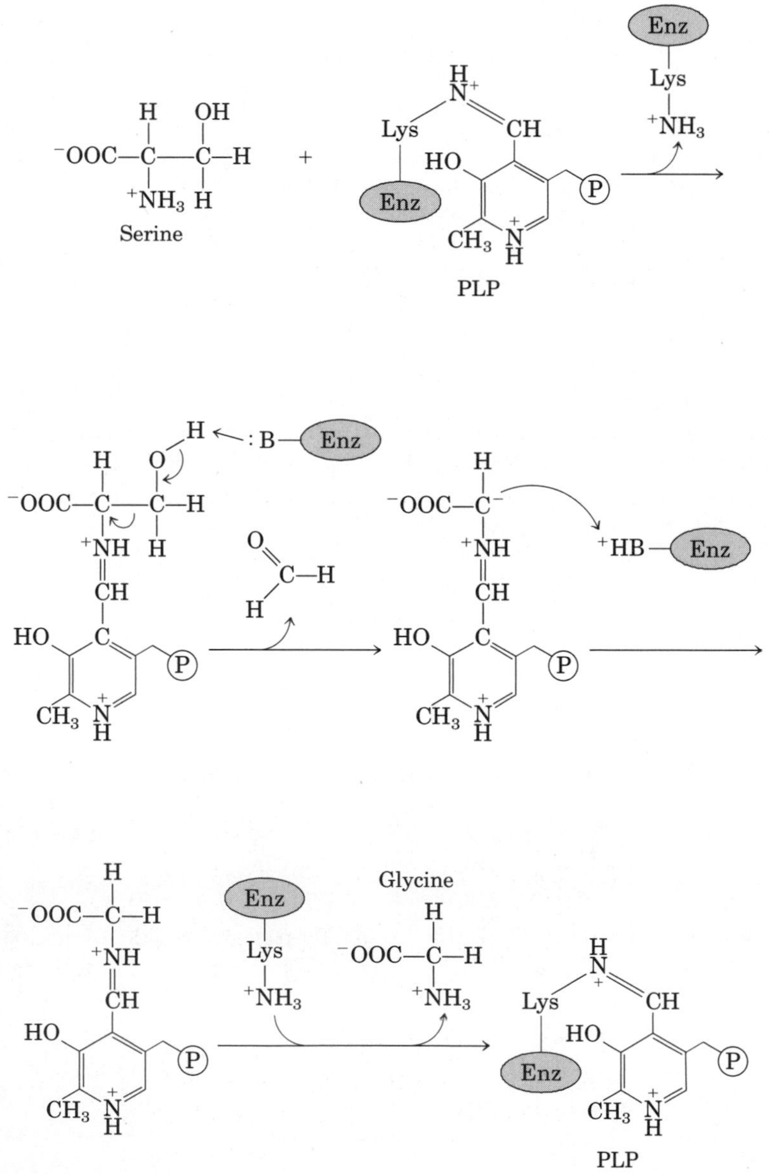

19. Parallel Pathways for Amino Acid and Fatty Acid Degradation The carbon skeleton of leucine is degraded by a series of reactions closely analogous to those of the citric acid cycle and β oxidation. For each reaction, **(a)** through **(f),** indicate its type, provide an analogous example from the citric acid cycle or β-oxidation pathway (where possible), and note any necessary cofactors.

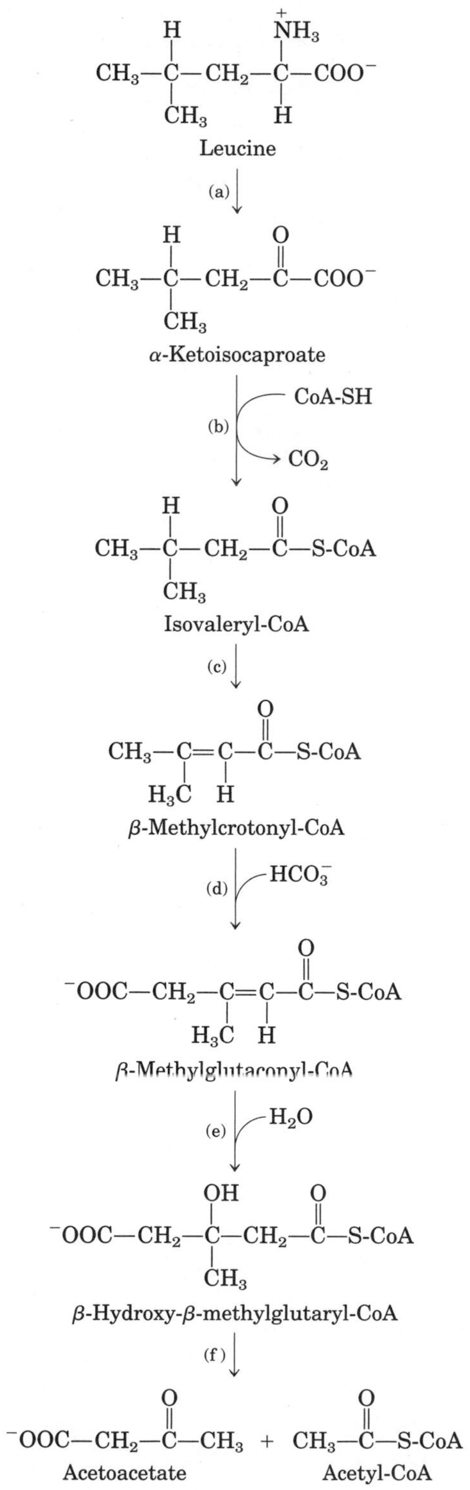

$CH_3-CH(CH_3)-CH_2-CH(\overset{+}{N}H_3)-COO^-$

Leucine

(a) ↓

$CH_3-CH(CH_3)-CH_2-C(=O)-COO^-$

α-Ketoisocaproate

(b) ↓ CoA-SH → CO_2

$CH_3-CH(CH_3)-CH_2-C(=O)-S\text{-}CoA$

Isovaleryl-CoA

(c) ↓

$CH_3-C(H_3C)=C(H)-C(=O)-S\text{-}CoA$

β-Methylcrotonyl-CoA

(d) ↓ HCO_3^-

$^-OOC-CH_2-C(H_3C)=C(H)-C(=O)-S\text{-}CoA$

β-Methylglutaconyl-CoA

(e) ↓ H_2O

$^-OOC-CH_2-C(OH)(CH_3)-CH_2-C(=O)-S\text{-}CoA$

β-Hydroxy-β-methylglutaryl-CoA

(f) ↓

$^-OOC-CH_2-C(=O)-CH_3 + CH_3-C(=O)-S\text{-}CoA$

Acetoacetate + Acetyl-CoA

Answer

(a) Transamination; no analogies in either pathway; requires PLP.

(b) Oxidative decarboxylation; analogous to oxidative decarboxylation of pyruvate to acetyl-CoA prior to entry into the citric acid cycle, and of α-ketoglutarate to succinyl-CoA in the citric acid cycle; requires NAD^+, FAD, lipoate, thiamine pyrophosphate.

(c) Dehydrogenation (oxidation); analogous to dehydrogenation of succinate to fumarate in the citric acid cycle and of fatty acyl–CoA to enoyl-CoA in β oxidation; requires FAD.

(d) Carboxylation; no analogous reaction in the citric acid cycle or β oxidation; requires ATP and biotin.

(e) Hydration; analogous to hydration of fumarate to malate in the citric acid cycle and of enoyl-CoA to 3-hydroxyacyl-CoA in β oxidation; no cofactors.

(f) Reverse aldol reaction; analogous to reverse of citrate synthase reaction in the citric acid cycle and identical to cleavage of β-hydroxy-β-methylglutaryl-CoA in formation of ketone bodies; no cofactors.

Data Analysis Problem

20. Maple Syrup Urine Disease Figure 18–28 shows the pathway for the degradation of branched-chain amino acids and the site of the biochemical defect that causes maple syrup urine disease. The initial findings that eventually led to the discovery of the defect in this disease were presented in three papers published in the late 1950s and early 1960s. This problem traces the history of the findings from initial clinical observations to proposal of a biochemical mechanism.

Menkes, Hurst, and Craig (1954) presented the cases of four siblings, all of whom died following a similar course of symptoms. In all four cases, the mother's pregnancy and the birth had been normal. The first 3 to 5 days of each child's life were also normal. But soon thereafter each child began having convulsions, and the children died between the ages of 11 days and 3 months. Autopsy showed considerable swelling of the brain in all cases. The children's urine had a strong, unusual "maple syrup" odor, starting from about the third day of life.

Menkes (1959) reported data collected from six more children. All showed symptoms similar to those described above, and died within 15 days to 20 months of birth. In one case, Menkes was able to obtain urine samples during the last months of the infant's life. When he treated the urine with 2,4-dinitrophenylhydrazone, which forms colored precipitates with keto compounds, he found three α-keto acids in unusually large amounts:

α-Ketoisocaproate	α-Ketoisovalerate	α-Keto-β-methyl-n-valerate
COO^-	COO^-	COO^-
$C{=}O$	$C{=}O$	$C{=}O$
CH_2	$CH_3{-}CH$	$CH_3{-}CH$
$CH_3{-}CH$	CH_3	CH_2
CH_3		CH_3

(a) These α-keto acids are produced by the deamination of amino acids. For each of the α-keto acids above, draw and name the amino acid from which it was derived.

Dancis, Levitz, and Westall (1960) collected further data that led them to propose the biochemical defect shown in Figure 18–28. In one case, they examined a patient whose urine first showed the maple syrup odor when he was 4 months old. At the age of 10 months (March 1956), the child was admitted to the hospital because he had a fever, and he showed grossly retarded motor development.

At the age of 20 months (January 1957), he was readmitted and was found to have the degenerative neurological symptoms seen in previous cases of maple syrup urine disease; he died soon after. Results of his blood and urine analyses are shown in the table below, along with normal values for each component.

	Urine (mg/24 h)			Plasma (mg/ml)	
	Normal	Patient		Normal	Patient
Amino acid(s)		Mar. 1956	Jan. 1957		Jan. 1957
Alanine	5–15	0.2	0.4	3.0–4.8	0.6
Asparagine and glutamine	5–15	0.4	0	3.0–5.0	2.0
Aspartic acid	1–2	0.2	1.5	0.1–0.2	0.04
Arginine	1.5–3	0.3	0.7	0.8–1.4	0.8
Cystine	2–4	0.5	0.3	1.0–1.5	0
Glutamic acid	1.5–3	0.7	1.6	1.0–1.5	0.9
Glycine	20–40	4.6	20.7	1.0–2.0	1.5
Histidine	8–15	0.3	4.7	1.0–1.7	0.7
Isoleucine	2–5	2.0	13.5	0.8–1.5	2.2
Leucine	3–8	2.7	39.4	1.7–2.4	14.5
Lysine	2–12	1.6	4.3	1.5–2.7	1.1
Methionine	2–5	1.4	1.4	0.3–0.6	2.7
Ornithine	1–2	0	1.3	0.6–0.8	0.5
Phenylalanine	2–4	0.4	2.6	1.0–1.7	0.8
Proline	2–4	0.5	0.3	1.5–3.0	0.9
Serine	5–15	1.2	0	1.3–2.2	0.9
Taurine	1–10	0.2	18.7	0.9–1.8	0.4
Threonine	5–10	0.6	0	1.2–1.6	0.3
Tryptophan	3–8	0.9	2.3	Not measured	0
Tyrosine	4–8	0.3	3.7	1.5–2.3	0.7
Valine	2–4	1.6	15.4	2.0–3.0	13.1

(b) The table includes taurine, an amino acid not normally found in proteins. Taurine is often produced as a by-product of cell damage. Its structure is:

$$H_3\overset{+}{N}-CH_2-CH_2-\overset{\overset{O}{\|}}{\underset{\underset{O}{\|}}{S}}-O^-$$

Based on its structure and the information in this chapter, what is the most likely amino acid precursor of taurine? Explain your reasoning.

(c) Compared with the normal values given in the table, which amino acids showed significantly elevated levels in the patient's blood in January 1957? Which ones in the patient's urine?

Based on their results and their knowledge of the pathway shown in Figure 18–28, Dancis and coauthors concluded: "although it appears most likely to the authors that the primary block is in the metabolic degradative pathway of the branched-chain amino acids, this cannot be considered established beyond question."

(d) How do the data presented here support this conclusion?

(e) Which data presented here do *not* fit this model of maple syrup urine disease? How do you explain these seemingly contradictory data?

(f) What data would you need to collect to be more secure in your conclusion?

Answer

(a) Leucine; valine; isoleucine

(b) Cysteine (derived from cystine). If cysteine were decarboxylated as shown in Figure 18–6, it would yield $H_3N^+—CH_2—CH_2—SH$, which could be oxidized to taurine.

(c) The January 1957 blood shows significantly elevated levels of isoleucine, leucine, methionine, and valine; the January 1957 urine, significantly elevated isoleucine, leucine, taurine, and valine.

(d) All patients had high levels of isoleucine, leucine, and valine in both blood and urine, suggesting a defect in the breakdown of these amino acids. Given that the urine also contained high levels of the keto forms of these three amino acids, the block in the pathway must occur after deamination but before dehydrogenation (as shown in Fig. 18–28).

(e) The model does not explain the high levels of methionine in blood and taurine in urine. The high taurine levels may be due to the death of brain cells during the end stage of the disease. However, the reason for high levels of methionine in blood are unclear; the pathway of methionine degradation is not linked with the degradation of branched-chain amino acids. Increased methionine could be a secondary effect of buildup of the other amino acids. It is important to keep in mind that the January 1957 samples were from an individual who was dying, so comparing blood and urine results with those of a healthy individual may not be appropriate.

(f) The following information is needed (and was eventually obtained by other workers): (1) The dehydrogenase activity is significantly reduced or missing in individuals with maple syrup urine disease. (2) The disease is inherited as a single-gene defect. (3) The defect occurs in a gene encoding all or part of the dehydrogenase. (4) The genetic defect leads to production of inactive enzyme.

References

Dancis, J., Levitz, M., & Westall, R. (1960) Maple syrup urine disease: branched-chain keto-aciduria. *Pediatrics* **25,** 72–79.

Menkes, J.H. (1959) Maple syrup disease: isolation and identification of organic acids in the urine. *Pediatrics* **23,** 348–353.

Menkes, J.H., Hurst, P.L., & Craig J.M. (1954) A new syndrome: progressive familial infantile cerebral dysfunction associated with an unusual urinary substance. *Pediatrics* **14,** 462-466.

Oxidative Phosphorylation and Photophosphorylation

chapter

19

1. **Oxidation-Reduction Reactions** The NADH dehydrogenase complex of the mitochondrial respiratory chain promotes the following series of oxidation-reduction reactions, in which Fe^{3+} and Fe^{2+} represent the iron in iron-sulfur centers, Q is ubiquinone, QH_2 is ubiquinol, and E is the enzyme:

   (1) $NADH + H^+ + E\text{-}FMN \longrightarrow NAD^+ + E\text{-}FMNH_2$

   (2) $E\text{-}FMNH_2 + 2Fe^{3+} \longrightarrow E\text{-}FMN + 2Fe^{2+} + 2H^+$

   (3) $2Fe^{2+} + 2H^+ + Q \longrightarrow 2Fe^{3+} + QH_2$

   Sum: $NADH + H^+ + Q \longrightarrow NAD^+ + QH_2$

   For each of the three reactions catalyzed by the NADH dehydrogenase complex, identify **(a)** the electron donor, **(b)** the electron acceptor, **(c)** the conjugate redox pair, **(d)** the reducing agent, and **(e)** the oxidizing agent.

   Answer Oxidation-reduction reactions require an electron donor and an electron acceptor. Recall that electron donors are reducing agents; electron acceptors are oxidizing agents.

   (1) NADH is the electron donor **(a)** and the reducing agent **(d);** E-FMN is the electron acceptor **(b)** and the oxidizing agent **(e);** NAD^+/NADH and E-FMN/E-$FMNH_2$ are conjugate redox pairs **(c).**

   (2) E-$FMNH_2$ is the electron donor **(a)** and reducing agent **(d);** Fe^{3+} is the electron acceptor **(b)** and oxidizing agent **(e);** E-FMN/E-$FMNH_2$ and Fe^{3+}/Fe^{2+} are redox pairs **(c).**

   (3) Fe^{2+} is the electron donor **(a)** and reducing agent **(d);** Q is the electron acceptor **(b)** and oxidizing agent **(e);** and Fe^{3+}/Fe^{2+} and Q/QH_2 are redox pairs **(c).**

2. **All Parts of Ubiquinone Have a Function** In electron transfer, only the quinone portion of ubiquinone undergoes oxidation-reduction; the isoprenoid side chain remains unchanged. What is the function of this chain?

   Answer The long isoprenoid side chain makes ubiquinone very soluble in lipids and allows it to diffuse in the semifluid membrane. This is important because ubiquinone transfers electrons from Complexes I and II to Complex III, all of which are embedded in the inner mitochondrial membrane.

3. Use of FAD Rather Than NAD^+ in Succinate Oxidation All the dehydrogenases of glycolysis and the citric acid cycle use NAD^+ (E'° for NAD^+/NADH is −0.32 V) as electron acceptor except succinate dehydrogenase, which uses covalently bound FAD (E'° for FAD/$FADH_2$ in this enzyme is 0.050 V). Suggest why FAD is a more appropriate electron acceptor than NAD^+ in the dehydrogenation of succinate, based on the E'° values of fumarate/succinate (E'° = 0.031), NAD^+/NADH, and the succinate dehydrogenase FAD/$FADH_2$.

Answer From the difference in standard reduction potential ($\Delta E'^\circ$) for each pair of half-reactions, we can calculate the $\Delta G'^\circ$ values for the oxidation of succinate using NAD^+ and oxidation using E-FAD.

For NAD^+:

$$\begin{aligned}\Delta G'^\circ &= -n\mathcal{F}\Delta E'^\circ \\ &= -2(96.5 \text{ kJ/V}\cdot\text{mol})(-0.32 \text{ V} - 0.031 \text{ V}) \\ &= \quad 68 \text{ kJ/mol}\end{aligned}$$

For E-FAD:

$$\begin{aligned}\Delta G'^\circ &= -2(96.5 \text{ kJ/V}\cdot\text{mol})(0.050 \text{ V} - 0.031 \text{ V}) \\ &= -3.7 \text{ kJ/mol}\end{aligned}$$

The oxidation of succinate by E-FAD is favored by the negative standard free-energy change, which is consistent with a K'_{eq} of >1. Oxidation by NAD^+ would require a large, positive, standard free-energy change and have a K'_{eq} favoring the synthesis of succinate.

4. Degree of Reduction of Electron Carriers in the Respiratory Chain The degree of reduction of each carrier in the respiratory chain is determined by conditions in the mitochondrion. For example, when NADH and O_2 are abundant, the steady-state degree of reduction of the carriers decreases as electrons pass from the substrate to O_2. When electron transfer is blocked, the carriers before the block become more reduced and those beyond the block become more oxidized (see Fig. 19–6). For each of the conditions below, predict the state of oxidation of ubiquinone and cytochromes *b*, c_1, *c*, and $a + a_3$.

(a) Abundant NADH and O_2, but cyanide added
(b) Abundant NADH, but O_2 exhausted
(c) Abundant O_2, but NADH exhausted
(d) Abundant NADH and O_2

Answer As shown in Figure 19–6, the oxidation-reduction state of the carriers in the electron-transfer system varies with the conditions.

(a) Cyanide inhibits cytochrome oxidase ($a + a_3$); all carriers become reduced.
(b) In the absence of O_2, no terminal electron acceptor is present; all carriers become reduced.
(c) In the absence of NADH, no carrier can be reduced; all carriers become oxidized.
(d) These are the usual conditions for an aerobic, actively metabolizing cell; the early carriers (e.g., Q) are somewhat reduced, while the late ones (e.g., cytochrome *c*) are oxidized.

5. Effect of Rotenone and Antimycin A on Electron Transfer Rotenone, a toxic natural product from plants, strongly inhibits NADH dehydrogenase of insect and fish mitochondria. Antimycin A, a toxic antibiotic, strongly inhibits the oxidation of ubiquinol.

(a) Explain why rotenone ingestion is lethal to some insect and fish species.
(b) Explain why antimycin A is a poison.
(c) Given that rotenone and antimycin A are equally effective in blocking their respective sites in the electron-transfer chain, which would be a more potent poison? Explain.

Answer

(a) The inhibition of NADH dehydrogenase by rotenone decreases the rate of electron flow through the respiratory chain, which in turn decreases the rate of ATP production. If this reduced rate is unable to meet its ATP requirements, the organism dies.

(b) Antimycin A strongly inhibits the oxidation of reduced Q in the respiratory chain, severely limiting the rate of electron transfer and ATP production.

(c) Electrons flow into the system at Complex I from the NAD^+-linked reactions and at Complex II from succinate and fatty acyl–CoA through FAD (see Figs. 19–8 and 19–16). Antimycin A inhibits electron flow (through Q) from *all* these sources, whereas rotenone inhibits flow only through Complex I. Thus, antimycin A is a more potent poison.

6. Uncouplers of Oxidative Phosphorylation In normal mitochondria the rate of electron transfer is tightly coupled to the demand for ATP. When the rate of use of ATP is relatively low, the rate of electron transfer is low; when demand for ATP increases, electron-transfer rate increases. Under these conditions of tight coupling, the number of ATP molecules produced per atom of oxygen consumed when NADH is the electron donor—the P/O ratio–is about 2.5.

(a) Predict the effect of a relatively low and a relatively high concentration of uncoupling agent on the rate of electron transfer and the P/O ratio.

(b) Ingestion of uncouplers causes profuse sweating and an increase in body temperature. Explain this phenomenon in molecular terms. What happens to the P/O ratio in the presence of uncouplers?

(c) The uncoupler 2,4-dinitrophenol was once prescribed as a weight-reducing drug. How could this agent, in principle, serve as a weight-reducing aid? Uncoupling agents are no longer prescribed because some deaths occurred following their use. Why might the ingestion of uncouplers lead to death?

Answer Uncouplers of oxidative phosphorylation stimulate the rate of electron flow but not ATP synthesis.

(a) At relatively low levels of an uncoupling agent, P/O ratios drop somewhat, but the cell can compensate for this by increasing the rate of electron flow; ATP levels can be kept relatively normal. At high levels of uncoupler, P/O ratios approach zero and the cell cannot maintain ATP levels.

(b) As amounts of an uncoupler increase, the P/O ratio decreases and the body struggles to make sufficient ATP by oxidizing more fuel. The heat produced by this increased rate of oxidation raises the body temperature. The P/O ratio is affected as noted in **(a).**

(c) Increased activity of the respiratory chain in the presence of an uncoupler requires the degradation of additional energy stores (glycogen and fat). By oxidizing more fuel in an attempt to produce the same amount of ATP, the organism loses weight. If the P/O ratio nears zero, the lack of ATP will be lethal.

7. Effects of Valinomycin on Oxidative Phosphorylation When the antibiotic valinomycin is added to actively respiring mitochondria, several things happen: the yield of ATP decreases, the rate of O_2 consumption increases, heat is released, and the pH gradient across the inner mitochondrial membrane increases. Does valinomycin act as an uncoupler or an inhibitor of oxidative phosphorylation? Explain the experimental observations in terms of the antibiotic's ability to transfer K^+ ions across the inner mitochondrial membrane.

Answer The observed effects are consistent with the action of an uncoupler—that is, an agent that causes the free energy released in electron transfer to appear as heat rather than in ATP. In respiring mitochondria, H^+ ions are translocated out of the matrix during electron transfer, creating a proton gradient and an electrical potential across the membrane. A significant portion of the free energy used to synthesize ATP originates from this electric potential. Valinomycin combines with K^+ ions to form a complex that passes through the inner mitochondrial membrane. So, as a proton is translocated out by electron transfer, a K^+ ion moves in, and the potential across the membrane is lost. This reduces the yield of ATP per mole of protons flowing through ATP synthase (F_oF_1). In other words, electron transfer and phosphorylation become uncoupled. In response to the decreased efficiency of ATP synthesis, the rate of electron transfer increases markedly. This results in an increase in the H^+ gradient, in oxygen consumption, and in the amount of heat released.

8. Mode of Action of Dicyclohexylcarbodiimide (DCCD) When DCCD is added to a suspension of tightly coupled, actively respiring mitochondria, the rate of electron transfer (measured by O_2 consumption) and the rate of ATP production dramatically decrease. If a solution of 2,4-dinitrophenol is now added to the preparation, O_2 consumption returns to normal but ATP production remains inhibited.

(a) What process in electron transfer or oxidative phosphorylation is affected by DCCD?

(b) Why does DCCD affect the O_2 consumption of mitochondria? Explain the effect of 2,4-dinitrophenol on the inhibited mitochondrial preparation.

(c) Which of the following inhibitors does DCCD most resemble in its action: antimycin A, rotenone, or oligomycin?

Answer

(a) DCCD inhibits ATP synthesis. In tightly coupled mitochondria, this inhibition leads to inhibition of electron transfer also.

(b) A decrease in electron transfer causes a decrease in O_2 consumption. 2,4-Dinitrophenol uncouples electron transfer from ATP synthesis, allowing respiration to increase. No ATP is synthesized and the P/O ratio decreases.

(c) DCCD and oligomycin inhibit ATP synthesis (see Table 19–4).

9. Compartmentalization of Citric Acid Cycle Components Isocitrate dehydrogenase is found only in the mitochondrion, but malate dehydrogenase is found in both the cytosol and mitochondrion. What is the role of cytosolic malate dehydrogenase?

Answer Malate dehydrogenase catalyzes the conversion of malate to oxaloacetate in the citric acid cycle, which takes place in the mitochondrion, and also plays a key role in the transport of reducing equivalents across the inner mitochondrial membrane via the malate-aspartate shuttle (Fig. 19–29). This shuttle requires the presence of malate dehydrogenase in the cytosol and the mitochondrial matrix.

10. The Malate–α-Ketoglutarate Transport System The transport system that conveys malate and α-ketoglutarate across the inner mitochondrial membrane (see Fig. 19–29) is inhibited by *n*-butylmalonate. Suppose *n*-butylmalonate is added to an aerobic suspension of kidney cells using glucose exclusively as fuel. Predict the effect of this inhibitor on **(a)** glycolysis, **(b)** oxygen consumption, **(c)** lactate formation, and **(d)** ATP synthesis.

Answer NADH produced in the cytosol cannot cross the inner mitochondrial membrane, but must be oxidized if glycolysis is to continue. Reducing equivalents from NADH enter the mitochondrion by way of the malate-aspartate shuttle. NADH reduces oxaloacetate to form malate and NAD^+, and the malate is transported into the mitochondrion. Cytosolic oxidation of glucose can continue, and the malate is converted back to oxaloacetate and NADH in the mitochondrion (see Fig. 19–29).

(a) If *n*-butylmalonate, an inhibitor of the malate–α-ketoglutarate transporter, is added to cells, NADH accumulates in the cytosol. This forces glycolysis to operate anaerobically, with reoxidation of NADH in the lactate dehydrogenase reaction.

(b) Because reducing equivalents from the oxidation reactions of glycolysis do not enter the mitochondrion, oxygen consumption slows and eventually ceases.

(c) The end product of anaerobic glycolysis, lactate, accumulates.

(d) ATP is not formed aerobically because the cells have converted to anaerobic glycolysis. Overall, ATP synthesis decreases drastically, to 2 ATP per glucose molecule.

11. Cellular ADP Concentration Controls ATP Formation Although ADP and P_i are required for the synthesis of ATP, the rate of synthesis depends mainly on the concentration of ADP, not P_i. Why?

Answer The steady-state concentration of P_i in the cell is much higher than that of ADP. As the ADP concentration rises as a result of ATP consumption, there is little change in $[P_i]$, so P_i cannot serve as a regulator.

12. Time Scales of Regulatory Events in Mitochondria Compare the likely time scales for the adjustments in respiratory rate caused by **(a)** increased [ADP] and **(b)** reduced pO_2. What accounts for the difference?

Answer In **(a),** respiratory control by ADP, the increase in respiratory rate is limited by the rate of diffusion of ADP, and the response would be expected to occur in fractions of a millisecond. The adjustment to **(b),** hypoxia mediated by HIF-1, requires a change in concentration of several proteins, the result of increased synthesis or degradation. The time scale for protein synthesis or degradation is typically many seconds to hours—much longer than the time required for changes in substrate concentration.

13. The Pasteur Effect When O_2 is added to an anaerobic suspension of cells consuming glucose at a high rate, the rate of glucose consumption declines greatly as the O_2 is used up, and accumulation of lactate ceases. This effect, first observed by Louis Pasteur in the 1860s, is characteristic of most cells capable of aerobic and anaerobic glucose catabolism.

(a) Why does the accumulation of lactate cease after O_2 is added?

(b) Why does the presence of O_2 decrease the rate of glucose consumption?

(c) How does the onset of O_2 consumption slow down the rate of glucose consumption? Explain in terms of specific enzymes.

Answer The addition of oxygen to an anaerobic suspension allows cells to convert from fermentation to oxidative phosphorylation as a mechanism for reoxidizing NADH and making ATP. Because ATP synthesis is much more efficient under aerobic conditions, the amount of glucose needed will decrease (the Pasteur effect). This decreased utilization of glucose in the presence of oxygen can be demonstrated in any tissue that is capable of aerobic and anaerobic glycolysis.

(a) Oxygen allows the tissue to convert from lactic acid fermentation to respiratory electron transfer and oxidative phosphorylation as the mechanism for NADH oxidation.

(b) Cells produce much more ATP per glucose molecule oxidized aerobically, so less glucose is needed.

(c) As [ATP] rises, phosphofructokinase-1 is inhibited, thus slowing the rate of glucose entry into the glycolytic pathway.

14. Respiration-Deficient Yeast Mutants and Ethanol Production Respiration-deficient yeast mutants (p^-; "petites") can be produced from wild-type parents by treatment with mutagenic agents. The mutants lack cytochrome oxidase, a deficit that markedly affects their metabolic behavior. One striking effect is that fermentation is not suppressed by O_2—that is, the mutants lack the Pasteur effect (see Problem 13). Some companies are very interested in using these mutants to ferment wood chips to ethanol for energy use. Explain the advantages of using these mutants rather than wild-type yeast for large-scale ethanol production. Why does the absence of cytochrome oxidase eliminate the Pasteur effect?

Answer The absence of cytochrome oxidase prevents these mutants from oxidizing the products of fermentation (ethanol, acetate, lactate, or glycerol) via the normal respiratory route. These mutants do not have a working citric acid cycle because they cannot reoxidize NADH through the O_2-dependent electron-transfer chain. Thus, catabolism of glucose stops at the ethanol stage, even in the presence of oxygen. The ability to carry out these fermentations in the presence of oxygen is a major practical advantage because completely anaerobic conditions are difficult to maintain. The Pasteur effect—the decrease in glucose consumption that occurs when oxygen is introduced—is not observed in the absence of an active citric acid cycle and electron-transfer chain.

15. Advantages of Supercomplexes for Electron Transfer There is growing evidence that mitochondrial Complexes I, II, III, and IV are part of a larger supercomplex. What might be the advantage of having all four complexes within a supercomplex?

Answer When electron-carrying complexes are bound together in a supercomplex, electron flow between complexes occurs in a solid state; this electron movement is kinetically favored compared with the situation in which electron flow depends on each complex diffusing to and colliding with the next complex in the chain.

16. How Many Protons in a Mitochondrion? Electron transfer translocates protons from the mitochondrial matrix to the external medium, establishing a pH gradient across the inner membrane (outside more acidic than inside). The tendency of protons to diffuse back into the matrix is the driving force for ATP synthesis by ATP synthase. During oxidative phosphorylation by a suspension of mitochondria in a medium of pH 7.4, the pH of the matrix has been measured as 7.7.

(a) Calculate $[H^+]$ in the external medium and in the matrix under these conditions.

(b) What is the outside-to-inside ratio of $[H^+]$? Comment on the energy inherent in this concentration difference. (Hint: see Eqn 11–4, p. 396)

(c) Calculate the number of protons in a respiring liver mitochondrion, assuming its inner matrix compartment is a sphere of diameter 1.5 μm.

(d) From these data, is the pH gradient alone sufficient to generate ATP?

(e) If not, suggest how the necessary energy for synthesis of ATP arises.

Answer

(a) Using the equation pH = $-\log [H^+]$, we can calculate external $[H^+] = 10^{-7.4} = 4.0 \times 10^{-8}$ M; and internal $[H^+] = 10^{-7.7} = 2.0 \times 10^{-8}$ M.

(b) From **(a),** the ratio is 2:1. We can calculate the free energy inherent in this *concentration* difference across the membrane. Assuming a temperature of 25 °C:

$$\begin{aligned}\Delta G &= RT \ln (C_2/C_1)\\ &= (2.48 \text{ kJ/mol}) \ln 2\\ &= -1.7 \text{ kJ/mol}\end{aligned}$$

(c) Given that the volume of the mitochondrion = $\frac{4}{3}\pi(0.75 \times 10^{-3} \text{ mm})^3$ and $[H^+] = 2.0 \times 10^{-8}$ M, the number of protons is

$$\frac{(1.33)(3.14)(0.75 \times 10^{-3} \text{ mm})^3(2.0 \times 10^{-8} \text{ mol/L})(6.02 \times 10^{23} \text{ protons/mol})}{(10^6 \text{ mm}^3\text{/L})} = 21 \text{ protons}$$

(d) No; the energy available from the H^+ concentration gradient, $2.3\Delta pH\, RT = 2.3(0.3)(2.48\text{ kJ/mol}) = 1.7\text{ kJ/mol}$, is insufficient to synthesize 1 mol of ATP.

(e) The total energy inherent in the pH gradient is the sum of the energy due to the concentration gradient and the energy due to the charge separation. The overall transmembrane electrical potential is the main factor in producing a sufficiently large ΔG_t (see Eqns 19–8 and 19–9).

17. Rate of ATP Turnover in Rat Heart Muscle Rat heart muscle operating aerobically fills more than 90% of its ATP needs by oxidative phosphorylation. Each gram of tissue consumes O_2 at the rate of 10.0 μmol/min, with glucose as the fuel source.

(a) Calculate the rate at which the heart muscle consumes glucose and produces ATP.

(b) For a steady-state concentration of ATP of 5.0 μmol/g of heart muscle tissue, calculate the time required (in seconds) to completely turn over the cellular pool of ATP. What does this result indicate about the need for tight regulation of ATP production? (Note: Concentrations are expressed as micromoles per gram of muscle tissue because the tissue is mostly water.)

Answer ATP turns over very rapidly in all types of tissues and cells.

(a) Glucose oxidation requires 6 mol of O_2 per mol of glucose. Therefore, glucose is consumed at the rate of (10.0 μmol/min · g)/6 = 1.7 μmol/min · g of tissue. If each glucose produces 32 ATP (see Table 19–5), the muscle produces ATP at the rate of (1.7 μmol glucose/min · g)(32 ATP/glucose) = 54 μmol/min · g, or 0.91 μmol/s · g.

(b) It takes (5.0 μmol/g)/(0.91 μmol/s · g) = 5.5 s to produce 5.0 μmol of ATP per gram, so the entire pool of ATP must be regenerated (turned over) every 5.5 s. In order to do this, the cell must regulate ATP synthesis precisely.

18. Rate of ATP Breakdown in Flight Muscle ATP production in the flight muscle of the fly *Lucilia sericata* results almost exclusively from oxidative phosphorylation. During flight, 187 mL of O_2/hr · g of body weight is needed to maintain an ATP concentration of 7.0 μmol/g of flight muscle. Assuming that flight muscle makes up 20% of the weight of the fly, calculate the rate at which the flight-muscle ATP pool turns over. How long would the reservoir of ATP last in the absence of oxidative phosphorylation? Assume that reducing equivalents are transferred by the glycerol 3-phosphate shuttle and that O_2 is at 25 °C and 101.3 kPa (1 atm).

Answer Using the gas laws ($PV = nRT$), we can calculate that 187 mL of O_2 contains

$$n = PV/RT = (1\text{ atm})(0.187\text{ L})/(0.08205\text{ L}\cdot\text{atm/mol}\cdot\text{K})(298\text{ K}) = 7650\ \mu\text{mol of } O_2$$

Thus, the rate of oxygen consumption by flight muscle is

$$(7650\ \mu\text{mol/hr})/(1\text{ g})(0.2)(3600\text{ s/hr}) = 10.6\ \mu\text{mol/s}\cdot\text{g}$$

Assuming a yield of 30 ATP per glucose (see Table 19–5; assume the use of the glycerol 3-phosphate shuttle), and given 6 O_2 consumed per glucose, the amount of ATP formed is

$$[(30\text{ ATP/glucose})/(6\ O_2\text{/glucose})](10.6\ \mu\text{mol } O_2\text{/s}\cdot\text{g}) = 53\ \mu\text{mol/s}\cdot\text{g}$$

Thus, a reservoir of 7.0 μmol/g would last (7.0 μmol/g)/(53 μmol/s · g) = 0.13 s.

19. Mitochondrial Disease and Cancer Mutations in the genes that encode certain mitochondrial proteins are associated with a high incidence of some types of cancer. How might defective mitochondria lead to cancer?

Answer Reactive oxygen species react with macromolecules, including DNA. If a mitochondrial defect leads to increased production of ROS, the nuclear genes that encode proto-oncogenes (pp. 473, 474) can be damaged, producing oncogenes and leading to unregulated cell division and cancer.

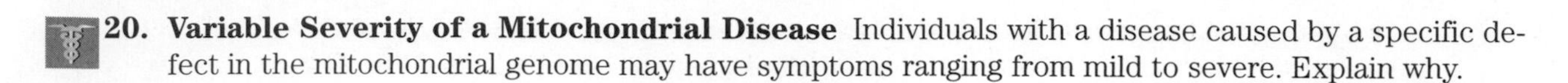

20. Variable Severity of a Mitochondrial Disease Individuals with a disease caused by a specific defect in the mitochondrial genome may have symptoms ranging from mild to severe. Explain why.

Answer The explanation is probably heteroplasmy. In some individuals, a mitochondrial mutation may affect only a small proportion of cells and tissues, because most mitochondria in these cells and tissues have normal genomes and the few mutant mitochondria do not significantly compromise the ability to produce ATP. In other individuals, the chance distribution of mitochondrial genomes during cell division has led to a high degree of heteroplasmy in which many cells have a majority of defective mitochondria, resulting in more severe symptoms.

21. Transmembrane Movement of Reducing Equivalents Under aerobic conditions, extramitochondrial NADH must be oxidized by the mitochondrial electron-transfer chain. Consider a preparation of rat hepatocytes containing mitochondria and all the cytosolic enzymes. If [4-^{3}H]NADH is introduced, radioactivity soon appears in the mitochondrial matrix. However, if [7-^{14}C]NADH is introduced, no radioactivity appears in the matrix. What do these observations reveal about the oxidation of extramitochondrial NADH by the electron-transfer chain?

^{3}H ^{3}H O C NH$_2$ N R   H H O ^{14}C NH$_2$ N R

[4-^{3}H]NADH   [7-^{14}C]NADH

Answer The malate-aspartate shuttle transfers electrons and protons from the cytoplasm into the mitochondrion. Neither NAD$^+$ nor NADH passes through the inner membrane, thus the labeled NAD moiety of [7-^{14}C]NADH remains in the cytosol. The ^{3}H on [4-^{3}H]NADH enters the mitochondrion via the malate-aspartate shuttle (see Fig. 19–29). In the cytosol, [4-^{3}H]NADH transfers its ^{3}H to oxaloacetate to form [^{3}H]malate, which enters the mitochondrion via the malate–α-ketoglutarate transporter, then donates the ^{3}H to NAD$^+$ to form [4-^{3}H]NADH in the matrix.

22. High Blood Alanine Level Associated with Defects in Oxidative Phosphorylation Most individuals with genetic defects in oxidative phosphorylation are found to have relatively high concentrations of alanine in their blood. Explain this in biochemical terms.

Answer In these individuals, the usual route for pyruvate metabolism—conversion to acetyl-CoA and entry into the citric acid cycle—is slowed by the decreased capacity for carrying electrons from NADH to oxygen. Accumulation of pyruvate in the tissues shifts the equilibrium for pyruvate-alanine transaminase, resulting in elevated concentrations of alanine in tissues and blood.

23. NAD Pools and Dehydrogenase Activities Although both pyruvate dehydrogenase and glyceraldehyde 3-phosphate dehydrogenase use NAD$^+$ as their electron acceptor, the two enzymes do not compete for the same cellular NAD pool. Why?

Answer Pyruvate dehydrogenase is located in the mitochondrion, and glyceraldehyde 3-phosphate dehydrogenase in the cytosol. Because the mitochondrial and cytosolic pools of NAD are separated by the inner mitochondrial membrane, the enzymes do not compete for the same NAD pool. However, reducing equivalents are transferred from one nicotinamide coenzyme pool to the other via shuttle mechanisms (see Problem 21).

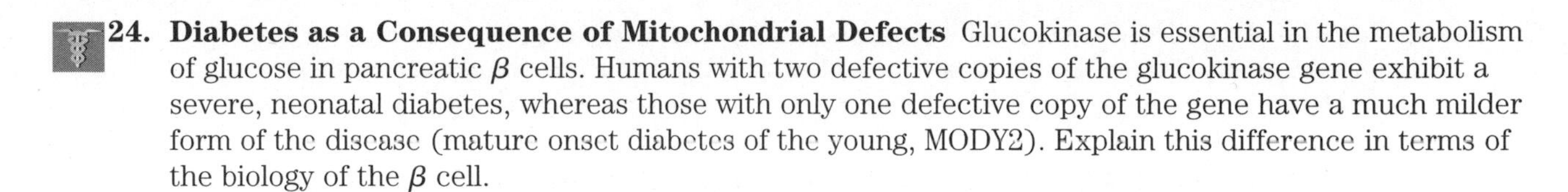

24. Diabetes as a Consequence of Mitochondrial Defects Glucokinase is essential in the metabolism of glucose in pancreatic β cells. Humans with two defective copies of the glucokinase gene exhibit a severe, neonatal diabetes, whereas those with only one defective copy of the gene have a much milder form of the disease (mature onset diabetes of the young, MODY2). Explain this difference in terms of the biology of the β cell.

Answer In β cells in which both copies of glucokinase are defective, the rate of glycolytic ATP production does not increase when blood glucose rises, and thus these cells cannot produce high enough concentrations of ATP to affect the ATP-dependent K^+ channel that indirectly regulates insulin secretion. With one functional copy of glucokinase, β cells can respond to very high glucose concentrations by producing suprathreshold concentrations of ATP, triggering insulin release.

25. Effects of Mutations in Mitochondrial Complex II Single nucleotide changes in the gene for succinate dehydrogenase (Complex II) are associated with midgut carcinoid tumors. Suggest a mechanism to explain this observation.

Answer Defects in Complex II result in increased production of ROS, damage to DNA, and mutations that lead to unregulated cell division (cancer). It is not clear why the cancer tends to occur in the midgut.

26. Photochemical Efficiency of Light at Different Wavelengths The rate of photosynthesis, measured by O_2 production, is higher when a green plant is illuminated with light of wavelength 680 nm than with light of 700 nm. However, illumination by a combination of light of 680 nm and 700 nm gives a higher rate of photosynthesis than light of either wavelength alone. Explain.

Answer Plants have two photosystems. Photosystem I absorbs light maximally at 700 nm and catalyzes cyclic photophosphorylation and $NADP^+$ reduction (see Fig. 19–56). Photosystem II absorbs light maximally at 680 nm, splits H_2O to O_2 and H^+, and donates electrons and H^+ to PSI. Therefore, light of 680 nm is better in promoting O_2 production, but maximum photosynthetic rates are observed only when plants are illuminated with light of both wavelengths.

27. Balance Sheet for Photosynthesis In 1804 Theodore de Saussure observed that the total weights of oxygen and dry organic matter produced by plants is greater than the weight of carbon dioxide consumed during photosynthesis. Where does the extra weight come from?

Answer Because the general reaction for plant photosynthesis is

$$CO_2 + H_2O \longrightarrow O_2 + \text{organic matter}$$

the extra weight must come from the water consumed in the overall reaction.

28. Role of H_2S in Some Photosynthetic Bacteria Illuminated purple sulfur bacteria carry out photosynthesis in the presence of H_2O and $^{14}CO_2$, but only if H_2S is added and O_2 is absent. During the course of photosynthesis, measured by formation of [^{14}C]carbohydrate, H_2S is converted to elemental sulfur, but no O_2 is evolved. What is the role of the conversion of H_2S to sulfur? Why is no O_2 evolved?

Answer Purple sulfur bacteria use H_2S as a source of electrons and protons:

$$H_2S \longrightarrow S + 2H^+ + 2e^-$$

The electrons are "activated" by a light energy–capturing photosystem. These cells produce their ATP by photophosphorylation and their NADPH from H_2S oxidation. Because H_2O is not split, O_2 is not evolved (photosystem II is absent).

29. Boosting the Reducing Power of Photosystem I by Light Absorption When photosystem I absorbs red light at 700 nm, the standard reduction potential of P700 changes from 0.40 V to about −1.2 V. What fraction of the absorbed light is trapped in the form of reducing power?

Answer For a change in standard reduction potential, ΔE, of 0.4 V − (−1.2 V), the free-energy change per electron is

$$\begin{aligned} \Delta G'^{\circ} &= n\mathcal{F}\Delta E'^{\circ} \\ &= -(96.48 \text{ kJ/V} \cdot \text{mol})(-1.6 \text{ V}) \\ &= 150 \text{ kJ/mol} \end{aligned}$$

Two photons are absorbed per electron elevated to a higher energy level, which for 700 nm light is equivalent to 2(170 kJ/mol) = 340 kJ/mol (see Fig. 19–46). Thus, the fraction of light energy trapped as reducing power is

$$(150 \text{ kJ/mol})/(340 \text{ kJ/mol}) = 0.44, \text{ or } 44\%$$

30. Electron Flow through Photosystems I and II Predict how an inhibitor of electron passage through pheophytin would affect electron flow through **(a)** photosystem II and **(b)** photosystem I. Explain your reasoning.

Answer

(a) Electron flow through PSII would stop, as there would be no way for electrons to move from PSII to the cytochrome b_6f complex and all the electron acceptors in PSII would very quickly be reduced.

(b) Electron flow through PSI would probably slow. The supply of electrons from PSII would be blocked, but with operation of the cyclic pathway, some electron flow through PSI could continue.

31. Limited ATP Synthesis in the Dark In a laboratory experiment, spinach chloroplasts are illuminated in the absence of ADP and P_i, then the light is turned off and ADP and P_i are added. ATP is synthesized for a short time in the dark. Explain this finding.

Answer Illumination of chloroplasts in the absence of ADP and P_i sets up a proton gradient across the thylakoid membrane. When ADP and P_i are added, ATP synthesis is driven by the gradient. In the absence of continuous illumination, the gradient soon becomes exhausted and ATP synthesis stops.

32. Mode of Action of the Herbicide DCMU When chloroplasts are treated with 3-(3,4-dichlorophenyl)-1,1-dimethylurea (DCMU, or diuron), a potent herbicide, O_2 evolution and photophosphorylation cease. Oxygen evolution, but not photophosphorylation, can be restored by addition of an external electron acceptor, or Hill reagent. How does DCMU act as a weed killer? Suggest a location for the inhibitory action of this herbicide in the scheme shown in Figure 19–56. Explain.

Answer DCMU must inhibit the electron-transfer system linking photosystem II and photosystem I at a position ahead of the first site of ATP production. DCMU competes with PQ_B for electrons from PQ_A (Table 19–4). Addition of a Hill reagent allows H_2O to be split and O_2 to be evolved, but electrons are pulled out of the system before the point of ATP synthesis and before the production of NADPH. DCMU kills plants by inhibiting ATP production.

33. Effect of Venturicidin on Oxygen Evolution Venturicidin is a powerful inhibitor of the chloroplast ATP synthase, interacting with the CF_o part of the enzyme and blocking proton passage through the CF_oCF_1 complex. How would venturicidin affect oxygen evolution in a suspension of well-illuminated chloroplasts? Would your answer change if the experiment were done in the presence of an uncou pling reagent such as 2,4-dinitrophenol (DNP)? Explain.

Answer Oxygen evolution requires continuing passage of electrons through PSII. Electrons will continue to flow through PSII and the cytochrome b_6f complex until the energetic cost of pumping a proton across the thylakoid membrane exceeds the energy available from absorption of a photon. This point is soon reached when proton flow through CF_oCF_1 is blocked by venturicidin, and oxygen evolution ceases. Addition of an uncoupling agent provides a route for protons to move through the thylakoid membrane, dissipating the energy of the proton gradient. Electrons can now continue to move through PSII and the cytochrome b_6f complex, and oxygen is produced in the water-splitting reaction.

34. Bioenergetics of Photophosphorylation The steady-state concentrations of ATP, ADP, and P_i in isolated spinach chloroplasts under full illumination at pH 7.0 are 120.0, 6.0, and 700.0 μM, respectively.

(a) What is the free-energy requirement for the synthesis of 1 mol of ATP under these conditions?

(b) The energy for ATP synthesis is furnished by light-induced electron transfer in the chloroplasts. What is the minimum voltage drop necessary (during transfer of a pair of electrons) to synthesize ATP under these conditions? (You may need to refer to Eqn 13–7, p. 515.)

Answer

(a) $$\Delta G = \Delta G'^\circ + RT \ln \frac{[\text{ATP}]}{[\text{ADP}][\text{P}_i]}$$

$$= 30.5 \text{ kJ/mol} + (2.48 \text{ kJ/mol}) \ln \frac{1.2 \times 10^{-4}}{(6.0 \times 10^{-6})(7.0 \times 10^{-4})}$$

$$= 30.5 \text{ kJ/mol} + 25.4 \text{ kJ/mol} = 55.9 \text{ kJ/mol}$$

$$= 56 \text{ kJ/mol (two significant figures)}$$

(b) $$\Delta G = -n\mathcal{J}\Delta E$$

$$\Delta E = -\Delta G/n\mathcal{J}$$

$$= \frac{-56 \text{ kJ/mol}}{-2(96.48 \text{ kJ/V} \cdot \text{mol})}$$

$$= 0.29 \text{ V}$$

35. Light Energy for a Redox Reaction Suppose you have isolated a new photosynthetic microorganism that oxidizes H_2S and passes the electrons to NAD^+. What wavelength of light would provide enough energy for H_2S to reduce NAD^+ under standard conditions? Assume 100% efficiency in the photochemical event, and use E'° of −243 mV for H_2S and −320 mV for NAD^+. See Figure 19–46 for the energy equivalents of wavelengths of light.

Answer First, calculate the standard free-energy change ($\Delta G'^\circ$) of the redox reaction

$$\text{NAD}^+ + \text{H}_2\text{S} \longrightarrow \text{NADH} + \text{S}^- + \text{H}^+$$

Because $\Delta E'^\circ = -320 \text{ mV} - (-243 \text{ mV}) = -\ 77 \text{ mV}$,

$$\Delta G'^\circ = -n\mathcal{J}\Delta E'^\circ$$

$$= (-2)(96.48 \text{ kJ/V} \cdot \text{mol})(-0.077 \text{ V}) = 15 \text{ kJ/mol}$$

This is the minimum energy needed to drive the reduction of NAD^+ by H_2S. Inspection of Figure 19–46 shows that the energy in a "mole" of photons (an einstein) in the visible part of the spectrum ranges from 170 to 300 kJ. Any visible light should have sufficient energy to drive the reduction of NADH by H_2S. In principle, and assuming 100% efficiency, even infrared light should have enough energy to drive this reaction.

36. Equilibrium Constant for Water-Splitting Reactions The coenzyme $NADP^+$ is the terminal electron acceptor in chloroplasts, according to the reaction

$$2H_2O + 2NADP^+ \longrightarrow 2NADPH + 2H^+ + O_2$$

Use the information in Table 19–2 to calculate the equilibrium constant for this reaction at 25 °C. (The relationship between K'_{eq} and $\Delta G'^\circ$ is discussed on p. 492.) How can the chloroplast overcome this unfavorable equilibrium?

Answer Using standard reduction potentials from Table 19–2, $\Delta E'^\circ$ for the reaction is $-0.324\ V - 0.816\ V = -1.140\ V$.

$$\begin{aligned}\Delta G'^\circ &= -n\mathcal{F}\Delta E'^\circ \\ &= -4(96.48\ kJ/V \cdot mol)(-1.140\ V) \\ &= 440\ kJ/mol\end{aligned}$$

(Note that $n = 4$ because 4 electrons are required to produce 1 mol of O_2.)

$$\begin{aligned}\Delta G'^\circ &= -RT \ln K'_{eq} \\ \ln K'_{eq} &= -\Delta G'^\circ/RT \\ &= (-440\ kJ/mol)/(2.48\ kJ/mol) \\ &= -177 \\ K'_{eq} &= e^{-177} = 1.35 \times 10^{-77}\end{aligned}$$

The equilibrium is clearly very unfavorable. In chloroplasts, the input of light energy overcomes this barrier.

37. Energetics of Phototransduction During photosynthesis, eight photons must be absorbed (four by each photosystem) for every O_2 molecule produced:

$$2H_2O + 2NADP^+ + 8\ \text{photons} \longrightarrow 2NADPH + 2H^+ + O_2$$

Assuming that these photons have a wavelength of 700 nm (red) and that the absorption and use of light energy are 100% efficient, calculate the free-energy change for the process.

Answer From Problem 36, $\Delta G'^\circ$ for the production of 1 mol of O_2 in this reaction is 440 kJ/mol. A light input of 8 photons (700 nm) is equivalent to (8)(170 kJ/einstein) = 1360 kJ/einstein (see Fig. 19–46). (An einstein is a "mole" of photons.) The overall standard free-energy change of the reaction is

$$\Delta G'^\circ = (440 - 1360)\ kJ/mol = -920\ kJ/mol$$

38. Electron Transfer to a Hill Reagent Isolated spinach chloroplasts evolve O_2 when illuminated in the presence of potassium ferricyanide (a Hill reagent), according to the equation

$$2H_2O + 4Fe^{3+} \longrightarrow O_2 + 4H^+ + 4Fe^{2+}$$

where Fe^{3+} represents ferricyanide and Fe^{2+}, ferrocyanide. Is NADPH produced in this process? Explain.

Answer No NADPH is produced. Artificial electron acceptors can remove electrons from the photosynthetic system and stimulate O_2 production. Ferricyanide competes with the cytochrome b_6f complex for electrons and removes them from the system. Consequently, P700 (of photosystem I) does not receive any electrons that can be activated for $NADP^+$ reduction. However, O_2 is evolved because all components of photosystem II are oxidized (see Fig. 19–56).

39. How Often Does a Chlorophyll Molecule Absorb a Photon? The amount of chlorophyll *a* (M_r 892) in a spinach leaf is about 20 μg/cm^2 of leaf. In noonday sunlight (average energy reaching the leaf is 5.4 J/cm^2 · min), the leaf absorbs about 50% of the radiation. How often does a single chlorophyll molecule absorb a photon? Given that the average lifetime of an excited chlorophyll molecule in vivo is 1 ns, what fraction of the chlorophyll molecules are excited at any one time?

Answer The leaf absorbs light in units of photons that vary in energy from 170 to 300 kJ/einstein, depending on wavelength (see Fig. 19–46). The leaf absorbs light energy at the rate of 0.5(5.4 J/cm^2 · min) = 2.7 J/cm^2 · min. Assuming an average energy of 270 kJ/einstein, this rate of light absorption is

$$(2.7 \times 10^{-3}\ \text{kJ/cm}^2 \cdot \text{min})/(270\ \text{kJ/einstein}) = 1 \times 10^{-5}\ \text{einstein/cm}^2 \cdot \text{min}$$

The concentration of chlorophyll in the leaf is

$$(20 \times 10^{-6}\ \text{g/cm}^2)/(892\ \text{g/mol}) = 2 \times 10^{-8}\ \text{mol/cm}^2$$

Thus, 1 mol of chlorophyll absorbs 1 einstein of photons every

$$(2 \times 10^{-8}\ \text{mol/cm}^2)/(1 \times 10^{-5}\ \text{einstein/cm}^2 \cdot \text{min}) = 2 \times 10^{-3}\ \text{min} = 0.1\ \text{s}$$

Because excitation lasts about 1 ns = 1×10^{-9} s, the fraction of chlorophylls excited at any one time is $(1 \times 10^{-9}\ \text{s})/(0.1\ \text{s}) = 1 \times 10^{-8}$, or one in every 10^8 molecules.

40. Effect of Monochromatic Light on Electron Flow The extent to which an electron carrier is oxidized or reduced during photosynthetic electron transfer can sometimes be observed directly with a spectrophotometer. When chloroplasts are illuminated with 700 nm light, cytochrome *f*, plastocyanin, and plastoquinone are oxidized. When chloroplasts are illuminated with 680 nm light, however, these electron carriers are reduced. Explain.

Answer Light at 700 nm activates electrons in P700 and $NADP^+$ is reduced (see Fig. 19–56). This drains all the electrons from the electron-transfer system between photosystems II and I, because light at 680 nm is not available to replace electrons by activating PSII. When light at 680 nm activates PSII (but not PSI), all the carriers between the two systems become reduced because no electrons are excited in PSI.

41. Function of Cyclic Photophosphorylation When the [NADPH]/[$NADP^+$] ratio in chloroplasts is high, photophosphorylation is predominantly cyclic (see Fig. 19–56). Is O_2 evolved during cyclic photophosphorylation? Is NADPH produced? Explain. What is the main function of cyclic photophosphorylation?

Answer Neither O_2 nor NADPH is produced. At high [NADPH]/[$NADP^+$] ratios, electron transfer from reduced ferredoxin to $NADP^+$ is inhibited and the electrons are diverted into the cytochrome b_6f complex. These electrons return to P700 and ATP is synthesized by photophosphorylation. Because electrons are not lost from P700, none are needed from PSII. Thus, H_2O is not split and O_2 is not produced. In addition, NADPH is not formed because the electrons return to P700. The function of cyclic photophosphorylation is to produce ATP.

Data Analysis Problem

42. Photophosphorylation: Discovery, Rejection, and Rediscovery In the 1930s and 1940s, researchers were beginning to make progress toward understanding the mechanism of photosynthesis. At the time, the role of "energy-rich phosphate bonds" (today, "ATP") in glycolysis and cellular respiration was just becoming known. There were many theories about the mechanism of photosynthesis, especially about the role of light. This problem focuses on what was then called the "primary photochemical process"—that is, on what it is, exactly, that the energy from captured light produces in the photosynthetic cell. Interestingly, one important part of the modern model of photosynthesis was proposed early on, only to be rejected, ignored for several years, then finally revived and accepted.

In 1944, Emerson, Stauffer, and Umbreit proposed that "the function of light energy in photosynthesis is the formation of 'energy-rich' phosphate bonds" (p. 107). In their model (hereafter, the "Emerson model"), the free energy necessary to drive both CO_2 fixation *and* reduction came from these "energy-rich phosphate bonds" (i.e., ATP), produced as a result of light absorption by a chlorophyll-containing protein.

This model was explicitly rejected by Rabinowitch (1945). After summarizing Emerson and coauthors' findings, Rabinowitch stated: "Until more positive evidence is provided, we are inclined to consider as more convincing a general argument against this hypothesis, which can be derived from energy considerations. Photosynthesis is eminently a problem of energy *accumulation*. What good can be served, then, by converting light quanta (even those of red light, which amount to about 43 kcal per Einstein) into 'phosphate quanta' of only 10 kcal per mole? This appears to be a start in the wrong direction—toward *dissipation* rather than toward accumulation of energy" (Vol. I, p. 228). This argument, along with other evidence, led to the abandonment of the Emerson model until the 1950s, when it was found to be correct—albeit in a modified form.

For each piece of information from Emerson and coauthors' article presented in (**a**) through (**d**) below, answer the following three questions:

1. How does this information support the Emerson model, in which light energy is used directly by chlorophyll *to make ATP*, and the ATP then provides the energy to drive CO_2 fixation and reduction?
2. How would Rabinowitch explain this information, based on his model (and most other models of the day), in which light energy is used directly by chlorophyll *to make reducing compounds*? Rabinowitch wrote: "Theoretically, there is no reason why *all* electronic energy contained in molecules excited by the absorption of light should not be available for oxidation-reduction" (Vol. I, p. 152). In this model, the reducing compounds are then used to fix and reduce CO_2, and the energy for these reactions comes from the large amounts of free energy released by the reduction reactions.
3. How is this information explained by our modern understanding of photosynthesis?

(a) Chlorophyll contains a Mg^{2+} ion, which is known to be an essential cofactor for many enzymes that catalyze phosphorylation and dephosphorylation reactions.

(b) A crude "chlorophyll protein" isolated from photosynthetic cells showed phosphorylating activity.

(c) The phosphorylating activity of the "chlorophyll protein" was inhibited by light.

(d) The levels of several different phosphorylated compounds in photosynthetic cells changed dramatically in response to light exposure. (Emerson and coworkers were not able to identify the specific compounds involved.)

As it turned out, the Emerson and Rabinowitch models were both partly correct and partly incorrect.

(e) Explain how the two models relate to our current model of photosynthesis.

In his rejection of the Emerson model, Rabinowitch went on to say: "The difficulty of the phosphate storage theory appears most clearly when one considers the fact that, in weak light, eight or ten quanta of light are sufficient to reduce one molecule of carbon dioxide. If each quantum should produce one molecule of high-energy phosphate, the accumulated energy would be only 80–100 kcal per Einstein—while photosynthesis requires *at least* 112 kcal per mole, and probably more, because of losses in irreversible partial reactions" (Vol. 1, p. 228).

(f) How does Rabinowitch's value of 8 to 10 photons per molecule of CO_2 reduced compare with the value accepted today? You need to consult Chapter 20 for some of the information required here.

(g) How would you rebut Rabinowitch's argument, based on our current knowledge about photosynthesis?

Answer

(a) (1) The presence of Mg^{2+} supports the hypothesis that chlorophyll is directly involved in catalysis of the phosphorylation reaction: ADP + $P_i \rightarrow$ ATP. (2) Many enzymes (or other proteins) that contain Mg^{2+} are not phosphorylating enzymes, so the presence of Mg^{2+} in chlorophyll does not prove its role in phosphorylation reactions. (3) The presence of Mg^{2+} is essential to chlorophyll's photochemical properties: light absorption and electron transfer.

(b) (1) Enzymes catalyze reversible reactions, so an isolated enzyme that can, under certain laboratory conditions, catalyze removal of a phosphoryl group could probably, under different conditions (such as in cells), catalyze addition of a phosphoryl group. So it is plausible that chlorophyll could be involved in the phosphorylation of ADP. (2) There are two possible explanations: the chlorophyll protein is a phosphatase only and does not catalyze ADP phosphorylation under cellular conditions, or the crude preparation contains a contaminating phosphatase activity that is unconnected to the photosynthetic reactions. (3) It is likely that the preparation was contaminated with a nonphotosynthetic phosphatase activity.

(c) (1) This light inhibition is what one would expect if the chlorophyll protein catalyzed the reaction ADP + P_i + light $\rightarrow$ ATP. Without light, the reverse reaction, a dephosphorylation, would be favored. In the presence of light, energy is provided and the equilibrium would shift to the right, reducing the phosphatase activity. (2) This inhibition must be an artifact of the isolation or assay methods. (3) It is unlikely that the crude preparation methods in use at the time preserved intact chloroplast membranes, so the inhibition must be an artifact.

(d) (1) In the presence of light, ATP is synthesized and other phosphorylated intermediates are consumed. (2) In the presence of light, glucose is produced and is metabolized by cellular respiration to produce ATP, with changes in the levels of phosphorylated intermediates. (3) In the presence of light, ATP is produced and other phosphorylated intermediates are consumed.

(e) Light energy is used to produce ATP (as in the Emerson model) *and* is used to produce reducing power (as in the Rabinowitch model).

(f) The approximate stoichiometry for photophosphorylation (Chapter 19) is that 8 photons yield 2 NADPH and about 3 ATP. Two NADPH and 3 ATP are required to reduce 1 CO_2 (Chapter 20). Thus, at a minimum, 8 photons are required per CO_2 molecule reduced. This is in good agreement with Rabinowitch's value.

(g) Because the energy of light is used to produce *both* ATP and NADPH, each photon absorbed contributes more than just 1 ATP for photosynthesis. The process of energy extraction from light is more efficient than Rabinowitch supposed, and plenty of energy is available for this process—even with red light.

Reference

Emerson, R.L., Stauffer, J.F., & Umbreit, W.W. (1944) Relationships between phosphorylation and photosynthesis in *Chlorella. Am. J. Botany* **31,** 107–120.

Rabinowitch, E.I. (1945) *Photosynthesis and Related Processes,* Interscience Publishers, New York.

chapter

20 Carbohydrate Biosynthesis in Plants and Bacteria

1. **Segregation of Metabolism in Organelles** What are the advantages to the plant cell of having different organelles to carry out different reaction sequences that share intermediates?

   Answer Within organelles, reaction intermediates and enzymes can be maintained at different levels from those in the cytosol and in other organelles. For example, the ATP/ADP ratio is lower in mitochondria than in the cytosol because the role of adenine nucleotides in the mitochondrial matrix is to accept a phosphoryl group, whereas the role in the cytosol is to donate a phosphoryl group. Similarly, different NADH/NAD^+ and NADPH/$NADP^+$ ratios reflect the reductive (biosynthetic) functions of the cytosol and the oxidative (catabolic) functions of the mitochondrial matrix. By segregating reaction sequences that share intermediates, the cell can regulate catabolic and anabolic processes separately.

2. **Phases of Photosynthesis** When a suspension of green algae is illuminated in the absence of CO_2 and then incubated with $^{14}CO_2$ in the dark, $^{14}CO_2$ is converted to [^{14}C]glucose for a brief time. What is the significance of this observation with regard to the CO_2-assimilation process, and how is it related to the light reactions of photosynthesis? Why does the conversion of $^{14}CO_2$ to [^{14}C]glucose stop after a brief time?

   Answer This observation suggests that photosynthesis occurs in two phases: (1) a light-dependent phase that generates ATP and NADPH, which are essential for CO_2 fixation, and (2) a light-independent (dark) phase, in which these energy-rich components are used for synthesis of glucose. In the absence of additional illumination, the supplies of NADPH and ATP become exhausted and CO_2 fixation ceases.

3. **Identification of Key Intermediates in CO_2 Assimilation** Calvin and his colleagues used the unicellular green alga *Chlorella* to study the carbon-assimilation reactions of photosynthesis. They incubated $^{14}CO_2$ with illuminated suspensions of algae and followed the time course of appearance of ^{14}C in two compounds, X and Y, under two sets of conditions. Suggest the identities of X and Y, based on your understanding of the Calvin cycle.

   (a) Illuminated *Chlorella* were grown with unlabeled CO_2, then the light was turned off and $^{14}CO_2$ was added (vertical dashed line in graph (a)). Under these conditions, X was the first compound to become labeled with ^{14}C; Y was unlabeled.

   (b) Illuminated *Chlorella* cells were grown with $^{14}CO_2$. Illumination was continued until all the $^{14}CO_2$ had disappeared (vertical dashed line in graph (b)). Under these conditions, X became labeled quickly but lost its radioactivity with time, whereas Y became more radioactive with time.

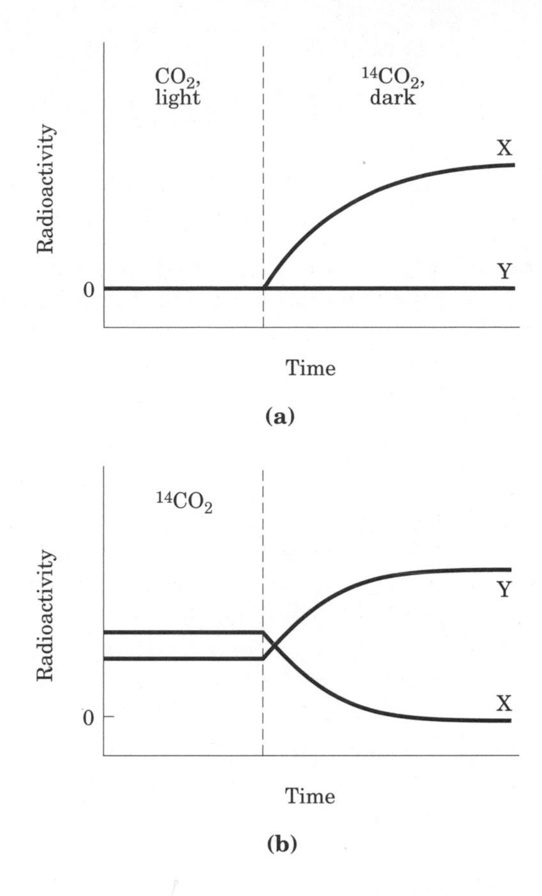

Answer Compound X is 3-phosphoglycerate and compound Y is ribulose 1,5-bisphosphate (see Fig. 20–4).

(a) Illumination of *Chlorella* in the presence of unlabeled CO_2 gives rise to steady-state levels of ribulose 1,5–bisphosphate, 3–phosphoglycerate, ATP, and NADPH. When the light is turned off, the production of ATP and NADPH ceases, but the Calvin cycle continues briefly until the residual ATP and NADPH are exhausted. Once this occurs, conversion of 3-phosphoglycerate (Stage 2 in Fig. 20–4) to hexoses, which depends on ATP and NADPH, is blocked. Thus, $^{14}CO_2$ added at the time the light is turned off is transformed primarily to 3-phosphoglycerate, but not to other intermediates of the cycle such as ribulose 1,5-bisphosphate.

(b) Illumination of *Chlorella* in the presence of $^{14}CO_2$ gives rise to steady-state levels of ^{14}C-labeled 3-phosphoglycerate and ribulose 1,5-bisphosphate. If the concentration of CO_2 is rapidly decreased, none is available for the ribulose 1,5-bisphosphate carboxylase reaction, and this constitutes a block at the fixation step (Stage 1 in Fig. 20–4). Because this experiment is carried out under conditions of constant illumination, the steps requiring ATP and NADPH are not blocked and all labeled 3-phosphoglycerate (X) can be converted to labeled ribulose 1,5-bisphosphate (Y). This results in a decrease in labeled X and a commensurate increase in labeled Y.

4. **Regulation of the Calvin Cycle** Iodoacetate reacts irreversibly with the free —SH groups of Cys residues in proteins.

$$\text{Cys}(\text{NAD}^+)(\text{SH}) + \text{ICH}_2\text{—C(=O)O}^- \xrightarrow{\text{HI}} \text{Cys}(\text{NAD}^+)(\text{S—CH}_2\text{—C(=O)O}^-)$$

Iodoacetate — Inactive enzyme

Predict which Calvin cycle enzyme(s) would be inhibited by iodoacetate, and explain why.

Answer Ribulose 5-phosphate kinase, fructose 1,6-bisphosphatase, sedoheptulose 1,7-bisphosphatase, and glyceraldehyde 3-phosphate dehydrogenase would be inhibited. All have mechanisms requiring activation by reduction of a critical disulfide bond to a pair of —SH groups. Iodoacetate reacts irreversibly with free —SH groups.

5. **Thioredoxin in Regulation of Calvin Cycle Enzymes** Motohashi and colleagues used thioredoxin as a hook to fish out from plant extracts the proteins that are activated by thioredoxin. To do this, they prepared a mutant thioredoxin in which one of the reactive Cys residues was replaced with a Ser. Explain why this modification was necessary for their experiments. [Motohashi, K., Kondoh, A., Stumpp, M.T., & Hisabori, T. (2001) Comprehensive survey of proteins targeted by chloroplast thioredoxin. *Proc. Natl. Acad. Sci. USA* **98,** 11,224-11,229.]

Answer The several Calvin cycle enzymes that are regulated by the effect of light on ferredoxin reduction are activated when reduced ferredoxin passes electrons to thioredoxin, reducing a disulfide bond to two —SH groups. For thioredoxin to play its catalytic role in the disulfide exchange reaction that activates the Calvin cycle enzymes, it must have both of its reactive —SH groups. The mutant thioredoxin, in which one of the reactive Cys residues is converted to a Ser (which lacks an —SH group), cannot form an internal disulfide, as it has only one —SH group. When this mutant thioredoxin interacts with one of the enzymes regulated by disulfide reduction, it forms a disulfide bond with that enzyme, but the next step, the reduction of the pair of Cys residues in the enzyme, cannot occur because the thioredoxin cannot form a disulfide. Consequently, thioredoxin remains covalently attached to the enzyme, and when the thioredoxin is isolated, the enzyme comes along, "hooked" on thioredoxin.

6. **Comparison of the Reductive and Oxidative Pentose Phosphate Pathways** The *reductive* pentose phosphate pathway generates a number of intermediates identical to those of the *oxidative* pentose phosphate pathway (Chapter 14). What role does each pathway play in cells where it is active?

Answer The *reductive* pentose phosphate pathway regenerates ribulose 1,5-bisphosphate from triose phosphates produced during photosynthesis, in a series of reactions involving sugars of three, four, five, six, and seven carbons and the enzymes transaldolase and transketolase. The oxidative pentose phosphate pathway plays a different metabolic role: it provides NADPH for reductive biosynthesis and pentose phosphates for nucleotide synthesis.

7. **Photorespiration and Mitochondrial Respiration** Compare the oxidative photosynthetic carbon cycle (C_2 cycle), also called *photorespiration,* with the *mitochondrial respiration* that drives ATP synthesis. Why are both processes referred to as respiration? Where in the cell do they occur, and under what circumstances? What is the path of electron flow in each?

Answer Both processes are called "respiration" because both consume O_2 and produce CO_2. Mitochondrial respiration, or cellular respiration, occurs only in mitochondria, providing most of the ATP required by the cell. In the mitochondrion, electrons derived from oxidation of various fuels in the matrix are passed through a chain of carriers in the inner mitochondrial membrane to O_2 (see Fig. 19–16); the proton gradient thus generated provides the energy for ATP synthesis. Photorespiration takes place only in plants. It is the long series of reactions, involving enzymes in the chloroplasts and peroxisomes, in which the glycolate produced when rubisco "fixes" molecular oxygen is converted to phosphoglycerate. In this reaction series (see Fig. 20–21), one-fourth of the carbons in glycolate are released as CO_2, and O_2 is consumed in the reaction catalyzed by glycolic acid oxidase. This amount of O_2 consumption is added to that occurring in the rubisco reaction. Photorespiration is an expensive side reaction of photosynthesis, resulting from the lack of specificity of rubisco.

8. **Rubisco and the Composition of the Atmosphere** N. E. Tolbert has argued that the dual specificity of rubisco for CO_2 and O_2 is not simply a leftover from evolution in a low-oxygen environment. He suggests that the relative activities of the carboxylase and oxygenase activities of rubisco actually have set, and now maintain, the ratio of CO_2 to O_2 in the earth's atmosphere. Discuss the pros and cons of this hypothesis, in molecular terms and in global terms. How does the existence of C_4 organisms bear on the hypothesis? [Tolbert, N.E. (1994) The role of photosynthesis and photorespiration in regulating atmospheric CO_2 and O_2. In *Regulation of Atmospheric* CO_2 *and* O_2 *by Photosynthetic Carbon Metabolism* (Tolbert, N.E. & Preiss, J., eds), pp. 8–33, Oxford University Press, New York.]

Answer This hypothesis assumes that the rubisco reaction is the most important chemical reaction in determining the composition of the earth's atmosphere. This may once have been true, but in the modern world, many other processes, such as the burning of fossil fuels and destruction of tropical forests (by burning and other means), probably have at least as large an impact on atmospheric composition. To the extent that C4 plants contribute to global CO_2 fixation, they also tend to counterbalance the effect of rubisco on the balance of CO_2 and O_2 in the atmosphere.

9. **Role of Sedoheptulose 1,7-Bisphosphatase** What effect on the cell and the organism might result from a defect in sedoheptulose 1,7-bisphosphatase in **(a)** a human hepatocyte and **(b)** the leaf cell of a green plant?

Answer

(a) In a human hepatocyte, NADPH produced by the pentose phosphate pathway—for which sedoheptulose 1,7-bisphosphatase activity is essential—is needed for the reductive steps in the synthesis of fatty acids, triacylglycerols, and sterols. Without this source of NADPH, all these processes would be inhibited. Cells would be unable to synthesize lipids and other reduced products.

(b) The lack of sedoheptulose 1,7-bisphosphatase would prevent the functioning of the Calvin cycle; the enzyme is essential in the series of reactions that convert triose phosphates to ribulose 1,5-bisphosphate. Without generation of ribulose 1,5-bisphosphate, the Calvin cycle is effectively blocked; there is no acceptor for CO_2 in the first step in that pathway. The leaf cell would therefore be incapable of fixing CO_2.

10. **Pathway of CO_2 Assimilation in Maize** If a maize (corn) plant is illuminated in the presence of $^{14}CO_2$, after about 1 second, more than 90% of all the radioactivity incorporated in the leaves is found at C-4 of malate, aspartate, and oxaloacetate. Only after 60 seconds does ^{14}C appear at C-1 of 3-phosphoglycerate. Explain.

Answer In maize, CO_2 is fixed by the C_4 pathway elucidated by Hatch and Slack. Phosphoenolpyruvate is rapidly carboxylated to oxaloacetate, some of which undergoes transamination to aspartate but most of which is reduced to malate in the mesophyll cells. Only after subsequent decarboxylation of labeled malate does $^{14}CO_2$ enter the Calvin cycle for conversion to glucose. The rate of entry into the cycle is limited by the rate of the rubisco-catalyzed reaction.

11. **Identifying CAM Plants** Given some $^{14}CO_2$ and all the tools typically present in a biochemistry research lab, how would you design a simple experiment to determine whether a plant was a typical C_4 plant or a CAM plant?

Answer The distinguishing features of CAM metabolism are the initial fixation of CO_2 at night and the storage of this fixed CO_2 in the vacuole until the next morning, when photosynthesis begins again. One could therefore measure the amount of $^{14}CO_2$ fixed in leaves during an hour of darkness, and the amount fixed during an hour of bright illumination. The CAM plant will take up much more CO_2 at night than a typical C_4 plant. One could also measure the concentration of organic acids in the vacuoles by titrating an extract of leaves; the CAM plant will have (in darkness) a much higher level of titratable acidity due to the malic acid stored in the vacuoles.

12. Chemistry of Malic Enzyme: Variation on a Theme Malic enzyme, found in the bundle-sheath cells of C_4 plants, carries out a reaction that has a counterpart in the citric acid cycle. What is the analogous reaction? Explain your choice.

Answer Malic enzyme catalyzes oxidative decarboxylation of a hydroxycarboxylic acid in the C_4 pathway:

$$\underset{\text{Malate}}{^{-}OOC\text{—}CH(OH)\text{—}CH_2\text{—}COO^{-}} + NADP^{+} \longrightarrow \underset{\text{Pyruvate}}{^{-}OOC\text{—}CO\text{—}CH_3} + CO_2 + NADPH + H^{+}$$

which is analogous to the reaction catalyzed in the citric acid cycle by the enzyme isocitrate dehydrogenase:

$$\text{Isocitrate} + NAD^{+} \longrightarrow \alpha\text{-ketoglutarate} + NADH + H^{+}$$

13. The Cost of Storing Glucose as Starch Write the sequence of steps and the net reaction required to calculate the cost, in ATP molecules, of converting a molecule of cytosolic glucose 6-phosphate to starch and back to glucose 6-phosphate. What fraction of the maximum number of ATP molecules available from complete catabolism of glucose 6-phosphate to CO_2 and H_2O does this cost represent?

Answer The reactions for formation of starch from glucose 6-phosphate are:
(1) Glucose 6-phosphate → glucose 1-phosphate
(2) ATP + glucose 1-phosphate → ADP-glucose + PP_i
(3) ADP-glucose + starch$_n$ → ADP + starch$_{n+1}$
The reactions for breakdown of starch are:
(4) Starch$_{n+1}$ + P_i → starch$_n$ + glucose 1-phosphate
(5) Glucose 1-phosphate → glucose 6-phosphate
The sum of reactions (1) through (5) is
(6) ATP + P_i → PP_i + ADP
Because it takes one ATP to convert ADP to ATP, we need to add
(7) ADP + ATP → ATP + ADP
The net reaction is the sum of reactions (6) and (7):
(8) ATP + P_i → ADP + PP_i
So, 1 mol of ATP is expended to store 1 mol of glucose 6-phosphate as starch. A mole of glucose 6-phosphate, when fully oxidized via glycolysis and the citric acid cycle, yields 31 to 33 mol of ATP, so the cost of storage is 1/33, or about 3.3% of the energy available.

14. Inorganic Pyrophosphatase The enzyme inorganic pyrophosphatase contributes to making many biosynthetic reactions that generate inorganic pyrophosphate essentially irreversible in cells. By keeping the concentration of PP_i very low, the enzyme "pulls" these reactions in the direction of PP_i formation. The synthesis of ADP-glucose in chloroplasts is one reaction that is pulled in the forward direction by this mechanism. However, the synthesis of UDP-glucose in the plant cytosol, which produces PP_i, is readily reversible in vivo. How do you reconcile these two facts?

Answer [PP_i] is high in the plant cell cytosol because the cytosol lacks inorganic pyrophosphatase, the enzyme that degrades PP_i to 2 P_i. Plants are unique in this regard. Animal cells have pyrophosphatase in their cytosol, and [PP_i] is therefore kept too low for PP_i to be a useful phosphoryl group donor.

15. Regulation of Starch and Sucrose Synthesis Sucrose synthesis occurs in the cytosol and starch synthesis in the chloroplast stroma, yet the two processes are intricately balanced. What factors shift the reactions in favor of **(a)** starch synthesis and **(b)** sucrose synthesis?

Answer

(a) Low levels of P_i in the cytosol and high levels of triose phosphate in the chloroplast favor formation of starch.

(b) High levels of triose phosphate in the cytosol favor formation of sucrose.

16. Regulation of Sucrose Synthesis In the regulation of sucrose synthesis from the triose phosphates produced during photosynthesis, 3-phosphoglycerate and P_i play critical roles (see Fig. 20–26). Explain why the concentrations of these two regulators reflect the rate of photosynthesis.

Answer 3-Phosphoglycerate is the primary product of photosynthesis; [P_i] rises when light-driven synthesis of ATP from ADP and P_i slows. The concentrations of these two metabolites thus provide clues to the energetic state of the leaf cell. When photosynthesis is occurring at a high rate, [P_i] drops and [3-phosphoglycerate] rises; in the dark, [P_i] rises and [3-phosphoglycerate] falls. The rate-limiting step in sucrose synthesis is the formation of ADP-glucose from glucose 1-phosphate and ATP, and this step is inhibited by P_i and activated by 3-phosphoglycerate (see Fig. 20–28). Sucrose synthesis therefore occurs at a high rate when photosynthesis is occurring, [P_i] is low, and [3-phosphoglycerate] is high.

17. Sucrose and Dental Caries The most prevalent infection in humans worldwide is dental caries, which stems from the colonization and destruction of tooth enamel by a variety of acidifying microorganisms. These organisms synthesize and live within a water-insoluble network of dextrans, called dental plaque, composed of ($\alpha 1 \rightarrow 6$)-linked polymers of glucose with many ($\alpha 1 \rightarrow 3$) branch points. Polymerization of dextran requires dietary sucrose, and the reaction is catalyzed by a bacterial enzyme, dextran-sucrose glucosyltransferase.

(a) Write the overall reaction for dextran polymerization.

(b) In addition to providing a substrate for the formation of dental plaque, how does dietary sucrose also provide oral bacteria with an abundant source of metabolic energy?

Answer

(a) $$\text{Sucrose} + \underset{\text{dextran}_n}{(\text{glucose})_n} \longrightarrow \underset{\text{dextran}_{n+1}}{(\text{glucose})_{n+1}} + \text{fructose}$$

(b) Fructose generated in the synthesis of dextran is readily taken up by the bacteria and metabolized to acidic compounds.

18. Differences between C_3 and C_4 Plants The plant genus *Atriplex* includes some C_3 and some C_4 species. From the data in the plots below (species 1, upper curve; species 2, lower curve), identify which is a C_3 plant and which is a C_4 plant. Justify your answer in molecular terms that account for the data in all three plots.

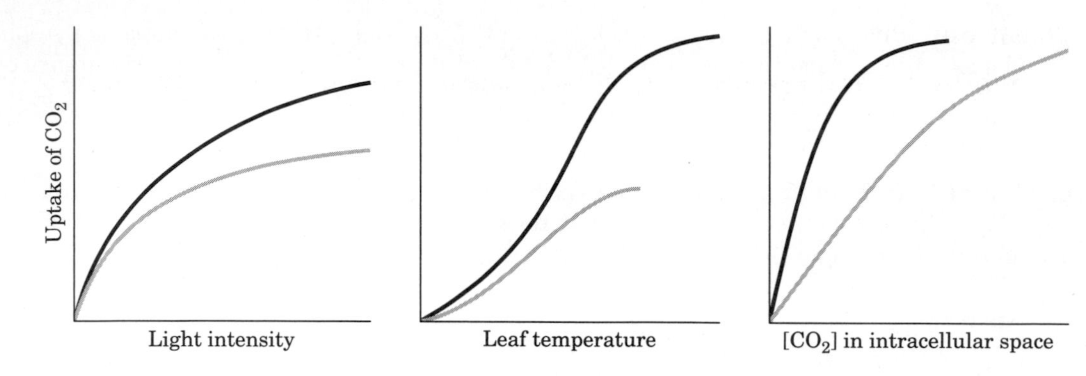

Answer Species 1 is the C_4 variety; species 2, the C_3 variety. The "mistaken" use of O_2 instead of CO_2 by rubisco is greater at higher light intensities (higher photosynthetic rates) and at higher temperatures, so under these conditions the C_4 plant has the advantage. In the rightmost panel, the fixation of CO_2 is measured as a function of $[CO_2]$, and again the C_4 plant has the advantage. The lower K_m for CO_2 of PEP carboxylase (relative to rubisco) gives the C_4 plant the advantage at lower CO_2 concentrations.

19. C_4 Pathway in a Single Cell In typical C_4 plants, the initial capture of CO_2 occurs in one cell type, and the Calvin cycle reactions occur in another (see Fig. 20–23). Voznesenskaya and colleagues have described a plant, *Bienertia cycloptera*—which grows in salty depressions of semidesert in Central Asia—that shows the biochemical properties of a C_4 plant but unlike typical C_4 plants does not segregate the reactions of CO_2 fixation into two cell types. PEP carboxylase and rubisco are present in the same cell. However, the cells have two types of chloroplasts, which are localized differently, as shown in the micrograph. One type, relatively poor in grana (thylakoids), is confined to the periphery; the more typical chloroplasts are clustered in the center of the cell, separated from the peripheral chloroplasts by large vacuoles. Thin cytosolic bridges pass through the vacuoles, connecting the peripheral and central cytosol. [Voznesenskaya, E.V., Fraceschi, V.R., Kiirats, O., Artyusheva, E.G., Freitag, H., & Edwards, G.E. (2002) Proof of C_4 photosynthesis without Kranz anatomy in *Bienertia cycloptera* (Chenopodiaceae). *Plant J.* **31,** 649–662.]

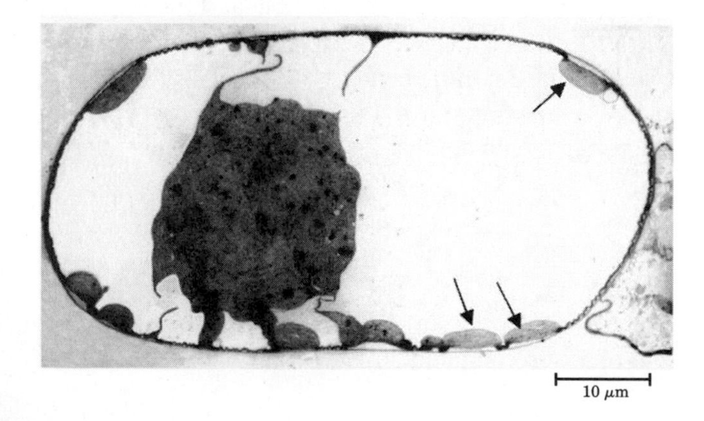

In this plant, where would you expect to find **(a)** PEP carboxylase, **(b)** rubisco, and **(c)** starch granules? Explain your answers with a model for CO_2 fixation in these C_4 cells.

Answer In this organism, the initial capture of CO_2 and the final CO_2 fixation by rubisco take place not in different cell types but in two different regions of the same cell. The enzymes for initial fixation of CO_2 into oxaloacetate are at the periphery, where the concentrations of CO_2 and O_2 are relatively high; rubisco is in the cluster of chloroplasts at the center of the cell, where the O_2 concentration is low, limited by diffusion into the cell.

(a) Periphery of the cell.

(b) Chloroplasts at center of the cell.

(c) Chloroplasts at center of the cell.

Data Analysis Problem

20. Rubisco of Bacterial Endosymbionts of Hydrothermal Vent Animals Undersea hydrothermal vents support remarkable ecosystems. At these extreme depths there is no light to support photosynthesis, yet thriving vent communities are found. Much of their primary productivity occurs through chemosynthesis carried out by bacterial symbionts that live in specialized organs (trophosomes) of certain vent animals.

Chemosynthesis in these bacteria involves a process that is virtually identical to photosynthesis. Carbon dioxide is fixed by rubisco and reduced to glucose, and the necessary ATP and NADPH are produced by electron-transfer processes similar to those of the light-dependent reactions of photosynthesis. The key difference is that in chemosynthesis, the energy driving electron transfer comes from a highly exergonic chemical reaction rather than from light. Different chemosynthetic bacteria use different reactions for this purpose. The bacteria found in hydrothermal vent animals typically use the oxidation of H_2S (abundant in the vent water) by O_2, producing elemental sulfur. These bacteria also use the conversion of H_2S to sulfur as a source of electrons for chemosynthetic CO_2 reduction.

(a) What is the overall reaction for chemosynthesis in these bacteria? You do not need to write a balanced equation; just give the starting materials and products.

(b) Ultimately, these endosymbiotic bacteria obtain their energy from sunlight. Explain how this occurs.

Robinson and colleagues (2003) explored the properties of rubisco from the bacterial endosymbiont of the giant tube worm *Riftia pachyptila*. Rubisco, from any source, catalyzes the reaction of either CO_2 (Fig. 20–7) or O_2 (Fig. 20–20) with ribulose 1,5-bisphosphate. In general, rubisco reacts more readily with CO_2 than O_2. The degree of selectivity (Ω) can be expressed as

$$\frac{V_{\text{carboxylation}}}{V_{\text{oxygenation}}} = \Omega \frac{[CO_2]}{[O_2]}$$

where V is the reaction velocity.

Robinson and coworkers measured the Ω value for the rubisco of the bacterial endosymbionts. They purified rubisco from tube-worm trophosomes, reacted it with mixtures of different ratios of O_2 and CO_2 in the presence of [1-^{3}H]ribulose 1,5-bisphosphate, and measured the ratio of [^{3}H]phosphoglycerate to [^{3}H]phosphoglycolate.

(c) The measured ratio of [^{3}H]phosphoglycerate to [^{3}H]phosphoglycolate is equal to the ratio $V_{\text{carboxylation}}/V_{\text{oxygenation}}$. Explain why.

(d) Why would [5-^{3}H]ribulose 1,5-bisphosphate not be a suitable substrate for this assay?

The Ω for the endosymbiont rubisco had a value of 8.6 ± 0.9.

(e) The atmospheric (molar) concentration of O_2 is 20% and that of CO_2 is about 380 parts per million. If the endosymbiont were to carry out chemosynthesis under these atmospheric conditions, what would be the value of $V_{\text{carboxylation}}/V_{\text{oxygenation}}$?

(f) Based on your answer to **(e)**, would you expect Ω for the rubisco of a terrestrial plant to be higher than, the same as, or lower than 8.6? Explain your reasoning.

Two stable isotopes of carbon are commonly found in the environment: the more abundant ^{12}C and the rare ^{13}C. All rubisco enzymes catalyze the fixation of $^{12}CO_2$ faster than that of $^{13}CO_2$. As a result, the carbon in glucose is slightly enriched in ^{12}C compared with the isotopic composition of CO_2 in the environment. Several factors are involved in this "preferential" use of $^{12}CO_2$, but one factor is the fundamental physics of gases. The temperature of a gas is related to the kinetic energy of its molecules. Kinetic energy is given by $\frac{1}{2}mv^2$, where m is molecular mass and v is velocity. Thus, at the same temperature (same kinetic energy), the molecules of a lighter gas will be moving faster than those of a heavier gas.

(g) How could this contribute to rubisco's "preference" for $^{12}CO_2$ over $^{13}CO_2$?

Some of the first convincing evidence that the tube-worm hosts were obtaining their fixed carbon from the endosymbionts was that the ^{13}C/^{12}C ratio in the animals was much closer to that of the bacteria than that of nonvent marine animals.

(h) Why is this more convincing evidence for a symbiotic relationship than earlier studies that simply showed the presence of rubisco in the bacteria found in trophosomes?

Answer

(a) By analogy to the oxygenic photosynthesis carried out by plants ($H_2O + CO_2 \rightarrow$ glucose $+ O_2$), the reaction would be $H_2S + O_2 + CO_2 \rightarrow$ glucose $+ H_2O + S$. This is the sum of the reduction of CO_2 by H_2S ($H_2S + CO_2 \rightarrow$ glucose $+ S$) and the energy input ($H_2S + O_2 \rightarrow S + H_2O$).

(b) The H_2S and CO_2 are produced chemically in deep-sea sediments, but the O_2, like the vast majority of O_2 on Earth, is produced by photosynthesis, which is driven by light energy.

(c) In the assay used by Robinson et al., ^3H labels the C-1 of ribulose 1,5-bisphosphate, so reaction with CO_2 yields one molecule of [^3H]3-phosphoglycerate and one molecule of unlabeled 3-phosphoglycerate; reaction with O_2 produces one molecule of [^3H]2-phosphoglycolate and one molecule of unlabeled 3-phosphoglycerate. Thus the ratio of [^3H]3-phosphoglycerate to [^3H]2-phosphoglycolate equals the ratio of carboxylation to oxygenation.

(d) If the ^3H labeled C-5, *both* oxygenation and carboxylation would yield [^3H]3-phosphoglycerate and it would be impossible to distinguish which reaction had produced the labeled product; the reaction could not be used to measure Ω.

(e) Substituting $\frac{[CO_2]}{[O_2]} = \frac{0.00038}{0.2} = 0.0019$ into $\frac{V_{carboxylation}}{V_{oxygenation}} = \Omega \frac{[CO_2]}{[O_2]}$ gives

$$\frac{V_{carboxylation}}{V_{oxygenation}} = (8.6)(0.0019) = 0.016$$

Therefore, the rate of oxygenation would be roughly 60 times the rate of carboxylation!

(f) If terrestrial plants had $\Omega = 8.6$, carboxylation would occur at a much lower rate than oxygenation. This would be extremely inefficient, so one would expect the rubisco of terrestrial plants to have an Ω substantially higher than 8.6. In fact, Ω values for land plants vary between 10 and 250. Even with these values, the expected rate of the oxygenation reaction is still very high.

(g) The rubisco reaction occurs with CO_2 as a gas. At the same temperature, $^{13}CO_2$ molecules diffuse more slowly than the lighter $^{12}CO_2$ molecules, and thus $^{13}CO_2$ will enter the active site (and become incorporated into substrate) more slowly than $^{12}CO_2$.

(h) For the relationship to be truly symbiotic, the tube worms must be getting a substantial amount of their carbon from the bacteria. The presence of rubisco in the endosymbionts simply shows that they are capable of chemosynthesis, not that they are supplying the host with a significant fraction of its carbon. On the other hand, showing that the $^{13}C : {}^{12}C$ ratio in the host is more similar to that in the endosymbiont than that in other marine animals strongly suggests that the tube worms are getting the majority of their carbon from the bacteria.

Reference

Robinson, J.J., Scott, K.M., Swanson, S.T., O'Leary, M.H., Horken, K., Tabita, F.R., & Cavanaugh, C.M. (2003) Kinetic isotope effect and characterization of form II RubisCO from the chemoautotrophic endosymbionts of the hydrothermal vent tubeworm *Riftia pachyptila*. *Limnol. Oceanogr.* **48,** 48–54.

chapter

21

Lipid Biosynthesis

1. **Pathway of Carbon in Fatty Acid Synthesis** Using your knowledge of fatty acid biosynthesis, provide an explanation for the following experimental observations:
   (a) Addition of uniformly labeled [^{14}C]acetyl-CoA to a soluble liver fraction yields palmitate uniformly labeled with ^{14}C.
   (b) However, addition of a *trace* of uniformly labeled [^{14}C]acetyl-CoA in the presence of an excess of unlabeled malonyl-CoA to a soluble liver fraction yields palmitate labeled with ^{14}C only in C-15 and C-16.

   Answer Recall that "loading" of the acyl carrier protein requires an initial addition of acetyl-CoA, followed by the addition of malonyl-CoA. Malonyl-CoA is normally produced by the addition of CO_2 to acetyl-CoA by acetyl-CoA carboxylase.
   (a) In the presence of an excess of [^{14}C]acetyl-CoA, the metabolic pool of malonyl-CoA becomes labeled at C-1 and C-2. This results in the formation of uniformly labeled palmitate.
   (b) If a *trace* of [^{14}C]acetyl-CoA is introduced in the presence of a large excess of unlabeled malonyl-CoA, the metabolic pool of malonyl-CoA does not become labeled. The trace of [^{14}C]acetyl-CoA is loaded onto the acyl carrier protein (to become C-15 and C-16 of palmitate) rather than transformed into malonyl-CoA (a slow, rate-controlling process). In addition, any labeled malonyl-CoA is diluted by the presence of excess unlabeled malonyl-CoA.

2. **Synthesis of Fatty Acids from Glucose** After a person has ingested large amounts of sucrose, the glucose and fructose that exceed caloric requirements are transformed to fatty acids for triacylglycerol synthesis. This fatty acid synthesis consumes acetyl-CoA, ATP, and NADPH. How are these substances produced from glucose?

   Answer Glucose and fructose are degraded to pyruvate via glycolysis in the cytosol. Pyruvate enters the mitochondrion and is oxidatively decarboxylated to acetyl-CoA, some of which enters the citric acid cycle. The reducing equivalents (NADH, $FADH_2$) produced in the citric acid cycle are used for ATP production by oxidative phosphorylation. The remainder of the acetyl-CoA is exported to the cytosol (via the acetyl group shuttle) for fatty acid synthesis (see Fig. 21–10). Some NADPH is produced in the pentose phosphate pathway (Fig. 14–21) in the cytosol, using glucose as a substrate, and some is produced by the action of malic enzyme on cytoplasmic malate.

3. **Net Equation of Fatty Acid Synthesis** Write the net equation for the biosynthesis of palmitate in rat liver, starting from mitochondrial acetyl-CoA and cytosolic NADPH, ATP, and CO_2.

   Answer Most of the acetyl-CoA used in fatty acid synthesis is formed from the oxidation of pyruvate in mitochondria. The inner mitochondrial membrane is impermeable to acetyl-CoA, and transfer across the membrane occurs via the acetyl group shuttle (see Fig. 21–10). This process requires the input of 1 ATP per acetyl-CoA

$$\text{Acetyl-CoA}_{(mit)} + \text{ATP} + \text{H}_2\text{O} \longrightarrow \text{acetyl-CoA}_{(cyt)} + \text{ADP} + \text{P}_i + \text{H}^+$$

or, for palmitate,

$$8\ \text{Acetyl-CoA}_{(mit)} + 8\text{ATP} + 8\text{H}_2\text{O} \longrightarrow 8\ \text{acetyl-CoA}_{(cyt)} + 8\text{ADP} + 8\text{P}_i + 8\text{H}^+$$

Cytosolic acetyl-CoA ("the primer") is condensed with malonyl-CoA, then reduced, hydrated, and reduced again. This process is repeated seven times, each time adding an additional acetyl unit to the fatty acyl–CoA. The equation for this process is

$$8\ \text{Acetyl-CoA} + 14\text{NADPH} + 6\text{H}^+ + 7\text{ATP} + \text{H}_2\text{O} \longrightarrow \text{palmitate} + 8\text{CoA} + 14\text{NADP}^+ + 7\text{ADP} + 7\text{P}_i$$

(Note that this differs from Eqn 21–3 in the text because the 7 H_2O for hydrolysis of 7 ATP were omitted from Eqn 21–1, as were 7 H^+ in the products of that reaction.)

The net equation for the overall process in the cytosol is

$$8\ \text{Acetyl-CoA}_{(mit)} + 15\text{ATP} + 14\text{NADPH} + 9\text{H}_2\text{O} \longrightarrow \text{palmitate} + 8\text{CoA} + 15\text{ADP} + 15\text{P}_i + 14\text{NADP}^+ + 2\text{H}^+$$

This may be viewed as the sum of two reactions:

$$8\ \text{Acetyl-CoA} + 14\text{NADPH} + 13\text{H}^+ \longrightarrow \text{CH}_3(\text{CH}_2)_{14}\text{COO}^- + 8\text{CoA} + 14\text{NADP}^+ + 6\text{H}_2\text{O}$$

$$15\text{ATP} + 15\text{H}_2\text{O} \longrightarrow 15\text{ADP} + 15\text{P}_i + 15\text{H}^+$$

4. **Pathway of Hydrogen in Fatty Acid Synthesis** Consider a preparation that contains all the enzymes and cofactors necessary for fatty acid biosynthesis from added acetyl-CoA and malonyl-CoA.
   (a) If [2-^{2}H]acetyl-CoA (labeled with deuterium, the heavy isotope of hydrogen)

   $$^-\text{OOC—C(}^2\text{H)}_2\text{—C(=O)S-CoA}$$

   and an excess of unlabeled malonyl-CoA are added as substrates, how many deuterium atoms are incorporated into every molecule of palmitate? What are their locations? Explain.
   (b) If unlabeled acetyl-CoA and [2-^{2}H]malonyl-CoA

   $$^-\text{OOC—C(}^2\text{H)}_2\text{—C(=O)S-CoA}$$

   are added as substrates, how many deuterium atoms are incorporated into every molecule of palmitate? What are their locations? Explain.

Answer

(a) Three deuteriums per palmitate; all located on C-16; all other two-carbon units are derived from unlabeled malonyl-CoA. (Note that the unlabeled malonyl-CoA is in excess.)

(b) Seven deuteriums per palmitate; one on each even-numbered carbon except C-16. One deuterium is lost from each labeled even-numbered carbon at the dehydration step (see Fig. 21–6).

5. **Energetics of β-Ketoacyl-ACP Synthase** In the condensation reaction catalyzed by β-ketoacyl-ACP synthase (see Fig. 21–6), a four-carbon unit is synthesized by the combination of a two-carbon unit and a three-carbon unit, with the release of CO_2. What is the thermodynamic advantage of this process over one that simply combines two two-carbon units?

Answer If a four-carbon unit were synthesized by combining two two-carbon units, the reaction would be reversible—for example, the reaction catalyzed by thiolase in the oxidation of fatty acids (see Fig. 17–8).

$$\text{CoA} + \text{CH}_3\text{—}\overset{\text{O}}{\overset{\|}{\text{C}}}\text{—CH}_2\text{—}\overset{\text{O}}{\overset{\|}{\text{C}}}\text{—CoA} \rightleftharpoons \text{CH}_3\text{—}\overset{\text{O}}{\overset{\|}{\text{C}}}\text{—CoA} + \text{CH}_3\text{—}\overset{\text{O}}{\overset{\|}{\text{C}}}\text{—CoA}$$

By using the three-carbon unit malonyl-CoA, the activated form of acetyl-CoA (recall that malonyl-CoA synthesis requires the input of energy from ATP), metabolite flow is driven in the direction of fatty acid synthesis by the exergonic release of CO_2.

6. **Modulation of Acetyl-CoA Carboxylase** Acetyl-CoA carboxylase is the principal regulation point in the biosynthesis of fatty acids. Some of the properties of the enzyme are described below.
   (a) Addition of citrate or isocitrate raises the V_{max} of the enzyme as much as 10-fold.
   (b) The enzyme exists in two interconvertible forms that differ markedly in their activities:

$$\text{Protomer (inactive)} \rightleftharpoons \text{filamentous polymer (active)}$$

   Citrate and isocitrate bind preferentially to the filamentous form, and palmitoyl-CoA binds preferentially to the protomer.
   Explain how these properties are consistent with the regulatory role of acetyl-CoA carboxylase in the biosynthesis of fatty acids.

Answer The rate-limiting step in the biosynthesis of fatty acids is the carboxylation of acetyl-CoA, catalyzed by acetyl-CoA carboxylase. High levels of citrate and isocitrate indicate that conditions are favorable for fatty acid synthesis: an active citric acid cycle is providing a plentiful supply of ATP, reducing equivalents, and acetyl-CoA. Citrate or isocitrate stimulates (increases the V_{max} of) the rate-limiting enzymatic step in fatty acid biosynthesis **(a).** Furthermore, because citrate and isocitrate bind more tightly to the filamentous form of the enzyme (the active form), the presence of citrate or isocitrate drives the protomer $\rightleftharpoons$ polymer equilibrium in the direction of the active (polymer) form **(b).** In contrast, palmitoyl-CoA (the end product of fatty acid biosynthesis) drives the equilibrium in the direction of the inactive (protomer) form. Hence, when the end product of fatty acid biosynthesis builds up, fatty acid biosynthesis is slowed down.

7. **Shuttling of Acetyl Groups across the Mitochondrial Inner Membrane** The acetyl group of acetyl-CoA, produced by the oxidative decarboxylation of pyruvate in the mitochondrion, is transferred to the cytosol by the acetyl group shuttle outlined in Figure 21–10.
   (a) Write the overall equation for the transfer of one acetyl group from the mitochondrion to the cytosol.
   (b) What is the cost of this process in ATPs per acetyl group?
   (c) In Chapter 17 we encountered an acyl group shuttle in the transfer of fatty acyl–CoA from the cytosol to the mitochondrion in preparation for β oxidation (see Fig. 17–6). One result of that shuttle was separation of the mitochondrial and cytosolic pools of CoA. Does the acetyl group shuttle also accomplish this? Explain.

Answer

(a) The reactions involved in the transfer of one acetyl group are

$$\text{Acetyl-CoA}_{(mit)} + \text{oxaloacetate}_{(mit)} \longrightarrow \text{citrate}_{(mit)} + \text{CoA}_{(mit)}$$
$$\text{Citrate}_{(mit)} \longrightarrow \text{citrate}_{(cyt)}$$
$$\text{Citrate}_{(cyt)} + \text{ATP} + \text{CoA}_{(cyt)} \longrightarrow \text{oxaloacetate}_{(cyt)} + \text{ADP} + \text{P}_i + \text{acetyl-CoA}_{(cyt)}$$
$$\text{Oxaloacetate}_{(cyt)} + \text{NADH} + \text{H}^+ \longrightarrow \text{malate}_{(cyt)} + \text{NAD}^+$$
$$\text{Malate}_{(cyt)} \longrightarrow \text{malate}_{(mit)}$$
$$\text{Malate}_{(mit)} + \text{NAD}^+ \longrightarrow \text{oxaloacetate}_{(mit)} + \text{NADH} + \text{H}^+$$

The overall equation is

$$\text{Acetyl-CoA}_{(\text{mit})} + \text{ATP} + \text{CoA}_{(\text{cyt})} \longrightarrow \text{acetyl-CoA}_{(\text{cyt})} + \text{ADP} + \text{P}_\text{i} + \text{CoA}_{(\text{mit})}$$

(b) The transfer of one acetyl group from the mitochondrial matrix to the cytosol (one turn of the acetyl group shuttle) is accompanied by the hydrolysis of ATP.

(c) The transfer requires the input of one cytosolic CoA, and one mitochondrial CoA is released. Thus, the shuttle keeps the cytosolic and mitochondrial pools of CoA separate.

8. Oxygen Requirement for Desaturases The biosynthesis of palmitoleate (see Fig. 21–12), a common unsaturated fatty acid with a cis double bond in the Δ^9 position, uses palmitate as a precursor. Can this be carried out under strictly anaerobic conditions? Explain.

Answer No; oxygen is required. The double bond in palmitoleic acid is introduced by an oxidation catalyzed by fatty acyl–CoA oxygenase (or fatty acyl–CoA desaturase; see Fig. 21–14), a mixed-function oxidase that requires molecular oxygen as a cosubstrate.

9. Energy Cost of Triacylglycerol Synthesis Use a net equation for the biosynthesis of tripalmitoylglycerol (tripalmitin) from glycerol and palmitate to show how many ATPs are required per molecule of tripalmitin formed.

Answer The reactions involved in conversion of glycerol and palmitate to tripalmitin are

$\text{ATP} + \text{glycerol} \longrightarrow \text{glycerol 3-phosphate} + \text{ADP} + \text{H}^+$
$\text{3 Palmitate} + \text{3ATP} + \text{3CoA} \longrightarrow \text{3 palmitoyl-CoA} + \text{3AMP} + 3\text{PP}_\text{i} + 3\text{H}^+$
$3\text{H}_2\text{O} + 3\text{PP}_\text{i} \longrightarrow 6\text{P}_\text{i} + 3\text{H}^+$
$\text{2 Palmitoyl-CoA} + \text{glycerol 3-phosphate} \longrightarrow \text{dipalmitoylglycerol 3-phosphate} + \text{CoA}$
$\text{Dipalmitoylglycerol 3-phosphate} + \text{H}_2\text{O} \longrightarrow \text{dipalmitoylglycerol} + \text{P}_\text{i}$
$\text{Dipalmitoylglycerol} + \text{palmitoyl-CoA} \longrightarrow \text{tripalmitin} + \text{CoA}$

The overall reaction is

$$\text{3 Palmitate} + \text{glycerol} + \text{4ATP} + 4\text{H}_2\text{O} \longrightarrow \text{tripalmitin} + \text{3AMP} + \text{ADP} + 7\text{P}_\text{i} + 7\text{H}^+$$

Including 3AMP + 3ATP → 6ADP, the overall reaction becomes

$$\text{3 Palmitate} + \text{glycerol} + \text{7ATP} + 4\text{H}_2\text{O} \longrightarrow \text{tripalmitin} + \text{7ADP} + 7\text{P}_\text{i} + 7\text{H}^+$$

10. Turnover of Triacylglycerols in Adipose Tissue When [^{14}C]glucose is added to the balanced diet of adult rats, there is no increase in the total amount of stored triacylglycerols, but the triacylglycerols become labeled with ^{14}C. Explain.

Answer In adult rats, stored triacylglycerols are maintained at a steady-state level through a balance of the rates of degradation and biosynthesis. Hence, the depots of fats are continually turned over, explaining the incorporation of ^{14}C from dietary [^{14}C]glucose.

11. Energy Cost of Phosphatidylcholine Synthesis Write the sequence of steps and the net reaction for the biosynthesis of phosphatidylcholine by the salvage pathway from oleate, palmitate, dihydroxyacetone phosphate, and choline. Starting from these precursors, what is the cost (in number of ATPs) of the synthesis of phosphatidylcholine by the salvage pathway?

Answer The sequence of steps required to carry out phosphatidylcholine biosynthesis is

$\text{Dihydroxyacetone phosphate} + \text{NADH} + \text{H}^+ \longrightarrow \text{glycerol 3-phosphate} + \text{NAD}^+$
$\text{Palmitate} + \text{ATP} + \text{CoA} \longrightarrow \text{palmitoyl-CoA} + \text{AMP} + \text{PP}_\text{i}$
$\text{Oleic acid} + \text{ATP} + \text{CoA} \longrightarrow \text{oleoyl-CoA} + \text{AMP} + \text{PP}_\text{i}$
$2\text{PP}_\text{i} + 2\text{H}_2\text{O} \longrightarrow 2\text{P}_\text{i}$
$\text{Glycerol 3-phosphate} + \text{palmitoyl-CoA} + \text{oleoyl-CoA} \longrightarrow \text{L-phosphatidate} + \text{2CoA}$

L-Phosphatidate + $H_2O \longrightarrow$ 1,2-diacylglycerol + P_i
ATP + choline $\longrightarrow$ ADP + phosphocholine
CTP + phosphocholine $\longrightarrow$ CDP-choline + PP_i
$PP_i + H_2O \longrightarrow 2P_i$
CDP-choline + 1,2-diacylglycerol $\longrightarrow$ CMP + phosphatidylcholine

The net reaction is

Dihydroxyacetone phosphate + NADH + palmitate + oleate + 3ATP + CTP + choline + $4H_2O \longrightarrow$ phosphatidylcholine + NAD^+ + 2AMP + ADP + CMP + $5P_i$ + H^+

This can also be viewed as the sum of two equations:

Dihydroxyacetone phosphate + palmitate + oleate + choline + NADH + $2H^+ \longrightarrow$ phosphatidylcholine + NAD^+ + $3H_2O$

3ATP + CTP + $7H_2O \longrightarrow$ 2AMP + ADP + CMP + $5P_i$ + $3H^+$

Note that the net equation shows the production of three nucleoside monophosphates (2 AMP + CMP). Recycling of these monophosphates to nucleoside diphosphates requires the input of 3 ATP.

2AMP + 2ATP $\longrightarrow$ 4ADP
CMP + ATP $\longrightarrow$ CDP + ADP

Furthermore, if we consider that CTP is energetically equivalent to ATP, then the synthesis of one phosphatidylcholine requires a total of 7 ATP.

12. Salvage Pathway for Synthesis of Phosphatidylcholine A young rat maintained on a diet deficient in methionine fails to thrive unless choline is included in the diet. Explain.

Answer The rat has two pathways for synthesizing phosphatidylcholine: the de novo pathway and the salvage pathway. The de novo pathway requires the transfer of a methyl group from *S*-adenosylmethionine (adoMet; see Fig. 21–27) to phosphatidylethanolamine. When the diet is deficient in methionine (an essential amino acid), the biosynthesis of adoMet and phosphatidylcholine is severely impaired. The salvage pathway, on the other hand, does not require adoMet but utilizes available choline. Thus, phosphatidylcholine can be synthesized when the diet is deficient in methionine, as long as choline is available.

13. Synthesis of Isopentenyl Pyrophosphate If 2-[^{14}C]acetyl-CoA is added to a rat liver homogenate that is synthesizing cholesterol, where will the ^{14}C label appear in Δ^3-isopentenyl pyrophosphate, the activated form of an isoprene unit?

Answer Three acetyl units are required for the synthesis of an isoprene unit (see Fig. 21–33). The ^{14}C label at C-2 of acetyl-CoA ends up in three places in the activated isoprene unit:

$^{14}CH_2$
C—$^{14}CH_2$—CH_2—
$^{14}CH_3$

14. Activated Donors in Lipid Synthesis In the biosynthesis of complex lipids, components are assembled by transfer of the appropriate group from an activated donor. For example, the activated donor of acetyl groups is acetyl-CoA. For each of the following groups, give the form of the activated donor: **(a)** phosphate; **(b)** D-glucosyl; **(c)** phosphoethanolamine; **(d)** D-galactosyl; **(e)** fatty acyl; **(f)** methyl; **(g)** the two-carbon group in fatty acid biosynthesis; **(h)** Δ^3-isopentenyl.

Answer **(a)** ATP; **(b)** UDP-glucose; **(c)** CDP-ethanolamine; **(d)** UDP-galactose; **(e)** fatty acyl–CoA; **(f)** *S*-adenosylmethionine; **(g)** malonyl-CoA; **(h)** Δ^3-isopentenyl pyrophosphate.

15. Importance of Fats in the Diet When young rats are placed on a totally fat-free diet, they grow poorly, develop a scaly dermatitis, lose hair, and soon die—symptoms that can be prevented if linoleate or plant material is included in the diet. What makes linoleate an essential fatty acid? Why can plant material be substituted?

Answer Linoleate is an unsaturated fatty acid required in the synthesis of prostaglandins. Animals lack the enzymes required to introduce double bonds into fatty acids beyond the Δ^9 position and therefore are unable to transform oleate to linoleate. Accordingly, linoleate is an essential fatty acid in animals. Plants can transform oleate to linoleate, and thus they provide animals with the required linoleate (Fig. 21–12).

16. Regulation of Cholesterol Biosynthesis Cholesterol in humans can be obtained from the diet or synthesized de novo. An adult human on a low-cholesterol diet typically synthesizes 600 mg of cholesterol per day in the liver. If the amount of cholesterol in the diet is large, de novo synthesis of cholesterol is drastically reduced. How is this regulation brought about?

Answer The rate-determining step in the biosynthesis of cholesterol is the synthesis of mevalonate, catalyzed by HMG-CoA reductase. This enzyme is subject to feedback regulation by the negative modulator cholesterol. High intracellular [cholesterol] also reduces transcription of the gene that encodes HMG-CoA reductase.

17. Lowering Serum Cholesterol Levels with Statins Patients treated with a statin drug generally exhibit a dramatic lowering of serum cholesterol. However, the amount of the enzyme HMG-CoA reductase present in cells can increase substantially. Suggest an explanation for this effect.

Answer When cholesterol levels decline because of treatment with a statin, cells attempt to compensate by increasing expression of the gene encoding HMG-CoA reductase. The statins are good competitive inhibitors of HMG-CoA reductase activity nonetheless and reduce overall production of cholesterol.

18. Roles of Thiol Esters in Cholesterol Biosynthesis Draw a mechanism for each of the three reactions shown in Figure 21–34, detailing the pathway for the synthesis of mevalonate from acetyl-CoA.

Answer Note: There are several plausible alternatives that a student might propose in the absence of a detailed knowledge of the literature on this enzyme. *Thiolase reaction:* begins with nucleophilic attack of an active-site Cys residue on the first acetyl-CoA substrate, displacing —S-CoA and forming a covalent thioester link between Cys and the acetyl group. A base on the enzyme then extracts a proton from the methyl group of the second acetyl-CoA, leaving a carbanion that attacks the carbonyl carbon of the thioester formed in the first step. The sulfhydryl of the Cys residue is displaced, creating the product acetoacetyl-CoA.

Enz
CoA-SH
Acetyl-CoA
Enz
Acetoacetyl-CoA

HMG-CoA synthase reaction: begins in the same way, with a covalent thioester link formed between the enzyme's Cys residue and the acetyl group of acetyl-CoA, with displacement of the —S-CoA. The —S-CoA dissociates as CoA-SH, and acetoacetyl-CoA binds to the enzyme. A proton is abstracted from the methyl group of the enzyme-linked acetyl, forming a carbanion that attacks the ketone carbonyl of the acetoacetyl-CoA substrate. The carbonyl is converted to a hydroxyl ion in this reaction, and this is protonated to create —OH. The thioester link with the enzyme is then cleaved hydrolytically to generate the HMG-CoA product.

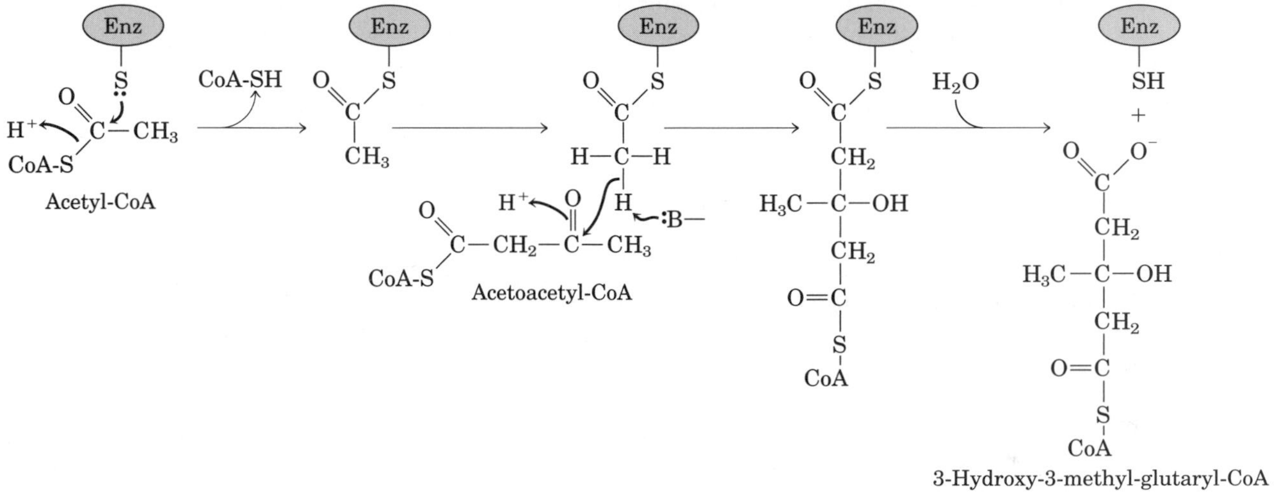

Acetyl-CoA

Acetoacetyl-CoA

3-Hydroxy-3-methyl-glutaryl-CoA

HMG-CoA reductase reaction: two successive hydride ions derived from NADPH first displace the —S-CoA, and then reduce the aldehyde to a hydroxyl group.

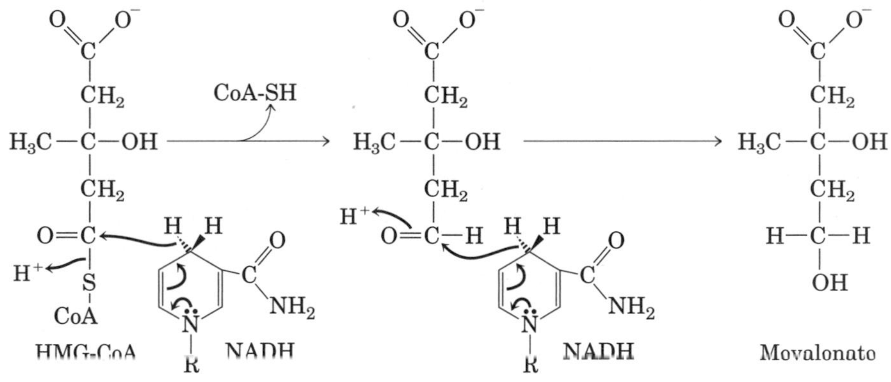

HMG-CoA NADH NADH Mevalonate

19. Potential Side Effects of Treatment with Statins Although clinical trials have not yet been carried out to document benefits or side effects, some physicians have suggested that patients being treated with statins also take a supplement of coenzyme Q. Suggest a rationale for this recommendation.

Answer Statins inhibit HMG-CoA reductase, an enzyme in the pathway to the synthesis of activated isoprenes, which are precursors of cholesterol and a wide range of isoprenoids, including coenzyme Q (ubiquinone). Hence, statins might reduce the levels of coenzyme Q available for mitochondrial respiration. Ubiquinone is obtained in the diet as well as by direct biosynthesis, but it is not yet clear how much is required and how well dietary sources can substitute for reduced synthesis. Reductions in the levels of particular isoprenoids may account for some side effects of statins.

Data Analysis Problem

20. Engineering *E. coli* to Produce Large Quantities of an Isoprenoid There is a huge variety of naturally occurring isoprenoids, some of which are medically or commercially important and produced industrially. The production methods include in vitro enzymatic synthesis, which is an expensive and low-yield process. In 1999, Wang, Oh, and Liao reported their experiments to engineer the easily grown bacterium *E. coli* to produce large amounts of astaxanthin, a commercially important isoprenoid.

Astaxanthin is a red-orange carotenoid pigment (an antioxidant) produced by marine algae. Marine animals such as shrimp, lobster, and some fish that feed on the algae get their orange color from the ingested astaxanthin. Astaxanthin is composed of eight isoprene units; its molecular formula is $C_{40}H_{52}O_4$:

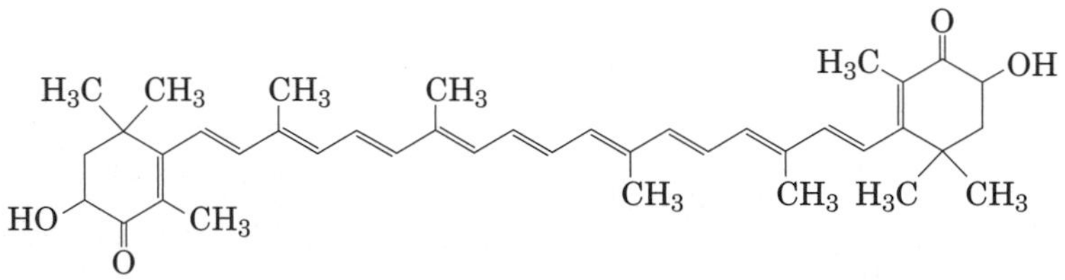

Astaxanthin

(a) Circle the eight isoprene units in the astaxanthin molecule. Hint: Use the projecting methyl groups as a guide.

Astaxanthin is synthesized by the pathway shown on the following page, starting with Δ^3-isopentenyl pyrophosphate (IPP). Steps ① and ② are shown in Figure 21–36, and the reaction catalyzed by IPP isomerase is shown in Figure 21–35.

(b) In step ④ of the pathway, two molecules of geranylgeranyl pyrophosphate are linked to form phytoene. Is this a head-to-head or a head-to-tail joining? (See Figure 21–36 for details.)

(c) Briefly describe the chemical transformation in step ⑤.

(d) The synthesis of cholesterol (Fig. 21–37) includes a cyclization (ring closure) that involves a net oxidation by O_2. Does the cyclization in step ⑥ of the astaxanthin synthetic pathway require a net oxidation of the substrate (lycopene)? Explain your reasoning.

E. coli does not make large quantities of many isoprenoids, and does not synthesize astaxanthin. It is known to synthesize small amounts of IPP, DMAPP, geranyl pyrophosphate, farnesyl pyrophosphate, and geranylgeranyl pyrophosphate. Wang and colleagues cloned several of the *E. coli* genes that encode enzymes needed for astaxanthin synthesis in plasmids that allow their overexpression. These genes included *idi,* which encodes IPP isomerase, and *ispA,* which encodes a prenyl transferase that catalyzes steps ① and ②.

To engineer an *E. coli* capable of the complete astaxanthin pathway, Wang and colleagues cloned several genes from other bacteria into plasmids that would allow their overexpression in *E. coli.* These genes included *crtE* from *Erwinia uredovora,* which encodes an enzyme that catalyzes step ③; and *crtB, crtI, crtY, crtZ,* and *crtW* from *Agrobacterium aurantiacum,* which encode enzymes for steps ④, ⑤, ⑥, ⑦, and ⑧, respectively.

The investigators also cloned the gene *gps* from *Archaeoglobus fulgidus*, overexpressed this gene in *E. coli*, and extracted the gene product. When this extract was reacted with [^{14}C]IPP and DMAPP, or geranyl pyrophosphate, or farnesyl pyrophosphate, only ^{14}C-labeled geranylgeranyl pyrophosphate was produced in all cases.

Δ^3-Isopentenyl pyrophosphate (IPP)

IPP isomerase

(1) Dimethylallyl pyrophosphate (DMAPP)

Geranyl pyrophosphate (C_{10}) + PP_i

(2) IPP

Farnesyl pyrophosphate (C_{15}) + PP_i

(3) IPP

Geranylgeranyl pyrophosphate (C_{20}) + PP_i

(4) Geranylgeranyl pyrophosphate

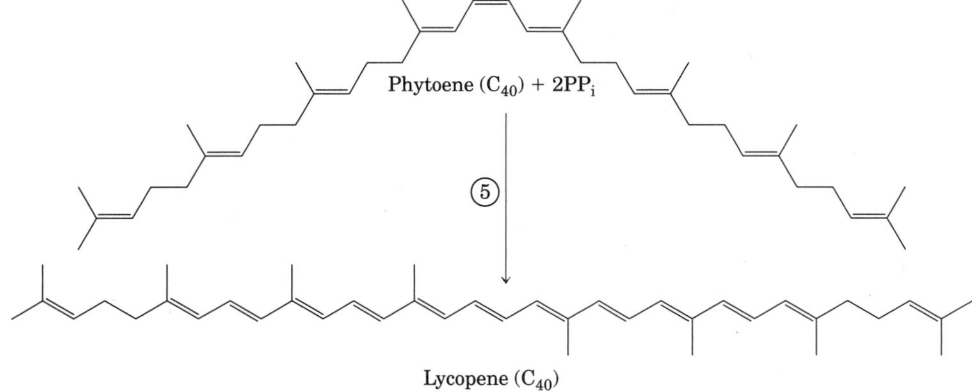

Phytoene (C_{40}) + $2PP_i$

(5)

Lycopene (C_{40})

(6)

β-Carotene (C_{40})

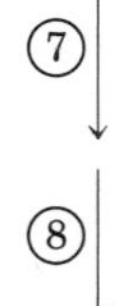

(7)

(8)

Astaxanthin (C_{40})

(e) Based on these data, which step(s) in the pathway are catalyzed by the enzyme encoded by *gps*? Explain your reasoning.

Wang and coworkers then constructed several *E. coli* strains overexpressing different genes and measured the orange color of the colonies (wild-type *E. coli* colonies are off-white) and the amount of astaxanthin produced. Their results are shown below.

Strain	Gene(s) overexpressed	Orange color	Astaxanthin yield (μg/g dry weight)
1	*crtBIZYW*	–	ND
2	*crtBIZYW, ispA*	–	ND
3	*crtBIZYW, idi*	–	ND
4	*crtBIZYW, idi, ispA*	–	ND
5	*crtBIZYW, crtE*	+	32.8
6	*crtBIZYW, crtE, ispA*	+	35.3
7	*crtBIZYW, crtE, idi*	++	234.1
8	*crtBIZYW, crtE, idi, ispA*	+++	390.3
9	*crtBIZYW, gps*	+	35.6
10	*crtBIZYW, gps, idi*	+++	1,418.8

Note: ND, not determined

(f) Comparing the results for strains 1 through 4 with those for strains 5 through 8, what can you conclude about the expression level of an enzyme capable of catalyzing step ③ of the astaxanthin synthetic pathway in wild-type *E. coli?* Explain your reasoning.

(g) Based on the data above, which enzyme is rate-limiting in this pathway, IPP isomerase or the enzyme encoded by *idi*? Explain your reasoning.

(h) Would you expect a strain overexpressing *crtBIZYW*, *gps*, and *crtE* to produce low (+), medium (++), or high (+++) levels of astaxanthin, as measured by its orange color? Explain your reasoning.

Answer

(a)

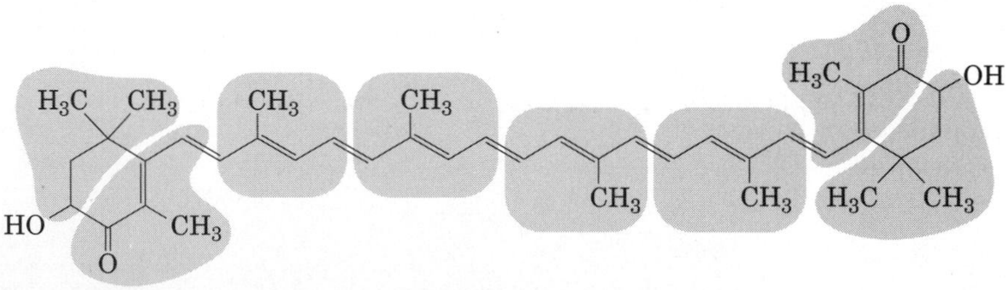

H_3C CH_3 CH_3 CH_3 H_3C O OH

HO O CH_3 CH_3 CH_3 H_3C CH_3

(b) Head-to-head. There are two ways to look at this. First, the "tail" of geranylgeranyl pyrophosphate has a branched dimethyl structure, as do both ends of phytoene. Second, no free —OH is formed by the release of PP_i, indicating that the two —O—Ⓟ—Ⓟ "heads" are linked to form phytoene.

(c) Four rounds of dehydrogenation convert four single bonds to double bonds.

(d) No. A count of single and double bonds in the following reaction shows that one double bond is replaced by two single bonds—so, there is no net oxidation or reduction.

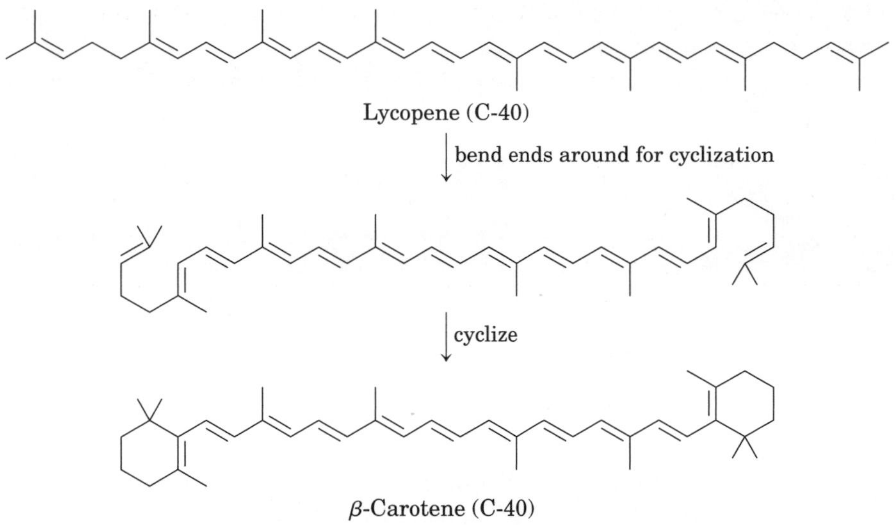

Lycopene (C-40)

bend ends around for cyclization

cyclize

β-Carotene (C-40)

(e) Steps ① through ③. The enzyme can convert IPP and DMAP to geranylgeranyl pyrophosphate, but catalyzes no further reactions in the pathway, as confirmed by results with the other substrates.

(f) Strains 1 through 4 lack *crtE* and have much lower astaxanthin production than strains 5 through 8, all of which overexpress *crtE*. Thus, overexpression of *crtE* leads to a substantial increase in astaxanthin production. Wild-type *E. coli* has some step ③ activity, but this conversion of farnesyl pyrophosphate to geranylgeranyl pyrophosphate is strongly rate-limiting.

(g) IPP isomerase. Comparing strains 5 and 6 shows that adding *ispA*, which catalyzes steps ① and ②, has little effect on astaxanthin production, so these steps are not rate-limiting. However, comparing strains 5 and 7 shows that adding *idi* substantially increases astaxanthin production, so IPP isomerase must be the rate-limiting step when *crtE* is overexpressed.

(h) A low (+) level, comparable to that of strains 5, 6, and 9. Without overexpression of *idi*, production of astaxanthin is limited by low IPP isomerase activity and the resulting limited supply of IPP.

Reference

Wang, C.-W., Oh, M.-K., & Liao, J.C. (1999) Engineered isoprenoid pathway enhances astaxanthin production in *Escherichia coli*. *Biotechnol. Bioeng.* **62,** 235–241.

chapter

22 Biosynthesis of Amino Acids, Nucleotides, and Related Molecules

1. **ATP Consumption by Root Nodules in Legumes** Bacteria residing in the root nodules of the pea plant consume more than 20% of the ATP produced by the plant. Suggest why these bacteria consume so much ATP.

   Answer Bacteria in the root nodules maintain a symbiotic relationship with the plant: the plant supplies ATP and reducing power, and the bacteria supply ammonium ion by reducing atmospheric nitrogen. This reduction requires large quantities of ATP.

2. **Glutamate Dehydrogenase and Protein Synthesis** The bacterium *Methylophilus methylotrophus* can synthesize protein from methanol and ammonia. Recombinant DNA techniques have improved the yield of protein by introducing into *M. methylotrophus* the glutamate dehydrogenase gene from *E. coli.* Why does this genetic manipulation increase the protein yield?

   Answer The synthesis of protein requires the synthesis of amino acids. The transfer of nitrogen from an ammonium ion to carbon skeletons—that is, amino acid synthesis—can be carried out in two ways: (1) combination of the NH_3 with glutamate to form glutamine, catalyzed by glutamine synthetase and (2) reductive amination of α-ketoglutarate to form glutamate, catalyzed by glutamate dehydrogenase. The latter process, which is promoted by the introduction of the *E. coli* enzyme, is especially important because glutamate is the amino group donor in all transamination reactions.

3. **PLP Reaction Mechanisms** Pyridoxal phosphate can help catalyze transformations one or two carbons removed from the α carbon of an amino acid. The enzyme threonine synthase (see Fig. 22–15) promotes the PLP-dependent conversion of phosphohomoserine to threonine. Suggest a mechanism for this reaction.

   Answer A link between enzyme-bound PLP and the phosphohomoserine substrate is first formed, with rearrangement to generate the ketimine at the α carbon of the substrate. This activates the β carbon for proton abstraction, leading to displacement of the phosphate and formation of a double bond between the β and γ carbons. A rearrangement (beginning with proton abstraction at the pyridoxal carbon adjacent to the substrate amino nitrogen) moves the double bond between the α and β carbons, and converts the ketimine to the aldimine form of PLP. Attack of water at the β carbon is then facilitated by the linked pyridoxal, followed by hydrolysis of the imine link between PLP and the product, to generate threonine.

4. **Transformation of Aspartate to Asparagine** There are two routes for transforming aspartate to asparagine at the expense of ATP. Many bacteria have an asparagine synthetase that uses ammonium ion as the nitrogen donor. Mammals have an asparagine synthetase that uses glutamine as the nitrogen donor. Given that the latter requires an extra ATP (for the synthesis of glutamine), why do mammals use this route?

Answer Recall that the ammonium ion is highly toxic to higher animals, especially to brain tissue. As NH_4^+ ions are transformed to glutamine, circulating NH_4^+ levels are reduced and toxic levels avoided.

5. **Equation for the Synthesis of Aspartate from Glucose** Write the net equation for the synthesis of aspartate (a nonessential amino acid) from glucose, carbon dioxide, and ammonia.

Answer We can approach this problem by working "backward" from aspartate to glucose as follows. Aspartate is synthesized from oxaloacetate by transamination from glutamate; glutamate is synthesized from α-ketoglutarate by glutamate dehydrogenase:

$$\text{Oxaloacetate} + \text{glutamate} \longrightarrow \text{aspartate} + \alpha\text{-ketoglutarate}$$

$$\alpha\text{-Ketoglutarate} + NH_3 + 2H^+ + \text{NADH} \longrightarrow \text{glutamate} + \text{NAD}^+ + H_2O$$

The sum of these reactions is

$$\text{Oxaloacetate} + NH_3 + 2H^+ + \text{NADH} \longrightarrow \text{aspartate} + \text{NAD}^+ + H_2O$$

Recall from Chapter 16 that oxaloacetate is synthesized from pyruvate by pyruvate carboxylase, and from Chapter 14 that pyruvate is produced from glucose via glycolysis:

$$\text{Pyruvate} + CO_2 + \text{ATP} + H_2O \longrightarrow \text{oxaloacetate} + \text{ADP} + P_i + 2H^+$$

$$\text{Glucose} + 2\text{NAD}^+ + 2\text{ADP} + 2P_i \longrightarrow 2\text{ pyruvate} + 2\text{NADH} + 2H^+ + 2\text{ATP} + 2H_2O$$

Thus, we can write the net equation for aspartate synthesis:

$$\text{Glucose} + 2CO_2 + 2NH_3 \longrightarrow 2\text{ aspartate} + 2H^+ + 2H_2O$$

6. **Asparagine Synthetase Inhibitors in Leukemia Therapy** Mammalian asparagine synthetase is a glutamine-dependent amidotransferase. Efforts to identify an effective inhibitor of human asparagine synthetase for use in chemotherapy for patients with leukemia has focused not on the amino-terminal glutaminase domain but on the carboxyl-terminal synthetase active site. Explain why the glutaminase domain is not a promising target for a useful drug.

Answer The amino-terminal glutaminase domain is quite similar in *all* glutamine amidotransferases. A drug that targeted this active site would probably inhibit many enzymes and thus be prone to producing many more side effects than a more specific inhibitor targeting the unique carboxyl-terminal synthetase active site.

7. **Phenylalanine Hydroxylase Deficiency and Diet** Tyrosine is normally a nonessential amino acid, but individuals with a genetic defect in phenylalanine hydroxylase require tyrosine in their diet for normal growth. Explain.

Answer In animals, tyrosine is synthesized from phenylalanine by phenylalanine hydroxylase. If this enzyme is defective, the biosynthetic route to tyrosine is blocked and this amino acid must be obtained from the diet.

8. **Cofactors for One-Carbon Transfer Reactions** Most one-carbon transfers are promoted by one of three cofactors: biotin, tetrahydrofolate, or *S*-adenosylmethionine (Chapter 18). *S*-Adenosylmethionine is generally used as a methyl group donor; the transfer potential of the methyl group in N^5-methyltetrahydrofolate is insufficient for most biosynthetic reactions. However, one example of the use of N^5-methyltetrahydrofolate in methyl group transfer is in methionine formation by the methionine synthase reaction (step ⑨ of Fig. 22–15); methionine is the immediate precursor of *S*-adenosylmethionine (see Fig. 18–18). Explain how the methyl group of *S*-adenosylmethionine can be derived from N^5-methyltetrahydrofolate, even though the transfer potential of the methyl group in N^5-methyltetrahydrofolate is one one-thousandth of that in *S*-adenosylmethionine.

Answer The transfer potential of the methyl group of N^5-methyltetrahydrofolate is quite sufficient for the synthesis of methionine, which has an even lower methyl group transfer potential. The methyl group of methionine is activated by addition of the adenosyl group from ATP, converting methionine to *S*-adenosylmethionine (see Fig. 18–18). Recall that adoMet synthesis is one of only two known biochemical reactions in which triphosphate is released from ATP. Hydrolysis of the triphosphate renders the reaction thermodynamically more favorable.

9. Concerted Regulation in Amino Acid Biosynthesis The glutamine synthetase of *E. coli* is independently modulated by various products of glutamine metabolism (see Fig. 22–6). In this concerted inhibition, the extent of enzyme inhibition is greater than the sum of the separate inhibitions caused by each product. For *E. coli* grown in a medium rich in histidine, what would be the advantage of concerted inhibition?

Answer Because the regulatory mechanism is concerted, the amount of inhibition caused by saturating concentrations of histidine is limited—that is, a large excess of one amino acid does not shut down the flow of glutamate to glutamine. Metabolite flow continues, albeit at a reduced rate. If the inhibition of glutamine synthase were *not* concerted, saturating concentrations of histidine would shut down the enzyme and cut off production of glutamine, which the bacterium needs to synthesize other products.

10. Relationship between Folic Acid Deficiency and Anemia Folic acid deficiency, believed to be the most common vitamin deficiency, causes a type of anemia in which hemoglobin synthesis is impaired and erythrocytes do not mature properly. What is the metabolic relationship between hemoglobin synthesis and folic acid deficiency?

Answer Folic acid is a precursor of the coenzyme tetrahydrofolate (Fig. 18–16), which is required in the biosynthesis of glycine (Fig. 22–12). Because glycine is a precursor of porphyrins, the heme component of hemoglobin, a folic acid deficiency results in an impairment of hemoglobin synthesis, especially if the diet is also low in glycine.

11. Nucleotide Biosynthesis in Amino Acid Auxotrophic Bacteria Wild-type *E. coli* cells can synthesize all 20 common amino acids, but some mutants, called amino acid auxotrophs, are unable to synthesize a specific amino acid and require its addition to the culture medium for optimal growth. Besides their role in protein synthesis, some amino acids are also precursors for other nitrogenous cell products. Consider the three amino acid auxotrophs that are unable to synthesize glycine, glutamine, and aspartate, respectively. For each mutant, what nitrogenous products other than proteins would the cell fail to synthesize?

Answer Glycine, glutamine, and aspartate are required for the de novo synthesis of purine nucleotides; aspartate for the de novo synthesis of UMP; aspartate and glutamine for the de novo synthesis of CTP. Thus, glycine auxotrophs would fail to synthesize adenine and guanine nucleotides. Glutamine auxotrophs would fail to synthesize adenine, guanine, and cytosine nucleotides. Aspartate auxotrophs would fail to synthesize adenine, guanine, cytosine, and uridine nucleotides.

12. Inhibitors of Nucleotide Biosynthesis Suggest mechanisms for the inhibition of **(a)** alanine racemase by L-fluoroalanine and **(b)** glutamine amidotransferases by azaserine.

Answer

(a) See Figure 18–6, step ②, for the reaction mechanism of amino acid racemization. The F atom of fluoroalanine is an excellent leaving group. Fluoroalanine causes irreversible (covalent) inhibition of alanine racemase. One plausible mechanism is:

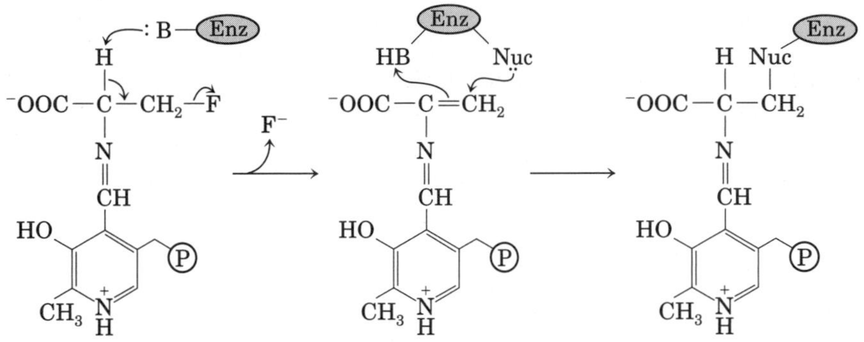

Nuc denotes any nucleophilic amino acid side chain in the enzyme active site.

(b) Azaserine (see Fig. 22–48) is an analog of glutamine. The diazoacetyl group is highly reactive and forms covalent bonds with nucleophiles at the active site of glutamine amidotransferases.

13. Mode of Action of Sulfa Drugs Some bacteria require *p*-aminobenzoate in the culture medium for normal growth, and their growth is severely inhibited by the addition of sulfanilamide, one of the earliest sulfa drugs. Moreover, in the presence of this drug, 5-aminoimidazole-4-carboxamide ribonucleotide (AICAR; see Fig. 22–33) accumulates in the culture medium. These effects are reversed by addition of excess *p*-aminobenzoate.

p-Aminobenzoate   Sulfanilamide

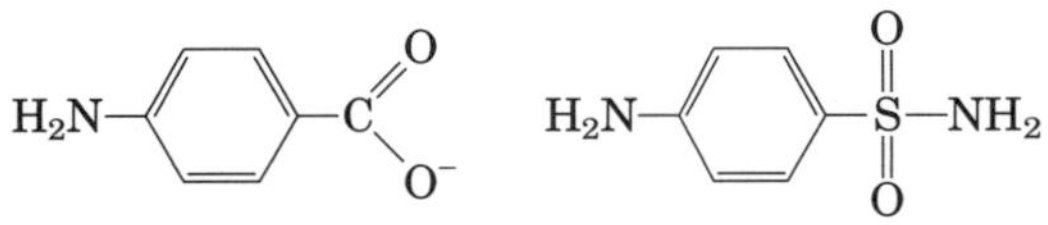

(a) What is the role of *p*-aminobenzoate in these bacteria? (Hint: see Fig. 18–16.)

(b) Why does AICAR accumulate in the presence of sulfanilamide?

(c) Why are the inhibition and accumulation reversed by addition of excess *p*-aminobenzoate?

Answer

(a) *p*-Aminobenzoate is a component of tetrahydrofolate (see Fig. 18–16) and its derivative, N^5,N^{10}-methylenetetrahydrofolate, the cofactor involved in the transfer of one-carbon units.

(b) Sulfanilamide is a structural analog of *p*-aminobenzoate. In the presence of sulfanilamide, bacteria are unable to synthesize tetrahydrofolate, a cofactor necessary for the transformation of AICAR to *N*-formylaminoimidazole-4-carboxamide ribonucleotide (FAICAR) by the addition of —CHO; thus, AICAR accumulates.

(c) Excess *p*-aminobenzoate reverses the growth inhibition and ribonucleotide accumulation by competing with sulfanilamide for the active site of the enzyme involved in tetrahydrofolate biosynthesis. The competitive inhibition by sulfanilamide is overcome by the addition of excess substrate (*p*-aminobenzoate).

14. Pathway of Carbon in Pyrimidine Biosynthesis Predict the locations of ^{14}C in orotate isolated from cells grown on a small amount of uniformly labeled [^{14}C]succinate. Justify your prediction.

Answer

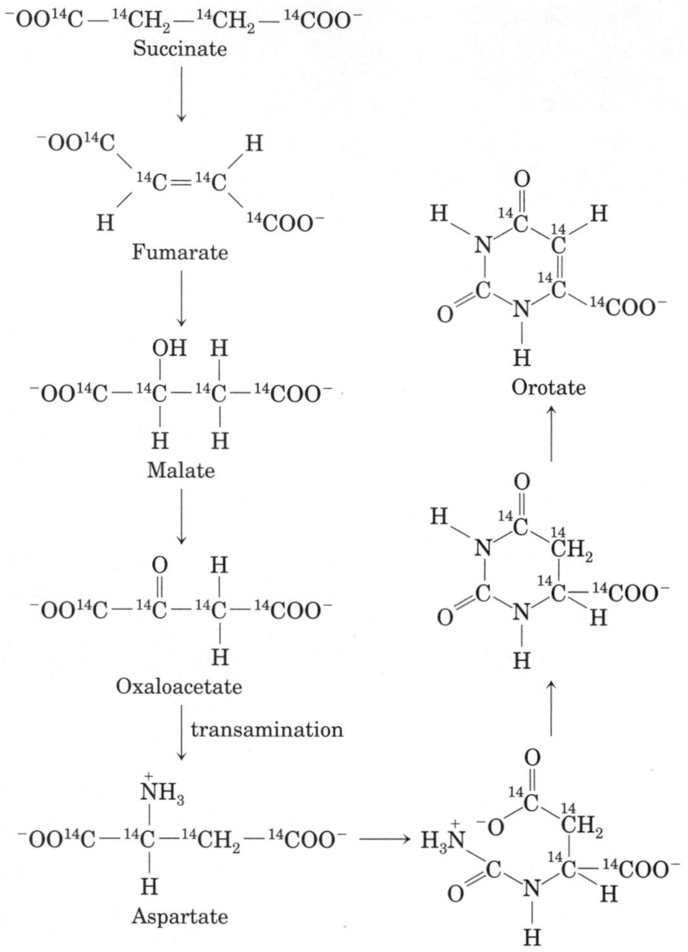

$^{-}OO^{14}C—^{14}CH_2—^{14}CH_2—^{14}COO^{-}$

Succinate

↓

$^{-}OO^{14}C(H)^{14}C{=}^{14}C(H)^{14}COO^{-}$

Fumarate

↓

$^{-}OO^{14}C—^{14}C(OH)(H)—^{14}C(H)(H)—^{14}COO^{-}$

Malate

↓

$^{-}OO^{14}C—^{14}C(=O)—^{14}C(H)(H)—^{14}COO^{-}$

Oxaloacetate

↓ transamination

$^{-}OO^{14}C—^{14}C(\overset{+}{N}H_3)(H)—^{14}CH_2—^{14}COO^{-}$

Aspartate

⟶ $H_3\overset{+}{N}$, ^{-}O, O, ^{14}C, $^{14}CH_2$, $^{14}C—^{14}COO^{-}$, H, C, N, H, O

↑

H, N, ^{14}C, O, $^{14}CH_2$, $^{14}C—^{14}COO^{-}$, H, C, N, H, O

↑

H, N, ^{14}C, O, ^{14}C, H, ^{14}C, $^{14}COO^{-}$, C, N, H, O

Orotate

15. Nucleotides as Poor Sources of Energy Under starvation conditions, organisms can use proteins and amino acids as sources of energy. Deamination of amino acids produces carbon skeletons that can enter the glycolytic pathway and the citric acid cycle to produce energy in the form of ATP. Nucleotides, on the other hand, are not similarly degraded for use as energy-yielding fuels. What observations about cellular physiology support this statement? What aspect of the structure of nucleotides makes them a relatively poor source of energy?

Answer Organisms do not store nucleotides to be used as fuel, and they do not completely degrade them, but rather hydrolyze them to release the bases, which can be recovered in salvage pathways. The low C:N ratio of nucleotides makes them poor sources of energy.

16. Treatment of Gout Allopurinol (see Fig. 22–47), an inhibitor of xanthine oxidase, is used to treat chronic gout. Explain the biochemical basis for this treatment. Patients treated with allopurinol sometimes develop xanthine stones in the kidneys, although the incidence of kidney damage is much lower than in untreated gout. Explain this observation in the light of the following solubilities in urine: uric acid, 0.15 g/L; xanthine, 0.05 g/L; and hypoxanthine, 1.4 g/L.

Answer Treatment with allopurinol has two biochemical consequences. (1) It inhibits conversion of hypoxanthine to uric acid, causing accumulation of hypoxanthine, which is more soluble than uric acid and more readily excreted. This alleviates the clinical problems associated with AMP degradation. (2) It inhibits conversion of guanine to uric acid, causing accumulation of xanthine, which is even less soluble than uric acid. This is the source of xanthine stones. Because less GMP than AMP is degraded, kidney damage caused by xanthine stones is less than that caused by untreated gout.

17. Inhibition of Nucleotide Synthesis by Azaserine The diazo compound *O*-(2-diazoacetyl)-L-serine, known also as azaserine (see Fig. 22–48), is a powerful inhibitor of glutamine amidotransferases. If growing cells are treated with azaserine, what intermediates of nucleotide biosynthesis will accumulate? Explain.

Answer In the de novo pathway of purine biosynthesis, the first step that requires glutamine is the conversion of 5-phosphoribosyl-1-pyrophosphate (PRPP) to 5-phospho-β-D-ribosylamine. In the presence of azaserine, which inhibits this conversion, PRPP accumulates.

Data Analysis Problem

18. Use of Modern Molecular Techniques to Determine the Synthetic Pathway of a Novel Amino Acid Most of the biosynthetic pathways described in this chapter were determined before the development of recombinant DNA technology and genomics, so the techniques were quite different from those that researchers would use today. Here we explore an example of the use of modern molecular techniques to investigate the pathway of synthesis of a novel amino acid, (2*S*)-4-amino-2-hydroxybutyrate (AHBA). The techniques mentioned here are described in various places in the book; this problem is designed to show how they can be integrated in a comprehensive study.

AHBA is a γ-amino acid that is a component of some aminoglycoside antibiotics, including the antibiotic butirosin. Antibiotics modified by the addition of an AHBA residue are often more resistant to inactivation by bacterial antibiotic-resistance enzymes. As a result, understanding how AHBA is synthesized and added to antibiotics is useful in the design of pharmaceuticals.

In an article published in 2005, Li and coworkers describe how they determined the synthetic pathway of AHBA from glutamate.

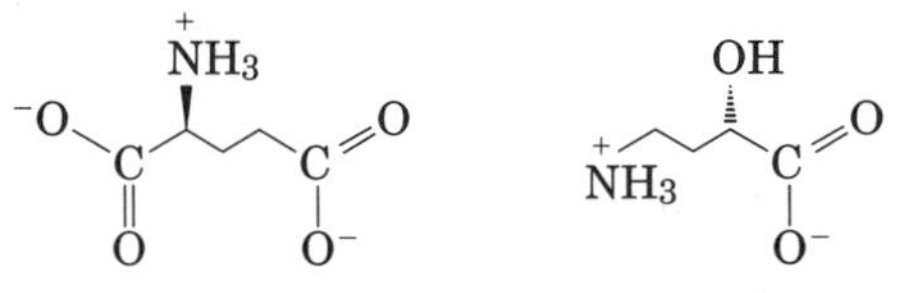

$\overset{+}{N}H_3$ OH

Glutamate   AHBA

(a) Briefly describe the chemical transformations needed to convert glutamate to AHBA. At this point, don't be concerned about the *order* of the reactions.

Li and colleagues began by cloning the butirosin biosynthetic gene cluster from the bacterium *Bacillus circulans*, which makes large quantities of butirosin. They identified five genes that are essential for the pathway: *btrI*, *btrJ*, *btrK*, *btrO*, and *btrV*. They cloned these genes into *E. coli* plasmids that allow overexpression of the genes, producing proteins with "histidine tags" (see p. 314) fused to their amino termini to facilitate purification.

The predicted amino acid sequence of the BtrI protein showed strong homology to known acyl carrier proteins (see Fig. 21–5). Using mass spectrometry (see Box 3–2), Li and colleagues found a molecular mass of 11,812 for the purified BtrI protein (including the His tag). When the purified BtrI was incubated with coenzyme A and an enzyme known to attach CoA to other acyl carrier proteins, the majority molecular species had an M_r of 12,153.

(b) How would you use these data to argue that BtrI can function as an acyl carrier protein with a CoA prosthetic group?

Using standard terminology, Li and coauthors called the form of the protein lacking CoA apo-BtrI and the form with CoA (linked as in Fig. 21–5) holo-BtrI. When holo-BtrI was incubated with glutamine, ATP, and purified BtrJ protein, the holo-BtrI species of M_r 12,153 was replaced with a species of M_r 12,281, corresponding to the thioester of glutamate and holo-BtrI. Based on these data, the authors proposed the following structure for the M_r 12,281 species (γ-glutamyl-*S*-BtrI):

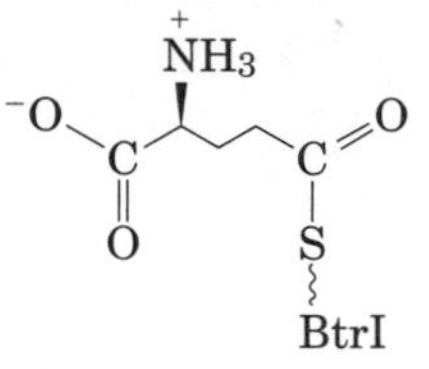

γ-Glutamyl-*S*-BtrI

(c) What other structure(s) is (are) consistent with the data above?

(d) Li and coauthors argued that the structure shown here (γ-glutamyl-*S*-BtrI) is likely to be correct because the α-carboxyl group must be removed at some point in the synthetic process. Explain the chemical basis of this argument. (Hint: See Fig. 18–6c.)

The BtrK protein showed significant homology to PLP-dependent amino acid decarboxylases, and BtrK isolated from *E. coli* was found to contain tightly bound PLP. When γ-glutamyl-*S*-BtrI was incubated with purified BtrK, a molecular species of M_r 12,240 was produced.

(e) What is the most likely structure of this species?

(f) Interestingly, when the investigators incubated glutamate and ATP with purified BtrI, BtrJ, and BtrK, they found a molecular species of M_r 12,370. What is the most likely structure of this species? Hint: Remember that BtrJ can use ATP to γ-glutamylate nucleophilic groups.

Li and colleagues found that BtrO is homologous to monooxygenase enzymes (see Box 21–1) that hydroxylate alkanes, using FMN as a cofactor, and BtrV is homologous to an NAD(P)H oxidoreductase. Two other genes in the cluster, *btrG* and *btrH*, probably encode enzymes that remove the γ-glutamyl group and attach AHBA to the target antibiotic molecule.

(g) Based on these data, propose a plausible pathway for the synthesis of AHBA and its addition to the target antibiotic. Include the enzymes that catalyze each step and any other substrates or cofactors needed (ATP, NAD, etc.).

Answer

(a) The α-carboxyl group is removed and an —OH is added to the γ carbon.

(b) BtrI has sequence homology with acyl carrier proteins. The molecular weight of BtrI increases when incubated under conditions in which CoA could be added to the protein. Adding CoA to a Ser residue would replace an —OH (formula weight (FW) 17) with a 4′-phosphopantetheine group (see Fig. 21–5, p. 809). This group has the formula $C_{11}H_{21}N_2O_7PS$ (FW 356). Thus, $11{,}182 - 17 + 356 = 12{,}151$, which is very close to the observed M_r of 12,153.

(c) The thioester could form with the α-carboxyl group.

(d) In the most common reaction for removing the α-carboxyl group of an amino acid (see Fig. 18–6, ©, p. 679), the carboxyl group must be free. Furthermore, it is difficult to imagine a decarboxylation reaction starting with a carboxyl group in its thioester form.

(e) $12{,}240 - 12{,}281 = 41$, close to the M_r of CO_2 (44). Given that BtrK is probably a decarboxylase, its most likely structure is the decarboxylated form:

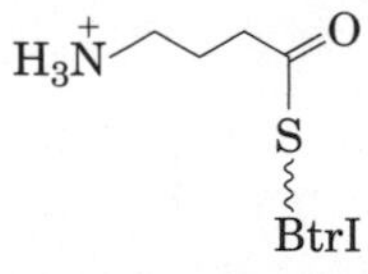

(f) 12,370 − 12,240 = 130. Glutamic acid ($C_5H_9NO_4$; M_r 147), minus the —OH (FW 17) removed in the glutamylation reaction, leaves a glutamyl group of FW 130; thus, γ-glutamylating the molecule above would add 130 to its M_r. BtrJ is capable of γ-glutamylating other substrates, so it may γ-glutamylate the structure above. The most likely site for this is the free amino group, giving the following structure:

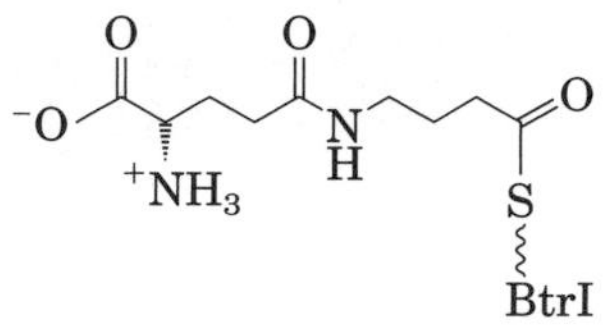

O O
$^-$O N O
$^+NH_3$ H S
BtrI

(g)

$\overset{+}{N}H_3$ $\overset{+}{N}H_3$ CO_2
$^-$O O BtrJ $^-$O O
O O$^-$ ATP ADP O S BtrK
+BtrI BtrI

Glutamate
ATP ADP
$H_3\overset{+}{N}$ O
BtrJ
S
BtrI

O O O_2 H_2O
$^-$O N O BtrO
$^+NH_3$ H S FMNH_2 FMN
BtrI BtrV
NAD^+ NADH

O O OH Antibiotic OH
$^-$O N O BtrG + BtrH $\overset{+}{N}H_3$ O
$^+NH_3$ H S Glutamate + BtrI O Antibiotic
BtrI

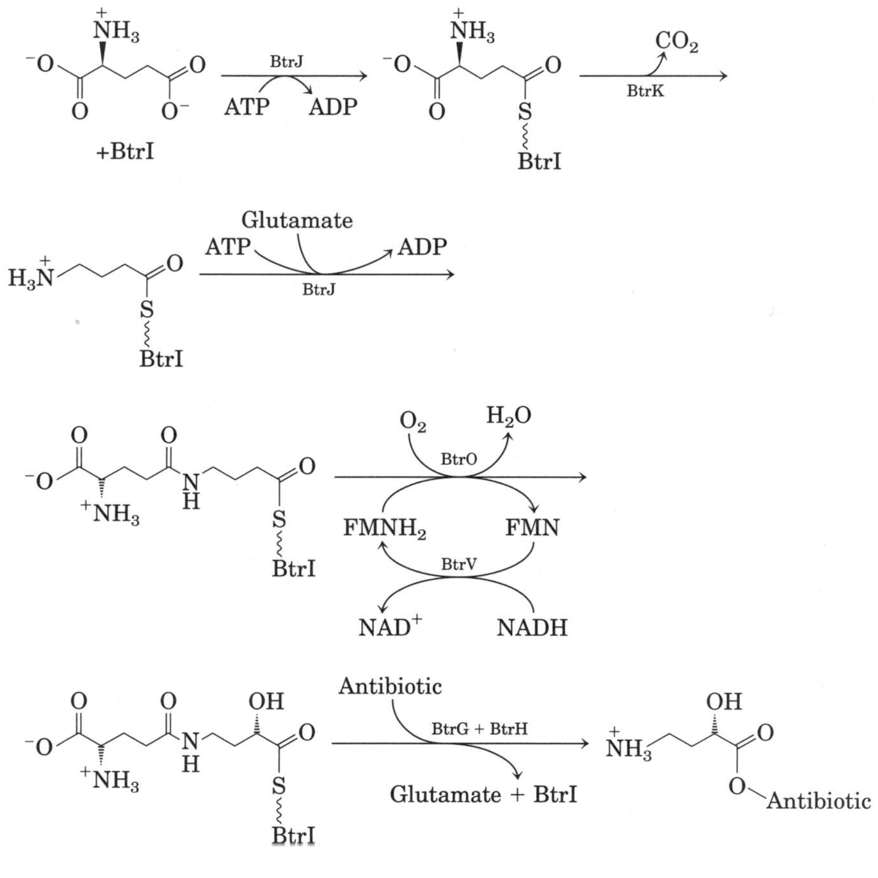

Reference

Li, Y., Llewellyn, N.M., Giri, R., Huang, F., & Spencer, J.B. (2005) Biosynthesis of the unique amino acid side chain of butirosin: possible protective-group chemistry in an acyl carrier protein–mediated pathway. *Chem. Biol.* **12,** 665–675.

chapter

23 Hormonal Regulation and Integration of Mammalian Metabolism

1. **Peptide Hormone Activity** Explain how two peptide hormones as structurally similar as oxytocin and vasopressin can have such different effects (see Fig. 23–10).

   Answer Oxytocin and vasopressin are recognized by different receptors, typically found in different cell types. These receptors are coupled to different downstream effects in their target cells.

2. **ATP and Phosphocreatine as Sources of Energy for Muscle** During muscle contraction, the concentration of phosphocreatine in skeletal muscle drops while the concentration of ATP remains fairly constant. However, in a classic experiment, Robert Davies found that if he first treated muscle with 1-fluoro-2,4-dinitrobenzene (p. 94), the concentration of ATP declined rapidly while the concentration of phosphocreatine remained unchanged during a series of contractions. Suggest an explanation.

   Answer Muscle contraction results in a net hydrolysis of ATP. Although the amount of ATP in muscle is very small, the supply can be rapidly replenished by phosphoryl group transfer from the phosphocreatine reservoir, catalyzed by creatine kinase:

   $$\text{Phosphocreatine} + \text{ADP} \rightleftharpoons \text{creatine} + \text{ADP}$$

   Because this reaction is rapid relative to the use of ATP by muscle, [ATP] remains in a steady state. The effect of pretreatment with fluoro-2,4-dinitrobenzene suggests that this is an effective inhibitor of creatine kinase. Under working conditions, the small amount of muscle ATP is quickly depleted and cannot be replenished.

3. **Metabolism of Glutamate in the Brain** Brain tissue takes up glutamate from the blood, transforms it into glutamine, then releases it into the blood. What is accomplished by this metabolic conversion? How does it take place? The amount of glutamine produced in the brain can actually exceed the amount of glutamate entering from the blood. How does this extra glutamine arise? (Hint: you may want to review amino acid catabolism in Chapter 18; recall that NH_3 is very toxic to the brain.)

   Answer Ammonia is very toxic to nervous tissue, especially the brain. Excess NH_3 in brain cells is removed by the transformation of glutamate to glutamine, catalyzed by glutamine synthetase. Glutamine is then exported to the blood and travels to the liver, where it is transformed to urea.

   The additional glutamine in the brain arises from the action of aminotransferases that transfer amino groups from amino acids to α-ketoglutarate (a citric acid cycle intermediate), forming glutamate, which is then converted to glutamine.

4. **Proteins as Fuel during Fasting** When muscle proteins are catabolized in skeletal muscle during a fast, what are the fates of the amino acids?

Answer Glucogenic amino acids are used to make glucose for the brain; others are oxidized in mitochondria via the citric acid cycle.

5. **Absence of Glycerol Kinase in Adipose Tissue** Glycerol 3-phosphate is required for the biosynthesis of triacylglycerols. Adipocytes, specialized for the synthesis and degradation of triacylglycerols, cannot use glycerol directly because they lack glycerol kinase, which catalyzes the reaction

$$\text{Glycerol} + \text{ATP} \longrightarrow \text{glycerol 3-phosphate} + \text{ADP}$$

How does adipose tissue obtain the glycerol 3-phosphate necessary for triacylglycerol synthesis?

Answer Adipose tissue converts glucose to the glycolytic intermediate dihydroxyacetone phosphate, which is reduced to glycerol 3-phosphate by the NADH-requiring enzyme glycerol 3-phosphate dehydrogenase:

$$\text{Glucose} \xrightarrow{\text{glycolysis}} \text{dihydroxyacetone phosphate}$$
$$\text{Dihydroxyacetone phosphate} + \text{NADH} + \text{H}^+ \rightleftharpoons \text{glycerol 3-phosphate} + \text{NAD}^+$$

6. **Oxygen Consumption during Exercise** A sedentary adult consumes about 0.05 L of O_2 in 10 seconds. A sprinter, running a 100 m race, consumes about 1 L of O_2 in 10 seconds. After finishing the race, the sprinter continues to breathe at an elevated (but declining) rate for some minutes, consuming an extra 4 L of O_2 above the amount consumed by the sedentary individual.
(a) Why does the need for O_2 increase dramatically during the sprint?
(b) Why does the demand for O_2 remain high after the sprint is completed?

Answer
(a) Increased muscle activity raises the demand for ATP, which is met by increased activity of the citric acid cycle enzymes and increased flow of electrons through the electron-transfer chain. This results in greater O_2 consumption.
(b) During the sprint, muscle transforms some glycogen to lactate (anaerobic glycolysis). After the sprint, lactate is transported to the liver, where it is converted back to glucose and glycogen. This process requires ATP and thus requires O_2 consumption above the resting rate.

7. **Thiamine Deficiency and Brain Function** Individuals with thiamine deficiency show some characteristic neurological signs and symptoms, including loss of reflexes, anxiety, and mental confusion. Why might thiamine deficiency be manifested by changes in brain function?

Answer Glucose is the primary fuel of the brain, and the brain is particularly sensitive to any change in the availability of glucose for energy production. A key reaction in glucose catabolism is the thiamine pyrophosphate–dependent oxidative decarboxylation of pyruvate to acetyl-CoA. Thus, a thiamine deficiency reduces the rate of glucose catabolism.

8. **Potency of Hormones** Under normal conditions, the human adrenal medulla secretes epinephrine ($C_9H_{13}NO_3$) at a rate sufficient to maintain a concentration of 10^{-10} M in circulating blood. To appreciate what that concentration means, calculate the diameter of a round swimming pool, with a water depth of 2.0 m, that would be needed to dissolve 1.0 g (about 1 teaspoon) of epinephrine to a concentration equal to that in blood.

Answer The concentration of epinephrine in blood is 10^{-10} M = 10^{-10} mol/L. The molecular weight of epinephrine is 183.21 g/mol. Thus, the mass of epinephrine (in grams) needed for a concentration of 10^{-10} mol/L is

$$(10^{-10} \text{ mol/L})(183.21 \text{ g/mol}) = 1.8 \times 10^{-8} \text{ g/L}$$

If ν is the volume in liters needed to dissolve 1 g of epinephrine to this concentration,

$$1.8 \times 10^{-8} \text{ g/L} = 1 \text{ g}/\nu$$
$$\nu = (1 \text{ g})/(1.8 \times 10^{-8} \text{ g/L}) = 5.5 \times 10^{7} \text{ L} = 5.5 \times 10^{10} \text{ cm}^3$$

If r is the radius of the pool, and h is the height of the water, the volume of water is $\nu = \pi r^2 h$ (volume of a cylinder). Rearranging to solve for r^2:

$$r^2 = \nu/\pi h$$

and substituting for ν, π, and h:

$$r^2 = (5.5 \times 10^{10} \text{ cm}^3)/3.14(200 \text{ cm}) = 8.8 \times 10^{7} \text{ cm}^2$$

Taking the square root:

$$r = 9{,}400 \text{ cm} = 94 \text{ m}$$

The diameter of the pool = $2r$ = 190 m. For comparison, an Olympic-size swimming pool is 50 m long.

9. **Regulation of Hormone Levels in the Blood** The half-life of most hormones in the blood is relatively short. For example, when radioactively labeled insulin is injected into an animal, half of the labeled hormone disappears from the blood within 30 min.
   (a) What is the importance of the relatively rapid inactivation of circulating hormones?
   (b) In view of this rapid inactivation, how is the level of circulating hormone kept constant under normal conditions?
   (c) In what ways can the organism make rapid changes in the level of a circulating hormone?

Answer

(a) Inactivation provides a rapid means to change hormone concentrations.

(b) A constant insulin level is maintained by equal rates of synthesis and degradation.

(c) Other means of varying hormone concentration include changes in rate of release from storage, rate of transport, and rate of conversion from prohormone to active hormone.

10. **Water-Soluble versus Lipid-Soluble Hormones** On the basis of their physical properties, hormones fall into one of two categories: those that are very soluble in water but relatively insoluble in lipids (e.g., epinephrine) and those that are relatively insoluble in water but highly soluble in lipids (e.g., steroid hormones). In their role as regulators of cellular activity, most water-soluble hormones do not enter their target cells. The lipid-soluble hormones, by contrast, do enter their target cells and ultimately act in the nucleus. What is the correlation between solubility, the location of receptors, and the mode of action of these two classes of hormones?

Answer Because of their low solubility in lipids, water-soluble hormones cannot penetrate the plasma membrane; they bind to receptors on the outer surface of the cell. In the case of epinephrine, this receptor is an enzyme that catalyzes the formation of a second messenger (cAMP) inside the cell. In contrast, lipid-soluble hormones readily penetrate the hydrophobic core of the plasma membrane. Once inside the cell, they can act on their target molecules or receptors directly.

11. Metabolic Differences between Muscle and Liver in a "Fight or Flight" Situation When an animal confronts a "fight or flight" situation, the release of epinephrine promotes glycogen breakdown in the liver, heart, and skeletal muscle. The end product of glycogen breakdown in the liver is glucose; the end product in skeletal muscle is pyruvate.

(a) What is the reason for the different products of glycogen breakdown in the two tissues?

(b) What is the advantage to an animal that must fight or flee of these specific glycogen breakdown routes?

Answer

(a) Glycogen breakdown in hepatocytes in response to epinephrine produces glucose 6-phosphate, which can be dephosphorylated by glucose 6-phosphatase; the glucose is then exported into the bloodstream. Heart and skeletal muscle lack glucose 6-phosphatase. Any glucose 6-phosphate produced in these tissues enters the glycolytic pathway and is metabolized to pyruvate. Under O_2-deficient conditions (actively contracting muscle), pyruvate is converted to lactate to regenerate NAD^+ for continued glycolysis.

(b) In a "fight or flight" situation, the concentration of glycolytic precursors in muscle tissue needs to be high in preparation for activity. Phosphorylated intermediates, such as glucose 6-phosphate, cannot escape from the cell because the membrane is not permeable to charged species; and glucose 6-phosphate is not exported by the glucose transporter. Thus, glucose 6-phosphate is retained in the cell and enters glycolysis. The liver, by contrast, does export glucose, which is necessary to maintain the blood glucose level. The glucose is transported in the bloodstream for uptake and use by muscle.

12. Excessive Amounts of Insulin Secretion: Hyperinsulinism Certain malignant tumors of the pancreas cause excessive production of insulin by the β cells. Affected individuals exhibit shaking and trembling, weakness and fatigue, sweating, and hunger.

(a) What is the effect of hyperinsulinism on the metabolism of carbohydrates, amino acids, and lipids by the liver?

(b) What are the causes of the observed symptoms? Suggest why this condition, if prolonged, leads to brain damage.

Answer

(a) Overproduction of insulin leads to excessive uptake of blood glucose by the liver and conversion to glycogen, leading to hypoglycemia. In addition, a general shutdown of amino acid and fatty acid catabolism occurs.

(b) Under conditions of hyperinsulinism, little circulating fuel is available to meet the requirements for ATP production, so the functions of energy-demanding tissues, such as brain and muscle, are compromised. This accounts for the shaking and trembling, weakness and fatigue. Excess insulin affects the mechanisms that regulate the autonomic nervous system, which in turn controls sweating. Finally, brain damage can result from lack of glucose because glucose is the main source of fuel for the brain.

13. Thermogenesis Caused by Thyroid Hormones Thyroid hormones are intimately involved in regulating the basal metabolic rate. Liver tissue of animals given excess thyroxine shows an increased rate of O_2 consumption and increased heat output (thermogenesis), but the ATP concentration in the tissue is normal. Different explanations have been offered for the thermogenic effect of thyroxine. One is that excess thryroxine causes uncoupling of oxidative phosphorylation in mitochondria. How could such an effect account for the observations? Another explanation suggests that the thermogenesis is due to an increased rate of ATP utilization by the thyroxine-stimulated tissue. Is this a reasonable explanation? Why?

Answer The observations are consistent with the thesis that thyroxine acts as an uncoupler of oxidative phosphorylation. Uncouplers lower the P/O ratio of tissues, and thus the tissue must increase respiration to meet the normal ATP demands. The observed thermogenesis could also be due to the increased rate of ATP utilization by the thyroxine-stimulated tissue, as the increased ATP demands are met by increased oxidative phosphorylation and thus respiration.

14. Function of Prohormones What are the possible advantages in the synthesis of hormones as prohormones?

Answer Because prohormones are inactive, they can be stored in quantity in secretory granules. Rapid activation is achieved by enzymatic cleavage in response to an appropriate signal.

15. Sources of Glucose during Starvation The typical human adult uses about 160 g of glucose per day, 120 g of which is used by the brain. The available reserve of glucose (~20 g of circulating glucose and ~190 g of glycogen) is adequate for about one day. After the reserve has been depleted during starvation, how would the body obtain more glucose?

Answer In animals, glucose can be synthesized from a variety of precursors by gluconeogenesis (Fig. 14–15). In humans, the principal precursors (obtained from storage depots) are glycerol from triacylglycerols and glucogenic amino acids from proteins. Oxaloacetate formed from CO_2 and pyruvate by pyruvate carboxykinase is also a source of glucose.

16. Parabiotic *ob/ob* mice By careful surgery, researchers can connect the circulatory systems of two mice so that the same blood circulates through both animals. In these **parabiotic** mice, products released into the blood by one animal reach the other animal via the shared circulation. Both animals are free to eat independently. If an *ob/ob* mouse (both copies of the *OB* gene are defective) and a normal *OB/OB* mouse (two good copies of the *OB* gene) were made parabiotic, what would happen to the weight of each mouse?

Answer The mutant (*ob/ob*) mouse is obese because it does not produce functional leptin, the signal to eat less and be more physically active. When the circulatory system of this mouse is joined with that of the normal mouse, leptin produced in the normal mouse reaches the hypothalamus of the obese mouse, triggering the "eat less" response; the obese mouse therefore loses weight. The normal *OB/OB* mouse continues to have enough leptin in its blood to maintain the signal to eat moderately, and it retains its normal body weight.

17. Calculation of Body Mass Index A portly biochemistry professor weighs 260 lb (118 kg) and is 5 feet 8 inches (173 cm) tall. What is his body mass index? How much weight would he have to lose to bring his body mass index down to 25 (normal)?

Answer BMI = [weight in kg]/[(height in m)^2]. The professor's current BMI = $118/(1.73)^2 = 39.5$. To achieve a BMI of 25, his weight would need to be $25(1.73)^2 = 74.8$ kg. He must lose current weight (118 kg) − desired weight (74.8 kg) = 43.2 kg (95.0 lb).

18. Insulin Secretion Predict the effects on insulin secretion by pancreatic β cells of exposure to the potassium ionophore valinomycin (p. 404). Explain your prediction.

Answer Valinomycin is a K^+ ionophore; it makes the membrane permeable to K^+. Valinomycin has the same effect as opening the K^+ channel, allowing exit of K^+ and consequent hyperpolarization of the membrane. This keeps the voltage-gated Ca^{2+} channel closed and reduces secretion of insulin from the cells. See Figure 23–28, p. 924.

19. Effects of a Deleted Insulin Receptor A strain of mice specifically lacking the insulin receptor of liver is found to have mild fasting hyperglycemia (blood glucose = 132 mg/dL, vs. 101 mg/dL in controls) and a more striking hyperglycemia in the fed state (glucose = 363 mg/dL, vs. 135 mg/dL in controls). The mice have higher than normal levels of glucose 6-phosphatase in the liver and elevated levels of insulin in the blood. Explain these observations.

Answer In these mice the hepatocytes lack the insulin receptor and so are insensitive to the blood insulin level. The cells continue to have high levels of glucose 6-phosphatase and high rates of gluconeogenesis, thus increasing blood glucose both during a fast and after a glucose-containing meal. The elevated blood glucose concentration triggers insulin release from pancreatic β cells. This accounts for the observed high level of insulin in the blood.

20. Decisions on Drug Safety The drug Avandia (rosiglitazone) is effective in lowering blood glucose in patients with type 2 diabetes, but also seems to carry an increased risk of heart attack. If it were your responsibility to decide whether this drug should remain on the market (labeled with suitable warnings of its side effects) or should be withdrawn, what factors would you weigh in making your decision?

Answer Some things to consider: What is the frequency of heart attack attributable to the drug? How does this frequency compare with the number of individuals spared the long-term consequences of type 2 diabetes? Are other, equally effective treatment options, with fewer adverse effects, available?

21. Type 2 Diabetes Medication The drugs acarbose (Precose) and miglitol (Glyset), used in the treatment of type 2 diabetes mellitus, inhibit α-glucosidases in the brush border of the small intestine. These enzymes degrade oligosaccharides derived from glycogen or starch to monosaccharides. Suggest a possible mechanism for the salutary effect of these drugs on individuals with diabetes. What side effects, if any, would you expect from these drugs. Why? (Hint: Review lactose intolerance, p. 545).

Answer Without intestinal glucosidase activity, absorption of glucose from dietary glycogen and starch is reduced, blunting the usual rise in blood glucose after the meal. The undigested oligosaccharides are fermented by bacteria in the large intestine, and the gases released cause intestinal discomfort.

Data Analysis Problem

22. Cloning the Pancreatic β-Cell Sulfonylurea Receptor Glyburide, a member of the sulfonylurea family of drugs shown on p. 925, is used to treat type 2 diabetes. It binds to and closes the ATP-gated K^+ channel shown in Figures 23–28 and 23–29.

(a) Given the mechanism shown in Figure 23–28, would treatment with glyburide result in increased or decreased insulin secretion by pancreatic β cells? Explain your reasoning.

(b) How does treatment with glyburide help reduce the symptoms of type 2 diabetes?

(c) Would you expect glyburide to be useful for treating type 1 diabetes? Why or why not?

Aguilar-Bryan and coauthors (1995) cloned the gene for the sulfonylurea receptor (SUR) portion of the ATP-gated K^+ channel from hamsters. The research team went to great lengths to ensure that the gene they cloned was in fact the SUR-encoding gene. Here we explore how it is possible for researchers to demonstrate that they have actually cloned the gene of interest rather than another gene.

The first step was to obtain pure SUR protein. As was already known, drugs such as glyburide bind SUR with very high affinity ($K_d < 10$ nM), and SUR has a molecular weight of 140 to 170 kDa. Aguilar-Bryan and coworkers made use of the high-affinity glyburide binding to tag the SUR protein with a radioactive label that would serve as a marker to purify the protein from a cell extract. First, they made a radiolabeled derivative of glyburide, using radioactive iodine (^{125}I):

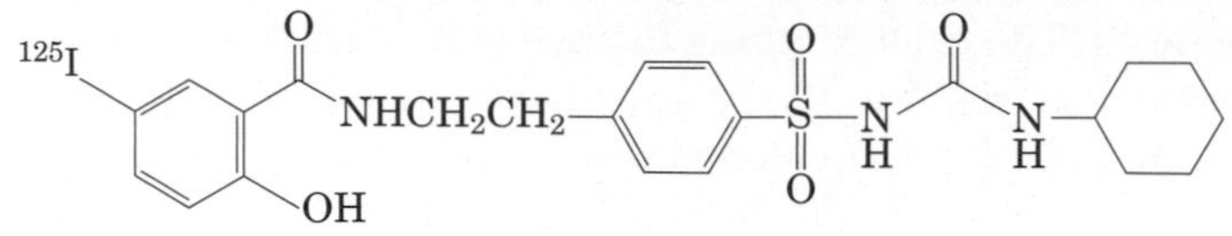

[^{125}I]5-Iodo-2-hydroxyglyburide

(d) In preliminary studies, the ^{125}I-labeled glyburide derivative (hereafter, [^{125}I]glyburide) was shown to have the same K_d and binding characteristics as unaltered glyburide. Why was it necessary to demonstrate this (what alternative possibilities did it rule out)?

Even though [^{125}I]glyburide bound to SUR with high affinity, a significant amount of the labeled drug would probably dissociate from the SUR protein during purification. To prevent this, [^{125}I]glyburide had to be covalently cross-linked to SUR. There are many methods for covalent cross-linking; Aguilar-Bryan and coworkers used UV light. When aromatic molecules are exposed to short-wave UV, they enter an excited state and readily form covalent bonds with nearby molecules. By cross-linking the radiolabeled glyburide to the SUR protein, the researchers could simply track the ^{125}I radioactivity to follow SUR through the purification procedure.

Aguilar-Bryan and colleagues treated hamster HIT cells (which express SUR) with [^{125}I]glyburide and UV light, purified the ^{125}I-labeled 140 kDa protein, and sequenced its amino-terminal 25 amino acid segment; they found the sequence PLAFCGTENHSAAYRVDQGVLNNGC. The investigators then generated antibodies that bound to two short peptides in this sequence, one that bound to PLAFCGTE and the other to HSAAYRVDQGV, and showed that these antibodies bound the purified ^{125}I-labeled 140 kDa protein.

(e) Why was it necessary to include this antibody-binding step?

Next, the researchers designed PCR primers based on the sequences above, and cloned a gene from a hamster cDNA library that encoded a protein that included these sequences (see Chapter 9 on biotechnology methods). The cloned putative *SUR* cDNA hybridized to an mRNA of the appropriate length that was present in cells known to contain SUR. The putative *SUR* cDNA did not hybridize to any mRNA fraction of the mRNAs isolated from hepatocytes, which do not express SUR.

(f) Why was it necessary to include this putative *SUR* cDNA–mRNA hybridization step?

Finally, the cloned gene was inserted into and expressed in COS cells, which do not normally express the *SUR* gene. The investigators mixed these cells with [^{125}I]glyburide with or without a large excess of unlabeled glyburide, exposed the cells to UV light, and measured the radioactivity of the 140 kDa protein produced. Their results are shown in the table.

Experiment	Cell type	Added putative SUR cDNA?	Added excess unlabeled glyburide?	^{125}I label in 140 kDa protein
1	HIT	No	No	+++
2	HIT	No	Yes	−
3	COS	No	No	−
4	COS	Yes	No	+++
5	COS	Yes	Yes	−

(g) Why was no ^{125}I-labeled 140 kDa protein found in experiment 2?

(h) How would you use the information in the table to argue that the cDNA encoded SUR?

(i) What other information would you want to collect to be more confident that you had cloned the *SUR* gene?

Answer

(a) Closing the ATP-gated K^+ channel would depolarize the membrane, leading to increased insulin release.

(b) Type 2 diabetes results from decreased sensitivity to insulin, not a deficit of insulin production; increasing circulating insulin levels will reduce the symptoms associated with this disease.

(c) Individuals with type 1 diabetes have deficient pancreatic β cells, so glyburide will have no beneficial effect.

(d) Iodine, like chlorine (the atom it replaces in the labeled glyburide), is a halogen, but it is a larger atom and has slightly different chemical properties. It is possible that the iodinated glyburide would not bind to SUR. If it bound to another molecule instead, the experiment would result in cloning of the gene for this other, incorrect protein.

(e) Although a protein has been "purified," the "purified" preparation might be a mixture of several proteins that co-purify under those experimental conditions. In this case, the amino acid sequence could be that of a protein that co-purifies with SUR. Using antibody binding to show that the peptide sequences are present in SUR excludes this possibility.

(f) Although the cloned gene does encode the 25 amino acid sequence found in SUR, it could be a gene that, coincidentally, encodes the same sequence in another protein. In this case, this other gene would most likely be expressed in different cells than the *SUR* gene. The mRNA hybridization results are consistent with the putative *SUR* cDNA actually encoding SUR.

(g) The excess unlabeled glyburide competes with labeled glyburide for the binding site on SUR. As a result, there is significantly less binding of labeled glyburide, so little or no radioactivity is detected in the 140 kDa protein.

(h) In the absence of excess unlabeled glyburide, labeled 140 kDa protein is found only in the presence of the putative *SUR* cDNA. Excess unlabeled glyburide competes with the labeled glyburide, and no ^{125}I-labeled 140 kDa protein is detected. This shows that the cDNA produces a glyburide-binding protein of the same molecular weight as SUR—strong evidence that the cloned gene encodes the SUR protein.

(i) Several additional steps are possible, such as: (1) Express the putative *SUR* cDNA in CHO (Chinese hamster ovary) cells and show that the transformed cells have ATP-gated K^+ channel activity. (2) Show that HIT cells with mutations in the putative *SUR* gene lack ATP-gated K^+ channel activity. (3) Show that experimental animals or human patients with mutations in the putative *SUR* gene are unable to secrete insulin.

Reference

Aguilar-Bryan, L., Nichols, C.G., Wechsler, S.W., Clement, J.P. IV, Boyd, A.E. III, González, G., Herrera-Sosa, H., Nguy, K., Bryan, J., & Nelson, D.A. (1995) Cloning of the β cell high-affinity sulfonylurea receptor: a regulator of insulin secretion. *Science* **268,** 423–426.

chapter

24 Genes and Chromosomes

1. **Packaging of DNA in a Virus** Bacteriophage T2 has a DNA of molecular weight 120×10^6 contained in a head about 210 nm long. Calculate the length of the DNA (assume the molecular weight of a nucleotide pair is 650) and compare it with the length of the T2 head.

   Answer The T2 DNA molecule has $(120 \times 10^6)/650 = 1.8 \times 10^5$ nucleotide pairs. Recall from Chapter 8 that a base pair occupies a length of 3.4 Å, so the length of the T2 DNA molecule is

   $$(1.8 \times 10^5 \text{ bp})(3.4 \text{ Å/bp}) = 6.1 \times 10^5 \text{ Å} = 6.1 \times 10^4 \text{ nm}$$

   Thus, the DNA molecule is 6.1×10^4 nm/210 nm = 290 times the length of the viral head. This nicely illustrates the necessity of compact packaging of DNA in viruses (see Fig. 24–1 and Table 24–1 for the relative lengths of DNA and viral head for other bacteriophages).

2. **The DNA of Phage M13** The base composition of phage M13 DNA is A, 23%; T, 36%; G, 21%; C, 20%. What does this tell you about the DNA of phage M13?

   Answer The complementarity between A and T, and between G and C, in the two strands of a duplex DNA underlies Chargaff's rule that the sum of pyrimidine nucleotides equals that of purine nucleotides in DNAs from (virtually) all species: that is, A = T, G = C, and A + G = C + T for duplex DNA. In M13 DNA, the percentage of A (23%) does not equal that of T (36%), nor does that of G (21%) equal that of C (20%); and A + G = 44%, whereas C + T = 56%. This lack of equality between purine and pyrimidine nucleotides shows that M13 DNA is single-stranded, *not* double-stranded; the relationships expected from complementarity between two strands of a duplex DNA are not seen. The M13 DNA is double-stranded only when replicating in the host cell.

3. **The *Mycoplasma* Genome** The complete genome of the simplest bacterium known, *Mycoplasma genitalium,* is a circular DNA molecule with 580,070 bp. Calculate the molecular weight and contour length (when relaxed) of this molecule. What is Lk_0 for the *Mycoplasma* chromosome? If $\sigma = -0.06$, what is Lk?

   Answer The molecular weight of a single nucleotide pair is 650 (see Problem 1), so the approximate molecular weight of the *Mycoplasma* DNA molecule is

   $$(580{,}070 \text{ bp})(650/\text{bp}) = 3.8 \times 10^8$$

   Given a length of 3.4 Å per base pair, the contour length is

   $$(580{,}070 \text{ bp})(3.4 \text{ Å/bp}) = 2.0 \times 10^6 \text{ Å} = 200 \ \mu\text{m}$$

   For relaxed circular DNA, the linking number is the number of base pairs divided by 10.5. For *M. genitalium,*

   $$Lk_0 = (580{,}070 \text{ bp})/(10.5 \text{ bp}) = 55{,}200$$

If $\sigma = -0.06$,

$$\Delta Lk = \sigma(Lk_0) = (-0.06)(55{,}200) = -3{,}310$$

Thus, $Lk - Lk_0 - 3{,}310 - 55{,}200 - 3{,}310 = 51{,}900$.

4. **Size of Eukaryotic Genes** An enzyme isolated from rat liver has 192 amino acid residues and is coded for by a gene with 1,440 bp. Explain the relationship between the number of amino acid residues in the enzyme and the number of nucleotide pairs in its gene.

   Answer Each amino acid is encoded by a triplet of 3 bp, so the 192 amino acids are encoded by 576 bp. The gene is in fact longer (1,440 bp). The additional 864 bp could be in introns (noncoding DNA, interrupting a coding segment) or could code for a signal sequence (or leader peptide). In addition, as discussed in Chapter 26, eukaryotic mRNAs have untranslated segments before and after the region coding for the polypeptide chain, which also contribute to the "extra" size of genes.

5. **Linking Number** A closed-circular DNA molecule in its relaxed form has an *Lk* of 500. Approximately how many base pairs are in this DNA? How is the linking number altered (increases, decreases, doesn't change, becomes undefined) when **(a)** a protein complex is bound to form a nucleosome, **(b)** one DNA strand is broken, **(c)** DNA gyrase and ATP are added to the DNA solution, or **(d)** the double helix is denatured by heat?

   Answer In relaxed DNA, the linking number (*Lk*) is equivalent to the number of turns in the DNA helix. *Lk* is a topological property and so does not vary when duplex DNA is twisted or deformed—as long as both DNA strands remain intact. *Lk* changes only if one or both strands are broken and rejoined. If a DNA strand remains broken, the molecule is no longer topologically constrained (the strands can unravel) and *Lk* is undefined.

   Because relaxed B-form DNA has about 10.5 bp per turn, a DNA with $Lk = 500$ has approximately (500 turns)(10.5 bp/turn) = 5,000 bp.

   (a) Doesn't change; the DNA strands are not cleaved and rejoined.

   (b) Becomes undefined; one of the strands remains broken.

   (c) Decreases; in the presence of ATP, gyrase (a type II topoisomerase that introduces negative supercoils) underwinds the DNA.

   (d) Doesn't change (assuming that the DNA strands are not broken and rejoined in the heating process).

6. **DNA Topology** In the presence of a eukaryotic condensin and a type II topoisomerase, the *Lk* of a relaxed closed-circular DNA molecule does not change. However, the DNA becomes highly knotted.

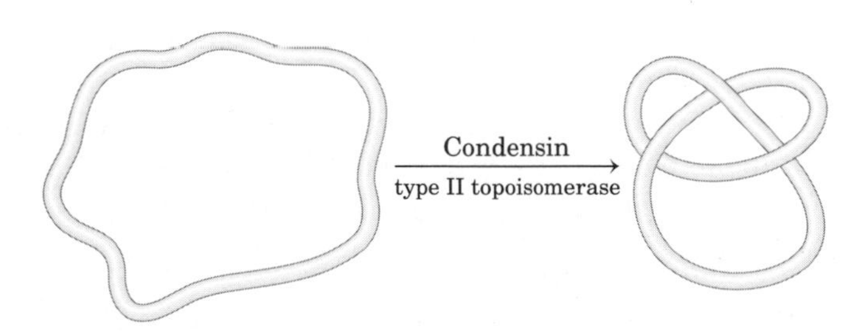

Formation of the knots requires breakage of the DNA, passage of a segment of DNA through the break, and religation by the topoisomerase. Given that every reaction of the topoisomerase would be expected to result in a change in linking number, how can *Lk* remain the same?

Answer For *Lk* to remain unchanged, the topoisomerase must introduce the same number of positive and negative supercoils.

7. **Superhelical Density** Bacteriophage λ infects *E. coli* by integrating its DNA into the bacterial chromosome. The success of this recombination depends on the topology of the *E. coli* DNA. When the superhelical density (σ) of the *E. coli* DNA is greater than −0.045, the probability of integration is <20%; when σ is less than −0.06, the probability is >70%. Plasmid DNA isolated from an *E. coli* culture is found to have a length of 13,800 bp and an *Lk* of 1,222. Calculate σ for this DNA and predict the likelihood that bacteriophage λ will be able to infect this culture.

Answer Superhelical density, also known as specific linking difference, is given by

$$\sigma = (Lk - Lk_0)/Lk_0$$

In this example, $Lk_0 = (13{,}800 \text{ bp})/(10.5 \text{ bp}) = 1{,}310$, and

$$\sigma = (1{,}222 - 1{,}310)/1{,}310$$
$$= -0.067$$

Because −0.067 is less than −0.06, there is a greater than 70% probability that the phage DNA will be incorporated into the *E. coli* DNA. [For more information about the relationship between the probability of recombination and the extent of DNA supercoiling, see Zechiedrich, E.L., Khodursky, A.B., & Cozzarelli, N.R. (1997) Topoisomerase IV, not gyrase, decatenates products of site-specific recombination in *Escherichia coli*. *Genes Dev.* ***11***, 2580–2592, especially Fig. 4 on p. 2585.]

8. **Altering Linking Number** **(a)** What is the *Lk* of a 5,000 bp circular duplex DNA molecule with a nick in one strand? **(b)** What is the *Lk* of the molecule in **(a)** when the nick is sealed (relaxed)? **(c)** How would the *Lk* of the molecule in **(b)** be affected by the action of a single molecule of *E. coli* topoisomerase I? **(d)** What is the *Lk* of the molecule in **(b)** after eight enzymatic turnovers by a single molecule of DNA gyrase in the presence of ATP? **(e)** What is the *Lk* of the molecule in **(d)** after four enzymatic turnovers by a single molecule of bacterial type I topoisomerase? **(f)** What is the *Lk* of the molecule in **(d)** after binding of one nucleosome?

Answer

(a) *Lk* is undefined; this is true for any DNA molecule in which the strands can be physically separated from each other. Here, the circular DNA is nicked, so one strand can be unwound and completely separated from the other; thus, there is no *Lk*.

(b) *Lk* = number of base pairs in the circle divided by number of base pairs per turn in relaxed B-form DNA (10.5). *Lk* = 5,000/10.5 = 476.

(c) Because the DNA is already relaxed, the topoisomerase I does not cause a net change. *Lk* remains at 476.

(d) DNA gyrase, in the presence of ATP, increases negative superhelicity, reducing the *Lk* in increments of 2. *Lk* declines by 8 × 2 = 16, so the final *Lk* = 460.

(e) Eukaryotic type I topoisomerases increase the *Lk* of underwound or negatively supercoiled DNA in increments of 1, so *Lk* in this case increases to 464.

(f) Nucleosome binding does not break any DNA strands and thus cannot change *Lk*. *Lk* remains at 460.

9. Chromatin Early evidence that helped researchers define nucleosome structure is illustrated by the agarose gel below, in which the thick bands represent DNA. It was generated by briefly treating chromatin with an enzyme that degrades DNA, then removing all protein and subjecting the purified DNA to electrophoresis. Numbers at the side of the gel denote the position to which a linear DNA of the indicated size would migrate. What does this gel tell you about chromatin structure? Why are the DNA bands thick and spread out rather than sharply defined?

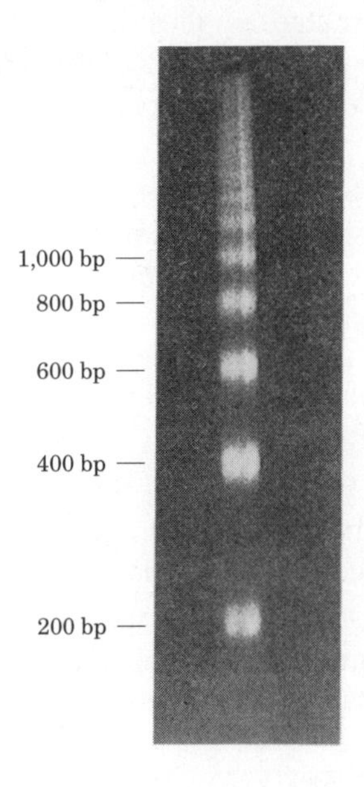

Answer The bands have a periodicity of about 200 bp (200, 400, 600 bp, etc.), showing that the chromatin is protected from nuclease digestion at regular intervals of 200 bp. This suggests that the nucleosomal cores (146 bp) provide the protection; this was verified in numerous subsequent investigations. Thus, the nucleosomes themselves are in a fairly regular array, occurring about once every 200 bp. The nuclease cuts the regions of double-stranded DNA that link but do not bind to the nucleosome cores—specifically, the spacer regions of about 60 bp. These regions are not always digested to completion; consequently, some bands correspond to the DNA from single nucleosomes (200 bp), others to two nucleosomes (400 bp), and so forth. If the nucleosomes were randomly distributed in the chromatin, a large number of differently sized DNA fragments would be generated by nuclease cleavage; in this case, a heterogeneous population of DNA fragments would have smeared through the gel.

The bands are thick because the spacer is fairly long (60 bp in this example) relative to the nucleosomal core (146 bp). The nuclease can cut essentially anywhere in the spacer, so the band corresponding to, for example, mononucleosomes has DNAs ranging in length from 146 to 206 bp.

10. DNA Structure Explain how the underwinding of a B-DNA helix might facilitate or stabilize the formation of Z-DNA.

Answer First, the shift from B- to Z-DNA requires a shift from a right-handed helix to a left-handed helix. Underwinding aids in the transition by providing a local region of unwound DNA that allows the shift to Z-DNA. (See Chapter 8, p. 281, for a description of sequences that permit the formation of Z-DNA.) Second, the left-handed helix of Z-DNA has a negative *Lk,* and underwinding of the DNA lowers the *Lk.* Third, underwinding essentially stores some free energy in the DNA that can be used to facilitate the B to Z transition.

11. Maintaining DNA Structure **(a)** Describe two structural features required for a DNA molecule to maintain a negatively supercoiled state. **(b)** List three structural changes that become more favorable when a DNA molecule is negatively supercoiled. **(c)** What enzyme, with the aid of ATP, can generate negative superhelicity in DNA? **(d)** Describe the physical mechanism by which this enzyme acts.

Answer

(a) A negatively supercoiled state implies that a DNA molecule is under some structural deformation, an energetically unfavorable condition. To maintain this state, both strands must be covalently closed, and the molecule must be either circular or (if linear) constrained at both ends.

(b) When a DNA molecule is negatively supercoiled, it is underwound. Any structural change involving DNA strand separation becomes more likely in negatively supercoiled DNA. This would include the formation of cruciforms and the direct unwinding of the DNA. Negative supercoiling also makes establishment of left-handed Z-DNA in a segment of DNA more likely because any underwinding of a right-handed helix aids the formation of a left-handed helix. Because the underwound DNA in a supercoiled molecule is placed under some structural deformation, the formation of plectonemic or solenoidal supercoils is favored as a means of relieving the strain.

(c) *E. coli* DNA topoisomerase II, or DNA gyrase, promotes the underwinding of DNA to introduce negative supercoils, when ATP is present.

(d) DNA gyrase binds the DNA at a point where it crosses on itself, cleaves both strands of one of the crossing segments, passes the other segment through the break, then reseals the break. The strand passage is directional and results in a change in Lk of -2.

12. Yeast Artificial Chromosomes (YACs) YACs are used to clone large pieces of DNA in yeast cells. What three types of DNA sequences are required to ensure proper replication and propagation of a YAC in a yeast cell?

Answer Stable artificial chromosomes require only three components: a centromere, telomeres at the ends, and an autonomous replicating sequence (replication origin).

13. Nucleoid Structure in Bacteria In bacteria, the transcription of a subset of genes is affected by DNA topology, with expression increasing or (more often) decreasing when the DNA is relaxed. When a bacterial chromosome is cleaved at a specific site by a restriction enzyme (one that cuts at a long, and thus rare, sequence), only nearby genes (within 10,000 bp) exhibit either an increase or decrease in expression. The transcription of genes elsewhere in the chromosome is unaffected. Explain. (Hint: See Fig. 24–37.)

Answer The bacterial nucleoid is organized into domains approximately 10,000 bp long. Cleavage by a restriction enzyme relaxes the DNA within a domain, but not outside the domain. Any gene in the cleaved domain for which expression is affected by DNA topology will be affected by the cleavage; genes outside the domain will not.

14. DNA Topoisomers When DNA is subjected to electrophoresis in an agarose gel, shorter molecules migrate faster than longer ones. Closed-circular DNAs of the same size but different linking number also can be separated on an agarose gel: topoisomers that are more supercoiled, and thus more condensed, migrate faster through the gel—from top to bottom in the gels shown on page S-279. A dye, chloroquine, was added to these gels. Chloroquine intercalates between base pairs and stabilizes a more underwound DNA structure. When the dye binds to a relaxed, closed-circular DNA, the DNA is underwound where the dye binds, and unbound regions take on positive supercoils to compensate. In the experiment shown here, topoisomerases were used to make preparations of the same DNA circle with different superhelical densities (σ). Completely relaxed DNA migrated to the position labeled N (for *n*icked), and highly supercoiled DNA (above the limit where individual topoisomers can be distinguished) to the position labeled X.

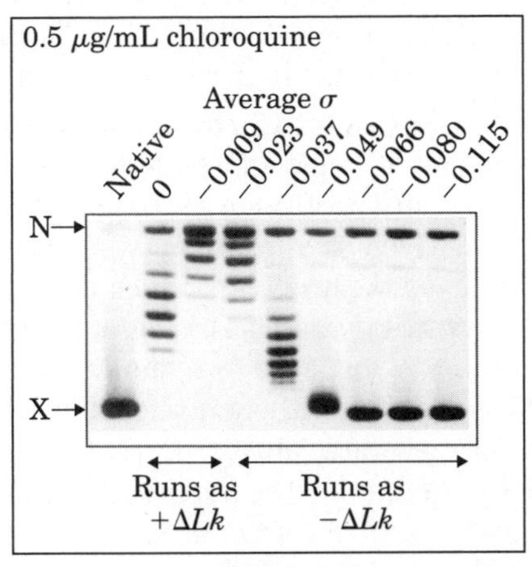

Gel A

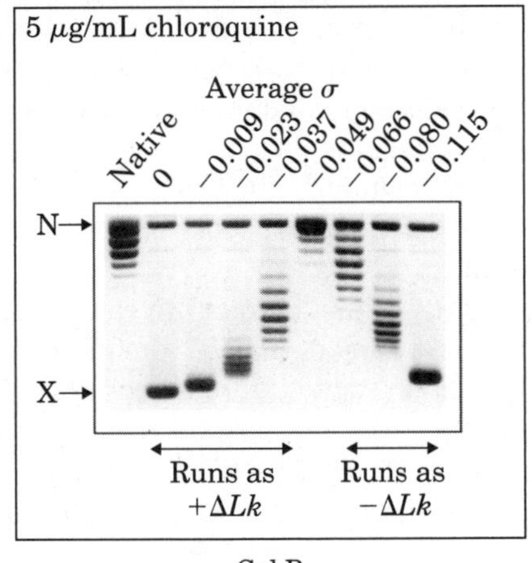

Gel B

(a) In gel A, why does the $\sigma = 0$ lane (i.e., DNA prepared so that $\sigma = 0$, on average) have multiple bands?

(b) In gel B, is the DNA from the $\sigma = 0$ preparation positively or negatively supercoiled in the presence of the intercalating dye?

(c) In both gels, the $\sigma = -0.115$ lane has two bands, one a highly supercoiled DNA and one relaxed. Propose a reason for the presence of relaxed DNA in these lanes (and others).

(d) The native DNA (leftmost lane in each gel) is the same DNA circle isolated from bacterial cells and untreated. What is the approximate superhelical density of this native DNA?

Answer

(a) When DNA ends are sealed to create a relaxed, closed circle, some DNA species are completely relaxed but others are trapped in slightly under- or overwound states. This gives rise to a distribution of topoisomers centered on the most relaxed species.

(b) Positively supercoiled.

(c) The DNA that is relaxed despite the addition of dye is DNA with one or both strands broken. DNA isolation procedures inevitably introduce small numbers of strand breaks in some of the closed-circular molecules.

(d) Approximately –0.05. This is determined by simply comparing native DNA with samples of known σ. In both gels, the native DNA migrates most closely with the sample of $\sigma = -0.049$.

Data Analysis Problem

15. Defining the Functional Elements of Yeast Chromosomes Figure 24–9 shows the major structural elements of a chromosome of baker's yeast (*Saccharomyces cerevisiae*). Heiter, Mann, Snyder, and Davis (1985) determined the properties of some of these elements. They based their study on the finding that in yeast cells, plasmids (which have genes and an origin of replication) act differently from chromosomes (which have these elements plus centromeres and telomeres) during mitosis. The plasmids are not manipulated by the mitotic apparatus and segregate randomly between daughter cells. Without a selectable marker to force the host cells to retain them (see Fig. 9–4), these plasmids are rapidly lost. In contrast, chromosomes, even without a selectable marker, are manipulated by the mitotic apparatus and are lost at a very low rate (about 10^{-5} per cell division).

Heiter and colleagues set out to determine the important components of yeast chromosomes by constructing plasmids with various parts of chromosomes and observing whether these "synthetic chromosomes" segregated properly during mitosis. To measure the rates of different types of failed chromosome segregation, the researchers needed a rapid assay to determine the number of copies of synthetic chromosomes present in different cells. This assay took advantage of the fact that wild-type yeast colonies are white whereas certain adenine-requiring (ade^-) mutants yield red colonies on nutrient media. Specifically, *ade2*$^-$ cells lack functional AIR carboxylase (the enzyme of step ⓺a in Figure 22–33) and accumulate AIR (5-aminoimidazole ribonucleotide) in their cytoplasm. This excess AIR is converted to a conspicuous red pigment. The other part of the assay involved the gene *SUP11*, which encodes an ochre suppressor (a type of nonsense suppressor; see Box 27–4) that suppresses the phenotype of some *ade2*$^-$ mutants.

Heiter and coworkers started with a diploid strain of yeast homozygous for *ade2*$^-$; these cells are red. When the mutant cells contain one copy of *SUP11*, the metabolic defect is partly suppressed and the cells are pink. When the cells contain two or more copies of *SUP11*, the defect is completely suppressed and the cells are white.

The researchers inserted one copy of *SUP11* into synthetic chromosomes containing various elements thought to be important in chromosome function, and then observed how well these chromosomes were passed from one generation to the next. These pink cells were plated on nonselective media, and the behavior of the synthetic chromosomes was observed. Specifically, Heiter and coworkers looked for colonies in which the synthetic chromosomes segregated improperly at the first division after plating, giving rise to a colony that is half one genotype and half the other. Because yeast cells are nonmotile, this will be a sectored colony, with one half one color and the other half another color.

(a) One way for the mitotic process to fail is *nondisjunction*: the chromosome replicates but the sister chromatids fail to separate, so both copies of the chromosome end up in the same daughter cell. Explain how nondisjunction of the synthetic chromosome would give rise to a colony that is half red and half white.

(b) Another way for the mitotic process to fail is *chromosome loss:* the chromosome does not enter the daughter nucleus or is not replicated. Explain how loss of the synthetic chromosome would give rise to a colony that is half red and half pink.

By counting the frequency of the different colony types, Heiter and colleagues could estimate the frequency of these aberrant mitotic events with different types of synthetic chromosome. First, they explored the requirement for centromeric sequences by constructing synthetic chromosomes with different-sized DNA fragments containing a known centromere. Their results are shown below.

Synthetic chromosome	Size of centromere-containing fragment (kbp)	Chromosome loss (%)	Nondisjunction (%)
1	none	—	>50
2	0.63	1.6	1.1
3	1.6	1.9	0.4
4	3.0	1.7	0.35
5	6.0	1.6	0.35

(c) Based on these data, what can you conclude about the size of the centromere required for normal mitotic segregation? Explain your reasoning.

(d) Interestingly, all the synthetic chromosomes created in these experiments were circular and lacked telomeres. Explain how they could be replicated more-or-less properly.

Heiter and colleagues next constructed a series of linear synthetic chromosomes that included the functional centromeric sequence and telomeres, and measured the total mitotic error rate (% loss + % nondisjunction) as a function of size:

Synthetic chromosome	Size (kbp)	Total error rate (%)
6	15	11.0
7	55	1.5
8	95	0.44
9	137	0.14

(e) Based on these data, what can you conclude about the chromosome size required for normal mitotic segregation? Explain your reasoning.

(f) Normal yeast chromosomes are linear, range from 250 kbp to 2,000 kbp in length, and have a mitotic error rate of about 10^{-5} per cell division. Extrapolating the results from **(e)**, do the centromeric and telomeric sequences used in these experiments explain the mitotic stability of normal yeast chromosomes, or must other elements be involved? Explain your reasoning. (Hint: A plot of log (error rate) vs. length will be helpful.)

Answer

(a) In nondisjunction, one daughter cell and all of its descendants get two copies of the synthetic chromosome and are white; the other daughter cell and all of its descendants get no copies of the synthetic chromosome and are red. This gives rise to a half-white, half-red colony.

(b) In chromosome loss, one daughter cell and all of its descendants get one copy of the synthetic chromosome and are pink; the other daughter and all its descendants get no copies of the synthetic chromosome and are red. This gives rise to a half-pink, half-red colony.

(c) The minimum functional centromere must be smaller than 0.63 kbp, since all fragments of this size or larger confer relative mitotic stability.

(d) Telomeres are required to fully replicate only linear DNA; a circular molecule can replicate without them.

(e) The larger the chromosome, the more faithfully it is segregated. The data show neither a minimum size below which the synthetic chromosome is completely unstable, nor a maximum size above which stability no longer changes.

(f)

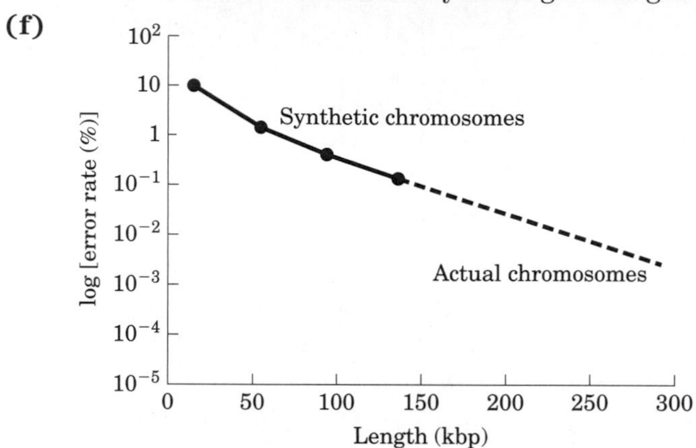

As shown in the graph, even if the synthetic chromosomes were as long as the normal yeast chromosomes, they would not be as stable. This suggests other, as yet undiscovered, elements are required for stability.

Reference

Heiter, P., Mann, C., Snyder, M., & Davis, R.W. (1985) Mitotic stability of yeast chromosomes: a colony color assay that measures nondisjunction and chromosome loss. *Cell* **40,** 381–392.

chapter

25

DNA Metabolism

1. **Conclusions from the Meselson-Stahl Experiment** The Meselson-Stahl experiment (see Fig. 25–2) proved that DNA undergoes semiconservative replication in *E. coli.* In the "dispersive" model of DNA replication, the parent DNA strands are cleaved into pieces of random size, then joined with pieces of newly replicated DNA to yield daughter duplexes. Explain how the results of Meselson and Stahl's experiment ruled out such a model.

   Answer If random, dispersive replication takes place, the density of the first-generation DNA in the Meselson-Stahl experiment would be the same as was actually observed: a single band midway between heavy and light DNA. In the second generation, however, all the DNA would again have the same density and would appear as a single band, midway between the band observed in the first generation and that of light DNA; this was not observed. *Two bands* were obtained in the experiment, ruling out the dispersive model.

2. **Heavy Isotope Analysis of DNA Replication** A culture of *E. coli* growing in a medium containing $^{15}NH_4Cl$ is switched to a medium containing $^{14}NH_4Cl$ for three generations (an eightfold increase in population). What is the molar ratio of hybrid DNA (^{15}N–^{14}N) to light DNA (^{14}N–^{14}N) at this point?

   Answer This experiment is an extension of the Meselson-Stahl experiment, which demonstrates that replication in *E. coli* is semiconservative. After three generations the molar ratio of ^{15}N–^{14}N DNA to ^{14}N–^{14}N DNA is 2/6 = 0.33.

3. **Replication of the *E. coli* Chromosome** The *E. coli* chromosome contains 4,639,221 bp.
   (a) How many turns of the double helix must be unwound during replication of the *E. coli* chromosome?
   (b) From the data in this chapter, how long would it take to replicate the *E. coli* chromosome at 37 °C if two replication forks proceeded from the origin? Assume replication occurs at a rate of 1,000 bp/s. Under some conditions *E. coli* cells can divide every 20 min. How might this be possible?
   (c) In the replication of the *E. coli* chromosome, about how many Okazaki fragments would be formed? What factors guarantee that the numerous Okazaki fragments are assembled in the correct order in the new DNA?

   Answer
   (a) During DNA replication, the complementary strands must unwind completely to allow the synthesis of a new strand on each template. Given the 10.5 bp/turn in B-DNA, and approximating the *E. coli* chromosome as 4.64×10^6 bp,

$$\text{the number of helical turns} = \frac{\text{number of base pairs}}{\text{number of base pairs per helical turn}}$$

$$= \frac{4.64 \times 10^6 \text{ bp}}{10.5 \text{ bp/turn}} = 4.42 \times 10^5 \text{ turns}$$

(b) Chromosomal DNA replication in *E. coli* starts at a fixed origin and proceeds bidirectionally. Each replication fork travels $(4.64 \times 10^6 \text{ bp})/2 = 2.32 \times 10^6$ bp during replication. If we assume a replication rate of 1,000 bp/s, the time required for the completion of DNA synthesis in each replication fork is

$$(2.32 \times 10^6 \text{ bp})/[(1000 \text{ bp/s})(60 \text{ s/min})] = 40 \text{ min}$$

This is about twice the time required for cell division. One possible explanation is that replication of an *E. coli* chromosome starts from two origins, each proceeding bidirectionally to yield four replication forks. In this mode, it would take 20 min to complete the replication of the chromosome. However, we know there is only *one* replication origin in the *E. coli* chromosome.

An alternative explanation is that a new round of replication begins before the previous one is completed: for cells dividing every 20 min, a replicative cycle is initiated every 20 min, and each daughter cell receives a chromosome that is half-replicated. This latter mode has been experimentally verified.

(c) The Okazaki fragments in *E. coli* are 1,000 to 2,000 nucleotides long, so $(4.64 \times 10^6$ nucleotides)/(2000 nucleotides) to $(4.64 \times 10^6$ nucleotides)/(1000 nucleotides) = 2,000 to 5,000 Okazaki fragments are formed. The fragments are firmly bound to the template strand by base pairing, and each fragment is quickly joined to the lagging strand by the successive action of DNA polymerase I and DNA ligase, thus preserving the correct order of the fragments. A mixed pool of Okazaki fragments, detached from their template, does *not* form during normal replication.

4. Base Composition of DNAs Made from Single-Stranded Templates Predict the base composition of the total DNA synthesized by DNA polymerase on templates provided by an equimolar mixture of the two complementary strands of bacteriophage ϕX174 DNA (a circular DNA molecule). The base composition of one strand is A, 24.7%; G, 24.1%; C, 18.5%; and T, 32.7%. What assumption is necessary to answer this problem?

Answer The sequence of a strand of duplex DNA is complementary to that of the other strand, as determined by Watson-Crick base pairing (A with T, and G with C). The DNA strand made from the given template strand has A, 32.7%; G, 18.5%; C, 24.1%; T, 24.7%. The DNA strand made from this complementary template strand has A, 24.7%; G, 24.1%; C, 18.5%; T, 32.7%. Thus, the composition of the *total* DNA synthesized is calculated as, for A, (32.7% + 24.6%)/2 = 28.7%; similarly, G = 21.3%; C = 21.3%; T = 28.7%. We are assuming that both template strands are completely replicated.

5. DNA Replication Kornberg and his colleagues incubated soluble extracts of *E. coli* with a mixture of dATP, dTTP, dGTP, and dCTP, all labeled with ^{32}P in the α-phosphate group. After a time, the incubation mixture was treated with trichloroacetic acid, which precipitates the DNA but not the nucleotide precursors. The precipitate was collected, and the extent of precursor incorporation into DNA was determined from the amount of radioactivity present in the precipitate.

(a) If any one of the four nucleotide precursors were omitted from the incubation mixture, would radioactivity be found in the precipitate? Explain.

(b) Would ^{32}P be incorporated into the DNA if only dTTP were labeled? Explain.

(c) Would radioactivity be found in the precipitate if ^{32}P labeled the β or γ phosphate rather than the α phosphate of the deoxyribonucleotides? Explain.

Answer

(a) No. Incorporation of ^{32}P into DNA results from the synthesis of new DNA, which requires the presence of *all four* nucleotide precursors.

(b) Yes. Although all four nucleotide precursors must be present for DNA synthesis, only one of them has to be radioactive in order for radioactivity to appear in the new DNA.

(c) No. No radioactivity would be incorporated if the ^{32}P label were not at the α phosphate because DNA polymerase, which catalyzes this reaction, cleaves off pyrophosphate—that is, the β and γ phosphate groups.

6. The Chemistry of DNA Replication All DNA polymerases synthesize new DNA strands in the 5′→3′ direction. In some respects, replication of the antiparallel strands of duplex DNA would be simpler if there were also a second type of polymerase, one that synthesized DNA in the 3′→5′ direction. The two types of polymerase could, in principle, coordinate DNA synthesis without the complicated mechanics required for lagging strand replication. However, no such 3′→5′-synthesizing enzyme has been found. Suggest two possible mechanisms for 3′→5′ DNA synthesis. Pyrophosphate should be one product of both proposed reactions. Could one or both mechanisms be supported in a cell? Why or why not? (Hint: You may suggest the use of DNA precursors not actually present in extant cells.)

Answer *Mechanism 1:* 3′-OH of an incoming dNTP attacks the α phosphate of the triphosphate at the 5′ end of the growing DNA strand, displacing pyrophosphate. This mechanism uses normal dNTPs, and the growing end of the DNA always has a triphosphate on the 5′ end.

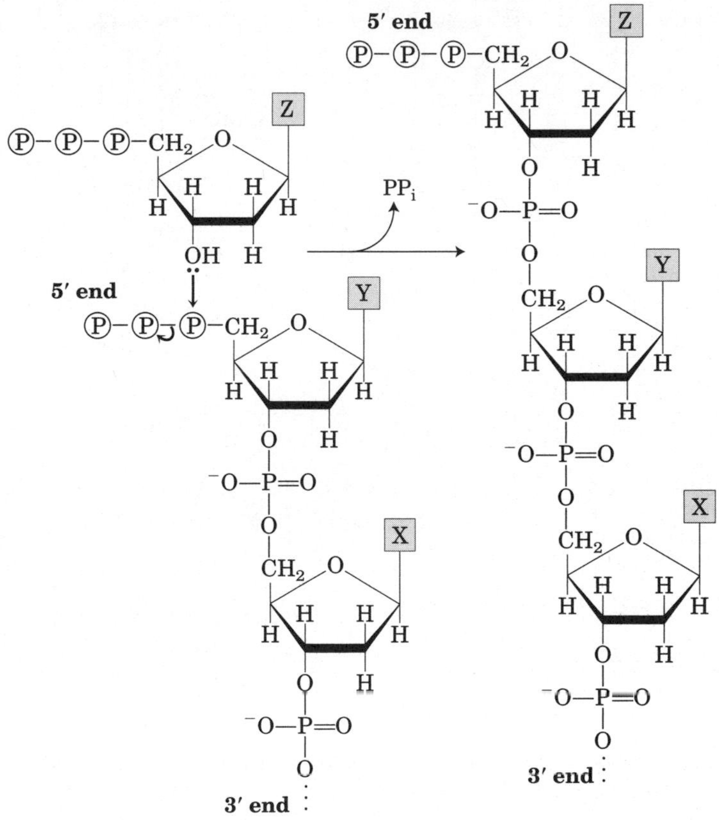

Mechanism 2: This uses a new type of precursor, nucleotide 3′-triphosphates. The growing end of the DNA strand has a 5′-OH, which attacks the α phosphate of an incoming deoxynucleotide 3′-triphosphate, displacing pyrophosphate. Note that this mechanism would require the evolution of new metabolic pathways to supply the needed deoxynucleotide 3′-triphosphates.

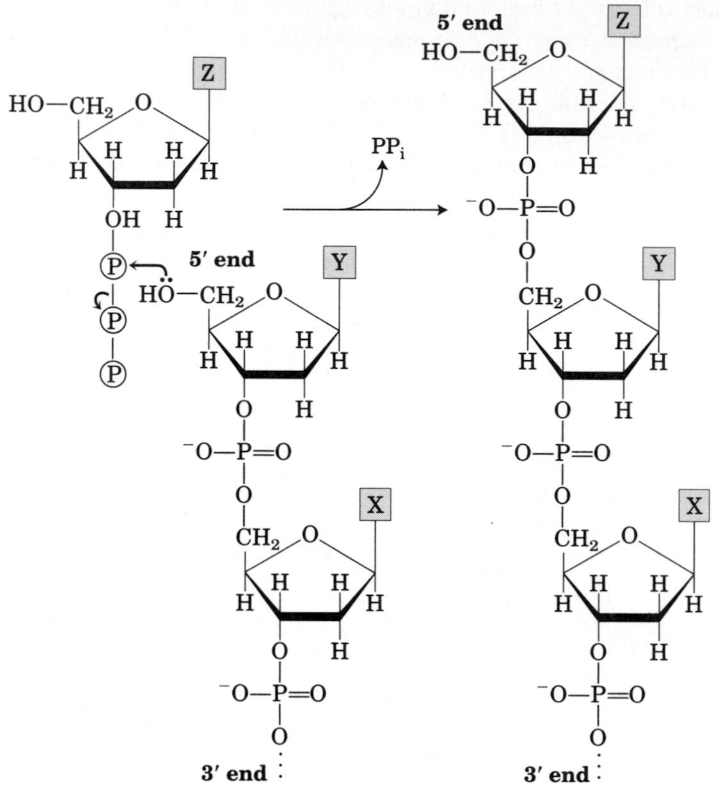

7. **Leading and Lagging Strands** Prepare a table that lists the names and compares the functions of the precursors, enzymes, and other proteins needed to make the leading strand versus the lagging strand during DNA replication in *E. coli.*

Answer In DNA replication, the *leading strand* is produced by continuous replication of the DNA template strand in the 5′ $\rightarrow$ 3′ direction. The *lagging strand* is synthesized in the form of short Okazaki fragments, which are then spliced together. The following table lists the participants in DNA replication in *E. coli.*

Component	**Function**	**Leading strand**	**Lagging strand**
dNTPs (dATP, dGTP, dGTP, dTTP)	Source of new nucleotides for new DNA strand; ATP energy source	X	X
Template	Parent strand provides information for identities of incoming nucleotides; stabilizes association of nucleotides in growing strand by base-pairing reactions.	X	X
DNA primer	Provides a free 3′ —OH, the point of attachment for incoming nucleotides	X	
DNA helicase	Unwinds double-stranded DNA just ahead of the replication fork; requires ATP	X	X
DNA gyrase	Topoisomerase that favors unwinding of the DNA at the replication fork by twisting the DNA; requires ATP	X	X
Single-stranded DNA–binding proteins	Prevents base pairing of unwound DNA strands; stabilizes single-stranded DNA	X	X
DNA polymerase III	Elongates the new DNA strand by adding nucleotides; requires the cofactors Mg^{2+} and Zn^{2+}	X	X
Pyrophosphatase	Hydrolyzes the PP_i released by polymerase activity; helps "pull" the reaction in the forward direction.	X	X
RNA primer	Same as DNA primer; used to start each Okazaki fragment		X
Primase	Synthesizes RNA primer		X
NTPs (ATP, GTP, CTP, UTP)	Used in synthesis of RNA primer		X
DNA polymerase I	Exonuclease that removes RNA primer by replacing NMPs with dNMPs in Okazaki fragments; requires the cofactors Mg^{2+} and Zn^{2+}		X
DNA ligase	Joins the Okazaki fragments in the lagging strand by catalyzing the formation of a phosphodiester bond; requires NAD^+ as energy source		X

8. Function of DNA Ligase Some *E. coli* mutants contain defective DNA ligase. When these mutants are exposed to ^{3}H-labeled thymine and the DNA produced is sedimented on an alkaline sucrose density gradient, two radioactive bands appear. One corresponds to a high molecular weight fraction, the other to a low molecular weight fraction. Explain.

Answer During replication of the DNA duplex, the leading strand is replicated continuously and the lagging strand is replicated in short fragments (Okazaki fragments), which are then spliced together by DNA ligase. Mutants with defective DNA ligase produce a DNA "duplex" in which one of the strands remains fragmented. Consequently, when this duplex is denatured by the alkaline conditions of the sucrose gradient, sedimentation results in one fraction containing the intact single strand (the high molecular weight band) and one fraction containing the unspliced fragments (the low molecular weight band).

9. Fidelity of Replication of DNA What factors promote the fidelity of replication during the synthesis of the leading strand of DNA? Would you expect the lagging strand to be made with the same fidelity? Give reasons for your answers.

Answer Fidelity of replication is ensured by Watson-Crick base pairing between the template and leading strand, and proofreading and removal of wrongly inserted nucleotides by the 3′-exonuclease activity of DNA polymerase III. The same fidelity would be expected in the lagging strand—perhaps. The factors ensuring fidelity of replication are operative in the leading and the lagging strands, but the greater number of distinct chemical operations involved in making the lagging strand might provide a greater opportunity for errors to arise.

10. Importance of DNA Topoisomerases in DNA Replication DNA unwinding, such as that occurring in replication, affects the superhelical density of DNA. In the absence of topoisomerases, the DNA would become overwound ahead of a replication fork as the DNA is unwound behind it. A bacterial replication fork will stall when the superhelical density (σ) of the DNA ahead of the fork reaches +0.14 (see Chapter 24).

Bidirectional replication is initiated at the origin of a 6,000 bp plasmid in vitro, in the absence of topoisomerases. The plasmid initially has a σ of −0.06. How many base pairs will be unwound and replicated by each replication fork before the forks stall? Assume that each fork travels at the same rate and that each includes all components necessary for elongation except topoisomerase.

Answer About 1,200 bp are unwound, or about 600 in each direction. The DNA is initially negatively supercoiled, with an Lk of about 537 and Lk_0 of about 571 (see Chapter 24). Unwinding 571 − 537 = 34 turns of DNA relaxes the DNA, and further unwinding adds positive supercoils. Unwinding another 79 turns (or 113 in all) brings the superhelical density in the remaining wound DNA to about +0.14—the stalling point. These 113 turns are equivalent to 113 turns × 10.5 bp/turn = 1,190 bp, or about 1,200 bp.

11. The Ames Test In a nutrient medium that lacks histidine, a thin layer of agar containing ~10^9 *Salmonella typhimurium* histidine auxotrophs (mutant cells that require histidine to survive) produces ~13 colonies over a two-day incubation period at 37 °C (see Fig. 25–21). How do these colonies arise in the absence of histidine? The experiment is repeated in the presence of 0.4 μg of 2-aminoanthracene. The number of colonies produced over two days exceeds 10,000. What does this indicate about 2-aminoanthracene? What can you surmise about its carcinogenicity?

Answer Occasionally, some of the histidine-requiring mutants spontaneously undergo back-mutation and regain their capacity to synthesize histidine, and thus can grow in a medium lacking histidine. The observation that only 13 of about 10^9 bacteria produce colonies indicates that the rate of back-mutation is quite low. The addition of 2-aminoanthracene increases the rate of back-mutations more than 800-fold, indicating that 2-aminoanthracene is mutagenic. Because about 90% of 300 known carcinogens are mutagenic, these observations suggest that 2-aminoanthracene is likely to be carcinogenic.

12. DNA Repair Mechanisms Vertebrate and plant cells often methylate cytosine in DNA to form 5-methylcytosine (see Fig. 8–5a). In these same cells, a specialized repair system recognizes G–T mismatches and repairs them to G≡C base pairs. How might this repair system be advantageous to the cell? (Explain in terms of the presence of 5-methylcytosine in the DNA.)

Answer This is very similar to Problem 6 of Chapter 8. Spontaneous deamination of 5-methylcytosine produces thymine, and thus a G–T mismatched pair. Such G–T pairs are among the most common mismatches in the DNA of eukaryotes. The specialized repair system restores the G≡C pair.

13. DNA Repair in People with Xeroderma Pigmentosum The condition known as xeroderma pigmentosum (XP) arises from mutations in at least seven different human genes (see Box 25–1). The deficiencies are generally in genes encoding enzymes involved in some part of the pathway for human nucleotide-excision repair. The various types of XP are denoted A through G (XPA, XPB, etc.), with a few additional variants lumped under the label XP-V.

Cultures of fibroblasts from healthy individuals and from patients with XPG are irradiated with ultraviolet light. The DNA is isolated and denatured, and the resulting single-stranded DNA is characterized by analytical ultracentrifugation.

(a) Samples from the normal fibroblasts show a significant reduction in the average molecular weight of the single-stranded DNA after irradiation, but samples from the XPG fibroblasts show no such reduction. Why might this be?

(b) If you assume that a nucleotide-excision repair system is operative in fibroblasts, which step might be defective in the cells from the patients with XPG? Explain.

Answer

(a) Ultraviolet irradiation of skin fibroblast DNA results in formation of pyrimidine dimers. In normal fibroblasts, the damaged DNA is repaired by excision of the pyrimidine dimer. One step in this process is cleavage of the damaged strand by a special excinuclease. Thus, the denatured single-stranded DNA isolated from normal cells after irradiation contains the many fragments caused by the cleavage, and the average molecular weight is lowered. These fragments of single-stranded DNA are absent from the XPG samples, as indicated by the unchanged average molecular weight.

(b) The absence of fragments in the single-stranded DNA from XPG cells after irradiation suggests that the special repair excinuclease is defective or missing.

14. Holliday Intermediates How does the formation of Holliday intermediates in homologous genetic recombination differ from their formation in site-specific recombination?

Answer During homologous genetic recombination, a Holliday intermediate may be formed almost anywhere within the two paired, homologous chromosomes. Once formed, the branch point of the intermediate can move extensively by branch migration. In site-specific recombination, the Holliday intermediate is formed between two specific sites, and branch migration is generally restricted by heterologous sequences on either side of the recombination sites.

15. A Connection between Replication and Site-Specific Recombination Most wild strains of *Saccharomyces cerevisiae* have multiple copies of the circular plasmid 2μ (named for its contour length of about 2 μm), which has ~6,300 bp of DNA. For its replication the plasmid uses the host replication system, under the same strict control as the host cell chromosomes, replicating only once per cell cycle. Replication of the plasmid is bidirectional, with both replication forks initiating at a single, well-defined origin. However, one replication cycle of a 2μ plasmid can result in more than two copies of the plasmid, allowing amplification of the plasmid copy number (number of plasmid copies per cell) whenever plasmid segregation at cell division leaves one daughter cell with fewer than the normal complement of plasmid copies. Amplification requires a site-specific recombination system encoded by the plasmid, which serves to invert one part of the plasmid relative to the other. Explain how a site-specific inversion event could result in amplification of the plasmid copy number. (Hint: Consider the situation when replication forks have duplicated one recombination site but not the other.)

Answer Once replication has proceeded from the origin to a point where one recombination site has been replicated but the other has not, site-specific recombination not only inverts the DNA between the recombination sites but also changes the direction of one replication fork relative to the other. The forks will chase each other around the DNA circle, generating many tandem copies of the plasmid. The multimeric circle can be resolved to monomers by additional site-specific recombination events.

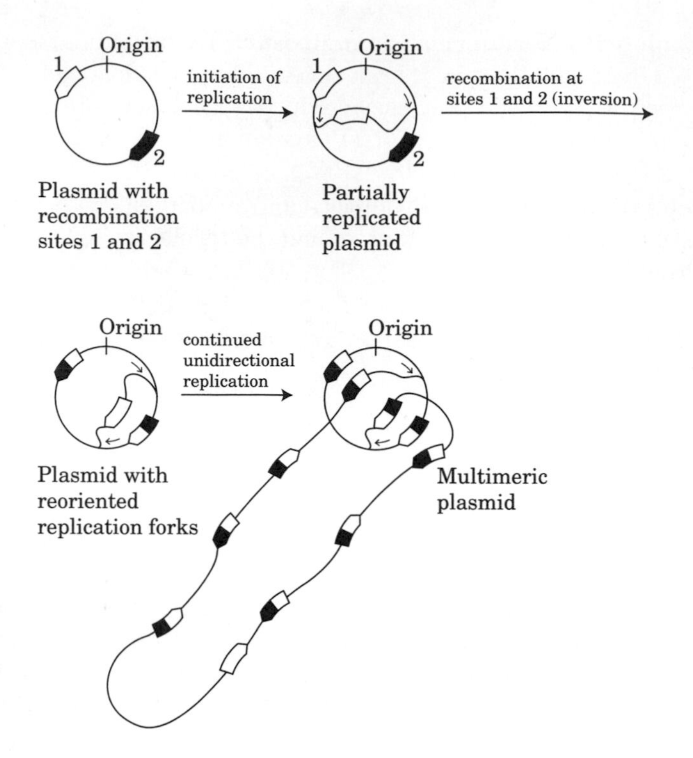

Data Analysis Problem

16. Mutagenesis in *Escherichia coli* Many mutagenic compounds act by alkylating the bases in DNA. The alkylating agent R7000 (7-methoxy-2-nitronaphtho[2,1-*b*]furan) is an extremely potent mutagen.

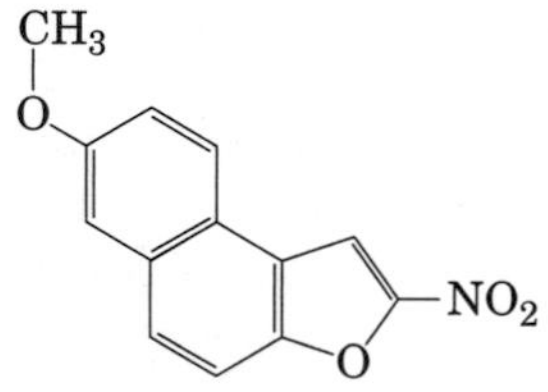

CH_3
O
NO_2
O

R7000

In vivo, R7000 is activated by the enzyme nitroreductase, and this more reactive form covalently attaches to DNA—primarily, but not exclusively, to G≡C base pairs.

In a 1996 study, Quillardet, Touati, and Hofnung explored the mechanisms by which R7000 causes mutations in *E. coli*. They compared the genotoxic activity of R7000 in two strains of *E. coli*: the wild-type (uvr^+) and mutants lacking *uvrA* activity (uvr^-; see Table 25–6). They first measured rates of mutagenesis. Rifampicin is an inhibitor of RNA polymerase (see Chapter 26). In its presence, cells will not grow unless certain mutations occur in the gene encoding RNA polymerase; the appearance of rifampicin-resistant colonies thus provides a useful measure of mutagenesis rates.

The effects of different concentrations of R7000 were determined, with the results shown in the graph following.

Rifampicin-resistant mutants produced
1,000
100
10
1
0 0.2 0.4 0.6 0.8 1
R7000 (μg/mL)
uvr$^-$
uvr$^+$

(a) Why are some mutants produced even when no R7000 is present?

Quillardet and colleagues also measured the survival rate of bacteria treated with different concentrations of R7000.

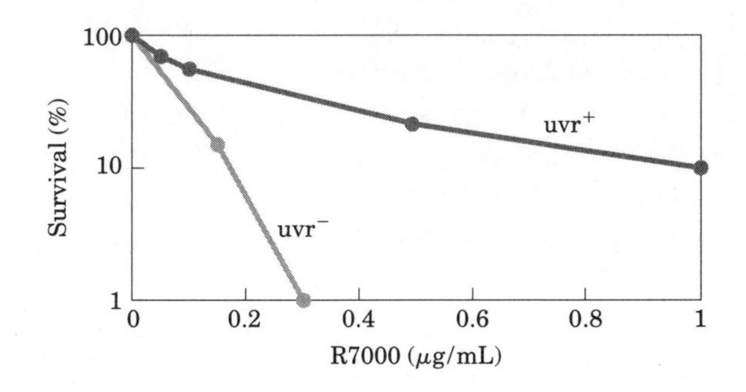

(b) Explain how treatment with R7000 is lethal to cells.

(c) Explain the differences in the mutagenesis curves and in the survival curves for the two types of bacteria, uvr$^+$ and uvr$^-$, as shown in the graphs.

The researchers then went on to measure the amount of R7000 covalently attached to the DNA in uvr$^+$ and uvr$^-$ *E. coli*. They incubated bacteria with [^{3}H]R7000 for 10 or 70 minutes, extracted the DNA, and measured its ^{3}H content in counts per minute (cpm) per μg of DNA.

	^{3}H in DNA (cpm/μg)	
Time (min)	uvr$^+$	uvr$^-$
10	76	159
70	69	228

(d) Explain why the amount of ^{3}H drops over time in the uvr$^+$ strain and rises over time in the uvr$^-$ strain.

Quillardet and colleagues then examined the particular DNA sequence changes caused by R7000 in the uvr$^+$ and uvr$^-$ bacteria. For this, they used six different strains of *E. coli*, each with a different point mutation in the *lacZ* gene, which encodes β-galactosidase (this enzyme catalyzes the same reaction as lactase; see Fig. 14–10). Cells with any of these mutations have a nonfunctional β-galactosidase and are unable to metabolize lactose (i.e., a Lac$^-$ phenotype). Each type of point mutation required a specific reverse mutation to restore *lacZ* gene function and Lac$^+$ phenotype. By plating cells on a medium containing lactose as the sole carbon source, it was possible to select for these reverse-mutated, Lac$^+$ cells. And by counting the number of Lac$^+$ cells following mutagenesis of a particular strain the researchers could measure the frequency of each type of mutation.

First, they looked at the mutation spectrum in uvr$^-$ cells. The following table shows the results for the six strains, CC101 through CC106 (with the point mutation required to produce Lac$^+$ cells indicated in parentheses).

	Number of Lac$^+$ cells (average ± SD)					
R7000 (μg/mL)	CC101 (A═T to C≡G)	CC102 (G≡C to A═T)	CC103 (G≡C to C≡G)	CC104 (G≡C to T═A)	CC105 (A═T to T═A)	CC106 (A═T to G≡C)
0	6 ± 3	11 ± 9	2 ± 1	5 ± 3	2 ± 1	1 ± 1
0.075	24 ± 19	34 ± 3	8 ± 4	82 ± 23	40 ± 14	4 ± 2
0.15	24 ± 4	26 ± 2	9 ± 5	180 ± 71	130 ± 50	3 ± 2

(e) Which types of mutation show significant increases above the background rate due to treatment with R7000? Provide a plausible explanation for why some have higher frequencies than others.

(f) Can all of the mutations you listed in **(e)** be explained as resulting from covalent attachment of R7000 to a G≡C base pair? Explain your reasoning.

(g) Figure 25–28b shows how methylation of guanine residues can lead to a G≡C to A═T mutation. Using a similar pathway, show how a G–R7000 adduct could lead to the G≡C to A═T or T═A mutations shown above. Which base pairs with the G—R7000 adduct?

The results for the uvr$^+$ bacteria are shown in the table below.

	Number of Lac$^+$ cells (average ± SD)					
R7000 (μg/mL)	CC101 (A═T to C≡G)	CC102 (G≡C to A═T)	CC103 (G≡C to C≡G)	CC104 (G≡C to T═A)	CC105 (A═T to T═A)	CC106 (A═T to G≡C)
0	2 ± 2	10 ± 9	3 ± 3	4 ± 2	6 ± 1	0.5 ± 1
1	7 ± 6	21 ± 9	8 ± 3	23 ± 15	13 ± 1	1 ± 1
5	4 ± 3	15 ± 7	22 ± 2	68 ± 25	67 ± 14	1 ± 1

(h) Do these results show that all mutation types are repaired with equal fidelity? Provide a plausible explanation for your answer.

Answer

(a) Even in the absence of an added mutagen, background mutations occur due to radiation, cellular chemical reactions, and so forth.

(b) If the DNA is sufficiently damaged, a substantial fraction of gene products are nonfunctional and the cell is nonviable.

(c) Cells with reduced DNA repair capability are more sensitive to mutagens. Because they less readily repair lesions caused by R7000, uvr$^-$ bacteria have an increased mutation rate and increased chance of lethal effects.

(d) In the uvr$^+$ strain, the excision-repair system removes DNA bases with attached [^{3}H]R7000, decreasing the ^{3}H in these cells over time. In the uvr$^-$ strain, the DNA is not repaired and ^{3}H level increases as [^{3}H]R7000 continues to react with the DNA.

(e) All mutations listed in the table except A═T to G≡C show significant increases over background. Each type of mutation results from a different type of interaction between R7000 and DNA. Because different types of interactions are not equally likely (due to differences in reactivity, steric constraints, etc.), the resulting mutations occur with different frequencies.

(f) No. Only those that start with a G≡C base pair are explained by this model. Thus A═T to C≡G and A═T to T═A must be due to R7000 attaching to an A or a T.

(g) R7000-G pairs with A. First, R7000 adds to G≡C to give R7000-G≡C. (Compare this with what happens with the CH_3-G in Fig. 25–28b, p. 1000.) If this is not repaired, one strand is replicated as R7000-G═A, which is repaired to T═A. The other strand is wild-type. If the replication produces R7000-G═T, a similar pathway leads to an A═T base pair.

(h) No. Compare data in the two tables, and keep in mind that different mutations occur at different frequencies.

A═T to C≡G: moderate in both strains; but better repair in the uvr^+ strain

G≡C to A═T: moderate in both; no real difference

G≡C to C≡G: higher in uvr^+; certainly less repair!

G≡C to T═A: high in both; no real difference

A═T to T═A: high in both; no real difference

A═T to G≡C: low in both; no real difference

Certain adducts may be more readily recognized by the repair apparatus than others and thus are repaired more rapidly and result in fewer mutations.

Reference

Quillardet, P., Touati, E., & Hofnung, M. (1996) Influence of the *uvr*-dependent nucleotide excision repair on DNA adducts formation and mutagenic spectrum of a potent genotoxic agent: 7-methoxy-2-nitronaphtho[2,1-*b*]furan (R7000). *Mutat. Res.* **358,** 113–122.

chapter

26 RNA Metabolism

1. **RNA Polymerase** **(a)** How long would it take for the *E. coli* RNA polymerase to synthesize the primary transcript for the *E. coli* genes encoding the enzymes for lactose metabolism (the 5,300 bp *lac* operon, considered in Chapter 28)? **(b)** How far along the DNA would the transcription "bubble" formed by RNA polymerase move in 10 seconds?

   Answer

   (a) Elongation of the RNA transcript in *E. coli* proceeds at about 50 to 90 nucleotides per second. Thus, the time required to produce the primary transcript in this case is

$$\frac{5{,}300 \text{ nucleotides}}{50 \text{ to } 90 \text{ nucleotides/s}} = 60 \text{ to } 100 \text{ s}$$

   (b) Because elongation occurs at 50 to 90 nucleotides per second, in 10 seconds the bubble travels

$$(10 \text{ s})(50 \text{ to } 90 \text{ nucleotides/s}) = 500 \text{ to } 900 \text{ nucleotides}$$

2. **Error Correction by RNA Polymerases** DNA polymerases are capable of editing and error correction, whereas the capacity for error correction in RNA polymerases seems to be quite limited. Given that a single base error in either replication or transcription can lead to an error in protein synthesis, suggest a possible biological explanation for this difference.

   Answer For each gene, RNA polymerase produces many RNA transcripts, so an error in any one transcript would result in only a small fraction of protein with an incorrect amino acid residue. This defective protein will probably be degraded quickly. An error in the mRNA does not pass to subsequent generations of cells because the mRNA itself is degraded. For DNA replication, however, errors would be transmitted to the next generation of cells.

3. **RNA Posttranscriptional Processing** Predict the likely effects of a mutation in the sequence (5′)AAUAAA in a eukaryotic mRNA transcript.

   Answer A key signal for cleavage and 3′-polyadenylation of the primary mRNA transcript is the sequence (5′)AAUAAA. After RNA polymerase II has transcribed beyond this sequence, an endonuclease cleaves the transcript at a position about 25 to 30 nucleotides 3′ to the AAUAAA, and polyadenylate polymerase adds a string of 20 to 250 A residues to the 3′ end, generating the 3′ poly(A) tail. The mutation would prevent this cleavage and polyadenylation. A transcript that is not polyadenylated is unstable, so the steady-state level of this mRNA would be very low and little or no protein would be produced.

4. Coding versus Template Strands The RNA genome of phage Qβ is the nontemplate strand, or coding strand, and when introduced into the cell it functions as an mRNA. Suppose the RNA replicase of phage Qβ synthesized primarily template-strand RNA and uniquely incorporated this, rather than nontemplate strands, into the viral particles. What would be the fate of the template strands when they entered a new cell? What enzyme would have to be included in the viral particles for successful invasion of a host cell?

Answer A template-strand RNA does not encode proteins, and by itself would not cause a productive infection. However, if an RNA-dependent RNA polymerase were included in the viral particle, it could copy the template strand to form a nontemplate strand after entry into the host cell. The nontemplate strand could serve as mRNA for synthesis of viral proteins, leading to a productive infection.

5. The Chemistry of Nucleic Acid Biosynthesis Describe three properties common to the reactions catalyzed by DNA polymerase, RNA polymerase, reverse transcriptase, and RNA replicase. How is the enzyme polynucleotide phosphorylase similar to and different from these four enzymes?

Answer These enzymes have at least four properties in common. (1) All are template directed, synthesizing a sequence complementary to the template. (2) Synthesis occurs in the $5' \rightarrow 3'$ direction. (3) All catalyze the addition of a nucleotide by the formation of a phosphodiester bond. (4) All use (deoxy)ribonucleoside triphosphates as substrates, and release pyrophosphate as a product.

Polynucleotide phosphorylase differs from these polymerases in points (1) and (4) (p. 1049). It does not use a template, but adds ribonucleotides to an RNA in a highly reversible reaction. The substrates (in the direction of synthesis) are ribonucleoside *diphosphates,* which are added to the polymer with the release of *phosphate* as product. In the cell, this enzyme probably catalyzes the reverse reaction to degrade RNAs.

6. RNA Splicing What is the minimum number of transesterification reactions needed to splice an intron from an mRNA transcript? Explain.

Answer A minimum of two transesterification steps are required. The mechanism for removal of introns from pre-mRNAs involves the formation of a lariat intermediate after the reaction is initiated. Each of the cleavage and rejoining reactions is a transesterification, in which a new phosphodiester bond is formed for every one that is broken. The first step is initiated by the attack of the 2′-hydroxyl group of an A residue within the intron on the bond linking the 3′ end of the first exon with the 5′ end of the intron. This generates a 3′-hydroxyl on the nucleotide at the 3′ end of the first exon, and effectively takes the intron out of the series of transesterifications by forming a lariat structure (see Fig. 26–16). The 3′ nucleotide of the first intron can then link to the first nucleotide of the second exon, again by a transesterification. The result of this second step is the union of the first and second exons and release of the intron as a lariat intermediate.

7. RNA Processing If the splicing of mRNA in a vertebrate cell is blocked, the rRNA modification reactions are also blocked. Suggest a reason for this.

Answer Many snoRNAs, required for rRNA modification reactions, are encoded in introns. If splicing does not occur, snoRNAs are not produced.

8. RNA Genomes The RNA viruses have relatively small genomes. For example, the single-stranded RNAs of retroviruses have about 10,000 nucleotides and the Qβ RNA is only 4,220 nucleotides long. Given the properties of reverse transcriptase and RNA replicase described in this chapter, can you suggest a reason for the small size of these viral genomes?

Answer Neither reverse transcriptase nor RNA replicase has a proofreading function, and hence these enzymes are much more error-prone than DNA polymerases. The smaller the genome, the fewer are the mutable sites and thus the lower the likelihood of accumulation of a lethal mutational load (i.e., one or a group of mutations that will inactivate the virus).

9. **Screening RNAs by SELEX** The practical limit for the number of different RNA sequences that can be screened in a SELEX experiment is 10^{15}. **(a)** Suppose you are working with oligonucleotides 32 nucleotides long. How many sequences exist in a randomized pool containing every sequence possible? **(b)** What percentage of these can be screened in a SELEX experiment? **(c)** Suppose you wish to select an RNA molecule that catalyzes the hydrolysis of a particular ester. From what you know about catalysis, propose a SELEX strategy that might allow you to select the appropriate catalyst.

Answer

(a) For the four different nucleotides, the number of possible sequences in a nucleic acid of length n residues is 4^n. Thus, for an oligonucleotide with 32 residues, the number of possible sequences is $4^{32} \approx 1.8 \times 10^{19}$.

(b) In an experiment limited to a sampling of 10^{15} sequences, $(100\%)(10^{15})/(1.8 \times 10^{19}) = 0.006\%$ of these could be sampled.

(c) For the "unnatural selection" step, use a chromatographic resin to which is bound a transition-state analog of the ester hydrolysis reaction (e.g., an appropriate phosphonate compound; see Box 6–3). RNAs that bind tightly to the transition-state analog may catalyze the corresponding reaction.

10. **Slow Death** The death cap mushroom, *Amanita phalloides,* contains several dangerous substances, including the lethal α-amanitin. This toxin blocks RNA elongation in consumers of the mushroom by binding to eukaryotic RNA polymerase II with very high affinity; it is deadly in concentrations as low as 10^{-8} M. The initial reaction to ingestion of the mushroom is gastrointestinal distress (caused by some of the other toxins). These symptoms disappear, but about 48 hours later, the mushroom-eater dies, usually from liver dysfunction. Speculate on why it takes this long for α-amanitin to kill.

Answer The turnover rates of mRNAs and their protein products vary, based on the rates of synthesis and degradation. Though RNA synthesis is quickly halted by the α-amanitin toxin, it takes several days for the critical mRNAs and proteins in the liver to degrade; once degraded they are not replenished, and the system fails.

11. **Detection of Rifampicin-Resistant Strains of Tuberculosis** Rifampicin is an important antibiotic used to treat tuberculosis as well as other mycobacterial diseases. Some strains of *Mycobacterium tuberculosis,* the causative agent of tuberculosis, are resistant to rifampicin. These strains become resistant through mutations that alter the *rpoB* gene, which encodes the β subunit of the RNA polymerase. Rifampicin cannot bind to the mutant RNA polymerase and so is unable to block the initiation of transcription. DNA sequences from a large number of rifampicin-resistant *M. tuberculosis* strains have been found to have mutations in a specific 69 bp region of *rpoB.* One well-characterized rifampicin-resistant strain has a single base pair alteration in *rpoB* that results in a His residue being replaced by an Asp residue in the β subunit.

(a) Based on your knowledge of protein chemistry, suggest a technique that would allow detection of the rifampicin-resistant strain containing this particular mutant protein.

(b) Based on your knowledge of nucleic acid chemistry, suggest a technique to identify the mutant form of *rpoB.*

Answer

(a) After cell lysis and partial purification of the protein extract, it could be subjected to isoelectric focusing. The β subunit protein could be detected by an antibody-based assay. The difference in amino acid residues between the normal β subunit and the mutated form (i.e., the different charges on the amino acids) would alter the electrophoretic mobility of the mutant β subunit protein (relative to that of the protein from a nonresistant strain) in an isoelectric focusing gel.

(b) Direct DNA sequencing (by the Sanger method) of the *rpoB* gene would allow detection of this particular mutation.

Biochemistry on the Internet

12. The Ribonuclease Gene Human pancreatic ribonuclease has 128 amino acid residues.

(a) What is the minimum number of nucleotide pairs required to code for this protein?

(b) The mRNA expressed in human pancreatic cells was copied with reverse transcriptase to create a "library" of human DNA. The sequence of the mRNA coding for human pancreatic ribonuclease was determined by sequencing the complementary DNA (cDNA) from this library that included an open reading frame for the protein. Use the Entrez database system (www.ncbi.nlm.nih.gov/Entrez) to find the published sequence of this mRNA (search the nucleotide database for accession number D26129). What is the length of this mRNA?

(c) How can you account for the discrepancy between the size you calculated in **(a)** and the actual length of the mRNA?

Answer

(a) Individual amino acids in proteins are coded for by a triplet of nucleotides. Therefore, encoding a protein with 128 amino acid residues will require $128 \times 3 = 384$ nucleotide pairs in the DNA.

(b) The GenBank report indicates that the cDNA derived from the mature mRNA contains 1,620 nucleotide pairs.

(c) The mRNA for pancreatic ribonuclease is clearly much larger than is necessary to code for the protein itself. (To see this rather dramatically, display the information graphically by changing the "Display" option from "GenBank" to "Graph.") Most of the nucleotides are untranslated regions at the $3'$ and $5'$ ends of the mRNA. The mRNA encodes a primary transcript longer than the mature ribonuclease; a signal sequence (see Chapter 27) is cleaved off to produce the mature and functional protein.

Data Analysis Problem

13. A Case of RNA Editing The AMPA (α-amino-3-hydroxy-5-methyl-4-isoxazolepropionic acid) receptor is an important component of the human nervous system. It is present in several forms, in different neurons, and some of this variety results from posttranscriptional modification. This problem explores research on the mechanism of this RNA editing.

An initial report by Sommer and coauthors (1991) looked at the sequence encoding a key Arg residue in the AMPA receptor. The sequence of the cDNA (see Fig. 9–14) for the AMPA receptor showed a CGG (Arg; see Fig. 27–7) codon for this amino acid. Surprisingly, the genomic DNA showed a CAG (Gln) codon at this position.

(a) Explain how this result is consistent with posttranscriptional modification of the AMPA receptor mRNA.

Rueter and colleagues (1995) explored this mechanism in detail. They first developed an assay to differentiate between edited and unedited transcripts, based on the Sanger method of DNA sequencing (see Fig. 8–33). They modified the technique to determine whether the base in question was an A (as in CAG) or not. They designed two DNA primers based on the genomic DNA sequence of this region of the AMPA gene. These primers, and the genomic DNA sequence of the nontemplate strand for the relevant region of the AMPA receptor gene, are shown below; the A residue that is edited is indicated by an asterisk.

```
                                       *
(5′)...  GTCTCTGGTTTTCCTTGGGTGCCTTTATGCAGCAAGGATGCGATATTTCGCCAAG...
Primer 1:                               CGTTCCTACGCTATAAAGCGGTTC (5′)
Primer 2:       (5′) CCTTGGGTGCCTTTA
```

To detect whether this A was present or had been edited to another base, Rueter and coworkers used the following procedure:

1. Prepared cDNA complementary to the mRNA, using primer 1, reverse transcriptase, dATP, dGTP, dCTP, and dTTP.
2. Removed the mRNA.
3. Annealed ^{32}P-labeled primer 2 to the cDNA, and reacted this with DNA polymerase, dGTP, dCTP, dTTP, and ddATP (dideoxy ATP; see Fig. 8–33).
4. Denatured the resulting duplexes and separated them with polyacrylamide gel electrophoresis (PAGE; p. 88).
5. Detected the ^{32}P-labeled DNA species with autoradiography.

They found that edited mRNA produced a 22 nucleotide [^{32}P]DNA, whereas unedited mRNA produced a 19 nucleotide [^{32}P]DNA.

(b) Using the sequences above, explain how the edited and unedited mRNAs resulted in these different products.

Using the same procedure, to measure the fraction of transcripts edited under different conditions, the researchers found that extracts of cultured epithelial cells (a common cell line called HeLa) could edit the mRNA at a high level. To determine the nature of the editing machinery, they pretreated an active HeLa cell extract as described in the table and measured its ability to edit AMPA mRNA. Proteinase K degrades only proteins; micrococcal nuclease, only DNA.

Sample	Pretreatment	% mRNA edited
1	None	18
2	Proteinase K	5
3	Heat to 65 °C	3
4	Heat to 85 °C	3
5	Micrococcal nuclease	17

(c) Use these data to argue that the editing machinery consists of protein. What is a key weakness in this argument?

To determine the exact nature of the edited base, Rueter and colleagues used the following procedure:

1. Produced mRNA, using [α-^{32}P]ATP in the reaction mixture.
2. Edited the labeled mRNA by incubating with HeLa extract.
3. Hydrolyzed the edited mRNA to single nucleotide monophosphates with nuclease P1.
4. Separated the nucleotide monophosphates with thin-layer chromatography (TLC; see Fig. 10–24).
5. Identified the resulting ^{32}P-labeled nucleotide monophosphates with autoradiography.

In unedited mRNA, they found only [^{32}P]AMP; in edited mRNA, they found mostly [^{32}P]AMP with some [^{32}P]IMP (inosine monophosphate; see Fig. 22–34).

(d) Why was it necessary to use [α-^{32}P]ATP rather than [β-^{32}P]ATP or [γ-^{32}P]ATP in this experiment?

(e) Why was it necessary to use [α-^{32}P]ATP rather than [α-^{32}P]GTP, [α-^{32}P]CTP, or [α-^{32}P]UTP?

(f) How does the result exclude the possibility that the entire A nucleotide (sugar, base, and phosphate) was removed and replaced by an I nucleotide during the editing process?

The researchers next edited mRNA that was labeled with [2,8-^{3}H]ATP, and repeated the above procedure. The only ^{3}H-labeled mononucleotides produced were AMP and IMP.

(g) How does this result exclude removal of the A base (leaving the sugar-phosphate backbone intact) followed by replacement with an I base as a mechanism of editing? What, then, is the most likely mechanism of editing in this case?

(h) How does changing an A to an I residue in the mRNA explain the Gln to Arg change in protein sequence in the two forms of AMPA receptor protein? (Hint: See Fig. 27–8).

Answer

(a) cDNA is produced by reverse transcription of mRNA; thus, the mRNA sequence is probably CGG. Because the genomic DNA transcribed to make the mRNA has the sequence CAG, the primary transcript most likely has CAG, which is posttranscriptionally modified to CGG.

(b) The unedited mRNA sequence is the same as that of the DNA (except for U replacing T). Unedited mRNA has the sequence (* indicates site of editing)

```
                                      *
(5′)...GUCUCUGGUUUUCCUUGGGUGCCUUUAUGCAGCAAGGAUGCGAUAUUUCGCCAAG...(3′)
```

In step 1, primer 1 anneals as shown:

```
                                      *
(5′)...GUCUCUGGUUUUCCUUGGGUGCCUUUAUGCAGCAAGGAUGCGAUAUUUCGCCAAG...(3′)
                                       |||||||||||||||||||||||
                   Primer 1: (3′)–CGTTCCTACGCTATAAAGCGGTTC–(5′)
```

cDNA (underlined) is synthesized from right to left:

```
                                      *
(5′)...GUCUCUGGUUUUCCUUGGGUGCCUUUAUGCAGCAAGGAUGCGAUAUUUCGCCAAG...(3′)
       |||||||||||||||||||||||||||||||||||||||||||||||||||||
(3′)...CAGAGACCAAAAGGAACCCACGGAAATACGTCGTTCCTACGCTATAAAGCGGTTC–(5′)
```

Then step 2 yields just the cDNA:

```
                                      *
(3′)...CAGAGACCAAAAGGAACCCACGGAAATACGTCGTTCCTACGCTATAAAGCGGTTC–(5′)
```

In step 3, primer 2 anneals to the cDNA:

```
      Primer 2: (5′)–CCTTGGGTGCCTTTA–(3′)
                     |||||||||||||||
(3′)...CAGAGACCAAAAGGAACCCACGGAAATACGTCGTTCCTACGCTATAAAGCGGTTC–(5′)
                                      *
```

DNA polymerase adds nucleotides to the 3′ end of this primer. Moving from left to right, it inserts T, G, C, and A. However, because the A from ddATP lacks the 3′ —OH needed to attach the next nucleotide, the chain is not elongated past this point. This A is shown in *italic;* the new DNA is underlined:

```
      Primer 2: (5′)–CCTTGGGTGCCTTTATGCA
                     |||||||||||||||||||
(3′)...CAGAGACCAAAAGGAACCCACGGAAATACGTCGTTCCTACGCTATAAAGCGGTTC–(5′)
                                      *
```

This yields a 19 nucleotide fragment for the unedited transcript. In the edited transcript, the *A is changed to G; in the cDNA this corresponds to C. At the start of step 3:

```
Primer 2:    (5′)–CCTTGGGTGCCTTTA–(3′)
                  |||||||||||||||
(3′)...CAGAGACCAAAAGGAACCCACGGAAATACGCCGTTCCTACGCTATAAAGCGGTTC–(5′)
                                     *
```

In this case, DNA polymerase can elongate past the edited base and will stop at the next T in the cDNA. The dideoxy A is *italic*; the new DNA is underlined.

```
Primer 2:    (5′)–CCTTGGGTGCCTTTATGCGGCA
                  ||||||||||||||||||||||
(3′)...CAGAGACCAAAAGGAACCCACGGAAATACGCCGTTCCTACGCTATAAAGCGGTTC–(5′)
```

This gives the 22 nucleotide product.

(c) Treatments (proteases, heat) known to disrupt protein function inhibit the editing activity, whereas treatments (nuclease) that do not affect proteins have little or no effect on editing. A key weakness of this argument is that the protein-disrupting treatments do not completely abolish editing. There could be some background editing or degradation of the mRNA even without the enzyme, or some of the enzyme might survive the treatments.

(d) Only the α phosphate of NTPs is incorporated into polynucleotides. If the researchers had used the other types of [^{32}P]NTPs, none of the products would have been labeled.

(e) Because only an A is being edited, only the fate of any A in the sequence is of interest.

(f) Given that only ATP was labeled, if the entire nucleotide were removed all radioactivity would have been removed from the mRNA, so only unmodified [^{32}P]AMP would be present on the chromatography plate.

(g) If the base were removed and replaced, one would expect to see only [^{3}H]AMP. The presence of [^{3}H]IMP indicates that the A to I change occurs without removal of H at positions 2 and 8. The most likely mechanism is chemical modification of A to I by hydrolytic deamination (see Fig. 22–34, p. 885).

(h) CAG is changed to CIG. This codon is read as CGG.

References

Rueter, S.M., Burns, C.M., Coode, S.A., Mookherjee, P., & Emesont, R.B. (1995) Glutamate receptor RNA editing in vitro by enzymatic conversion of adenosine to inosine. *Science* **267,** 1491–1494.

Sommer, B., Köhler, M., Sprengel, R., & Seeburg, P.H. (1991) RNA editing in brain controls a determinant of ion flow in glutamate-gated channels. *Cell* **67,** 11–19.

chapter

27

Protein Metabolism

1. **Messenger RNA Translation** Predict the amino acid sequences of peptides formed by ribosomes in response to the following mRNA sequences, assuming that the reading frame begins with the first three bases in each sequence.
   (a) GGUCAGUCGCUCCUGAUU
   (b) UUGGAUGCGCCAUAAUUUGCU
   (c) CAUGAUGCCUGUUGCUAC
   (d) AUGGACGAA

   Answer The genetic code is nonoverlapping, unpunctuated, and in triplet (see Fig. 27–7).
   (a) Gly–Gln–Ser–Leu–Leu–Ile
   (b) Leu–Asp–Ala–Pro (UAA is a stop codon)
   (c) His–Asp–Ala–Cys–Cys–Tyr
   (d) Met–Asp–Glu in eukaryotes; fMet–Asp–Glu in prokaryotes

2. **How Many Different mRNA Sequences Can Specify One Amino Acid Sequence?** Write all the possible mRNA sequences that can code for the simple tripeptide segment Leu–Met–Tyr. Your answer will give you some idea about the number of possible mRNAs that can code for one polypeptide.

   Answer The genetic code is degenerate, meaning that a given amino acid may be specified by more than one codon (see Table 27–3 and Fig. 27–7). Leu is specified by six different codons: UUA, UUG, CUU, CUC, CUA, CUG. Met, when not used as an initiation codon, is specified by AUG; Tyr is specified by two codons: UAC, UAU. Thus, $6 \times 1 \times 2 = 12$ possible mRNA sequences can code for a tripeptide segment Leu–Met–Tyr:
   UUAAUGUAU, UUAAUGUAC, UUGAUGUAU, UUGAUGUAC,
   CUUAUGUAU, CUUAUGUAC, CUCAUGUAU, CUCAUGUAC,
   CUAAUGUAU, CUAAUGUAC, CUGAUGUAU, CUGAUGUAC

3. **Can the Base Sequence of an mRNA Be Predicted from the Amino Acid Sequence of Its Polypeptide Product?** A given sequence of bases in an mRNA will code for one and only one sequence of amino acids in a polypeptide, if the reading frame is specified. From a given sequence of amino acid residues in a protein such as cytochrome *c,* can we predict the base sequence of the unique mRNA that coded it? Give reasons for your answer.

   Answer No. Because nearly every amino acid has more than one codon, any given polypeptide can be coded for by a number of different base sequences (see Problem 2). However, because some amino acids are encoded by only one codon and those with multiple codons often share the same nucleotide at two of the three positions, *certain parts* of the mRNA sequence encoding a protein of known amino acid sequence can be predicted with high certainty.

4. Coding of a Polypeptide by Duplex DNA The template strand of a segment of double-helical DNA contains the sequence

(5′)CTTAACACCCCTGACTTCGCGCCGTCG(3′)

(a) What is the base sequence of the mRNA that can be transcribed from this strand?

(b) What amino acid sequence could be coded by the mRNA in **(a)**, starting from the 5′ end?

(c) If the complementary (nontemplate) strand of this DNA were transcribed and translated, would the resulting amino acid sequence be the same as in **(b)**? Explain the biological significance of your answer.

Answer The template strand serves as the template for RNA synthesis; the nontemplate strand is identical in sequence to the RNA transcribed from the gene, with U in place of T.

(a) (5′)CGACGGCGCGAAGUCAGGGGUGUUAAG(3′)

(b) Arg–Arg–Arg–Glu–Val–Arg–Gly–Val–Lys

(c) No. The base sequence of mRNA transcribed from the nontemplate strand would be (5′)CUUAACACCCCUGACUUCGCGCCGUCG. This mRNA, when translated, would result in a different peptide from **(b)**. It would have the amino acid sequence

Leu–Asn–Thr–Pro–Asp–Phe–Ala–Pro–Ser

The complementary antiparallel strands in double-helical DNA do not have the same base sequence in the 5′→3′ direction. RNA is transcribed from only one specific strand of duplex DNA. The RNA polymerase must therefore recognize and bind to the correct strand.

5. Methionine Has Only One Codon Methionine is one of two amino acids with only one codon. How does the single codon for methionine specify the initiating residue and interior Met residues of polypeptides synthesized by *E. coli*?

Answer There are two tRNAs for methionine: $tRNA^{fMet}$, which is the initiating tRNA, and $tRNA^{Met}$, which can insert a Met residue in interior positions in a polypeptide. The $tRNA^{fMet}$ reacts with methionine to yield Met-$tRNA^{fMet}$, promoted by methionine aminoacyl-tRNA synthetase. The amino group of its Met residue is then formylated by N^{10}-formyltetrahydrofolate to yield fMet-$tRNA^{fMet}$. Free methionine or Met-$tRNA^{Met}$ cannot be formylated. Only fMet-$tRNA^{fMet}$ is recognized by the initiation factor IF-2 and is aligned with the initiating AUG positioned at the ribosomal P site in the initiation complex. AUG codons in the interior of the mRNA are eventually positioned at the ribosomal A site and can bind and incorporate only Met-$tRNA^{Met}$.

6. Synthetic mRNAs The genetic code was elucidated with polyribonucleotides synthesized either enzymatically or chemically in the laboratory. Given what we now know about the genetic code, how would you make a polyribonucleotide that could serve as an mRNA coding predominantly for many Phe residues and a small number of Leu and Ser residues? What other amino acid(s) would be coded for by this polyribonucleotide, but in smaller amounts?

Answer Polynucleotide phosphorylase is template-independent and does not require a primer. The base composition of the RNA formed by this enzyme reflects the relative concentrations of the nucleoside 5′-diphosphates in the reaction mixture. To prepare the required polyribonucleotide, allow polynucleotide phosphorylase to act on a mixture of UDP and CDP in which UDP has, say, five times the concentration of CDP. The result would be a synthetic RNA polymer with many UUU triplets (coding for Phe), a smaller number of UUC (also Phe), UCU (Ser), and CUU (Leu), a yet smaller number of UCC (also Ser), CUC (also Leu), and CCU (Pro), and smallest number of CCC (Pro).

7. Energy Cost of Protein Biosynthesis Determine the minimum energy cost, in terms of ATP equivalents expended, required for the biosynthesis of the β-globin chain of hemoglobin (146 residues), starting from a pool including all necessary amino acids, ATP, and GTP. Compare your answer with the direct energy cost of the biosynthesis of a linear glycogen chain of 146 glucose residues in (α1→4) linkage, starting from a pool including glucose, UTP, and ATP (Chapter 15). From your data, what is the *extra* energy cost of making a protein, in which all the residues are ordered in a specific sequence, compared with the cost of making a polysaccharide containing the same number of residues but lacking the informational content of the protein?

In addition to the direct energy cost for the synthesis of a protein, there are indirect energy costs—those required for the cell to make the necessary enzymes for protein synthesis. Compare the magnitude of the indirect costs to a eukaryotic cell of the biosynthesis of linear (α1→4) glycogen chains and the biosynthesis of polypeptides, in terms of the enzymatic machinery involved.

Answer The number of ATP equivalents required for the synthesis of a *polypeptide* with n residues is:

$2n$	for the charging of tRNA (ATP → AMP + PP_i; PP_i → $2P_i$)
1	for initiation (GTP → GDP + P_i)
$n - 1$	for the formation of $n - 1$ peptide bonds (GTP → GDP + P_i)
$n - 1$	for the $n - 1$ translocation steps (GTP → GDP + P_i)

Total for polypeptide = $4n - 1$

The number of ATP equivalents required for the synthesis of a linear *glycogen* chain of n glucose residues is:

n	for the phosphorylation of glucose (Glucose + ATP → ADP + glucose 6-phosphate)
0	for the conversion of glucose 6-phosphate to glucose 1-phosphate
$2n$	for the activation of glucose 1-phosphate to UDP-glucose (Glucose 1-phosphate + UTP → UDP-glucose + PP_i; PP_i → $2P_i$)
$-n$	(i.e., n generated) for formation of polymer from UDP-glucose (UDP-glucose + glycogen → UDP + glycogen-glucose)

Total for linear glycogen = $2n$

Thus, the total number of ATP equivalents required for the synthesis of one molecule of β-globin is $(4 \times 146) - 1 = 583$. The total number for the synthesis of a linear chain of 146 glucose residues is $2 \times 146 = 292$. The extra energy cost for the synthesis of β-globin is $583 - 292 = 291$ ATP equivalents; this reflects the cost of the information contained in the protein.

In order to synthesize a protein from amino acids, at least 20 aminoacyl-tRNA synthetases (activating enzymes), 70 ribosomal proteins, 4 rRNAs, 32 or more tRNAs, an mRNA, and 10 or more auxiliary enzymes must be made by the eukaryotic cell. Synthesis of these proteins and RNA molecules is energetically expensive. In contrast, the synthesis of an (α1→4) chain of glycogen from glucose requires only four or five enzymes (see Chapter 15).

8. Predicting Anticodons from Codons Most amino acids have more than one codon and attach to more than one tRNA, each with a different anticodon. Write all possible anticodons for the four codons of glycine: (5′)GGU, GGC, GGA, and GGG.

(a) From your answer, which of the positions in the anticodons are primary determinants of their codon specificity in the case of glycine?

(b) Which of these anticodon-codon pairings has/have a wobbly base pair?

(c) In which of the anticodon-codon pairings do all three positions exhibit strong Watson-Crick hydrogen bonding?

Answer All the anticodons for the four Gly codons have the sequence (5′)XCC. The first position of each anticodon is determined by the specific codon it interacts with, and by the wobble hypothesis (see Table 27–4). For example, the wobble position of the codon GGU is U, which can be recognized by either A, G, or I. So this codon has three possible anticodons: ACC, GCC, and ICC. By the same token, the anticodons for the GGC codon are GCC and ICC; the anticodons for the GGA codon are UCC and ICC; and the anticodons for the GGG codon are CCC and UCC.

(a) The 3′ and the middle position. The 5′ position is the wobble position.

(b) Anticodons GCC, ICC, and UCC each recognize more than one codon by virtue of wobble base pairing. Anticodons ACC and CCC each recognize only one codon and thus are not involved in wobble base pairing.

(c) The pairing of anticodons ACC and CCC with their respective codons involves Watson-Crick base pairing in all three positions, A pairing with U, and C pairing with G.

9. Effect of Single-Base Changes on Amino Acid Sequence Much important confirmatory evidence on the genetic code has come from assessing changes in the amino acid sequence of mutant proteins after a single base has been changed in the gene that encodes the protein. Which of the following amino acid replacements would be consistent with the genetic code if the replacements were caused by a single base change? Which cannot be the result of a single-base mutation? Why?

(a) Phe→Leu
(b) Lys→Ala
(c) Ala→Thr
(d) Phe→Lys
(e) Ile→Leu
(f) His→Glu
(g) Pro→Ser

Answer For each part of this problem, "yes" indicates that the replacement is consistent with a single base change. The various codons for each amino acid are listed. The single base changes (where they exist) that would cause the mutation are underlined.

(a) Yes.
Phe: UUU UUC UUA UUG
Leu: CUU CUC CUA CUG

(b) No single-base mutation could convert the codons for Lys to the codons for Ala.
Lys: AAA AAG
Ala: GCU GCC GCA GCG

(c) Yes.
Ala: GCU GCC GCA GCG
Thr: ACU ACC ACA ACG

(d) No single-base mutation could convert the codons for Phe to the codons for Lys.
Phe: UUU UUC UUA UUG
Lys: AAA AAG

(e) Yes.
Ile: AUU AUC AUA
Leu: CUU CUC CUA

(f) No single-base mutation could convert the codons for His to the codons for Glu.
His: CAU CAC
Glu: GAA GAG

(g) Yes.
Pro: CCU CCC CCA CCG
Ser: UCU UCC UCA UCG

10. Basis of the Sickle-Cell Mutation Sickle-cell hemoglobin has a Val residue at position 6 of the β-globin chain, instead of the Glu residue found in normal hemoglobin A. Can you predict what change took place in the DNA codon for glutamate to account for replacement of the Glu residue by Val?

Answer The two DNA codons for Glu are GAA and GAG, and the four DNA codons for Val are GTT, GTC, GTA, and GTG. A single base change in GAA to form GTA or in GAG to form GTG could account for the Glu→Val replacement in sickle-cell hemoglobin. Much less likely are two-base changes from GAA to GTG, GTT, or GTC; or from GAG to GTA, GTT, or GTC.

11. Proofreading by Aminoacyl-tRNA Synthetases The isoleucyl-tRNA synthetase has a proofreading function that ensures the fidelity of the aminoacylation reaction, but the histidyl-tRNA synthetase lacks such a proofreading function. Explain.

Answer Isoleucine is similar in structure to several other amino acids, particularly valine. Distinguishing between valine and isoleucine in the aminoacylation process requires the second filter of a proofreading function. Histidine has a structure unlike that of any other amino acid, and this structure provides opportunities for binding specificity adequate to ensure accurate aminoacylation of the cognate tRNA.

12. Importance of the "Second Genetic Code" Some aminoacyl-tRNA synthetases do not recognize and bind the anticodon of their cognate tRNAs but instead use other structural features of the tRNAs to impart binding specificity. The tRNAs for alanine apparently fall into this category.

(a) What features of $tRNA^{Ala}$ are recognized by Ala-tRNA synthetase?

(b) Describe the consequences of a C→G mutation in the third position of the anticodon of $tRNA^{Ala}$.

(c) What other kinds of mutations might have similar effects?

(d) Mutations of these types are never found in natural populations of organisms. Why? (Hint: Consider what might happen both to individual proteins and to the organism as a whole.)

Answer

(a) The only nucleotides of $tRNA^{Ala}$ required for recognition by Ala-tRNA synthetase are those of the G^3–U^{70} base pair in the amino acid arm (see Fig. 27–18a).

(b) Mutations in the anticodon region would produce a $tRNA^{Ala}$ capable of recognizing and binding to codons for amino acids other than alanine. However, changes in the anticodon of $tRNA^{Ala}$ would *not* affect the specificity of the charging reaction by Ala-tRNA synthetase.

There are four Ala codons, GCU, GCC, GCA, and GCG. The third position of each $tRNA^{Ala}$ should be a C because this position interacts with the first position of the Ala codons, which is a G in all four. Thus, changing the C in the third position of $tRNA^{Ala}$ to a G would allow the mutant $tRNA^{Ala}$ to recognize CCU, CCC, CCA, and CCG, all of which specify Pro. Ala residues would be inserted at sites coding for Pro.

(c) Another mutation with similar effects would be, for example, a mutation in the $tRNA^{Ala}$ synthetase such that it recognized proline instead of alanine.

(d) The Ala→Pro substitution resulting from these mutations would render most of the proteins in the cell inactive, making these mutations lethal; hence, their effects would not be observed.

13. Maintaining the Fidelity of Protein Synthesis The chemical mechanisms used to avoid errors in protein synthesis are different from those used during DNA replication. DNA polymerases use a 3′→5′ exonuclease proofreading activity to remove mispaired nucleotides incorrectly inserted into a growing DNA strand. There is no analogous proofreading function on ribosomes and, in fact, the identity of an amino acid attached to an incoming tRNA and added to the growing polypeptide is never checked. A proofreading step that hydrolyzed the previously formed peptide bond after an incorrect amino acid had been inserted into a growing polypeptide (analogous to the proofreading step of DNA polymerases) would be impractical. Why? (Hint: Consider how the link between the growing polypeptide and the mRNA is maintained during elongation; see Figs. 27–29 and 27–30.)

Answer The amino acid most recently added to a growing polypeptide chain is the only one covalently attached to a tRNA and hence is the only link between the polypeptide and the mRNA that encodes it. A proofreading activity that severed this link would halt synthesis of the polypeptide and release it from the mRNA.

14. Predicting the Cellular Location of a Protein The gene for a eukaryotic polypeptide 300 amino acid residues long is altered so that a signal sequence recognized by SRP occurs at the polypeptide's amino terminus and a nuclear localization signal (NLS) occurs internally, beginning at residue 150. Where is the protein likely to be found in the cell?

Answer The protein would be directed into the endoplasmic reticulum, and from there the targeting depends upon additional signals. SRP binds the amino-terminal signal early in protein synthesis and directs the nascent polypeptide and ribosome to receptors in the ER. Because the altered protein is translocated into the ER lumen as it is synthesized, the NLS is never accessible to the proteins involved in nuclear targeting.

15. Requirements for Protein Translocation across a Membrane The secreted bacterial protein OmpA has a precursor, ProOmpA, which has the amino-terminal signal sequence required for secretion. If purified ProOmpA is denatured with 8 M urea and the urea is then removed (such as by running the protein solution rapidly through a gel filtration column) the protein can be translocated across isolated bacterial inner membranes in vitro. However, translocation becomes impossible if ProOmpA is first allowed to incubate for a few hours in the absence of urea. Furthermore, the capacity for translocation is maintained for an extended period if ProOmpA is first incubated in the presence of another bacterial protein called trigger factor. Describe the probable function of this factor.

Answer Trigger factor is a molecular chaperone that stabilizes an unfolded and translocation-competent conformation of ProOmpA.

16. Protein-Coding Capacity of a Viral DNA The 5,386 bp genome of bacteriophage ϕX174 includes genes for 10 proteins, designated A to K, with sizes given in the table below. How much DNA would be required to encode these 10 proteins? How can you reconcile the size of the ϕX174 genome with its protein-coding capacity?

Protein	Number of amino acid residues
A	455
B	120
C	86
D	152
E	91
F	427
G	175
H	328
J	38
K	56

Answer The minimum number of base pairs = 3 × the number of amino acids.

Protein	Number of amino acid residues	Minimum base pairs
A	455	1,365
B	120	360
C	86	258
D	152	456
E	91	273
F	427	1,281
G	175	525
H	328	984
J	38	114
K	56	168

Summing the last column, the minimum number of base pairs required = 5,784 bp. Thus, the 5,386 bp genome of bacteriophage ϕX174 would not be large enough to code for the ten amino acid sequences, not to mention the necessary regulatory elements and promoters, *unless* some of the coding sequences for these viral proteins are nested or overlap.

Data Analysis Problem

17. Designing Proteins by Using Randomly Generated Genes Studies of the amino acid sequence and corresponding three-dimensional structure of wild-type or mutant proteins have led to significant insights into the principles that govern protein folding. An important test of this understanding would be to *design* a protein based on these principles and see whether it folds as expected.

Kamtekar and colleagues (1993) used aspects of the genetic code to generate random protein sequences with defined patterns of hydrophilic and hydrophobic residues. Their clever approach combined knowledge about protein structure, amino acid properties, and the genetic code to explore the factors that influence protein structure.

They set out to generate a set of proteins with the simple four-helix bundle structure shown below, with α helices (shown as cylinders) connected by segments of random coil. Each α helix is amphipathic—the R groups on one side of the helix are exclusively hydrophobic (light gray) and those on the other side are exclusively hydrophilic (dark gray). A protein consisting of four of these helices separated by short segments of random coil would be expected to fold so that the hydrophilic sides of the helices face the solvent.

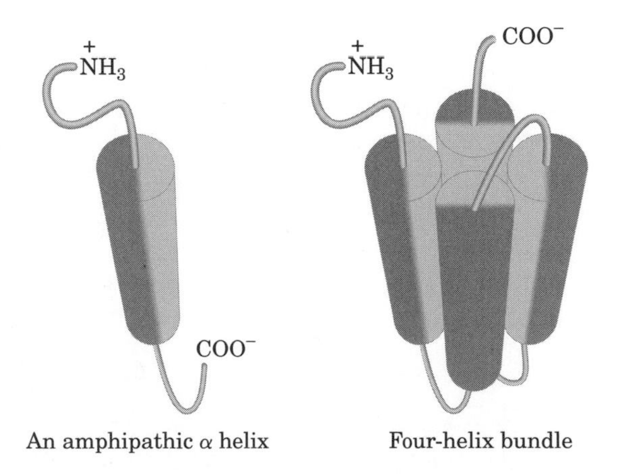

An amphipathic α helix

Four-helix bundle

(a) What forces or interactions hold the four α helices together in this bundled structure?

Figure 4–4a shows a segment of α helix consisting of 10 amino acid residues. With the gray central rod as a divider, four of the R groups (purple spheres) extend from the left side of the helix and six extend from the right.

(b) Number the R groups in Figure 4–4a, from top (amino terminus; 1) to bottom (carboxyl terminus; 10). Which R groups extend from the left side and which from the right?

(c) Suppose you wanted to design this 10 amino acid segment to be an amphipathic helix, with the left side hydrophilic and the right side hydrophobic. Give a sequence of 10 amino acids that could potentially fold into such a structure. There are many possible correct answers here.

(d) Give one possible double-stranded DNA sequence that could encode the amino acid sequence you chose for **(c)**. (It is an internal portion of a protein, so you do not need to include start or stop codons.)

Rather than designing proteins with specific sequences, Kamtekar and colleagues designed proteins with partially random sequences, with hydrophilic and hydrophobic amino acid residues placed in a controlled pattern. They did this by taking advantage of some interesting features of the genetic code to construct a library of synthetic DNA molecules with partially random sequences arranged in a particular pattern.

To design a DNA sequence that would encode random hydrophobic amino acid sequences, the researchers began with the degenerate codon NTN, where N can be A, G, C, or T. They filled each N position by including an equimolar mixture of A, G, C, and T in the DNA synthesis reaction to generate a mixture of DNA molecules with different nucleotides at that position (see Fig. 8–35). Similarly, to encode random polar amino acid sequences, they began with the degenerate codon NAN and used an equimolar mixture of A, G, and C (but in this case, no T) to fill the N positions.

(e) Which amino acids can be encoded by the NTN triplet? Are all amino acids in this set hydrophobic? Does the set include *all* the hydrophobic amino acids?

(f) Which amino acids can be encoded by the NAN triplet? Are all of these polar? Does the set include *all* the polar amino acids?

(g) In creating the NAN codons, why was it necessary to leave T out of the reaction mixture?

Kamtekar and coworkers cloned this library of random DNA sequences into plasmids, selected 48 that produced the correct patterning of hydrophilic and hydrophobic amino acids, and expressed these in *E. coli*. The next challenge was to determine whether the proteins folded as expected. It would be very time-consuming to express each protein, crystallize it, and determine its complete three-dimensional structure. Instead, the investigators used the *E. coli* protein-processing machinery to screen out sequences that led to highly defective proteins. In this initial screening, they kept only those clones that resulted in a band of protein with the expected molecular weight on SDS polyacrylamide gel electrophoresis (see Fig. 3–18).

(h) Why would a grossly misfolded protein fail to produce a band of the expected molecular weight on electrophoresis?

Several proteins passed this initial test, and further exploration showed that they had the expected four-helix structure.

(i) Why didn't all of the random-sequence proteins that passed the initial screening test produce four-helix structures?

Answer

(a) The helices associate through hydrophobic and van der Waals interactions.

(b) R groups 3, 6, 7, and 10 extend to the left; 1, 2, 4, 5, 8, and 9 extend to the right.

(c) One possible sequence is

```
  1   2   3   4   5   6   7   8   9   10
N–Phe–Ile–Glu–Val–Met–Asn–Ser–Ala–Phe–Gln–C
```

(d) One possible DNA sequence for the amino acid sequence in **(c)** is

```
Nontemplate strand
(5′)–TTTATTGAAGTAATGAATAGTGCATTCCAG–(3′)
     ||||||||||||||||||||||||||||||
(3′)–AAATAACTTCATTACTTATCACGTAAGGTC–(5′)
Template strand
```

(e) Phe, Leu, Ile, Met, and Val. All are hydrophobic, but the set does not include *all* the hydrophobic amino acids; Trp, Pro, Ala, and Gly are missing.

(f) Tyr, His, Gln, Asn, Lys, Asp, and Glu. All of these are hydrophilic, although Tyr is less hydrophilic than the others. The set does not include *all* the hydrophilic amino acids; Ser, Thr, and Arg are missing.

(g) Omitting T from the mixture excludes codons starting or ending with T—thus excluding Tyr, which is not very hydrophilic, and, more importantly, excluding the two possible stop codons (TAA and TAG). No other amino acids in the NAN set are excluded by omitting T.

(h) Misfolded proteins are often degraded in the cell. Therefore, if a synthetic gene has produced a protein that forms a band on the SDS gel, it is likely that this protein is folded properly.

(i) Protein folding depends on more than hydrophobic and van der Waals interactions. There are many reasons why a synthesized random-sequence protein might not fold into the four-helix structure. For example, hydrogen bonds between hydrophilic side chains could disrupt the structure. Also, not all sequences have an equal propensity to form an α helix.

Reference

Kamtekar, S., Schiffer, J.M., Xiong, H., Babik, J.M., & Hecht, M.H. (1993) Protein design by binary patterning of polar and nonpolar amino acids. *Science* **262,** 1680–1685.

chapter

28 Regulation of Gene Expression

1. **Effect of mRNA and Protein Stability on Regulation** *E. coli* cells are growing in a medium with glucose as the sole carbon source. Tryptophan is suddenly added. The cells continue to grow, and divide every 30 min. Describe (qualitatively) how the amount of tryptophan synthase activity in the cells changes under the following conditions:
   (a) The *trp* mRNA is stable (degraded slowly over many hours).
   (b) The *trp* mRNA is degraded rapidly, but tryptophan synthase is stable.
   (c) The *trp* mRNA and tryptophan synthase are both degraded rapidly.

   Answer The mRNA from the *trpEDCBA* operon encodes several enzymes for tryptophan biosynthesis; the *trpB* and *trpA* genes encode tryptophan synthase. The complete *trp* mRNA is synthesized only when the concentration of tryptophan (actually, that of charged Trp-tRNA) is low. This is the result of repression and attenuation (see Problem 6). There is no strong regulation at translation, so when *trp* mRNA is present, tryptophan synthase is produced. Although the regulation of tryptophan biosynthesis, like most biosynthetic pathways, is fine-tuned by feedback control, feedback inhibition by tryptophan is exerted at the branch point anthranilate synthase (the product of the *trpE* and *trpD* genes; see Fig. 28–19), so the activity of tryptophan synthase is not strongly affected by [tryptophan].

   (a) If the *trp* mRNA is stable relative to cell generation time, it persists in the population of bacteria even after tryptophan has been added, and tryptophan synthase continues to be synthesized and active. In a simple model, we would expect to see the enzyme activity per cell roughly halved for each generation (30 min); that is, the activity would be slowly diluted out by the increasing numbers of cells.

   (b) Again, if the enzyme is stable relative to the generation time, it persists in the population, even after the addition of tryptophan, and remains active.

   (c) If the mRNA and enzyme are unstable (degraded rapidly relative to cell generation time), attenuation of transcription of the *trp* operon caused by addition of tryptophan leads to an abrupt decrease in levels of *trp* mRNA and tryptophan synthase.

2. **Negative Regulation** Describe the probable effects on gene expression in the *lac* operon of a mutation in **(a)** the *lac* operator that deletes most of O_1; **(b)** the *lacI* gene that inactivates the repressor; and **(c)** the promoter that alters the region around position -10.

   Answer The *lac* operon is negatively regulated by a repressor, the product of the *lacI* gene. The Lac repressor binds to specific DNA sequences called the operators (O_1 and pseudo-operators O_2 and O_3). Binding of the repressor prevents efficient initiation of transcription by RNA polymerase from the promoter. An inducer (allolactose or an analog) binds to the repressor and prevents its binding to the operator, thereby relieving the repression and allowing transcription of the *lac* operon.

(a) This mutation in the operator, the binding site for repressor, would lower the affinity for the repressor and hence reduce repressor binding. This would allow continued transcription (thus expression) of the *lac* operon even in the absence of inducer—referred to as constitutive expression.

(b) A mutation in the *lacI* gene that produced a repressor unable to bind to the operator would lead to constitutive expression (no repression in the absence of inducer). A mutation that prevented binding of the repressor to the inducer without affecting the ability to bind to the operator would lead to a noninducible phenotype (constant repression).

(c) The region of the *lac* gene around position -10 is the promoter region. The *lac* promoter is not particularly strong. Mutations have been observed that either increase or decrease its efficiency of initiating transcription. Base substitutions that make the promoter sequence more similar to the consensus generate a stronger promoter (promoter "up" mutations); those that make the promoter less similar to the consensus generate a weaker promoter (promoter "down" mutations). An "up" mutation would make the *lac* operon independent of positive regulation by the cAMP-CAP complex (when the operon is induced). A "down" mutation would not allow expression even in the derepressed state (presence of inducer) and hence would produce a noninducible phenotype.

3. Specific DNA Binding by Regulatory Proteins A typical bacterial repressor protein discriminates between its specific DNA binding site (operator) and nonspecific DNA by a factor of 10^4 to 10^6. About 10 molecules of repressor per cell are sufficient to ensure a high level of repression. Assume that a very similar repressor existed in a human cell, with a similar specificity for its binding site. How many copies of the repressor would be required to elicit a level of repression similar to that in the bacterial cell? (Hint: The *E. coli* genome contains about 4.6 million bp; the human haploid genome has about 3.2 billion bp.)

Answer The discrete DNA-binding domains of transcriptional regulatory proteins form specific complexes with defined sequences of DNA. Their affinity for these defined sequences is about 10^4 to 10^6 greater than their affinity for other sequences.

Using the example of the Lac repressor, the binding site (operator) is 22 bp long. Ten molecules of the Lac repressor are sufficient to keep this operator in a bound state, even in the context of 4.6×10^6 bp of nonspecific DNA (the rest of the *E. coli* genome). This amounts to finding one specific site in a sea of $(4.6 \times 10^6 \text{ bp})/(22 \text{ bp}) = 2.1 \times 10^5$ nonspecific sites. For the hypothetical repressor in a human cell, let us use the same size binding site (22 bp), although this is larger than most sites so far characterized. The human repressor must find its specific site within $(3.2 \times 10^9 \text{ bp})/(22 \text{ bp}) = 1.5 \times 10^8$ nonspecific sites. Thus, the ratio of nonspecific to specific sites is $(1.5 \times 10^8)/(2.1 \times 10^5) = 710$ times greater in the human cell. Extrapolating from the Lac repressor information, we can estimate that $(710)(10 \text{ molecules}) = 7{,}000$ molecules of repressor are needed per cell.

4. Repressor Concentration in *E. coli* The dissociation constant for a particular repressor-operator complex is very low, about 10^{-13} M. An *E. coli* cell (volume 2×10^{-12} mL) contains 10 copies of the repressor. Calculate the cellular concentration of the repressor protein. How does this value compare with the dissociation constant of the repressor-operator complex? What is the significance of this answer?

Answer Repressor concentration is

$$10 \text{ molecules} \times \frac{1 \text{ mol}}{6.02 \times 10^{23} \text{ molecules}} \times \frac{10^3 \text{ mL}}{\text{L}} \times \frac{1}{2 \times 10^{-12} \text{ mL}} = 8 \times 10^{-9} \text{ M}$$

This is $10^{-13} \text{ M}/8 \times 10^{-9} \text{ M} \approx 10^5$ times greater than the dissociation constant. Because the dissociation constant represents the concentration of repressor needed for half the operator sites to be filled, we can conclude that, with 10 copies of active repressor in the cell, the operator site is always bound by a repressor molecule.

5. Catabolite Repression *E. coli* cells are growing in a medium containing lactose but no glucose. Indicate whether each of the following changes or conditions would increase, decrease, or not change the expression of the *lac* operon. It may be helpful to draw a model depicting what is happening in each situation.

(a) Addition of a high concentration of glucose

(b) A mutation that prevents dissociation of the Lac repressor from the operator

(c) A mutation that completely inactivates β-galactosidase

(d) A mutation that completely inactivates galactoside permease

(e) A mutation that prevents binding of CRP to its binding site near the *lac* promoter

Answer Each condition would *decrease* the expression of *lac* operon genes.

(a) Increased [glucose] results in a decrease in [cAMP]. A cAMP-CRP complex is required for strong binding of the RNA polymerase to the *lac* promoter. As [cAMP] declines, so does expression of the *lac* operon.

(b) A repressor that binds irreversibly to the *lac* operator blocks the binding of RNA polymerase, even if lactose is present, thus preventing expression of the *lac* operon.

(c) If β-galactosidase is not active, no lactose is converted to allolactose, and hence the normal inducer of the operon is absent. Expression of the operon decreases.

(d) A mutation that inactivates the galactoside permease reduces transport of lactose into the cell, even when extracellular levels of lactose are high. Little or no allolactose is produced to induce expression of the *lac* operon, so expression decreases.

(e) In the presence of cAMP, CRP binds to a site near the *lac* promoter and significantly enhances transcription of the operon by stabilizing the open complex of the RNA polymerase. Mutations that prevent CRP binding to the DNA decrease expression of the *lac* operon.

6. Transcription Attenuation How would transcription of the *E. coli trp* operon be affected by the following manipulations of the leader region of the *trp* mRNA?

(a) Increasing the distance (number of bases) between the leader peptide gene and sequence 2

(b) Increasing the distance between sequences 2 and 3

(c) Removing sequence 4

(d) Changing the two Trp codons in the leader peptide gene to His codons

(e) Eliminating the ribosome-binding site for the gene that encodes the leader peptide

(f) Changing several nucleotides in sequence 3 so that it can base-pair with sequence 4 but not with sequence 2

Answer The *trp* operon is subject to regulation by repression and attenuation. Attenuation depends on the tight coupling between transcription and translation in bacteria. When [tryptophan] is high, the *trp* leader is completely translated and the ribosome blocks sequence 2. This allows the transcribed sequences 3 and 4 to form the stem-and-loop attenuator structure (see Fig. 28–21). Formation of the 3:4 loop, which resembles a ρ-independent transcription terminator, terminates transcription of the *trp* operon before genes *E, D, C, B,* and *A* are transcribed, and the enzymes for tryptophan biosynthesis are not produced. When [tryptophan] is low, translation of the *trp* leader stalls at two Trp codons. In this position, the ribosome does not cover sequence 2, and sequence 2 is free to base-pair with sequence 3 in an alternative secondary structure. Formation of the 2:3 stem-and-loop precludes formation of the 3:4 attenuator loop, and transcription proceeds through the *trpEDCBA* genes. Thus, when [tryptophan] is low, the biosynthetic genes are expressed and more tryptophan is synthesized.

(a) Increasing the distance between sequence 1 (encoding the *trp* leader peptide) and sequence 2 results in constitutive expression of the *trp* operon. Whether the ribosome stalls at the leader sequence because of lack of tryptophan, or translates through the leader sequence in the presence of tryptophan, sequence 2 is free to form the 2:3 stem and loop, preventing formation of the 3:4 attenuator.

(b) A large increase in the distance between sequences 2 and 3 discourages formation of the 2:3 stem and loop. Thus, at low [tryptophan], even though the ribosome stalls, the 3:4 attenuator is more likely to form than the 2:3 stem and loop, and expression of the *trp* operon decreases.

(c) Because sequence 4 is required to form the 3:4 attenuator stem and loop, no attenuation can occur in its absence.

(d) If the Trp codons in the leader sequence were changed to His codons, expression of the operon would be regulated by the availability of histidine rather than tryptophan.

(e) Eliminating the ribosome-binding site from the leader sequence would prevent the synthesis of the leader protein, which is the tryptophan sensor in this system.

(f) Changing sequence 3 so that it can base-pair with sequence 4, but not with sequence 2, would mean that only the 3:4 attenuator structure can form. Thus there would be an inappropriate decrease in *trp* operon expression when tryptophan levels were low.

7. Repressors and Repression How would the SOS response in *E. coli* be affected by a mutation in the *lexA* gene that prevented autocatalytic cleavage of the LexA protein?

Answer Induction of the SOS response could not occur, making the cells more sensitive to high levels of DNA damage.

8. Regulation by Recombination In the phase variation system of *Salmonella,* what would happen to the cell if the Hin recombinase became more active and promoted recombination (DNA inversion) several times in each cell generation?

Answer *Salmonella* switches expression between two different flagellin genes, *H1* and *H2*, about once every 1,000 generations, in order to evade the immune system of the host organism. This switch in expression is accomplished by a site-specific recombination system, requiring the action of the Hin recombinase on *hix* sites that flank the *H2* gene. *H2* is expressed and *H1* is repressed in one orientation of *hin* (when the *H2* promoter is oriented toward the *H2* gene), whereas *H1* is expressed and *H2* is inactive in the other orientation (see Fig. 28–28).

More rapid switching due to a more active Hin recombinase would lead to a mixed population of *Salmonella,* some with H1 flagellin and others with H2 flagellin. The host immune system, faced with both types of flagella, would mount an attack on both, greatly reducing the numbers of *Salmonella.* In other words, the protective advantage (to *Salmonella*) of phase variation would be lost.

9. Initiation of Transcription in Eukaryotes A new RNA polymerase activity is discovered in crude extracts of cells derived from an exotic fungus. The RNA polymerase initiates transcription only from a single, highly specialized promoter. As the polymerase is purified, its activity declines, and the purified enzyme is completely inactive unless crude extract is added to the reaction mixture. Suggest an explanation for these observations.

Answer The observation that the new RNA polymerase loses activity as it is purified, and that this activity is restored by addition of crude extract, suggests that some factor remaining in the crude extract is necessary for activity under the assay conditions. Any of several factors could have dissociated and been eliminated during purification: (1) an activator such as cAMP-CRP; (2) a loosely associated subunit of the protein that is necessary for activity; or (3) a specificity factor (similar to the σ subunit of the *E. coli* RNA polymerase) necessary for efficient transcription from its promoter. Finally, proteolysis could be responsible for loss of activity during successive stages of purification. Small amounts of polymerase remaining in the crude extract would then supply the only activity in the fraction. In the absence of quantitative information (a purification table), it is impossible to determine whether this explanation is feasible.

10. **Functional Domains in Regulatory Proteins** A biochemist replaces the DNA-binding domain the yeast Gal4 protein with the DNA-binding domain from the Lac repressor, and finds that the en neered protein no longer regulates transcription of the *GAL* genes in yeast. Draw a diagram of th different functional domains you would expect to find in the Gal4 protein and in the engineered pro Why does the engineered protein no longer regulate transcription of the *GAL* genes? What might l done to the DNA-binding site recognized by this chimeric protein to make it functional in activatin transcription of *GAL* genes?

    Answer Transcriptional activators have at least two domains that frequently function sep rately: the DNA-binding domain and the activation domain. The DNA-binding domain is re quired for the sequence-specific binding of the protein to DNA. A different portion of the tein is responsible for activation; this domain may directly interact with the RNA polymer or it may facilitate the action of coactivators or other proteins that stimulate transcription

    Gal4 protein

Gal4 DNA-binding domain	Gal4 activator domain

    Engineered protein

Lac repressor DNA-binding domain	Gal4 activator domain

    The biochemist has done part of a domain-swap experiment; the activation domain of Gal4 protein is fused to the DNA-binding domain of the Lac repressor. This new hybrid no long recognizes the Gal4 binding site in DNA (called UAS_G) because that DNA-binding domain no longer present. However, it can bind to the repressor-binding site in the operator portio the *lac* operon. Therefore, replacement of UAS_G with the *lac* operator (binding site for the Lac repressor) in the *GAL* operon should allow the Lac repressor-Gal4 hybrid protein to fu tion as a transcriptional activator of *GAL* genes in yeast.

11. **Nucleosome Modification during Transcriptional Activation** To prepare genomic regions fo transcription, certain histones in the resident nucleosomes are acetylated and methylated at speci locations. Once transcription is no longer needed, these modifications need to be reversed. In mam mals, the methylation of Arg residues in histones is reversed by peptidylarginine deiminases (PAD The reaction promoted by these enzymes does not yield unmethylated arginine. Instead, it produc citrulline residues in the histone. What is the other product of the reaction? Suggest a mechanism this reaction.

    Answer Methylamine. The reaction proceeds with attack of water on the guanidinium car of the modified arginine.

12. **Inheritance Mechanisms in Development** A *Drosophila* egg that is bcd^-/bcd^- may develop no mally, but as an adult will not be able to produce viable offspring. Explain.

    Answer The *bicoid (bcd)* gene product is a major anterior morphogen. The mRNA from *bicoid* gene is synthesized by nurse cells in female flies and deposited in the unfertilized e near its anterior pole. The Bicoid protein produced by the maternal mRNA diffuses throug the egg cell, creating a concentration gradient that establishes the location of the anterior of the embryo. The amounts of Bicoid protein present in various parts of the embryo affec the subsequent expression of a number of other genes in a threshold-dependent manner. L of Bicoid protein results in development of an embryo with two abdomens but neither hea nor thorax; this embryo dies before it hatches as a larva. If a female fly that is heterozygou for the *bicoid* gene (bcd^+/bcd^-) is mated with a male that has a *bicoid* deficiency, the fen will be able to produce *bicoid* mRNA, allowing for the development of phenotypically normal female progeny that lack the *bicoid* gene (bcd^-/bcd^-). The female progeny from this cross, however, will not be able to produce viable offspring because their nurse cells are incapable of producing *bicoid* mRNA for deposition in their egg cells.

Biochemistry on the Internet

13. TATA-Binding Protein and the TATA Box To examine the roles of hydrogen bonds and hydrophobic interactions between transcription factors and DNA, go to FirstGlance in Jmol at http://firstglance.jmol.org and enter the PDB ID 1TGH. This file models the interactions between a human TATA-binding protein and a segment of double-stranded DNA. Once the structure loads, click the "Spin" button to stop the molecule from rotating. When the molecule has reloaded, click the "Contacts" link. With the radio button for "Chains" selected, click on any part of the protein (Chain A, displays in blue) to select it as the target. Click "Show Atoms Contacting Target" and, in the list of contacts to display, check only "Show putatively hydrogen-bonded non-water" to display hydrogen bonds between the protein and the TATA box DNA. Then click on the rightmost option for viewing images (Maximum detail: Target & Contacts Balls and Sticks, Colored by Element). With this view you should be able to use the zoom and rotate controls and mouse clicks to answer the following questions.

(a) Which of the base pairs in the DNA form hydrogen bonds with the protein? Which of these contribute to the specific recognition of the TATA box by this protein? (Hydrogen-bond length between hydrogen donor and hydrogen acceptor ranges from 2.5 to 3.3 Å.)

(b) Which amino acid residues in the protein interact with these base pairs?

(c) What is the sequence of the DNA in this model and which portions of the sequence are recognized by the TATA-binding protein?

(d) Examine the hydrophobic interactions in this complex. Are they rare or numerous? To answer this question, click on "Return to contacts" and check the option to "Show hydrophobic (apolar van der Waals) interactions."

Answer

(a) The nucleotides with backbone elements involved in hydrogen bonding include A106 and A110 of chain B and A118, T119, and A122 of chain C. The nucleotides with bases involved in hydrogen bonding include A106 and T107 of chain B, and A118 and T119 of chain C. It is the interactions between the protein and the individual bases (not the backbone) that contribute directly to DNA sequence recognition.

(b)

	Chain B	Chain C
Base	A106|||Asn^{253} T107|||Asn^{253}	A118|||Asn^{163} T119|||Asn^{163}
Backbone	A106|||Arg^{290} A110|||Ser^{212}	A118|||Arg^{199} T119|||Arg^{204} A122|||Ser^{303}

In the TATA-binding protein, the residues most often hydrogen bonded to the DNA are Arg and Asn. In general, the amino acid residues typically hydrogen bonded through their side chains to bases in DNA include Asn, Gln, Glu, Lys, and Arg.

(c) Chain B: TATATATA (residues 103 to 110)
Chain C: ATATATAT (residues 122 to 115)
The TATA-binding protein makes specific contact with the base pairs A106/T119 and T107/A118 in the middle of this sequence and interacts with the backbone at each end, at A104/T121 and T109/A116.

(d) The hydrophobic interactions are numerous. Many binding interactions of this type involve the burying of large amounts of hydrophobic surface.

Data Analysis Problem

14. Engineering a Genetic Toggle Switch in *Escherichia coli* Gene regulation is often described as an "on or off" phenomenon—a gene is either fully expressed or not expressed at all. In fact, repression and activation of a gene involve ligand-binding reactions, so genes can show intermediate levels of expression when intermediate levels of regulatory molecules are present. For example, for the *E. coli lac* operon, consider the binding equilibrium of the Lac repressor, operator DNA, and inducer (see Fig. 28–7). Although this is a complex, cooperative process, it can be approximately modeled by the following reaction (R is repressor; IPTG is the inducer isopropyl-β-D-thiogalactoside):

$$\text{R} + \text{IPTG} \xrightleftharpoons{K_d = 10^{-4}\,\text{M}} \text{R} \cdot \text{IPTG}$$

Free repressor, R, binds to the operator and prevents transcription of the *lac* operon; the R • IPTG complex does not bind to the operator and thus transcription of the *lac* operon can proceed.

(a) Using Equation 5–8, we can calculate the relative expression level of the proteins of the *lac* operon as a function of [IPTG]. Use this calculation to determine over what range of [IPTG] the expression level would vary from 10% to 90%.

(b) Describe qualitatively the level of *lac* operon proteins present in an *E. coli* cell before, during, and after induction with IPTG. You need not give the amounts at exact times—just indicate the general trends.

Gardner, Cantor, and Collins (2000) set out to make a "genetic toggle switch"—a gene-regulatory system with two key characteristics of a light switch. (A) *It has only two states:* it is either fully on or fully off; it is not a dimmer switch. In biochemical terms, the target gene or gene system (operon) is either fully expressed or not expressed at all; it cannot be expressed at an intermediate level. (B) *Both states are stable:* although you must use a finger to flip the light switch from one state to the other, once you have flipped it and removed your finger, the switch stays in that state. In biochemical terms, exposure to an inducer or some other signal changes the expression state of the gene or operon, and it remains in that state once the signal is removed.

(c) Explain how the *lac* operon lacks both characteristics A and B.

To make their "toggle switch," Gardner and coworkers constructed a plasmid from the following components:

OP_{lac} The operator-promoter region of the *E. coli lac* operon

OP_{λ} The operator-promoter region of λ phage

lacI The gene encoding the *lac* repressor protein, LacI. In the absence of IPTG, this protein strongly represses OP_{lac}; in the presence of IPTG, it allows full expression from OP_{lac}.

rep^{ts} The gene encoding a temperature-sensitive mutant λ repressor protein, rep^{ts}. At 37 °C this protein strongly represses OP_{λ}; at 42 °C it allows full expression from OP_{λ}.

GFP The gene for green fluorescent protein (GFP), a highly fluorescent reporter protein (see Fig. 9–15)

T Transcription terminator

The investigators arranged these components (see figure below) so that the two promoters were reciprocally repressed: OP_{lac} controlled expression of *rep*^{ts}, and OP_{λ} controlled expression of *lacI*. The state of this system was reported by the expression level of *GFP*, which was also under the control of OP_{lac}.

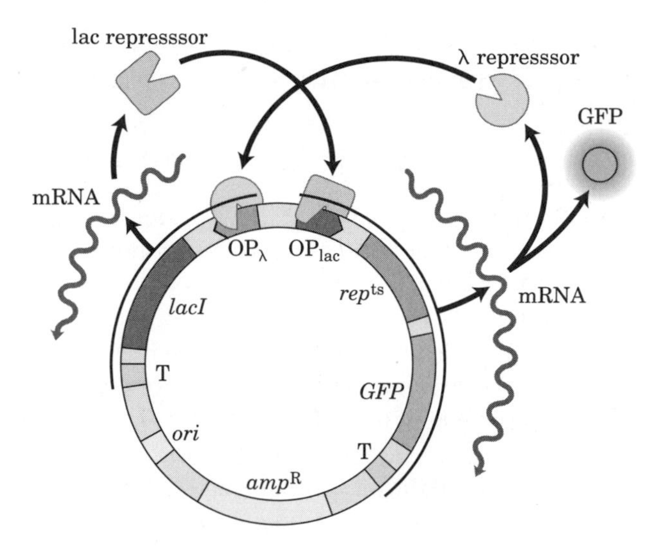

(d) The constructed system has two states: GFP-on (high level of expression) and GFP-off (low level of expression). For each state, describe which proteins are present and which promoters are being expressed.

(e) Treatment with IPTG would be expected to toggle the system from one state to the other. From which state to which? Explain your reasoning.

(f) Treatment with heat (42 °C) would be expected to toggle the system from one state to the other. From which state to which? Explain your reasoning.

(g) Why would this plasmid be expected to have characteristics A and B as described above?

To confirm that their construct did indeed exhibit these characteristics, Gardner and colleagues first showed that, once switched, the GFP expression level (high or low) was stable for long periods of time (characteristic B). Next, they measured GFP level at different concentrations of the inducer IPTG, with the following results.

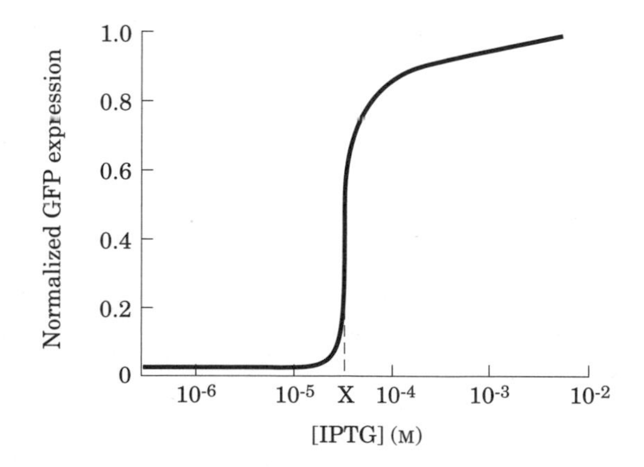

They noticed that the average GFP expression level was intermediate at concentration X of IPTG. However, when they measured the GFP expression level *in individual cells* at [IPTG] = X, they found either a high level or a low level of GFP—no cells showed an intermediate level.

(h) Explain how this finding demonstrates that the system has characteristic A. What is happening to cause the bimodal distribution of expression levels at [IPTG] = X?

Answer

(a) For 10% expression (90% repression), 10% of the repressor has bound inducer and 90% is free and available to bind the operator. The calculation uses Equation 5–8 (p. 156), with $\theta = 0.1$ and $K_d = 10^{-4}$ M.

$$\theta = \frac{[\text{IPTG}]}{[\text{IPTG}] + K_d} = \frac{[\text{IPTG}]}{[\text{IPTG}] + 10^{-4}\text{M}}$$

$$0.1 = \frac{[\text{IPTG}]}{[\text{IPTG}] + 10^{-4}\text{M}} \quad \text{so } 0.9\,[\text{IPTG}] = 10^{-5} \text{ or } [\text{IPTG}] = 1.1 \times 10^{-5}\,\text{M}$$

For 90% expression, 90% of the repressor has bound inducer, so $\theta = 0.9$. Entering the values for θ and K_d in Equation 5–8 gives [IPTG] $= 9 \times 10^{-4}$ M. Thus, gene expression varies 10-fold over a roughly 10-fold [IPTG] range.

(b) You would expect the protein levels to be low before induction, rise during induction, and then decay as synthesis stops and the proteins are degraded.

(c) As shown in **(a)**, the *lac* operon has more levels of expression than just on or off; thus it does not have characteristic A. As shown in **(b)**, expression of the *lac* operon subsides once the inducer is removed; thus it lacks characteristic B.

(d) *GFP-on:* rep^{ts} and GFP are expressed at high levels; rep^{ts} represses OP_{λ}, so no LacI protein is produced. *GFP-off:* LacI is expressed at a high level; LacI represses OP_{lac}, so rep^{ts} and GFP are not produced.

(e) IPTG treatment switches the system from GFP-off to GFP-on. IPTG has an effect only when LacI is present, so affects only the GFP-off state. Adding IPTG relieves the repression of OP_{lac} allowing high-level expression of rep^{ts}, which turns off expression of LacI, and high-level expression of GFP.

(f) Heat treatment switches the system from GFP-on to GFP-off. Heat has an effect only when rep^{ts} is present, so affects only the GFP-on state. Heat inactivates rep^{ts} and relieves the repression of OP_{λ}, allowing high-level expression of LacI. LacI then acts at OP_{lac} to repress synthesis of rep^{ts} and GFP.

(g) *Characteristic A:* the system is not stable in the intermediate state. At some point, one repressor will act more strongly than the other due to chance fluctuations in expression; this shuts off expression of the other repressor and locks the system in one state. *Characteristic B:* once one repressor is expressed, it prevents the synthesis of the other; thus the system remains in one state even after the switching stimulus has been removed.

(h) At no time does any cell express an intermediate level of GFP—this is a confirmation of characteristic A. At the intermediate concentration (X) of inducer, some cells have switched to GFP-on while others have not yet made the switch and remain in the GFP-off state; none are in between. The bimodal distribution of expression levels at [IPTG] = X is caused by the mixed population of GFP-on and GFP-off cells.

Reference

Gardner, T.S., Cantor, C.R., & Collins, J.J. (2000) Construction of a genetic toggle switch in *Escherichia coli*. *Nature* **403,** 339–342.